ZERO to INFINITY

The Foundations of Physics

K&E Series on Knots and Everything – Vol. 41

ZERO to INFINITY

The Foundations of Physics

Peter Rowlands

University of Liverpool, UK

World Scientific

NEW JERSEY • LONDON • SINGAPORE • BEIJING • SHANGHAI • HONG KONG • TAIPEI • CHENNAI

Published by

World Scientific Publishing Co. Pte. Ltd.

5 Toh Tuck Link, Singapore 596224

USA office: 27 Warren Street, Suite 401-402, Hackensack, NJ 07601

UK office: 57 Shelton Street, Covent Garden, London WC2H 9HE

Library of Congress Cataloging-in-Publication Data

Rowlands, Peter.

Zero to infinity : the foundations of physics / by Peter Rowlands.

p. cm. -- (Series on knots and everything ; v. 41)

Includes bibliographical references and index.

ISBN-13: 978-981-270-914-1 (hardcover)

ISBN-10: 981-270-914-2 (hardcover)

1. Physics. 2. Mathematical physics. I. Title.

QC21.3.R69 2007

530--dc22

2007022004

cover image: courtesy of Vanessa Hill

copyright of the designs of the cover belongs to Vanessa Hill

British Library Cataloguing-in-Publication Data

A catalogue record for this book is available from the British Library.

First published 2007

Reprinted 2008

Printed in Singapore by B & JO Enterprise

To the memory of my parents, Ernest William Rowlands
and Theresa Mary Rowlands

Preface

Physics appears to be the only source of fundamental knowledge about the natural world. No other system of thought or methodology has shown any of its systematic explanatory or predictive power. This success has been achieved by a continued attempt at minimalism and reductionism, and it appears that the greatest success has been achieved from the simplest possible foundations. Yet, because we have been obliged to approach the subject by an inductive method, working back from complicated observations to simple explanations, we have still to discover the ultimate foundations on which this whole conceptual scheme has been based. We know that the ultimate theory must be simple, probably extremely simple, but, because it must also be unique, we have no precedent which would help us to make the discovery. Yet the belief that the discovery is possible remains, and has led to many approaches towards a 'unified theory of physics' or a 'theory of everything', none of which seems to be close to success.

Obviously, no one expects to succeed instantly with a theory that will simply explain everything. What we would hope to do is to find a *process*, a systematic way of proceeding with strong indications that we were on the right track. This is what is being aimed at in this book. Positions that are rejected from the outset in the search include model-dependent theories of any kind; the aim of the work is resolutely abstract. One of the particular approaches avoided is the restructuring of particle physics in terms of multidimensional space-time strings or membranes. We can, of course, do this as a mathematical *representation*, and the procedure for doing so is sketched out in chapters 4, 15 and 18, but a string theory would not be a unified theory, even if we should chance to find out the 'correct' one from the many thousands of possible alternatives. A unified theory has to explain the *concept* of dimension, as well as the number (why 10? why 11?); it also has to explain space and time, their similarities and differences, and even the use of mathematics to explain physics. But there are also intrinsic difficulties with the approach which explain why it does not offer a true unity.

Even the first stage of combining space and time in the simplest way, by adding the single time dimension to space's three in special relativity, causes us problems when we look at quantum mechanics, a theory which appears to be an essential starting-point for all foundational work in physics. Time, unlike space, is simply not an observable in quantum mechanics, though the two quantities are assumed to have identical status in the relativistic 'space-time' concept. When we make the space-time part of an even-more complicated structure, as in general relativity, where 'curvature' is used to eliminate mass or gravity, the problems multiply, and we find singularities, nonlinearity, unrenormalizable infinities, and the violation of fundamental physical laws. We solve some problems, but create others.

Again, we must reject the idea that a single cosmic creation event has structured the laws of physics in a particular way, and that they could have been different in different circumstances. The idea could, in principle, be true, but then we would have no abstract subject of physics, no generality, no absolute mathematics, and no meaningful concept of conservation, the process which makes physics universal. The very idea that we could discover a *unified* theory of physics is impossible in such a context. Physics is fractured in the very act of creation. In addition, such explanations have the habit of becoming self-fulfilling prophecies. We simply refer difficulties to special conditions that occurred in the 'early universe', and deprive ourselves of understanding fundamental physical phenomena which ought to be valid at all places in all epochs. This does not, of course, mean that we cannot discover historical evolution over time for structures such as stars and galaxies, and galactic clusters. What it does mean is that physics, if it is to be a truly unified subject, should not be determined by cosmic history, whether or not this turns out to be true. The laws of physics cannot be the result of an accident.

Even the very successful approach to physics using symmetry groups, as employed extensively, for example, in particle physics, should be treated with caution. It is assumed that, if we find a group structure which accommodates all four known physical interactions, then we will have solved the problem of their relationship, and we will have a 'Grand Unified' theory of particle interactions. Of course, such a result would be a very significant step, and the idea is discussed in some detail in chapter 15, but we would not have solved the problem from a fundamental point of view unless we could explain *why* we have this particular group structure, and, indeed, why we have a group structure at all. Obviously, we have to proceed in understanding nature by stages, but a group structure will never be an end stage, nor will *any* structure. As long as anything complicated remains unexplained in our theory, we will not be able to describe it as a theory of *everything*.

So it is far from obvious how we would construct such a theory, but there is one important clue as to where we should start. One fundamental idea, and one only, has the necessary simplicity and intrinsic inexplicability to be the foundation for everything else. This is *nothing*, zero in mathematical terms. We could imagine creating a theory of everything if it was also a theory of nothing. The question is: can it be done? Can we start from zero, and use it to structure nature as we understand it today? The proposition would seem to be impossible, but, in fact, it is not, and it is the aim of the present research programme to justify this statement.

It would probably be impossible to do this by purely logical development from first principles, though, in a sense, mathematics attempts such an approach. Mathematics certainly provides a very powerful formalism, of which physics makes extensive use, but its own logical foundations, as Gödel proved, remain insecurely based on a seemingly empirical process of counting. Computing provides another alternative, and Wolfram and others have seen the development of complexity from simplicity in systems governed by cellular automata;[1] with the further assumption that the 'right' complex structures will somehow finally emerge, but, again, the empirical counting process is assumed, along with the idea that only discrete concepts matter. Why discreteness is to be privileged and what discreteness actually *is* remain unexplained. Physics allows us a very different route to the foundations, through the application of inductive methodology to masses of empirical data, and a ruthless Darwinian selection of the only formalisms which work, and it is in seeing what mathematical structures are essential to physics *at its very foundations* that we see what structures are also essential to mathematics and computing. By finding the common origin of mathematics, physics and computing and the way they deal with zero totality, using the dual processes of induction and deduction, we can finally track down the route through which 'everything' finally comes from nothing.

The structure of the book reflects this process. The first chapter develops a computing analogy to see how a zero totality can be used to create a *universal rewrite system*, which then allows us to structure mathematics without first assuming the number system or discreteness. This exercise leads us to a very definite mathematical structure, *with zero conceptual totality*, which we can then work towards in a deductive context. The next chapter is inductive, and takes physics as far as we can towards its ultimate foundations by analysing the most fundamental concepts that we are capable of imagining in a physical context. The procedures used in the mathematical structure can then be seen to correspond with the ones we have derived by induction as the basic components of physics. Further analysis of the physical context then shows us, in chapter 3, that the most convenient packaging of the mathematical structure is the one that provides the shortest route to zero totality, at the same time as presenting us with the fundamental equation that drives the whole of physics. Most of the remaining chapters then present the working out of the consequences of chapters 1, 2 and 3 in all the detail necessary to show that the structure is sufficient to generate the results which are considered foundational to physics, even to the point of numerical detail. Chapters 19 and 20, however, stand apart in showing that the kind of information processing structures that make physics and mathematics 'spontaneously' emerge in nature also apply (in a fractal sense) to biological and other large-scale systems. Only by creating the most efficient information processing possible could these large-scale systems be created against the natural tendency to disorder or increased entropy, and it is difficult to imagine that any information processing could be more efficient than the one produced by Nature's own rewrite code.

If this process is true to any considerable degree, then it will be of significance to everyone, scientist and nonscientist alike, with or without mathematical training. So, the book has been written in such a way that there are long sections of conceptual argument, which should appeal to the general reader as well as the professional scientist. However, there is no disguising the fact that mathematics lies at the heart of this book, and that credible results, in many areas, can only be achieved by using the full mathematical formalisms. So, these are also given in full detail where required. The idea has been to develop all the ideas from as foundational a position as possible, but the presentation concentrates generally on results which may be considered original in some respect, and only makes use of established work where it is absolutely necessary to the argument. Since the mathematics of quaternions and multivariate vectors is essential to the argument, an appendix is included at the end of the first chapter giving an elementary treatment of these algebras.

Despite the obvious novelty of the fundamental position, and some of the specific formalisms employed, most of the results generated certainly support 'orthodox' or 'mainstream' science, where existing work is available for comparison. Of course, there are new results and predictions, but there is no challenge here to the bases of quantum mechanics, classical physics, particle physics, or anything else now universally accepted. New ideas are certainly put forward in areas which are still highly speculative, such as certain aspects of cosmology, and some of the *physical* interpretations (for example, in relation to the emergent nature of fractional charges in quarks, or the gravitational-inertial explanation of general relativity) differ from the usual (though not exclusively accepted) ones, while retaining the overall mathematical formalisms which really define the theories. However, nothing here proposes deviations from the experimental evidence as now understood, though some new predicted results are available for testing, and some have already been confirmed since they were first predicted. In addition, many speculative concepts now in the literature would be ruled out by the analysis, while a few might be vindicated. A complete reading of the book should indicate that every position adopted is founded on the results incorporated in chapters 1, 2 and 3. The theoretical position is put forward as an organic whole, and every statement within it, in a sense, reinforces every other. Important links are shown through a system of cross-referencing.

To aid the reader, there is a synopsis of the contents of the chapters at the beginning of each, and a summary of the entire argument at the end. These can be used to get a general idea of the argument where details prove troublesome or appear to require too much specialised knowledge. As with all presentations of novel results, readers will have to make up their own minds about the thesis being proposed, but the book is intended to contain the minimum of 'speculation', in the ordinary sense of that word. Only a few of the mass calculations and the section at the end dealing with the derivation of the cosmic background radiation are consciously put forward as speculative proposals, and, in the latter case, it could be argued that Ockham's razor ought to favour an argument that leaves fewer unexplained facts than any known alternative. Many parts of the book (chapters 5, 6, 10 and 11, especially) are the working out of the consequences of new formalisms with the appropriate degree of mathematical rigour, the physical consequences following on directly from the mathematics, while the ideas on algebra and rewrite alphabets in chapters 1 and 3 leave plenty of scope for further development in the direction of practical application. Elsewhere, it is hoped that the sheer simplicity of the basic ideas, and their apparent ability to explain a great number of seemingly diverse facts, will recommend them to the reader's attention.

The project has a long history. I can write here in detail only of my own trajectory; those of my colleagues would, of course, be different and would have different emphases. The germ of several significant ideas began with student speculations.[2] The essential philosophy was developed between the last years at school and the first years at university. The group symmetry of space, time, mass and charge was in place by the mid-1970s, along with the first particle physics ideas based on charge structures, and some of the gravitational and cosmological ideas outlined in chapters 18 and 21. The first publications came at the end of the 1970s and the beginning of the 1980s, but success in 'respected' publication outlets was a long time coming.

Lee Smolin has described how the 'philosophical way of doing theoretical physics' of the 1920s 'gradually lost out to a more pragmatic, hard-nosed style of research', which was 'completed when the center of gravity of physics moved to the United States in the 1940s'.[3] By the 1970s, at the exact moment when the first ideas in this book were being developed, 'the transition was complete'. Smolin reports that: 'As a graduate student, I was told by my teachers that it was impossible to make a career working on problems in the foundations of physics. My mentors pointed out that there were no interesting new experiments in that area, whereas particle physics was driven by a continuous stream of new experimental discoveries.' The experiments, of course, led to the establishment of the Standard Model of particle physics, around 1973, but, since that time, no really new unifying principle seems to have been discovered, and the abandonment of research into the foundations of the subject has made such a discovery increasingly unlikely. The remarkable thing is that the pattern that has set in over the last half century or so has made many physicists seemingly unable to conceive of the *concept* of researching the foundations. Work of this kind seems to create bafflement in many and downright hostility in others.

Serious publication outlets were certainly minimal during the 1980s, but the situation improved towards the end of the decade when the PIRT series of conferences (Physical Interpretations of Relativity) were started in London, by Michael Duffy. I first attended in 1990, and, as a result of these meetings (which are also now held in Moscow, Calcutta and Budapest), I came into contact with the Vigier conferences (Toronto-Berkeley-Paris), organised from 1995 by Geoffrey Hunter, Stanley Jeffers and Richard Amoroso, the ANPA conferences at Cambridge, organised by Keith Bowden and Arleta Ford, which I attended from 1998, and the CASYS conferences at Liège, organised by Daniel Dubois, which I first attended in 2003. All these meetings, in their different ways, have been concerned with the foundations of the subject, and with tackling important questions in a freely inquiring spirit; and it is largely through contacts made through these and related events that I first met the collaborators, who are named on the title page of this book. During these years, also, I published three books, summarising my work from the 1980s, and consolidated the new view of relativistic quantum mechanics I had been developing as a result of my theories of symmetry. I managed to publish the first paper on this topic in 1994, and set about relating the particle physics consequences with my earlier work on this subject.

John Cullerne was my first real collaborator, and worked with me for many hours, principally during 1997-2000, on Dirac algebra and the derivation of the Standard Model and other aspects of particle physics. (See **5**, **6**, **14**, **15**.) Our intense weekly discussions were always a source of great and mutual intellectual stimulation. After this time, John's other commitments took hold and our collaboration became less intense, although it still continues on an occasional basis. In particular, John has acted as adviser on parts of chapters **10-13**, the orthodox making a potent combination with the unorthodox. A new departure was the universal rewrite system, which was the result of my collaboration with my computer science colleague, Bernard Diaz, from about 1997 (see **1**, **3**, **20**, **Appendix B**). Because of its extremely fundamental nature, this has proved a difficult area in which to work, one needing endless examination and re-examination of the concepts. Brian Koberlein, whom I met through ANPA, worked with me intensely for a few days in Cambridge, and then via email, on groups and dual systems, and, separately, on the comparison of the nilpotent and idempotent versions of quantum mechanics (see **4**, **7**, **15**). Presentations by Brian also stimulated the work which appears in **18.9**.

Peter Marcer (see **20**) was another ANPA contact and we have now had a wide-ranging collaboration for many years, on many subjects, beginning as informal discussion, and continuing under the auspices of the British Computer Society's Cybernetics Machine Group and the CASYS conferences in Liège. Through Peter, I have also had a fruitful interaction with Edgar Mitchell and Walter Schempp, our co-authors on the ground-breaking paper, 'Zenergy' (see **20**). Finally, through a London frontiers meeting, organized by Simon Daniel, as a result of an earlier Vigier meeting in Paris, I met the biologist Vanessa Hill, and we soon realised that we had a potentially powerful collaboration on applying algebraic and geometric concepts in biology (see **19**). Both of these collaborations (as recorded in chapters **19** and **20**) are developing rapidly and expanding into areas that we had not previously connected with the project.

Apart from these formal collaborations, I have had stimulating discussions and contacts with many other researchers, including Ruggero Santilli, Erik Trell, Stein Johansen, Jeremy Dunning-Davies, Clive Kilmister, Ted Bastin, Lou Kauffman, Dan Kurth, Mark Curtis, Sarah Bell, Cynthia Whitney, John Spencer, John Valentine, Mark Stuckey, Jose Almeida, Otto van Nieuwnehuijze, Tolga Yarman, Tuomo Suntola, Sergey Siparov, Vladimir Gladyshev and Tatyana Gladysheva.[4] Besides these there are a huge number of people to whom I and my collaborators are indebted. In particular, there are the organizers and participants of PIRT, Vigier, ANPA and CASYS, for many stimulating presentations and discussions; the Swansea / Bristol / Keele group (Viv and Mary Pope, Alan Winfield, Anthony Osborne); my colleagues David Edwards, Mike Houlden, Dominic Dickson, John Fry and Christos Touramanis, for their support and interest over many years. Mike, in particular, has been an endless source of new problems for me to challenge, and his advice and encouragement has been without parallel. Apart from my collaborators, he is the person of all to whom I am most indebted. The British Computer Society have been generous in their financial support for the Cybernetics Machine Group's activities; and I have been a beneficiary on several occasions, along with Peter and Vanessa. I am also grateful for funding to Dmitri Pavlov, and to the British Council, as well as to the University's Physics and Computer Departments.

Peter Rowlands
Oliver Lodge Laboratory
University of Liverpool

Contents

Preface vii

1. Zero 1
   1.1 An Origin for Everything 2
   1.2 The Genesis of Number 3
   1.3 The Genesis of Algebra 9
   1.4 Group Representations 14
   1.5 Rewriting Nature 18
   1.6 Quaternions and Vectors 24

2. Why Does Physics Work? 32
   2.1 A Foundational Level 33
   2.2 The Origin of Abstraction 35
   2.3 Symmetry 36
   2.4 The Meaning of the Conservation Laws 39
   2.5 The Mathematical Structure of Physical Quantities 42
   2.6 Where Does Dimensionality Come From? 45
   2.7 A Group of Order 4 50
   2.8 Noether's Theorem Revisited 54
   2.9 Analytic Versus Synthetic 55
   2.10 The Power of Analogy 58
   2.11 The Nature of Reality 60

3. The Emergence of Physics 63
   3.1 The Mathematical Character of Physics 63
   3.2 The Algebra of Space, Time, Mass and Charge 64
   3.3 The Dirac Algebra 69
   3.4 The Creation of the Dirac State 72
   3.5 The Nilpotent Dirac Equation 77
   3.6 Uniqueness, Qubits and Quantum Computing 81
   3.7 The Completeness of Mathematical Physics 82
   3.8 Theoretical Computation 86

4. Groups and Representations 88
4.1 The Dirac Equation and Quantum Field Theory 88
4.2 Reversals of Properties 91
4.3 The Dual Group and Higher Symmetries 93
4.4 A Broken Octonion 96
4.5 A Hierarchy of Dualities 99
4.6 Dimensionality 101
4.7 Symmetry Hierarchy 103
4.8 Colour Representation 105
4.9 3-D (Vector) Representation 107
4.10 Tetrahedral Representation 109

5. Breaking the Dirac Code 111
5.1 Singularities and Redundancy 111
5.2 Redundancy in the Dirac Equation 113
5.3 Defragmenting the Dirac Equation 115
5.4 The Dirac 4-Spinor 118
5.5 The 4-Component Differential Operator 123
5.6 C-Linear Maps and Lifts 125
5.7 The Quaternion Form Derived from a Matrix Representation 127
5.8 Bilinear Covariants and the Dirac Lagrangian 131
5.9 Removing Redundancies in Relativistic Quantum Mechanics 133
5.10 Orthonormality of the Nilpotent Solutions of the Dirac Equation 134

6. The Dirac Nilpotent 137
6.1 Spin 137
6.2 Helicity 139
6.3 Fermions and Bosons 141
6.4 Vacuum 146
6.5 *CPT* Symmetry 149
6.6 Baryons 151
6.7 Gluons and Exotic States 154
6.8 Parities of Bosons and Baryons 156
6.9 Supersymmetry and Renormalization 157
6.10 Annihilation and Creation Operators 160
6.11 The Quantum Field 161
6.12 The Nilpotent State 163
6.13 Nonlocality 164
6.14 BRST Quantization 166

7. Nonrelativistic Quantum Mechanics and the Classical Transition 168
7.1 The Bispinor Form of the Dirac Equation 168
7.2 The Schrödinger Approximation 171
7.3 The Heisenberg Formulation of Quantum Mechanics 174

7.4 Heisenberg v. Schrödinger 175
7.5 The Quantum-Classical Transition 177
7.6 The Classical Limit 179
7.7 The Dirac Nilpotent Using Discrete Differentiation 182
7.8 Idempotent and Nilpotent Versions of Quantum Mechanics 185
7.9 A Fundamental Quantum Mechanical Duality 186

8. The Classical and Special Relativistic Approximations 191
8.1 Linear Versus Orbital Dynamics 191
8.2 Scaling Relations 192
8.3 Special Relativity 195
8.4 The Significance of the Proper Time 199
8.5 The Nature of Classical Physics 203
8.6 Constructed Quantities 206
8.7 Classical Mechanics 207
8.8 Classical Electromagnetic Theory 212

9. The Resolution of Paradoxes 218
9.1 Paradoxes Relating to Conservation and Nonconservation 218
9.2 Paradoxes Relating to Continuity and Discontinuity 222
9.3 Irreversibility and Causality 225
9.4 The Mass Frame and Zero-Point Energy 228
9.5 Two Versions of Relativity 232
9.6 Thermodynamics and the Arrow of Time 235

10. Electric, Strong and Weak Interactions 239
10.1 The Dirac Equation in the Coulomb Field 240
10.2 Condensed Matter: The Kronig-Penney Model 246
10.3 The Helium Atom 248
10.4 $SU(3)$ 250
10.5 The Quark-Antiquark and Three-Quark Interactions 252
10.6 Angular Momentum 258
10.7 The Weak Filled Vacuum 260
10.8 The Origin of the Higgs Mechanism 263
10.9 $SU(2)_L \times U(1)$ 264
10.10 The Weak Interaction and the Dirac Formalism 266
10.11 The Higgs Mechanism for $U(1)$ and $SU(2)_L$ 270
10.12 The Spherical Harmonic Oscillator 273
10.13 The Weak Interaction as a Harmonic Oscillator 277
10.14 A Strong-Electroweak Solution of the Dirac Equation 280

11. QED and its Analogues 285
11.1 A Perturbation Expansion of the Dirac Equation for QED 285
11.2 Integral Solutions of the Dirac Equation 289

11.3 Renormalization 290
11.4 Green's Function Solution 293
11.5 The Propagator Method in Lowest Order 298
11.6 Electron Scattering 300
11.7 Strong and Weak Analogues 305
11.8 QFD Using Nilpotents 307
11.9 The Success of the Nilpotent Method 309

12. Vacuum 310
12.1 Physics and Observables 310
12.2 Zero-Point Energy 312
12.3 The Weak Vacuum 313
12.4 The Strong Vacuum 315
12.5 The Electric Vacuum 316
12.6 The Gravitational Vacuum 318
12.7 The Casimir Effect 319
12.8 Berry's Geometric Phase 321

13. Fermion and Boson Structures 324
13.1 The Charge Structures of Quarks and Leptons 324
13.2 A Unified Representation for Quarks / Leptons 327
13.3 Conservation of Charge Type and Conservation of Angular Momentum 329
13.4 Phase Diagrams for Charge Conservation 330
13.5 Quark and Lepton Charge Structures in Tabular Form 332
13.6 Mesons and Baryons 338
13.7 The Standard Model 339
13.8 A Pentad Structure for Charges and their Transitions 345
13.9 Lepton-Like Quarks 346

14. A Representation of Strong and Weak Interactions 349
14.1 Charge Occupancy 349
14.2 Symmetries in a Matrix Representation 353
14.3 Constructing a Baryon 355
14.4 Constructing a Meson 358
14.5 Lepton Structures 359
14.6 The Electroweak Interaction Mechanism 360
14.7 The Production of Leptons 363
14.8 Electroweak Mixing 366
14.9 *SU*(2) Transitions 367
14.10 The Higgs Coupling 369
14.11 The Mass Gap for Any Gauge Group 371

15. Grand Unification and Particle Masses 374
15.1 A Dirac Equation for Charge 375
15.2 *SU*(5) Symmetry 378
15.3 The Grand Unification Group Generators 379
15.4 The Dirac Algebra Operators and *SU*(5) Generators 382
15.5 Superspace and Higher Symmetries 386
15.6 Grand Unification and the Planck Mass 388
15.7 The Generation of Mass 394
15.8 The Higgs Model for Fermions 397
15.9 The Masses of Baryons and Bosons 398
15.10 The Masses of Fermions 405
15.11 The CKM Mixing 408
15.12 A Summary of the Mass Calculations 411

16. The Factor 2 and Duality 414
16.1 Duality and Physics 414
16.2 Kinematics and the Virial Theorem 415
16.3 Relativity 418
16.4 Spin and the Anomalous Magnetic Moment 422
16.5 The Linear Harmonic Oscillator 424
16.6 The Heisenberg Uncertainty Principle 426
16.7 Fermions and Bosons 426
16.8 Radiation Reaction 428
16.9 Supersymmetry and the Berry Phase 431
16.10 Physics and Duality 435
16.11 The Factor 2 and Electroweak Mixing 439
16.12 Alternative Dualities 440
16.13 Mathematical Doubling and the Self-Duality of the Dirac Nilpotent 442

17. Gravity and Inertia 444
17.1 The Continuity of Mass-Energy 445
17.2 The Speed of Gravity 448
17.3 What is General Relativity About? 449
17.4 General Relativity and Quantum Mechanics 452
17.5 The Schwarzschild Solution 455
17.6 Gravitational Redshift 457
17.7 The Gravitational Deflection of Electromagnetic Radiation 459
17.8 The Gravitational Time-Delay of Electromagnetic Radiation 461
17.9 Perihelion and Periastron Precession 463
17.10 The Inertial Correction 467
17.11 The Aberration of Space 470
17.12 Gravomagnetic Effects 472
17.13 A Linear Interpretation of the Gravitational Field 477

18. Dimensionality, Strings and Quantum Gravity 484
- 18.1 Discreteness and Dimensionality 484
- 18.2 Dimensionality and Chirality 486
- 18.3 '4-Dimensional' Space-Time 488
- 18.4 Proper Time and Causality 489
- 18.5 The Klein Bottle Analogy 490
- 18.6 A String Theory Without Strings 491
- 18.7 Twistor Representations 493
- 18.8 Quantum Gravitational Inertia 495
- 18.9 Calculation of Quantized Gravitational Inertia 499

19. Nature's Code 502
- 19.1 The Dirac Nilpotent as the Origin of Symmetry-Breaking 502
- 19.2 The Significance of the Pseudoscalar Term 505
- 19.3 Spin and Aggregation of Matter 509
- 19.4 Self-Organization of Matter 511
- 19.5 The Filled Weak Vacuum and the One-Handed Bias in Nature 513
- 19.6 The Idea of 3-Dimensionality 514
- 19.7 Application to Biology: DNA and RNA Structure 515
- 19.8 Transcription 516
- 19.9 Translation and Triplet Codons 517
- 19.10 Triplet Codons and the Dirac Algebra 518
- 19.11 The Five Platonic Solids 527
- 19.12 Fibonacci Numbers 529
- 19.13 Application of Geometrical Structures to DNA and Genetic Coding 532
- 19.14 Pentagonal Symmetry Within DNA 539
- 19.15 The Cube and the Harmonic Oscillator 550
- 19.16 The Rewrite Process as Nature's Code 552
- 19.17 The Unification of Physics and Biology 555

20. Nature's Rules 556
- 20.1 A Semantic Model of Computation 557
- 20.2 Scientific Perspectives on Computation 559
- 20.3 The Nilpotent Structure of the Universal Grammar 562
- 20.4 General Relativity and NQM Semantic Description 564
- 20.5 Analysis over the Surreals 565
- 20.6 The Heaviside Operator 567
- 20.7 Wheeler's Meaning Circuit 569
- 20.8 Anticipatory Computation and Other Ideas Supporting the NUCRS 571
- 20.9 A Boundary Condition and the Holographic Principle 573
- 20.10 Quantum Holography 574
- 20.11 The Bra and Ket Notation 577
- 20.12 The Universe as a Quantum Thermodynamic Engine 578
- 20.13 The Riemann Zeta Function 581

20.14 Galactic Structure 583
20.15 Quantum Thermodynamics and Evolution 585
20.16 DNA as a Rewrite System 590
20.17 Brains as Quantum Carnot Engines 595
20.18 Language and Universal Grammar 596
20.19 Nature's Process 599

21. Infinity 600
21.1 A Version of Mach's Principle 601
21.2 Gravity and Inertia 606
21.3 Cosmology and Physics 608
21.4 Information Loss and Radiation 610
21.5 A Numerological Coincidence? 611
21.6 Vacuum Acceleration and Radiation 613
21.7 The Concept of Creation 619

Appendix A Summary and Predictions 623
A.1 Summary of the Main Argument 623
A.2 Predictions 629

Appendix B The Infinite Square Roots of –1 633

References 639

Index 669

Chapter 1

Zero

The bulk of this chapter (**1.2** – **1.5**) is the result of a collaboration with **BERNARD DIAZ**, and incorporates the results of a number of co-authored papers. The aim of the chapter is to show that mathematics, and in particular the mathematics needed for physics, can be generated from the idea of zero. Mathematics is shown to be derivable from a zero totality, without assuming numbers, by structuring it by analogy with a computer rewrite system. Here, we assume that any deviation from the zero state (a 'creation') forces a continued attempt to recover the original zero totality. The fundamental rule has a zero-totality alphabet or set of states, which generates itself by acting on any subalphabet, necessarily generating a new zero-totality alphabet by acting on itself. Then, simply by assuming a totally undefined nonzero state, we produce a conjugate (or 'negative') and a series of complex forms (not yet associated with numbers), which act on each other to produce further complex forms or their conjugates. We can identify operations which we describe as commutative, which can be continued indefinitely, and others which are anticommutative, which are restricted to the two forms and their operation on each other. The requirement of uniqueness forces us to default on the second, with the consequence that we generate an ordinal series of discrete, dimensional systems, which we can use as a number series, and so introduce this concept for the first time. The link between divisibility or discreteness and dimensionality, is now established as the fundamental process by which the idea of discreteness, and, even that of number, actually enters into mathematics. We can now generate an infinite series by the repeated application of three processes, described as conjugation, complexification and dimensionalization, with important consequences for physics. The overall strategy of the chapter is to derive the general process and its important application to mathematics in sections 1.2 to 1.4, before developing the more specifically technical details of the computer rewrite interpretation in section 1.5.

1.1 An Origin for Everything

What is the most fundamental possible idea? What is the simplest thing we can possibly start with to obtain the universe as we know it today? Is there some point at which we could say that the origin of everything had been established because it could not possibly have been imagined otherwise? If we believe that there really are answers to these questions, then we must imagine that they must come in a form so obvious that no special argument will be needed to present them. It is almost inconceivable, also, that such answers would not also lead to the long-desired 'unified theory of physics'.

Physics has long been recognised as the most fundamental physical science, perhaps even the most fundamental possible structure for the whole of human knowledge. It certainly gives the appearance of being the most fundamental way possible of accessing knowledge about what we call the 'natural world'. Our current understanding of physics has been accumulated as the result of a succession of inductive inferences derived from information which is ultimately empirical in origin. We have learnt from experience that certain patterns of thought produce successful outcomes, that relatively simple generalisations appear to have a surprisingly wide currency, and that, as Galileo expressed it: 'The Book of Nature is written in the language of mathematics.' The idea that physics may be, in some sense, 'unified' in a way which has not so far been discovered has a sound basis in the way that the subject has developed over the last five hundred years.

A 'unified' theory, however, means something very particular. It means a theory which derives all its results from a single source, not simply one in which we put all our known results into a single comprehensive package. This is a crucial distinction which is not always made in discussions of the subject; but, in principle, a truly unified theory cannot come from an act of unification; that is, we cannot, for example, create a truly unified theory simply by combining quantum mechanics and general relativity in a new mathematical superstructure. Such attempts have always failed in the past, and will continue to do so in the future, because the concept of unification as combination is invalid. Unification is really about finding *descent from a common origin.* Creating a sophisticated mathematical superstructure will not provide answers to the fundamental questions that we would expect from a truly unified theory, a theory of 'everything'.

Many physicists believe that we can assume concepts such as space, time and matter to be fundamental and even inexplicable. Many also believe that we must assume that the fundamental ideas will be expressed in a mathematical language that already exists independently of the physical principles we are expressing. Others again are prepared to believe that the unifying mathematical structure, whatever it is, will be a sophisticated one which only a few will be able to truly understand. *I do not believe in any of these things.* If physics is to prove itself the most fundamental possible way of understanding the 'natural world', then it must *explain* space, time and matter, as well as use them, and it must *generate* the mathematical structures it uses; it must also show how all things that are sophisticated arise from things that are much more simple. In principle, all possible complications must be removed from the ultimate starting point. It has to be intrinsically simple and absolutely single.

So, how do we construct such a theory? What would be left after we have removed all the complications? What could be our single and simple starting point? There is only one possible answer: zero, absolutely nothing, or perhaps 'no thing'. This is the only thing in all our experience which has no structure and includes no complication. So, zero must be our starting point. It must also be where we finish, for nothing, as we all know, comes from nothing – *nihil ex nihilo fit*. The idea that the universe may be a totality of 'nothing', in some sense, has been discussed for quite a number of years, especially in the context of the 'big bang' theory of creation, where the total energies of matter (positive) and gravitation (negative) could cancel each other out. This is the kind of reasoning behind the statement of the science-writer, Peter Atkins, that 'the seemingly something is elegantly reorganized nothing, and … the net content of the universe is … nothing'.[1] However, to find the most fundamental possible theory, we must go even further, and state that the universe and everything we can possible experience or conceive is also *conceptually* nothing. No other position is extreme or uncompromising enough to be able to explain *everything*.

1.2 The Genesis of Number

As we have said, many scientists today believe that the concept of nothing has a fundamental role to play in an understanding of the universe and its contents. It is, for example, often stated that the universe itself could have emerged as a quantum fluctuation from an essentially zero state, and that the total zero energy is maintained by the total positive mass energy being equated numerically to the negative gravitational energy generated by this mass through a special application of Mach's principle (the idea that the inertia or mass of physical objects is determined by their interactions with the rest of the matter in the universe). However, the concept of nothing may be even more powerful than this, for it may be that the nothingness applies also to the entire conceptual scheme of which matter and the universe are merely components. Nothing is unique among conceptual ideas in being infinitely degenerate (infinitely capable of reinterpretation) and it may be that this infinite degeneracy is the key to an understanding of its special power.

However, since 'nothing comes from nothing', we are left with the question of how we preserve the total nothingness in the presence of the seemingly 'something' which we call 'the universe'. Both mathematics and physics suggest that the answer lies in the concept of *duality*, the idea that to every fundamental concept there exists some kind of partnership or relationship with another that is its 'dual'. As Nicholas Young has noted, 'the idea of duality pervades mathematics',[2] the pairing of positive and negative numbers being only the most obvious example. So we might well expect that the scheme required for preserving the conceptual nothingness of mathematical physics, is essentially a dualistic one. (This will be discussed in the next chapter.) But we still have to identify the origin of such processes, and it is here that we will need the concept of infinite degeneracy, for we have a logical analogy in the *rewrite* systems which are fundamental to computing.

Obviously, we cannot generate a fundamental theory by imagining in advance that computers actually exist or that the universe is structured like a computer, but the 'rewrite' concept is independent of any connection with software or hardware, and can be conceived as a purely abstract process. If we could develop a universal system which endlessly rewrote itself, *superveniently* (that is, without any sense of temporal progression), we would also have a system which was, by definition, infinitely degenerate; and, if at every rewriting, the dualistic principle (or, in the first instance, zero totality) applied, we would have a universal process relevant to mathematics, physics and all such attempts at a fundamental description of ultimate 'reality'.

In a recent paper,[3] Deutsch *et al* state that, 'Though the truths of logic and pure mathematics are objective and independent of any contingent facts or laws of nature, our *knowledge* of these truths depends entirely on our knowledge of the laws of physics.' According to these authors we have been forced by 'recent progress in the theory of computation', 'to abandon the classical view that computation, and hence mathematical proof, are purely logical notions independent of that of computation as a physical process'. Mathematical structures, however autonomous, 'are revealed to us only through the physical world'. We can go further and state that that mathematical structure which is most fundamental in understanding the physical world is also likely to be the structure which is most fundamental to understanding mathematics itself.

The application to physics is particularly important because it is a strong test of the worth of a fundamental idea. Mathematics can be structured on fundamental principles in a large variety of ways, but physics has to survive the test of observation and experiment under many different conditions. Physics thus appears to tell us not only what is necessary to a foundational structure for mathematics, but also what is the most efficient such structure. The mathematical foundations of physics may also be expected to provide a route to understanding the principles important in the foundations of quantum computing. So, how then do we generate mathematics, and, in particular, the mathematics especially relevant to physics, purely from the concept of zero?

Well, we might think that mathematics begins with the natural numbers or positive integers, 1, 2, 3, ..., as we are conditioned into believing that counting is an elementary process. So, it has become a standard procedure to derive mathematical structures from the natural numbers, and then progress by successively extending the set to incorporate negative, rational, algebraic, real, and complex numbers, before proceeding to higher algebraic structures involving, say, quaternions, vectors, Grassmann and Clifford algebras, Hilbert spaces, and even higher structures (including transfinite ones). However, to begin mathematics with the integers, though natural to our human perceptions, is to start from a position already beyond the beginning. The integers are loaded with a mass of assumptions about mathematics. They are not fundamentally simple but already contain packaged information about things beyond the integer series itself. This makes them a convenient codification of mathematics, but not a simplified starting-point. The number 1 is not the most obvious initial step from 0 because it contains, for example, the notion of discreteness, as well as ordinality (or ordering). Further, although there are many ways in which mathematics can be generated from foundational elements, physics seems to indicate that discreteness at the foundational level is always associated with dimensionality (see chapter 2), and hence is not a truly primary concept. In addition, there is no obvious route of progression from natural numbers to reals, which then appear as a nonlogical extension. It would seem to be more logical, in terms of rewrite procedures, to consider the real or noncountable 'numbers' as anterior to the integers, though it is even more logical to suppose that even real numbers are not the ultimate starting point.

Suppose, then, that we make no prior mathematical assumptions, in particular any conventional concept of mathematics / numbering. We assume instead that zero totality is the only possible state and that any deviation from this state, such as the assumption of a nonzero state, say R, generates an automatic mechanism for recovering it. Zero totality, however, is also infinitely degenerate; and to express this we propose that, while each state is a zero totality state, it is not a *unique* zero. Our rewriting procedure will then consist of defining a zero totality 'system', otherwise undefined, to contain nothing new within it (symbolized by $\rightarrow$) and a new zero totality outside it (symbolized by $\Rightarrow$). For convenience, though without any assumption as to form, we will describe the totality, as so far understood, as an 'alphabet'. Then an examination – here represented by a 'concatenation' or placing together, with no algebraic significance – of the alphabet with respect to anything other than itself (a 'subalphabet') will always yield the alphabet, because anything less than the totality will generate nothing new. On the other hand, an examination of the alphabet with respect to itself will necessarily generate a new zero totality alphabet, because a finite alphabet cannot represent a unique zero state. To summarise:

(subalphabet) (alphabet) $\rightarrow$ (alphabet) *i.e. there is nothing new*
(alphabet) (alphabet) $\Rightarrow$ (new alphabet) *i.e. the zero totality is not unique*

These are the only conditions. However, as we shall see, the maintenance of a total zero state will require duality at all times. All terms are necessarily paired terms, which are, in principle, indistinguishable individually. In addition, the nature of the new alphabet produced by $\Rightarrow$ will always be determined by the need to satisfy $\rightarrow$ in all possible cases. That is, we can't work out what a new alphabet will look like till we have worked out all the ways in which concatenation with its subalphabets will yield only itself.

Suppose, then, that we assume a nonzero R. This will not only be the first subalphabet, but it will also need the zero-creating conjugate, say R^*, to make it a conjugated (zero) alphabet. That is:

$$(R)\,(R) \Rightarrow (R, R^*)$$

We now apply $\rightarrow$ to this alphabet to show that, within itself, it produces nothing new.

$$(R)\,(R, R^*) \rightarrow (R, R^*) \;\; ; \;\; (R^*)\,(R, R^*) \rightarrow (R^*, R) \equiv (R, R^*)$$

We note here that no concept of 'ordering' is required by concatenation. So, all possible concatenations have produced only the alphabet itself. From these rules it is easy to show that

$$(R)\,(R) \rightarrow (R)\,;\,(R^*)\,(R) \rightarrow (R^*)\,;\,(R)\,(R^*) \rightarrow (R^*)\,;\;(R^*)\,(R^*) \rightarrow (R)$$

But, of course, the zero-totality alphabet (R, R^*) cannot be unique, and concatenation with itself must produce a new conjugated alphabet, whose additional terms must be chosen in such a way that the subalphabets yield nothing new. So we try

$$(R, R^*)\,(R, R^*) \Rightarrow (R, R^*, A, A^*)$$

and find that applying $\rightarrow$ to this new alphabet means that

$$\begin{aligned}
(R)\,(R, R^*, A, A^*) &\rightarrow (R, R^*, A, A^*) \equiv (R, R^*, A, A^*)\\
(R^*)\,(R, R^*, A, A^*) &\rightarrow (R^*, R, A^*, A) \equiv (R, R^*, A, A^*)\\
(A)\,(R, R^*, A, A^*) &\rightarrow (A, A^*, R^*, R) \equiv (R, R^*, A, A)\\
(A^*)\,(R, R^*, A, A^*) &\rightarrow (A^*, A, R, R^*) \equiv (R, R^*, A, A^*)
\end{aligned}$$

It will be apparent that, to maintain an unchanged alphabet (and to specify that R, R^*, A, A^* remain distinct), we have to arrange that each term cycles into another. From these rules, it would appear that we may also derive

$$(R)\,(A) \rightarrow (A)\,;\,(R)\,(A^*) \rightarrow (A^*)\,;\,(R^*)\,(A) \rightarrow (A^*)\,;\,(R^*)\,(A^*) \rightarrow (A);$$

$$(A)\,(A) \rightarrow (R^*)\,;\,(A^*)\,(A^*) \rightarrow (R^*)\,;\,(A)\,(A^*) \rightarrow (R).$$

Of course, absolute duality of terms like A and A^* would seem to imply that wc could write down expressions such as

$$(A)\,(A) \rightarrow (R)\,;\,(A^*)\,(A^*) \rightarrow (R^*)\,;\,(A)\,(A^*) \rightarrow (R^*)$$

instead of the ones chosen. However, since R and A would now be indistinguishable, this would, in effect, be equivalent to not extending the alphabet.

At the next stage, the process of ensuring that the new alphabet only produces itself when concatenated with its subalphabets ($\rightarrow$) requires that we now introduce concatenated *terms*, such as AB, AB^* into the alphabet. So:

$$(R, R^*, A, A^*)\,(R, R^*, A, A^*) \Rightarrow (R, R^*, A, A^*, B, B^*, AB, AB^*).$$

As in all previous cases, if we successively perform the $\rightarrow$ operation with (R^*), (A), (A^*), (B), (B^*), (AB), (AB^*), or any combination of these that is less than the full alphabet, the answer will be unchanged.

$$(R)\,(R, R^*, A, A^*, B, B^*, AB, AB^*) \rightarrow (R, R^*, A, A^*, B, B^*, AB, AB^*)$$

$$(R^*)\,(R, R^*, A, A^*, B, B^*, AB, AB^*) \rightarrow (R^*, R, A^*, A, B^*, B, AB^*, AB)$$

$$(A)\ (R, R^*, A, A^*, B, B^*, AB, AB^*) \rightarrow (A, A^*, R^*, R, AB, AB^*, B, B^*)$$
$$(A^*)\ (R, R^*, A, A^*, B, B^*, AB, AB^*) \rightarrow (A^*, A, R, R^*, AB^*, AB, B^*, B)$$
$$(B)\ (R, R^*, A, A^*, B, B^*, AB, AB^*) \rightarrow (B, B^*, AB, AB^*, R^*, R, A, A^*)$$
$$(B^*)\ (R, R^*, A, A^*, B, B^*, AB, AB^*) \rightarrow (B^*, B, AB^*, AB, R, R^*, A^*, A)$$
$$(AB)\ (R, R^*, A, A^*, B, B^*, AB, AB^*) \rightarrow (AB, AB^*, B, B^*, A, A^*, R^*, R)$$
$$(AB^*)\ (R, R^*, A, A^*, B, B^*, AB, AB^*) \rightarrow (AB^*, AB, B^*, B, A^*, A, R, R^*)$$

This is, of course, identical in our formalism to:

$$(R)\ (R, R^*, A, A^*, B, B^*, AB, AB^*) \rightarrow (R, R^*, A, A^*, B, B^*, AB, AB^*)$$
$$(R^*)\ (R, R^*, A, A^*, B, B^*, AB, AB^*) \rightarrow (R, R^*, A, A^*, B, B^*, AB, AB^*)$$
$$(A)\ (R, R^*, A, A^*, B, B^*, AB, AB^*) \rightarrow (R, R^*, A, A^*, B, B^*, AB, AB^*)$$
$$(A^*)\ (R, R^*, A, A^*, B, B^*, AB, AB^*) \rightarrow (R, R^*, A, A^*, B, B^*, AB, AB^*)$$
$$(B)\ (R, R^*, A, A^*, B, B^*, AB, AB^*) \rightarrow (R, R^*, A, A^*, B, B^*, AB, AB^*)$$
$$(B^*)\ (R, R^*, A, A^*, B, B^*, AB, AB^*) \rightarrow (R, R^*, A, A^*, B, B^*, AB, AB^*)$$
$$(AB)\ (R, R^*, A, A^*, B, B^*, AB, AB^*) \rightarrow (R, R^*, A, A^*, B, B^*, AB, AB^*)$$
$$(AB^*)\ (R, R^*, A, A^*, B, B^*, AB, AB^*) \rightarrow (R, R^*, A, A^*, B, B^*, AB, AB^*)$$

However, with both A and B in the alphabet, we start generating apparent ambiguities, with regard to AB and AB^*. Thus, there would seem to be no absolute way of determining the last two terms in the last two expressions. If the concatenated (AB), (AB^*), etc., are to be considered as valid terms, then something must disappear when we take $(AB)\ (AB)$, under $\rightarrow$, as the only new terms allowed are those within the alphabet. In fact, we find that we have two options:

$$(AB)\ (AB) \quad \rightarrow \quad (R) \qquad (\textit{commutative})$$
$$(AB)\ (AB) \quad \rightarrow \quad (R^*) \qquad (\textit{anticommutative})$$

with conjugates, of course, producing the appropriately conjugated results.

Both options would appear to be possible, but their effects will be very different. The anticommutative option, will not be repeatable when the alphabet is extended to incorporate new terms, such as (C), (D), etc., whereas the commutative option, can be repeated indefinitely. The anticommutative option effectively produces a closed 'cycle' with components (A, B, AB) and their conjugates, which excludes any further $C, D \ldots$-type term of anticommuting with them. All other such terms will commute.

But there is another fundamental difference between the two options. If we choose the commutative option, we have a series of terms, such as *A*, *B*, *C* ..., which are completely indistinguishable. In other words, we wouldn't know whether or not we had created something new. We would not be extending the alphabet. If, however, we choose the *anticommutative* option (as we are, in fact, obliged to do if we want to generate something new), each term will always be distinguishable from all the others, because its anticommutative partner must be unique. That is, if *B* is the anticommutative partner to *A*, then *C*, *D*, ... etc., are not. So we can always identify a term by its anticommutative partner. Uniqueness is only possible because partnership is involved.

In fact, as we will find, the particular choice that makes both mathematics and physics possible as we know them, is also the most 'efficient', in requiring the minimum of choice or decision-making. That is, we automatically default at the anticommutative option whenever this is possible. The result is that the alphabets generated by $\Rightarrow$ incorporate a regular series of identically structured closed cycles, each of which commutes with all others. The structure of this is identical to that which is familiar to us as the infinite series of finite (binary) integers of conventional mathematics. The closed cycles form an infinite ordinal sequence, establishing for the first time the meaning of both the number 1 and the binary symbol 1 as it appears in classical Boolean logic. The logical 1 becomes potentially a conjugation state of 0, that is, a subset alphabet defined within the system; and the alphabets structure themselves as an infinite series of binary digits.

The necessity of introducing anticommutativity has, in fact, simultaneously created the concepts of *discreteness* and *dimensionality* (specifically 3-dimensionality). As we will see from physics, there is a key connection between 3-dimensionality and discreteness in physical systems: one cannot exist without the other. Here, 3-dimensionality, i.e. anticommutativity, becomes the ultimate source of discreteness in a zero totality universe, because an anticommutative system has the property of *closure*, whereas a commutative system remains open to infinity. Geometry is logically prior to algebra and arithmetic.

1.3 The Genesis of Algebra

The process we have generated so far is based on the idea that zero totality is a *universal* requirement. It is not intrinsically algebraic, or even mathematical. It has not made the prior assumption that mathematical structures actually exist, and, in principle, it is independent of any concept of observers or physical entities. In addition, although we have used a kind of 'symbolic' representation for *R*, *A*, *B*, *C*, etc., these terms are not symbolic in the conventional sense, and do not, *a priori*, represent algebraic or any other entities.

However, the process has now reached a stage where we have developed an equivalent of numbering, ordinality or counting via a system of integers; and the concept of discreteness has emerged along with dimensionality. To maintain *R*, *A*, *B*, *C*, etc., as non-integral, non-countable and non-discrete, and yet establish their relationship with respect to the new number system which they generate, we describe them as real, non-denumerable and continuous, where these terms imply only the absence of those properties which we associate with the integers. In this sense *R*, which is simply a category of things which are unspecified and undefined, and not even definable as a set, can, by using the new concept of numbering, be re-structured as the set of reals ($\Re$), defined by the Cantor continuum (with cardinality $\aleph_1$, as opposed to the cardinality $\aleph_0$ of the integers). If we regard 1 as a 'unit' within the set, then we can proceed to find in $\Re$ all the number systems that can be constructed as countable with a 1:1 relation with the integers: rational numbers, algebraic numbers, etc., without completely specifying the set, and all of these, of course, become subject to standard arithmetical operations.

At the same time, we can see that the conjugate set R^* can be interpreted, in this re-structuring, as the set of negative reals; while *A* and A^* become the continuous sets based on the complex units *i* and $-i$, which each square to -1. *B* and B^*, with the necessary introduction of anticommutativity and dimensionality, convert the units of *A*, A^*, *B*, B^* into those of the *quaternions*, $\mathbf{i}$, $-\mathbf{i}$, $\mathbf{j}$, $-\mathbf{j}$, with the units of *AB* and AB^* becoming equivalent to $\mathbf{ij} = \mathbf{k}$ and $-\mathbf{ij} = -\mathbf{k}$. Further extensions to the alphabet will then be accomplished via an infinite series of new quaternion systems; and a new type of dimensional or constructible 'real' numbers will emerge to represent terms such as *AC*, *BC*, *ABC* (with countable units squaring to 1), which will be equivalent to those of Robinson's non-standard analysis or Skolem's non-standard arithmetic. It will also be possible to develop new types of mathematics by combining different aspects of the overall structure (now definable as a *rewrite* system) in novel ways, as has been the usual procedure in mathematics, and we may conjecture that all branches of mathematics that can conceivably exist may be generated by procedures internal to this structure. In particular, we may note that, once we have the concept of number, and the additional principle of complexity, we may structure John Conway's development of the surreal number system in terms of a rewrite process, with a defined alphabet.

Alternative systems of units will certainly be possible, where they can related by a 1:1 mapping to the overall structure, for example the negative unit (or iso-Minkowskian) system developed by Santilli, with its powerful applications in both physics and pure mathematics.[4] Again, since the whole generating process is defined through an initial lack of specificity, with constraints only applied when zero totality or infinite degeneracy require them, then there is no need to even consider such concepts as associativity in the context of defining a universal rewrite system; however, once numbering is introduced, then we are certainly free to introduce alternative relations between defined and undefined quantities, with particular specificities now introduced – for example, the associativity which we break in defining octonions.

We can, also use the integral ordinal sequence established with dimensionalization to restructure the subset alphabets as a series of finite groups, the order of which doubles at every stage, producing an ordinal binary enumeration. In effect, we use the process to create the concept of finite group, though it is significant that the group concept generally requires the avoidance of specific limitations on the overall system. With the group structure, the 'multiplication' and 'squaring' of elements, in addition to identity and inversion, become operations which are fundamental to the principles of duality and zero totality.

In terms of 'units' (once we have established their existence), we can express the developing structures in the form:

order 2 ± 1

order 4 $\pm 1, \pm \boldsymbol{i}_1$

order 8 $\pm 1, \pm \boldsymbol{i}_1, \pm \boldsymbol{j}_1, \pm \boldsymbol{i}_1\boldsymbol{j}_1$

order 16 $\pm 1, \pm \boldsymbol{i}_1, \pm \boldsymbol{j}_1, \pm \boldsymbol{i}_1\boldsymbol{j}_1, \pm \boldsymbol{i}_2, \pm \boldsymbol{i}_2\boldsymbol{i}_1, \pm \boldsymbol{i}_2\boldsymbol{j}_1, \pm \boldsymbol{i}_2\boldsymbol{i}_1\boldsymbol{j}_1$

order 32 $\pm 1, \pm \boldsymbol{i}_1, \pm \boldsymbol{j}_1, \pm \boldsymbol{i}_1\boldsymbol{j}_1, \pm \boldsymbol{i}_2, \pm \boldsymbol{i}_2\boldsymbol{i}_1, \pm \boldsymbol{i}_2\boldsymbol{j}_1, \pm \boldsymbol{i}_2\boldsymbol{i}_1\boldsymbol{j}_1,$
$\pm \boldsymbol{j}_2, \pm \boldsymbol{j}_2\boldsymbol{i}_1, \pm \boldsymbol{j}_2\boldsymbol{j}_1, \pm \boldsymbol{j}_2\boldsymbol{i}_1\boldsymbol{j}_1, \pm \boldsymbol{j}_2\boldsymbol{i}_2, \pm \boldsymbol{j}_2\boldsymbol{i}_2\boldsymbol{i}_1, \pm \boldsymbol{j}_2\boldsymbol{i}_2\boldsymbol{j}_1, \pm \boldsymbol{j}_2\boldsymbol{i}_2\boldsymbol{i}_1\boldsymbol{j}_1$

order 64 $\pm 1, \pm \boldsymbol{i}_1, \pm \boldsymbol{j}_1, \pm \boldsymbol{i}_1\boldsymbol{j}_1, \pm \boldsymbol{i}_2\boldsymbol{i}_1, \pm \boldsymbol{i}_2\boldsymbol{i}_1, \pm \boldsymbol{i}_2\boldsymbol{j}_1, \pm \boldsymbol{i}_2\boldsymbol{i}_1\boldsymbol{j}_1,$
$\pm \boldsymbol{j}_2, \pm \boldsymbol{j}_2\boldsymbol{i}_1, \pm \boldsymbol{j}_2\boldsymbol{j}_1, \pm \boldsymbol{j}_2\boldsymbol{i}_1\boldsymbol{j}_1, \pm \boldsymbol{j}_2\boldsymbol{i}_2, \pm \boldsymbol{j}_2\boldsymbol{i}_2\boldsymbol{i}_1, \pm \boldsymbol{j}_2\boldsymbol{i}_2\boldsymbol{j}_1, \pm \boldsymbol{j}_2\boldsymbol{i}_2\boldsymbol{i}_1\boldsymbol{j}_1$
$\pm \boldsymbol{i}_3, \pm \boldsymbol{i}_3\boldsymbol{i}_1, \pm \boldsymbol{i}_3\boldsymbol{j}_1, \pm \boldsymbol{i}_3\boldsymbol{i}_1\boldsymbol{j}_1, \pm \boldsymbol{i}_3\boldsymbol{i}_2, \pm \boldsymbol{i}_3\boldsymbol{i}_2\boldsymbol{i}_1, \pm \boldsymbol{i}_3\boldsymbol{i}_2\boldsymbol{j}_1, \pm \boldsymbol{i}_3\boldsymbol{i}_2\boldsymbol{i}_1\boldsymbol{j}_1,$
$\pm \boldsymbol{i}_3\boldsymbol{j}_2, \pm \boldsymbol{i}_3\boldsymbol{j}_2\boldsymbol{i}_1, \pm \boldsymbol{i}_3\boldsymbol{j}_2\boldsymbol{j}_1, \pm \boldsymbol{i}_3\boldsymbol{j}_2\boldsymbol{i}_1\boldsymbol{j}_1, \pm \boldsymbol{i}_3\boldsymbol{j}_2\boldsymbol{i}_2, \pm \boldsymbol{i}_3\boldsymbol{j}_2\boldsymbol{i}_2\boldsymbol{i}_1,$
$\pm \boldsymbol{i}_3\boldsymbol{j}_2\boldsymbol{i}_2\boldsymbol{j}_1, \pm \boldsymbol{i}_3\boldsymbol{j}_2\boldsymbol{i}_2\boldsymbol{i}_1\boldsymbol{j}_1$

We begin with a unit integer (1) and then imagine finding an infinite series of 'duals' to this unit. We suppose that the dualling process must be carried out with respect to all previous duals, so that the entire set of characters generated becomes the new 'unit', and ensure that the total result is zero at every stage. The first dual then becomes –1, generating a new 'unit' consisting of (1, –1). Following this, we have a series of terms to which we can give symbols such as $\boldsymbol{i}_1$, $\boldsymbol{j}_1$, etc. Usually, of course, $\boldsymbol{i}_1\boldsymbol{j}_1$ would be written $\boldsymbol{k}_1$, but no new independent unit is created by this notation. An alternative expression could be in terms of multiplying factors:

order 2	$(1, -1)$
order 4	$(1, -1) \times (1, \boldsymbol{i}_1)$
order 8	$(1, -1) \times (1, \boldsymbol{i}_1) \times (1, \boldsymbol{j}_1)$
order 16	$(1, -1) \times (1, \boldsymbol{i}_1) \times (1, \boldsymbol{j}_1) \times (1, \boldsymbol{i}_2)$
order 32	$(1, -1) \times (1, \boldsymbol{i}_1) \times (1, \boldsymbol{j}_1) \times (1, \boldsymbol{i}_2) \times (1, \boldsymbol{j}_2)$
order 64	$(1, -1) \times (1, \boldsymbol{i}_1) \times (1, \boldsymbol{j}_1) \times (1, \boldsymbol{i}_2) \times (1, \boldsymbol{j}_2) \times (1, \boldsymbol{i}_3)$,

with the series repeating for an endless succession of indistinguishable $\boldsymbol{i}_n$ and $\boldsymbol{j}_n$ values. To define the character sets as true 'units', we require that the product of any unit with itself, or with any subunit, generates only the unit. So, for example, at order 8, we will have the products:

$$(\pm 1) \times (\pm 1, \pm \boldsymbol{i}_1, \pm \boldsymbol{j}_1, \pm \boldsymbol{i}_1\boldsymbol{j}_1) = (\pm 1, \pm \boldsymbol{i}_1, \pm \boldsymbol{j}_1, \pm \boldsymbol{i}_1\boldsymbol{j}_1)$$
$$(\pm \boldsymbol{i}_1) \times (\pm 1, \pm \boldsymbol{i}_1, \pm \boldsymbol{j}_1, \pm \boldsymbol{i}_1\boldsymbol{j}_1) = (\pm 1, \pm \boldsymbol{i}_1, \pm \boldsymbol{j}_1, \pm \boldsymbol{i}_1\boldsymbol{j}_1)$$
$$(\pm \boldsymbol{j}_1) \times (\pm 1, \pm \boldsymbol{i}_1, \pm \boldsymbol{j}_1, \pm \boldsymbol{i}_1\boldsymbol{j}_1) = (\pm 1, \pm \boldsymbol{i}_1, \pm \boldsymbol{j}_1, \pm \boldsymbol{i}_1\boldsymbol{j}_1)$$
$$(\pm \boldsymbol{i}_1\boldsymbol{j}_1) \times (\pm 1, \pm \boldsymbol{i}_1, \pm \boldsymbol{j}_1, \pm \boldsymbol{i}_1\boldsymbol{j}_1) = (\pm 1, \pm \boldsymbol{i}_1, \pm \boldsymbol{j}_1, \pm \boldsymbol{i}_1\boldsymbol{j}_1)$$
$$(\pm 1, \pm \boldsymbol{i}_1) \times (\pm 1, \pm \boldsymbol{i}_1, \pm \boldsymbol{j}_1, \pm \boldsymbol{i}_1\boldsymbol{j}_1) = (\pm 1, \pm \boldsymbol{i}_1, \pm \boldsymbol{j}_1, \pm \boldsymbol{i}_1\boldsymbol{j}_1)$$
$$(\pm 1, \pm \boldsymbol{j}_1) \times (\pm 1, \pm \boldsymbol{i}_1, \pm \boldsymbol{j}_1, \pm \boldsymbol{i}_1\boldsymbol{j}_1) = (\pm 1, \pm \boldsymbol{i}_1, \pm \boldsymbol{j}_1, \pm \boldsymbol{i}_1\boldsymbol{j}_1)$$
$$(\pm 1, \pm \boldsymbol{i}_1, \pm \boldsymbol{j}_1) \times (\pm 1, \pm \boldsymbol{i}_1, \pm \boldsymbol{j}_1, \pm \boldsymbol{i}_1\boldsymbol{j}_1) = (\pm 1, \pm \boldsymbol{i}_1, \pm \boldsymbol{j}_1, \pm \boldsymbol{i}_1\boldsymbol{j}_1)$$
$$(\pm 1, \pm \boldsymbol{i}_1, \pm \boldsymbol{j}_1, \pm \boldsymbol{i}_1\boldsymbol{j}_1) \times (\pm 1, \pm \boldsymbol{i}_1, \pm \boldsymbol{j}_1, \pm \boldsymbol{i}_1\boldsymbol{j}_1) = (\pm 1, \pm \boldsymbol{i}_1, \pm \boldsymbol{j}_1, \pm \boldsymbol{i}_1\boldsymbol{j}_1), \text{ etc.}$$

For this to be always true, the terms $\boldsymbol{i}_1, \boldsymbol{j}_1, \boldsymbol{i}_2, \boldsymbol{j}_2, \boldsymbol{i}_3, \boldsymbol{j}_3$, etc. are required to have the properties of imaginary units, or square roots of –1, while the products, such as $\boldsymbol{i}_1\boldsymbol{j}_1$, must be imaginary or real units, that is, square roots of either –1 or +1. The two possibilities lead to entirely different consequences, for we can generate an unlimited number of complex products which are square roots of 1, but, for any complex number, such as $\boldsymbol{i}_1$, there is *only a single complex product* of the form $\boldsymbol{i}_1\boldsymbol{j}_1$, which is itself complex. So, if $\boldsymbol{i}_1\boldsymbol{j}_1$ is complex, then $\boldsymbol{i}_1$, $\boldsymbol{j}_1$, and $\boldsymbol{i}_1\boldsymbol{j}_1$ form a *closed system* – equivalent to the cyclic quaternion system of complex numbers,

$\boldsymbol{i}, \boldsymbol{j}, \boldsymbol{k}$. The system is closed with the dimensionality fixed at 3, and no other option is available. As we have seen, the choice is not arbitrary within a universal system; to create something new, we are obliged to choose the first option as default, generating an infinite number of identically structured closed systems.

Of course, the potentially infinite sequence of $\boldsymbol{i}_n$ values, with commutativity between $\boldsymbol{i}_m$ and $\boldsymbol{i}_n$ or $\boldsymbol{j}_n$ $(m \neq n)$, creates the possibility of an infinite-dimensional vector-like algebra, while the anticommutativity between $\boldsymbol{i}_n$ and $\boldsymbol{j}_n$ ensures the finite- and, specifically, three-dimensionality of each of the quaternion systems. The commutativity of $\boldsymbol{i}_m$ and $\boldsymbol{i}_n$ is equivalent to defining $(\boldsymbol{i}_m\boldsymbol{i}_n)^2$ as 1, while the anticommutativity of $\boldsymbol{i}_n$ and $\boldsymbol{j}_n$ defines $(\boldsymbol{i}_n\boldsymbol{j}_n)^2$ as the conjugate, or -1. If $\boldsymbol{i}_1$ forms a quaternion system with $\boldsymbol{j}_1$, and the complex product $\boldsymbol{i}_1\boldsymbol{j}_1$, then no product of $\boldsymbol{i}_1$ with any other complex number of the form $\boldsymbol{i}_1, \boldsymbol{i}_2, \boldsymbol{i}_3, \boldsymbol{i}_4, \ldots$ or $\boldsymbol{j}_2, \boldsymbol{j}_3, \boldsymbol{j}_4, \ldots$ will itself be complex. If we take $\boldsymbol{i}_1, \boldsymbol{j}_1$, and $\boldsymbol{i}_1\boldsymbol{j}_1$ as a quaternion system $(\boldsymbol{i}, \boldsymbol{j}, \boldsymbol{k})$, then any further complexification, to produce, say, $\boldsymbol{i}_2\boldsymbol{i}_1$, $\boldsymbol{i}_2\boldsymbol{j}_1$, and $\boldsymbol{i}_2\boldsymbol{i}_1\boldsymbol{j}_1$, will produce a system equivalent to the multivariate vectors, complexified quaternions, or Pauli matrices $(\mathbf{i}, \mathbf{j}, \mathbf{k})$, which square to positive scalar units. (The mathematical argument is detailed in the appendix to the chapter.)

As a result of this, we can define two separate processes of dualling through complex numbers: the ordinary process of complexification (i.e. multiplying by a complex number of the form $A = \boldsymbol{i}$), which can be continued to infinity (here symbolized by using terms of the form $\boldsymbol{i}_n$); and the restricted process of dimensionalization, which introduces a complementary complex factor of the form $B = \boldsymbol{j}$ (here symbolized by terms of the form $\boldsymbol{j}_n$ for each $\boldsymbol{i}_n$), which applies separately, and uniquely, to every complex operator, converting the $\boldsymbol{i}_n$ into an element of a quaternion set. The processes of *complexification* and *dimensionalization*, with their respective open and closed algebras, become simply alternative forms of duality, along with *conjugation*, or the introduction of alternative signs, + and –. It is notable that there is no such thing, in principle, as a pure complex number, only an incomplete representation of a quaternion set.

While further applications of conjugation have no effect on the structure of the group elements, repeated applications of complexification and dimensionalization take the sequence through the infinite series of quaternionic structures. Since the repeated application of conjugation makes no change to the structure, the series follows the pattern:

order 2	conjugation	$\times$	$(1, -1)$
order 4	complexification	$\times$	$(1, \boldsymbol{i}_1)$
order 8	dimensionalization	$\times$	$(1, \boldsymbol{j}_1)$

order 16	complexification	×	$(1, \boldsymbol{i}_2)$
order 32	dimensionalization	×	$(1, \boldsymbol{j}_2)$
order 64	complexification	×	$(1, \boldsymbol{i}_3)$

We will show later that the three processes, taken together, and repeated indefinitely, provide the entire structure of mathematical duality required for physical application.

There is also a further sense in which such separate processes can also be seen as aspects of a single process of duality. This is evident in the fact that we can switch between the processes in our *representations* of them. For example, we can use the discrete process of dimensionalization to represent the alternative (and intrinsically non-discrete) processes of conjugation and complexification. We can even represent them on the same 2-dimensional (Argand) diagram, where, the x-y plot uses the space to the right and left of the origin to represent + and – values, and the combination of x and y coordinates to represent complex numbers. (It is significant here that we can *only* represent dual processes through discreteness, though this does not mean that the processes themselves must be discrete.) Discreteness also enables us to find more convenient ways of describing the emerging algebra.

1.4 Group Representations

The structure, as so far defined, is closely related to that of Clifford or geometrical algebra, $Cl(m, n) \equiv G(m, n)$, which is structured on m units squaring to 1 and n to –1. It also has elements of Grassmann algebra. It is interesting that these fundamentally dualistic algebras arise as aspects of simply defining a zero totality. The same applies to the group representations, which further emphasize the link with mathematical duality, and are particularly relevant to mathematical physics. Though the discussion is confined to the finite or discrete groups, the incorporation of fundamentally undefined terms as the original basis of the algebra means that it applies to continuous or Lie groups as well.

We may begin with the C_2 group, which can now be represented by 1 and –1; a dual system will extend this to four elements, producing an equivalent to $C_2 \times C_2$, and we choose the only way of extending a group including 1 and –1 to encompass four elements, by making the unknown elements (hitherto represented by the generic A and A^*) acquire the characters that we describe by the algebraic symbols i and $-i$. The group of $1, -1, i, -i$ is not, of course, $C_2 \times C_2$, or D_2, but C_4.

However, it contains the same *information* as $C_2 \times C_2$, for we can write this information in the form of the complex ordered pairs: 1, i; 1, $-i$; -1, i; -1, $-i$, which *is* of the form $C_2 \times C_2$, and is the only domain in which $\pm\, i$ can exist.

If we are now required to dual the C_4 group, the most efficient and ordinally-structured way of retaining elements equivalent to 1, -1, i, $-i$ in an extended group of order eight, is by supposing that we can expand i, $-i$ into the necessarily *cyclic* and noncommutative operators $\boldsymbol{i}$, $-\boldsymbol{i}$, $\boldsymbol{j}$, $-\boldsymbol{j}$, $\boldsymbol{k}$, $-\boldsymbol{k}$, which we describe as quaternions. The definition of the quaternion group Q_4, with elements 1, -1, $\boldsymbol{i}$, $-\boldsymbol{i}$, $\boldsymbol{j}$, $-\boldsymbol{j}$, $\boldsymbol{k}$, $-\boldsymbol{k}$, is simply a statement of the fact that the complex C_4 group has been dualistically extended on the basis that $\boldsymbol{ij}$ ($= \boldsymbol{k}$) has the same kind of properties as $\boldsymbol{i}$ and $\boldsymbol{j}$, with $(\boldsymbol{ij})(\boldsymbol{ij}) = -1$. Again, we can represent the same information by a C_2 multiplication, using a group of the form $C_2 \times C_2 \times C_2$. The cyclic nature of the quaternions is significant here, because the eight possible ($C_2 \times C_2 \times C_2$) combinations of $\pm\, \boldsymbol{i}$, $\pm\, \boldsymbol{j}$, $\pm\, \boldsymbol{k}$ become sufficient to generate the entire information produced by the elements of Q_4. In effect, describing a set of operators, such as $\boldsymbol{i}$, $\boldsymbol{j}$, $\boldsymbol{k}$, as 'cyclic' means reducing the amount of independent information they contain by a factor 2, because $\boldsymbol{k}$, for example, arises purely from the product $\boldsymbol{ij}$. It could even be argued that the necessity of maintaining the equivalence of the Q_4 and $C_2 \times C_2 \times C_2$ representations is the determining factor in making the quaternion operators cyclic. In addition, the cyclicity prevents the definition of further complex terms, such as $\boldsymbol{I}$, where $(\boldsymbol{iI})(\boldsymbol{iI}) = -1$, though there are an unlimited number of $\boldsymbol{I}$ terms such that $(\boldsymbol{iI})(\boldsymbol{iI}) = 1$.

The process can be continued further using terms of this kind. We dual Q_4 by complexifying it to the complex quaternion or multivariate 'vector' group 1, -1, i, $-i$, $\boldsymbol{i}$, $-\boldsymbol{i}$, $\boldsymbol{j}$, $-\boldsymbol{j}$, $\boldsymbol{k}$, $-\boldsymbol{k}$, $i\boldsymbol{i}$, $-i\boldsymbol{i}$, $i\boldsymbol{j}$, $-i\boldsymbol{j}$, $i\boldsymbol{k}$, $-i\boldsymbol{k}$, of order 16, which has a related $C_2 \times C_2 \times C_2 \times C_2$ formulation, and which may also be written 1, -1, i, $-i$, $i\mathbf{i}$, $i\mathbf{i}$, $i\mathbf{j}$, $-i\mathbf{j}$, $i\mathbf{k}$, $-i\mathbf{k}$, $\mathbf{i}$, $-\mathbf{i}$, $\mathbf{j}$, $-\mathbf{j}$, $\mathbf{k}$, $-\mathbf{k}$, where a complex quaternion, such as $i\boldsymbol{i}$ becomes the equivalent of the multivariate vector $\mathbf{i}$. (It is significant, here, that a possible alternative dualling of quaternions to octonions, with sixteen components, would fail to maintain the group structure, as octonions are nonassociative.) We then expand the complex terms to a three-dimensional status, to produce a double quaternion group, say 1, -1, $\boldsymbol{I}$, $-\boldsymbol{I}$, $\boldsymbol{J}$, $-\boldsymbol{J}$, $\boldsymbol{K}$, $-\boldsymbol{K}$, $\boldsymbol{i}$, $-\boldsymbol{i}$, $\boldsymbol{j}$, $-\boldsymbol{j}$, $\boldsymbol{k}$, $-\boldsymbol{k}$, of order 32, which has a related $C_2 \times C_2 \times C_2 \times C_2 \times C_2$ formulation. Then we complexify again, to produce a multivariate vector-quaternion group 1, -1, i, $-i$, $i\mathbf{i}$, $-i\mathbf{i}$, $i\mathbf{j}$, $-i\mathbf{j}$, $i\mathbf{k}$, $-i\mathbf{k}$, $\mathbf{i}$, $-\mathbf{i}$, $\mathbf{j}$, $-\mathbf{j}$, $\mathbf{k}$, $-\mathbf{k}$, $\boldsymbol{i}$, $-\boldsymbol{i}$, $\boldsymbol{j}$, $-\boldsymbol{j}$, $\boldsymbol{k}$, $-\boldsymbol{k}$, $i\boldsymbol{i}$, $-i\boldsymbol{i}$, $i\boldsymbol{j}$, $-i\boldsymbol{j}$, $i\boldsymbol{k}$, $-i\boldsymbol{k}$, and 36 real and complex combinations of vectors and quaternions, forming a group of 64, with a related $C_2 \times C_2 \times C_2 \times C_2 \times C_2 \times C_2$ formulation. Because of the reduction of information involved in defining both multivariate vectors and quaternions as cyclic, and in one producing complex, and the other real, products, the $C_2 \times C_2 \times C_2 \times C_2 \times C_2 \times C_2$ formulation can be expressed by the 64 possible combinations of $\pm\,\mathbf{i}$, $\pm\,\mathbf{j}$, $\pm\,\mathbf{k}$, $\pm\,\boldsymbol{i}$, $\pm\,\boldsymbol{j}$, $\pm\,\boldsymbol{k}$. Further dualling is possible on the same basis, but it is clear that only three fundamental principles are required to continue the dualling to infinity – opposite signs (or equivalent), the distinction between real and imaginary components, and the introduction of cyclic dimensionality – and to establish every conceivable combination of these, that is to establish every type of dualling, requires a group of 64 elements.

C_2	C_2	± 1	conjugate
C_4	$C_2 \times C_2$	$\pm 1, \pm i$	complexify
Q_4	$C_2 \times C_2 \times C_2$	$\pm 1, \pm \boldsymbol{i}, \pm \boldsymbol{j}, \pm \boldsymbol{k}$	dimensionalize
V_{16}	$C_2 \times C_2 \times C_2 \times C_2$	$\pm 1, \pm i, \pm \boldsymbol{i}, \pm \boldsymbol{j}, \pm \boldsymbol{k}$	complexify
QQ_{32}	$C_2 \times C_2 \times C_2 \times C_2 \times C_2$	$\pm 1, \pm \boldsymbol{I}, \pm \boldsymbol{J}, \pm \boldsymbol{K}, \pm \boldsymbol{i}, \pm \boldsymbol{j}, \pm$	dimensionalize
VQ_{64}	$C_2 \times C_2 \times C_2 \times C_2 \times C_2 \times C_2$	$\pm 1, \pm i, \pm \boldsymbol{I}, \pm \boldsymbol{J}, \pm \boldsymbol{K}, \pm \boldsymbol{i}, \pm \boldsymbol{j}, \pm \boldsymbol{k}$	complexify

The process becomes entirely repetitive at the level of V_{16}, while VQ_{64} is what we obtain by combining C_2, C_4, Q_4, and V_{16} as independent elements, establishing conjugation, complexification, dimensionalization and repetition.[5] Beyond this stage, we can consider the sequence proceeding through an infinite series of quaternionic structures by repeated processes of complexification and dimensionalization, creating an algebra of infinite dimensions, whose units are each quaternionic. Repetition necessarily sets in as soon as we establish the principle of closure, and closure, as we shall see, allows us an immediate procedure for returning to zero.

The order 16 group, as we have seen, is of special interest as creating what is effectively a 'real' dimensional structure of the kind observed in normal 3-dimensional vector space. The order 16 group of complex quaternions is notably equivalent to the 'real' dimensional structure of 3-dimensional multivariate vector space, as used in the geometrical algebra of Hestenes and others, and applied by them to the algebra of physical space and time, to generate electron spin as a natural consequence of spatial three-dimensionality.[6,7] The components, $\pm 1, \pm \boldsymbol{i}_1, \pm \boldsymbol{j}_1, \pm \boldsymbol{i}_1\boldsymbol{j}_1, \pm \boldsymbol{i}_2, \pm \boldsymbol{i}_2\boldsymbol{i}_1, \pm \boldsymbol{i}_2\boldsymbol{j}_1, \pm \boldsymbol{i}_2\boldsymbol{i}_1\boldsymbol{j}_1$, can be rearranged and written in the form ± 1, $\pm i$, $\pm \mathbf{i}$, $\pm \mathbf{j}$, $\pm \mathbf{k}$, $\pm i\mathbf{i}$, $\pm i\mathbf{j}$, $\pm i\mathbf{k}$, where ± 1, $\pm i$, become the respective scalar and pseudoscalar, and $\mathbf{i}$, $\mathbf{j}$, $\mathbf{k}$, and $i\mathbf{i}$, $i\mathbf{j}$, $i\mathbf{k}$ the respective vector and pseudovector terms of this algebra. In this algebra, which is also isomorphic to that of Pauli matrices, the 'total' product of two multivariate vectors **a** and **b** is of the form **a.b** + *i* **a** × **b**, and the 'total' products of the vector units is of the form **ii** = **jj** = **kk** = 1; and **ij** = –**ji** = *i***k**; **jk** = –**kj** = *i***i**; and **ki** = –**ik** = *i***j**. In summary, quaternions follow the multiplication rules:

$$i^2 = j^2 = k^2 = -1$$
$$ij = -ji = k$$
$$jk = -kj = i$$
$$ki = -ik = j$$
$$ijk = -1,$$

while those for complexified quaternions are:

$$(ii)^2 = (ij)^2 = (ik)^2 = 1$$
$$(ii)(ij) = -(ij)(ii) = i(ik)$$
$$(ij)(ik) = -(ik)(ij) = i(ii)$$
$$(ik)(ii) = -(ii)(ik) = i(ij)$$
$$(ii)(ij)(ik) = i,$$

which are isomorphic to those for multivariate vectors, or Pauli matrices:

$$\mathbf{i}^2 = \mathbf{j}^2 = \mathbf{k}^2 = 1$$
$$\mathbf{ij} = -\mathbf{ji} = i\mathbf{k}$$
$$\mathbf{jk} = -\mathbf{kj} = i\mathbf{i}$$
$$\mathbf{ki} = -\mathbf{ik} = i\mathbf{j}$$
$$\mathbf{ijk} = i.$$

The succession, then, allowing for conjugation (±) within each group, becomes:

order 2	real scalar
order 4	complex scalar (real scalar plus pseudoscalar)
order 8	quaternions
order 16	complex quaternions or multivariate 4-vectors
order 32	double quaternions
order 64	complex double quaternions or multivariate vector quaternions

The order 16 group (if we are to retain the maximum indistinguishability by avoiding octonion-type nonassociativity) is also the point at which the extension of the sequence becomes one of repetition. So, a complete specification of an interative generating procedure could be made by using the groups of order 2, 4, 8 and 16. Taken as independent entities, these may be combined minimally in the group of order 64, using the symbols ± 1, ± *i*, ± ***i***, ± ***j***, ± ***k***, ± **i**, ± **j**, ± **k**, to represent the respective units required by the scalar, pseudoscalar, quaternion and multivariate vector groups. This will take on physical significance when we realize that the algebra of this group is that of the gamma matrices used in the Dirac equation – the quantum equation determining the behaviour of the most fundamental components of matter – and that these matrices may be represented as the terms ***k***, ***ii*i**, ***ii*j**, ***ii*k**, ***ij***, whose binomial combinations, as we will show, are sufficient to generate the entire group.

1.5 Rewriting Nature

So far, we have concentrated on showing that a requirement of non-unique zero totality automatically generates a structure which has the characteristics of an algebra, including countable numbers, even without the prior assumption of the existence of mathematics. The structure effectively 'rewrites' itself in the way that occurs in the formal rewrite systems used in computing, examples of which are given by von Koch (1905)[8], Chomsky (1956)[9], Naur *et al* (1960)[10], Mandelbrot (1982)[11], Wolfram (1985)[12], and Prusinkiewicz and Lindenmayer (1990)[13], among others, and, as previously suggested, we may include the work of Conway here. The system we have proposed, however, unlike those, is a *universal* one, which effectively rewrites all known rewrite systems, and so extends the applicability and power of the entire concept to any sphere of computing. In this sense, as well as providing a foundation for mathematics, it also provides a foundation for computing, in either its classical or quantum forms.

Now, rewrite systems (or production systems) are synonymous with computing in the sense that most software is written in a language that must be rewritten as symbols for some hardware to interpret. Formal rewrite, or production, systems are pieces of software that take an object usually represented as a string of characters and using a set of *rewrite rules* generate a new string representing an altered state of the object. If required, a second *realisation system* takes the string and produces a visualisation or manifestation of the objects being represented. However, using the concept of zero totality, we can show that we can define a rewrite system from which other rewrite systems may be constructed at a basic level. Of course, while such a system can be encapsulated, for convenience, in a computer program written in a high level language, that program must be recognised as being different from the rewriting mechanism which it represents. It would simply be a way of realising this using existing knowledge, while the originating mechanism would represent 'computer language' at a much more fundamental level.

Rewriting in the computing sense always begins with an initial state. In this case, it is a string representation of 0. We begin with the idea that only 0 is unique. Everything that is not 0 is undefined. In rewriting, we start with an argument denying that we have a non-0 starting-point. We assume that we are not entitled to posit anything other than 0, and that we are forced to rewrite when we start from any other position. In the process we observe the significance of the concept of hierarchy, and of the difference between *recursion* (where the entire set of objects and relations is already in existence) and *iteration* (where the objects and relations are successively built up in a process which starts from the simplest).

Traditionally computer rewrite systems involve objects defined in terms of symbols representing characters drawn from a finite alphabet, and a series of states. To move from state to state we apply a finite set of rules – rewrite rules or productions – to a string of the symbols that represents the current state of the complex object. Some stopping mechanism is defined to identify the end of one state and the start of the next (for example we can define that for each symbol or group of symbols in a string, and working in a specific order, we will apply every rule that applies). It is usual in such systems to halt the execution of the entire system if there is no change in the string generated or if the changes are cycling, or after a specified number of iterations. Differing stopping mechanisms determine different families of rewrite systems, and in each family, alternative rules and halting conditions may result in strings representing differing species of object. Allowing new rules to be added dynamically to the existing set and allowing rules to be invoked in a stochastic fashion are means whereby more complexity may be introduced.

To explain how an apparent 'something' results from an ultimate nothing we need to show how a universal alphabet that encompasses duality and nothingness can be developed using a universal rewriting system. There are, in fact, two methods by which the elements of this alphabet may be discovered. One of these methods yields an infinite number of subset alphabets each of which has properties that can be exploited, for example using further rewrite systems based on the subset alphabet. As we have seen from the algebraic development, at this stage, a powerful simplification becomes apparent, reducing the procedure to three fundamental processes.

If we relax the rules regarding the finiteness of the characters in the alphabet(s) and the number of states, but continue to assume the rest of the constraints described above, a more universal rewrite system is defined. Such a system has alphabets as its complex objects and subset alphabets (all the symbols so far delivered) as states. For this to remain a rewrite system an initial state (that can be re-written) must exist, and for it to be universal there must be, we may conjecture, a minimum of two rewrite rules (productions). One of these, *create,* delivers a new symbol at each invocation. The term 'symbol' is used here because what is delivered may be a single character of the alphabet, a subset alphabet, or indeed an entire alphabet. The second rule, *conserve*, examines all symbols currently in existence to ensure that no anomalies exist as a consequence of bringing the new one into existence.

With such a minimum universal rewrite system, the initial state (usually called the Δ-state) must contain at least one symbol that we can use to identify that the universe is empty. However, any symbol we choose is immediately (and simultaneously) a symbol, a character of the final alphabet, a subset alphabet and full alphabet in its own right. We may choose, arbitrarily, the single symbol 0 (zero), and set it as the string representing the complex object in the Δ-state {0}. We are obliged to make an arbitrary choice here because we cannot use *create* without the Δ-state – the minimum rewrite system condition for a universal system. If we were to use *conserve* now it would simply return that 0 is unique, fixed, and consistent and no change from the Δ-state would be generated. We must therefore now invoke *create* supplying the Δ-state as parameter, or source, string.

If we presume that *create* is an algorithm with stopping criteria, it returns a result target string containing a new symbol. If the paradigm for the algorithm were recursive, the resulting symbol (E is used here) would represent every character of the alphabet at the first step. To create any refining character, a specific e_x, using the *recursive* paradigm would be impractical because of the implied infinity and storage requirement. We may not use an iterative paradigm at this stage because we would have to supply an upper limit and / or need to identify which of the infinite characters we are creating. Both of these actions require a character not yet in the character set (alphabet) we have so far defined.

The pair of symbols, the string {0, E} is our new object (alphabet) which we now submit to *conserve* which examines every combination of symbols (Table 1.1):

Table 1.1

	0	E
0	00	0E
E	E0	EE

We note that 00, the 'transition' from 0 to 0, conserves 0. The combination 0E is the transition from 0 to E and is balanced, for all E, by its conjugate partner E0 which is the transition back from E to 0, thereby conserving 0. The combination EE, the transition from every symbol E to every other, is anomalous and must be returned by *conserve* as unexplained or 'inconsistent' as it does not appear to conserve 0. However, at infinity, all transitions represented by EE will have been examined, EE will be declared 'nilpotent' (i.e. 'squaring' to 0) in that it delivers 0, and we will be left with three generic combinations:

(00, 0E, E0)

However, it is impractical to use the recursive version of *conserve* to examine further the elements of E because of the implied infinite number of iterations.

We return to the *create* process and accept that we must postulate symbols Δ_a, Δ_b, … Δ_n drawn from E such that they are in an arbitrary ordinal sequence. We note that there is an infinite number of such sequences because choice of Δ_a is arbitrary. However, we may now use an iterative paradigm for *create* and because *n* is specified, an iterative (or recursive) *conserve* can be constructed. At the end of each invocation we are now presented with a symmetrical table of transitions that represent the simplest set of properties for the current set of *n* symbols (Table 1.2).

Table 1.2

	0	Δ_a	Δ_b	Δ_c	…	Δ_n
0	00	$0\Delta_a$	$0\Delta_b$	$0\Delta_c$		$0\Delta_n$
Δ_a	Δ_a0	$\Delta_a\Delta_a$	$\Delta_a\Delta_b$	$\Delta_a\Delta_c$		$\Delta_a\Delta_n$
Δ_b	Δ_b0	$\Delta_b\Delta_a$	$\Delta_b\Delta_b$	$\Delta_b\Delta_c$		$\Delta_b\Delta_n$
Δ_c	Δ_c0	$\Delta_c\Delta_a$	$\Delta_c\Delta_b$	$\Delta_c\Delta_c$		$\Delta_c\Delta_n$
:						
Δ_n	Δ_n0	$\Delta_n\Delta_a$	$\Delta_n\Delta_b$	$\Delta_n\Delta_c$		$\Delta_n\Delta_n$

The Δ_a row and Δ_a column illustrate the conjugate pair structure observed earlier. The remaining cells of Table 1.2 identify explicitly each Δ symbol to Δ symbol transition observed generically in Table 1.1. Off diagonal there are symmetrical conjugate pairs; there are, for example, three such cancelling pairs when $n = b$ and six when $n = c$. The diagonal cells of the table contain transitions from each symbol to itself and do not cancel out in this way.

We now invoke the *conserve* process, noting that it does not define the transition property but merely identifies those novel transition combinations that appear not to conserve 0. When $n = a$, the symbol Δ_a is added to the alphabet and the transition $0\Delta_a$ is introduced. We need $\Delta_a 0$ (and the idea that this is a conjugate form) to conserve 0. However, this leaves the combination $\Delta_a\Delta_a$ unexplained (i.e. novel) and to conserve 0 we must conjecture that, whatever it is, is balanced by whatever is to come – or both are 'nilpotent' in the sense introduced above. To discover this we invoke *create* to add a new symbol to the alphabet which then defines (arbitrarily) the $n = b$ row and column. At $n = b$ (in *conserve*) we continue to require the conjugate explanation for all off-diagonal elements in the table. In addition, we have non-0 to non-0 symbol transitions, each of which has a cancelling conjugate, and which must ultimately yield a symbol already in the alphabet. However, when these transitions are explained we still have $\Delta_b\Delta_b$ as novel, and require the method of explaining the novelty used earlier. We see that at every invocation of *conserve* we define the need for an additional symbol, delivered by *create* – it is inherent that both processes are obligatory. Other processes may now be conjectured within the rewrite system that impart meaning to 'transition' and also to each transition from Δ_n to Δ_n; however, in each case all of what is to come must balance the $\Delta_n\Delta_n$ in the diagonal position. 'Balance' in this explanation assumes that the 00 transition yields 0; however, we could consider it to yield a conjugate of some form. Where this is the case we may consider each newly created diagonal element as 'balancing' that conjugate by delivering the unconjugated form. In each case the new symbol created carries the entire subset alphabet.

The properties and symbols emerge from the application of the two rewrite rules and would have been equally valid for any of the infinite alternative selections. Significantly, since the ultimate aim is to recover the zero state through an infinite series of processes, the emergence should be seen as being of a *supervenient* nature, that is, without temporal connotation. Furthermore, the symbol delivered at each step has all the properties of all the symbols previously delivered, and in a hierarchical and orthogonal fashion.

Finally, we note that the symbol 0, the existence of the Δ-state, and the processes *create* and *conserve* are outside the rewrite system in that they must exist before the system can function. If we can allow these assumptions, we may also presume the existence of some natural machine that will deliver, for a set of appropriate rewrite rules, a corresponding alphabet where the symbols themselves map to specific rules. In terms of the rewrite procedures we have adopted, the assumption of any non-zero category (previously expressed as *R*) must immediately lead to the return to zero, which, in mathematical terms, becomes equivalent to supposing a 'negative' category or 'conjugate' (R^*) corresponding to the original assumption; the combination must then produce the desired zero state by some process (not yet specified as 'addition'). In terms of Table 1.2, this is the recognition that $\Delta_a\Delta_a$ leads to the creation of the new symbol Δ_b. A process of 'self-referencing', correlation or 'combination', which can be expressed in the form *RR* (again without assuming a specific mathematical interpretation), has produced an extended categorization, from (*R*) to (*R*, –*R*). At this point we have created ordinality (through the existence of an implied + and an explicit –), though not yet counting, as there is no discreteness or anything fixed involved in the procedure. (The same applies in the Dedekind 'cut', used to define real numbers in conventional mathematics; despite its name, this is a definition of ordinality without a prior assumption of discreteness.)

However, discreteness emerges naturally as a result of extending the alphabet, and it is the next application of the create procedure ($\Delta_b\Delta_b \rightarrow \Delta_c$) which leads to the number system as we know it, for now we have an undifferentiated 'set' of possible origins for the 'negative' ordinal category or conjugate. We describe these as complex forms, and each must have its own conjugate. In mathematical terms, the complex category remains completely undefined in respect to the real category, and has no ordinal relation to it. There are infinitely possible or indefinitely possible systems that are represented by the mathematical *A*, even for a seemingly specified real category. It is only when we express this fact in the next creation stage that we are able to begin to extend ordinality towards enumeration, for this stage leads to what become mathematical 'combinations' of complex categories. We find here that to every conceivable *A*, e.g. *A*, *B*, *C*, ..., there are indefinitely possible (commutative) combinations leading to the original real category (e.g. $AC\,AC = R$), but very definite and restricted (anticommutative) ones leading to the conjugate (e.g. $AB\,AB = -R$).

These alternative possibilities provide the respective foundations for infinite- and finite-dimensional algebras. The infinite-dimensional algebra, as we will see in chapter 3, leads to the infinite Hilbert vector space of quantum mechanics, while the Hamilton or 3-dimensional algebra is responsible for the cyclic system of quaternions. It is the cyclicity of the latter which introduces discreteness or closure, and the concept of 'unity', and it is thus no coincidence that discreteness in physics is invariably associated with dimensionality (see chapter 2). As we have seen in 1.3, in our generation of the associated algebraic structure, we can choose the default position of taking the conjugate combination to create a regular ordinal sequence. We now find that only 'one' independent A-type concept (say B) is associated with each conceivable A, and we can sequence the terms ordinally by choosing indistinguishability between the As in every conceivable respect. So the sequence, although arbitrary, becomes a series of integral binary enumerations, which we can also apply to ordinality in the real categories. Defining 1, at all, in this way, automatically creates a *dual* system, with −1, etc., equivalent to requiring $1 + 1 = 2$, and generating the Peano idea of 'successor'. With the reals, integers, and complexity as fundamental aspects of the system, the remaining mathematical number categories (and higher algebras) can be defined by applying the ordinality condition in a variety of ways, as in conventional mathematics. No new principle is required.

In effect, the hierarchical and orthogonal mathematical structure suggested by the rewrite mechanism is the following:

R	undefined	Δ_a
R, R^*	conjugation	Δ_b
R, R^*, A, A^*	complexification	Δ_c
$R, R^*, A, A^*, B, B^*, AB, AB^*$	dimensionalization	Δ_d
$R, R^*, A, A^*, B, B^*, AB, AB^*,$ $C, C^*, AC, AC^*, BC, BC^*, ABC, ABC^*$	repetition	Δ_e

As before, all the character sets are generated by the simple rule that no non-zero character set or alphabet is complete, so every self-referencing yields a new or extended alphabet, which is necessarily self-conjugated; and we can, of course, extend the process to infinity, as in section 1.3. The significance of this system for both classical and quantum computing will become apparent after we have first examined the foundations of physics.

1.6 Quaternions and Vectors

Because of their great importance in physics, it will be convenient at this point to provide a technical appendix, giving a short account of the relationship between quaternions and vectors. Quaternions were finally discovered by William Rowan Hamilton in 1843, but he had actually started work, many years earlier, with the Argand diagram, used to represent complex numbers ($a + bi$), in which the x-axis represented the real system of numbers (a), and the y-axis, drawn orthogonally to it, represented the imaginary numbers (bi). This, he thought, could be seen as representing two-dimensional space, and he sought to extend it to provide a 3-dimensional version using an extension of complex numbers. There is, of course only one set of *real* numbers, but Hamilton reasoned that, with the true nature of imaginary numbers being unknown, there was no reason why there could not be two sets of imaginary numbers, which could be drawn orthogonally to each other, say bi and cj. However, he found very quickly that such a system of 'triplets' would not work, for the system did not exhibit 'closure'; that is, the product ij had no meaning within the system, and had to represent some other quantity.

The breakthrough came when he realised that if he invented a *third* system of imaginary numbers (dk), he could define a system that exhibited closure, though he had to sacrifice the principle of commutativity (the idea that ab always equals ba) to do it. In this system, $\boldsymbol{i}, \boldsymbol{j}$ and $\boldsymbol{k}$, the three square roots of -1, are related by the formulae:

$$\boldsymbol{i}^2 = \boldsymbol{j}^2 = \boldsymbol{k}^2 = \boldsymbol{ijk} = -1.$$

However, the products of the units are noncommutative, since

$$\boldsymbol{ij} = -\boldsymbol{ji} = \boldsymbol{k}$$

and so on. This is obvious from consideration of $\boldsymbol{ij}\,\boldsymbol{ji} = -\boldsymbol{i}^2 = 1$. The full rules become:

$$\begin{aligned} \boldsymbol{i}^2 = \boldsymbol{j}^2 = \boldsymbol{k}^2 &= -1 \\ \boldsymbol{ij} = -\boldsymbol{ji} &= \boldsymbol{k} \\ \boldsymbol{jk} = -\boldsymbol{kj} &= \boldsymbol{i} \\ \boldsymbol{ki} = -\boldsymbol{ik} &= \boldsymbol{j} \\ \boldsymbol{ijk} &= -1, \end{aligned}$$

Hamilton suspected what Frobenius later proved (in 1878) that the system was unique; no other extension of complex algebra was possible that maintained the same algebraic rules. He was convinced also that he had therefore found the true explanation for the three-dimensionality of space. But, as with ordinary complex numbers, the three imaginary parts had to be accompanied by a real part, making a four-part number, which he described as a *quaternion*. But, if the three imaginary parts represented the three dimensions of space, what did the real part mean? Almost immediately, he postulated that it must be time. Now, quaternions were used enthusiastically in the mid-nineteenth century by mathematical physicists like P. G. Tait and James Clerk Maxwell. But, in the end, several factors caused them to be discarded in favour of a new form of vector algebra, promoted strongly by Willard Gibbs and Oliver Heaviside. The first was that quaternions, in normal usage, gave the wrong signs when squared for Pythagorean addition, the answers being negative where positive ones would have been expected; while the second was the fact that quaternions were complicated structures made up of real and imaginary parts, and only the imaginary (or three-dimensional) part was needed in most physical applications. So vector algebra extracted this imaginary part from the complete quaternion and changed it from an imaginary to a real quantity, devising a restricted set of rules for its usage, which were derived ultimately from the more extensive quaternion algebra.

This system proved to be of such immediate utility that, despite the efforts of Tait and others, vector algebra eventually succeeded quaternion algebra in all major applications. However, it has long been established that quaternions were at least the *parent* of vector theory. From the very beginning, Hamilton had realised that quaternions involved a vector part (the imaginary one) and a scalar part (the real one), and he even defined equivalents of the modern vector and scalar products, and the modern vector differential operator, 'del', while Maxwell, using quaternions, developed a great deal of modern vector calculus; the terms 'vector' and 'scalar', used in the modern sense, are also Hamilton's own. In principle, the vector algebraists did little more, in mathematical terms, than separate out vector and scalar parts, and vector and scalar products, change them over from real to imaginary, or imaginary to real, and devise a neater and more consistent notation, though vector algebra still uses Hamilton's **i**, **j**, **k** as well as his ∇.

However, quaternions are, in themselves, and not just in their application to the development of vectors, one of the most powerful tools ever presented to the physicist – many people, in fact, use them routinely without realising that it is quaternions that they are actually using. In fact, if they had been used in the correct way early on, they would have pre-empted many later physical developments that came about much more tortuously.[14] For example, Einstein developed, in 1905, a simple kinematical theory, based on only two physical postulates, for explaining a whole series of facts concerning the 'Electrodynamics of Moving Bodies', but, within two and a half years, this theory (the special theory of relativity) was, in a way, superseded by an even simpler theory, due to Hermann Minkowski, which was based on mathematical, rather than physical, assumptions.

In classical vector space, Pythagoras' theorem required the length of a line element r to be invariant to arbitrary changes in the components from x, y, z to x', y', z' according to the rule:

$$r^2 = x'^2 + y'^2 + z'^2 = x^2 + y^2 + z^2.$$

To this equation, Minkowski added a term representing time, multiplied by a unit-conversion factor numerically equal to the velocity of light. In Einstein's words, written after his conversion to the new scheme: 'Mathematically, we can characterize the generalized Lorentz transformation thus: it expresses x', y', z', t' in terms of linear homogeneous functions of x, y, z, t of such a kind that the relation

$$x'^2 + y'^2 + z'^2 - c^2t'^2 = x^2 + y^2 + z^2 - c^2t^2$$

is satisfied identically.'[15] Minkowski thus replaced the ordinary vector described by three real parts (x, y, z), representing Euclidean space, with a space-time *4-vector* with three real parts and one imaginary (x, y, z, ict); and proceeded to apply the same principle to every part of physics previously described by vectors.

Minkowski's argument, though very striking, was not entirely original. Its mathematics had been almost anticipated by Poincaré, who had effectively introduced the Minkowski metric in his paper of July 1905, without fully realising the relativistic implications, and the idea of four dimensions had had a long prehistory. But the parallel with quaternions and with Hamilton's view of the connection of space and time are obvious, and it was soon pointed out to Minkowski himself that his space-time had a basically quaternion-type structure. A. W. Conway, for example, used quaternions in a simple and elegant way to derive the Lorentz transformations and all the other relativistic equations.[16] The interest in this issue is not just historical, for, even now, it is usually considered mathematically convenient, to write 4-vector line elements in what is effectively a quaternion form,

$$ds^2 = c^2\,dt^2 - dr^2 = c^2dt^2 - dx^2 - dy^2 - dz^2,$$

whose very structure is equivalent to making the time element real and the space elements imaginary.

Many applications of vector theory, of course, don't require the time component, but, even in these cases, a proper attention to quaternion theory shows that what seems arbitrary and awkward in vector theory has a natural and simple explanation in terms of quaternions. Vector theory, as it stands, breaks most of the rules of ordinary algebra – namely, commutativity, associativity, the law of moduli, unambiguous division and closure – and it has two kinds of 'product', neither of which is a product in the ordinary algebraic sense. The rules of vector algebra are designed purely to fit physical requirements and seem to make very little mathematical sense. However, they do make sense if seen in a quaternionic context.

The most convenient way of showing this is to use one of the 'Clifford algebras', named after William K. Clifford, who discovered its basic principles in the 1870s. The term is generically used to cover all the algebras (also called geometric algebras) which combine such elements as real and complex numbers, vectors and quaternions, but the type which we need for our immediate purposes is the Cl (3, 1) algebra, an algebra isomorphic to complex quaternions, which is, in effect, the complete reverse of the Cl (1, 3) quaternion algebra. ($Cl\ (m, n) \equiv G\ (m, n)$ means an algebra founded on m square roots of 1 and n square roots of –1.) This is the algebra which Hestenes has called 'multivariate' vector algebra, and it effectively subjects 4-vectors to the full quaternionic rules of multiplication.

First of all, we look at quaternion multiplication. If we take two quaternions, with zero real parts:

$$\boldsymbol{a} = x\boldsymbol{i} + y\boldsymbol{j} + z\boldsymbol{k}$$

and

$$\boldsymbol{a}' = x'\boldsymbol{i} + y'\boldsymbol{j} + z'\boldsymbol{k},$$

and multiply them, we obtain

$$\boldsymbol{aa}' = -(xx' + yy' + zz') + \boldsymbol{i}\,(yz' - zy') + \boldsymbol{j}\,(zx' - xz') + \boldsymbol{k}\,(xy' - yx').$$

This, as Hamilton originally showed, divides into a scalar product $-(xx' + yy' + zz')$ and a vector product $\boldsymbol{i}\,(yz' - zy') + \boldsymbol{j}\,(zx' - xz') + \boldsymbol{k}\,(xy' - yx')$, and we can recognise, almost immediately, that the first term is virtually identical to the normal scalar product of two vectors (except for the negative sign) and the second term is the same as the normal vector product – except that the unit imaginary quaternions ***i***, ***j***, ***k*** replace the unit vectors **i**, **j**, **k**. (It is convenient to use the convention of bold italic script for quaternions and bold for vectors.) The important thing about this multiplication, however, is that it obeys the ordinary rules of algebra (except for commutativity). It doesn't require any special definition or any special rules. In particular, the product is a quaternion, like the multiplicands, and so we have 'closure'.

In ordinary vector algebra, by contrast, there are two methods of multiplication and neither method exhibits 'closure': the dot product of two vectors produces a scalar and the cross product a pseudovector, and both products break other basic algebraic rules. What we really need is a *single* method of multiplication of vectors which exhibits closure and which maintains all normal algebraic properties except commutativity, a method which *combines* dot and cross products. By *direct analogy with quaternions*, it is easily shown that this 'full' product of vectors **a** and **a'** is of the form:

$$\mathbf{aa'} = \mathbf{a.a'} + i\,\mathbf{a} \times \mathbf{a'}.$$

This is the basic rule of the Clifford algebra *Cl* (3,1). Assuming that (as with quaternions) the basic quantity is a combination of vector and scalar, the product exhibits closure and all the other basic algebraic properties except commutativity. The imaginary sign before the cross product is necessary to establish complete symmetry with the quaternion system, and the introduction of imaginary quantities into vector theory gives us an easy explanation of the difference between scalars and pseudoscalars, and vectors and pseudovectors.

Pseudoscalars are simply imaginary scalars, that is, ordinary (real) scalars which have been multiplied by *i*, the square root of –1; and pseudovectors are ordinary (real) vectors multiplied by the same factor. Both types of term arise naturally from the algebra. If we take two orthogonal vectors, **a** and **b**, then the scalar product is 0, and the total product is a pseudovector

$$\mathbf{ab} = i\,\mathbf{a} \times \mathbf{b}.$$

An example is the area of a rectangle. If **a** and **b** are parallel, on the other hand, then the total product is a real scalar

$$\mathbf{ab} = \mathbf{a.b}.$$

The product of a pseudovector $i\,\mathbf{a} \times \mathbf{b}$ and parallel real vector **c** will be an imaginary scalar quantity, and, therefore, a pseudoscalar. For example, if **a**, **b** and **c** are the vectors representing the sides of a cube, then the volume is given by the total product

$$\mathbf{abc} = i\,\mathbf{a} \times \mathbf{b.c}.$$

The product of a pseudovector $i\,\mathbf{a} \times \mathbf{b}$ and a perpendicular real vector **c**, however, which involves the multiplication of two imaginary terms, will be a real quantity, $i\,(i\,\mathbf{a} \times \mathbf{b}) \times \mathbf{c}$ and, therefore, a true vector. An example of this occurs in the formula

$$\mathbf{F} = e\,\mathbf{v} \times \mathbf{B},$$

where we are effectively multiplying a vector e **v** by an orthogonal pseudovector **B**. We also obtain a real vector if we multiply a pseudovector like area (**A**) by a pseudoscalar like pressure (P), as in the formula

$$\mathbf{F} = P\mathbf{A},$$

and we obtain a real scalar, work (PV), if we take the product of the pseudoscalars, pressure and volume. Once again, the imaginary operators are removed by multiplication.

A great advantage of this system is that we have no need to define separate vector and scalar products for the unit vectors **i**, **j** and **k**. We simply define total products of the form

$$\mathbf{ii} = \mathbf{i.i} + i\,\mathbf{i} \times \mathbf{i} = \mathbf{i.i} = 1,$$

and $$\mathbf{ij} = \mathbf{i.j} + i\,\mathbf{i} \times \mathbf{j} = i\,\mathbf{k},$$

with $$\mathbf{ji} = \mathbf{j.i} + i\,\mathbf{j} \times \mathbf{i} = -i\,\mathbf{k}.$$

Products of two vectors are called 'bivectors', and products of three vectors are 'trivectors'. (They are literally 'triple products'.) For unit vectors **i** and **j**, the bivector (or 'area' element) is i**k**, while the trivector (or 'volume' element) is i. The full rules are written:

$$\mathbf{i}^2 = \mathbf{j}^2 = \mathbf{k}^2 = 1$$
$$\mathbf{ij} = -\mathbf{ji} = i\mathbf{k}$$
$$\mathbf{jk} = -\mathbf{kj} = i\mathbf{i}$$
$$\mathbf{ki} = -\mathbf{ik} = i\mathbf{j}$$
$$\mathbf{ijk} = i. \qquad (1.1)$$

Physicists, of course, have long used this algebra, in the form of the so-called 'Pauli matrices', which are employed in applying spin to the Schrödinger equation. These are defined as:

$$\sigma_x = \begin{pmatrix} 0 & 1 \\ 1 & 0 \end{pmatrix} \qquad \sigma_y = \begin{pmatrix} 0 & -i \\ i & 0 \end{pmatrix} \qquad \sigma_z = \begin{pmatrix} 1 & 0 \\ 0 & -1 \end{pmatrix},$$

and follow the rules:

$$\sigma_x\sigma_y = -\,\sigma_y\sigma_x = i\sigma_x$$
$$\sigma_y\sigma_z = -\,\sigma_z\sigma_y = i\sigma_x$$
$$\sigma_z\sigma_x = -\,\sigma_x\sigma_z = i\sigma_y$$

and
$$\sigma_x\sigma_x = \sigma_y\sigma_y = \sigma_z\sigma_z = \begin{pmatrix} 1 & 0 \\ 0 & 1 \end{pmatrix} = \mathbf{I}.$$

Using the algebraic operators, in fact, makes it completely unnecessary to use any matrices at all, and, by applying the full multiplication properties of vectors to quantum mechanics, we can show that the spin of the electron can be derived from the nonrelativistic Schrödinger equation (by adding a cross product term), just as readily as it can from the relativistic equation of Dirac. So, if the true 'quaternion'-like nature of vector quantities had been recognised at the time, it would have been possible to derive the spin of the electron directly from nonrelativistic quantum mechanics, without having to put it into the equation ad hoc. This also tells us that spin has nothing to do with relativity or 4-vectors; it involves only the vector part of the quaternion, and, as we might expect, analogies have been found in classical physics, where vectors are equally important.

Another way of obtaining the same algebra is to use complexified quaternions, so that
$$\mathbf{i} = i\boldsymbol{i}; \qquad \mathbf{j} = i\boldsymbol{j}; \qquad \mathbf{k} = i\boldsymbol{k}.$$

Then
$$(i\boldsymbol{i})^2 = (i\boldsymbol{j})^2 = (i\boldsymbol{k})^2 = 1$$
$$(i\boldsymbol{i})(i\boldsymbol{j}) = -(i\boldsymbol{j})(i\boldsymbol{i}) = i(i\boldsymbol{k})$$
$$(i\boldsymbol{j})(i\boldsymbol{k}) = -(i\boldsymbol{k})(i\boldsymbol{j}) = i(i\boldsymbol{i})$$
$$(i\boldsymbol{k})(i\boldsymbol{i}) = -(i\boldsymbol{i})(i\boldsymbol{k}) = i(i\boldsymbol{j})$$
$$(i\boldsymbol{i})(i\boldsymbol{j})(i\boldsymbol{k}) = i.$$

These rules are clearly the same as those defined in (1.1).[17] The multivariate vectors are, notably, as anticommutative as the parent quaternions. An anticommutative, associative algebra of this kind is necessarily 3-dimensional, but an infinite vector (Grassmann) algebra can be defined by making the terms commutative.

Chapter 2

Why Does Physics Work?

At this point, it is necessary to temporarily set aside our synthetic development of physics from mathematics, to investigate physics, as we know it, from an inductive viewpoint. This will vastly simplify the structure that we will need to derive, in the next chapter, from first principles. An inductive investigation into the foundations of physics suggests that the subject is structured in such a way that it avoids characterizing nature. This is the origin of symmetry and duality, and it explains the separation between the system and observation in both classical and quantum physics. The most fundamental symmetry that can be discovered inductively appears to be a group-symmetry between the four parameters space, time, mass and charge, which divides them between three sets of opposing properties: conserved / nonconserved; divisible / indivisible; real / imaginary. Each parameter has one property in common with each other, with the two remaining properties in absolute opposition, producing a zero conceptual totality. Dimensionality appears to be linked to divisibility, the two divisible quantities, space and charge, also being dimensional. The dimensions of charge are the respective sources of electromagnetic, strong and weak interactions, which, it is presumed, would be identical under some set of idealised conditions yet to be discovered. In addition, mass, time, space and charge take on the respective mathematical characters associated with real scalars, pseudoscalars, multivariate vectors and quaternions. The binary operations between the group elements incorporate Pythagorean addition and universal 'interactions' between all mass and all charge elements. If the absoluteness of the symmetry is accepted, then many facts concerning the parameters are immediately explained and various paradoxes eliminated. We can also predict new mathematical theorems purely on symmetry grounds, including a direct connection between the conservation of type of charge and conservation of angular momentum.

2.1 A Foundational Level

In chapter 1, we saw that mathematics could be regarded as an emergent property of a system privileging zero totality. The question we now need to ask is: is physics equally emergent? This, I intend to show, relates, in a very fundamental way, to another important question: why does physics work? That is, why is physics, or knowledge ultimately based on the assumption of the validity of physics, the only system of thought which has produced a reproducible description of reality capable of endless extension from within itself? How is it that out of the many possible systems of thought which have tried to grasp at fundamental concerns, one and one only has been successful, and that one is physics? Can a philosophical approach help us to find the answer, and can we use this knowledge to tackle important physical questions? Can a theory of knowledge be used both to explain physics and to suggest possibilities for its future development? Will the much sought-after 'unified theory' only be possible when we have located the structure which makes physics such a uniquely successful means of describing the processes of nature?

I believe that, to answer these questions, we must look at areas in which physics appears to be counter-intuitive. Physics could be described as the science of measurement (at least in its classical form), but its structure does not suggest that it is based on measurement convenience. There is, in fact, only one process of measurement known, the counting of spatial intervals, using a fixed scale of units defined arbitrarily. All measurements, even those supposedly representing other quantities, such as time, are really measurements of space. However, no attempt to structure the whole of physics on the basis of space alone has been so far successful. Physics seems to demand the existence of other fundamental quantities, which cannot be directly measured. The measurement process required by physics does not define the whole of nature. It would seem, therefore, that we have a problem, for concepts like countability, measurability and observability are the very foundation of physics, and its particular distinguishing feature. How do we explain this seemingly strange fact?

Of course, the fundamental principles of physics have been developed on the basis of a long step-by-step process of conceptualizing and testing by experiment, and there are theories such as classical mechanics, electromagnetic theory and quantum mechanics which have a vast range of applicability in describing the processes of nature. It is the requirement that the experimental justification must always accompany the theoretical developments that has led to their success, but this does not tell us why these particular theoretical structures are required by nature. Another unexplained fact is that many of the principles used, such as the various conservation laws, and the irreversibility of time, appear to be *intrinsically simple*, and suggest that simpler theories are more likely to be inherently true than complex ones. Is there some more fundamental principle of knowledge which determines these structures and which privileges simplicity over complexity? These questions need answering at the basic level before we proceed to uncovering the more esoteric aspects of the subject.

Certain aspects of physics, in fact, suggest the existence of a core level of basic information which is completely independent of any hypothesis or model-building. This information is concerned with the definition of the fundamental parameters of measurement and how they are structured, though, as we know from quantum mechanics, the measurement process is not concerned only with measurable quantities. Here, we are not so much describing nature itself as specifying the characteristics of the simplest categories needed to make such a description possible. It is, of course, sometimes argued that physics should not necessarily be concerned with the simplest possible ideas because nature or 'reality' may well not be simple in principle, and the simplicity may be ours rather than nature's. This argument, however, is based on a misconception. Physics as we know it has evolved because it has created a set of simple categories which have been successful in devising the human construct that we call the 'description of nature'. Whether or not such simplicity is truly characteristic of some hypothetical 'external' reality is a question outside the realm of physics.

Now, the purpose of isolating a foundational level for physics would be to give a simple account of those important facts which are purely concerned with our own processes of measurement or description, and not with, say, the nature of matter or the structure of the universe. Such information, if isolated, could be dealt with more efficiently than if linked, unnecessarily, with more specific or more complicated theories, and could lead to the creation of wholly new sources of foundational information, maybe to the extent of explaining the things originally excluded. In addition, if we pitch our foundational theories at a high level of sophistication, as we are often tempted to do, we cut ourselves off from understanding their origins. Many truly fundamental results have often turned out to be simple in principle, even when it has taken a sophisticated approach to find them. The simple bases are often found only after a prolonged struggle with more complicated ideas, but it is possible that we could discover some of them more easily by a direct analysis.

In attempting to reach the foundational level, then, we need to separate out the truly fundamental ideas from the mass of sophistications which inevitably accompany them, but the really basic ideas may not be particularly difficult to find. Obviously, for example, space and time are basic; it is impossible to conceive a theory of physics without space and time. A combined space-time, however, whether 'curved' (as in general relativity) or otherwise, would not be basic, even discounting the fact that such a combination remains problematic, in any case, at least in areas of quantum mechanics. It is the separate identities of space and time, which are important at the fundamental level, and combinations should always be regarded as a sophistication, as something to be discovered after we have investigated the separate identities of the components.

2.2 The Origin of Abstraction

It is interesting to look at the historical background of the development of physics as we know it today. This began in the late Middle Ages (c 1300 – 1450), with the attempt to explain processes, such as free fall and uniformly accelerated motion, through dcfining quantities based on the variations of space and time, the most abstract ideas then known, and still the only means of observing variation in nature. The theories then developed encompassed various states of uniform and nonuniform velocity, and uniform and nonuniform acceleration, which were expressed in both mathematical and graphical forms. The authors of these theories were not scientists in the modern sense, but theologians, more concerned with comprehending the mind of an unknowable God than with explaining physical observations. Hence, their explanations were pitched to the highest point of abstraction and to the greatest level of simplicity then attainable.

The success of these mathematical theologians is attributable to the extremism of their ideas, their relentless application of Ockham's razor (the principle that the idea based on the minimum assumptions is the most likely to be true) resulting in the creation of abstract concepts of a universal generality, rather than specific explanations of particular physical phenomena, as in the previous Aristotelian model. The method was adopted and extended by both Galileo and Newton, who were educated at universities which were still mediaeval in outlook, but Newton found it necessary to incorporate a third concept, mass, on the same level as space and time, to develop a more general system of dynamics. The significance of mass was that it established the principle that fundamental physical laws could be built around the fact that some concepts were conserved quantities while others were not. The Newtonian procedure also established the fact that the mathematical system used to describe fundamental physical laws, based as it was on differential equations and an infinite number of interacting particles, was not a direct description of nature, but an idealised abstraction which could never be observed in terms of direct physical measurements.

In principle, this separation between observables and the mathematical system carries over into quantum mechanics, and is extremely counter-intuitive. Clearly, we have not developed the present mathematical structure of physics for *convenience*, as is often maintained; it has rather evolved by natural selection because no other structure will work. None of the great abstract theories, such as classical mechanics, electromagnetic theory, or quantum mechanics, has been accepted wholeheartedly, even by physicists, as a self-evident fundamental truth, precisely because such theories run counter to our starting point of structuring the world of nature on direct observation and measurement. Yet no theory based on alternative principles has had the slightest success. Also, successful theories of this kind have led to the belief that theories explaining new phenomena must be similar in kind, and can be derived by analogy and symmetry.

This happened after the development of Newtonian mechanics and gravitational theory, when a two-hundred-year search for the laws of electricity, magnetism and optics was based on the starting assumption that analogies must exist with the Newtonian system. Through a long series of experimental and theoretical developments, the analogy was eventually successful through the creation of the mass-like parameter charge, and the additional component of the relativistic connection of space and time (which accounted for the magnetic aspect), which was then applied, by analogy, to the original Newtonian theory. In more recent times, it has been widely assumed that the two new physical forces discovered through radioactivity and particle physics, the strong and weak interactions, must be in some way analogous to the electric force, under ideal conditions; and that some process must exist which breaks an otherwise perfect symmetry between the three forces.

2.3 Symmetry

Symmetry (or analogy) has been the driving force of much of theoretical particle physics, as it was, previously, of classical physics, and physicists seem to *expect* to find symmetries in nature. There are many classic theorems which invoke symmetry, for example, Noether's theorem, which allows us to relate the conservation laws of energy, momentum and angular momentum to the translation and translation-rotation symmetries of time and space. It is this fact which gives us an important clue as to what really makes physics work. The other important clue is provided by the fact already stated that the work of mediaeval theologians aimed at expressing the unknowability of God has been successfully transported into physics and used with the totally different objective of describing the natural world in terms of abstract mathematics.

These clues suggest that the important philosophical principle determining the structure of physics is that nature cannot be characterized. It is neither measurable nor unmeasurable; and in fact has no single defining characteristic. Any system which assumes a special characteristic of any kind limits us to the asymptotic discovery of our initial assumption. To create a universally applicable system, we have to avoid any such limiting assumption; and, to make physics work, we have had to incorporate concepts which we would not have chosen on first principles, but which introduce a systematic contradiction of our starting assumptions. For example, though we started with space and time as variable quantities, we found that nature threw up quantities which we could not exclude which turned out to be *in*variable, such as mass and charge; and, though we started by assuming that we could measure everything directly (through space), we quickly found that we had to structure physics to incorporate quantities which could not be directly measured (time, mass and charge).

The in-built procedure which allows us to have these opposite characteristics within the same system is symmetry. If we apply a perfectly symmetrical structure to the explanation of nature, we ensure that, whatever characteristic we choose in one part of the system will be countered by its exactly symmetrical opposite in another. Symmetry thus allows us to do the apparently impossible, that it is to characterize nature using a measurement process – because the system as a whole is not measurable – while, at the same time allowing us to reduce our number of fundamental assumptions. By a process of choosing the only methodology which produces a match with experimental results, we have adopted a structure for physics which is perfectly symmetrical in its foundations, though we have not yet recognized this fact. However, if we do begin to recognize it, we will be able to design new physics according to the specifications which have, in this way, been forced upon us.

But what symmetry? There are many in physics, and we must decide which are the most important and fundamental. I will argue that there are just four basic parameters and that we can use such techniques as dimensional analysis to show that all other physical parameters arise from compounded versions of these elementary ones. We can also show that it is precisely these parameters which are assumed to be the elementary ones in the statement of the CPT theorem (that the laws of physics are unchanged under simultaneous changes in the signs of charges and space and time coordinates). Our historical analysis has, in addition, suggested that, once we have selected space and time as our 'variables', there is only one other type of information that is likely to fit the description 'basic', for the fundamental bases of the whole of physics are undoubtedly the four known interactions. Once we truly understand these, and their sources, mass and three types of charge, then we will also understand physics. The symmetry, then, which I believe to underlie all the other symmetries in physics is that between the fundamental parameters space, time, mass and charge, where charge is a general term representing the sources of the three nongravitational interactions. Essentially, space is characterized by the properties required of a parameter of measurement: it is real, nonconserved, and countable. Nature, however, forces us to include, at the same level within our system, parameters which have directly opposing properties. These turn out to be those which we describe as time, mass and charge; and the symmetry between the four parameters appears to be absolute and representable in terms of a group structure.

It is perhaps slightly surprising that physicists have been so often prepared to tackle fundamental questions without taking proper account of the ideas which appear to be most basic. Although we can't hope to *analyse* really basic ideas, we can learn a great deal by setting one off against the other. It would surely be profitable to examine the properties of these parameters as closely as possible and look for patterns, or symmetries of one sort or another, that would help to clarify their meaning and uses. Symmetry has been such a powerful tool in understanding particle physics and the fundamental interactions that we have every reason to expect to find it here. We should, at any rate, examine the properties of our parameters to see whether or not it exists. In fact, as we shall show, classical mechanics, electrodynamics, relativity, quantum mechanics, and particle physics, can be readily accommodated within this structure. Many theorems can be established as consequences of the emerging symmetrical pattern; and new physical facts can be predicted, even in quantitative terms. The structure is at once more simple and powerful than any of the principles derived from it.

2.4 The Meaning of the Conservation Laws

One of the most important aspects of Newtonian physics is its introduction of the parameter mass on the same fundamental level as space and time. Introducing mass also introduces the idea of conservation, which can be described in absolute terms, and leads to a whole series of conservation principles, which are invariably the bases of dynamical systems. Physical equations then become statements that one quantity is conserved while another is not. So, the conserved property of mass also shows up the contrasting, nonconserved natures of space and time, which are incorporated into physics in the differential forms with which these quantities are associated in fundamental equations.

In Newton's original formulation, the quantity force was defined as a product of the conserved mass and the differential forms of the nonconserved space and time (dx, dt, with the *second* differential, d^2x / dt^2, being used for reasons which will become apparent), and the conservation property of mass was then established through the 'third law of motion', which effectively stated that force was zero within a conservative system. The method was powerful because it was extreme. It had to be true in all cases. Later versions of dynamics, associated with Euler, Maupertuis, Lagrange, and Hamilton, were structured on exactly the same pattern: a quantity was defined which incorporated mass and the differentials of space and time, and was then shown to be subject to extreme behaviour. The quantity was set to zero, or a constant value, or a maximum or minimum. And the extremum principles so derived, in particular the conservation laws of linear momentum, angular momentum and energy, were, in principle, nothing other than the fact that mass was conserved against all possible variations in space and time.

Subsequent developments in electrodynamics suggested that electric charge had precisely the same property, and the conservation laws of mass and electric charge are considered among the most fundamental in physics (although we now, of course, interpret the meaning of 'mass' in terms of energy). It is almost inconceivable to imagine any circumstance in which they would be violated. Again, it is almost certain that some kind of conservation law applies also to the sources of the weak and strong nuclear interactions, which have in the last two decades increasingly been referred to under the same generic label of 'charge'.[1] Lepton and baryon conservation are obvious consequences, baryons being the only particles with strong, as well as weak, components, and leptons being the only particles with weak, but no strong, components. Of particular importance is the fact that the conservation laws of mass and charge are not merely global, applying to the total amount of each quantity in the universe, but also *local*, applying to the amount of each quantity at a given place in a given time. Classically, we give each element of mass or charge an *identity* which it retains throughout all interactions, subject only, in the case of charge, to its annihilation by an element with the opposite sign; and, although the identity of individual particles is not maintained in quantum mechanics, we still retain the requirement of local conservation.

Strikingly, when we look at *nonconservation*, as manifested by space and time, we immediately see that it is the *exact opposite* of conservation and it is just as definite a property, though it manifests itself in many different ways. For example, if the elements of mass and charge have individual, specific and permanent identities, those of space and time have no identity whatsoever. One manifestation of this is the property of *translation symmetry* which applies to both space and time. This implies that every element of space and time is exactly like every other, and is not only indistinguishable in practice, but *must be stated to be indistinguishable* when we write down physical equations. The translation property has profound physical consequences, as Noether's theorem – the mathematical result, already mentioned, which states that, for every global transformation preserving the Lagrangian density (a constructed quantity involving mass and the differentials of space and time, cf 8.7), there exists a conserved quantity – shows us that the translation symmetry of time is precisely identical to the conservation of energy, and that the translation symmetry of space is precisely identical to the conservation of linear momentum.

The third conservation principle, that of angular momentum, is explained by Noether's theorem as a result of space's extra three-dimensional property of rotation symmetry, meaning that there is no more identity for spatial *directions* than there is for spatial locations. In addition to having no unique elements, space also lacks a unique set of dimensions. One direction in space is identical to any other; this is the fact that is responsible for space's *affine* structure, the infinite number of possible resolutions of a vector into dimensional components. In particular, it is always possible to define a 3-component vector-space (x, y, z) in terms of a coordinate system with nonzero values in a single resultant dimension (r) and zero values elsewhere, while a single dimension can always be resolved into a 3-component vector-space.

The meaning of the conservation laws now becomes more apparent. They are not only the absolute defining principles of all aspects of fundamental physics, but must also be accompanied by absolute *nonconservation* principles. We could, in fact, illustrate the exactly opposite nature of conservation and nonconservation by expressing the identity or uniqueness properties of mass and charge in terms of 'translation' *a*symmetries. Translation asymmetry then means that one element of mass or charge cannot be 'translated to' (or exchanged for) any other within a system, however similar. Each element of mass and charge is untranslatable to any other. Such conserved quantities can, of course, only be defined with respect to changes in the nonconserved quantities, and what we define as interactions can often be thought of as a statement of all those possible changes in the nonconserved quantities which will maintain the value of the conserved ones. We look at what remains invariant (conserved) under certain groups of transformations (nonconserved). A similar concept of rotation asymmetry might be expected to apply to charge, but we will return to this later.

But there are also other manifestations of nonconservation in space and time. The absoluteness of the nonconservation properties is also apparent in the *gauge invariance* used in both classical and quantum physics. In classical or quantum electrodynamics, electric and magnetic field terms remain invariant under arbitrary changes in the vector and scalar potentials, or phase changes in the quantum mechanical wavefunction, brought about, essentially, by translations (or rotations) in the space and time coordinates. (It is significant that the aspects of identical particle wavefunctions which are truly interchangeable in quantum mechanics are the nonconserved space and time coordinates, not the conserved charges.) Gauge invariance is simply a way of expressing the fact that a system will remain conservative under arbitrary changes in the coordinates which do not produce changes in the values of conserved quantities such as charge, energy, momentum and angular momentum. In other words, we cannot know the absolute phase or value of potential because we cannot choose to fix values of coordinates which are subject to absolute and arbitrary change. Even more significantly, in the Yang-Mills principle used in particle physics, the arbitrary phase changes are specifically *local*, rather than global. Nonconservation, therefore, must be local in exactly the same way as conservation.

In general terms, the whole of physics is based on defining systems in which conserved quantities remain fixed while nonconserved quantities vary absolutely. A conserved quantity can only be defined with respect to changes in a nonconserved quantity. The conserved quantities, of course, include composite ones such as energy, momentum and angular momentum, as well as the more elementary ones, mass and charge, but the conservation of the composite ones depends directly on the conservation of the more elementary ones, and each of the composite conservation laws relates also to one of the aspects of the *nonconservation* of space or time. The behaviour of *all* physical systems is then described entirely in terms of a combination of conservation and nonconservation principles. In effect, we look at how mass and charge, and such quantities as energy, momentum, force or action remain constant, or zero, or a maximum or a minimum, because of the more fundamental requirements involving mass and charge, while the space and time coordinates alter arbitrarily. The alteration of space and time is expressed by describing them in terms of differentials, and the very fact that we have based physics on differential equations and the definition of systems involving conservation requirements is an expression of the presence of both absolutely conserved and absolutely nonconserved terms in nature. Quantum mechanics, rather than being intrinsically strange, is really more fundamental and more logical than classical physics in simply making this more explicit.

2.5 The Mathematical Structure of Physical Quantities

Ancient knowledge tells us that space is three-dimensional. Three independent spatial axes can be drawn at right angles. In other words, the dimensions of space may be combined 'vectorially', or by the Pythagorean addition of their squared values. Special relativity, which emerged from electrodynamics in the early twentieth century indicated, however, that time, in many respects, could be treated as a fourth dimension of space, at the price of making the time component an imaginary (or 'non-orderable') number, with negative squared values, the whole combination being described by the Minkowski space-time 4-vector ($\mathbf{i}x$; $\mathbf{j}y$; $\mathbf{k}z$; it). Many people, of course, refer to this as a mere mathematical convenience, or 'trick', but we still have to explain why such a convenient 'trick' actually works. It is instructive, therefore, to look at the equivalent representation of mass and charge.

Here, we may imagine, as is generally believed, that the electric, strong and weak source terms (say e, s, w) would, in some idealised regime, be of a similar nature, suggesting that we could describe them, by analogy with space, as 'dimensions' of a single quantity, generically known as 'charge'. This could be justified in terms of the fact that the Newton and Coulomb inverse-square force laws effectively square mass and charge terms, in the same way as space and time terms are squared in Pythagorean addition. Now, we have the intriguing fact, long known but never explained, that forces between like masses are attractive, whereas forces between like charges (of all kinds) are repulsive; that is, the forces between like masses and like charges have opposite signs. However, if we choose to represent charges by imaginary numbers and masses by real ones, we then have a symmetrical representation for the Newton and Coulomb force laws:

$$F = -\frac{Gm_1m_2}{r^2}$$

$$F = -\frac{iq_1iq_2}{4\pi\varepsilon_0 r^2}$$

The three types of source would, of course, have to be distinguished from each other in some way, and so would each require a different imaginary number, or square root of −1. However, the mathematics required for such a situation is already available and has been well-known for a hundred and fifty years. This is the *quaternion* system, which we have already encountered, in which $\boldsymbol{i}, \boldsymbol{j}$ and $\boldsymbol{k}$, the three square roots of −1, are related by the formulae:

$$\boldsymbol{i}^2 = \boldsymbol{j}^2 = \boldsymbol{k}^2 = \boldsymbol{ijk} = -1.$$

Effectively the quaternion components are the reverse of the 4-vectors used in Minkowski space-time: three imaginary parts and one real one (an ordinary real number or scalar), as opposed to three real vector parts and one imaginary pseudoscalar. Their special significance is that they are unique; no other *associative* extension of ordinary complex algebra involving imaginary dimensions is possible. A system with two imaginary parts is impossible, and there are no systems with four, five or six imaginary parts. A system with seven imaginary parts is possible (octonions), but this requires breaking the rule of associativity (that $(ab)c = a(bc)$); and there are no systems with more than seven.

It seems that, for dimensions determined by Pythagorean addition, the number three has a special significance. The three imaginary parts, in this representation, would be associated with the three components of charge (say, $\boldsymbol{i}s, \boldsymbol{j}e, \boldsymbol{k}w$), leaving the real fourth part to represent mass. Space and time would then become a three real- and one imaginary-part system by *symmetry*, and the necessary mathematical connection between space and time would be explained as a consequence of the necessary mathematical connection between charge and mass:

space-time	$\mathbf{i}x$	$\mathbf{j}y$	$\mathbf{k}z$	it
mass-charge	$\boldsymbol{i}s$	$\boldsymbol{j}e$	$\boldsymbol{k}w$	m

There is, of course, a complication in that quaternions have different rules of multiplication to vectors, there being no such thing as a 'full product' (**ab**) between two vectors, **a** and **b**, such as exists between two quaternions. However, if we were to postulate that the symmetry between space-time and mass-charge should be an exact one, we could extend the vector property of space to incorporate a quaternionic-like 'full' product between two vectors, combining the scalar product with *i* times the vector product, so that

$$\mathbf{ab} = \mathbf{a.b} + i\,\mathbf{a} \times \mathbf{b}.$$

This procedure has been fully justified by several decades of mathematical development, and it turns out that the extra pseudovector terms in the full product are just those required to explain the otherwise 'mysterious' spin property in quantum mechanics.

It is here that, now we understand the meaning of 'dimensionality' in the case of charge, we can return to the subject of its rotation asymmetry. If charge is absolutely conserved, then we can expect conservation in dimension as well as in quantity, which is what we mean by rotation *a*symmetry. That is, the sources of the electromagnetic, weak and strong interactions should be separately conserved, and incapable of interconversion. We could consider charge units to be arranged along axes which are fundamentally irrotational, so that one type of charge (say, electric) can never be converted into another (say, weak or strong). Immediately, this tells us that the proton, which has a strong charge measured by its baryon number, cannot decay to products like the positron and neutral pion, which have none. (The Weinberg-Salam unification of electromagnetic and weak forces is not, of course, affected because this theory is a statement of the identity of effect in the two interactions, under ideal conditions, not of identity of the sources; the three quaternion operators ***i***, ***j*** and ***k*** are different sources, though identical in effect.) In fact, separate conservation laws should lead immediately, as we have said, to baryon and lepton conservation. We will see later how this produces a new manifestation of Noether's theorem.

Dimensionality, however, is not the only advantage of an imaginary representation for charge, for imaginary numbers have yet another important property. This is the fact that equal representation must be given to positive and negative values of imaginary quantities. Neither positive nor negative imaginary values may be privileged in algebraic equations; every equation which has a positive solution also has an algebraically indistinguishable negative solution (the complex conjugate). Consequently, all our charges (but not necessarily real masses) must exist in both positive and negative states. This is exactly what we require to explain the existence of antiparticles, for even those particles, such as the neutron and neutrino, which have no electric charge still have antiparticles because they have strong and / or weak charges whose signs may be changed under the process of charge conjugation. In addition to this, imaginary charge of any type is inaccessible except in the presence of a field (that is, of other charges of the same type) – it can only be accessed as a *squared* quantity, while real mass, though likewise accessible as a squared quantity through the gravitational field, can also be detected directly through inertia.

Even imaginary time, however, has other 'convenient' properties, outside of its representation in the Minkowski 4-vector, for an imaginary representation also makes uniform velocity (= dx / dt) imaginary, while acceleration, with its squaring of time (d^2x / dt^2), remains real; this, of course, would explain why time 'measurement' is only possible through force and acceleration, and not through uniform motion, time being only only accessible or 'orderable' as a squared quantity. There are similar advantages in quantum mechanics, where imaginary numbers seemingly appear for no reason in wavefunctions for time-related quantities, but become completely valid in a context where the imaginary representation is a fundamental requirement.

2.6 Where Does Dimensionality Come From?

Dimensionality has two aspects – the fact of multidimensionality, and the Pythagorean addition of the multiple dimensions by squares. Assuming these two aspects, we can solve the problem of *three*-dimensionality, for no other dimensionality can be represented by an associative division algebra, but dimensionality itself appears to be too complex to be a basic property. It must be explicable on some more fundamental grounds. It is necessary, here, to compare the dimensional (or multidimensional) parameters, space and charge, with the nondimensional (or one-dimensional) parameters, time and mass.

Though space and time are associated mathematically in the Minkowski formalism, there is good evidence to believe that they are fundamentally different. Space, for example, is *always* used in direct measurement; it is, in fact, impossible to measure directly any other quantity. The principle of *simulation* (for example, by film, hologram, photograph or sound recording) lies in allowing changes of space to represent supposed changes in time, mass and charge, and would be altogether impossible if we could measure all four parameters directly. So-called 'time'-measuring devices, such as pendulums, mechanical clocks, and crystal and atomic oscillators, all use some concept of repetition of a spatial interval. Special conditions have to be used to set up such measurements, whereas any object whatsoever can be used to measure space. Space also is reversible – and it is this reversibility which is used in the measurement of time – but time is not. In quantum mechanics, time, though a variable, is not even an observable – a fact which causes a major problem for any attempt at quantizing general relativity.

There also seems to be a deep philosophical problem with the infinite divisibility of time, as Zeno's ancient paradoxes suggest. In the well-known argument about the race between Achilles and the Tortoise, Achilles, in any number of time intervals, should never catch up with the Tortoise, to whom he has given a lead, because, each time he thinks he has caught up, he finds the Tortoise has already moved further ahead, even if only by an ever smaller amount. Another example is the Dichotomy Paradox, in which an object moving over any distance can never get started because it must cover half the distance before it covers the whole, and a quarter of the distance before it covers half, and so on; to go any distance in a finite amount of time, it must already have been involved in an infinite number of operations. Of course, these paradoxes are, in a sense, 'answered' by the use of limits or infinite series, but Whitrow, who has made the most extensive and influential recent study, thinks that the answers still leave the problem incomplete.[2]

On the basis of such paradoxes, Whitrow writes: 'One can, therefore, conclude that the idea of the infinite divisibility of time must be rejected, or ... one must recognize that it is ... a logical fiction.' And the science writers, Coveney and Highfield, conclude that: 'Either one can seek to deny the notion of 'becoming', in which case time assumes essentially space-like properties; or one must reject the assumption that time, like space, is infinitely divisible into ever smaller portions.'[2] Achilles, for example, never catches the Tortoise because we have assumed that the time for the race can be divided up into finite intervals. The paradoxes seem to show, according to Whitrow, that motion is 'impossible if time (and, correlatively, space) is divisible ad infinitum'.

Whitehead thought that the paradoxes showed an 'instant of time' to be 'nonsense',[3] while Bergson, according to Whitrow, 'enthusiastically adopted the view' that time 'is wholly indivisible', 'as a means of escaping the difficulties raised by Zeno, concerning both temporal continuity and atomicity, without abandoning belief in the reality of time. ... Unfortunately, in attacking the geometrization (or spatialization) of time he went too far and argued that, because time is essentially different from space, therefore it is fundamentally irreducible to mathematical terms.' According to our analysis, there is good evidence that one cannot simply assume that time can be indefinitely subdivided like space. There is every reason to believe, in fact, that time, unlike space, is an absolute continuum. There is no infinite succession of measurable instants in time, as supposed in the paradoxes, because there are no instants. Time cannot actually be divided. In more contemporary jargon, space is digital, time is analogue – and we have both concepts in nature because we have both parameters.

The space-time distinction has profound consequences for both mathematics and physics, and, if we believe (as I do) that physics is the inherent creator of mathematics, and not merely the employer of its techniques, then we will say that the mathematics which is possible is created because it is possible physically. We can, for example, say that time is the set of reals with the standard topology superimposed, and is nonalgorithmic; space is the set of reals without the topology, and is algorithmic. Abraham Robinson, in his *Non-Standard Analysis*,[4] has successfully treated infinitesimals as though they had the properties of real numbers, and has shown that proofs of many theorems become much simpler by this method, although all non-standard proofs may be duplicated by standard ones (and vice versa). Non-standard analysis is also closely related to Skolem's non-standard arithmetic of 1934, with its denumerable model of the reals, and the so-called non-Archimedean geometry, which relates this to space. These versions of non-standard mathematics are a reflection of a discreteness in space while 'standard' results (based on limits) rely on the continuity of time.

Though we often describe real numbers in terms of a continuous line in space, the 'continuity' which we attributed to space because of its indefinite divisibility is not what is meant by the absolute continuity of time. Absolute continuity cannot be visualised and any process used to describe it would deny continuity. The property which space has that is often referred to as 'continuity' is indefinite elasticity, its 'continual' recountability or its unending divisibility. But it is this very divisibility of space which denies it *absolute* continuity; and the elastic nature of the divisibility comes from the entirely different property of nonconservation. A nonconserved quantity necessarily has nonfixed units, but they are units nonetheless. The whole process of measurement depends crucially on the divisibility of space, or creation and recreation of discontinuities within it. Thus the entire problem of Zeno's paradoxes disappears as soon as we accept that we can have discontinuities or divisibility in space, but not in time.

The discontinuities in space are in both quantity and direction; it can be reversed and changed in orientation; and, without both of these properties, measurement would be impossible. Time, however, cannot be reversed, precisely because it is absolutely continuous. Any reversal of time would require discontinuity. For the same reason, time cannot be multidimensional, or, in our terminology, 'dimensional', as branching of any kind would require the discontinuity of an origin. It is also equally impossible for a discrete quantity, like space, to be nondimensional, for one cannot demonstrate discreteness in a one-dimensional system. Though we think of a line as one-dimensional, it is in fact no such thing: it is a one-dimensional construction within a two-dimensional one.[5] If our space was truly one-dimensional we would only have a point with no extension. We couldn't demonstrate discreteness, and certainly not discreteness with variability, as we demand of space. Interestingly, it is dimensionality as such, rather than any particular level of dimensionality which is responsible for creating the additional level of discreteness required by the introduction of algebraic numbers, and even of transcendental numbers such as π and e, two independent dimensions are sufficient to create the required level of incommensurability at the rational number level, and the introduction of a third dimension requires no qualitatively new type of number.

The distinction in status between space and time is even responsible for the fundamental fact that time, in the definition of velocity and acceleration, the basic quantities used in dynamics, is the independent variable, whereas space is the dependent variable. So we use dx / dt, rather than dt / dx. This situation arises because time, as a continuous quantity, unlike space, is not susceptible to measurement – a situation which is even more obvious in quantum mechanics than it is in classical physics. We have no control over the variation of time, and so its variation is necessarily independent.

The existence of both 'standard' and 'nonstandard' versions of analysis and arithmetic are consequences of the prior existence of space and time. The mathematical options that are available, here and elsewhere, are almost certainly a reflection of the availability of physical options. Continuity and discontinuity, finiteness and infinity, and so on, probably exist as mathematical categories because they are also physical categories. For example, the discrete process of differentiation (using infinitesimals) is essentially modelled on variation in space; while the continuous process (using limits) is modelled on variation in time. Each is a valid option, as differentiation is a property linked to nonconservation, and not concerned, in principle, with the difference between absolute continuity and indefinite divisibility. (It is significant that the solutions of Zeno's paradoxes which invoke the concept of limit tacitly assume the time-based definition of differentiation.[6]) In arithmetical terms, the Cantorian definition of an absolutely continuous set of real numbers has equal validity with the idea of an infinitely constructible, though not absolutely continuous, set of real numbers based on algorithmic processes. Of significance here is the Löwenheim-Skolem theorem, that any consistent finite, formal theory has a denumerable model, with the elements of its domain in a one-to-one correspondence with the positive integers.

A further significant aspect of nonstandard analysis is that it has been a key ingredient in developing *topos* theory, which provides a view of space as arising from some kind of mathematical structure and possibly including dynamism, the points of the space being infinitesimal (but smeared) and nilpotent, or square roots of zero. *Topos* theory offers very attractive possibilities for formalizing some of the points made here. Space, in the present theory, arises from the process of counting or measurement; dimensionality is a necessary consequence of discreteness (or smearing); dynamism, with time, mass and charge emerging automatically from the symmetry, is in this sense in-built; and, as we will see later, the nilpotent wavefunction which we will construct for the pointlike charges could be related to the nilpotents needed to construct the points in space, especially as the nilpotent wavefunction creates an infinite series of further nilpotents in the continuous vacuum.

The continuous-discrete distinction also occurs, as we would expect, between mass and charge. Mass (in the sense that it incorporates fields and energy) is an absolute continuum present in all systems and at every point in space; this is why it is unipolar, unlike charge, for negative mass would necessarily require a break in the continuum, as would any multidimensionality. The property is crucial to the Higgs mechanism, which provides rest masses for the fundamental particles, and to quantum mechanics in general, as it implies a continuum of mass-energy, or filled vacuum. Charge, on the other hand, is divisible and observed in units. Naturally, because charge is also a conserved quantity, unlike space, these units must be fixed, unlike those of space. Again, charge as a noncontinuous quantity is also dimensional.

Our analysis will, further, allow us to deal with two well-known *physical* paradoxes. One is the 'reversibility paradox', where time, according to the laws of physics, whether classical or quantum, is reversible in mathematical sign, when it is clearly not reversible in physical consequences. Time, however, is characterised by imaginary numbers, and imaginary numbers are not privileged according to sign. Thus, it is quite possible to have a time which has equal positive and negative mathematical solutions because it is imaginary, but which has only one physical direction because it is continuous. The corresponding unipolarity, or single sign, of mass is the reason why we have a *CPT*, rather than an *MCPT*, theorem, *C* standing for charge conjugation, *P* for space reflection and *T* for time reversal, all of which have two mathematical sign options.

The other apparent paradox is wave-particle duality. This arises from the fact that, when we mathematically combine space and time in Minkowski's 4-vector formalism, as symmetry apparently requires us to do, we have two options: we can either make time space-like (or discrete) or space time-like (or continuous). Using the discrete options, we obtain particles, special relativity and Heisenberg's quantum mechanics. Using the continuous options, we obtain waves, Lorentzian relativity and Schrödinger's wave mechanics. Heisenberg makes everything discrete, so mass becomes charge-like quanta in quantum mechanics; the parallel combination of mass and charge in the quaternion structure then produces the discrete concept of rest mass, as opposed to the continuous source of gravity. Schrödinger, by contrast, makes everything continuous, so charge becomes mass-like wavefunctions in wave mechanics. In measurement, the true situations are restored, for Heisenberg's uncertainty principle and virtual vacuum effectively reintroduce continuous mass, while the collapse of the wavefunction in Schrödinger's formulation (see 7.2, 9.3) is an effective restoration of discreteness to particle states involving 'charge'.

2.7 A Group of Order 4

We have shown that the four basic parameters may be distributed between three sets of opposing paired categories: real / imaginary (alternatively, orderable / nonorderable), conserved / nonconserved (alternatively, with elements unique / nonunique), divisible / indivisible (alternatively, discrete / continuous, countable / noncountable, dimensional / nondimensional), with each parameter paired off with a different partner in each of the categories, according to the following scheme:

space	real	nonconserved	divisible
time	imaginary	nonconserved	indivisible
mass	real	conserved	indivisible
charge	imaginary	conserved	divisible

The symmetry appears to be exact, and is probably the exclusive source of physical information at the fundamental level. The properties where they match, seem to be exactly identical, and where they oppose, to be in exact opposition. Each pair forms an abstract group of order 2 (C_2):

e	a
a	e

The exactness of the symmetry, with a perfect distribution between properties and 'antiproperties', creates the possibility of a total conceptual nothingness or uncharacterizability in nature.

Overall, the scheme incorporates a group of order 4, in which any parameter can be the identity element and each is its own inverse. (The duality of space-time elements and their inverses is, interestingly, a feature of string theory.[7]) We can easily generate an algebraic representation by denoting the properties of space (real, nonconserved, divisible) by, say, x, y, z, with the opposing antiproperties (imaginary, conserved, indivisible) denoted by $-x$, $-y$, $-z$ (though this representation is not, of course, unique). The algebraic summation here neatly represents the required zero totality. The group now becomes:

space	x	y	z
time	$-x$	y	$-z$
mass	x	$-y$	$-z$
charge	$-x$	$-y$	z

With group multiplication rules of the form:

$$x * x = -x * -x = x$$

$$x * -x = -x * x = -x$$

$$x * y = y * -x = 0$$

and similarly for y and z, we can establish a group multiplication table of the form:

*	space	time	mass	charge
space	space	time	mass	charge
time	time	space	charge	mass
mass	mass	charge	space	time
charge	charge	mass	time	space

This is the characteristic multiplication table of the Klein-4 or D_2 group, with space as the identity element and each element its own inverse. However, there is no reason to privilege space with respect to the other parameters, since the symbols x and $-x$, y and $-y$, z and $-z$ are arbitrarily selected, and any of the other three parameters may be made the identity by defining its properties as x, y, z. For example, if mass is made the identity element, then the group properties and antiproperties may be represented by:

space	x	$-y$	$-z$
time	$-x$	$-y$	z
mass	x	y	z
charge	$-x$	y	$-z$

and the multiplication table becomes:

*	mass	charge	time	space
mass	mass	time	charge	space
charge	time	mass	space	charge
time	charge	space	mass	time
space	space	time	charge	mass

Various further representations are possible, and seem to be relevant, in particular, to the mathematical structure of the Dirac equation, which we will introduce in chapter 3. For example, the identity element, say mass, could be represented by the scalar part of a quaternion (1) and the other three terms by the imaginary operators ***i***, ***j***, ***k***, if we choose only the modular values, and ignore the + and – signs:

*	1M	***i***C	***j***T	***k***S
1M	1M	***i***C	***j***T	***k***S
iC	***i***C	(–)1M	***k***S	(–)***j***T
jT	***j***T	(–)***k***S	(–)1M	(–)***i***C
kS	***k***S	***j***T	***i***C	(–)1M

With the + and – signs added, we would require the full (and now cyclic) quaternion group structure of eight components. It is important to recognise here that the quaternion operators are extrinsically derived and not an integral component of the parameters space, time, mass and charge. Though the addition of these operators creates a new group structure, this structure is a relation between new mathematical constructs and not between the parameters themselves; it also presupposes the validity of the original symmetry between the parameters.[8]

The group representation that we choose for the fundamental parameters will depend on what we are actually looking at; purely 'qualitative' conceptions of a parameter's nature put into symbolic form will produce different representations to those derived using the internal algebras generated by the parameters themselves, and it appears that the 'canonical' representation is D_2. That D_2 is more fundamental than C_2 may seem at first a surprising result, as C_2 is the simplest possible group, a direct description of the 'something from nothing'-type duality which we might use to avoid characterizing nature. However, physics is ultimately structured on the need to define a concept of *measurement* (a 'probe') and a single fundamental C_2 would not allow this. By definition, a perfect C_2 symmetry would simply create an unrecognizable negating 'response' to a measurement probe; and, while a perfect C_2 would be meaningless, an 'imperfect' C_2 (allowing oppositeness in only a limited number of characteristics) would be incomplete. The simplest perfect symmetry (in the physical sense) based on the C_2 principle, which allows recognition of the response, is then D_2.

We have shown in chapter 1 how this group structure both arises in the most fundamental aspects of mathematics, and, in chapter 3, we will show how it is realised in fundamental physics as the basis of the Dirac or fermionic (particle) state, the single structure which incorporates the entire information available to physics. Only two members of the original group, say space and time, are needed to define the structure. With a third, say mass or charge, we can create quantum mechanics.

Various binary operations link the group's elements, as with all groups, but the simple operation of algebraic multiplication, which arises from the Dirac state (see chapter 3), and the necessary relations between the respective mathematical representations of mass, time, charge and space as scalar, pseudoscalar, quaternion or vector, seems to require the existence of fundamental constants linking the parameters to each other, and to their inverses, and returns us to the squaring process which is the origin of dimensionality. The existence of the binary operation of squaring within the group, or multiplication of a unit of any parameter by an identically-valued unit of the same parameter, seems to be linked to the same operation being responsible for the 4-dimensionality of space-time and mass-charge.

The method of 'squaring', however, reflects the status of parameters as either conserved or nonconserved. The nonconserved quantities, space and time, have nonidentifiable units, and so squaring produces only nonidentifiable squared versions of these units. The conserved quantities, on the other hand, have units with individual identities, and so 'squaring' involves the multiplication of each eligible unit with each other, as in m_1m_2 and q_1q_2, for mass and charge. Such 'squaring' must be a universal operation between any units of mass and charge, no individual unit being privileged. It will be convenient to give this process the name of 'interaction', and it will be recognised that 'interaction' in this sense is universal and nonlocal.

2.8 Noether's Theorem Revisited

The symmetries we have discussed are prescriptive as well as descriptive. New results may be derived on symmetry grounds even before we have the means of working out their mathematical or physical consequences. One example is associated with an extension of Noether's theorem. This theorem, as we have seen, requires the translation symmetry of time to be linked to the conservation of energy. Of course, since energy is related to mass by the equation $E = mc^2$, then the translation symmetry of time is also linked to the conservation of mass (that is, mass in the general sense, not rest mass). To put it another way, the nonconservation of time is responsible for the conservation of mass. This result could have been derived from symmetry alone, as it is inherent in the Klein-4 group structure. In fact, Noether's theorem itself, linking conserved quantities with transformations, is really an expression of the fact that every fundamental conserved quantity must be symmetrical to a fundamental nonconserved quantity. And so, extending the analogy, we can link the conservation of the quantity of charge with the nonconservation, or translation symmetry of space; and since the latter is already linked with the conservation of linear momentum, we can propose a theorem in which the conservation of linear momentum is responsible for the conservation of the quantity of charge (of any type). By the same kind of reasoning, we can make the conservation of *type* of charge linked to the rotation symmetry of space, and so to the conservation of angular momentum, as in the following scheme:

symmetry	**conserved quantity**	**linked conservation**
space translation	linear momentum	value of charge
time translation	energy	value of mass
space rotation	angular momentum	type of charge

Even on existing knowledge, it is possible to give some special cases of the applications of these new general theorems. Thus, the conservation of electric charge within a system has been known, since Fritz London's work of 1927, to be identical to invariance under transformations of the electrostatic potential by a constant representing changes of phase, and the phase changes are of the kind involved in the conservation of linear momentum. Since, in a conservative system, electrostatic potential varies only with the spatial coordinates, this is, in effect, a statement of the principle that the quantity of electric charge is conserved because the spatial coordinates are not. The result is exactly what we would expect from the symmetry of the parameter group; it is a special case of the first theorem, though we could also extend it to weak and strong charges.

In the second, and more significant case, the relation between spin and statistics observed in fundamental particles could be explained by saying that fermions and bosons have different values of spin angular momentum (respectively, ½-integer and integer multiples of Planck's constant $\hbar$), and they also differ in that fermions carry weak units of charge, where bosons do not. In some way, then, the presence of a particular *type* of charge determines the angular momentum state of the particle, so conservation of this type of charge is linked with the value of angular momentum. In later chapters we will show that the conservation of angular momentum does indeed require the separate conservation of weak, strong and electric charges, through the conservation of the separate properties of orientation with respect to the linear momentum, direction, and magnitude; and that the general theorem is of crucial importance for the understanding of particle structures and of symmetry-breaking between the three interactions.

2.9 Analytic Versus Synthetic

The question that we began with has been resolved to the extent that we can identify a group of fundamental parameters that forms the basis of all physical experience and all physical laws, but it will be necessary, before we attempt specific applications, and a derivation of the even deeper foundations, to look at some more general philosophical issues, and to return to a more detailed description of the historical process which created the subject as we know it today. Though the question of why physics works should certainly be approached with the aim of commonsense comprehensibility, this does not mean that we must expect reality to be structured in a commonsense way. In fact, a serious analysis quickly shows that it is not, and that at least one particular metaphysical proposition must be accepted before we can make sense of the fundamentals of physics. In a sense, the proposition is an ancient one, but it has never perhaps been applied with the rigour it deserves, and certainly not in a physical context. The proposition states simply that reality cannot be characterized. We cannot choose, for example, to say that 'reality' is countable, measurable, or observable. We have no fundamental right to suppose that it is any of these things simply because it would be convenient for us if it were. Though common sense would seem to require it, and the attempt indeed has often been made, it is, in fact, philosophically absurd to insist that 'nature' or 'reality' is intrinsically susceptible to measurement, or, that it is, say, intrinsically continuous or discrete.

This proposition, has several significant consequences for the way physics is structured. Our examination of the structure of fundamental physics, for example, has revealed that symmetry is an essential ingredient in a successful theory. It has also suggested that symmetries at the fundamental level must be absolute. Such absoluteness cannot be achieved at any but the most abstract or analytic level. The major fundamental ideas in physics have always been analytic rather than synthetic. Strangely, such analytic ideas have always been resisted, even by other physicists, especially when their bases are abstract rather than concretely realisable. The Newtonian method is a classic instance. Like quantum mechanics in our own time, it was opposed *in principle* by nearly all leading scientists of the time, because it postulated an abstract concept of gravitational force independent of any known mechanism which could produce it. It was used only because it worked. That is, it succeeded ultimately, not on account of its fundamental analytical validity, but because it could be applied *synthetically* to a wide range of physical phenomena.

Newton's method separated the abstract *system* from physical *measurement.* The system had a perfection that could never be physically realised. The principle underlying all his work was that there were a few certain types of information which were more fundamental than others and that these were abstract and could be defined precisely in an abstract way without regard to any model of nature based on concrete terms. He believed that 'to derive two or three general Principles of Motion from Phaenomena, and afterwards to tell us how the Properties and Actions of all corporeal Things follow from these manifest Principles, would be a very great step in Philosophy, though the Causes of those Principles were not yet discover'd'.[9] For Newton, though not for his mechanistically-inclined contemporaries, the ultimate causes of things were abstract rather than mechanical. The laws describing the system did not depend on any physical hypotheses. As he said in Query 28 of the *Opticks* of 1717: 'the main Business of natural Philosophy is to argue from Phaenomena without feigning Hypotheses, and to deduce Causes from Effects, till we come to the very first Cause, which certainly is not mechanical ...'[10]

According to this way of doing physics, universal laws are abstract definitions and do not primarily describe nature. Scientific knowledge is not organised around hypotheses, mechanistic or otherwise, but around fundamental abstract principles which are not derived directly from experiment, but which are the basic channels through which experimental information is organised. Since they only concern details and since assumptions of any kind can be made in hypotheses, experimental tests are not tests of the validity of fundamental laws.

All other physical laws than the universal ones are solutions of general equations which are ultimately approximate or local. Again, the universal law cannot describe any particular physical system, but is rather a totally abstract statement of a relationship between fundamental parameters of measurement such as mass, space and time. Mathematically, universal laws are expressed by differential equations of which there is no exact solution. The solution always involves an approximation, which does not directly relate to the equation. Differential equations are expressed, not in terms of algebraic relations between the quantities themselves, but in terms of their rates of change (as in dx / dt), or rates of rates of change (as in d^2x / dt^2). To convert from relations involving rates of change to simple and direct relations between the original quantities (which is described as 'solving' the equations), one has to reduce the general equation to a particular case by imposing 'boundary conditions' and this is essentially a process of approximation. In effect, general and exact laws cannot give us direct knowledge; to obtain the latter, we have to reduce the infinite number of possible solutions to a particular and individual case using some kind of approximation.

In Newton's own system, the universal law of gravitation, everything attracting everything else, meant that the system was fundamentally indeterminate. Perfection certainly existed in the inverse-square law of attraction between all particles, but the very universality of this law made it impossible to have perfection in a system of such particles in motion. The motion of every particle depended on an infinite number of interactions; strictly speaking, it was not even possible to specify the motion of a particle unless the effect of all these interactions was known. There was no perfection in observed nature, only in the abstract system.

It is significant that mass, in Newton's theory, is something other than the mere quantity of matter. He even describes it as a kind of 'force', the 'impressed force', and gives it force-like properties. The relation between matter and force in the theory allows something outside of matter and opposing it, which can be treated abstractly. Newton, with his theological inclinations, called it 'spirit', which brought him into conflict with the more strictly materialist followers of Descartes. However, in later physics, there is always something other than matter, which has this characteristic. In the nineteenth century, it might be the field concept or aether; in the twentieth century it might be energy or vacuum. A version of it comes into particle physics with the distinction between fermions and bosons. The ultimate origin, in my view, is the distinction between mass and charge (which roughly correspond to the nineteenth-century ideas of aether and matter), but it is important that physicists have always found the need for something in opposition to the matter concept. Physicists cannot be pure 'materialists'.

2.10 The Power of Analogy

It will be evident from the previous sections that the concept of 'charge' has a fundamental role to play in physics, though this has not always been fully recognised. Its historical development is also an instance of the power of analogy. The idea of 'charge' developed as the eighteenth century struggled to establish the electrostatic force as inverse-square, like gravity, while Kant showed that this was a natural result of 3-dimensional space. The whole thrust of this work was to produce an analogy between the different physical forces. Though current electricity and electromagnetism complicated the picture, Maxwell, eventually set down, in mathematical form, all the laws that were known to be valid for electric and magnetic fields – in particular, those of Coulomb, Ampère and Faraday – and in doing so noted that there was an asymmetry which could be corrected by the addition of another term to the law of Ampère. This term was the so-called displacement current – a current which he supposed must exist when static charge was supplied to a parallel plate capacitor.

Even though there was no physical justification for such a current except by analogy, the assumption had a remarkable effect, for Maxwell was immediately able to generate wave equations whose velocity was exactly that of light.

The objection to Maxwell's theory of the electromagnetic field for many years was its intrinsically abstract nature. William Thomson (Lord Kelvin) famously declared: 'I never satisfy myself until I can make a mechanical model of a thing. If I can make a mechanical model I can understand it. As long as I cannot make a mechanical model all the way through I cannot understand; and that is why I cannot understand the electromagnetic theory.'[11] However, following the special theory of relativity, the whole of electromagnetic theory became explicable as an extension of Coulomb's inverse-square law by the addition of a fourth dimension onto that of space in Minkowski's space-time (which did away with the need even for Einstein's simplified kinematics). The simple parallel between electromagnetism and gravity (or between mass and charge) had at last been established, and it was natural to assume (as in general relativity) that the 4-dimensional space-time connection applied to the gravitational force, as well as the electromagnetic.

The method of analogy presupposes the more fundamental concept of symmetry, and this would seem, as we have seen, to be the magic ingredient which makes physics work. Symmetry allows us to do what Newton and other analytic physicists have wished to do: to define an abstract, unknowable reality, combined with a process of observation or measurement of its parts. Symmetry is not really, as we might imagine, a measure of similarity, but a measure of difference, or, to be more specific, absolute oppositeness or negation. Symmetry between two concepts means absolute identity in most respects, combined with absolute opposition in one. So symmetry allows us to characterize a part of reality without characterizing the whole. Only through symmetry can unity result in diversity. And physics works in such a way that when you characterize a part of reality in a certain way, you are necessarily characterizing the rest as different (i.e. opposite). This is what explains Newton's success in introducing mass as a conserved quantity opposed to the variables space and time, and also the success of his opposition of matter and what he called 'spirit' (ultimately resolving itself as charge and mass). But he didn't consciously set out to do this; he developed by the ruthless application of analytical techniques the only procedures that would work.

The same applies to the creators of quantum mechanics. Though the abstract aspects of the Newtonian and Maxwellian theories were long resisted on account of their intrinsically abstract nature, quantum mechanics has forced modern physicists into the same abstract positions. When Werner Heisenberg introduced his new mechanics, strongly influenced by the formalized dynamical tradition dating back to Lagrange, in which relations were expressed only between observable quantities, he abandoned the reality of Bohr's physical electron orbits and the concept of orbital radius, in order to retain the measurable quantity of frequency as a fundamental observable. This led to Bohr's Copenhagen interpretation, in which the abstract system was effectively separated from the physical measuring apparatus. The subsequent development of the ideas of nonlocality and entangled states, backed up by strong experimental evidence, has led physics back to the indeterminate infinity of interacting states required even in the classical Newtonian theory, but ignored by his successors.

Quantum mechanics has left many people puzzled. It is clearly a highly successful theory, which can make predictions to eleven places of decimals in the case of the magnetic moment of the electron, but why does it imply that there is no fundamental 'naïve' reality in which real particles with real positions and real momentum states interact with each other with real forces? The answer ought to be simple, and I believe that it is. We have no right to believe that nature can be described according to the principles of measurement, or according to those of 'naïve' realism. Quantum mechanics, in fact, merely takes to an extreme the principles of conservation and nonconservation which underlie the more classical areas of physics. We simply construct conservation equations for charge and mass which allow for the *complete* variation of the nonconserved parameters space and time which is part of their original specification. It describes the ultimate abstract system based on symmetry.

2.11 The Nature of Reality

There is no such thing as 'reality'. Physics has been constructed in such a way that it avoids creating any such concept. The power and the generality of the subject originates entirely in this. However, circumscribed beings like ourselves cannot avoid thinking in 'realistic' terms, and so we have created a system in which apparent reality in one aspect is countered by total nonreality in another. Hence, at the most fundamental level, physics is described by an abstract system, whose relationship to the original concept of measurement is only ever indirect, though it must always be present. Measurement is a component of the system, but it cannot describe it completely. In addition, the mathematical structures which we employ are not a separate system which we *apply* to physics but an integral component of it. Valid mathematical structures have an ultimately physical origin.

Abstraction is certainly one of the key features of this approach, as is simplicity (which is effectively the same thing). But simplicity and abstraction alone are not enough, for fundamental ideas, however simple and abstract, and however few, are still a characterization of reality. They also do not explain why simplicity, though no doubt 'convenient', provides explanations for systems that are clearly complex; but, according to our reasoning, this is because physics has a principle of symmetry which makes it possible to overcome the problem of characterization. If our fundamental ideas have the built-in opposition which symmetry imposes, then they will never uniquely characterize reality. The idea, as we have said, has ancient roots, and has appeared in various systems of belief and philosophies in many guises. As the poet Coleridge expressed it: 'Every power in nature and in spirit, must evolve an opposite, as the sole means and condition of its manifestation: and all opposition is a tendency to re-union. This is the universal Law of Polarity or essential Duality.'[12] If to every concept there is an exactly symmetrical opposite, then we never have to specify reality if we use both at the same time.

Symmetry can also be exact, in a way that no specific idea can, and it helps us to reduce our starting assumptions without reducing our range of options. In also helps to explain the necessity of simplicity, for it is, in fact, impossible to conceive of one without also requiring the other. But symmetry does *not* mean identity. Space and time are not identical, but *typically symmetrical* concepts: they have some points of absolute identity and others which are different, and indeed symmetrically opposite – facts which, as we have seen, ultimately explain wave-particle duality and Zeno's paradoxes. Clearly, to obtain these exact identities and opposites, we cannot have true symmetry until we have stripped down knowledge to the simplest possible way of thinking. If our symmetries are to be absolute, they must also be simple. It follows also that the programme to reduce everything to an aspect of space in multiple dimensions is fundamentally misconceived because physics needs its symmetrical opposites for its success. Unity requires *oppositeness* as well as similarity.

It is clear that the search for a specifically *unified* theory, however secular it has now become, is essentially at one with the originally theologically-inspired project of the fourteenth century, subsequently continued into the seventeenth century by Galileo and Newton, and we now have a better understanding of what such a theory would actually look like. It would certainly be characterized by abstraction, simplicity and symmetry. It would also be, in principle, extreme, no compromise being allowed for an ultimate theory. There would be no mathematics, other than that derived through symmetry principles, no model-dependent structures of any kind, and no arbitrary succession of inexplicable events, as supposed in some versions of cosmology. It would certainly look different from any theory yet devised for the various particular aspects of physics, yet these would all be ultimately deducible from it. Our analysis of such structures as we consider to be at the heart of physics suggests that they are not there either for mathematical or for measurement convenience. Physics does not work because it provides a simple or convenient description of 'reality'. Physics works because it has successfully, and uniquely, avoided characterizing nature.

Chapter 3

The Emergence of Physics

Certain sections of this chapter (**3.2** and **3.6** to **3.8**) continue the collaboration with **BERNARD DIAZ** begun in chapter 1. Having established inductively the fundamental symmetry which is the basis of physics, we can now proceed to derive this on a deductive basis. The algebra needed to combine the respective real scalar, pseudoscalar, vector and quaternion operators of mass, time, space and charge, as independent units, and the processes defined in the rewrite procedure as conjugation, complexification and dimensionalization, is the 32-part Dirac algebra of the gamma matrices, the simplest way of generating which is via five composite units. Mathematically, this must be accomplished by taking one of the two three-dimensional parameters and superimposing each of the dimensional parts separately onto one of the other three parameters. If we do this using charge, we create a conserved and quantized composite which we call the *Dirac state*. It is easy to show that the Dirac state has an infinite number of *nilpotent* solutions, that is ones which square to zero, and we can exploit this by matching the conserved Dirac state with the nonconserved partner which we know must exist (the differential operator) to create the *Dirac equation*, which, in this form, becomes the most fundamental equation in physics. Of course, the Dirac state is only the most convenient way of packaging the information needed to create a fundamental physical unit, and the system requires that the units continue to infinity, which is accomplished by making them components of an infinite vector space. The nilpotent nature, however, ensures that they are *unique* units, and the idea has major significance for theoretical computing, in addition to physics and mathematics.

3.1 The Mathematical Character of Physics

Physics, we can be assured, has not adopted a mathematical character by accident, nor is it described in mathematical terms for 'convenience'. It is not at all convenient that the mathematics required to define general physical laws (calculus) is essentially incompatible with the mathematics required for observation (counting). Physics has adopted a mathematical character because it is a description of exactly the same thing as mathematics. Each is meaningless without the other, and each works to preserve the characterlessness of what it describes. Their fundamental concepts simply use different terminology to describe exactly the same thing. The apparent differences of approach are purely historical in origin. The two methodologies have arrived at the same point by taking different routes – inductive empiricism in the case of physics, deductive rationalism in the case of mathematics. The fact that contrasting approaches yield related information is entirely to be expected from the dualistic nature of the thing being investigated, for discovery of fundamental dualities is the result of all attempts at scientific investigation.

There is no evidence to suppose that what we describe as 'reality', either in the concrete sense of something directly perceived, or in the sense of an ontological concept beyond immediate perception (another essential dualism!), has any defining characteristic. In fact it would appear that 'reality' or 'nature' goes out of its way to avoid being characterized. Physicists have learned to deal with this by default, effectively by a natural selection of the only method which works. This is to insert a probe into nature, observe how the response denies validity to the probe, and then incorporate both probe and response under the guise of 'symmetry'. Inserting a probe is the process we define as 'measurement' or observation. Measurability depends on discreteness. The simplest discrete thing is a point. So we start with the creation of a point or point 'particle'. A single particle has to have inherent symmetries, so bringing in the dualities that measurement apparently denies. The most fundamental are the ones we have defined as belonging to the four parameters space, time, mass and charge, and these turn out to be the same symmetries or dualities that lie at the heart of the foundations of mathematics.

3.2 The Algebra of Space, Time, Mass and Charge

Each of the processes involved in the generation of the sequence of mathematical structures by the rewrite mechanism – conjugation, complexification, and dimensionalization – would appear to have a realization in physics, which seemingly contrives to use the minimum possible structure for returning to zero without privileging any of the component processes. The structure previously proposed as foundational to physics suggests that the only truly fundamental parameters are space, time, mass(-energy) and charge, which also have an internal group symmetry, which, for the purposes of this discussion, can be expressed in the following form:

space	nonconjugated	real	dimensional
time	nonconjugated	complex	nondimensional
mass	conjugated	real	nondimensional
charge	conjugated	complex	dimensional

corresponding to the already proposed structure:

space	nonconserved	real	dimensional / discrete
time	nonconserved	imaginary	nondimensional / continuous
mass	conserved	real	nondimensional / continuous
charge	conserved	imaginary	dimensional / discrete

Conjugated here is equivalent to conserved, so a positive charge (or source of mass-energy) cannot be created without also creating a negative one. The original C_2 duality (R / R^*) originates from the act of creating 'something from nothing' (R from 0). The creation is the very definition of *nonconservation*, as is the concept of 'successor' which it implies, while the real / imaginary and noncountable / countable distinctions, which we have derived inductively from observed physical characteristics of the parameters, are identical to the further C_2 distinctions which extend the original C_2 duality into complexity and cyclic dimensionality.

Particularly significant, here, is the fact, ascertained from the mathematics, that countability or discreteness is a necessary requirement for cyclic multidimensionality, for unidimensionality is an obviously necessary property of a continuous or noncountable quantity – it can't have an origin. Only the (3-) dimensional quantities, space and charge, are countable, and dimensional quantities are firmly identified with discrete ones. Multidimensionality is, clearly, also identified as a *necessary* property of discreteness, which has to have a reference or origin, and, physically, one cannot imagine a mechanism for dividing the units in a single dimension. As we have previously stated, it is because this is impossible for time that it becomes physically irreversible, and for the same reason mass-energy becomes physically unipolar, with only one sign – and profound consequences for quantum mechanics and particle physics. Neither quantity allows a dimensional discontinuity or origin representing a zero state.

In addition, the mathematical processes which allow for the continual recreation of new non-integral structures in 1-to-1 correspondence with the integers would be inconceivable in a system without dimensionality; while, as previously outlined, two versions of the 'real' numbers are required: the uncountable ones of the Cantor continuum and standard analysis (for mass), and the countable ones of the Löwenheim-Skolem arithmetic and Robinson's non-standard analysis (for space). Spatial real numbers may be expected in the quantum context, but, in classical terms, where 'measurement' may be defined as a linear, integral, process, rational and algebraic numbers emerge from the dimensional apparatus, but the transition to real numbers comes from using the variability condition to define space in terms of a locus, or variation under certain fixed conditions. The fixity here determines the countability, against the non-countability of time.

It can be seen, in addition, that the parameters not only encode the three processes involved in mathematical dualling on an equal basis, but also represent stages in the emergent algebra that it creates:

order 2	real scalar	1	mass
order 4	pseudoscalar	i	time
order 8	quaternions	$\boldsymbol{i}, \boldsymbol{j}, \boldsymbol{k}$	charge
order 16	multivariate vectors	$\mathbf{i}, \mathbf{j}, \mathbf{k}$	space

And, as has already become apparent, if we put these four mathematical structures together in a single algebra, they constitute the complex double quaternion or multivariate vector quaternion algebra, which occurs at order 64, and which is the algebra required to represent the Dirac state. The Dirac state, as we will show, is, simply, the combined state of space, time, mass and charge, putting the four parameters onto an equal overall footing in a single mathematical representation. Its behaviour as a unit also shows how the group of space, time, mass and charge has all the elements required to extend physical duality to infinity. Here, *conjugation* introduces opposite algebraic signs; *complexification* multiplies throughout by a single imaginary term; dimensionalization multiplies again by a new imaginary term completing the quaternion set:

order 2	conjugation	×	$(1, -1)$
order 4	complexification	×	$(1, \boldsymbol{i}_1)$
order 8	dimensionalization	×	$(1, \boldsymbol{j}_1)$
order 16	complexification	×	$(1, \boldsymbol{i}_2)$
order 32	dimensionalization	×	$(1, \boldsymbol{j}_2)$
order 64	complexification	×	$(1, \boldsymbol{i}_3)$

At the fourth stage, significantly, the cycle begins to repeat. So we have:

order 2	real scalar
order 4	complex scalar (real scalar plus pseudoscalar)
order 8	quaternions
order 16	complex quaternions or multivariate 4-vectors
order 32	double quaternions
order 64	complex double quaternions or multivariate vector quaternions

To incorporate the first four as independent units of a single universal system we need order 64 (the Dirac algebra), and we can recognize the four as *introducing* the units of the physical quantities which, as we will show, make up the nilpotent state vectors constituting the Dirac state.

order 2	mass	real scalar
order 4	time	pseudoscalar
order 8	charge	quaternion
order 16	space	multivariate vector

The mathematical structure we have generated using the rewrite system is related closely to Clifford algebra, and it is possible to see that this can be taken as an origin for the conventional mathematics based on defining a number system. The conversion to a specifically physical application occurs when we apply the whole structure (i.e. all possible zero totality alphabets) at once, while using the fact that a repeating structure can be recognised after the first four alphabets. This requires the constraint that the information contained within the alphabets allows an immediate return to zero within a universal perspective.[1]

Significantly, the Dirac state uses a compactified form of the Dirac algebra, with the components projected onto a 3-dimensional operator. This enables the terms of the first two alphabets to be *extrinsically* structured according to the discrete numbering system which is introduced only with the third, while retaining their *intrinsically* continuous (unstructured) character, and so preserves the interpretation of higher order alphabets as extensions of lower order ones.[2] That is, the alphabets incorporating time, space and mass, become extrinsically structured using the one associated with charge. So, if we begin with

time	space	mass	charge
i	$\mathbf{i}\ \mathbf{j}\ \mathbf{k}$	1	$\boldsymbol{i}\ \boldsymbol{j}\ \boldsymbol{k}$

the first three terms, when associated with their undefined or real number 'solutions' (say, t, x, y, z, m), and their sign options (see 3.5), become:

$$\pm\, it \qquad \pm\, \mathbf{i}x \pm \mathbf{j}y \pm \mathbf{k}z \qquad 1m$$

Using the charge units as extrinsic structuring, we obtain new composite physical units:

$$ik \qquad \mathit{i}\mathbf{i} \;\; \mathit{i}\mathbf{j} \;\; \mathit{i}\mathbf{k} \qquad j$$

with new real number values (E, p_x, p_y, p_z, m), which may now be named for the first time as *energy* (E), *momentum* ($\mathbf{p}$) and *rest mass* (m), together with a collective unit of *angular momentum*, as will be explained in 3.4:

$$E \qquad \mathbf{p} \qquad m$$

$$\pm ikE \qquad \pm i\mathbf{i}p_x \;\pm i\mathbf{j}p_y \;\pm i\mathbf{k}p_z \qquad jm$$

Here, the sign options of the real number solutions are those available from the parent parameters and their associated mathematical structures. (As with the parameters, 4 sets of sign options are also the minimum needed to zero all possible versions of 3 concepts.) Connections between the terms in the combined (Dirac) state now automatically introduce the ideas of the quantum state and relativity without requiring an extrinsic physical origin.

In thus applying a unit or numbering structure (derived from 3-dimensionality) to the original alphabetic categories, we specify that they must be algebraically related. Thus, we force an algebraic connection between the units $\mathbf{i}, \mathbf{j}, \mathbf{k}$, 1 and i, and the only structure which can accommodate this connection is itself 3-dimensional. So we end with an interlocking of 3 × 3-D systems, one of which remains incomplete, leading to two of the four parameters remaining unconjugated (nonconserved). The Dirac state, therefore, is not complete in itself. It is only completely specified in partnership with the 'rest of the universe'. In principle, this must create zero totality. In the special case where ($\pm ikE \pm i\mathbf{p} + jm$) is a *nilpotent*, squaring to zero, the conjugate or dual state, $-(\pm ikE \pm i\mathbf{p} + jm)$, becomes the 'rest of the universe' (or, as we will later describe it, *vacuum*) because both the superposition, $(\pm ikE \pm i\mathbf{p} + jm) - (\pm ikE \pm i\mathbf{p} + jm)$, and the combination $-(\pm ikE \pm i\mathbf{p} + jm)(\pm ikE \pm i\mathbf{p} + jm) = 0$ for a nilpotent state.[3] Summarising, we could say that the requirement of simultaneous validity for the first four alphabetic structures creates problems in representing discrete quantities with complete quaternionic structures and continuous quantities with incomplete ones. In effect, we can only create zero totality for all the first four alphabetic structures simultaneously if they are packaged in such a way that we obtain nilpotent solutions for the combined state ($\pm ikE \pm i\mathbf{p} + jm$), with E, $\mathbf{p}$, m real. The packaging of the four physical components at order 64 in the Dirac nilpotent, however, allows us to introduce a new level of closure, again determined by anticommutativity and 3-dimensionality, with an immediate (potential) return to zero. With E, $\mathbf{p}$ and m in the Dirac state represented by real numbers (from the parent quantities, time, space and mass), we can define solutions for which the state is unique (0 if squared). However, the infinite variation within the universal rewrite system determines that there are infinitely many quaternionic systems beyond the three used in creating the Dirac state.

How do we incorporate these into the physical picture? The answer is that these are commutative to those used in the Dirac state, so there will be infinitely many Dirac states, each with different commutative coefficients. So the infinite variation within the universal rewrite system determines that each new state introduces a coefficient which is necessarily commutative with those of the nilpotent system. The recipe here is for a quantum mechanical universe based on nilpotent states within an infinite-dimensional Hilbert space (which is conveniently defined by a Grassmann algebra). Only in such a system can the uniqueness of zero, as an infinitely degenerate concept, be maintained.In mathematical form, individual antisymmetric nilpotents $\psi_1 = (\pm ikE_1 \pm i\mathbf{p}_1 + jm_1)$, ψ_2, ψ_3, ..., have coefficients which are unrepeated but arbitrary units or strings of commutative units. This generates an infinite-dimensional Grassmann algebra, with successive outer products defined by the Slater determinant, with $\psi_1 \wedge \psi_1 = 0$

$$\text{and} \qquad \psi_1 \wedge \psi_2 = -\,\psi_2 \wedge \psi_1 \qquad \text{etc.}$$

The nilpotent units ψ_n must be both nilpotent and antisymmetric, and each must be unique to avoid the trivial case of $R = 0$ (an immediate return to zero). This infinite-dimensional algebra is equivalent to the complex Hilbert space of conventional quantum theory, with algebraic and nonlocal superposition of fermionic or Dirac states throughout the entire universe. A possible consequence of an infinity of Dirac nilpotent states with commutative coefficients, constructing a 'universe', is that, from within such a universe, we would have no idea of the internal structure available within the external commutative coefficients. We could even suppose that they contain 'parallel universes', though the idea is probably meaningless.

3.3 The Dirac Algebra

In the parameter group, not only are the properties dual, but so is the distribution between the parameters. This is why the minimum representation of the full duality is the Dirac algebra, of order $C_2 \times C_2 \times C_2 \times C_2 \times C_2 \times C_2$, which is produced by the 64 possible combinations of the 'vector quaternion', $\pm \mathbf{i}$, $\pm \mathbf{j}$, $\pm \mathbf{k}$,

$\pm \boldsymbol{i}, \pm \boldsymbol{j}, \pm \boldsymbol{k}$. Hidden within this representation, but expressive of the cyclic nature of the operators, are the respective pseudoscalar and scalar terms, $\pm i$ and ± 1 (we could, alternatively, use the 'double vector', $\pm$ **i**, $\pm$ **j**, $\pm$ **k**, $\pm\, ii$, $\pm\, ij$, $\pm\, ik$). The fact that the Dirac algebra can be derived from a combination of two three-dimensional operators now suggests a further possibility.[4] This is that one of the two three-dimensional parameters may be mapped on to the other three parameters, represented as the 'dimensions', and, in fact, the smallest set of units from which the full algebra can be derived comes from exactly such a mapping.

In the Dirac algebra, the various combinations of the 8 basic units, 1, i, **i**, j, **k**, $\boldsymbol{i}, \boldsymbol{j}, \boldsymbol{k}$, generate the entire set of 32 parts, composed of:

2 complex numbers	$(1, i)$
6 complex unit vectors	$(1, i) \times (\mathbf{i}, \mathbf{j}, \mathbf{k})$
6 complex unit quaternions	$(1, i) \times (\boldsymbol{i}, \boldsymbol{j}, \boldsymbol{k})$
18 complex vector quaternions	$(1, i) \times (\mathbf{i}, \mathbf{j}, \mathbf{k}) \times (\boldsymbol{i}, \boldsymbol{j}, \boldsymbol{k})$,

each with + and – parts. Alternatively, we can specify 1 real scalar, 1 imaginary scalar, 3 real vectors, 3 imaginary vectors, 3 quaternions, 3 imaginary quaternions, 9 real vector quaternions and 9 imaginary vector quaternions.

However, 32 parts can also be derived from the binomial combinations of 5 quantities, so we can also generate the entire structure from a *pentad* set, equivalent to the gamma matrices, such as $i\boldsymbol{k}$; $\mathbf{i}\boldsymbol{i}$; $\mathbf{j}\boldsymbol{i}$; $\mathbf{k}\boldsymbol{i}$; $\boldsymbol{j}$. In effect, generating the algebra from 5 units could be taken as 'simpler' and more efficient than generating it from 8 units, and so the compactified composite set could be taken, in some senses, as more mathematically 'fundamental' than the original basic set.

Operationally, the pentad set is isomorphic to the one defined by the 5 γ matrices, traditionally used in conventional quantum mechanics, and we can make correlations of the form:

$$\begin{aligned} \gamma^0 &= -i\boldsymbol{i} \\ \gamma^1 &= \mathbf{i}\boldsymbol{k} \\ \gamma^2 &= \mathbf{j}\boldsymbol{k} \\ \gamma^3 &= \mathbf{k}\boldsymbol{k} \\ \gamma^5 &= i\boldsymbol{j} \end{aligned} \qquad \text{or} \qquad (3.1)$$

$$\begin{aligned} \gamma^0 &= i\boldsymbol{k} \\ \gamma^1 &= \mathbf{i}\boldsymbol{i} \\ \gamma^2 &= \mathbf{j}\boldsymbol{i} \\ \gamma^3 &= \mathbf{k}\boldsymbol{i} \\ \gamma^5 &= i\boldsymbol{j}. \end{aligned} \qquad (3.2)$$

Here, the γ terms and their equivalents all anticommute, while $\gamma^5 = i\,\gamma^0\gamma^1\gamma^2\gamma^3$. The squares of γ^0 and γ^5 are 1, while the squares of the other terms are –1. The full 32 parts can be derived as follows, with terms of the opposite sign produced by reversing the order of multiplication:

$$1$$
$$\gamma^0 = i\boldsymbol{k},\ \gamma^1 - i\mathbf{i},\ \gamma^2 - i\mathbf{j},\ \gamma^3 = i\mathbf{k},\ \gamma^5 = i\boldsymbol{j},$$
$$\gamma^0\gamma^1 = i\boldsymbol{j}\mathbf{i},\ \gamma^0\gamma^2 = i\boldsymbol{j}\mathbf{j},\ \gamma^0\gamma^3 = i\boldsymbol{j}\mathbf{k},\ \gamma^0\gamma^5 = \boldsymbol{i},\ \gamma^1\gamma^2 = -i\mathbf{k},$$
$$\gamma^1\gamma^3 = i\mathbf{j},\ \gamma^1\gamma^5 = i\boldsymbol{k}\mathbf{i},\ \gamma^2\gamma^3 = -i\mathbf{i},\ \gamma^2\gamma^5 = i\boldsymbol{k}\mathbf{j},\ \gamma^3\gamma^5 = i\boldsymbol{k}\mathbf{k},$$
$$\gamma^0\gamma^1\gamma^2 = \boldsymbol{k}\mathbf{k},\ \gamma^0\gamma^1\gamma^3 = -\boldsymbol{k}\mathbf{j},\ \gamma^0\gamma^1\gamma^5 = \mathbf{i},\ \gamma^0\gamma^2\gamma^3 = \boldsymbol{k}\mathbf{i},\gamma^0\gamma^2\gamma^5 = \mathbf{j},$$
$$\gamma^0\gamma^3\gamma^5 = \mathbf{k},\ \gamma^1\gamma^2\gamma^3 = -i\boldsymbol{i},\ \gamma^1\gamma^2\gamma^5 = \boldsymbol{j}\mathbf{k},\ \gamma^1\gamma^3\gamma^5 = -\boldsymbol{j}\mathbf{j},\ \gamma^2\gamma^3\gamma^5 = \boldsymbol{j}\mathbf{i},$$
$$\gamma^0\gamma^1\gamma^2\gamma^3 = \boldsymbol{j},\ \gamma^0\gamma^1\gamma^2\gamma^5 = -i\boldsymbol{i}\mathbf{k},\ \gamma^0\gamma^1\gamma^3\gamma^5 = i\boldsymbol{i}\mathbf{j},\ \gamma^0\gamma^2\gamma^3\gamma^3 = -i\boldsymbol{i}\mathbf{i},\ \gamma^1\gamma^2\gamma^3\gamma^5 = \boldsymbol{k},$$
$$\gamma^0\gamma^1\gamma^2\gamma^3\gamma^5 = -i.$$

A set of five units of this kind, or pentad, will always generate the entire Dirac algebra. The 32 parts turn out to be 1 and *i*, and six Dirac pentads, three based on the quaternion operators and three on the vector operators (cf the table in 15.4). Any of the pentad sets can be used as the basis for the five gamma matrices in the Dirac equation and as generators for the Dirac group, but it is most convenient to choose a set based on the quaternions, as here, because charge is a conserved quantity, and the mathematical structure then has a convenient physical interpretation.

The Dirac pentad is the largest possible anti-commuting set, and the six such sets are always generated in the same way, by mapping one 'triad' or 3-dimensional operator (vector or quaternion) onto another. We can always map the gamma algebra to a double-quaternion algebra by defining a triad of anti-commuting numbers (there are 60 such sets) as a quaternion set, then calculating the conjugate triad (another quaternion set) which commutes with the original triad (there are 30 such triad pairs). Such a pair is then the double-quaternion (or quaternion / vector) set, which gives the 3 quaternion, 3 vector, 9 product algebra. It can be shown that, no matter how such a triad pair is chosen, *every* pentad must contain two members of one triad, while the remaining three terms are the product of the third member of the triad with each member of the conjugate triad.[5]

The mathematical and physical symmetry breaking, which is characteristic of the formation of the Dirac state, occurs when the conjugate triads are chosen. In the general algebra, one is never required to define any triads as privileged, so the algebra has 5-dimensional global symmetry. However, once a triad is chosen, the conjugate triad is also uniquely defined. It is this choice which breaks the global symmetry, and allows one to define an *SU*(3) component within the overall structure, as is found in fundamental physics.

3.4 The Creation of the Dirac State

Space, time, mass and charge are separate structures governed by different algebras, but physics has found a way of combining them into a single compactified package, which provides a dramatic short-cut back to the zero origin. This is what we call the Dirac state, and its origins may be found in the algebra, and its most efficient generation via a pentad. There are, as we have seen, many ways of constructing a pentad to generate the Dirac algebra, but all involve taking the components of one of the two 3-dimensional parameters (space or charge) and superimposing one on the units of each of the 3 other parameters. In principle, we could use space, but the conservation property makes it more convenient to use charge. In the case of vector space, the components are not uniquely determined, because the quantity is nonconserved, and can even be arbitrarily reduced to a single one. However, in the case of charge, the conservation property is directly related to the conservation of angular momentum, and so brings in the spatial rotation simultaneously, as becomes evident in the full explanation of symmetry-breaking.

Physically, of course, if the basic units really do represent those of space, time, mass and charge, the new, composite units must represent entirely new physical parameters, produced by the combinations, and, if we choose to perform the 'compactification' using the units of charge we will create composite units that incorporate the properties that are characteristic of charge, namely conservation and discrete quantization. We begin with:

time	space	mass	charge
i	$\mathbf{i}\ \mathbf{j}\ \mathbf{k}$	1	$\boldsymbol{i}\ \boldsymbol{j}\ \boldsymbol{k}$

Then, taking, each of the charge units onto one of the algebraic expressions representing time, mass or space,

i	$\mathbf{i}\ \mathbf{j}\ \mathbf{k}$	1	$\boldsymbol{i}\ \boldsymbol{j}\ \boldsymbol{k}$
$\boldsymbol{k}$	$\boldsymbol{i}$	$\boldsymbol{j}$	

we obtain the following combinations:

$i\boldsymbol{k}$	$\boldsymbol{i}\mathbf{i}\ \ \boldsymbol{i}\mathbf{j}\ \ \boldsymbol{i}\mathbf{k}$	$\boldsymbol{j}$

For mathematical convenience, and for compatibility with the conventional way of writing the Dirac algebra, we will often write this in the form:

$\boldsymbol{k}$	$i\boldsymbol{i}\mathbf{i}\ \ i\boldsymbol{i}\mathbf{j}\ \ i\boldsymbol{i}\mathbf{k}$	$i\boldsymbol{j}$

The new composite quantities produced by the application of the conserved and quantized units of charge to the parameters time, space and mass naturally combine the characteristics of their parent quantities. Physically, by putting quantized charge components onto time, space and mass units, we introduce quantization to the composite terms, and, since the charges are also conserved quantities, we create a *quantum state* with fixed composite quantities E, $\mathbf{p}$, m. The charge input makes them all conserved and quantized, with the act of imposing charge's three-dimensional structure onto the original time, space and mass being identical to the act of quantization; but the Dirac energy (E), the Dirac momentum ($\mathbf{p}$) and the Dirac rest mass (m) also retain the respective pseudoscalar, multivariate vector, and real scalar properties of time, space and mass.

$$\begin{array}{ccc} ik & i\mathbf{i}\ i\mathbf{j}\ i\mathbf{k} & j \\ E & \mathbf{p} & m \end{array}$$

The concept of 'rest mass' emerges only in this act of 'quantization'. Another conserved and quantized quantity, the Dirac angular momentum, relates to the directional properties of the vector term, and, in some sense, to the Dirac state as a whole. It is often stated that Dirac himself, on the basis of the quantization of angular momentum incorporated in the Dirac equation, predicted that a magnetic monopole could exist with charge automatically quantized in integral multiples of fundamental constants, and that the existence of one such monopole anywhere in the universe would explain charge quantization. However, as we see here, the fundamentally quantized nature of charge is what *explains* the quantization of angular momentum, and other quantities, in the Dirac state. So, the position is actually reversed.

The combination, however, has another important physical consequence, as the quaternion units, i, j, k, are changed from being symmetrical and indistinguishable representations of independent charges into composite units whose symmetry is broken, by being associated with quantities with different mathematical properties (pseudoscalar, vector and real scalar); and, from the composition of ik, the combined ($i\mathbf{i}$, $i\mathbf{j}$, $i\mathbf{k}$), and j, it is possible to derive the respective $SU(2)$, $SU(3)$ and $U(1)$ symmetries associated with the weak, strong and electric charges. The symmetry between the weak, strong and electromagnetic interactions is thus broken in the creation of the Dirac state.

$$\begin{array}{ccc} ik & i\mathbf{i}\ i\mathbf{j}\ i\mathbf{k} & j \\ w & s & e \end{array}$$

With this built-in degree of symmetry-breaking, we are on the way to understanding important aspects of fundamental physics.

Significantly, the three components E, $\mathbf{p}$, and m, of the Dirac state, which we represent in the form ($\pm\, \boldsymbol{k}E \pm \boldsymbol{i}i\mathbf{p} + \boldsymbol{i}jm$) or ($\pm\, i\boldsymbol{k}E \pm \boldsymbol{i}\mathbf{p} + \boldsymbol{j}m$), are, from the fundamental properties of their parent-parameters time, space, and mass (-energy), specified by unrestricted real number values (though space's are countable in the Löwenheim-Skolem sense). Thus, it is possible, using the anticommuting properties of the quaternion and vector operators, and the presence of at least one complex term, to find values of the state, which square to a *zero numerical solution*. These, in turn, become the units of an infinite higher algebra (Hilbert space), which provides the basic parameterisation that we describe as physics. In effect, the presence of anticommutativity allows physics to create a more direct route to the zeroing or conjugation of an act of 'creation', at the level of the 64-element Dirac algebra, for, in parameterizing the physical world using this algebra, we create a structure which zeros itself by being a nilpotent or square root of zero, so producing a cyclicity at a higher level which incorporates the whole range of procedures required for the rewrite mechanism. The next stage is then simply to make infinitely or indefinitely many applications of this closed system or 'unit' structure to construct the entire physical universe, in the same way as we iterate applications of the quaternion system to construct a system of mathematics.

In terms of the individual Dirac state, the process of creating a conserved state is paralleled by a description in terms of the equivalent process of *nonconservation*. In a nonconserved form this produces the respective quantum (or differential) operators:

$i\boldsymbol{k}$	$\boldsymbol{i}\mathbf{i}$ $\boldsymbol{i}\mathbf{j}$ $\boldsymbol{i}\mathbf{k}$	$\boldsymbol{j}$
$\partial/\partial t$	∇	m

which act on a variable component in the *state vector* or wavefunction, representing vacuum or the rest of the universe.

Treating the momentum term as a single quantity, the free-fermion Dirac state vector becomes $\psi = (\pm\, i\boldsymbol{k}E \pm \boldsymbol{i}\mathbf{p} + \boldsymbol{j}m)\, e^{-i(Et\, -\, \mathbf{p.r.})}$, where ($\pm\, i\boldsymbol{k}E \pm \boldsymbol{i}\mathbf{p} + \boldsymbol{j}m$) expresses the absolute conservation of charge and mass(-energy), and the exponential term, operated upon by ($\mp\, \boldsymbol{k}\partial/\partial t \pm \boldsymbol{i}i\nabla + \boldsymbol{j}m$), the absolute nonconservation of space and time. As a nilpotent, the state vector becomes a precise expression of the fundamentally dualistic process of returning 'something' back to 'nothing' through a squaring operation or self-interaction. The exponential or 'wave' term is a mathematical representation of the group of space and time translations and rotations, providing the maximal variation (or 'nonconservation') for space and time coordinates in the idealised free particle state. It is defined in such a way that, application of the nonconservation or differential operator, $(\mp k\partial/\partial t \pm ii\nabla + jm)$, produces an *eigenvalue* or result which is identical to the expression for the Dirac state $(\pm ikE \pm i\mathbf{p} + jm)$. This process, which we describe as the *Dirac equation*, thus expresses the fact that our fundamental duality has been represented in terms of conservation and nonconservation, and that the effect of applying both is to maintain the zero totality.

The form of the function term (henceforth described as the 'phase' or 'phase factor') is, in fact, totally determined by the nature of the differential operator. In the case of a free particle, the exponential is the only function whose form remains unchanged under differentiation, while a complex exponential shows no progressive change over time; the terms **p.r** and *Et*, in any case, each have each one component (**p**, *t*) which is imaginary. The conventional definition of a fundamental 'particle', of course, assumes an irreducible representation of the Poincaré group, or the group of space and time translations and rotations compatible with Lorentz or special relativistic invariance (i.e. with the representation of space and time as components of a 4-vector), but it can be seen here that such translations and rotations are essentially identical to the conservation properties related to charge and rest mass which define a particle in the present theory.

Both $(\mp k\partial/\partial t \pm ii\nabla + jm)$ and $(\pm ikE \pm i\mathbf{p} + jm)$, here, which each contain four possible sign combinations, are expressed most conveniently as row or column vectors with four components; this is yet another 4-vector mapping, and one which can be accomplished with the four quaternion components, 1, ***i***, ***j***, ***k***, if required. For a non-free (or interacting) state, the phase would, of course, be different but the eigenvalue format would be the same, so that

$$(\pm ikE \pm i\mathbf{p} + jm)\,(\pm ikE \pm i\mathbf{p} + jm) = E^2 - p^2 - m^2 = 0$$

would always be true. In these terms, we can use the symbols E and $\mathbf{p}$ to mean either the operators $i\partial/\partial t$ and $-i\nabla$ or the eigenvalues which they produce by acting on ψ. For comparison with established practice in both classical and quantum physics, we can also write the operator $E = i\partial/\partial t$ as $\mathcal{H}$ (the Hamiltonian, or total energy operator). It is easy to show that, for one spatial dimension x, the operation $[x, \mathbf{p}] = x\mathbf{p} - \mathbf{p}x$, acting on ψ, gives $i\psi$, which can be expressed by the *commutation relation*: $[x, \mathbf{p}] = i$.

It is of deep significance, here, however, that the application of quaternion operators in an expression such as $(\pm ikE \pm i\mathbf{p} + jm)$ does not in itself create the Dirac state – the same algebraic expression could have been used in a purely mathematical factorization of the classical special relativistic energy-momentum expression. It is the act of *equating these operators to the three fundamental charge units*, with their properties of quantization and conservation, that creates the Dirac state by restructuring the meaning of the terms to which they are applied as quantized and conserved ones. The creation of the nilpotent operator is thus equivalent to the process of quantization of E, $\mathbf{p}$ and m, which thus become 'dimensionalized' while, the quaternions specifying the weak, strong and electric charges ($\boldsymbol{k}$, $\boldsymbol{i}$, $\boldsymbol{j}$) become distinguished by being attached, respectively, to scalar, multivariate vector and pseudoscalar operators. It is, of course, relevant here that it is the anticommutative aspect of the quaternion algebra, which introduces discreteness, enumeration, or countability, through cyclicity, and so introduces the discrete quantum state as a unit.

This process, as we shall see, implies the direct relation between conservation of charge and conservation of angular momentum which we have predicted must exist purely on the grounds of symmetry. The same act also establishes direct and inverse numerical relationships between the units E and $\mathbf{p}$, and between those of t and $\mathbf{r}$, leading to the introduction of the constants $\hbar$ and c, and the equations of special relativity (see chapter 8). A third constant, G, is required when we involve m. These constants, as has long been known, have no intrinsic meaning; they are simply the inevitable consequence of creating a composite state. (They also allow an alternative nilpotent expression for relativity, $(\pm\, \boldsymbol{ik}t \pm \boldsymbol{i}\mathbf{r} + \boldsymbol{j}\tau)$, with the conjugate parameters, time (t), space ($\mathbf{r}$) and proper time (τ) replacing energy, momentum and rest mass.) With the explicit introduction of $\hbar$, the operator $\mathcal{H} = E$ becomes $i\hbar\partial/\partial t$, while the operator $\mathbf{p}$ becomes $-i\hbar\nabla$, with an anticommutation relation $[x, \mathbf{p}] = i\hbar$, though the usual convention is to choose units such that $\hbar = 1$ and $c = 1$.[6] (That E and $\mathbf{p}$ are the 'energy' and 'momentum' terms that carry over into the classical transition will be shown in chapter 8.)

The identities of the three 'charge' operators are preserved, even in the combinations of $(\pm\, \boldsymbol{ik}E \pm \boldsymbol{i}\mathbf{p} + \boldsymbol{j}m)$, but they now become discriminated into ones with timelike (weak), spacelike (strong) and masslike (electric) properties, and the effects can be distinguished physically by the aspects of angular momentum conservation to which they relate. The Dirac algebra, which produces the simplest possible combination of all the dualistic properties required by space, time, mass and charge, generates a broken symmetry in the manifestations of the charges' interactions, though it also suggests their idealised unification in an overall $SO(10)$, $SU(5)$ or $U(5)$ group structure.

3.5 The Nilpotent Dirac Equation

The nilpotent Dirac equation can be derived purely from the requirements of a fundamental rewrite algebra, but it is important to show that this form of the equation can also be derived in a more conventional manner. Thus, applying eq. (3.1), from section 3.3, directly to the conventional form of the Dirac equation,

$$\left(\gamma^{\mu}\partial_{\mu}+im\right)\psi=\left(\gamma^{0}\frac{\partial}{\partial t}+\gamma^{1}\frac{\partial}{\partial x}+\gamma^{2}\frac{\partial}{\partial y}+\gamma^{3}\frac{\partial}{\partial y}+im\right)\psi=0\,,$$

we obtain
$$\left(-\boldsymbol{ii}\frac{\partial}{\partial t}+\boldsymbol{ki}\mathbf{i}\frac{\partial}{\partial x}+\boldsymbol{kj}\frac{\partial}{\partial y}+\boldsymbol{kk}\frac{\partial}{\partial y}+im\right)\psi=0\,.$$

Multiplying the equation from the left by $\boldsymbol{j}$ then alters the algebraic representation to (3.2) and the Dirac equation becomes:

$$\left(\boldsymbol{ik}\frac{\partial}{\partial t}+\boldsymbol{i}\mathbf{i}\frac{\partial}{\partial x}+\boldsymbol{i}\mathbf{j}\frac{\partial}{\partial y}+\boldsymbol{i}\mathbf{k}\frac{\partial}{\partial y}+\boldsymbol{ij}m\right)\psi=0\,, \qquad (3.3)$$

or, in more compact form,

$$\left(\boldsymbol{ik}\frac{\partial}{\partial t}+\boldsymbol{i}\nabla+\boldsymbol{ij}m\right)\psi=0\,. \qquad (3.4)$$

This is a symmetrical equation in which all terms are equally operated on by a γ term and all the γ operators have equivalent status.

If we now apply a free-particle solution, such as

$$\psi = A\,e^{-i(Et\,-\,\mathbf{p.r})},$$

to equation (3.3), we find that:

$$(\boldsymbol{k}E + \boldsymbol{ii}\mathbf{i}p_x + \boldsymbol{ii}\mathbf{j}p_y + \boldsymbol{ii}\mathbf{k}p_x + \boldsymbol{ij}\,m)\,A\,e^{-i(Et\,-\,\mathbf{p.r})} = 0,$$

or, in a more compact form,

$$(\boldsymbol{k}E + \boldsymbol{ii}\,\mathbf{p} + \boldsymbol{ij}\,m)\,A\,e^{-i(Et\,-\,\mathbf{p.r})} = 0,$$

where $\mathbf{p}$ is a multivariate vector. Allowing for two signs of E and two of $\mathbf{p}$, we obtain:

$$(\pm\,\boldsymbol{k}E \pm \boldsymbol{ii}\,\mathbf{p} + \boldsymbol{ij}\,m)\,A\,e^{-i(Et\,-\,\mathbf{p.r})} = 0.$$

The equation is only valid when A is a multiple of $(\pm\,\boldsymbol{k}E \pm \boldsymbol{ii}\,\mathbf{p} + \boldsymbol{ij}\,m)$. In principle, this means that A, and hence ψ, must be a nilpotent or square root of zero. Here, of course, we rely on the fact, that, for a multivariate $\mathbf{p}$, the product $\mathbf{pp}$ becomes identical to the product of the scalar magnitudes $pp = p^2$. It is, additionally, identical to the product of the helicities ($\boldsymbol{\sigma}.\mathbf{p}$) ($\boldsymbol{\sigma}.\mathbf{p}$) (where $\boldsymbol{\sigma}$ is effectively equivalent to the vector term $-\mathbf{1}$), indicating that the multivariate vector (or equivalent Pauli matrix) representation of $\mathbf{p}$ automatically incorporates the concept of spin (see 6.3). (It is, of course, possible to use either $\mathbf{p}$ or $\boldsymbol{\sigma}.\mathbf{p}$ in the quaternion state vector.)

The nilpotent version of the Dirac state vector seems to have a more fundamental status than the conventional one, but the two may be easily related. In an equation such as:

$$(\boldsymbol{i}E - i\boldsymbol{k}\ \mathbf{p} - im)\ (\boldsymbol{i}E - i\boldsymbol{k}\ \mathbf{p} + im) = 0, \tag{3.5}$$

which is just one version of the usual energy-mass-momentum relation. The two bracketed terms here are clearly different, and so are not square roots of zero or nilpotents. However, if we multiply from the left by $-\boldsymbol{j}$ and from the right by $\boldsymbol{j}$ (in the process changing the Clifford algebra of the operators from $Cl_{4,1}$ to $Cl_{2,3}$),[7] we get:

$$-\boldsymbol{j}(\boldsymbol{i}E - i\boldsymbol{k}\ \mathbf{p} - im)\ (\boldsymbol{i}E - i\boldsymbol{k}\ \mathbf{p} + im)\boldsymbol{j} = 0, \tag{3.6}$$

which becomes the familiar

$$(\boldsymbol{k}E + i\boldsymbol{i}\ \mathbf{p} + i\boldsymbol{j}\ m)\ (\boldsymbol{k}E + i\boldsymbol{i}\ \mathbf{p} + i\boldsymbol{j}\ m) = 0.$$

The two bracketed terms are now identical, and so each is a nilpotent. Equation (3.5) is, in effect, the usual way of writing the Dirac equation, with the left-hand bracket becoming the differential operator and the right-hand bracket the wavefunction, but, although theoretically equivalent, it is, in fact, much less powerful than (3.6).

In principle, the Dirac equation, as used by Dirac, requires wavefunctions to be *ideals*, based on idempotents (quantities remaining unchanged when multiplied by themselves), and incorporating only a restricted part of the algebra. However, as the analysis in chapters 5 and 6 will show, limiting the wavefunction to an ideal limits drastically the physical interpretation that one can derive from the algebra. In other words, a restriction of the wavefunction to *part* of the algebra is a consequence of the fact that the traditional form of the Dirac equation is incomplete – a fact which results in the necessity of second quantization. However, an additional process of second quantization is, in fact, unnecessary. In incorporating the m term directly into the γ matrix structure, (3.6) effectively goes beyond the conventional Dirac wavefunction towards a quantum field interpretation, though it retains all the physical interpretation available to the conventional term. It is also the only form of the equation which gives all the parameters equal status.

One of the strengths of the nilpotent version here defined is that, both the column vector and each of its individual terms is a nilpotent, which can be converted to an idempotent on pre- or post-multiplication by a single quaternionic operator. A minimal left ideal is a column vector of 'primitive' idempotents multiplied from the left, which is a subgroup of some larger group. The ($\pm$ ***k****E* $\pm$ *ii*p + *ijm*) form is a column vector, multiplied from the left, composed of idempotents before multiplication by ***j***, and nilpotents after multiplication, and it is, in effect, a subgroup of a larger group. It is an extreme case of an ideal.

The nilpotent formalism is so intrinsically powerful that it seems to imply a foundational status. At the quantum level, the physical universe appears to be composed entirely of nilpotent fermionic or antifermionic wavefunctions of this kind or of combinations of them. Antifermionic wavefunctions merely reverse or conjugate the sign of ***k****E* in fermion wavefunctions, producing equally nilpotent terms such as ($\mp$ ***k****E* $\pm$ *ii* **p** + *ijm*), while bosonic wavefunctions are nothing other than combinations of the two. The 'universe', as described by physicists, is essentially an entanglement of all possible nilpotent states. Significantly, no nilpotent can be identical to any other; each must be unique, with *E*, *p* and *m* being unspecified real numbers. Any individual nilpotent wavefunction structure of the form ($\pm$ *i****k****E* $\pm$ ***i*****p** + ***j****m*) must be unique because a product of identical ones would zero the wavefunction of the entire set of fermionic states, so Pauli exclusion becomes obvious.

This specification of uniqueness requires instant correlation, at the same time as the 4-vector nature of the operator connecting *E* and **p** (or, equivalently, space and time) requires time-delayed action between discrete sources. It is also a reflection of the uniqueness or local conservation of individual charge components. The nonzero Berry phase, as manifested in the quantum Hall, Aharonov-Bohm or Jahn-Teller effects, is, as will be shown in later chapters, effectively a realisation of the equivalent antifermionic wavefunction in a fermion's physical 'environment', and is, in this sense, a unique signature. Ordinality is preserved, but enumeration is reserved for the nilpotent units rather than their component parts.

It will be shown in chapters 5 and 6 that the Dirac equation in the nilpotent form becomes the most fundamental equation in physics, incorporating in a compactified structure all the conservation and nonconservation principles which make up classical and quantum physics. All other physical principles are in some sense defined in relation to it, and can be discovered through exploring its many consequences. Through the equation, for example (as alluded to in the previous section), the conserved / nonconserved pairings of E and t, and $\mathbf{p}$ and $\mathbf{r}$, become *conjugate variables*, that is, ones which exchange statements about conservation into equivalent statements about nonconservation, and vice versa. Information about one of the conjugate partners can always be exchanged for information about the other. Much of physics, both quantum and classical, is constructed in this way (e.g. via Poisson brackets), and the exchange can be established by the mathematical technique of *Fourier transformation*.

In connection with this we observe that energy is a pseudoscalar and is conserved in quantity and individual element (i.e. is translation asymmetric) in precisely the same way as the pseudoscalar parameter time is not conserved in quantity and individual element (i.e. is translation symmetric). It may be regarded as the link between time and the real scalar quantity mass (the gravitational source), as the conservation of energy is directly linked to the conservation of mass. Linear and angular momentum are, respectively, vector and pseudovector, and are conserved in quantity and individual element (i.e. are translation and rotation asymmetric) in precisely the same way that the vector parameter space is not conserved in quantity and individual element (i.e. is translation and rotation symmetric). They may be regarded as the respective links between space and the quantitative values of the different charge types and the quaternion operators applied to them.

In addition, as suggested previously, the existence of four 'solutions', that is, four sets of sign variations for the Dirac state, becomes an obvious consequence of the derivation of the Dirac terms E, $\mathbf{p}$ and m from the original parameters time, mass and space. The two signs for the E term derive from the two signs for the imaginary time parameter, while the two signs for $\mathbf{p}$ result from its dimensionality and countability. However, this makes the negative versions of E fundamentally different in character from the negative versions of $\mathbf{p}$, being 'mathematical' rather than 'physical' in origin. Thus, negative $\mathbf{p}$ terms are of equal status to positive ones, but negative E terms, like negative time, do not exist in the ground state of the universe. The m term, in addition, remains positive in all cases because the parent parameter, mass, is unipolar, and, in the transition from quantum to classical systems, m, as a positive scalar, remains unchanged. These distinctions are significant in understanding such fundamental physical processes as the Higgs mechanism.

3.6 Uniqueness, Qubits and Quantum Computing

The nilpotent algebra used in the Dirac formalism, together with its infinite Hilbert space expansion, provides a mathematics of uniqueness previously unexplored. Mathematics is normally structured on the notion that its units are capable of repeated application, but the infinite nilpotent algebra is structured on the idea that its units, though variable, cannot be repeated. This is because a combination of any two identical nilpotents, such as the Hilbert space formalism requires, will be automatically zero. The only way to make a nonzero universe out of these is for each to be unique. This is manifested physically as Pauli exclusion.[8]

By using a series of operators that can be repeated, conventional mathematics loses some of the information which is potentially available; an algebraic structure based on unique operators may be expected to produce an entirely new set of mathematical results. The uniqueness of fermionic nilpotents reflects the uniqueness of charges, just as the superposition of the wavefunctions in the infinite Hilbert space reflects their interactions via mathematical multiplication, and the instant correlation of $\psi_m \psi_n$, etc. (cf chapter 17) reflects the universal gravitational interaction of $E_m E_n$, etc. Significantly, the uniqueness, though the result of an infinite superposition, is manifested within any finite set of nilpotent operators. The nilpotent algebra thus allows us to use the iterative procedure to represent a recursive system. Though we can only examine a finite number of elements, we can identify that they have exact positions within an infinite set. So both iterative and recursive requirements are satisfied at once.[9]

The formalism is only possible because the terms E, $\mathbf{p}$ and m, like the original parameters time, space and mass, from which they were derived, have the full range of real number values. In principle, then, each individual nilpotent can be unique; and must be if, as we believe, the entire universe can be structured as a superposition of fermionic states, with any nonuniqueness in the components producing immediate zeroing. The generating algebra which we have created by our rewrite mechanism can then be extended to infinity, through the physical property of fermionic wavefunctions being nonlocally connected throughout the entire universe. In principle, it is the mathematical interconnectedness of the nilpotent operators that allows us to group its components as a 'unit' of an even higher algebra, which may be in the form either of the conventional complex Hilbert space or, alternatively, the equivalent geometric algebra as demonstrated by Matzke (or even a complex version of the latter).[10] We may, as previously suggested, consider the nilpotents ψ_1, ψ_2, ψ_3, …, with coefficients which are unrepeated but arbitrary units or strings of units of the form $\boldsymbol{i}_s$, as forming an infinite-dimensional Grassmann algebra, with successive outer products defined by the Slater determinant, and so requiring $\psi_1 \wedge \psi_1 = 0$ and $\psi_1 \wedge \psi_2 = - \psi_2 \wedge \psi_1$, etc. To create such an algebra, it would seem, the state vector units ψ_n must be both nilpotent and antisymmetric.[11]

Such algebras create the doubling mechanism provided in our foundational algebraic structure by terms of the form $(1, \boldsymbol{i}_n)$. In effect, the fermionic nilpotents become isomorphic to the fundamental unit of quantum information, or qubit, composed of two orthogonal vectors and their superposition states. So, taking the tensor product of every qubit expands the space exponentially, exactly as in our mathematics of duality. In principle, also, such fermionic qubits would be uniquely labelled, as required for quantum computing, and, theoretically, the fermionic states could be identified by a manifestation of the Berry phase, such as the quantum Hall effect (see chapter 12). (Alternatively, an ideal Bose-Einstein condensate would consist of fermionic nilpotents differing only by their opposite spin states, or sign of $\mathbf{p}$ in the nilpotent formalism.) Deutsch, in his classic foundational paper on quantum computing,[12] states that any physical process can be modelled perfectly by a quantum computer. This, according to our understanding, is because a quantum computer is ultimately described in terms of the same units as real physical processes. The physical universe is the set of all possible quantum computers.

3.7 The Completeness of Mathematical Physics

The description of fundamental processes in chapters 2 and 3 has been aimed at capturing the way that mathematics and physics, in particular, operate at a more fundamental level, and, by doing so, to gain an extra power of understanding and manipulation, but the formalism has significance for the whole of constructible knowledge beyond its immediate utility. Central to this is the idea that the various approaches to fundamental knowledge are aspects of the same overall package, and have deeper connections than has previously been realised. Mathematics, physics, theoretical computation, and even philosophy, emerge together out of the basic idea of duality, though the concept applied is more fundamental than that name, with its connotations of a necessary discreteness, would imply.

Mathematics has been shown to be constructible using this mechanism, with an order which is more coherent than one produced by starting with integers, and it is not something merely 'applied' to physics for 'convenience'. It is, as we have said, extremely *inconvenient*, as the mathematical laws of physics are general differential equations, which have to be reinterpreted ('solved', using different boundary conditions) every time a measurement is taken. Observation and theory, in physics, necessarily use incompatible types of mathematics because observation depends only on one member of the parameter group (space), while theory sets up the properties of the other members in opposition. So, mathematics is required only because it is a fundamental component of physics, and the structure of mathematics itself seems to suggest physical boundaries to the type of ideas which can be made mathematically useful (though obviously not in the form of a purely one-to-one correspondence). In fact, as we have demonstrated, physics can become a kind of test of the ultimate value of mathematical structures at the fundamental level.

For example, physics appears to insist on the fact that all discrete quantities must be dimensional. This would not be required of a mathematical theory based, as most are, on the primacy of the integer series. However, if we begin mathematics with the integer series, then we have major problems in accommodating the reals – there is no natural progression – and, if we assume, like most axiomatists, that the most fundamental proposition in mathematics is $1 + 1 = 2$, we will come up against the problems that Gödel identified, with axiomatic theories or 'rigidly logical systems' which are intrinsically incomplete.

However, in physical terms, we may suppose that the integer series is not primary and that arithmetic, although the most psychologically familiar, is not the most fundamental branch of mathematics; and, further, that, the moment we assume that the number 1 (or even number at all) is the most basic concept in mathematics (or indeed in human thought), we have at the same time brought in a whole package of information that we will never be able to establish from first principles. Physics, in fact, tells us that integers and discrete numbering are not primary; they are associated with dimensionality, and dimensionality only has a meaning in the context of complexity. The integers are really a *codification* of a multiplicity of prior stages in mathematical evolution. To begin with them will necessarily produce an incompleteness in our logical procedures, with key steps appearing merely as assumptions in a circular argument. But, if we begin at the true primary stage, with a zero end product at every level, we effectively remove the incompleteness in our axiomatization. We also reach a primary stage in which even the word 'dual' loses its meaning, although its convenience for the later stages makes it worth retaining if separated from its numerical associations.

The very applicability of the concept of 'duality' to the process of returning from 'something' to 'nothing' implies that the actual processes of counting and generating numbers are created, along with 'addition', 'squaring', and other arithmetical procedures, at the same time as the categories of conjugation, complexification, and dimensionalization are separated from their dualistic counterparts. Defining the integers as an ordinal set within a much more fundamental process allows us to create new mathematical processes in which this ordinal set is applied in other ways, and so we can create types of mathematics where the relation to physical categories is less direct, but the ultimate 'physical' or 'dual' origin will remain.[13]

By rejecting the 'loaded information' that the integers represent, and basing our mathematics on an immediate zero totality, it appears that we are able to produce a mathematical structure which has the potential of avoiding the incompleteness indicated by Gödel's theorem, in contrast to the conventional approaches, based on the primacy of the number system, which have necessarily led to the discovery that a more primitive structure cannot be recovered than the one initially assumed. From this mathematical structure, we have been able to develop an insight into how physics works, and, using this, to suggest a process that leads naturally to a formulation for quantum computation.

From a purely physical point of view, the Dirac nilpotent would appear to be the perfect way of producing something from nothing; its structure, as we will show in chapter 4, also effectively incorporates or generates all the discrete and continuous groups of interest in fundamental physics, from C_2 to E_8, while the infinite imaging of the fermion state in the vacuum (see 6.5) and the infinite entanglement of all nilpotent fermion states extends the dualling to infinity, as required. At the other end of the scale, the later chapters will show how this concept applies to the structure of fundamental particles and the four fundamental physical interactions. The conservation laws incorporated into the nilpotent operator ($\pm\, \boldsymbol{k}E \pm \boldsymbol{i}i\, \mathbf{p} + \boldsymbol{i}\boldsymbol{j}m$) include those of mass-energy and the three types of charge, information on the latter being carried by the orientation, direction and magnitude of the angular momentum. It is these conservation laws, defined against the nonconservation or variation of space and time, which determine the behaviour of physical systems.

The state vector, in this form, is seen to have a clear physical meaning, as containing the entire available information on the energy, momentum and mass of the particle, while the Dirac equation, in any of its major forms, becomes a direct consequence of energy-momentum-mass conservation, combined with space and time translation and rotation, or nonconservation. The use of a differential operator, acting as the direct mechanism for reducing the totality to zero, allows us to directly vary space and time, while conserving mass. The equation itself thus expresses the fundamental duality of our view of 'nature', for the left-hand term (the differential operator) specifies the nonconserved aspects, and the right-hand term (the wavefunction) the conserved aspects; and the combination of state vector plus operator establishes the principle that conservation (of mass, energy, momentum) cannot be defined without defining what is at the same time *not* conserved. While classical physics does this by defining itself in terms of differential equations which vary space and time, while keeping the mass, energy and momentum fixed within a system, the use of the state vector or wavefunction in quantum mechanics allows a more complete incorporation of the complete set of translations and rotations of space and time coordinates within which the conservation rules operate.

The definition of the Dirac nilpotent suggests that this is the most efficient way of parameterizing nature while ensuring its total 'nothingness'. It may be possible to relate this to the aims of *topos* theory, as mentioned previously, in using a nilpotent Pythagorean structure to create a 'parameter space' which contains within itself dynamical and other physical possibilities.[14] The uniqueness of the individual Dirac nilpotents, together with their necessary entanglement with each other and their infinite interaction with the vacuum, suggest that this is a real number space, with the numbers countable in the Robinson or Löwenheim-Skolem sense. Through the Dirac equation, the nilpotents are then interpreted simultaneously in terms of conservation (the eigenvalue) and nonconservation (the operator). Quantum physics thus becomes a natural consequence of the fundamental meaning of conservation and nonconservation, and its separation from the physics of measurement (classical physics) becomes obvious. It is still necessary, however, to make sense of the classical transition, and also of the relationship between gravity and the other forces. Considerations of such ideas may also suggest the origin of the classical laws of thermodynamics (see chapters 7 and 8).

3.8 Theoretical Computation

In addition to its relevance to physics and mathematics, the structure devised in chapters 2 and 3 may be found relevant also to aspects of theoretical computation especially abstract machine specification where notation and the needs of rewriting (substitution) languages are explicitly required.[15] The universal rewrite system that is proposed may be mapped to a Turing machine, very close to Turing's original assumptions, where every operation 'consists of some change in the physical system consisting of the computer and his tape',[16] and every subset alphabet can be used in such an environment. For example at a simple level the subset alphabet with conjugation alone, when appropriately wrapped, provides an exact mapping to a Boolean encoding and, when a symbolism for the conjugate character is added, maps to a ternary encoding. (For an approach towards using the universal rewrite system directly for digital computing, see Appendix B.)

A physical universe composed of a potentially infinite series of unique (but changeable) nilpotents, originating in the supervenient dualistic processes needed to maintain the zero total state, has itself all the characteristics of a Turing machine. The description of physical systems in these terms allows a mapping of Turing systems to other physical processes and suggests a novel approach to investigating such systems. Here the algebraic and rewrite structure that underlies the mapping can be used to simulate and demonstrate such systems.

In addition, infinitely parallel and serial systems are posited by the method in the process of generating algebras with infinite and finite dimensions. Though the system developed here is parallel in the first instance, and ideal for quantum computation, we have options in what we can select out from the mathematical structure, and could also have chosen a serial representation. Indeed, to propose the universal alphabet in a representation that encompasses physics as we know it we are required to follow both a serial and an iterative procedure. The structure presented has all the properties required of a universal rewrite system that can generate its own alphabet.[17]

Finally, in addition to its immediate relevance to quantum computation and theoretical computation, to mathematics, and to physics, the approach appears to have possible practical application in parallel computation. This is especially the case when cast as parallel agents having autonomous actions mediated by message passing within a well defined spatial and temporal set of constraints. The required properties of this processing environment are captured by the concept of a subset alphabet, and process steps and communication mechanisms are represented as rewrite rules. It is likely that this sort of parallel processing environment will have immediate application to our understanding of the complexity of biological and biotechnological systems (see chapter 19). In addition to its providing a program for unravelling the structure of physics, the universal alphabet and rewrite system proposed in these chapters seems to have virtually unlimited potential for direct application.

Chapter 4

Groups and Representations

Some parts of **4.3** to **4.7** are based on collaborative work done with **BRIAN KOBERLEIN**. The Dirac nilpotent and group representations allow many mathematical reformulations of the basic structures, using reversals of properties and a potentially infinite hierarchy of dualities. The formal representations can be extended to include all the group structures of interest in physics. At the same time, there are at least three significant visual representations, which show the interchangeability of the dual processes of conjugation, complexification and dimensionalization outlined in chapters 1 and 2, and allow the possibility of incorporating the already extensive results of mathematical topology into the description of fundamental physics processes.

4.1 The Dirac Equation and Quantum Field Theory

The Dirac equation, in its nilpotent form, allows an explicit treatment of mass as a 'fifth' dimension, on a par with the four of space and time. The five-dimensional nature can even be represented in the compact '5-vector' form

$$\gamma^{\mu} D_{\mu} \, \psi = 0,$$

where $\mu = 0, 1, 2, 3, 5$, and

$$D_{\mu} = \frac{\partial}{\partial t} + \nabla + m \, .$$

In quaternion terms, however, mass is effectively only a third dimension, the three of space being reduced to one in the product $\boldsymbol{\gamma}.\boldsymbol{\nabla}$. In effect, the equation presents us with a three-dimensional system, whose units are represented by the three imaginary quaternion operators $\boldsymbol{i}, \boldsymbol{j}, \boldsymbol{k}$. As these terms refer to the energy, momentum and mass operators, the equation, in a sense, incorporates three degrees of freedom associated with the parameter mass. Mass conservation is shown with respect to time and space variation, and with respect to mass itself.

So, the equation seems to be suggesting a quasi-'three-dimensionality' in mass (and even, indirectly, for a quantum system, in time, cf 8.3[1]), which is somehow linked to the existence of three independent parameters of measurement.

It has been frequently noted that, in effectively squaring the time derivative of the free particle nonrelativistic Schrödinger equation

$$i\partial\psi/\partial t = -(\hbar^2/2m)\,\nabla^2\psi,$$

to achieve a relativistic invariance similar to that of the classical energy-momentum-rest mass relation, the Klein-Gordon equation

$$-(1/c^2)\,\partial^2\psi/\partial t^2 + \nabla^2\psi = 0,$$

the relativistic equation which applies to bosons as well as fermions, makes time spacelike, while, in taking the 'square root' of the space derivative, the conventional Dirac equation

$$(\gamma^\mu\partial_\mu + im)\,\psi = 0,$$

which applies only to fermions, makes space timelike. The timelike aspect of the space terms can be seen in the fact that the operators applied to the space derivative in the Dirac equation are square roots of –1, just as they are for the *time* operators in conventional 4-vectors, while the operator applied to the time derivative in the equation is a square root of 1, just like those we employ for conventional space operators. The metric is thus reversed from – + + + to + – – –. (For the even more complicated distortion of the metric which occurs with the use of γ matrices, see 5.3.)

In fact, the symmetry between space-time and mass-charge is so exact that any reversal of role between space and time is likely also to produce a corresponding reversal of role between mass and charge. An example of this effect is seen in the prediction of negative energy or mass states by the Dirac equation, which subsequent theory has to *interpret* as referring to opposite charge states (or antiparticles), with the assumption that the negative energy states are all filled, a possibility which is not present in the equation itself. In fact, the Dirac *theory* fails to accommodate antiparticles, though Dirac himself actually predicted them, after overcoming some initial conceptual difficulties. The regular interpretation of the Dirac equation assumes, in effect, that a reversal has taken place. The correct interpretation, however, provided by quantum field theory, with a filled vacuum, or, more comprehensively, by the nilpotent version of the Dirac equation, where charge is explicit, requires the return of mass and charge to their original status, though the possibility of *mathematical* transformation remains. So we have:

Quantum Field Theory	**Dirac's Theory**
Space: spacelike	Space: timelike
Time: timelike	Time: spacelike
Mass: masslike	Mass: chargelike
(+ energy)	(+ and – energy)
Charge: chargelike	Charge: masslike
(antiparticles exist)	(no antiparticles)

The logical difficulties of the conventional Dirac formulation result from the fact that the wavefunction, as written, is nonquantum; quantization is confined to the differential operator. However, quantum field theory makes explicit use of the principle that charge is conserved locally rather than globally, and that a particle may be 'created' or 'annihilated' at a point if the corresponding antiparticle is also created or annihilated simultaneously. The theory also avoids the problems introduced by the existence of two incompatible covariant one-particle equations (Klein-Gordon and Dirac) by making all energy states positive and all antiparticles result from conjugate charge states, thus restoring each of the parameters to its true status. It will be shown in chapter 6 that the nilpotent Dirac equation, unlike the conventional version, is itself a quantum field equation, and that quantum field theory can be derived entirely from first principles. For the moment, however, it will be convenient to outline the conventional version of quantum field theory.

In quantum field theory the particle wavefunctions are treated as quanta of excitation of a matter field with coefficients specifying the probabilities of the creation and destruction of the quanta. Assuming *CPT* symmetry, particles and antiparticles emerge simultaneously with positive energy only and there is no explicit use of an unobservable sea of negative energy states, as Dirac originally proposed. Negative energy particle solutions of the wave equations propagating backward in time become equivalent to positive energy antiparticle solutions propagating forward in time. However, the filled 'sea' of negative energy states is still necessary to establish the absolute continuity of mass-energy, and quantum field theory alone does not explain the universal asymmetry between fermions and antifermions, or the difference between matter and vacuum.

The direct quantization of the field energy is referred to as a 'second' quantization of matter, the 'first' quantization being an application of wave properties to the field. Second quantization means that the fermion is taken as a quantum of excitation of a fermion field extending over all space. In fact, because fermions and antifermions have equal status in the theory (though not in the universe), they may both be incorporated into a unified description of a common fermion-antifermion field. The quantum mechanical wavefunction for the particle is now replaced by a *field operator*, which is a Fourier sum over individual wavefunctions each multiplied by a coefficient representing the probability of the creation or destruction of a quantum of that momentum or wave number (reciprocal wavelength) at any given point. The fact that the quanta are localised excitations is an expression of the principle that charge is conserved locally because individual charges are unique.

Mathematically, quantum field theory describes the creation and annihilation of fermions and antifermions and is specified in terms of *creation* and *annihilation operators* ($a^\dagger(\mathbf{p})$, $b^\dagger(\mathbf{p})$ and $a(\mathbf{p})$, $b(\mathbf{p})$), which operate on the wavefunction's $e^{-i(Et\,-\,\mathbf{p.r})}$ and in effect add to or remove from a basis state one particle or field quantum with momentum $\mathbf{p}$ or wave number $\mathbf{k}$; and the probabilities of the occurrence of these events may be expressed in terms of the coefficients $u(\mathbf{p})$ and $v(\mathbf{p})$. Incorporating terms representing both the two possible orientations for spin along the direction of momentum, or right- and left-handed helicities, the positive and negative 'energy' (i.e. wave number) components of the field operators for electrons and positrons (or antielectrons) are of the form:

$$\Psi^+(\mathbf{r},t) = \int u(\mathbf{p})a(\mathbf{p})_{(R+L)}e^{-i(Et-\mathbf{p.r})}d^3\mathbf{p}$$

and

$$\Psi^-(\mathbf{r},t) = \int v(-\mathbf{p})b(\mathbf{p})^\dagger_{(R+L)}e^{-i(Et-\mathbf{p.r})}d^3\mathbf{p}\,.$$

The total field operator $\Psi(\mathbf{r}, t)$, or sum of the two components, is, in effect, a quantized wavefunction which may be substituted for the ordinary wavefunction in the Dirac equation to define a new *field equation*. In chapter 6, it will be shown that the nilpotent Dirac state vectors are equivalent to quantum field operators acting on vacuum. For the present, however, we will regard quantum field theory as an interesting example of the 'reversal or properties' available in a dual system, and return to the similar reversal in the Dirac equation involving mass and charge.

4.2 Reversals of Properties

Of course, mass, within itself, has no capacity for being made dimensional, and the extra degrees of freedom provided by the Dirac state vector have to be derived from its relationship with the independent fundamental parameters space and time. It is intriguing (though explicable within the context of fundamental duality) that the number of parameters available to 'replace' the parameter charge should be exactly equal to the number of 'dimensions' available to charge itself. However, because the three terms are connected by an exact algebraic relationship (in principle, because of conservation of mass), one of them is, in a sense, redundant, and this is why it is possible to have alternative descriptions of the energy equation with the third term represented, arbitrarily, by a quaternion operator or an imaginary scalar, as we choose between the nilpotent and conventional versions of the Dirac equation.

Reversals and partial reversals of the characteristics of parameters like space, time, mass and charge are permitted because of the fundamental symmetries which exist between them. Usually, they occur naturally or by accident, as in the development of the Dirac equation, and have to be themselves reversed to produce a meaningful physical picture, but there is no reason, in fact, why we should not employ them consciously for mathematical convenience, as several previous authors have done.[2-5] For example, the quaternions and 4-vectors of the equation

$$\left(i\boldsymbol{k}\frac{\partial}{\partial t}+\boldsymbol{i}\nabla+\boldsymbol{j}m\right)\psi=0$$

could be reversed to produce

$$\left(\mathbf{k}\frac{\partial}{\partial t}+\mathbf{i}\nabla_q+i\mathbf{j}m\right)\psi=0\,,$$

an equation, which could be used with vector, or scalar plus vector, wavefunctions. Here, the quaternion operators in the quaternion version of ∇ have disappeared in the product $\gamma^\mu\partial_\mu$. Much of the use of quaternion operators ∇_q and quaternion wavefunctions by previous authors has been in this spirit, and quite different to that necessitated by the Dirac algebra.[2-12] Through various reversals of parameter characteristics, the Dirac equation offers multiple possibilities for the employment of quaternions.

Reversal of properties is a general option resulting from duality, and it can be seen occurring in both relativity and quantum mechanics. Although Minkowski, in 1908, in mathematically uniting space and time coordinates in what we now call a 4-vector, proclaimed that: 'From now on, space by itself, and time by itself, are destined to sink into shadows, and only a kind of union of both to retain an independent existence',[13] the very structure of the 4-vector proves that the two concepts retain significant physical differences. Space and time, though combined mathematically in the 4-vector principle used in special relativity (in the limited form represented by $it + \mathbf{r}$, rather than the more fundamental $\boldsymbol{k}it + \boldsymbol{i}\mathbf{r}$), are very different physically, and the requirements of mathematical combination must do violence to the physical characteristics of one or the other. For example, in the process of combining them in *physical* measurement, as has been mentioned, we have to make a decision about whether time should be made discrete or space continuous; and the decision we make will also have consequences for the symmetrical parameters, mass and charge, and everything that depends on them. A spacelike combination assumes that all the significant physical quantities (space, time, mass, charge, energy, momentum, angular momentum) are discrete, and gives us particles, the discrete quantum concept, Heisenberg's quantum mechanics and Einstein's special relativity. A timelike combination assumes that the same quantities are all continuous, and gives us waves, stochastic electrodynamics, Schrödinger's wave mechanics and the Lorentz-Poincaré version of relativity. Of course, as in the reversals of the conventional Dirac theory, there is a price to pay for taking either of these unphysical options. Thus, Heisenberg's uncertainty principle restores continuity to the truly continuous quantities: time, mass and energy, while the so-called 'collapse of the wavefunction' in the Schrödinger theory restores discreteness to the parameters which are truly discrete: space, charge, momentum and angular momentum (cf 7.4).

4.3 The Dual Group and Higher Symmetries

Mathematically, it is possible to create a dual set or parameters, one form of which is seen in those versions of the Dirac theory in which certain characteristics of space and time, and mass and charge are reversed, for example the real / imaginary characteristics.

space*	nonconserved	imaginary	countable
time*	nonconserved	real	noncountable
mass*	conserved	imaginary	noncountable
charge*	conserved	real	countable

It has been suggested, previously, that these are the effective representations of space, time, mass and charge as used in the non-quantum-field version of the Dirac theory. The dual group may be represented symbolically by:

space*	$-x$	y	z
time*	x	y	$-z$
mass*	$-x$	$-y$	$-z$
charge*	$-x$	$-y$	z

In this case the multiplication rule is:

$$x * x = -x * -x = -x$$

$$x * -x = -x * x = x$$

$$x * y = x * -y = 0$$

and the group multiplication table becomes:

*	space*	time*	mass*	charge*
space*	mass*	charge*	space*	time*
time*	charge*	mass*	time*	space*
mass*	space*	time*	mass*	charge*
charge*	time*	space*	charge*	mass*

Here, mass* becomes the identity element, though, again this is arbitrary, and changing the signs of both x and y, for example, would make space* the identity element. The 'dual group' to space, time, mass and charge will, of course, have many manifestations, of which the Dirac representation is just one.[14] (Vacuum is another.) Such representations will involve mathematical reversals of physical properties. Thus, in incorporating both the explicit quantization of E-$\mathbf{p}$-m and the quaternion operators, the Dirac equation combines space* and time with an effective restructuring of mass with the properties of charge. Significantly, this has five components. The fact that only one direction of spin is well-defined is a consequence of using space* for space.[15]

However, it seems likely that any representation other than the canonical one (that is, the original group) will always be reduced to it under the process of measurement, as we have seen happens with the Heisenberg and Schrödinger formulations of quantum mechanics. The respective options here seem to be S, T^*, M^*, C, for Heisenberg, and S^*, T, M, C^*, for Schrödinger, with the true S, T, M, C being restored in the measurement processes.

Taking both the canonical D_2 or $C_2 \times C_2$ group of S, T, M, C, and the dual D_2 or $C_2 \times C_2$ group of S^*, T^*, M^*, C^*, the combined group is then extended to $C_2 \times C_2 \times C_2$, with a quaternion representation (Q_4) with both signs of scalar. This extends the mathematical representational space, but is not needed in the physical representational space, and, in the more sophisticated (quantum field) versions of Dirac theory, it becomes redundant.

In the canonical group structure, the pairs of property / antiproperty (divisible / indivisible; conserved / nonconserved; real / imaginary pair) each form C_2 groups of order 2. However, algebraically, the real / imaginary (or orderable / nonorderable) pair also forms a C_4 group (1, i, –1, –i) of order 4. The total group structure is then $C_4 \times C_2 \times C_2$ (or $C_4 \times D_2$), of order 16. In this representation, the dual parameters must possess the same $C_4 \times D_2$ structure; so the total group symmetry must be of order 32. Additionally, these dual structures must mutually anticommute. Since the dual of C_4 is itself, the complete group structure must be $C_4 \times Q_4$, where Q_4 is the quaternion group (of order 8). This structure is simply the complexified quaternions. The total group structure may therefore be represented by a quaternion vector pair.

There is a further distinction to be made between the representations of the parameter properties (e.g. real / imaginary) by existence / nonexistence conditions, as here; and the explicit representation of these properties by their explicit natures (e.g. vector / quaternion). The minimum representation in the latter case is of the order $C_2 \times C_2 \times C_2 \times C_2 \times C_2 \times C_2$, or the 64-term Dirac group. In the former case, there are at least two striking visual representations of the group relations, which bring out the significance of the C_2 distinctions and of the principle of cyclic dimensionality (cf sections 4.8 and 4.9). A group version of (S, T, M, C) × (S*, T*, M*, C*) would also require 64 terms.

Taken together, the complex quaternions in the $C_4 \times Q_4$ structure do not form a general group. It is clear, in fact, that *general object symmetry*, as required to describe a fundamental particle, cannot form a group, as the existence of nilpotents (such as these describe) is central to quantum duality. However, one is led to ask if the complex quaternion algebra is the product of some larger symmetry group, which appears in its current form via spontaneous symmetry breaking. There are a number of ways in which a supersymmetry can be created. Perhaps the simplest is to represent the state of an object as a vector-like term, as in

$$O \Rightarrow 1 + \mathbf{i} + \mathbf{j} + \mathbf{k} + i + i\mathbf{I} + i\mathbf{J} + i\mathbf{K},$$

where **i**, **j**, **k** and **I**, **J**, **K** represent commutative multivariate vector (or quaternion) systems.

In this form, the state of an object resembles a broken octonion. The dimensionality in this representation presupposes the concept of *rotation*, and the rotational symmetry of an octonion is the exceptional Lie group G_2. The above vector is then an octonion with one orientation (in this case *i*) constrained so as not to transform into other coordinates. The symmetry group of such a constrained octonion is $SU(3)$. Within this constrained symmetry, there exist two other obvious subgroups: the quaternion vector, with symmetry $SU(2)$, and the complex 'vector', with symmetry $U(1)$. It is clear then that object symmetry has the form $SU(3) \times SU(2) \times U(1)$, which can be generated from the spontaneous symmetry breaking of the octonion symmetry G_2. So the inherent symmetry of an abstract object, with no *a priori* internal structure, is required to have the internal structure of the Standard Model. The Standard Model therefore arises from the basic properties of space, time, mass and charge (cf chapters 13 and 14).

4.4 A Broken Octonion

An octonion symmetry is certainly one way of structuring the fundamental dualities required to define a particle or to parameterize nature; and the structure can be arrived at from more than one starting position. We could, for example, consider the original group, space, time, mass and charge, and the dual group, space*, time*, mass*, charge*, as the 'commutative' and 'noncommutative' ways of producing the abstract D_2 structure, and then combine the two into a larger group structure $C_2 \times D_2$ of order 8. The identity element, say mass, could then be represented by the scalar part of a quaternion (1) and the other three terms by the imaginary operators $\boldsymbol{i}, \boldsymbol{j}, \boldsymbol{k}$, while mass*, charge*, space*, time* are represented by -1, $-\boldsymbol{i}$, $-\boldsymbol{j}$, $-\boldsymbol{k}$. So, the multiplication table:

*	1	i	j	k	-1	$-i$	$-j$	$-k$
1	1	i	j	k	-1	$-i$	$-j$	$-k$
i	i	-1	k	$-j$	$-i$	1	$-k$	j
j	j	$-k$	-1	i	$-j$	k	1	$-i$
k	k	j	$-i$	-1	$-k$	$-j$	i	1
-1	-1	$-i$	$-j$	$-k$	1	i	j	k
$-i$	$-i$	1	$-k$	j	i	-1	k	$-j$
$-j$	$-j$	k	1	$-i$	j	$-k$	-1	i
$-k$	$-k$	$-j$	i	1	k	j	$-i$	-1

can be used to represent the group relations between M, C, S, T and M^*, C^*, S^*,

*T**:

*	*M*	*C*	*S*	*T*	*M**	*C**	*S**	*T**
M	*M*	*C*	*S*	*T*	*M**	*C**	*S**	*T**
C	*C*	*M**	*T*	*S**	*C**	*M*	*T**	*S*
S	*S*	*T**	*M**	*C*	*S**	*T*	*M*	*C**
T	*T*	*S*	*C**	*M**	*T**	*S**	*C*	*M*
*M**	*M**	*C**	*S**	*T**	*M*	*C*	*S*	*T*
*C**	*C**	*M*	*T**	*S*	*C*	*M**	*T*	*S**
*S**	*S**	*T*	*M*	*C**	*S*	*T**	*M**	*C*
*T**	*T**	*S**	*C*	*M*	*T*	*S*	*C**	*M**

If the 3-dimensionality of charge and space is directly involved, the overall structure requires a quaternion (±1, ±***i****s*, ±***j****e*, ±***k****w*) and a quaternion-like 4-vector (±*i*, ±**i***x*, ±**j***y*, ±**k***z*) (a double quaternion in total) within another overall quaternion-type arrangement. This can be accomplished using an octonion, with sixteen members (±1*m*, ±***i****s*, ±***j****e*, ±***k****w*, ±***e****t*, ±***f****x*, ±***g****y*, ±***h****z*) (as illustrated below). Although this is no longer a group, it combines two conjugate groups of order 8: *M*, *C*(3), *S*(3), *T* and *M**, *C*(3)*, *S*(3)*, *T**, and can be represented in a group structure through the use of left-product or right-product octonions. In addition, the nonassociativity of the dimensional terms in this octonion extension seems to be lost within terms which effectively cancel each other out, and are of no physical significance.

*	1	*i*	*j*	*k*	*e*	*f*	*g*	*h*
1	1	*i*	*j*	*k*	*e*	*f*	*g*	*h*
i	*i*	–1	*k*	–*j*	*f*	–*e*	–*h*	*g*
j	*j*	–*k*	–1	*i*	*g*	*h*	–*e*	–*f*
k	*k*	*j*	–*i*	–1	*h*	–*g*	*f*	–*e*
e	*e*	–*f*	–*g*	–*h*	–1	*i*	*j*	*k*
f	*f*	*e*	–*h*	*g*	–*i*	–1	–*k*	*j*
g	*g*	*h*	*e*	–*f*	–*j*	*k*	–1	–*i*
h	*h*	–*g*	*f*	*e*	–*k*	–*j*	*i*	–1

*	*m*	*s*	*e*	*w*	*t*	*x*	*y*	*z*
m	*m*	*s*	*e*	*w*	*t*	*x*	*y*	*z*
s	*s*	–*m*	*w*	–*e*	*x*	*t*	–*z*	*y*
e	*e*	–*w*	–*m*	*s*	*y*	*z*	–*t*	–*x*
w	*w*	*e*	–*s*	–*m*	*z*	–*y*	*x*	–*t*
t	*t*	–*x*	–*y*	–*z*	–*m*	*s*	*e*	*w*
x	*x*	*t*	–*z*	*y*	–*s*	–*m*	–*w*	*e*
y	*y*	*z*	*t*	–*x*	–*e*	*w*	–*m*	–*s*
z	*z*	–*y*	*x*	*t*	–*w*	–*e*	*s*	–*m*

A broken octonion symmetry, as we have already established, creates for us the algebra we need for quantum mechanics, with the 8-unit octonion splitting into the 4-vector part, representing space and time, with real vector units **i**, **j**, **k** and imaginary scalar *i*; and the quaternion part, representing charge and mass, with imaginary 'vector' units ***i***, ***j***, ***k*** and real scalar 1. It is significant, of course, that the octonion representation is simply that, and not a statement of the underlying algebra of physics. If we choose an octonion representation, we will necessarily obtain results which do not accord with nature alongside those which do, and we must have a more absolute way of distinguishing the results which really correspond to physically meaningful phenomena from those which are merely artefacts of the mathematical representation. The same applies to another possible representation: that of twistors or complex 4-D space. The 4 real units **i**, **j**, **k**, 1 and 4 imaginary units ***i***, ***j***, ***k***, *i*, certainly incorporate the concept of a twistor space, and, where the real / imaginary distinction is more significant than the dimensional / nondimensional, or conserved / nonconserved, the representation will be accurate.

If we take charge as the identity element, and represent it by a scalar, the remaining structure for time, space and mass (and, implicitly, the energy, momentum and mass operators) becomes that of the Dirac algebra, providing an alternative method of generating this formalism, and also of *SU*(5) or *U*(5) (as described later). Such representations do not determine the properties of the group members, space, time, mass and charge. They exist because the group has four components, and can, therefore, be represented by a 4-component structure like a quaternion, in which the link between elements is made by a binary operation (squaring); but the link between a group with four components and a 4-dimensional space-time or mass-charge may be in itself significant. In addition, the D_2 group is the group of rotations of the rectangle (identity and rotations about three spatial axes). This could be another way of linking the double 3 + 1 symmetry of the units of space, time, mass and charge with the structure of the group.

4.5 A Hierarchy of Dualities

We can, as we have seen, construct a D_2 group for space, time, mass and charge on the basis of assigning algebraic symbols to properties without concerning ourselves about the actual mathematics associated with the four parameters. However, as soon as one assigns such mathematics, the group relationship changes. The resulting structures form a hierarchy of dualities, analogous to the hierarchy of dualities that we can use to create the natural numbers in binary form, and which can be considered as a restricted form of the universal rewrite system. The hierarchy stems from the fact that any mathematical representation of a symmetry will necessarily be an imperfect or 'broken' symmetry, like the definition of a 'system' in the context of Newton's third law of motion, and will necessarily lead to a doubling process, tending to a more 'perfect' symmetry.

The doubling can be seen as a realization of the fact that everything has to have a dual or C_2 partner to make something from nothing, and the mathematics within the parameter structures has to be constructed so that the doubling process is actually possible. The real / imaginary distinction and 3-dimensionality are ways in which doubling can occur, but they act in subtly different ways. Both ideas are associated with the squaring process, which is a kind of doubling, just as square rooting is a kind of halving. This means that squaring via C_2 is as natural a way of producing 'something from nothing'-type duality as adding.

We can explain the duality hierarchy in a restricted sense in terms of a 'Fundamental Theorem of Symmetry', though the full explanation, of course, requires the universal rewrite system: For every symmetry, there must exist a corresponding asymmetry. As such, a simple duality can always be represented as C_2. If we define science as the measurement of a system to determine results, there will be three basic duals. The System Dual corresponds to conservation / nonconservation. A system consists of an object within an environment. The object and the environment are therefore dual. Mass and charge, here, are object properties, space and time are environment properties. The second or Measurement Dual corresponds to countability / noncountability (discreteness / continuity). For any system, counting is either possible, or it is not. Thus 'measurable' and 'nonmeasurable' (countable and noncountable) are dual. Space and charge are countable, time and mass are not. The third, or Observation Dual does not correspond directly to the real / imaginary distinction, but it creates it indirectly in the form of the duality between the 'canonical' D_2 group and its own dual, and it is central to the idea of 'probe' and 'response'. Any measurement consists of the observer and the observed. Thus observer / observed are dual.

Though these three dualities must exist within any scientific model, they also generate further dualities. The System / Observation also forms a duality, since it creates two possibilities: the object is the observer and the environment is the observed, or the environment is the observer and the object is the observed. It is this duality which is expressed when we reverse the physically incorrect assumptions of either the Schrödinger or the Heisenberg mathematical systems in the process of measurement.

The System / Observation Dual corresponds directly to the real / imaginary distinction since it is a two-fold dual. That is, there are four quantities (object-observer, object-observed, environment-observer, environment-observed), and they must have an internal duality. This is why the real / imaginary distinction becomes $C_4 = (1, -1, i, -i)$, where $(1, -1)$ and $(i, -i)$ have a dual symmetry. This last duality is also what generates the classical / quantum symmetry, and can explain why the i shows up in quantum mechanics. We can also say that the real / imaginary distinction is the principle responsible for doubling the original group to include the dual group. So the creation of a dual group is the same thing, in effect, as making the real / imaginary distinction explicit. In addition, without this distinction, there would be no nilpotent structure, as only complexity can create the required squaring to zero, and without the nilpotent structure we would have no extension of dualling to infinity. We can further suppose that any of the C_2 dualities in space, time, mass and charge will produce the factor 2, which occurs in many aspects of physics as a direct mathematical expression of the fundamental nature of duality (cf chapter 16), and that one duality can be translated into another. The C_2 'something from nothing' doubling is then extended to higher orders automatically, because of its own 'self-dynamism'.

The important aspect for the space, time, mass, charge model is that there is an asymmetry in the formulation. In this sense, even the fundamental parameter group of space, time, mass and charge is a *mathematically* (though not physically) broken symmetry, because the full $C_2 \times C_2 \times C_2$ (which includes the dual group) is not needed for physics at any given time, but only half of it. (We can relate it in some sense to the need to define vacuum along with the fermionic state.) The System, Observation, and System-Observation duals, also, form a closed set. The Measurement duality is not part of this set, but, rather, a kind of truth 'external' to the system. This means that the four duals do not form a group, but require a set of order 32, which is not a group, but which has the complex numbers as a subgroup. Thus, the result is the complexified quaternions. The higher doublings then occur through the recognition of this structure as a larger group structure, and lead, through the application of the rotation symmetries implicit in the multidimensional parameters, to the creation of the Lie algebras.

4.6 Dimensionality

We have seen that, bringing in the real / imaginary distinction doubles the order of the group; introducing 3-dimensionality doubles it again. The doubling effect is natural, due to the basic concept of duality, and the 3-dimensionality is itself related to the real / imaginary distinction, but the process of creating multidimensionality by doubling is very subtle. Though we know how to create multidimensionality through the universal rewrite system, it is worth considering the process in a more restricted way to gain a more 'physical' insight into its ultimate meaning.

The only way for a group to have some form of 'dimensionality' is for it to be non-Abelian or non-commutative in some form. It is not possible for a non-Abelian group to be the product of Abelian or commutative groups. Therefore, one cannot simply 'generate' dimensionality out of the 'doubling' effect alone. The smallest dimensional finite group is the quaternion group, which is of dimension 3 (not 4, since 'time' is commutative). This appears at order 8. Of the 5 finite groups of order 8, C_8, $C_2 \times C_4$, $C_2 \times C_2 \times C_2$, D_4, and Q_4, all are Abelian except for Q_4. So, somehow (and as we also know from the rewrite system), there must be an argument requiring Q_4, or at some higher order, a non-Abelian group of which Q_4 is a subgroup.

3-dimensionality is a source of the factor 2 in many aspects of physics. Dynamically and quantum mechanically, the factor 2 is associated with vector terms. Dynamically, it comes from action and reaction, or the virial theorem (as a direct product of 3-dimensionality); quantum mechanically, it comes from the noncommutativity of the vector terms in the spin angular momentum. In general, the factor 2 in physics frequently comes from the division between discreteness – the source of multidimensionality – and continuity. In the group structures it is also associated with the division between real and complex representations. The complex version of Dirac algebra (as described previously) has 32 terms, as opposed to the 'real' version with 16. It appears to be relevant, therefore, that it is the imaginary version of vector algebra, in the quaternion, which requires 3-dimensionality or 3 + 1-dimensionality.

In principle, the 'rule of doubling' means that the multidimensionality of quaternions has to be of such an order as would create the same number of elements as we would get by doubling, even though they will not be the same elements. That is, considerations of duality mean that we need a doubling of elements, but some other fundamental consideration requires that the perfection of the originally Abelian symmetry is violated, and the doubling occurs in a 'broken' way. Duality itself creates the spatial 3-dimensionality, and the 3 + 1 dimensionality of, say, space plus time, mass and charge (or space, time and mass plus charge) becomes the double 3 + 1 dimensionality of space plus time and mass plus charge when we double the grouping by including the dual group. This, however, is a requirement, not a prescription.

In the space, time, mass, charge set, 3-dimensionality becomes a necessary result, through symmetry, of extending the real / imaginary distinction from time to charge. Charge, being discrete, is necessarily multidimensional, and, being also imaginary, requires a description in the form of a multidimensional imaginary algebra. What seemingly happens here is that the group duality consideration, requiring a multiplication of elements such that the number of elements increases from 4 to 8, but allowing only 2 of the elements to acquire the property that would make this possible, requires 3-dimensionality, and this 3-dimensionality is expressed by what we happen to call quaternions, of order 8, which are noncommutative because of the rotation property. The real / imaginary division itself then requires a further multiplication of elements from 8 to 16, creating, in the process, the parallel 3-dimensionality of the real vector parameter, space.

As we have seen, one method of producing the quaternion group mathematically is to start with the three property / antiproperty dualities (conservation / nonconservation, countability / noncountability, real description / imaginary description), and treat them initially as simple C_2 symmetries, with two categories, exactly opposite. These multiply to $C_2 \times C_2 \times C_2$, of order 8. But if we then take the real / imaginary duality, although only contributing an order 2, as, mathematically, of the form C_4, we can, by symmetry, take each of the dualities to be of the same form. In other words, the group of order 8 will have 3 unique C_4 subgroups, since Q_4 has three C_4 subgroups as its only subgroups.

Thus, the order 8 group will be Q_4, automatically introducing dimensionality of order 3.

Like the direct doubling due to the real / imaginary distinction, which creates a mathematics of ordered pairs, it is evident that the dimensional doubling, as we know already from pure mathematics, is related to the introduction of imaginary numbers. Both are a product of the Pythagorean squaring process, which is simultaneously an expression of dimensionality and of the group relationships between the fundamental parameters, in addition to being the actual mechanism by which the group doubling process becomes related to the mathematical factor 2. In the case of multidimensional quantities this factor is derived by making dimensional multiplication noncommutative, and only 3-dimensional quantities may accomplish this in a way that preserves symmetrical closure. In addition, symmetry considerations determine that only two parameters may become multidimensional, and so only 2 have this property of noncommutativity.

From a purely physical point of view, the central origin of dimension can be taken as an equivalent to the particle / wave duality which comes directly from the discrete / continuous division, for this is certainly the origin of the division between multidimensionality and unidimensionality. Only discrete quantities can have more than one dimension, while at the same time the idea of discreteness makes multidimensionality a requirement; continuous quantities, by the same token, must necessarily be unidimensional. Particles express discreteness, though they have no dimension (or extension) themselves, while waves express continuity, though, as the opposite of particles, they must, in a sense, have dimension. Effectively, particles mark the discreteness of space, by being the divisions or zero points of it, and so, in this sense, are not discrete at all. Once again, this is similar to the Heisenberg-Schrödinger distinction. Each incorporates its opposite to be able to define itself. There is a dual aspect to everything.

4.7 Symmetry Hierarchy

Although we have demonstrated that the foundational approach is sufficiently general to encompass the Standard Model, we have not yet presented a complete picture of object symmetry. The foundational properties of space, time, mass, and charge require a direct connection to the full set of division algebras (real, complex, quaternion, octonion), as seen above. As a result, the foundational approach is closely related to a hierarchy of symmetry structures derived from these algebras. The most famous of these hierarchies is known as the Freudenthal-Tits Magic Square, which is a 4×4 array of groups, associated with the Jordan algebras of 3×3 Hermitian matrices. The result is derived from the groups $G \times G$, and is given by

	R	C	Q	O
R	$SO(3) = SU(2)$	$SU(3)$	$Sp(3)$	F_4
C	$SU(3)$	$SU(3) \times SU(3)$	$SU(6)$	E_6
Q	$Sp(3)$	$SU(6)$	$SO(12)$	E_7
O	F_4	E_6	E_7	E_8

It is clear from this table that products of the octonion algebra, together with the octonion symmetry G_2, form the set of exceptional Lie algebras which are of such great interest in higher dimensional models such as string theory. It is evident, therefore, that there is a fundamental connection between the foundational symmetries of space, time, mass, and charge, and the 'higher' symmetries of models such as string theory and the Standard Model. Clearly, there exists enough freedom to express these models in a foundational context. Furthermore, one can argue that the foundational approach is more powerful, as it requires the above symmetries as a consequence, rather than imposing them *a priori*. In fact, the problem with string theories in general is that the formalisms are too general, too all-encompassing, with the physical information buried too deep in the mathematical structure to be easily extracted. There are, literally, many thousands of possible approaches (and supposedly $\sim 10^{500}$ possible vacua!), with no obvious way of deciding which provides the closest fit to the physical picture.

The same is true, to a lesser extent, with abstract modelling based on groups. A group should be seen as a stage in the discovery of the physical principles it embodies, not as the object of discovery itself. An example of this is the $SU(3)_f$ structure which eventually led to the quark theory. It is the quark theory that is now considered to be important, not the $SU(3)_f$ structure, especially as this is now known to be only a component in a larger $SU(6)_f$. The use of groups can also be deceptive. As already stated, it would, for example, be possible by a process of analytical or inductive thinking to find an octonion structure as a fundamental component of physics – certainly, if one started from string theory; however, the significant aspect of octonions as used here is that the symmetry is broken. This alters the information provided by the octonion symmetry, at the same time as greatly reducing it. A pure octonion symmetry would give false and truc information at the same time, with no obvious way of deciding which aspects of it constituted real physics.

4.8 Colour Representation

The multiple duality of the parameter group has many strikingly simple visual representations using colours and geometric diagrams in 3 dimensions. These become possible because dimensionality is a direct result of duality, and a 3-colour system is a way of simulating 3-dimensionality. The four parameters, space, time, mass and charge, for example, may be represented by concentric circles, the parameter chosen as the identity element for the group occupying the centre circle. The division of the properties into three components is reflected by the division of the circles into three sectors. The properties (say, real, nonconserved, discrete) are represented in Figure 4.1, by shadings standing for primary colours (say, red, green, blue), and the 'antiproperties' (imaginary, conserved, continuous) by the complementary secondary 'colours' (cyan, magenta, yellow). All of these choices are individually arbitrary, as is the choice of secondary colours to represent the properties, and primary colours to represent the 'antiproperties' in Figure 4.2. The division between properties and antiproperties is also a completely free choice. Only the overall pattern is fixed. As configured, with space selected as the identity element, and the colour representation for the properties selected as indicated, the innermost circle represents space, the next charge, the next mass, and the outermost circle time. But this will be changed as soon as we redefine any of the colour representations or exchange the status of any or the property-antiproperty pairs. In addition to being an alternative representation of the main group, Figure 4.2 may also be used, simultaneously with Figure 4.1, as a representation of the dual group, which can be obtained (for example) by exchanging the status of real and imaginary quantities (as we have seen in some versions of the Dirac theory).

The nature of the fundamental parameter group is demonstrated in this representation by summing up the colour combinations in each of the circles. This results in a white inner circle for the identity element, and a sequence of the three primary colours (Figure 4.3) or secondary colours (Figure 4.4), which adds up to a white totality.

Figure 4.1

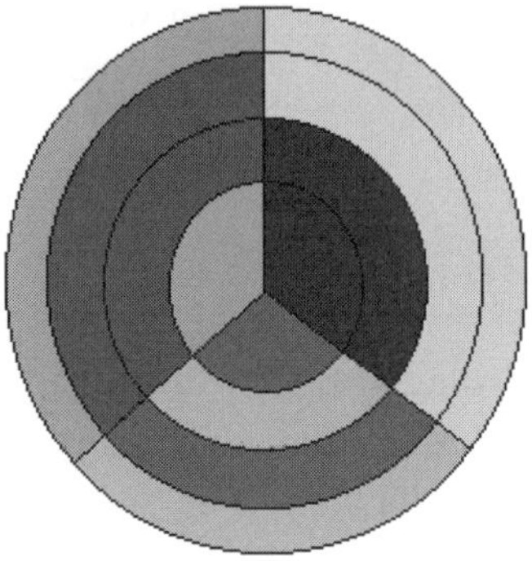

Figure 4.2

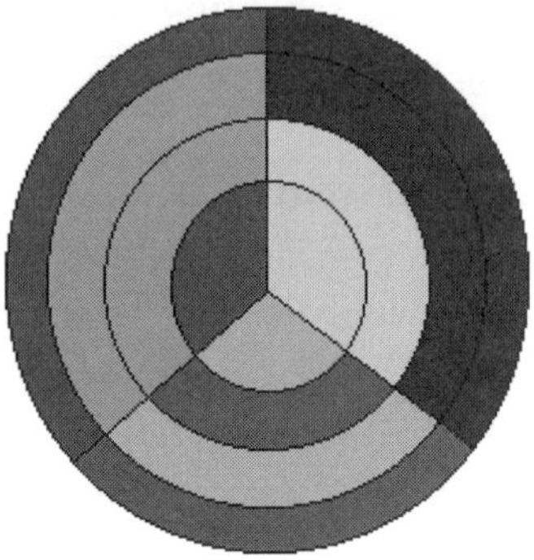

Figure 4.3

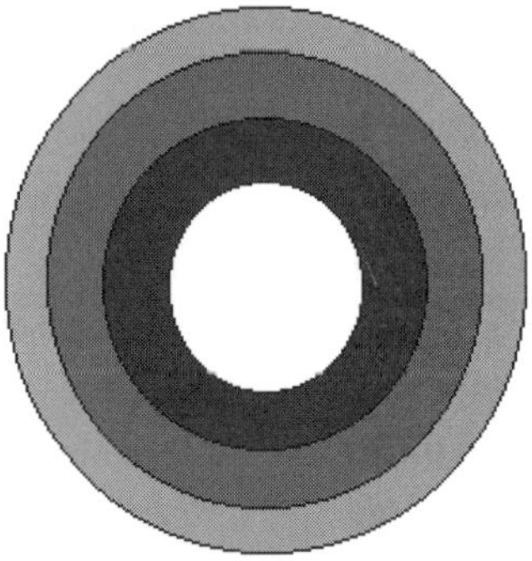

Figure 4.4

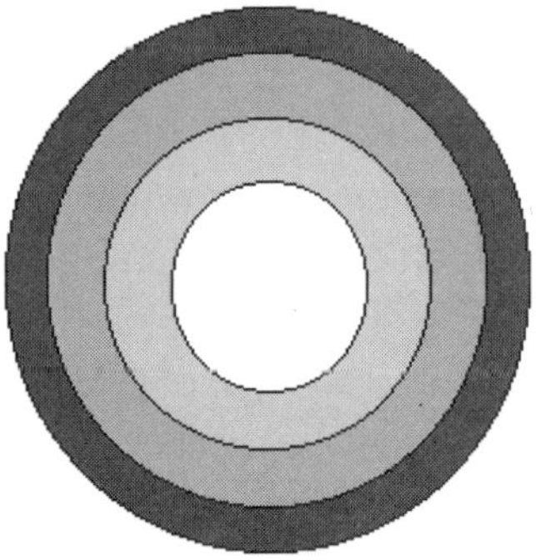

Adding up the property-antiproperty combinations in the sectors also results in a white totality for each sector, as expected – the representational equivalent of a total conceptual nothingness (Figure 4.5). Clearly, this colour representation derives its effectiveness from the fact that a three-colour system is a one-to-one simulation of three-dimensionality in an alternative vector or quaternion representation.[16] Such representations are also possible in a more direct form.

Figure 4.5

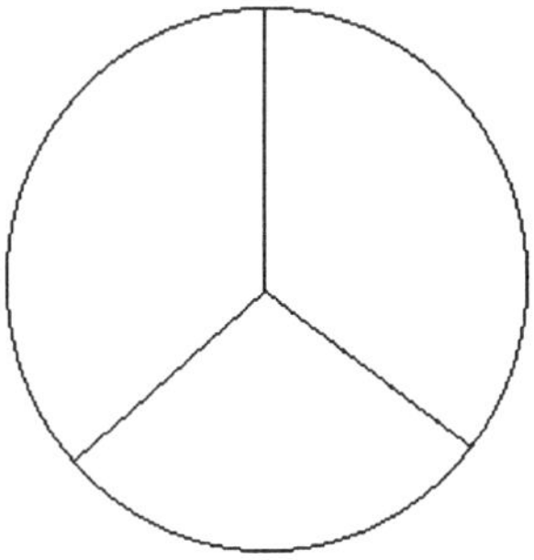

4.9 3-D (Vector) Representation

The alternative (vector or quaternion) representation (Figures 4.6 and 4.7) can be seen as a literal interpretation of the x, y and z, used in representing the canonical parameter group. The x, y and z directions represent the properties, and the $-x$, $-y$, and $-z$ directions the antiproperties (again, according to an arbitrary choice). The four 'red' lines (represented as solid), in Figurc 4.6, drawn from the origin of these 3-dimensional axes, then represent the four parameters, and the 'cyan' lines (represented as dotted) those of the dual group. Figure 4.7 shows the same representation as Figure 4.6, but without the axes. The red lines are reflections of each other in two planes. We can represent these as preserving the sign of the volume element (or identity), if the axes are taken in the same cyclic sense; and so they correspond to the parameter group. The red plus cyan lines are the reflections of each other in a single plane, and do not preserve the sign of the volume element; and so form the parameter group taken with its dual. The reflection of a line in three planes produces its exact dual.

It will be apparent that the representation of the dualities of the parameter group using either the three real spatial dimensions or the pseudo-dimensions of the three primary colours is a powerful way of bringing out the connections between duality and dimensionality, and the fact that all the individual dualities are, in effect, versions of the same mechanism. It is also a convenient way of showing how the parameter group can be used to represent a kind of 'super-duality' of all the elements, conveniently displayed using the particular duality of dimensionality. (The 3-D representation is also, incidentally, an extremely convenient diagram for illustrating the basis of the Bell inequalities, which is particularly significant if we consider that the angular momentum, in some form, contains the entire information about a fermionic or bosonic state.)

Figure 4.6

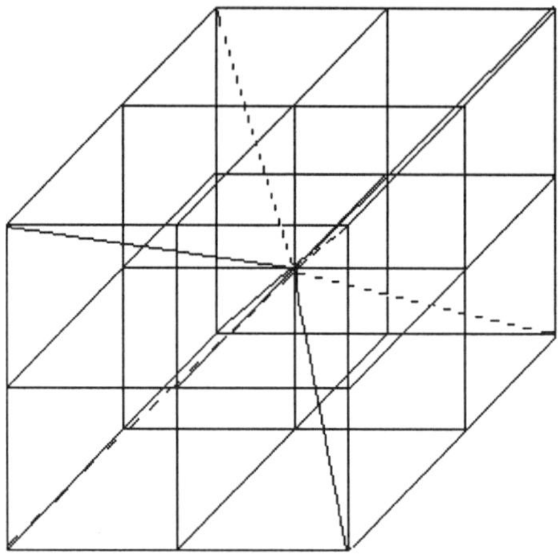

Figure 4.7

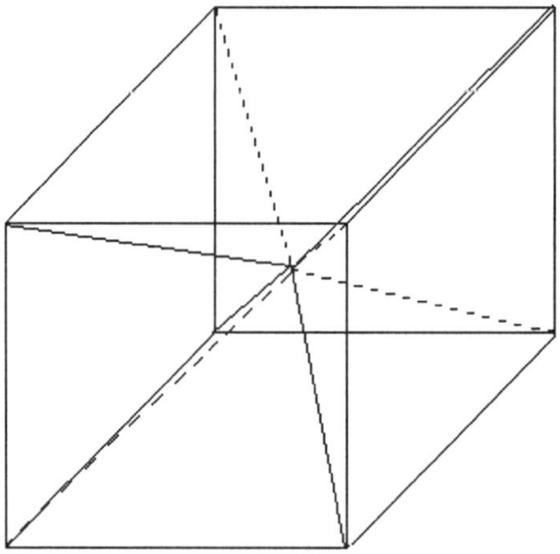

To map between the colour and 3-D representations, we could represent the positive x, y, z axes in Figure 4.6 by the primary colours, say red, green, blue, against a black background, and the corresponding negative directions by the complementary secondary colours, which, in this case, would be cyan, magenta, yellow. The four red vectors of Figure 4.6 would become respectively white, blue, red, green, as determined by the 'colour' coding in Figures 4.1 and 4.3, and the four cyan vectors of Figure 6 would become respectively white, yellow, cyan, magenta, as determined by the 'colour' coding in Figures 4.2 and 4.4.

4.10 Tetrahedral Representation

Yet another 3-D representation (Figure 4.8) would place the parameters at the vertices of a regular tetrahedron, with the six edges coloured to represent the properties and antiproperties as in Figure 4.8.

Figure 4.8

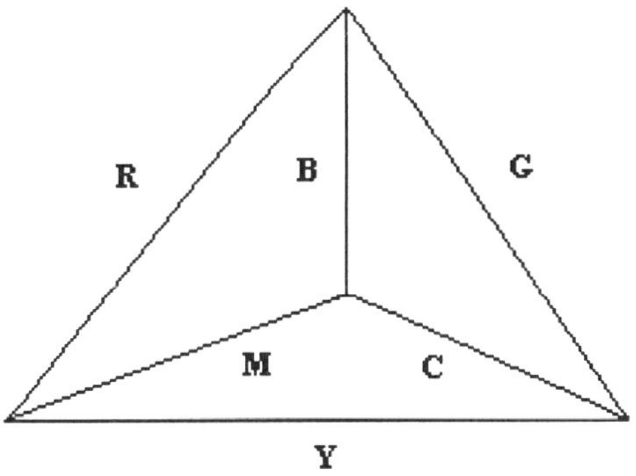

We can consider the faces of the tetrahedron to be the members of the dual group, and, clearly, an alternative representation would reverse primary and secondary colours (represented here by the labels R, B, G and M, Y, C) and / or the roles of faces and vertices. It might be possible to consider the tetrahedron as close-packed with inverted tetrahedra with complementary colour-representation in an all-white solid-space, which can be extended to infinity in a spiral or 5-fold symmetry, or as forming a star tetrahedron with its inverted dual.

An interesting possibility is that a structure like the one in Figure 4.8, if flattened out in a 2-dimensional space, could be considered as a 'dart' or 'kite' in a Penrose tiling pattern (the base line being optional to the connections between the vertices representing the parameters). (Here, we consider the tetrahedral arrangement as a concept in graph theory, leaving out the 3-D aspect.) Each of these 4-sided figures is an equilateral triangle, with one of the sides having a kink, either inwards or outwards, to produce the fourth vertex. Penrose tiling is, of course, a five-fold symmetry, introducing the Fibonacci sequence (see chapter 19), and the combination tends to produce figures with 5 or a multiple of 5 vertices on the outside.[17]

Typically, a group of five darts (or kites) will produce a star-shaped pattern with each of the darts joined with all the others at its apex, and with its two nearest neighbours along two of its edges. The star then has five inner and fiver outer vertices, and, surrounding a central star made of darts, we will have ten kites, each joined by two edges to two nearest neighbours, and by another edge to one of the ten outer edges of the star. If we assume that each of the 4 vertices of any dart (3 symmetric and 1 asymmetric) must represent one of space, time, mass and charge, and that joint vertices may only represent one parameter, then putting a 3-dimensional parameter like charge or space, at the centre of the star forces us to choose the inner and outer vertices in such a way that the other 3-dimensional parameter occurs three times in the five inner or outer vertices, while each of the other parameters occurs only once. It is interesting that that other five-fold structure (the Dirac algebra) emerging from a combination of two types of 3 + 1-component units (space-time 4-vectors and mass-charge quaternions) is also forced to 'privilege' one of its two 3-dimensional quantities, space or charge.

It is certainly possible that there is a connection between the kite plus dart generation of Penrose tiling and the eight parts of space, time, mass and charge becoming the five of Dirac energy-momentum-mass. The 7-D mapping with 7 possible neighbourhoods for the tiling patterns may relate to the imaginary part of an octonion-type structure. It may also be relevant that Penrose tiling is fractal, as is the infinite series of pentad vacuum 'images' produced by a fermion, and it is as unique and unperiodic as fermionic composition, with angles that are never repeated. Such visual representations not only provide a stimulating way of comprehending important physical symmetries, but suggest possibilities for incorporating the profound insights available within topology into the deepest structure of the subject.

Chapter 5

Breaking the Dirac Code

Several sections of this chapter (especially **5.4** to **5.6** and **5.10**) stem from collaborative work done with **JOHN CULLERNE**. It is proposed that the Dirac equation, as normally interpreted, incorporates intrinsic redundancies whose removal necessarily leads to an enormous gain in calculating power and physical interpretation. Streamlined versions of the Dirac equation can be developed which remove the redundancies and singularities from many areas of quantum physics while giving quantum representations to specific particle states. Effectively, the 'Dirac code' is broken; that is, the mysterious mathematical aspects of the conventional Dirac equation are replaced by an apparatus whose origin and meaning are transparently physical. Further physical meaning becomes apparent when we create formal transformations to and from the conventional representation. The technical developments include the derivation of bilinear covariants and a formal proof that the four Dirac solutions are orthonormal.

5.1 Singularities and Redundancy

The transformation of coordinate systems is a powerful device used in many areas of physics, but its most significant effects occur when the coordinate systems before and after the transformation are not absolutely equivalent. The particular choice of coordinate system will then often determine the subsequent development of the mathematical structure as a physical representation of the system being investigated; and whether or not the transformation is desirable will be determined by how 'correct' this physical representation is considered to be. Spherical polar coordinates, for example, are ideal for systems with point or radial sources – as with the hydrogen atom or the Schwarzschild solution in general relativity – and this is because they privilege one spatial dimension (the radial) over the others, as required. In these cases the source of the field can be considered as a singularity, but in other cases the choice of such a coordinate system can lead to a singularity appearing where none exists in reality. If we find that the mere choice of a coordinate system leads to singularities, then we have, of course, introduced a problem that needn't exist. But it isn't always only a single problem. Singularities can often be avoided or overcome by using special mathematical techniques or careful definition of the valid 'physical' limits of the system being investigated, but, sometimes, the singularity acts as a barrier (or 'attractor'), which separates the system into two seemingly unconnected halves, leading to immediate duplication of the information required; in addition, the existence of an uncrossable boundary can lead to *repeated* duplication of information as the system attempts to compensate in one way or another for the loss of a connectedness that ought to exist.

The choice of coordinate transformations to represent rotation has always been a particularly difficult problem, especially when matrix methods, such as Euler angle rotation in $SO(3)$, are used. The origin of the problem is a singularity at the point where the angle $\beta = \pi / 2$. As one author writes: 'Inherent in every minimal Euler angle rotation sequence in $SO(3)$ – the group whose elements are the Special Orthogonal matrices in R^3 – is at least one singularity.'[1] Essentially, the problem arises from the use of an intrinsically 2-dimensional mathematical structure to represent a 3-dimensional reality; we can, for example, show that the problem is immediately solved and the singularity removed when the intrinsically 3-dimensional quaternions are introduced. Exactly the same kind of reasoning can be applied to relativistic quantum mechanics, where problems emerge from the imposition of a matrix representation. Relativistic quantum mechanics as represented by the Dirac equation and quantum field theory produces at least one type of singularity that appears to be an artefact of the system – the infrared divergence (see 11.4). It also leads to infinities that have to be removed by renormalization, even in the ideal case of free particles where there is apparently no real source for the divergent terms. In addition, there appears to be a great deal of redundancy. For example, QCD calculations using Feynman diagrams derived from the standard gamma matrix representation require ten million calculations for a six gluon interaction, whereas the alternative algebraic approach using twistor space, originally proposed by Witten,[2] reduces the calculations required to only six. Even with this method, it is clear that redundancies are still visible; so the question we should ask is whether it is possible to find a coordinate system for the fermionic state which removes redundancy entirely.

5.2 Redundancy in the Dirac Equation

The Dirac equation, as conventionally written in matrix form,

$$(\gamma^{\mu}\partial_{\mu} + im)\ \psi = 0, \tag{5.1}$$

though apparently compact, in fact contains a large amount of redundancy. The ultimate source of this redundancy is a faultline in the matrix representation for the gamma operators which is most clearly manifested in the three momentum operators or spatial differentials. Here, we find a system which is not rotation symmetric, unlike physical momentum or 3-D space. The mathematical constraints brought about by using matrices force us into a physical representation which does not reflect reality, and which may therefore be inadvertently introducing a 'redundancy barrier' of the kind discussed in the previous section. (In fact, any mathematical representation which makes a 3-D system rotation asymmetric may be considered either the origin or the signature of a singularity.)

The gamma operators in matrix form are normally expressed using Pauli matrices. To understand the more fundamental algebra involved, however, we need first to look at the multivariate vectors, with multiplication rules:

$$\begin{aligned} \mathbf{i}^2 = \mathbf{j}^2 = \mathbf{k}^2 &= -i\mathbf{ijk} = 1 \\ \mathbf{ij} = -\mathbf{ji} &= i\mathbf{k} \\ \mathbf{jk} = -\mathbf{kj} &= i\mathbf{i} \\ \mathbf{ki} = -\mathbf{ik} &= i\mathbf{j}. \end{aligned} \tag{5.2}$$

Isomorphic to these, *but not fully equivalent*, are the Pauli matrices, σ_x, σ_y, σ_z, used to represent spin in the Schrödinger equation. Hestenes used the isomorphism, in 1966, to derive the origin of spin from the $i\mathbf{a} \times \mathbf{b}$ term in the full product. Among other things, Hestenes and his followers were able to show how, writing the Schrödinger equation in terms of a multivariate, rather than ordinary, vector ∇, automatically generates fermionic half-integral spin.[3-4] So, defining

$$\sigma_x = \begin{pmatrix} 0 & 1 \\ 1 & 0 \end{pmatrix} \quad \sigma_y = \begin{pmatrix} 0 & -i \\ i & 0 \end{pmatrix} \quad \sigma_z = \begin{pmatrix} 1 & 0 \\ 0 & -1 \end{pmatrix} \quad \text{with unit } I = \begin{pmatrix} 1 & 0 \\ 0 & 1 \end{pmatrix}$$

we have:

$$\begin{aligned} \sigma_x\sigma_y &= -\ \sigma_y\sigma_x = i\sigma_x \\ \sigma_y\sigma_z &= -\ \sigma_z\sigma_y = i\sigma_x \\ \sigma_z\sigma_x &= -\ \sigma_x\sigma_z = i\sigma_y\,. \end{aligned} \tag{5.3}$$

But there is a fundamental difference between the two isomorphic systems (5.2) and (5.3). Pauli matrices are not symmetric in three dimensions because they are based on a two-dimensional number system – the complex plane. The notable thing about Pauli matrices is the fact that the multiplication rules for their algebraic equivalents involve an extra pseudoscalar term (i) in the pseudovector products. This means that, at least one of the three matrices must have complex coefficients, creating an asymmetric relationship between them.

This complexity must also carry over to the gamma matrices, as these are conventionally defined in terms of Pauli matrix components.

$$\boldsymbol{\gamma} = \begin{pmatrix} 0 & \boldsymbol{\sigma} \\ -\boldsymbol{\sigma} & 0 \end{pmatrix} \qquad \gamma_0 = \begin{pmatrix} I & 0 \\ 0 & -I \end{pmatrix} \qquad \gamma_5 = \begin{pmatrix} 0 & -I \\ -I & 0 \end{pmatrix}$$

$$\gamma_1 = \begin{pmatrix} 0 & \sigma_x \\ -\sigma_x & 0 \end{pmatrix} \quad \gamma_2 = \begin{pmatrix} 0 & \sigma_y \\ -\sigma_y & 0 \end{pmatrix} \quad \gamma_3 = \begin{pmatrix} 0 & \sigma_z \\ -\sigma_z & 0 \end{pmatrix}$$

where $\gamma_5 = i\gamma_0\gamma_1\gamma_2\gamma_3$, leading to:

$$\gamma_1 = \begin{pmatrix} 0 & 0 & 0 & 1 \\ 0 & 0 & 1 & 0 \\ 0 & -1 & 0 & 0 \\ -1 & 0 & 0 & 0 \end{pmatrix} \gamma_2 = \begin{pmatrix} 0 & 0 & 0 & -i \\ 0 & 0 & i & 0 \\ 0 & i & 0 & 0 \\ -i & 0 & 0 & 0 \end{pmatrix} \gamma_3 = \begin{pmatrix} 0 & 0 & 1 & 0 \\ 0 & 0 & 0 & -1 \\ -1 & 0 & 0 & 0 \\ 0 & 1 & 0 & 0 \end{pmatrix}$$

$$\gamma_0 = \begin{pmatrix} 1 & 0 & 0 & 0 \\ 0 & 1 & 0 & 0 \\ 0 & 0 & -1 & 0 \\ 0 & 0 & 0 & -1 \end{pmatrix} \qquad I = \begin{pmatrix} 1 & 0 & 0 & 0 \\ 0 & 1 & 0 & 0 \\ 0 & 0 & 1 & 0 \\ 0 & 0 & 0 & 1 \end{pmatrix}$$

With these components, the 4×4 Dirac differential operator now becomes:

$$\begin{pmatrix} \frac{\partial}{\partial t} + im & 0 & \frac{\partial}{\partial z} & \frac{\partial}{\partial x} - i\frac{\partial}{\partial y} \\ 0 & \frac{\partial}{\partial t} + im & \frac{\partial}{\partial x} + i\frac{\partial}{\partial y} & -\frac{\partial}{\partial z} \\ -\frac{\partial}{\partial z} & -\frac{\partial}{\partial x} + i\frac{\partial}{\partial y} & -\frac{\partial}{\partial t} + im & 0 \\ -\frac{\partial}{\partial x} - i\frac{\partial}{\partial y} & \frac{\partial}{\partial z} & 0 & -\frac{\partial}{\partial t} + im \end{pmatrix}$$

from which we can derive two positive and two negative energy free-particle spinors for ψ, with respective phases exp $(-ip.x)$ and exp $(ip.x)$, where p and x are 4-vectors. The positive energy spinors become:

$$\left(\frac{E+m}{2m}\right)^{1/2}\begin{pmatrix}1\\0\\\frac{p_z}{E+m}\\\frac{p_x+ip_y}{E+m}\end{pmatrix} \quad \text{and} \quad \left(\frac{E+m}{2m}\right)^{1/2}\begin{pmatrix}0\\1\\\frac{p_x-ip_y}{E+m}\\\frac{-p_z}{E+m}\end{pmatrix}$$

and the negative energy spinors:

$$\left(\frac{E+m}{2m}\right)^{1/2}\begin{pmatrix}\frac{p_z}{E+m}\\\frac{p_x+ip_y}{E+m}\\1\\0\end{pmatrix} \quad \text{and} \quad \left(\frac{E+m}{2m}\right)^{1/2}\begin{pmatrix}\frac{p_x-ip_y}{E+m}\\\frac{-p_z}{E+m}\\0\\1\end{pmatrix}$$

Immediately, we note a potential problem: there is no such physical object as $p_x + ip_y$ or $p_x - ip_y$ (or even $E + m$). Obviously, using the complex notation makes p_x orthogonal to p_y, so acting as a kind of substitute vector addition, but it can't be extended to create a 3-dimensional operator on an equivalent basis, and it has the unwelcome consequence of making p_x physically different from p_y.

5.3 Defragmenting the Dirac Equation

In fact, though the Dirac equation, as conventionally written in (5.1), is the fundamental basis of particle physics, it is inconvenient in many ways in addition to the unphysical nature of the spinor solutions, with the main problems relating to its asymmetric structure. Thus the operator $(\gamma^\mu\partial_\mu + im)$ is a 4×4 matrix, as are each of the four terms γ^μ, while ψ is a 4-component (vector) spinor. There are many possible choices for the γ matrices, but, whatever choice is made, there will be mixing of the energy $(\gamma^0\partial_0)$ and mass terms, and a situation in which some of the momentum terms $(\boldsymbol{\gamma}.\nabla)$ have real matrix coefficients and others imaginary ones.

Essentially, there are four obvious problems with the matrices:

(1) They cause fragmentation of the equation, mixing up energy, momentum and

mass terms.

(2) They take up too much logical space, requiring 16 pieces of information for one operation.

(3) They lack symmetry. There are 5 terms in the equation, but only 4 have a γ matrix. Yet there is a fifth matrix (γ^5) in the algebra.

(4) Even more significantly, as we have seen, the momentum operators are made asymmetric; giving one of the momentum operators an imaginary representation means that our phase space has two spacelike and two timelike components (signature + + − −), rather than the 3 + 1 structure that we believe represents physical reality (meaning that the representation is not fully relativistic). Ultimately, this leads to singularities and redundancy on a massive scale, which cannot be fully realised until we have found an alternative formalism which removes them.

Fig. 5.1 Defragmentation, with separate 'bins' for energy, momentum and mass

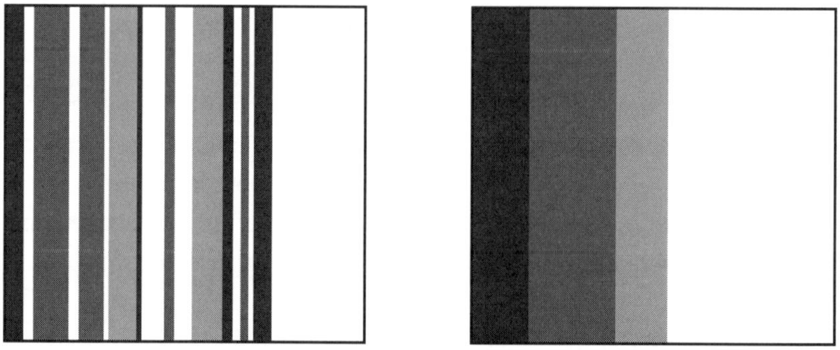

There is, of course, no need to use matrices at all, other than historical precedent; and, if we use simpler algebraic operators, we can *defragment* the equation, that is, separate energy, momentum and mass terms from each other, each in its own 'bin', in the same way as we defragment the real and imaginary parts in physical equations using ordinary complex numbers. We will also see that using such operators additionally solves the logical space problem by reducing the 16 operators of each 4 × 4 matrix to a single term. The only requirement is to find a system of five operators in which

$$(\gamma^0)^2 = (\gamma^5)^2 = 1 \quad (\gamma^1)^2 = (\gamma^2)^2 = (\gamma^3)^2 = -1$$

and *all* terms anticommute with each other, so that $\gamma^0\gamma^1 = -\gamma^1\gamma^0$, etc

The most elegant way of achieving this result might be to use geometrical or Clifford algebra. However, a more *physically* expressive option, as we have seen, is to use a combination of quaternions and multivariate vectors, as outlined in chapters 1 and 3. As discussed in chapter 3, we can easily relate the two algebras by making mappings of the form:

$$\begin{aligned} \gamma^0 &= -\boldsymbol{i}i \\ \gamma^1 &= \mathbf{i}\boldsymbol{k} \\ \gamma^2 &= \mathbf{j}\boldsymbol{k} \\ \gamma^3 &= \mathbf{k}\boldsymbol{k} \\ \gamma^5 &= \boldsymbol{i}\boldsymbol{j} \end{aligned} \qquad (5.4) \qquad \text{or} \qquad \begin{aligned} \gamma^0 &= i\boldsymbol{k} \\ \gamma^1 &= \mathbf{i}\boldsymbol{i} \\ \gamma^2 &= \mathbf{j}\boldsymbol{i} \\ \gamma^3 &= \mathbf{k}\boldsymbol{i} \\ \gamma^5 &= \boldsymbol{i}\boldsymbol{j} \end{aligned} \qquad (5.5)$$

Both of these mappings generate the full 32-part algebra; and both are relevant to the construction of a defragmented Dirac equation. If, for example, we substitute (5.4) into the component form of the Dirac equation,

$$\left(\gamma^0 \frac{\partial}{\partial t} + \gamma^1 \frac{\partial}{\partial x} + \gamma^2 \frac{\partial}{\partial y} + \gamma^3 \frac{\partial}{\partial z} + im\right)\psi = 0, \qquad (5.6)$$

we obtain:

$$\left(-\boldsymbol{i}i \frac{\partial}{\partial t} + \boldsymbol{k}\mathbf{i} \frac{\partial}{\partial x} + \boldsymbol{k}\mathbf{j} \frac{\partial}{\partial y} + \boldsymbol{k}\mathbf{k} \frac{\partial}{\partial z} + im\right)\psi = 0. \qquad (5.7)$$

A key move now is to multiply the equation from the left by $\boldsymbol{j}$, altering the representation to (5.5), and obtaining

$$\left(i\boldsymbol{k} \frac{\partial}{\partial t} + \boldsymbol{i}\mathbf{i} \frac{\partial}{\partial x} + \boldsymbol{i}\mathbf{j} \frac{\partial}{\partial y} + \boldsymbol{i}\mathbf{k} \frac{\partial}{\partial z} + i\boldsymbol{j}m\right)\psi = 0. \qquad (5.8)$$

This apparently trivial step has profound consequences. The equation is now fully symmetrical, and the quaternion operators provide the 3 separated 'bins' we require:

$\boldsymbol{k}$	$\boldsymbol{i}$	$\boldsymbol{j}$
energy	momentum	mass

Although they are mathematically compatible, the new equation (5.8) takes us well beyond the conventional one. We can see this by applying the plane wave solution for a free-particle

$$\psi = A\, e^{-i(Et - \mathbf{p.r})},$$

where A is the amplitude and $e^{-i(Et - \mathbf{p.r})}$ the phase. We then find that:

$$(\boldsymbol{k}E + i\boldsymbol{i}\mathbf{i}p_x + i\boldsymbol{i}\mathbf{j}p_y + i\boldsymbol{i}\mathbf{k}p_x + i\boldsymbol{j}\, m)\, A\, e^{-i(Et - \mathbf{p.r})} = 0,$$

which may be more conveniently written in the form,

$$(\boldsymbol{k}E + i\boldsymbol{i}\,\mathbf{p} + i\boldsymbol{j}\, m)\, A\, e^{-i(Et - \mathbf{p.r})} = 0, \qquad (5.9)$$

where **p** is a multivariate vector.

Here, ($\boldsymbol{k}E + \boldsymbol{ii}\,\mathbf{p} + \boldsymbol{ij}\,m$) has the property of being a *nilpotent*, or square root of zero, because

$$(\boldsymbol{k}E + \boldsymbol{ii}\,\mathbf{p} + \boldsymbol{ij}\,m)\,(\boldsymbol{k}E + \boldsymbol{ii}\,\mathbf{p} + \boldsymbol{ij}\,m) = -E^2 + p^2 + m^2 = 0. \qquad (5.10)$$

The only way in which this can be nontrivially accomplished in (5.9) is if A itself is the *same nilpotent*; and, because equation (5.8) was obtained from equation (5.5) only by multiplying from the left, the amplitude of the Dirac wavefunction, even within the conventional form of the Dirac equation, must be equally nilpotent, or nilpotent subject to multiplication by any factor from the right. The derivation, of course, relies on the fact, that, for a multivariate **p**, the product **pp** has a meaning identical to the product of the scalar magnitudes $pp = p^2$. It is also identical to the product of the helicities ($\boldsymbol{\sigma}.\mathbf{p}$) ($\boldsymbol{\sigma}.\mathbf{p}$), which in the case of a fermionic state with positive energy is equal to $(-p)(-p)$, indicating that the multivariate vectors (as equivalent to Pauli matrices) automatically incorporate the concept of spin (see 6.3). (However, while using a multivariate vector **p** means that we obtain the spin directly, without needing $\boldsymbol{\sigma}.\mathbf{p}$, in cases where, for mathematical convenience, we reduce it to an ordinary vector, the spin must be explicitly introduced.)

5.4 The Dirac 4-Spinor

Conventionally, the Dirac equation allows four solutions, corresponding to the four combinations of fermion and antifermion, and spin up and spin down, which may be arranged in a column vector, or Dirac 4-spinor.

$$\Psi = \begin{pmatrix} \psi_1 \\ \psi_2 \\ \psi_3 \\ \psi_4 \end{pmatrix}$$

In calculation, this is usually reduced to 2 × 2-spinors, one representing fermion states and the other antifermion states, each with spin up and spin down terms. So, the ψ we have used in the previous sections is not really a single term, but a 4-component spinor, which accommodates fermion and antifermion states, as well as spin up and spin down. However, identification of these is now easy:

fermion / antifermion	$\pm E$
spin up / down	$\pm \mathbf{p}$

Since a multivariate **p** or **∇** already incorporates fermionic spin, we may identify the solutions as those produced by the four combinations of ± E and ± **p** (or **σ.p**). Reversal of the sign of ***k***E produces the wavefunction for an antiparticle, while reversal of the sign of ***ii*****p** changes the direction of spin. In this form of the equation, the operator **p**, because it is multivariate and is always effectively multiplied by itself in the eigenvalue produced by the differential operator, can represent the momentum vector **p**, the scalar value of p, or the spin term **σ.p**. There is no need to multiply the wavefunction operators by any other term, such as an additional spinor, to convert momentum states to spin states; spin is built into the structure of the multivariate operator **p**.[5] It is also of no significance from a fundamental point of view whether or not we specify a preferred direction, as in p_x, p_y, or p_z. We can, in many instances, derive general expressions for **p** from expressions originally derived for such directional terms; the equations for the general and the particular are, in each case, identical. As in conventional theory, to obtain a scalar probability density $\psi\psi^*$, we need to use a combination of all four solutions.

We now need a 4-spinor with 4 amplitudes and 4 phases, with all 4 variations of ± E and ± **p** applied to $(\boldsymbol{k}E + \boldsymbol{ii}\,\mathbf{p} + \boldsymbol{ij}\,m)\,e^{-i(Et\,-\,\mathbf{p.r})}$. In our notation, we could write these terms in the form

$$\psi_1 = A_1\, e^{-i(Et\,-\,\mathbf{p.r})}$$
$$\psi_2 = A_2\, e^{-i(Et\,+\,\mathbf{p.r})}$$
$$\psi_3 = A_3\, e^{i(Et\,-\,\mathbf{p.r}}$$
$$\psi_4 = A_4\, e^{i(Et\,+\,\mathbf{p.r})}.$$

If we now substitute these expressions into the equation

$$\left(ik\frac{\partial}{\partial t} + \boldsymbol{i}\nabla + \boldsymbol{i}jm\right)\psi = 0\,,$$

and apply a single differential operator, we obtain the Dirac 4-spinor as a column vector with 4 components:

$$\psi_1 = (\boldsymbol{k}E + \boldsymbol{ii}\,\mathbf{p} + \boldsymbol{ij}\,m)\, e^{-i(Et\,-\,\mathbf{p.r})}$$
$$\psi_2 = (\boldsymbol{k}E - \boldsymbol{ii}\,\mathbf{p} + \boldsymbol{ij}\,m)\, e^{-i(Et\,+\,\mathbf{p.r})}$$
$$\psi_3 = (-\boldsymbol{k}E + \boldsymbol{ii}\,\mathbf{p} + \boldsymbol{ij}\,m)\, e^{i(Et\,-\,\mathbf{p.r})}$$
$$\psi_4 = (-\boldsymbol{k}E - \boldsymbol{ii}\,\mathbf{p} + \boldsymbol{ij}\,m)\, e^{i(Et\,+\,\mathbf{p.r})}.$$

each of which is operated on by $(\boldsymbol{ik}\partial/\partial t + \boldsymbol{i}\nabla + \boldsymbol{ij}\,m)$. (The orthonormality of these solutions is shown in section 5.10.)

However, there is one more trick we can play (which is analogous to the switch between Eulerian and Lagrangian representations in dynamics), and it is essential to obtaining a fully defragmented Dirac equation. The equation, as so far constructed, uses a single differential operator acting on all four components of the Dirac spinor. There is, however, a much more useful form of the equation which exploits the nilpotent properties of the wavefunction more directly. Since our differential operator has been reduced to a single term from the 16 in the original matrix, we now have the logical space to turn it into a 4-spinor, like the wavefunction, and reduce to a *single phase*. That is, we *transfer* the variation in the signs of E and $\mathbf{p}$ from the exponential to the differential operator, which, instead of being a single expression, now becomes a 4-term row vector, forming a scalar product with the four terms in the column vector representing the Dirac 4-spinor:

So we now have four differential operators, each corresponding exactly with one of the terms in the wavefunction, but each operating on the same exponential terms. In principle, we have implicitly incorporated the idea that an antifermion going backwards in time is equivalent to a fermion going forwards in time, without having to make the idea explicit. Arranging the four terms, for convenience, in double columns, we have:

$$\left(ik\frac{\partial}{\partial t}+\boldsymbol{i}\nabla+ijm\right)(\boldsymbol{k}E+i\boldsymbol{i}\mathbf{p}+ijm)e^{-i(Et-\mathbf{p.r})}=0$$

$$\left(ik\frac{\partial}{\partial t}-\boldsymbol{i}\nabla+ijm\right)(\boldsymbol{k}E-i\boldsymbol{i}\mathbf{p}+ijm)e^{-i(Et-\mathbf{p.r})}=0$$

$$\left(-ik\frac{\partial}{\partial t}+\boldsymbol{i}\nabla+ijm\right)(-\boldsymbol{k}E+i\boldsymbol{i}\mathbf{p}+ijm)e^{-i(Et-\mathbf{p.r})}=0$$

$$\left(-ik\frac{\partial}{\partial t}-\boldsymbol{i}\nabla+ijm\right)(-\boldsymbol{k}E-i\boldsymbol{i}\mathbf{p}+ijm)e^{-i(Et-\mathbf{p.r})}=0$$

The Dirac 4-spinor equation for a free particle can now be represented by a row vector of 4 differential operators acting on a column vector of 4 eigenstates. Using a compactified notation:

$$\left(\pm ik\frac{\partial}{\partial t}\pm\boldsymbol{i}\nabla+ijm\right)(\pm\boldsymbol{k}E\pm i\boldsymbol{i}\mathbf{p}+ijm)e^{-i(Et-\mathbf{p.r})}=0 \tag{5.11}$$

row *column*

This formalism automatically includes the Feynman representation of antistates with negative energy going backwards in time. It ensures that negative energy only occurs within negative time, and that all states defined in a time compatible with thermodynamics are positive energy states. It follows also that positive energy states require positive mass states, if we define mass as the 'proper energy', or energy within the fermion's inertial frame, in the same way as the (positive) proper time τ in the conjugate nilpotent expression ($\pm \boldsymbol{k}t \pm \boldsymbol{ii}\mathbf{r} + \boldsymbol{ij}\tau$), is defined as the time within the fermion's inertial frame ($\mathbf{r} = 0$).

If we had written (5.10) in spinor form, we could, of course, have derived (5.11) from it directly, by converting the E and $\mathbf{p}$ terms into quantum operators, and we could use it in the same way to derive the Klein-Gordon equation, which, in this formalism, becomes merely a branch of Dirac, with an additional conversion of the amplitude:

$$\left(\pm i\boldsymbol{k}\frac{\partial}{\partial t} \pm \boldsymbol{i}\nabla + \boldsymbol{ij}m\right)\left(\pm i\boldsymbol{k}\frac{\partial}{\partial t} \pm \boldsymbol{i}\nabla + \boldsymbol{ij}m\right)e^{-i(Et-\mathbf{p.r})} = 0 .$$

However, while this equation can be applied to a wavefunction of any kind, for example the scalar wavefunction of a boson, because the nilpotency is provided by the double differential operator, (5.11) can only be applied to a state which is a nilpotent.

An important qualification must be made, however, to the formalism as presented in (5.11). This is derived directly from the Dirac equation as normally presented. But, from a *physical* point of view, it is correct to write the nilpotent operator in the form used in chapter 3:

$$\left(\pm i\boldsymbol{k}E \pm \boldsymbol{i}\mathbf{p} + \boldsymbol{j}m\right)$$

in which the energy operator becomes a pseudoscalar, and the momentum and mass operators remain real. Here, significantly, the convention for positive and negative energy states is reversed from that used in the version directly from the conventional Dirac equation. Using this convention, (5.11) would be written in the alternative form:

$$\left(\mp i\boldsymbol{k}\frac{\partial}{\partial t} \pm \boldsymbol{i}\nabla + \boldsymbol{j}m\right)\left(\pm i\boldsymbol{k}E \pm \boldsymbol{i}\mathbf{p} + \boldsymbol{j}m\right)e^{-i(Et-\mathbf{p.r})} = 0 , \qquad (5.12)$$

with the sign of the $\partial / \partial t$ operator reversed. Both conventions will be used in the following chapters. In general, where compatibility with the conventional Dirac equation is convenient, especially where calculations are involved, the convention will be that of (5.11). However, where the emphasis is on physical interpretation, (5.12) will be preferred. Though the form ($\pm \boldsymbol{k}E \pm \boldsymbol{ii}\,\mathbf{p} + \boldsymbol{ij}\,m$) for the amplitude will be adopted where the convention is that of (5.11), it will always be assumed that, *physically*, energy, and not mass, is pseudoscalar.

Using either convention, it may be said, the equation in the defragmented nilpotent form has also reached its most perfect pitch of simplicity and symmetry, creating an irreducible form of quantum mechanics. Operator and amplitude are essentially identical, since $(\pm \boldsymbol{k}E \pm \boldsymbol{ii}\mathbf{p} + \boldsymbol{ij}m)$ is just

$$\left(\pm i\boldsymbol{k}\frac{\partial}{\partial t} \pm \boldsymbol{i}\nabla + \boldsymbol{ij}m \right)$$

in operator form. In fact, we can even do away with the equation altogether! All we need is to specify the operator meanings for the terms separated by the three quaternion 'bins':

$$(\pm \boldsymbol{k}E \pm \boldsymbol{ii}\,\mathbf{p} + \boldsymbol{ij}\,m)$$

That is, E and $\mathbf{p}$ don't need to be the differential operators for a free state. They can be covariant derivatives (preserving the form of a Lagrangian under a local gauge transformation) or incorporate field terms of any kind. In principle, they should contain the terms needed to express every interaction to which the fermion is subject. Of course, if E and $\mathbf{p}$ contain field terms, then the phase will no longer be a pure exponential $e^{-i(Et - \mathbf{p}.\mathbf{r})}$. It will be whatever function is needed to make the amplitude nilpotent. Both the amplitude and phase terms will be uniquely defined once the differential operator has been specified. And because there is only one phase term, analytical solutions will be easier to find. Again, although the operator is in principle a 4-component spinor, the terms are not independent, and all the information is contained in the lead term. The remaining terms then simply offer a fixed pattern of sign variations for the energy and momentum components; and the total effect of this variation is merely to ensure that fermion amplitudes (for states with equal values of E and $\mathbf{p}$) have only two possible products with each other after normalization: 0 if they are identical and 1 if they are not (almost like the reverse of a delta function). Essentially, then, although we will continue to use four components for the maximum clarity, the fermionic state, and all calculation related to it, can be reduced to a single-line operator, composed of an energy term, a momentum term and a mass term. Symbolically, we have $\boldsymbol{k}E + \boldsymbol{ii}\mathbf{p} + \boldsymbol{ij}m$ or $i\boldsymbol{k}E + \boldsymbol{i}\mathbf{p} + \boldsymbol{j}m$, where E and $\mathbf{p}$ are either operators or eigenvalues. Amplitude and phase are uniquely determined by the same operator and each is quantized in the same way. Formal second quantization is unnecessary. In this formulation, the differential operator and the amplitude are essentially identical, and are quantized in identical ways. The reason for writing one as a differential operator and one as an eigenvalue term is that we then have a simultaneous representation of the nonconserved and the conserved parts of the equation; the Dirac equation for a free particle can then be seen as a mathematical expression of the absolute conservation of E, $\mathbf{p}$ and m and the absolute *nonconservation*, or variability, of $\mathbf{r}$ and t. In this version of the equation, also, as we have seen, the four solutions exist at the same time and on the same footing – they differ only in signs of E and $\mathbf{p}$. Essentially, we have the basic requirements for automatic second quantization and a quantum field theory, the nilpotent expressions displaying the characteristics of full quantum field operators rather than wavefunctions in the more restricted sense; and we can show that quantum field integrals acting on vacuum produce the nilpotent state vector (see 6.13).

5.5 The 4-Component Differential Operator

Collecting all four terms into a single expression, we have

$$\left(\pm \boldsymbol{i}k\frac{\partial}{\partial t} \pm \boldsymbol{i}\nabla + \boldsymbol{i}\boldsymbol{j}m\right)(\boldsymbol{k}E \pm i\boldsymbol{i}\mathbf{p} + i\boldsymbol{j}m)e^{-i(Et-\mathbf{p}.\mathbf{r})} = 0 \tag{5.13}$$

as our version of the Dirac equation for a free particle. (Of course, as we will demonstrate later, this form of the differential operator and eigenvalue term can be used, not only for free particle states, but also for bound states, where the phase term is not a simple exponential, and where E and $\mathbf{p}$ also incorporate field terms.)

Equation (5.13) is of special significance, for, as we have seen, the differential operator and the eigenvalue part of the wavefunction, or amplitude, are essentially identical, and both are quantized. Reducing this to the eigenvalue form, applying the additional i factor (convention (5.12)), and multiplying out, produces the classical relativistic momentum-energy conservation equation:

$$(\pm\, i\boldsymbol{k}E \pm \boldsymbol{i}\,\mathbf{p} + \boldsymbol{j}\,m)\,(\pm\, i\boldsymbol{k}E \pm \boldsymbol{i}\,\mathbf{p} + \boldsymbol{j}\,m) = E^2 - p^2 - m^2 = 0. \tag{5.14}$$

If we assume, for a moment, that classical physics is logically prior to quantum mechanics (which, according to our understanding, is the opposite of the true position), we can 'derive' the free-particle Dirac equation very simply by the reverse process of factorizing (5.14), purely algebraically, into $(\pm\, i\boldsymbol{k}E \pm \boldsymbol{i}\,\mathbf{p} + \boldsymbol{j}\,m)$ $(\pm\, i\boldsymbol{k}E \pm \boldsymbol{i}\,\mathbf{p} + \boldsymbol{j}\,m)$, and then replacing the classical E and $\mathbf{p}$ terms with the standard quantum operators acting on the free-particle exponential. Using purely quantum mechanical procedures, we can multiply equation (5.13) from the left by the same differential operator to show that a Klein-Gordon condition applies to fermionic, as well as bosonic, states, as

$$\left(\pm \boldsymbol{k}\frac{\partial}{\partial t} \pm i\boldsymbol{i}\nabla + \boldsymbol{j}m\right)\left(\pm \boldsymbol{k}\frac{\partial}{\partial t} \pm i\boldsymbol{i}\nabla + \boldsymbol{j}m\right)(i\boldsymbol{k}E \pm \boldsymbol{i}\mathbf{p} + \boldsymbol{j}m)e^{-i(Et - \mathbf{p.r})} = 0$$

becomes
$$\left(\frac{\partial^2}{\partial t^2} - \nabla^2 + m^2\right)(i\boldsymbol{k}E \pm \boldsymbol{i}\mathbf{p} + \boldsymbol{j}m)e^{-i(Et - \mathbf{p.r})} = 0 .$$

The Dirac equation actually overspecifies its components. The algebra, as normally written, specifies the same information three times: in the E-$\mathbf{p}$-m terms, in the spinors, and in the exponentials. In the nilpotent formulation, the information is specified only once, in the first term of the quaternion state vector (QSV). The QSV then automatically selects the three remaining terms in sequence, incorporating all values of $\pm E \pm \mathbf{p}$. The phase factor for a free particle is algebraically the same for any state, and determined only by the values of E and $\mathbf{p}$. The same principle applies for a bound state, though here the E and $\mathbf{p}$ terms are determined as eigenvalues produced by the differential operator acting on the variable or nonquaternionic part of the wavefunction (which we again describe as the phase factor). The Hamiltonian is completely determined by the QSV, the differential operator having an eigenvalue identical to the state vector, when operating on the common phase. The nature of the phase is totally determined by whatever function is necessary to produce the correct nilpotent when acted on by the differential operator. Thus, the anticommuting pentad term

$$(i\boldsymbol{k}E + \boldsymbol{i}\mathbf{p} + i\boldsymbol{j}m)$$

completely defines a state of a free fermion, spin up, because the phase term is necessarily always $e^{-i(Et - \mathbf{p.r})}$, for *any* free state, and the complete specification of the state vector follows automatically, as:

$$\begin{pmatrix} i\boldsymbol{k}E + \boldsymbol{i}\mathbf{p} + \boldsymbol{j}m \\ i\boldsymbol{k}E - \boldsymbol{i}\mathbf{p} + \boldsymbol{j}m \\ -i\boldsymbol{k}E + \boldsymbol{i}\mathbf{p} + \boldsymbol{j}m \\ -i\boldsymbol{k}E - \boldsymbol{i}\mathbf{p} + \boldsymbol{j}m \end{pmatrix}.$$

The differential operator is, in fact, *identical* to the state vector, but with the E and $\mathbf{p}$ representing the relevant quantum operators with or without field terms, rather than eigenvalues. The first term of the state vector codifies *all* the information about a state. The exponential phase term in the free particle case is an aspect of the automatic application of the state to the vacuum. In defining states, in fact, including composite ones, we can avoid using the phase term altogether. We simply use vector multiplication of the quaternion state vectors.

It is significant that there are exactly four solutions to the Dirac equation. Both quaternion and complex operators, of course, require an equal representation for + and – signs, suggesting eight possible sign combinations for $\pm ikE \pm \boldsymbol{i}\,\mathbf{p} \pm \boldsymbol{j}\,m$; but only four of these will be independent, since the overall sign for the state vector is an arbitrary scalar factor. So the sign of one of ikE, $\boldsymbol{i}\,\mathbf{p}$ or $\boldsymbol{j}\,m$ must behave as if fixed. The compactification process producing the Dirac state effectively removes a sign degree of freedom. With only E and $\mathbf{p}$ terms represented in the phase, it becomes evident that the fixed term must be m. Four solutions also result from the fact that the quaternionic structure of the state vector can be related to the conventional 4×4 matrix formulation using quaternionic matrices (cf section 5.7), and the conventional formulation is itself uniquely determined by the 4-D space-time signature of the equation, a $2n$-D space-time requiring a $2^n \times 2^n$ matrix representation of the Clifford algebra. In the case of quaternionic matrices, it is also significant that the (hidden) quaternion operators $\boldsymbol{i}, \boldsymbol{j}, \boldsymbol{k}$ applied, along with 1, to the rows and columns, and also to the rows of the Dirac 4-spinor, are *identical* in meaning to the same operators applied to the terms in the nilpotent state vector, as one can be derived from the other.

5.6 *C*-Linear Maps and Lifts

We can use some rather technical arguments to show that the nilpotent and matrix formulations of the Dirac equation are interconvertible in both directions, though this does not mean, of course, that they are exactly equivalent, and both conversion processes use rather subtle properties of the formalisms, rather than the more obvious ones. Neither process is essential to the development of the nilpotent formulation, but the matrix $\rightarrow$ nilpotent transition, in particular, provides significant physical insights. To convert from the nilpotent representation to matrix form, it is convenient to use C-linear maps and lifts. In our expressions for the Dirac equation and the Dirac wavefunction, $\mathbf{p}$ is understood to be a multivariate momentum vector with the usual three components. The most usual representation for multivariate vectors is the set of Pauli matrices, σ_0, σ_x, σ_y, σ_z, which means that we may write the Dirac equation as follows:

$$\left(ik\frac{\partial}{\partial t} + \boldsymbol{i\sigma}.\nabla + \boldsymbol{ij}m \right)\psi = 0 ,$$

where **σ.∇** is understood now to mean the scalar product, $\sigma_x\partial_x + \sigma_y\partial_y + \sigma_z\partial_z$. This has an equivalent 2×2 matrix form:

$$\begin{pmatrix} \boldsymbol{k}E + \boldsymbol{ii}p_x + \boldsymbol{ij}m & \boldsymbol{ii}(p_x - ip_y) \\ \boldsymbol{ii}(p_x + ip_y) & \boldsymbol{k}E - \boldsymbol{ii}p_x + \boldsymbol{ij}m \end{pmatrix} \begin{pmatrix} \phi \\ \chi \end{pmatrix} = 0 ,$$

which leads to the coupled linear differential equations

$$(\boldsymbol{k}E + \boldsymbol{ii}\, p_z + \boldsymbol{ij}\, m)\phi + \boldsymbol{ii}\,(p_x - ip_y)\chi = 0,$$
$$\boldsymbol{ii}\,(p_x + ip_y)\phi + (\boldsymbol{k}E - \boldsymbol{ii}\, p_z + \boldsymbol{ij}\, m)\chi = 0.$$

Choosing the momentum to be along the z-axis ($p = p_z$), these differential equations decouple to leave

$$(\boldsymbol{k}E + \boldsymbol{ii}\, p + \boldsymbol{ij}\, m)\phi = 0,$$
$$(\boldsymbol{k}E - \boldsymbol{ii}\, p + \boldsymbol{ij}\, m)\chi = 0,$$

with

$$\phi = \psi_1 = (\boldsymbol{k}E + \boldsymbol{ii}\mathbf{p} + \boldsymbol{ij}m)\begin{pmatrix} 1 \\ 0 \end{pmatrix} e^{-i(Et - \mathbf{p}.\mathbf{r})},$$

$$\chi = \psi_2 = (\boldsymbol{k}E - \boldsymbol{ii}\mathbf{p} + \boldsymbol{ij}m)\begin{pmatrix} 0 \\ 1 \end{pmatrix} e^{-i(Et + \mathbf{p}.\mathbf{r})}.$$

These are the positive energy solutions; the quaternion representation allows a simple deduction of the negative energy solutions:

$$\psi_3 = (-\boldsymbol{k}E + \boldsymbol{ii}\mathbf{p} + \boldsymbol{ij}m)\begin{pmatrix} 1 \\ 0 \end{pmatrix} e^{i(Et + \mathbf{p}.\mathbf{r})},$$

$$\psi_4 = (-\boldsymbol{k}E - \boldsymbol{ii}\mathbf{p} + \boldsymbol{ij}m)\begin{pmatrix} 0 \\ 1 \end{pmatrix} e^{i(Et - \mathbf{p}.\mathbf{r})}.$$

To carry out quantum mechanical calculations, we need to define a scalar product for the wavefunctions. However, the wavefunctions in the quaternion / spinor form are effectively a tensor product of two representations. It is convenient to use a C-linear map **F**: $H^C \rightarrow C^2$ is a mapping from complex quaternions, H^C ($q_\mu \mathbf{i}_\mu$, $q_\mu \in C$) to C^2.

$$\mathbf{F}: H^C \rightarrow C^2, \quad \mathbf{Q} \rightarrow \mathbf{Q}\begin{pmatrix} 1 \\ 0 \end{pmatrix},$$

where **Q** is a complex quaternion, and

$$\mathbf{i}_0 = \begin{pmatrix} 1 & 0 \\ 0 & 1 \end{pmatrix}, \quad \mathbf{i}_1 = \boldsymbol{i} = \begin{pmatrix} 0 & -i \\ -i & 0 \end{pmatrix}, \quad \mathbf{i}_2 = \boldsymbol{j} = \begin{pmatrix} 0 & -1 \\ 1 & 0 \end{pmatrix}, \quad \mathbf{i}_3 = \boldsymbol{k} = \begin{pmatrix} -i & 0 \\ 0 & i \end{pmatrix}.$$

Applying this to ψ_1, we obtain

$$\psi_1 = \frac{1}{\sqrt{2}} \begin{pmatrix} -i \\ \kappa + i\varepsilon \end{pmatrix} \otimes \begin{pmatrix} 1 \\ 0 \end{pmatrix},$$

where $\kappa = p\ /\ E$, $\varepsilon = m\ /\ E$, and $\otimes$ is the tensor product between the two representations. Hence all four solutions may take the form

$$\psi_1 = \frac{1}{\sqrt{2}} \begin{pmatrix} -i \\ 0 \\ \kappa + i\varepsilon \\ 0 \end{pmatrix} \quad \psi_2 = \frac{1}{\sqrt{2}} \begin{pmatrix} 0 \\ -i \\ 0 \\ -\kappa + i\varepsilon \end{pmatrix} \quad \psi_3 = \frac{1}{\sqrt{2}} \begin{pmatrix} i \\ 0 \\ \kappa + i\varepsilon \\ 0 \end{pmatrix} \quad \psi_4 = \frac{1}{\sqrt{2}} \begin{pmatrix} 0 \\ i \\ 0 \\ -\kappa + i\varepsilon \end{pmatrix}$$

This mapping leads to the conventional Dirac equation for these states, which in the case of ψ_1 is

$$\begin{pmatrix} -iE & 0 & -im + p & 0 \\ 0 & -iE & 0 & -im - ip \\ im + p & 0 & iE & 0 \\ 0 & im - p & 0 & iE \end{pmatrix} \begin{pmatrix} -i \\ 0 \\ \kappa + i\varepsilon \\ 0 \end{pmatrix} = 0.$$

5.7 The Quaternion Form Derived from a Matrix Representation

The procedure outlined in the previous section shows that the quaternion nilpotent algebra can be used to generate the standard version of the Dirac spinors. The reverse process is also possible, but requires a more subtle argument which suggests why exactly four solutions are required (only four are, of course, *possible* if they are to be independent). It also suggests that a link may be found with Dirac algebras that structure the four solutions as components of a four-vector or quaternion set.

Here we interpret the matrices as dyadics formed from quaternion (or, alternatively, 4-vector) components, arranged by row and column. Starting with the conventional form of the equation:

$$(\boldsymbol{\alpha}.\mathbf{p} + \beta m - E)\ \psi = 0,$$

where $\beta = \gamma^0$ and $\boldsymbol{\alpha}\beta = \boldsymbol{\gamma}$ (so $\boldsymbol{\alpha} = -ij\mathbf{1}$), and taking (for convenience, without loss of generality) $p = p_y$, we obtain[6]

$$\alpha_y = \begin{pmatrix} 0 & 0 & 0 & -i \\ 0 & 0 & i & 0 \\ 0 & -i & 0 & 0 \\ i & 0 & 0 & 0 \end{pmatrix}.$$

Using[5]

$$\beta = \begin{pmatrix} 0 & 0 & i & 0 \\ 0 & 0 & 0 & i \\ -i & 0 & 0 & 0 \\ 0 & -i & 0 & 0 \end{pmatrix}$$

and applying the unit 4×4 matrix to E, the Dirac equation becomes

$$(\alpha.\mathbf{p} + \beta m - E)\psi = \begin{pmatrix} -E & 0 & im & -ip \\ 0 & -E & ip & im \\ im+p & 0 & iE & 0 \\ ip & -im & 0 & -E \end{pmatrix} \begin{pmatrix} \psi_1 \\ \psi_2 \\ \psi_3 \\ \psi_4 \end{pmatrix} = 0,$$

where the column vector is the usual 4-component spinor, and the terms E and $\mathbf{p}$ are the quantum operators which give the eigenvalues represented by these symbols when applied to the exponential term of the wavefunction. (It will be convenient here to refer to the components of the 4×4 matrix in terms of these eigenvalues rather than in terms of the operators which produce them.)

We can interpret the rows and columns of this matrix as having either 4-vector or quaternion coefficients. Let us choose the quaternion operators

$$\boldsymbol{j}, \boldsymbol{i}, \boldsymbol{k}, 1$$

as the respective coefficients of the 4 rows. The 4×4 matrix now becomes a single row bra matrix with the columns:

$$-\boldsymbol{j}E \;-\boldsymbol{i}km + ip$$
$$-\boldsymbol{i}E \;-\boldsymbol{i}kp - im$$
$$\boldsymbol{i}jm + \boldsymbol{i}ip - \boldsymbol{k}E$$
$$-\boldsymbol{i}jp + \boldsymbol{i}im - E\,.$$

If we multiply these terms from the left by the respective *column* coefficients

$$\boldsymbol{i}, -\boldsymbol{j}, 1, \boldsymbol{k},$$

we obtain in each case the expression

$$-\boldsymbol{k}E + i\boldsymbol{i}\,\mathbf{p} + i\boldsymbol{j}\,m,$$

which, when multiplied from the right by a wavefunction beginning with this term, gives a zero product. In order to show that this is equivalent to the Dirac equation in matrix form, we need to show that the four solutions ψ_1, ψ_2, ψ_3, ψ_4, multiplied by the appropriate quaternion row coefficients, each result in expressions beginning with this term.

Suppose, therefore, that we have the four solutions, as previously assumed:

$$\psi_1 = (\boldsymbol{k}E + i\boldsymbol{i}\,\mathbf{p} + i\boldsymbol{j}\,m)\,e^{-i(Et - \mathbf{p.r})}$$
$$\psi_2 = (\boldsymbol{k}E - i\boldsymbol{i}\,\mathbf{p} + i\boldsymbol{j}\,m)\,e^{-i(Et - \mathbf{p.r})}$$
$$\psi_3 = (-\boldsymbol{k}E + i\boldsymbol{i}\,\mathbf{p} + i\boldsymbol{j}\,m)\,e^{-i(Et - \mathbf{p.r})}$$
$$\psi_4 = (-\boldsymbol{k}E - i\boldsymbol{i}\,\mathbf{p} + i\boldsymbol{j}\,m)\,e^{-i(Et - \mathbf{p.r})}.$$

We can show that the terms

$$\boldsymbol{k}\,\psi_1\,\boldsymbol{k}$$
$$-\boldsymbol{j}\,\psi_2\boldsymbol{j}$$
$$1\,\psi_3\,1$$
$$\boldsymbol{i}\,\psi_4\,\boldsymbol{i}$$

each produce the expression

$$(-\boldsymbol{k}E + i\boldsymbol{i}\,\mathbf{p} + i\boldsymbol{j}\,m)\,e^{-i(Et - \mathbf{p.r})}.$$

So, multiplying each of the terms

$$\boldsymbol{k}\,\psi_1$$
$$-\boldsymbol{j}\,\psi_2$$
$$1\,\psi_3$$
$$\boldsymbol{i}\,\psi_4$$

from the left by $(-\boldsymbol{k}E + i\boldsymbol{i}\,\mathbf{p} + i\boldsymbol{j}\,m)$ results in a zero product.

For convenience, we may rearrange these to give a ket (or column) matrix of the form:

$$\begin{pmatrix} \psi_1 \\ \psi_2 \\ \psi_3 \\ \psi_4 \end{pmatrix} = \boldsymbol{i}\psi_4 - \boldsymbol{j}\psi_2 + 1\psi_3 + \boldsymbol{k}\psi_1,$$

where the row coefficients are identical to the column coefficients of the bra matrix (and even allow the elimination of the − sign before $-\boldsymbol{j}$). The resulting equation is identical to the quaternionic Dirac equation, which we have previously derived by direct means, with the four solutions representing the four possible combinations of $\pm\,E$ and of $\pm\,\mathbf{p}$ states.

The derivation demonstrates that the reason for the use of 4 × 4 matrices is, in fact, the fundamentally quaternionic nature of the Dirac wavefunction. Ultimately, this is because of the 4-vector space-time used in relativistic equations, which is symmetrical with the quaternion algebra for mass and charge. According to our understanding, the use of quaternionic operators to define the weak, strong and electromagnetic charges (*w*, *s*, *e*) maps directly onto the use of the same operators for the terms *E*, **p**, *m* in the wavefunction, the existence of these terms as independent units stemming ultimately from the separate identities of the fundamental parameters time, space and mass(-energy) (*T*, **S**, *M*). Because quaternion operators define the meaning of the rows and columns in the Dirac matrix, the only way we can map charges *w*, *s*, *e* onto the *E*, **p**, *m* terms via these operators is to have 4 × 4 matrices, and hence a 4-dimensional space-time signature in the equation.

It is significant that the application of quaternion operators to *w*, *s*, *e* and *E*, **p**, *m* – and, by implication, to the more fundamental parameters time, space and mass (*T*, **S**, *M*) – is incomplete, each of these groups of three terms requiring a fourth to complete it. We can identify the respective fourth terms, without difficulty, as mass (*M*), angular momentum (*J*) and charge (*Q*). The 4 × 4 Dirac matrix, in effect, *incorporates the fourth term as a zero quantity*. But, while the matrix requires four quantities to which the columns and rows apply, the 4-component spinor allows only four possible solutions from their combination. Interpreting the solutions in terms of the number of relative sign combinations of the component terms allows only *three* of the terms to be nonzero. Significantly, the excluded term in each case is an invariant, the system requiring only one invariant quantity (e.g. *m* or *M*) to demonstrate the variability of the others (*E*, **p**, or *T*, **S**).

In connection with this, we may note that the term (***k*** *E* + ***ii*** *p* + ***ij*** *m*) can be factorised as

$$(\boldsymbol{k}E + \boldsymbol{ij}m)\left(1 + \frac{\boldsymbol{ii}p}{\boldsymbol{k}E + \boldsymbol{ij}m}\right)$$

which is, effectively, the conventional expression for the full Dirac spinor:

$$\Psi = \begin{pmatrix} \phi \\ \chi \end{pmatrix} = \begin{pmatrix} \phi \\ \phi \dfrac{\boldsymbol{\sigma}.\mathbf{p}}{E+m} \end{pmatrix}.$$

Arranged in this way, the state vector is a four-component quantity (three of the components coming from **p**), but, as the vector element of **p** disappears in the product **σ.p**, it can also be seen, more fundamentally, as a three-component quaternion, illustrating the fact, which many authors have proclaimed and demonstrated, that spin has nothing to do with relativity or even with quantum physics. The fact that the wavefunction requires only the three imaginary quaternions suggests that a full quaternion (or equivalent scalar plus vector) representation will contain redundant information, leading to a zero term.

5.8 Bilinear Covariants and the Dirac Lagrangian

For a Dirac wavefunction of the form

$$\psi = (\boldsymbol{k}E + \boldsymbol{i}i\,\mathbf{p} + i\boldsymbol{j}\,m)\,e^{-i(Et\,-\,\mathbf{p.r})}$$

we can also define an Hermitian conjugate,

$$\psi^{\dagger} = \gamma^0 \psi \gamma^0 = \psi^* = (\boldsymbol{k}E - \boldsymbol{i}i\,\mathbf{p} - i\boldsymbol{j}\,m)\,e^{i(Et\,-\,\mathbf{p.r})}$$

and an adjoint wavefunction,

$$\overline{\psi} = \psi^{\dagger}\gamma^0 = \psi^{\dagger}\,i\boldsymbol{k} = (-\,i\boldsymbol{E} - \boldsymbol{j}\mathbf{p} + \boldsymbol{i}\,m)\,e^{i(Et\,-\,\mathbf{p.r})},$$

though it may be equally convenient to replace this with the same function multiplied from the left by $i\boldsymbol{k}$, so that

$$\overline{\psi} = (\boldsymbol{k}E + \boldsymbol{i}i\,\mathbf{p} + i\boldsymbol{j}\,m)\,e^{i(Et\,-\,\mathbf{p.r})},$$

as this function equally satisfies the adjoint equation

$$\left(\frac{\partial\overline{\psi}}{\partial t}i\boldsymbol{k} + \nabla\overline{\psi}\boldsymbol{i} - \overline{\psi}i\boldsymbol{j}m\right) = 0\,,$$

and is only ever used as a multiplier from the left.

From this, we may derive the bilinear covariants, and hence the current density terms (using the four solution sum and the normalising factor):

$$\overline{\psi}\,\gamma^0\psi = \overline{\psi}\,i\boldsymbol{k}\psi = i\boldsymbol{k}\,E^2/E^2 = i\boldsymbol{k}$$

$$\overline{\psi}\,\gamma^1\psi = \overline{\psi}\,\mathbf{i}\boldsymbol{i}\psi = -\mathbf{i}\boldsymbol{i}\,p^2/E^2$$

$$\overline{\psi}\,\gamma^2\psi = \overline{\psi}\,\mathbf{i}\boldsymbol{j}\psi = -\mathbf{i}\boldsymbol{j}\,p^2/E^2$$

$$\overline{\psi}\,\gamma^3\psi = \overline{\psi}\,\mathbf{i}\boldsymbol{k}\psi = -\mathbf{i}\boldsymbol{k}\,p^2/E^2$$

$$\overline{\psi}\,\gamma^5\psi = \overline{\psi}\,i\boldsymbol{j}\psi = -\,i\boldsymbol{j}\,m^2/E^2.$$

The first four quantities (the scalar and vector terms) are the components of the current-probability density 4-vector, $J^{\mu} = \overline{\psi}\,\gamma^{\mu}\psi$, where $\partial^{\mu}J^{\mu} = 0$ in the absence of external fields. Using the bra and ket notation for the respective row and column vectors, represented by $\bar{\psi}$ and ψ, we may also write this in the form $\langle\bar{\psi}|\gamma^{\mu}|\psi\rangle$.

Using the original form of the adjoint wavefunction, the scalar, vector and pseudoscalar terms become:

$$\bar{\psi}\,\gamma^0\psi = 1$$
$$\bar{\psi}\,\gamma^1\psi = -\boldsymbol{ij}\mathbf{i}\,p^2/E^2$$
$$\bar{\psi}\,\gamma^2\psi = -\boldsymbol{ij}\mathbf{j}\,p^2/E^2$$
$$\bar{\psi}\,\gamma^3\psi = -\boldsymbol{ij}\mathbf{k}\,p^2/E^2$$
$$\bar{\psi}\,\gamma^5\psi = \boldsymbol{i}\,m^2/E^2,$$

while the pseudovector terms take the values:

$$\bar{\psi}\,\gamma^5\gamma^0\psi = -\boldsymbol{j}\,p^2/E^2$$
$$\bar{\psi}\,\gamma^5\gamma^1\psi = -\mathbf{i}$$
$$\bar{\psi}\,\gamma^5\gamma^2\psi = -\mathbf{j}$$
$$\bar{\psi}\,\gamma^5\gamma^3\psi = -\mathbf{k},$$

and the tensor terms are zeroed.

We may also use the bilinear covariants to construct the zero Dirac Lagrangian, which results when the equations of motion are obeyed:

$$\mathcal{L} = \bar{\psi}\,(i\gamma^{\mu}\partial_{\mu} - m)\psi = 0.$$

The Dirac state (or operator) should, in principle, contain all the information about fermions (and bosons) that can exist. In this case, it should be a unique key to explaining fundamental physical facts. The nilpotent Dirac state is, of course, a quantum operator. So it is useful to show that it can do conventional quantum mechanics. Taking $\psi\psi^*$ as the scalar product of the column vector with terms ψ_1, ψ_2 ψ_3, ψ_4, with the complex *quaternion* conjugate row vector with terms, ψ_1^*, ψ_2^*, ψ_3^*, ψ_4^*, the resulting sum = $(\pm \boldsymbol{k}E \pm \boldsymbol{ii}\mathbf{p} + \boldsymbol{ij}m)(\mp \boldsymbol{k}E \pm \boldsymbol{ii}\mathbf{p} + \boldsymbol{ij}m)$ gives the positive definite value $8E^2$, which becomes 1 on application of the normalising factor $1/\sqrt{2}E$ to each of ψ_1, ψ_2 ψ_3, ψ_4, though, because of the nature of nilpotent mathematics, we will seldom need to make the normalisation explicit. That is, the product of the amplitude and its complex quaternion conjugate defines the probability density that is needed for ordinary quantum mechanics. The use of the complex quaternion conjugate here is a reflection of the fact that the state vector has acquired an extra complexity through pre-multiplication by the factor $\boldsymbol{j}$, and, for a strict definition of probability density, this is the conjugate required. Using this definition, the 'reciprocal' of ($\pm\, \boldsymbol{k}E \pm \boldsymbol{ii}\,\mathbf{p} + \boldsymbol{ij}\,m$) that occurs, for example, in the propagator, can be identified as $(\mp kE \pm ii\mathbf{p} + ijm)$.

5.9 Removing Redundancies in Relativistic Quantum Mechanics

Redundancy in relativistic quantum mechanics has been associated with the singularities produced when the momentum operator is treated mathematically in a way that makes it rotation asymmetric. This can be overcome by the use of an algebra which is fundamentally 3-dimensional. It then appears that the fragmentation of the momentum operator in the conventional representation is totally avoidable, leading to a much more coherent picture of the fermionic state. It is clear from this procedure that much of the mathematical apparatus associated with the Dirac equation is totally redundant. In the nilpotent version which emerges, there is only a one-line operator, which has the same form whether the fermion is free or interacting. There is no need, in principle, to define a wavefunction, as the phase and amplitude is determined uniquely by the operator. There is no need, in addition, to define either operator or wavefunction as a 4-component spinor, as none of the terms is totally independent of the others. All we need to specify is that any product of two quaternionic fermionic states will always result in a nonquaternionic value, this being the only contribution made by the '4-component' structure. In addition, there is only one phase, whether the system is fermionic, antifermionic or bosonic.[7] Physically, of course, the quaternion labels, ***k***, ***i***, and ***j***, which provide the additional 'solutions', have multiple functions, which now can be seen to be related in a fundamental way.[8] As we will show, they are, respectively, *T*, *P*, and *C* operators; and the generators of weak, strong, and electric vacua; they are also involved in the respective production of spin 1 bosons, Bose-Einstein condensates, and spin 0 bosons.

Many authors have applied geometrical and hypergeometrical algebras to the Dirac equation to make it easier to manipulate or more amenable to physical interpretation, with varying degrees of success; but there seems to be only one approach which gives a precise and exact solution to the fundamental problems of particle physics and relativistic quantum mechanics, and it requires all its elements in exactly the right place before it displays its full power. Ultimately, it would seem that the approach is also the simplest, its defragmentation of the Dirac equation making analytic calculation in significant examples relatively easy. Relativistic calculations turn out to be easier than nonrelativistic ones because all the elements are positioned in their correct places. Physical interpretations and explanations emerge purely from the nilpotent structure.

Preliminary work on the less tractable problems posed by aggregated matter suggests that it might have just as significant an effect in those areas as well.

5.10 Orthonormality of the Nilpotent Solutions of the Dirac Equation

It will be convenient here to show that the solutions in the nilpotent form can be described as orthonormal. Using the bra-ket notation we can write the eigenvalue equations for the operator:

$$\mathcal{D} = (\boldsymbol{k}\partial/\partial t + i\,\boldsymbol{i}\nabla + i\boldsymbol{j}\, m)$$

as

$$\mathcal{D}\,|\psi_1\rangle = q_1\,|\psi_1\rangle$$
$$\mathcal{D}\,|\psi_2\rangle = q_1\,|\psi_2\rangle$$
$$\mathcal{D}\,|\psi_3\rangle = q_1\,|\psi_3\rangle$$
$$\mathcal{D}\,|\psi_4\rangle = q_1\,|\psi_4\rangle,$$

where $|\psi_1\rangle = e^{-i(Et\ -\ \mathbf{p.r})}\,|1\rangle$, $|\psi_2\rangle = e^{-i(Et\ +\ \mathbf{p.r})}\,|2\rangle$, $|\psi_3\rangle = e^{i(Et\ +\ \mathbf{p.r})}\,|3\rangle$, $|\psi_4\rangle = e^{i(Et\ -\ \mathbf{p.r})}\,|4\rangle$, and $\{|\psi_1\rangle, |\psi_2\rangle, |\psi_3\rangle, |\psi_4\rangle\}$ are components of a basis, not necessarily orthonormal, to be determined.

One may find the Hermitian conjugate for $\mathcal{D}$ by appealing to the mappings for γ matrices which lead to the construction of $\mathcal{D}$ in quaternion form. We find that the Hermitian conjugate of $\mathcal{D}$ is easily obtained by making a complex-quaternion conjugation; that is, conjugating imaginary i as well as $\boldsymbol{i}, \boldsymbol{j}$ and $\boldsymbol{k}$. The operator $\mathcal{D}^\dagger$ is the Hermitian conjugate of $\mathcal{D}$ which is defined as:

$$\mathcal{D}^\dagger = (\boldsymbol{k}\partial/\partial t - i\,\boldsymbol{i}\nabla + i\boldsymbol{j}\, m).$$

Using this we may now construct matrix elements for the operator $\mathcal{D}$ in the basis $\{|\psi_1\rangle, |\psi_2\rangle, |\psi_3\rangle, |\psi_4\rangle\}$. The matrix element $\langle\psi_2|\mathcal{D}|\,\psi_1\rangle = q_1\langle\psi_2|\psi_1\rangle$, may also be constructed by applying $\mathcal{D}^\dagger$ to $|\psi_2\rangle$ first; that is, $\langle\psi_2|\mathcal{D}|\,\psi_1\rangle = (\mathcal{D}^\dagger|\psi_2\rangle)^\dagger|\psi_1\rangle = q_3\langle\psi_2|\psi_1\rangle$. This means that $q_1\langle\psi_2|\psi_1\rangle = q_3\langle\psi_2|\psi_1\rangle$, which can only be true if $\langle\psi_2|\psi_1\rangle = 0$; that is, if $|\psi_1\rangle$ and $|\psi_2\rangle$ are orthogonal. Similar calculations for $\langle\psi_3|\,\mathcal{D}\,|\,\psi_1\rangle$ and $\langle\psi_4|\,\mathcal{D}\,|\,\psi_1\rangle$ lead one to the conclusion that $\{|\psi_1\rangle, |\psi_4\rangle\}$ form a subspace that is necessarily orthogonal to $\{|\psi_2\rangle, |\psi_3\rangle\}$. $|\psi_1\rangle$ and $|\psi_4\rangle$ are not necessarily orthogonal and the same goes for $|\psi_2\rangle$ and $|\psi_3\rangle$. The most general basis one could write for $\{|\psi_1\rangle, |\psi_2\rangle, |\psi_3\rangle, |\psi_4\rangle\}$ is:

$$|\psi_1\rangle = \begin{pmatrix} \alpha_1 \\ 0 \\ 0 \\ \beta_1 \end{pmatrix} e^{-i(Et - \mathbf{p.r})}, \qquad |\psi_2\rangle = \begin{pmatrix} 0 \\ \alpha_2 \\ \beta_2 \\ 0 \end{pmatrix} e^{-i(Et + \mathbf{p.r})},$$

$$|\psi_3\rangle = \begin{pmatrix} 0 \\ \alpha_3 \\ \beta_3 \\ 0 \end{pmatrix} e^{i(Et + \mathbf{p.r})}, \qquad |\psi_4\rangle = \begin{pmatrix} \alpha_4 \\ 0 \\ 0 \\ \beta_4 \end{pmatrix} e^{i(Et - \mathbf{p.r})}.$$

However, with further consideration we find that these eigenfunctions may be orthogonalized using a Gram-Schmidt method leading to a basis of the following form:

$$|\psi_1\rangle = \begin{pmatrix} 1 \\ 0 \\ 0 \\ 0 \end{pmatrix} e^{-i(Et - \mathbf{p.r})}, \qquad |\psi_2\rangle = \begin{pmatrix} 0 \\ 1 \\ 0 \\ 0 \end{pmatrix} e^{-i(Et + \mathbf{p.r})},$$

$$|\psi_3\rangle = \begin{pmatrix} 0 \\ 0 \\ 1 \\ 0 \end{pmatrix} e^{i(Et + \mathbf{p.r})}, \qquad |\psi_4\rangle = \begin{pmatrix} 0 \\ 0 \\ 0 \\ 1 \end{pmatrix} e^{i(Et - \mathbf{p.r})}.$$

Since the application of $\not{\partial}$ onto the above states leads to solutions of the Dirac equation $(i\boldsymbol{k}\partial/\partial t + \boldsymbol{i}\nabla + i\boldsymbol{j}\, m)\,\psi = 0$, we find that the quaternion wave functions ψ_1, ψ_2, ψ_3, ψ_4, are given by:

$$(\boldsymbol{k}E + i\boldsymbol{i}\,\mathbf{p} + i\boldsymbol{j}\,m)\,|\psi_1\rangle \qquad \text{(a)}$$
$$(\boldsymbol{k}E - i\boldsymbol{i}\,\mathbf{p} + i\boldsymbol{j}\,m)\,|\psi_1\rangle \qquad \text{(b)}$$
$$(-\boldsymbol{k}E + i\boldsymbol{i}\,\mathbf{p} + i\boldsymbol{j}\,m)\,|\psi_1\rangle \qquad \text{(c)}$$
$$(-\boldsymbol{k}E - i\boldsymbol{i}\,\mathbf{p} + i\boldsymbol{j}\,m)\,|\psi_1\rangle \qquad \text{(d).}$$

If we consider an alignment of the momentum vector **p** in the direction of the z axis (p_x and p_y are both zero), we see that $\pm\,\mathbf{p}$, which appears in states (a) to (d), may be used to identify the eigenstates of helicity. This convention is chosen to coincide with the normal Dirac conventions for positive and negative energy states and spin orientations: (a) represents an electron with spin parallel to momentum, (b) an electron with spin anti-parallel to momentum, (c) represents a positron with spin parallel to momentum, (d) a positron with spin anti-parallel to momentum. This is consistent with what happens to these states on application of *T*, *C* or *P* operators Indeed, we now have a basis within which we can construct operators in matrix form, the matrix operators for *T*, *P* or *C* being as follows:

$$T = \begin{pmatrix} 0 & 0 & e^{2iEt} & 0 \\ 0 & 0 & 0 & e^{2iEt} \\ e^{-2iEt} & 0 & 0 & 0 \\ 0 & e^{-2iEt} & 0 & 0 \end{pmatrix}, \quad P = \begin{pmatrix} 0 & e^{-2ipz} & 0 & 0 \\ e^{-2ipz} & 0 & 0 & 0 \\ 0 & 0 & 0 & e^{2ipz} \\ 0 & 0 & e^{2ipz} & 0 \end{pmatrix},$$

$$C = \begin{pmatrix} 0 & 0 & 0 & e^{2i(Et-pz)} \\ 0 & 0 & e^{2i(Et+pz)} & 0 \\ 0 & e^{-2i(Et+pz)} & 0 & 0 \\ e^{-2i(Et-pz)} & 0 & 0 & 0 \end{pmatrix}.$$

Chapter 6

The Dirac Nilpotent

Some parts of this chapter (especially **6.4**, **6.10**, **6.11**) were developed and refined with **JOHN CULLERNE**. Here, we explore the rich formal structure of the Dirac nilpotent. The formulation appears to be so powerful that it immediately suggests physical possibilities. Fermions and bosons, baryons (with their 'quark' components), *CPT* symmetry, vacuum in all its manifestations (including weak, strong, and electric), supersymmetry, renormalization, the quantum field, annihilation and creation operators, and many other important concepts emerge out of the formalism alone, with no empirical or model-dependent input. More conventional exercises are also possible, such as the derivation of spin and helicity.

6.1 Spin

The Dirac nilpotent represents the most concise packaging of the dualistic information contained in the parameter group, the most complete way of parameterizing nature; and, as we have seen, the combination of all the desired physical elements, with all the inherent symmetries, into a single, parameterization of nature, is the same as the process of 'quantization' of energy, momentum (or angular momentum) and 'rest mass'. It seems that we get 'something' from 'nothing', not just in a physical way, by perfect symmetry between the parameters denying overall characterization, but also literally, by making the fundamental unit of our characterization a square root of zero, and that this becomes zero in the Dirac equation when we apply to it a differential operator, and generate an exact equal to it as an eigenvalue. Many significant aspects of physics are explained through a direct investigation of this rich formal structure.

In addition to such radical reformulations, various standard results can be accommodated into the new formalism by replacing the gamma matrices with quaternion operators. Particularly important is the derivation of fermion spin. In the conventional treatment of spin, we write

$$[\hat{\boldsymbol{\sigma}}, \mathcal{H}] = [\hat{\boldsymbol{\sigma}}, i\gamma_0\boldsymbol{\gamma}.\mathbf{p} + \gamma_0 m].$$

Here, $\hat{\sigma}_l = i\gamma_0\gamma_5\gamma_l$, with $l = 1, 2, 3$

and $i\gamma_0\boldsymbol{\gamma}.\mathbf{p} = i\gamma_0\gamma_1 p_1 + i\gamma_0\gamma_2 p_2 + i\gamma_0\gamma_3 p_3$

while $\gamma_0 = i\boldsymbol{k}$; $\gamma_1 = \mathbf{i}\boldsymbol{i}$; $\gamma_2 = \mathbf{j}\boldsymbol{i}$; $\gamma_3 = \mathbf{k}\boldsymbol{i}$; $\gamma_5 = \boldsymbol{ij}$.

So, $\hat{\sigma}_1 = -\mathbf{i}$; $\hat{\sigma}_2 = -\mathbf{j}$; $\hat{\sigma}_3 = -\mathbf{k}$

or $\hat{\boldsymbol{\sigma}} = -\mathbf{1}$,

and $\boldsymbol{\gamma} = i\mathbf{1}$,

where **1** is the unit (spin) vector.

Since $\gamma_0 m = \boldsymbol{ik}m$ has no vector term and $\hat{\boldsymbol{\sigma}}$ no quaternion, they commute, and we may derive the conventional

$$[\hat{\boldsymbol{\sigma}}, \gamma_0 m] = 0$$

and $[\hat{\boldsymbol{\sigma}}, \mathcal{H}] = [\hat{\boldsymbol{\sigma}}, i\gamma_0\boldsymbol{\gamma}.\mathbf{p}]$.

Now, $i\gamma_0\boldsymbol{\gamma}.\mathbf{p} = -\boldsymbol{j}\,(\mathbf{i}p_1 + \mathbf{j}p_2 + \mathbf{k}p_3)$.

So,
$$\begin{aligned}[\hat{\boldsymbol{\sigma}}, \mathcal{H}] &= 2\boldsymbol{j}\,(\mathbf{ij}p_2 + \mathbf{ik}p_3 + \mathbf{ji}p_1 + \mathbf{jk}p_3 + \mathbf{ki}p_1 + \mathbf{kj}p_2) \\ &= 2\boldsymbol{ij}\,(\mathbf{k}(p_2 - p_1) + \mathbf{j}(p_1 - p_3) + \mathbf{i}(p_3 - p_2)) \\ &= 2\boldsymbol{ij}\,\mathbf{1} \times \mathbf{p}. \qquad (6.1)\end{aligned}$$

In more conventional terms,

$$\begin{aligned}[\hat{\boldsymbol{\sigma}}, \mathcal{H}] &= 2i\boldsymbol{ki}\,(\mathbf{k}(p_2 - p_1) + \mathbf{j}(p_1 - p_3) + \mathbf{i}(p_3 - p_2)) \\ &= 2i\boldsymbol{k}\,\boldsymbol{\gamma} \times \mathbf{p} \\ &= 2\gamma_0\,\boldsymbol{\gamma} \times \mathbf{p}. \qquad (6.2)\end{aligned}$$

Simultaneously, if **L** is the orbital angular momentum $\mathbf{r} \times \mathbf{p}$,

$$[\mathbf{L}, \mathcal{H}] = [\mathbf{r} \times \mathbf{p}, i\gamma_0\boldsymbol{\gamma}.\mathbf{p} + \gamma_0 m] = [\mathbf{r} \times \mathbf{p}, i\gamma_0\boldsymbol{\gamma}.\mathbf{p}].$$

Taking out common factors,

$$\begin{aligned}[\mathbf{L}, \mathcal{H}] &= i\gamma_0\,[\mathbf{r}, \boldsymbol{\gamma}.\mathbf{p}] \times \mathbf{p} \\ &= -\boldsymbol{ki}\,[\mathbf{r}, \mathbf{1}.\mathbf{p}] \times \mathbf{p} \\ &= -\boldsymbol{j}\,[\mathbf{r}, \mathbf{1}.\mathbf{p}] \times \mathbf{p}.\end{aligned}$$

Now,

$$[\mathbf{r}, \mathbf{1.p}]\psi = -i\mathbf{i}\left(x\frac{\partial\psi}{\partial x} - \frac{\partial(x\psi)}{\partial x}\right) - i\mathbf{j}\left(y\frac{\partial\psi}{\partial y} - \frac{\partial(y\psi)}{\partial y}\right) - i\mathbf{k}\left(\frac{\partial\psi}{\partial z} - \frac{\partial(z\psi)}{\partial z}\right) = i\mathbf{1}\psi.$$

Hence, $$[\mathbf{L}, \mathcal{H}] = -\, \boldsymbol{ij}\, \mathbf{1} \times \mathbf{p}. \tag{6.3}$$

This, again, can be converted into conventional terms:

$$[\mathbf{L}, \mathcal{H}] = \boldsymbol{iki}\, \mathbf{1} \times \mathbf{p} = i\, \gamma_0 \boldsymbol{\gamma} \times \mathbf{p}. \tag{6.4}$$

Using either (6.1) and (6.3) or (6.2) and (6.4), we may write

$$[\mathbf{L} - \mathbf{1} / 2, \mathcal{H}] = 0$$

or $$[\mathbf{L} + \hat{\boldsymbol{\sigma}} / 2, \mathcal{H}] = 0.$$

Hence, $(\mathbf{L} - \mathbf{1} / 2)$ or $(\mathbf{L} + \hat{\boldsymbol{\sigma}} / 2)$ is a constant of the motion. The extra spin term $(\hat{\boldsymbol{\sigma}} / 2)$, with its property of requiring 4π rather than 2π rotation to complete a cycle, is thus necessary to maintaining the total conservation of angular momentum.

6.2 Helicity

The term

$$\hat{\boldsymbol{\sigma}}.\mathbf{p} = -\,p_1 - p_2 - p_3 = -\,p$$

is defined as helicity, and, since it has no vector or quaternion terms, and has only terms of the form $\partial / \partial x$, $\partial / \partial y$, and $\partial / \partial z$ in common with

$$i\gamma_0\boldsymbol{\gamma}.\mathbf{p} = -\boldsymbol{j}\,(\mathbf{i}p_1 + \mathbf{j}p_2 + \mathbf{k}p_3)$$

and also clearly commutes with $\gamma_0 m = \boldsymbol{ik}m$, then

$$[\hat{\boldsymbol{\sigma}}.\mathbf{p}, \mathcal{H}] = 0$$

and the helicity is a constant of the motion.

For a hypothetical particle with zero mass, the term $\boldsymbol{k}\, E + \boldsymbol{ii}\, p + \boldsymbol{ij}\, m$ reduces to $\boldsymbol{k}\, E + \boldsymbol{ii}\, p$, where p actually represents $\hat{\boldsymbol{\sigma}}.\mathbf{p}$. E also becomes equal to $\pm\, p$. For positive energy states,

$$E = \hat{\boldsymbol{\sigma}}.\mathbf{p}.$$

So the spin is aligned antiparallel to the momentum (has left-handed helicity). Then,

$$\boldsymbol{ij}\,(\boldsymbol{k}\, E + \boldsymbol{ii}\, p) = \boldsymbol{ij}\,(\boldsymbol{k} - \boldsymbol{ii}\,)\, E = (\boldsymbol{ii}\, - \boldsymbol{k})\, E$$

and the spinor wavefunction follows the rule:

$$\boldsymbol{ij}\, u_L = -\, u_L. \tag{6.5}$$

For negative energy states,

$$E = - \hat{\boldsymbol{\sigma}}.\mathbf{p}.$$

In this case, the spin is aligned parallel to the momentum (has right-handed helicity). Then,

$$\boldsymbol{ij}\ (\boldsymbol{k}\ E + \boldsymbol{ii}\ p) = \boldsymbol{ij}\ (\boldsymbol{k} + \boldsymbol{ii}\ )\ E = (\boldsymbol{ii}\ + \boldsymbol{k})\ E$$

and the spinor wavefunction follows the rule:

$$\boldsymbol{ij}\ u_R = u_R. \tag{6.6}$$

From (6.5) and (6.6), we may derive the relations

$$\left(\frac{1-\boldsymbol{ij}}{2}\right)u_L = \left(\frac{1-\gamma^5}{2}\right) = u_L = u_L$$

and

$$\left(\frac{1-\boldsymbol{ij}}{2}\right)u_R = \left(\frac{1-\gamma^5}{2}\right) = u_R = 0.$$

These are the equations which, in conventional theory, produce two sharply-defined helicity states ($\boldsymbol{\sigma}.\mathbf{p} / 2p = ½$ and $-½$), of which the right-handed state (p) is suppressed in the case of assumed massless fermions of positive energy (E) and the left-handed state ($-p$) in the case of massless fermions of 'negative energy' ($-E$), or antifermions, in which $E = -p$.[1] Helicity is a pseudoscalar and so changes sign under parity transformations (or reversals of the signs of spatial coordinates); this means that parity must be violated in interactions such as those involving massless fermions because the two helicity states do not then make equal contributions to the interaction. Parity violation, in this case, is made inevitable by the suppression of mass, and the fixing of the E / p ratio.

If we define the right- and left-handed components of the wavefunction, ψ_R and ψ_L by the expressions

$$\psi_R = \left(\frac{1+\gamma^5}{2}\right) = \left(\frac{1+\boldsymbol{ij}}{2}\right)\psi = \psi$$

and

$$\psi_L = \left(\frac{1-\gamma^5}{2}\right) = \left(\frac{1-\boldsymbol{ij}}{2}\right)\psi = \psi\,,$$

we may derive the explicit expressions

$$\left(\frac{1+\boldsymbol{ij}}{2}\right)(\boldsymbol{k}E + \boldsymbol{ii}p + \boldsymbol{ij}m) = \left(\frac{1+\boldsymbol{ij}}{2}\right)(\boldsymbol{k}(E+p)+m) \tag{6.7}$$

and $$\left(\frac{1-ij}{2}\right)(kE+iip+ijm)=\left(\frac{1-ij}{2}\right)\left(\frac{-km}{E+p}\right)(k(E+p)+m) \tag{6.8}$$

It is apparent from (6.7) that $\psi_R = 0$ when $m = 0$ and $E = -p = \hat{\boldsymbol{\sigma}}.\mathbf{p}$. Also,

$$\psi_L - \psi_R = (\boldsymbol{ii}\ E + \boldsymbol{k}\ p + m)\ e^{-i(Et\ -\ \mathbf{p.r})}$$

and $$\psi_L + \psi_R = (\boldsymbol{k}\ E + \boldsymbol{ii}\ p + \boldsymbol{ij}\ m)\ e^{-i(Et\ -\ \mathbf{p.r})},$$

as expected.

In the case where $m \neq 0$, we can still define wavefunctions with spinors

$$u = u_R + u_L,$$

where $$u_R = (\boldsymbol{k}\ E + \boldsymbol{ii}\ p + \boldsymbol{ij}\ m)$$

and $$u_L = (\boldsymbol{k}\ E - \boldsymbol{ii}\ p - \boldsymbol{ij}\ m),$$

but these no longer represent sharply-defined helicity states. For fermions with nonzero mass, but at rest, $p = 0$ and $E = m$. The wavefunctions must then contain terms of the form

$$(\boldsymbol{k}\ E \pm \boldsymbol{ij}\ m) = E\ (\boldsymbol{k} \pm \boldsymbol{ij}).$$

When multiplied by the conventional parity operator $\mathbf{P} = \gamma^0 = \boldsymbol{ii}$, these become

$$\boldsymbol{ii}\ E\ (\boldsymbol{k} \pm \boldsymbol{ij}) = -E\ (\boldsymbol{k} \pm \boldsymbol{ij}),$$

which is effectively a charge conjugation (or exchange of particle and antiparticle). This not only preserves *CP* invariance, but also suggests that fermions at rest have opposite intrinsic parities to their respective antifermions.

6.3 Fermions and Bosons

There are good reasons for believing that the nilpotent form of the Dirac equation is the most fundamental, all other forms yielding only a partial range of its consequences. It is automatically second quantized, with the differential operator and the state vector being essentially identical, fulfilling all the requirements of a quantum field theory; it removes (as will be shown in chapter 11) the infrared divergence in the fermion propagator, and the (ultraviolet) divergent loop calculation for the self-energy of the non-interacting fermion; and it introduces supersymmetry as a mathematical operation without the need for additional particles. It also allows an easy calculation of parity states, and a simple method of introducing *C*, *P* or *T* transformations. In addition, state vectors for fermions, antifermions, bosons and baryons have immediately recognizable forms. A fermion, for example, may be represented by a row (or column) vector, whose components are four creation (or annihilation) operators:

$$\begin{array}{ll}
(ikE+\boldsymbol{i}\,\mathbf{p}+\boldsymbol{j}\,m) & \text{fermion spin up} \\
(ikE-\boldsymbol{i}\,\mathbf{p}+\boldsymbol{j}\,m) & \text{fermion spin down} \\
(-ikE+\boldsymbol{i}\,\mathbf{p}+\boldsymbol{j}\,m) & \text{antifermion spin up} \\
(-ikE-\boldsymbol{i}\,\mathbf{p}+\boldsymbol{j}\,m) & \text{antifermion spin down.}
\end{array}$$

The antifermion then takes up the corresponding column (or row) vector:

$$\begin{array}{l}
(-ikE+\boldsymbol{i}\,\mathbf{p}+\boldsymbol{j}\,m) \\
(-ikE-\boldsymbol{i}\,\mathbf{p}+\boldsymbol{j}\,m) \\
(ikE+\boldsymbol{i}\,\mathbf{p}+\boldsymbol{j}\,m) \\
(ikE-\boldsymbol{i}\,\mathbf{p}+\boldsymbol{j}\,m).
\end{array}$$

The spin 1 boson produced by their combination is simply the scalar product:

$$\begin{array}{ll}
(ikE+\boldsymbol{i}\,\mathbf{p}+\boldsymbol{j}\,m) & (-ikE+\boldsymbol{i}\,\mathbf{p}+\boldsymbol{j}\,m) \\
(ikE-\boldsymbol{i}\,\mathbf{p}+\boldsymbol{j}\,m) & (-ikE-\boldsymbol{i}\,\mathbf{p}+\boldsymbol{j}\,m) \\
(-ikE+\boldsymbol{i}\,\mathbf{p}+\boldsymbol{j}\,m) & (ikE+\boldsymbol{i}\,\mathbf{p}+\boldsymbol{j}\,m) \\
(-ikE-\boldsymbol{i}\,\mathbf{p}+\boldsymbol{j}\,m) & (ikE-\boldsymbol{i}\,\mathbf{p}+\boldsymbol{j}\,m).
\end{array}$$

The spin 0 boson is obtained by reversing the **p** signs in either fermion or antifermion:

$$\begin{array}{ll}
(ikE+\boldsymbol{i}\,\mathbf{p}+\boldsymbol{j}\,m) & (-ikE-\boldsymbol{i}\,\mathbf{p}+\boldsymbol{j}\,m) \\
(ikE-\boldsymbol{i}\,\mathbf{p}+\boldsymbol{j}\,m) & (-ikE+\boldsymbol{i}\,\mathbf{p}+\boldsymbol{j}\,m) \\
(-ikE+\boldsymbol{i}\,\mathbf{p}+\boldsymbol{j}\,m) & (ikE-\boldsymbol{i}\,\mathbf{p}+\boldsymbol{j}\,m) \\
(-ikE-\boldsymbol{i}\,\mathbf{p}+\boldsymbol{j}\,m) & (ikE+\boldsymbol{i}\,\mathbf{p}+\boldsymbol{j}\,m).
\end{array}$$

Significantly, massless spin 0 particles (Goldstone bosons) are ruled out on purely algebraic grounds:

$$\begin{array}{ll}
(ikE+\boldsymbol{i}\,\mathbf{p}) & (-ikE-\boldsymbol{i}\,\mathbf{p})=0 \\
(ikE-\boldsymbol{ii}\,\mathbf{p}) & (-ikE+\boldsymbol{i}\,\mathbf{p})=0 \\
(-ikE+\boldsymbol{i}\,\mathbf{p}) & (ikE-\boldsymbol{i}\,\mathbf{p})\ \ =0 \\
(-ikE-\boldsymbol{i}\,\mathbf{p}) & (ikE+\boldsymbol{i}\,\mathbf{p})\ \ =0.
\end{array}$$

Effectively, this means also that a massless left-handed fermion cannot exist at the same time as a massless left-handed antifermion. The spin 1 boson has components that may be exclusively one-handed because if the spins of fermion and antifermion are aligned (in the **p** component), their helicities will be opposite because of their opposite signs of E. A spin 0 boson, however, has fermion and antifermion components that are in opposite spin (or **p**) alignment, so with states of the *same* helicity because of their opposite signs of E. The fact that nilpotency determines that spin 0 bosons must have nonzero mass ensures that this is equivalent to having fermion and antifermion states of the same helicity, while masslessness requires states of opposite helicity. As we have seen, the anticommutativity of the operations involved in defining $[\mathbf{L}, \mathcal{H}]$ ultimately ensure that the helicity term is antiparallel (or left-handed) for positive energy states and parallel (or right-handed) for negative energy states.

Massless spin 1 states, however, are allowed, since

$$\begin{array}{ll} (ikE+\boldsymbol{i}\,\mathbf{p}) & (-ikE+\boldsymbol{i}\,\mathbf{p}) \\ (ikE-\boldsymbol{i}\,\mathbf{p}) & (-ikE-\boldsymbol{i}\,\mathbf{p}) \\ (-ikE+\boldsymbol{i}\,\mathbf{p}) & (ikE+\boldsymbol{i}\,\mathbf{p}) \\ (-ikE-\boldsymbol{i}\,\mathbf{p}) & (ikE-\boldsymbol{i}\,\mathbf{p}) \end{array}$$

has a nonzero scalar sum. Pauli exclusion is also automatic, since:

$$\begin{array}{lll} (ikE+\boldsymbol{i}\,\mathbf{p}+\boldsymbol{j}\,m) & (ikE+\boldsymbol{i}\,\mathbf{p}+\boldsymbol{j}\,m) & =0 \\ (ikE-\boldsymbol{i}\,\mathbf{p}+\boldsymbol{j}\,m) & (ikE-\boldsymbol{i}\,\mathbf{p}+\boldsymbol{j}\,m) & =0 \\ (-ikE+\boldsymbol{i}\,\mathbf{p}+\boldsymbol{j}\,m) & (-ikE+\boldsymbol{i}\,\mathbf{p}+\boldsymbol{j}\,m) & =0 \\ (-ikE-\boldsymbol{i}\,\mathbf{p}+\boldsymbol{j}\,m) & (-ikE-\boldsymbol{i}\,\mathbf{p}+\boldsymbol{j}\,m) & =0. \end{array}$$

The Pauli exclusion principle in this form demands nonlocality. It also means that Pauli exclusion can be established for each individual fermion, without using the entire Slater determinant derived from the Grassmann algebra.

However, a boson-like state (the 'Bose-Einstein condensate') can be constructed from two fermions (or equivalent, e.g. fermion + flux line) if the spins (or, in the case of He^3, momentum states) are opposite

$$\begin{array}{ll} (ikE+\boldsymbol{i}\,\mathbf{p}+\boldsymbol{j}\,m) & (ikE-\boldsymbol{i}\,\mathbf{p}+\boldsymbol{j}\,m) \\ (ikE-\boldsymbol{i}\,\mathbf{p}+\boldsymbol{j}\,m) & (ikE+\boldsymbol{i}\,\mathbf{p}+\boldsymbol{j}\,m) \\ (-ikE+\boldsymbol{i}\,\mathbf{p}+\boldsymbol{j}\,m) & (-ikE-\boldsymbol{i}\,\mathbf{p}+\boldsymbol{j}\,m) \\ (-ikE-\boldsymbol{i}\,\mathbf{p}+\boldsymbol{j}\,m) & (-ikE+\boldsymbol{i}\,\mathbf{p}+\boldsymbol{j}\,m) \end{array}$$

This last state incorporates such physical manifestations as the Aharonov-Bohm effect, the Jahn-Teller effect, the quantum Hall effect, Cooper pairs, and even-even (spin 0) nuclei. Even electrons or other fermions with their spins aligned in a magnetic field can be considered as producing an overall state (fermion plus flux line) which acts in a bosonic way.

We notice here that the boson is structured as a unified state, with E, **p** and m

values common to the fermionic and antifermionic parts. We can, in fact, postulate that the signature of a completely interacting dynamical theory of composite particles is that the E, $\mathbf{p}$ and m values have meaning only in the context of the *entire state*. This will become especially significant with baryons. The *hyperentanglement* (entanglement in every degree of freedom), which is thus required and predicted by the nilpotent structure, has now been observed in photons.[2]

Because the state vector always represents four terms with the complete variation of signs in E and $\mathbf{p}$, the interaction vertex between any fermion / antifermion and any other

$$(\pm \, ikE_1 \pm i\mathbf{p}_1 + jm_1)(\pm \, ikE_2 \pm i\mathbf{p}_2 + jm_2)$$

will remove the quaternionic operators, leaving only scalars and vectors. (The exchange of energy and momentum at such a vertex is the characteristic feature of an interaction.) Of course, the standard definition of nilpotent wavefunctions as antisymmetric also still applies (with the antisymmetric nature due to the spin), since

$$\begin{aligned}(\pm \, ikE_1 \pm i\mathbf{p}_1 + jm_1)(\pm \, ikE_2 \pm i\mathbf{p}_2 + jm_2) - (\pm \, ikE_2 \pm i\mathbf{p}_2 + jm_2)(\pm \, ikE_1 \pm i\mathbf{p}_1 + jm_1) \\ = 4\mathbf{p}_2\mathbf{p}_1 - 4\mathbf{p}_1\mathbf{p}_2 = -8i\mathbf{p}_1 \times \mathbf{p}_2 = 8i\mathbf{p}_2 \times \mathbf{p}_1 \qquad (6.9)\end{aligned}$$

before normalization, while

$$\begin{aligned}(\pm \, ikE_1 \pm i\mathbf{p}_1 + jm_1)(\pm \, ikE_2 \pm i\mathbf{p}_2 + jm_2) + (\pm \, ikE_2 \pm i\mathbf{p}_2 + jm_2)(\pm \, ikE_1 \pm i\mathbf{p}_1 + jm_1) \\ = 4E_1E_2 + 4E_2E_1 - 4\mathbf{p}_1\mathbf{p}_2 - 4\mathbf{p}_2\mathbf{p}_1 - 4m_1m_2 - 4m_2m_1 = 0 \, .\end{aligned}$$

Because of the Grassmann algebra used, and the 'indistinguishability' condition for fermions, more complex systems require the Slater determinant.

The reduction of the $\psi_1\psi_2 - \psi_2\psi_1$ state, in (6.9), to a specification in terms of spin is only possible because the 4-component spinor structure of ψ ensures that all other terms are zeroed. (For scalar values of ψ_1 and ψ_2 the opposite rules would clearly apply.) It is only when the E, $\mathbf{p}$ and m values become numerically equal that the vertex can be defined as a new *combined* bosonic state. The phases will then also become identical. So, a boson is created at a vertex when the fermion and antifermion acquire the same phase. Now, the result $\psi_1\psi_2 - \psi_2\psi_1 = -8i\mathbf{p}_1 \times \mathbf{p}_2$ is remarkable, because this term is 0 unless $\mathbf{p}_1$ and $\mathbf{p}_2$ are pointing in different directions, irrespective of their scalar magnitudes. In other words, Pauli exclusion requires fermions *to have unique spin axes at any given instant*. Each fermion is distinguished from every other solely by the instantaneous relative direction (or angle) of its spin axis, which is effectively a *unique phase* for the $\mathbf{p}$

term's position in its cycle, even though this cannot be known absolutely, and corresponding to the unique phase relating to E and **p**.[3] Because of the nature of spin, expressing this uniqueness, **p** has only a single defined component, conventionally described as the 'third component' (cf the uniqueness of the angles in Penrose tiling).[4]

In principle, individual fermions are 1-dimensional in having only a single well-defined direction for spin; however, the combination of all possible spin directions, in an infinite universe, constitutes (and constructs) 3-dimensional Euclidean space, and it is with respect to this space that we define **p** as having rotation symmetry and a random 3-dimensional orientation (and, of course, any two fermions with non-parallel momentum vectors will automatically generate a 3-dimensionality, from $\mathbf{p}_1$, $\mathbf{p}_2$ and $\mathbf{p}_1 \times \mathbf{p}_2$). It is outside of the fermion and its view of the rest of the universe. A single fermion's view becomes part of the approaching Euclidean view of a more complex state. Euclidean space is that of the zero-point energy (sections 9.4 and 12.2). But, though the nonlocal connection of fermions with each other (via **p** terms, etc.) is through the mass continuum, not the singularities involved in electric, strong and weak interactions, it is still symmetrical with respect to a point source, and will be a Newton-Coulomb interaction, with $U(1)$ symmetry, *exactly as is required to represent the space of all possible spin axes*. (It is notable, here, that Pauli exclusion requires that the **p** direction must be unique both in this 'real' space and in the quaternionic space provided by the orthogonal components E, **p**, m.)

It seems likely also that the universal collection of spin axis directions at any given instant is not repeatable, giving a unique direction for time, and – including energy as the 'time' component – making fermionic world-lines unique as well. A further connection with irreversibility or a unique time direction is suggested by the fact that an interaction between fermions with differently oriented vectors $\mathbf{p}_1$ and $\mathbf{p}_2$ will produced a reduced magnitude of momentum within the system unless the even more 'organized' amount of rest mass is reduced. In either case, the system will become less coherent or more entropic, as required by the second law of thermodynamics, the only law of physics which requires an irreversibility of time.

It is obvious, of course, from the scalar structure of the wavefunction why the Dirac equation cannot apply to bosons. A nilpotent differential operator can only produce zero by operating on a nilpotent wavefunction. However, the Klein-Gordon differential operator is already equivalent to a zero eigenvalue, and so can be applied to either fermions or bosons.

6.4 Vacuum

Since the vacuum plays such a significant part in many fundamental processes, the formulation of a vacuum operator (or vacuum operators) will be necessary to a full theory. A fermion creation operator $(\boldsymbol{i}\boldsymbol{k}E + \mathbf{i}\mathbf{p} + \boldsymbol{j}m)$ is unaffected if postmultiplied by $\boldsymbol{k}\,(\boldsymbol{i}\boldsymbol{k}E + \mathbf{i}\mathbf{p} + \boldsymbol{j}m)$ (if we assume that scalar factors are removed by normalisation). That is,

$$(\boldsymbol{i}\boldsymbol{k}E + \boldsymbol{i}\,\mathbf{p} + \boldsymbol{j}m) = (\boldsymbol{i}\boldsymbol{k}E + \boldsymbol{i}\,\mathbf{p} + \boldsymbol{j}m)\ \boldsymbol{k}\,(\boldsymbol{i}\boldsymbol{k}E + \boldsymbol{i}\,\mathbf{p} + \boldsymbol{j}m)\ \boldsymbol{k}\,(\boldsymbol{i}\boldsymbol{k}E + \boldsymbol{i}\,\mathbf{p} + \boldsymbol{j}m)\ \ldots$$

The same applies if the operator is postmultiplied by $\boldsymbol{i}\,((\boldsymbol{i}\boldsymbol{k}E + \boldsymbol{i}\,\mathbf{p} + \boldsymbol{j}m)$ or $\boldsymbol{j}\,(\boldsymbol{i}\boldsymbol{k}E + \boldsymbol{i}\,\mathbf{p} + \boldsymbol{j}m)$. In effect, $\boldsymbol{k}\,(\boldsymbol{i}\boldsymbol{k}E + \boldsymbol{i}\,\mathbf{p} + \boldsymbol{j}m)$, $\boldsymbol{i}\,(\boldsymbol{i}\boldsymbol{k}E + \boldsymbol{i}\,\mathbf{p} + \boldsymbol{j}m)$ and $\boldsymbol{j}\,(\boldsymbol{i}\boldsymbol{k}E + \boldsymbol{i}\,\mathbf{p} + \boldsymbol{j}m)$ act as vacuum operators, leaving the fermion state unchanged. However,

$$(\boldsymbol{i}\boldsymbol{k}E + \boldsymbol{i}\,\mathbf{p} + \boldsymbol{j}m) = (\boldsymbol{i}\boldsymbol{k}E + \boldsymbol{i}\,\mathbf{p} + \boldsymbol{j}m)\ \boldsymbol{k}\,(\boldsymbol{i}\boldsymbol{k}E + \boldsymbol{i}\,\mathbf{p} + \boldsymbol{j}m)\ \boldsymbol{k}\,(\boldsymbol{i}\boldsymbol{k}E + \boldsymbol{i}\,\mathbf{p} + \boldsymbol{j}m)\ \ldots$$

can also be written as

$$(\boldsymbol{i}\boldsymbol{k}E + \boldsymbol{i}\,\mathbf{p} + \boldsymbol{j}\,m) = (\boldsymbol{i}\boldsymbol{k}E + \boldsymbol{i}\,\mathbf{p} + \boldsymbol{j}\,m)\,(-\boldsymbol{i}\boldsymbol{k}E + \boldsymbol{i}\,\mathbf{p} + \boldsymbol{j}\,m)\,(\boldsymbol{i}\boldsymbol{k}E + \boldsymbol{i}\,\mathbf{p} + \boldsymbol{j}\,m)\ \ldots$$

with alternate states implying antifermion creation; or with the whole operation implying alternate creations of fermion and boson. Taken to infinity, this defines a nonlocal process.

Assuming an appropriate normalization, we may construct a generalised nilpotent vacuum operator as a diagonal matrix, which may be premultiplied by a 4-component quaternion row state vector or postmultiplied by a 4-component quaternion column state vector, representing a fermion state. In the first case, we write:

$$((\boldsymbol{i}\boldsymbol{k}E + \mathbf{i}\mathbf{p} + \boldsymbol{j}m)\ (\boldsymbol{i}\boldsymbol{k}E - \mathbf{i}\mathbf{p} + \boldsymbol{j}m)\ (-\boldsymbol{i}\boldsymbol{k}E + \mathbf{i}\mathbf{p} + \boldsymbol{j}m)\ (-\boldsymbol{i}\boldsymbol{k}E - \mathbf{i}\mathbf{p} + \boldsymbol{j}m)) \times$$

$$\boldsymbol{k}\begin{pmatrix} \boldsymbol{i}\boldsymbol{k}E + \mathbf{i}\mathbf{p} + \boldsymbol{j}m & 0 & 0 & 0 \\ 0 & \boldsymbol{i}\boldsymbol{k}E - \mathbf{i}\mathbf{p} + \boldsymbol{j}m & 0 & 0 \\ 0 & 0 & -\boldsymbol{i}\boldsymbol{k}E + \mathbf{i}\mathbf{p} + \boldsymbol{j}m & 0 \\ 0 & 0 & 0 & -\boldsymbol{i}\boldsymbol{k}E - \mathbf{i}\mathbf{p} + \boldsymbol{j}m \end{pmatrix} e^{-i(Et - \mathbf{p}.\mathbf{r})} =$$

$$((\boldsymbol{i}\boldsymbol{k}E + \mathbf{i}\mathbf{p} + \boldsymbol{j}m)\ (\boldsymbol{i}\boldsymbol{k}E - \mathbf{i}\mathbf{p} + \boldsymbol{j}m)\ (-\boldsymbol{i}\boldsymbol{k}E + \mathbf{i}\mathbf{p} + \boldsymbol{j}m)\ (-\boldsymbol{i}\boldsymbol{k}E - \mathbf{i}\mathbf{p} + \boldsymbol{j}m))\ e^{-i(Et - \mathbf{p}.\mathbf{r})},$$

assuming the appropriate normalisation constants. The vacuum wavefunction operator (when applied to a row vector) is always $\boldsymbol{k}$ × matrix form of state vector × exponential term. The vacuum operator omits the exponential term. The order is reversed when applied to a column vector.

The vacuum operator, here, clearly leaves the original fermion state unchanged. The individual creation operators, or individual components of the row vector, $(\pm ikE \pm i\mathbf{p} + jm)$, which specify the complete fermion system, can be considered as being postmultiplied by $k\ (\pm ikE \pm i\mathbf{p} + jm)$ to return to their original state, after normalization. The process can be continued indefinitely, with the fermion acting continually on the vacuum to reproduce itself:

$(\pm ikE \pm i\mathbf{p} + jm)\ k\ (\pm ikE \pm i\mathbf{p} + jm)\ k\ (\pm ikE \pm i\mathbf{p} + jm)\ k\ (\pm ikE \pm i\mathbf{p} + jm) \ldots$

However, $k\ (\pm ikE \pm i\mathbf{p} + jm)\ k$ is the same as the antistate to $(\pm ikE \pm i\mathbf{p} + jm)$, or $(\mp ikE \pm i\mathbf{p} + jm)$, making this equivalent to

$$(\pm ikE \pm i\mathbf{p} + jm)\ (\mp ikE \pm i\mathbf{p} + jm)\ (\pm ikE \pm i\mathbf{p} + jm)\ (\mp ikE \pm i\mathbf{p} + jm) \ldots$$

Physically, the fermion can be considered to see in the vacuum its 'image' or virtual antistate, producing a kind of virtual bosonic combination, and leading to an infinite alternating series of virtual fermions and bosons. Each real fermion state creates a virtual antifermion mirror image of itself in the vacuum, while each real antifermion state creates a virtual fermion mirror image of itself. The combined real and virtual particle creates a virtual boson state. Real fermions and real antifermions, of course, provide *real* mirror images of each other. Taking the fermion state as a whole, this links up with the idea that the supersymmetry operator Q and its Hermitian conjugate $Q^\dagger$ (which convert bosons to fermions and fermions to bosons) are simply the respective fermion and antifermion operators, $(\pm ikE \pm i\mathbf{p} + jm)$ and $(\mp ikE \pm i\mathbf{p} + jm)$. In the context of renormalization, with this conception of vacuum, we could see an infinite succession of boson and fermion loops cancelling each other, without needing to generate a new set of supersymmetric partners. The bosons and fermions become their own supersymmetric partners.

The bosons, here, are assumed to be spin 1, created from a fermion-antifermion pair, with the same spin, but opposite helicities, like all known gauge bosons, but we could also imagine a vacuum of the form $j\ (\pm ikE \pm i\mathbf{p} + jm)$, or $-j\ (\pm ikE \pm i\mathbf{p} + jm)$, in which the bosonic state would be spin 0. For reasons which will quickly become apparent, premultiplication by k could be said to produce a 'weak' vacuum while premultiplication by j produces an 'electric' one.

$k\ (ikE + i\,\mathbf{p} + j\,m)$	weak vacuum	fermion creation
$i\ (ikE + i\,\mathbf{p} + j\,m)$	strong vacuum	gluon plasma
$j\ (ikE + i\,\mathbf{p} + j\,m)$	electric vacuum	$SU(2)$

In fact, the three vacuum coefficients, k, i and j, can be seen as originating in (or being responsible for) the concept of discrete (point-like) charge. In this interpretation, the charges act as a discrete partitioning of the continuous vacuum responsible for zero-point energy, with the separate conservation laws for weak, strong and electric charges implying that the three discrete partitions are entirely independent of each other. And the full fermion spinor ($\pm \boldsymbol{k}E \pm i\boldsymbol{i}\,\mathbf{p} + i\boldsymbol{j}\,m$) can be seen as being equivalent to fermion creation plus three vacuum 'reflections', corresponding to the charge states:

$(i\boldsymbol{k}E + \boldsymbol{i}\,\mathbf{p} + \boldsymbol{j}\,m)$	fermion creation
$(i\boldsymbol{k}E - \boldsymbol{i}\,\mathbf{p} + \boldsymbol{j}\,m)$	strong reflection
$(-i\boldsymbol{k}E + \boldsymbol{i}\,\mathbf{p} + \boldsymbol{j}\,m)$	weak reflection
$(-i\boldsymbol{k}E - \boldsymbol{i}\,\mathbf{p} + \boldsymbol{j}\,m)$	electric reflection

Significantly, the 'weak' reflection is equivalent to a simultaneous switch from fermion to antifermion and (by preserving the sign of **p**) from one helicity state to another. The 'electric' reflection, on the other hand is a full charge conjugation (with no change in helicity), while the 'strong' reflection is a spin reversal.[5]

The connection between the quaternionic operators applied to charge and the (hidden) ones used in the Dirac 4-spinor now gives us a new understanding of the physical meaning of charge as a vacuum generator: $\boldsymbol{i}$, $\boldsymbol{j}$, and $\boldsymbol{k}$ are, simultaneously, the respective operators applied to strong, electric and weak charges, and also the creators of the strong, electric and weak vacuum images of a real fermion (which itself may be presumed to be 'generated' by the 'mass' operator, 1). Charge is, in effect, a kind of vacuum state, linked to the quantum field nature of the state vector.

The quantum field, of course, interprets the $U(1)$ symmetry, or scalar phase (gauge) invariance, of the electromagnetic interaction as a process of absorption and emission by charged particles of virtual quanta of the electromagnetic field (photons), and the process is illustrated by the use of Feynman diagrams, in which a fermion, such as an electron (represented by a straight line), enters a vertex, from which it is shown emerging, after emitting the photon (which is represented by a wavy line, connected to another vertex, showing the absorption process) (cf sections 10.10 and 11.8). The quantum field allows pair production of fermion and antifermion from virtual (and real) bosons, in addition to the reverse process of mutual annihilation, and, in the case of the vacuum, extends the process to infinity, so that the virtual bosons produce further pairs in an infinite succession of 'loop diagrams', whose effects can only be eliminated in standard theory by the process of renormalization. The processes are evident in the nilpotent structures for fermions, bosons and vacuum, as outlined in this chapter, and similar ones apply to the $SU(2)$ gauge symmetry applicable to the weak interaction, which is mediated by the W and Z bosons, and the $SU(3)$ symmetry of the strong interaction, mediated by gluons. These processes are discussed in chapters 11-14.

It appears that the nilpotent state vector incorporates real and virtual components in the same way as mass and charge. *Zitterbewegung* (see 7.6) is a switching between them. This is why state vectors are supersymmetric. It is a quantum equivalent of action plus virtual reaction. To put it another way, the 'reflection' of a fermionic creation in a charged 'mirror' is equivalent to defining the rest of the universe for that creation, just as Newton's classical process of action and reaction is really between a body and the rest of the universe, rather than between two isolated bodies. In this sense, every time a new fermion state is created, the universe is (re)created as well. And, because we can only define a fermion by also defining the rest of the universe, the fermion itself is only half of the picture. This is what we mean by saying that a fermion has half-integral spin.[6] The fermion state is incomplete without its vacuum (and, indeed, supersymmetric) partner; they are analogous to the action and reaction sides of a steady-state potential energy equation, with the fermion state alone represented by kinetic energy; and it is even possible to apply a classical kinetic energy equation for magnetic moment in a magnetic field to produce the ½-integral value of spin. The real fermion and its set of dual vacuum images combine to produce a single-valued bosonic spin state, analogous to a conserved physical system, simultaneously incorporating both action and reaction sides of Newton's third law of motion or a virial doubling of the kinetic energy in a potential energy term. This is why fermion and antifermion state vectors have identical components, with only the *order* privileging either $+E$ or $-E$ states as the 'real' ones. In addition, fermions not only carry with them the virtual vacuum partners which describe their interactions with the rest of the universe, but they are also able, in appropriate circumstances, to realise them (singly) as real (mass-shell) partners, either through the creation of real boson or boson-like states or, in less compactified form, through applications of the Berry phase.

6.5 *CPT* Symmetry

There are three fundamental symmetry operations in particle physics:

P	Parity	reverses signs of space coordinates
T	Time reversal	reverses sign of time coordinate
C	Charge conjugation	exchanges particle and antiparticle

The laws of physics are not preserved under these transformations taken separately, but are preserved under all three operations taken together, in any order (*CPT*). *CPT* symmetry is another mathematical consequences of a nilpotent representation. Here, we can represent the component *P*, *T* and *C* operations on a nilpotent wavefunction by using a different operator to represent each type of transformation. So, the ***i***, ***k***, and ***j*** operators are applied to a nilpotent state vector to represent the respective *P*, *T*, and *C* transformations:

Parity (*P*):

$$\begin{aligned}
\boldsymbol{i}\,(ik E + \boldsymbol{i}\,\mathbf{p} + \boldsymbol{j}\,m)\,\boldsymbol{i} &= (ik E - \boldsymbol{i}\,\mathbf{p} + \boldsymbol{j}\,m)\\
\boldsymbol{i}\,(ik E - \boldsymbol{i}\,\mathbf{p} + \boldsymbol{j}\,m)\,\boldsymbol{i} &= (ik E + \boldsymbol{i}\,\mathbf{p} + \boldsymbol{j}\,m)\\
\boldsymbol{i}\,(-ik E + \boldsymbol{i}\,\mathbf{p} + ij\,m)\,\boldsymbol{i} &= (-ik E - \boldsymbol{i}\,\mathbf{p} + \boldsymbol{j}\,m)\\
\boldsymbol{i}\,(-ik E - \boldsymbol{i}\,\mathbf{p} + \boldsymbol{j}\,m)\,\boldsymbol{i} &= (-ik E + \boldsymbol{i}\,\mathbf{p} + \boldsymbol{j}\,m)
\end{aligned}$$

Time reversal (*T*):

$$\begin{aligned}
\boldsymbol{k}\,(ik E + \boldsymbol{i}\,\mathbf{p} + \boldsymbol{j}\,m)\,\boldsymbol{k} &= (-ik E + \boldsymbol{i}\,\mathbf{p} + \boldsymbol{j}\,m)\\
\boldsymbol{k}\,(ik E - \boldsymbol{i}\,\mathbf{p} + \boldsymbol{j}\,m)\,\boldsymbol{k} &= (-ik E - \boldsymbol{i}\,\mathbf{p} + \boldsymbol{j}\,m)\\
\boldsymbol{k}\,(-ik E + \boldsymbol{i}\,\mathbf{p} + ij\,m)\,\boldsymbol{k} &= (ik E + \boldsymbol{i}\,\mathbf{p} + \boldsymbol{j}\,m)\\
\boldsymbol{k}\,(-ik E - \boldsymbol{i}\,\mathbf{p} + \boldsymbol{j}\,m)\,\boldsymbol{k} &= (ik E - \boldsymbol{i}\,\mathbf{p} + \boldsymbol{j}\,m)
\end{aligned}$$

Charge conjugation (*C*):

$$\begin{aligned}
-\boldsymbol{j}\,(ik E + \boldsymbol{i}\,\mathbf{p} + \boldsymbol{j}\,m)\,\boldsymbol{j} &= (-ik E - \boldsymbol{i}\,\mathbf{p} + \boldsymbol{j}\,m)\\
-\boldsymbol{j}\,(ik E - \boldsymbol{i}\,\mathbf{p} + \boldsymbol{j}\,m)\,\boldsymbol{j} &= (-ik E + \boldsymbol{i}\,\mathbf{p} + \boldsymbol{j}\,m)\\
-\boldsymbol{j}\,(-ik E + \boldsymbol{i}\,\mathbf{p} + ij\,m)\,\boldsymbol{j} &= (ik E - \boldsymbol{i}\,\mathbf{p} + \boldsymbol{j}\,m)\\
-\boldsymbol{j}\,(-ik E - \boldsymbol{i}\,\mathbf{p} + \boldsymbol{j}\,m)\,\boldsymbol{j} &= (ik E + \boldsymbol{i}\,\mathbf{p} + \boldsymbol{j}\,m)
\end{aligned}$$

The last may also be written:

$$ij\left(\pm ikE \pm \boldsymbol{i}\mathbf{p} + \boldsymbol{j}m\right)ij = \left(\mp ikE \mp \boldsymbol{i}\mathbf{p} + \boldsymbol{j}m\right)$$

From these, we see immediately that:

$CP = T$: $\quad -\boldsymbol{j}\,(\boldsymbol{i}\,(ik E + \boldsymbol{i}\,\mathbf{p} + \boldsymbol{j}\,m)\,\boldsymbol{i})\,\boldsymbol{j} = \boldsymbol{k}\,(ik E + \boldsymbol{i}\,\mathbf{p} + \boldsymbol{j}\,m)\,\boldsymbol{k} = (-ik E + \boldsymbol{i}\,\mathbf{p} + \boldsymbol{j}\,m)$

$PT = C$: $\quad \boldsymbol{i}\,(\boldsymbol{k}\,(ik E + \boldsymbol{i}\,\mathbf{p} + \boldsymbol{j}\,m)\,\boldsymbol{k})\,\boldsymbol{i} = -\boldsymbol{j}\,(ik E + \boldsymbol{i}\,\mathbf{p} + \boldsymbol{j}\,m)\,\boldsymbol{j} = (-ik E - \boldsymbol{i}\,\mathbf{p} + \boldsymbol{j}\,m)$

$TC = P$ $\quad \boldsymbol{k}\,(-\boldsymbol{j}\,(ik E + \boldsymbol{i}\,\mathbf{p} + \boldsymbol{j}\,m)\,\boldsymbol{j})\,\boldsymbol{k} = \boldsymbol{i}\,(ik E + \boldsymbol{i}\,\mathbf{p} + \boldsymbol{j}\,m)\,\boldsymbol{i} = (ik E - \boldsymbol{i}\,\mathbf{p} + \boldsymbol{j}\,m)$

and that $TCP \equiv CPT \equiv$ identity, because:

$$\boldsymbol{k}\,(-\boldsymbol{j}\,(\boldsymbol{i}\,(\boldsymbol{i}kE + \boldsymbol{i}\,\mathbf{p} + \boldsymbol{j}\,m)\,\boldsymbol{i})\,\boldsymbol{j})\,\boldsymbol{k} = -\boldsymbol{kji}\,(\boldsymbol{i}kE + \boldsymbol{i}\,\mathbf{p} + \boldsymbol{j}\,m)\,\boldsymbol{ijk} = (\boldsymbol{i}kE + \boldsymbol{i}\,\mathbf{p} + \boldsymbol{j}\,m).$$

Using this formalism for the transformations, the correct intrinsic parities of ground-state baryons and bosons are easily recovered (cf section 6.8).

It is of interest, in connection with the violation of symmetries that occurs in the weak interaction, that no fundamental process can tell, in principle, whether the symmetry violated, along with charge conjugation, is P or T. We can only tell whether the violation is of one or two of these symmetries. By convention, and because it is easier to measure, we assume that the first symmetry violated is P, but the result does not depend on any fundamental justification.

A violation of charge conjugation, such as happens (at least partially) in the weak interaction, would effectively make the parameter m unable to assume a sign opposite to $\mathbf{p}$ and E. Violation of C would therefore lead to something like:

$$-\boldsymbol{j}\,\psi\boldsymbol{j} \rightarrow \psi$$

In such a case, since

$$-\boldsymbol{j}\,\psi\boldsymbol{j} = -\boldsymbol{j}\,(\boldsymbol{i}kE + \boldsymbol{i}\,\mathbf{p} + \boldsymbol{j}\,m)\,\boldsymbol{j} = (-\boldsymbol{i}kE - \boldsymbol{i}\,\mathbf{p} + \boldsymbol{j}\,m) \rightarrow \psi + 2\boldsymbol{j}m,$$

either that part of the matter quaternion which is involved in the weak interaction would require a term like $2\boldsymbol{j}m$ to be equivalent to zero, or we would need to add its equivalent value elsewhere. Similar arguments could be applied to define violations of P and T in this formalism.

We could, of course, construct the fermion operator in such a way as to transform the sign of the $\boldsymbol{j}m$ term; the convention here adopted (based on the rewrite formalism) is that such a transformation applies to the continuous vacuum state. In principle, however, the nilpotent operator could be made isodual according to the formalism introduced by Santilli to incorporate fermionic and antifermionic matter into physics on an equal basis.[7,8] Santilli has proposed that the nilpotent formalism generated in this book incorporates his iso-, geno- and hyper-Dirac structures as well as their isoduals.[9]

6.6 Baryons

We have already postulated an entangled system of two nilpotent states (fermion and antifermion) to describe bosons. Can we extend this idea to three nilpotent states to describe baryons? Conventionally, we consider a baryon to be made up of three fermionic components, to which we assign colour to overcome Pauli exclusion. Can we relate this concept of colour to the fundamental structure of nilpotents? Can we have a 3-component state vector? Obviously a combination involving identical fermions will be impossible, because

$$(\boldsymbol{i}kE + \boldsymbol{i}\,\mathbf{p} + \boldsymbol{j}\,m)\,(\boldsymbol{i}kE + \boldsymbol{i}\,\mathbf{p} + \boldsymbol{j}\,m)\,(\boldsymbol{i}kE + \boldsymbol{i}\,\mathbf{p} + \boldsymbol{j}\,m) = 0.$$

However, the products

$$(\boldsymbol{i}kE + \boldsymbol{i}\,\mathbf{p} + \boldsymbol{j}\,m)\,(\boldsymbol{i}kE + \boldsymbol{j}\,m)\,(\boldsymbol{i}kE + \boldsymbol{j}\,m) \rightarrow (\boldsymbol{i}kE + \boldsymbol{i}\,\mathbf{p} + \boldsymbol{j}\,m)$$
$$(\boldsymbol{i}kE + \boldsymbol{j}\,m)\,(\boldsymbol{i}kE + \boldsymbol{i}\,\mathbf{p} + \boldsymbol{j}\,m)\,(\boldsymbol{i}kE + \boldsymbol{j}\,m) \rightarrow (\boldsymbol{i}kE - \boldsymbol{i}\,\mathbf{p} + \boldsymbol{j}\,m)$$
$$(\boldsymbol{i}kE + \boldsymbol{j}\,m)\,(\boldsymbol{i}kE + \boldsymbol{j}\,m)\,(\boldsymbol{i}kE + \boldsymbol{i}\,\mathbf{p} + \boldsymbol{j}\,m) \rightarrow (\boldsymbol{i}kE + \boldsymbol{i}\,\mathbf{p} + \boldsymbol{j}\,m)$$

result in the characteristic fermionic structure, after normalization of the scalar factor $-p^2$. Also

$$(\boldsymbol{i}kE - \boldsymbol{i}\,\mathbf{p} + \boldsymbol{j}\,m)\,(\boldsymbol{i}kE + \boldsymbol{j}\,m)\,(\boldsymbol{i}kE + \boldsymbol{j}\,m) \rightarrow (\boldsymbol{i}kE - \boldsymbol{i}\,\mathbf{p} + \boldsymbol{j}\,m)$$
$$(\boldsymbol{i}kE + \boldsymbol{j}\,m)\,(\boldsymbol{i}kE - \boldsymbol{i}\,\mathbf{p} + \boldsymbol{j}\,m)\,(\boldsymbol{i}kE + \boldsymbol{j}\,m) \rightarrow (\boldsymbol{i}kE + \boldsymbol{i}\,\mathbf{p} + \boldsymbol{j}\,m)$$
$$(\boldsymbol{i}kE + \boldsymbol{j}\,m)\,(\boldsymbol{i}kE + \boldsymbol{j}\,m)\,(\boldsymbol{i}kE - \boldsymbol{i}\,\mathbf{p} + \boldsymbol{j}\,m) \rightarrow (\boldsymbol{i}kE - \boldsymbol{i}\,\mathbf{p} + \boldsymbol{j}\,m).$$

So it is possible to have a nonzero state vector if we use the *vector* properties of **p** and the arbitrary nature of its sign (+ or –). A state vector of the form, privileging the **p** components:

$$(\boldsymbol{i}kE \pm \boldsymbol{i}\,\mathbf{i}p_X + \boldsymbol{j}\,m)\,(\boldsymbol{i}kE \pm \boldsymbol{i}\,\mathbf{j}p_y + \boldsymbol{j}\,m)\,(\boldsymbol{i}kE \pm \boldsymbol{i}\,\mathbf{k}p_Z + \boldsymbol{j}\,m)$$

has six independent allowed phases, i.e. when

$$\mathbf{p} = \pm\,\mathbf{i}p_X\,,\ \mathbf{p} = \pm\,\mathbf{j}p_y\,,\ \mathbf{p} = \pm\,\mathbf{k}p_Z.$$

In these phases, only *one* of the three components of momentum, p_x, p_y, p_z, is nonzero and represents the total **p**, but the phases must be *gauge invariant*, i.e. indistinguishable, or all present at once. This requires an exact symmetry with an $SU(3)$ group structure, with eight generators, exactly comparable to the conventional symmetry of the coloured quark model, with three symmetric and three antisymmetric phases, and transitions mediated by eight massless spin 1 gluons.

Choosing the labels *B*, *G* and *R* to represent the **p** variation within the brackets, with the $+\,\boldsymbol{i}\mathbf{p}$ phases representing a positive or cyclic combination of the three, and the $-\,\boldsymbol{i}\mathbf{p}$ phases a negative or anticyclic combination, we can represent the total state vector, incorporating all six phases, via the Jacobi identity, as

$$\psi \sim (BGR - BRG + GRB - GBR + RBG - RGB).$$

This has exactly the same group structure as the standard 'coloured' baryon wavefunction made of *R*, *G* and *B* 'quarks', with the mappings:

$$(\boldsymbol{k}E + \boldsymbol{i}\,\mathbf{i}p_x + \boldsymbol{j}\,m)\,(\boldsymbol{k}E + \ldots + \boldsymbol{j}\,m)\,(\boldsymbol{k}E + \ldots + \boldsymbol{j}\,m) \qquad +RGB$$
$$(\boldsymbol{k}E - \boldsymbol{i}\,\mathbf{i}p_x + \boldsymbol{j}\,m)\,(\boldsymbol{k}E - \ldots + \boldsymbol{j}\,m)\,(\boldsymbol{k}E - \ldots + \boldsymbol{j}\,m) \qquad -RBG$$
$$(\boldsymbol{k}E + \ldots + \boldsymbol{j}\,m)\,(\boldsymbol{k}E + \boldsymbol{i}\,\mathbf{j}p_y + \boldsymbol{j}\,m)\,(\boldsymbol{k}E + \ldots + \boldsymbol{j}\,m) \qquad +BRG$$
$$(\boldsymbol{k}E - \ldots + \boldsymbol{j}\,m)\,(\boldsymbol{k}E - \boldsymbol{i}\,\mathbf{j}p_y + \boldsymbol{j}\,m)\,(\boldsymbol{k}E - \ldots + \boldsymbol{j}\,m) \qquad -GRB$$
$$(\boldsymbol{k}E + \ldots + \boldsymbol{j}\,m)\,(\boldsymbol{k}E + \ldots + \boldsymbol{j}\,m)\,(\boldsymbol{k}E + \boldsymbol{i}\,\mathbf{k}p_z + \boldsymbol{j}\,m) \qquad +GBR$$
$$(\boldsymbol{k}E - \ldots + \boldsymbol{j}\,m)\,(\boldsymbol{k}E - \ldots + \boldsymbol{j}\,m)\,(\boldsymbol{k}E - \boldsymbol{i}\,\mathbf{k}p_z + \boldsymbol{j}\,m) \qquad -BGR \qquad (6.10)$$

These are recognisably the same three cyclic and three anticyclic combinations as in the standard QCD representation of the baryon wavefunction. That is, it has an $SU(3)$ structure, with 8 generators. And, since it allows only one term to have a momentum component at any one time, it can accommodate both spin ½ baryons (where the signs of the **p** term in different components are different) and spin $^3/_2$ baryons (where the signs of the **p** term in different components are the same). Clearly,

$$(\boldsymbol{k}E \mp \boldsymbol{i}\,\mathbf{i}p_x + \boldsymbol{j}\,m)\,(\boldsymbol{k}E \pm \boldsymbol{i}\,\mathbf{j}p_y + \boldsymbol{j}\,m)\,(\boldsymbol{k}E \mp \boldsymbol{i}\,\mathbf{k}p_z + \boldsymbol{j}\,m) \qquad \text{spin } \tfrac{1}{2}$$

and $$(\boldsymbol{k}E \pm \boldsymbol{i}\,\mathbf{i}p_x + \boldsymbol{j}\,m)\,(\boldsymbol{k}E \pm \boldsymbol{i}\,\mathbf{j}p_y + \boldsymbol{j}\,m)\,(\boldsymbol{k}E \pm \boldsymbol{i}\,\mathbf{k}p_z + \boldsymbol{j}\,m) \qquad \text{spin } {}^3/_2$$

both result in the same final structures. With the spinor terms included, each of these is represented by a tensor product of three spinors, for example:

$$(\boldsymbol{k}E + \boldsymbol{j}\,m)\,(\boldsymbol{k}E + \boldsymbol{j}\,m)\,(\boldsymbol{k}E + \boldsymbol{i}\,\mathbf{p} + \boldsymbol{j}\,m) \rightarrow \left(\frac{1}{2}\right) \otimes \left(\frac{1}{2}\right) \otimes \left(\frac{1}{2}\right)$$

where $$\left(\frac{1}{2}\right) \otimes \left(\frac{1}{2}\right) \otimes \left(\frac{1}{2}\right) = \left(\frac{3}{2}\right) \oplus \left(\frac{1}{2}\right) \oplus \left(\frac{1}{2}\right).$$

The perfect gauge invariance that we obtain between the 'three quark' states in a baryon should apply additionally where a bosonic ('quark-antiquark') state can be defined in terms of the same varying directional properties of its **p** operator. The structure is determined solely by the nilpotent nature of the fermion wavefunction. Putting in an extra **p** into the brackets missing them, we immediately reduce to zero. Because there is only one spin term, the nilpotent representation also predicts that the spin, in the case of a baryon is a property of the baryon wavefunction as a whole, not of component quark wavefunctions.

Very significantly, the full symmetry between the 3 momentum components (which is a P transformation, using operator $\boldsymbol{i}$, exactly as supposed for the strong vacuum) can only apply if the momentum operators can be equally + or –. That is, the symmetry evident in (6.10) requires equivalent status for the **+p** and **–p** states associated with positive energy. With all phases of the interaction present at the same time (perfect gauge invariance), this is equivalent to saying fermionic states of both negative and positive helicity (or left- and right-handedness) must be present simultaneously in the baryon state. In other words, the baryonic state must have non-zero (positive) mass via the Higgs mechanism. In principle, this immediately solves the so-called mass gap problem (see section 14.11). At the same time, the requirement of unbroken gauge invariance, which is a consequence of the vector nature of **p**, requires that the mediators must be massless, and so spin 1.

'Colour' transitions can be seen as involving either an exchange of the components of **p** between the individual quarks, or as a relative switching of quark positions, so that the colours either move with the respective p_x, p_y, p_z components, or switch with them. In either model the effect is the same, and a sign reversal in **p** is an additional necessary result. One method of picturing the exact symmetry presented in (6.10) is to imagine an automatic mechanism of transfer between the phases. And, since the E and **p** terms in the state vector really represent time and space derivatives, we can replace these with the covariant derivatives needed for invariance under a local $SU(3)$ gauge transformation.

Another significant aspect of the $SU(3)$ symmetry or strong interaction is that, because it depends entirely on the nilpotency of the component state vectors, it is entirely nonlocal. That is, the exchange of momentum **p** involved is entirely independent of any spatial position of the 3 components of the baryon. We can suppose, therefore, that the rate of change of momentum (or 'force') is constant with respect to spatial positioning or separation. A constant force is equivalent to a potential which is linear with distance, exactly as is required for the conventional strong interaction.

As is the case with bosons, the baryon representation can only exist as a unified or entangled state. It is not really a representation of a combination of 3 independent fermions. It is equally significant that the representation is impossible in a conventional spinor formulation, with terms such as $p_x + ip_y$, or in any representation in which the momentum operators cannot show the full affine nature of the vector concept.

6.7 Gluons and Exotic States

Now, to maintain the $SU(3)$ group symmetry, in this representation, the mediators of the strong force will be eight spin 1 bosons or vertices constructed from

$$(\pm ikE \mp \boldsymbol{i}\,\mathbf{i}p_x)\,(\mp ikE \mp \boldsymbol{i}\,\mathbf{j}p_y) \qquad (\pm ikE \mp \boldsymbol{i}\,\mathbf{j}p_y)\,(\mp ikE \mp \boldsymbol{i}\,\mathbf{i}p_x)$$

$$(\pm ikE \mp \boldsymbol{i}\,\mathbf{j}p_y)\,(\mp ikE + \boldsymbol{i}\,\mathbf{k}p_z) \qquad (\pm ikE \mp \boldsymbol{i}\,\mathbf{k}p_z)\,(\top ikE \mp \boldsymbol{i}\,\mathbf{j}p_y)$$

$$(\pm ikE \mp \boldsymbol{i}\,\mathbf{i}p_z)\,(\mp ikE \mp \boldsymbol{i}\,\mathbf{i}p_x) \qquad (\pm ikE \mp \boldsymbol{i}\,\mathbf{i}p_x)\,(\mp ikE \mp \boldsymbol{i}\,\mathbf{i}p_z)$$

and two combinations from

$$(\pm ikE \mp \boldsymbol{i}\,\mathbf{i}p_x)\,(\mp ikE \mp \boldsymbol{i}\,\mathbf{i}p_x) \qquad (\pm ikE \mp \boldsymbol{i}\,\mathbf{j}p_y)\,(\mp ikE \mp \boldsymbol{i}\,\mathbf{j}p_v)$$

$$(\pm ikE \mp \boldsymbol{i}\,\mathbf{k}p_z)\,(\mp ikE \mp \boldsymbol{i}\,\mathbf{k}p_z)$$

These structures are, of course, identical to an equivalent set in which both brackets undergo a complete sign reversal

$$(\mp ikE \pm \boldsymbol{i}\,\mathbf{i}p_x)\,(\pm ikE \pm \boldsymbol{i}\,\mathbf{i}p_y) \text{ or } (\pm ikE \pm \boldsymbol{i}\,\mathbf{i}p_y)\,(\mp ikE \pm \boldsymbol{i}\,\mathbf{i}p_x), \text{ etc.}$$

The fact that $(ikE - \boldsymbol{i}\,\mathbf{i}p_x)\,(-ikE - \boldsymbol{i}\,\mathbf{i}p_y)$ easily transforms to $(ikE - \boldsymbol{i}\,\mathbf{i}p_x)\,(ikE + \boldsymbol{i}\,\mathbf{i}p_y)$ implies a 'strong' vacuum $\boldsymbol{i}\,(kE + \boldsymbol{i}\,\mathbf{i}p_x)$ or $\boldsymbol{i}\,(kE + \boldsymbol{i}\,\mathbf{i}p_y)$ with

$$(ikE + \boldsymbol{i}\,p_x)\,\boldsymbol{i}\,(ikE + \boldsymbol{i}\,p_y)\,\boldsymbol{i}\,(ikE + \boldsymbol{i}\,p_y)\,\boldsymbol{i}\,(ikE + \boldsymbol{i}\,p_x)\,\boldsymbol{i}\,(ikE + \boldsymbol{i}\,p_y)\;\ldots$$

or

$$(ikE + \boldsymbol{i}\,p_x)\,(ikE - \boldsymbol{i}\,p_y)\,(ikE + \boldsymbol{i}\,p_y)\,(ikE - \boldsymbol{i}\,p_x)\,(ikE - \boldsymbol{i}\,p_y)\;\ldots$$

producing the interactions of an (assumed) massless quark with the gluon sea. The important thing here is that applying any of these mediators will produce a sign change in the **p** component that leads to mass.

The nilpotent structure allows possible state vectors for exotic states, though it does not prove that they necessarily exist, for example, a spin 2 glueball:

$$(\pm ikE \pm ip_x + jm)\,(\mp ikE \pm ip_y + jm)\,(\pm ikE \pm ip_y + jm)\,(\mp ikE \pm ip_x + jm)$$

A spin 0 glueball would be represented by:

$$(\pm ikE \pm ip_x + jm)\,(\mp ikE \mp ip_x + jm)\,(\pm ikE \pm ip_y + jm)\,(\mp ikE \mp ip_y + jm),$$

or

$$(\pm ikE \mp ip_x + jm)\,(\mp ikE \pm ip_x + jm)\,(\pm ikE \mp ip_y + jm)\,(\mp ikE \pm ip_y + jm),$$

which, significantly, cannot be massless.

More exotic structures might include 'pentaquarks', or baryon-boson combinations, where the **p** phase of the 'baryon' component would not coincide with that of the 'boson', for example:

$$(ikE + \boldsymbol{j}\,m)\,(ikE \pm \boldsymbol{i}\,p_y + \boldsymbol{j}\,m)\,(ikE + \boldsymbol{j}\,m)\,(ikE + \boldsymbol{i}\,p_z + \boldsymbol{j}\,m)\,(-ikE - \boldsymbol{i}\,p_z + \boldsymbol{j}\,m).$$

Significantly, the existence of such a structure (which might be considered as a possible stage in the pion exchange process between baryons in nuclei) would necessitate identical (though not coincident) colour-phase transitions for both baryons and bosons.

6.8 Parities of Bosons and Baryons

Defining the parity transformation on ψ as $\boldsymbol{i}\,\psi\,\boldsymbol{i}$, we can now investigate the intrinsic parities of ground state bosons and baryons. Applying the transformation to a scalar boson, we obtain:

$$\boldsymbol{i}\,(\boldsymbol{ik}E + \boldsymbol{i}\,\mathbf{p} + \boldsymbol{j}\,m)\,(-\boldsymbol{ik}E - \boldsymbol{i}\,\mathbf{p} + \boldsymbol{j}\,m)\,\boldsymbol{i};$$
$$\boldsymbol{i}\,(\boldsymbol{ik}E - \boldsymbol{i}\,\mathbf{p} + \boldsymbol{j}\,m)\,(-\boldsymbol{ik}E + \boldsymbol{ii}\,\mathbf{p} + \boldsymbol{j}\,m)\,\boldsymbol{i};$$
$$\boldsymbol{i}\,(-\boldsymbol{ik}E + \boldsymbol{ii}\,\mathbf{p} + \boldsymbol{ij}\,m)\,(\boldsymbol{ik}E - \boldsymbol{i}\,\mathbf{p} + \boldsymbol{j}\,m)\,\boldsymbol{i};$$
$$\boldsymbol{i}\,(-\boldsymbol{ik}E - \boldsymbol{i}\,\mathbf{p} + \boldsymbol{j}\,m)\,(\boldsymbol{ik}E + \boldsymbol{i}\,\mathbf{p} + \boldsymbol{j}\,m)\,\boldsymbol{i}.$$

If we take the first term, and use $-\,\boldsymbol{i}\,\boldsymbol{i} = 1$, then the parity transformation produces:

$$-\boldsymbol{i}\,(\boldsymbol{ik}E + \boldsymbol{i}\,\mathbf{p} + \boldsymbol{j}\,m)\,\boldsymbol{i}\,\boldsymbol{i}\,(\boldsymbol{ik}E - \boldsymbol{i}\,\mathbf{p} + \boldsymbol{j}\,m)\,\boldsymbol{i}.$$

Now, we have a parity transformation on each bracket, with an additional – sign. This produces:

$$-\,(\boldsymbol{ik}E - \boldsymbol{i}\,\mathbf{p} + \boldsymbol{j}\,m)\,(-\boldsymbol{ik}E + \boldsymbol{ii}\,\mathbf{p} + \boldsymbol{ij}\,m).$$

Applying the transformation to each of the terms, we obtain:

$$-\,(\boldsymbol{ik}E - \boldsymbol{i}\,\mathbf{p} + \boldsymbol{j}\,m)\,(-\boldsymbol{ik}E + \boldsymbol{ii}\,\mathbf{p} + \boldsymbol{ij}\,m);$$
$$-\,(\boldsymbol{ik}E + \boldsymbol{i}\,\mathbf{p} + \boldsymbol{j}\,m)\,(-\boldsymbol{ik}E - \boldsymbol{i}\,\mathbf{p} + \boldsymbol{j}\,m);$$
$$-\,(-\boldsymbol{ik}E - \boldsymbol{i}\,\mathbf{p} + \boldsymbol{j}\,m)\,(\boldsymbol{ik}E + \boldsymbol{i}\,\mathbf{p} + \boldsymbol{j}\,m);$$
$$-\,(-\boldsymbol{ik}E + \boldsymbol{ii}\,\mathbf{p} + \boldsymbol{ij}\,m)\,(\boldsymbol{ik}E - \boldsymbol{i}\,\mathbf{p} + \boldsymbol{j}\,m).$$

The total transformed wavefunction $\boldsymbol{i}\psi\boldsymbol{i}$ thus becomes $-\psi$. The original wavefunction therefore has negative parity. For the vector meson, we obtain:

$$-\,(\boldsymbol{ik}E - \boldsymbol{i}\,\mathbf{p} + \boldsymbol{j}\,m)\,(-\boldsymbol{ik}E - \boldsymbol{i}\,\mathbf{p} + \boldsymbol{j}\,m);$$
$$-\,(\boldsymbol{k}E + \boldsymbol{ii}\,\mathbf{p} + \boldsymbol{ij}\,m)\,(-\boldsymbol{ik}E + \boldsymbol{ii}\,\mathbf{p} + \boldsymbol{ij}\,m);$$
$$-\,(-\boldsymbol{ik}E - \boldsymbol{i}\,\mathbf{p} + \boldsymbol{j}\,m)\,(\boldsymbol{ik}E - \boldsymbol{i}\,\mathbf{p} + \boldsymbol{j}\,m);$$
$$-\,(-\boldsymbol{ik}E + \boldsymbol{ii}\,\mathbf{p} + \boldsymbol{ij}\,m)\,(\boldsymbol{ik}E + \boldsymbol{i}\,\mathbf{p} + \boldsymbol{j}\,m).$$

Again, $\boldsymbol{i}\psi\boldsymbol{i}$ becomes $-\psi$, and the original wavefunction has negative parity.

Let us try the same operation on a baryon. Taking one of the terms:

$$(\boldsymbol{ik}E + \boldsymbol{j}\,m)\,(\boldsymbol{ik}E + \boldsymbol{j}\,m)\,(\boldsymbol{ik}E + \boldsymbol{i}\,\mathbf{p} + \boldsymbol{j}\,m),$$

we apply a parity transformation:

$$\boldsymbol{i}\,(\boldsymbol{ik}E + \boldsymbol{j}\,m)\,(\boldsymbol{ik}E + \boldsymbol{j}\,m)\,(\boldsymbol{ik}E + \boldsymbol{i}\,\mathbf{p} + \boldsymbol{j}\,m)\,\boldsymbol{i}.$$

This time, we can write it in the form:

$$\boldsymbol{i}\,(\boldsymbol{ik}E + \boldsymbol{j}\,m)\,\boldsymbol{i}\,\boldsymbol{i}\,(\boldsymbol{ik}E + \boldsymbol{j}\,m)\,\boldsymbol{i}\,\boldsymbol{i}\,(\boldsymbol{ik}E + \boldsymbol{i}\,\mathbf{p} + \boldsymbol{j}\,m)\,\boldsymbol{i},$$

with no sign change. This term becomes:

$$(ikE + \boldsymbol{j}\, m)\, (ikE + \boldsymbol{j}\, m)\ \ (ikE - \boldsymbol{i}\, \mathbf{p} + \boldsymbol{j}\, m).$$

Taken over all the terms (three with **p**, and three with – **p**), then:

$$(ikE + \boldsymbol{j}\, m)\, (ikE + \boldsymbol{j}\, m)\, (ikE - \boldsymbol{i}\, \mathbf{p} + \boldsymbol{j}\, m)$$
$$(ikE + \boldsymbol{j}\, m)\, (ikE + \boldsymbol{i}\, \mathbf{p} + \boldsymbol{j}\, m)\, (ikE + \boldsymbol{j}\, m)$$
$$(ikE + \boldsymbol{j}\, m)\, (ikE - \boldsymbol{i}\, \mathbf{p} + \boldsymbol{j}\, m)\, (ikE + \boldsymbol{j}\, m)$$
$$(ikE + \boldsymbol{j}\, m)\, (kE + ij\, m)\, (ikE + \boldsymbol{i}\, \mathbf{p} + \boldsymbol{j}\, m)$$
$$(ikE - \boldsymbol{i}\, \mathbf{p} + \boldsymbol{j}\, m)\, (ikE + \boldsymbol{j}\, m)\, (ikE + \boldsymbol{j}\, m)$$
$$(ikE + \boldsymbol{i}\, \mathbf{p} + \boldsymbol{j}\, m)\, (ikE + \boldsymbol{j}\, m)\, (ikE + \boldsymbol{j}\, m),$$

and
$$\boldsymbol{i}\, \psi \boldsymbol{i} = \psi.$$

The baryon wavefunction has positive parity.

These calculations, of course, apply to the ground state values only, because if extra angular momentum terms are added, then extra terms must be supplied to the wavefunctions, the sign of parity reversing for each additional term.

6.9 Supersymmetry and Renormalization

The fermion and antifermion state vectors are not only quantum field operators (removing the need for representation by quantum field integrals, cf section 6.11), but also supersymmetric operators, equivalent to Q and $Q^\dagger$, respectively converting boson to fermion and fermion to boson, and each being the Hermitian conjugate (i.e. vacuum 'image') of the other. With this conception of vacuum, we can imagine a renormalization process, involving an infinite succession of boson and fermion loops cancelling each other out, without needing to generate a new set of extra supersymmetric partners or encountering a hierarchy problem. The formalism also produces a perturbation expansion for a first-order QED coupling with a state vector of the form:

$$\Psi_1 = -e\, \Sigma\, [\boldsymbol{k}E + \boldsymbol{i}i\, \boldsymbol{\sigma}.(\mathbf{p} + \mathbf{k}) + \boldsymbol{i}jm]^{-1} (\boldsymbol{i}k\phi\ - \boldsymbol{i}\, \boldsymbol{\sigma}.\mathbf{A})\, (\boldsymbol{k}E + \boldsymbol{i}i\boldsymbol{\sigma}.\mathbf{p} + \boldsymbol{i}jm)\, e^{-i(Et - (\mathbf{p} + \mathbf{k}).\mathbf{r})},$$

which automatically becomes 0 for a self-interacting electron, and similar cases (cf section 11.1). Pure vacuum interactions in this formalism require no renormalization, although charge values vary with the strength of real interactions in the usual way, while a fermion propagator of the form

$$S_F(p) = \frac{1}{(\boldsymbol{k}E + \boldsymbol{i}i\boldsymbol{\sigma}.\mathbf{p} + \boldsymbol{i}jm)}$$

eliminates any infrared divergence by having a denominator which conjugates to a non-zero scalar using its vacuum 'image' (see section 11.4).

Of course, in some cases, the fermion (or antifermion) produces its 'image' in a real antifermionic (or fermionic) state. This is the origin of the nonzero Berry phase, Jahn-Teller effect, Aharonov-Bohm effect, quantum Hall effect, even-even nuclei, and many other similar phenomena. A Bose-Einstein condensate in He^4 or Cooper pairing in a normal superconductor is a slightly different way of producing a bosonic-type state, as it is composed of a fermion-fermion pairing with opposite spins (total spin 0), as in $(\pm ikE \pm \mathbf{ip} + jm)$ $(\pm ikE \mp \mathbf{ip} + jm)$. The 'vacuum' equivalent for this would be $\boldsymbol{i}$ $(\pm ikE \pm \mathbf{ip} + jm)$ or $\boldsymbol{i}$ $(\pm ikE \mp \mathbf{ip} + jm)$. If the component fermions are physically separate objects, as with atoms of He^3, then the momentum directions may be opposite (say, in a harmonic oscillator-type arrangement) while spin directions are the same, making $(\pm ikE \pm \mathbf{ip} + jm)$ $(\pm ikE \mp \mathbf{ip} + jm)$ represent spin 1. Superpositions of paired fermion states with different energies, such as $(\pm ikE_1 \pm \mathbf{ip}_1 + jm_1)$ $(\pm ikE_2 \pm \mathbf{ip}_2 + jm_2)$ are also, in a sense, bosonic. A product of an even number of fermionic quaternion state vectors will necessarily be bosonic, and a product of an odd number fermionic, as in conventional theory, and the products be similarly described by the use of Slater determinants to give equal status to all components, as in conventional quantum mechanics.

Another effect which is observed is the reverse coupling of a real boson of spin 0 or 1 to an 'environment' to produce a fermion-like state. Perhaps the Higgs mechanism occurs in this way, but a more immediate possibility is the coupling of gluons to the quark-gluon plasma to deliver the total spin of ½ or $^3/_2$ to a baryon. If a boson state is represented by $(\pm ikE \pm \mathbf{ip} + jm)$ $(\mp ikE \pm \mathbf{ip} + jm)$ or $(\pm ikE \pm \mathbf{ip} + jm)$ $(\mp ikE \mp \mathbf{ip} + jm)$ we may imagine the respective 'fermion' and 'antifermion' components, say $(\pm ikE \pm \mathbf{ip} + jm)$ and $(\mp ikE \pm \mathbf{ip} + jm)$, being respectively pre- and post-multiplied by terms like $(\pm ikE \pm \mathbf{ip} + jm)$ $\boldsymbol{k}$ and $\boldsymbol{k}$ $(\mp ikE \pm \mathbf{ip} + jm)$, so that the state becomes

$$\ldots (\pm ikE \pm \mathbf{ip} + jm)\, \boldsymbol{k}\, (\pm ikE \pm \mathbf{ip} + jm)\, \boldsymbol{k}\, (\pm ikE \pm \mathbf{ip} + jm)\, \boldsymbol{k}\, (\pm kE \pm i\mathbf{ip} + ijm)$$
$$(\mp ikE \pm \mathbf{ip} + m)\, \boldsymbol{k}\, (\mp ikE \pm \mathbf{ip} + m)\, \boldsymbol{k}\, (\mp ikE \pm \mathbf{ip} + m)\, \boldsymbol{k}\, (\mp ikE \pm \mathbf{ip} + m) \ldots$$

Alternatively, we can imagine the boson responding to only the left-multiplied (fermion) terms or the right-multiplied (antifermion) ones.

Vacuum fermions and vacuum antifermions have a similar relationship to real fermions and real antifermions, although both states, in this case, are virtual. The mirror image states of all possible fermion states constitute the zero point energy of the vacuum.[10] Each possible state provides a virtual vacuum energy of $\hbar\, \omega / 2$, like the ground state of a harmonic oscillator (which, of course, it is). To create a real fermion state, we excite a virtual vacuum state of $-\hbar\omega / 2$ up to the level $\hbar\omega / 2$, using a total energy quantum of $\hbar\omega$. Counting real and virtual particles, we have the same number of fermions and antifermions in the universe, but, in a universe with a non-symmetric ground state (such as we will demonstrate must exist), fermions will be predominantly real and antifermions predominantly virtual; and, counting real and virtual particles, and assigning $+E$ to fermions and $-E$ to fermions, we obtain a total energy of zero.

Real	Fermion	Antifermion
Vacuum	Antifermion	Fermion

The existence of mirror image vacuum states for all fermionic particles accounts for the structure of the Dirac quaternion state vector. We incorporate both real and virtual components and interpret the *zitterbewegung* as a switching between them. The four creation operators create both the real particle and its set of dual vacuum images. All fermion wavefunctions are, in this sense, single-valued, producing an effective combination analogous to a simultaneous consideration of the two sides of Newton's third law of motion or a virial doubling of the kinetic energy in a potential energy term. Fermion and antifermion state vectors thus have identical components; only the *order* privileges either $+E$ or $-E$ states as the 'real' ones, and a similar principle applies to the spin states.

The vacuum is really an expression of the continuous or noncountable nature of mass-energy ('mass', as the source of gravity). Continuity, as we have seen in chapter 2, automatically makes mass-energy unidimensional and unipolar. Since it is also real, it is therefore restricted to a single mathematical sign, which is usually taken as positive. We can interpret this as implying a non-symmetric ground state or a filled vacuum. The filled vacuum for the ground state is that of negative energy or antifermions. In the rewrite system, it has an exact counterpart in the way that binary negative units are defined as an infinite strings of 1s. So if we add 1 to ...111111 it becomes ...000000. (See Appendix B.) Ultimately, this is the result of losing a degree of freedom (that of $-m$) when we create the nilpotent structure. Compactification necessarily produces duplication.[11]

In physical terms, it manifests itself in the Higgs field, which breaks charge conjugation symmetry for the weak interaction, and gives rest masses to the fermions and weak gauge bosons. (The reaction half of the system, in this case, is equivalent to what Newton called the 'impressed force' or the inertia.) It is also responsible for quantum mechanical nonlocality and the instantaneous transmission of the static gravitational force – though not the acceleration-dependent inertial or GTR component, or the inertial reaction force that we actually measure in systems with localised mass (and with which gravity is often confused) (cf chapter 17). Significantly, gravitational potential energy is often represented as negative. Again, according to the arguments advanced in chapter 17, it has no dependence on *c*, and no interconversion with mass or other forms of potential energy; that is, it does not behave like energy in the ordinary sense, which would accord with its special status.

6.10 Annihilation and Creation Operators

Yet another significant aspect of the quaternion Dirac algebra using nilpotents is that it is already effectively second quantized. We have seen that the quaternion operator $(\boldsymbol{ik}\partial/\partial t + \boldsymbol{i}\nabla + \boldsymbol{ij}\, m)$ acting on a state

$$(\boldsymbol{k}E + \boldsymbol{ii}\,\mathbf{p} + \boldsymbol{ij}\, m)\, e^{-i(Et - \mathbf{p.r})}$$

leads to the equation:

$$(\boldsymbol{ik}\partial/\partial t + \boldsymbol{i}\nabla + \boldsymbol{ij}\, m)\,(\boldsymbol{k}E + \boldsymbol{ii}\,\mathbf{p} + \boldsymbol{ij}\, m)\, e^{-i(Et - \mathbf{p.r})} = 0.$$

This operation with operator $\mathcal{D} = (\boldsymbol{ik}\partial/\partial t + \boldsymbol{i}\nabla + \boldsymbol{ij}\, m)$ may be thought of as a creation operation acting on the single particle fermion state which is already filled. The result is therefore zero. We may obtain the corresponding annihilation operation by finding the Hermitian conjugate of $(\boldsymbol{ik}\partial/\partial t + \boldsymbol{i}\nabla + \boldsymbol{ij}\, m)$.

We know that:

$$\mathcal{D}\, e^{-i(Et - \mathbf{p.r})} = (\boldsymbol{k}E + \boldsymbol{ii}\,\mathbf{p} + \boldsymbol{ij}\, m)\, e^{-i(Et - \mathbf{p.r})}. \tag{6.11}$$

$\mathcal{D}$ is acting as though it were the creation operator $a^\dagger$ acting on the group of translations and rotations that we call vacuum. We can therefore write $a^\dagger$ as

$$a^\dagger = (1/2E)\,(\boldsymbol{k}E + \boldsymbol{ii}\,\mathbf{p} + \boldsymbol{ij}\, m).$$

Now, the Hermitian conjugate of expression (6.11) is:

$$\mathcal{D}^\dagger\, e^{-i(Et - \mathbf{p.r})} = (-\boldsymbol{k}E + \boldsymbol{ii}\,\mathbf{p} + \boldsymbol{ij}\, m)\, e^{i(Et - \mathbf{p.r})}.$$

Here, because it describes the creation of the anti-particle, $\mathcal{D}^\dagger$ is acting as though it were the annihilation operator a conjugate to $a^\dagger$, so that

$$a = (1/2E)\,(-\boldsymbol{k}E + \boldsymbol{ii}\,\mathbf{p} + \boldsymbol{ij}\, m).$$

The operators have been modified by a factor (1 / 2*E*) so that, when acting on a state, $a^\dagger a$ will reproduce the state up to a scalar multiplicative factor which does not affect the quaternion properties of the state.

It is easy to verify that these two operators have the commutation relations appropriate to fermion annihilation and creation operators:

$$aa\dagger + a\dagger a = 1.$$

We now need to find the vacuum state that $a^\dagger$ can act upon to lead to the single particle state:

$$(\boldsymbol{k}E + \boldsymbol{i}i\ \mathbf{p} + \boldsymbol{i}j\ m)\ e^{-i(Et - \mathbf{p.r})}.$$

We need only consider the following:

$$a\ (\boldsymbol{k}E + \boldsymbol{i}i\ \mathbf{p} + \boldsymbol{i}j\ m)\ e^{-i(Et - \mathbf{p.r})} = |0\rangle$$

$$= 2E\ (1 / 2E)\ (E - \boldsymbol{i}j\ \mathbf{p} + \boldsymbol{i}i\ m)\ e^{-i(Et - \mathbf{p.r})},$$

where $|0\rangle$ is vacuum. To check this we need only observe that

$$a^\dagger\ (a\ (\boldsymbol{k}E + \boldsymbol{i}i\ \mathbf{p} + \boldsymbol{i}j\ m)\ e^{-i(Et - \mathbf{p.r})} = \mathbf{a}^\dagger |0\rangle$$

$$= a^\dagger\ (1 / 2E)\ 2E\ (E - \boldsymbol{i}j\ \mathbf{p} + \boldsymbol{i}i\ m)\ e^{-i(Et - \mathbf{p.r})}$$

$$= (1 / 2E)\ 2E\ (\boldsymbol{k}E + \boldsymbol{i}i\ \mathbf{p} + \boldsymbol{i}j\ m)\ e^{-i(Et - \mathbf{p.r})}$$

$$= (\boldsymbol{k}E + \boldsymbol{i}i\ \mathbf{p} + \boldsymbol{i}j\ m)\ e^{-i(Et - \mathbf{p.r})}.$$

It is easy to verify that the further action of *a* onto vacuum leads to zero.

The interaction of a fermion with the (infinite) vacuum, or mass-energy continuum, produces an infinite succession of products or superpositions of Dirac nilpotent states. This extends the dualling processes to infinity. Each of the 'virtual' states produced also acts in the same way, producing a pattern of the same form as the Conway system of constructed real numbers. The requirement of infinite dualling ensures the entanglement of all states in the universe, although, as with classical interference, decoherence will make this virtually unobservable except in special cases.

6.11 The Quantum Field

The preceding sections have shown that the quaternion state vectors in this formulation have remarkable properties. They are nilpotents or square roots of 0, and they have the properties of quantum field operators, creation and annihilation operators, and even supersymmetry operators. Because the differential operators are, in effect, identical to the quaternion state vectors, the theory already has the second quantization required of quantum field theory, so it should be possible to show rigorously that they are identical to quantum field integrals acting on vacuum. Let us define the following:

The creation operator for an electron, spin up:

$$a^{\dagger}(\mathbf{p}) = (\boldsymbol{ik}E + \boldsymbol{i}\mathbf{p} + \boldsymbol{j}m) / 2E$$

The annihilation operator for an electron, spin up:

$$a(\mathbf{p}) = (-\boldsymbol{ik}E + \boldsymbol{i}\mathbf{p} + \boldsymbol{j}m) / 2E$$

The annihilation operator for an electron, spin down:

$$a^{\dagger}(-\mathbf{p}) = (\boldsymbol{ik}E - \boldsymbol{i}\mathbf{p} + \boldsymbol{j}m) / 2E$$

The creation operator for electron, spin up:

$$a(-\mathbf{p}) = (-\boldsymbol{ik}E - \boldsymbol{i}\mathbf{p} + \boldsymbol{j}m) / 2E$$

The annihilation operator for a positron, spin down:

$$b(\mathbf{p}) = (\boldsymbol{ik}E + \boldsymbol{i}\mathbf{p} + \boldsymbol{j}m) / 2E$$

The creation operator for a positron, spin down:

$$b^{\dagger}(\mathbf{p}) = (-\boldsymbol{ik}E + \boldsymbol{i}\mathbf{p} + \boldsymbol{j}m) / 2E$$

The annihilation operator for a positron, spin up:

$$b(-\mathbf{p}) = (\boldsymbol{ik}E - \boldsymbol{i}\mathbf{p} + \boldsymbol{j}m) / 2E$$

The creation operator for a positron, spin up:

$$b^{\dagger}(-\mathbf{p}) = (-\boldsymbol{ik}E - \boldsymbol{i}\mathbf{p} + \boldsymbol{j}m) / 2E$$

The anticommutators,

$$\{a^{\dagger}(\mathbf{p}_1), a(\mathbf{p}_2)\} = \delta^3(\mathbf{p}_1 - \mathbf{p}_2)$$

and

$$\{b^{\dagger}(\mathbf{p}_1), a(\mathbf{p}_2)\} = \delta^3(\mathbf{p}_1 - \mathbf{p}_2)$$

are, of course, only valid where $\mathbf{p}_1 = \mathbf{p}_2$.

Writing out the Fourier superpositions of all possible states, we have

$$\psi(x) = \int d^3\mathbf{p}\left\{\left(a(\mathbf{p})\begin{pmatrix}1\\0\end{pmatrix} + a(-\mathbf{p})\begin{pmatrix}0\\1\end{pmatrix}\right)e^{-ipx} + \left(b^{\dagger}(\mathbf{p})\begin{pmatrix}1\\0\end{pmatrix} + b^{\dagger}(-\mathbf{p})\begin{pmatrix}0\\1\end{pmatrix}\right)e^{ipx}\right\}$$

$$\bar{\psi}(x) = \int d^3\mathbf{p}\left\{\left(a^{\dagger}(\mathbf{p})\begin{pmatrix}1\\0\end{pmatrix} + a^{\dagger}(-\mathbf{p})\begin{pmatrix}0\\1\end{pmatrix}\right)e^{ipx} + \left(b(\mathbf{p})\begin{pmatrix}1\\0\end{pmatrix} + b(-\mathbf{p})\begin{pmatrix}0\\1\end{pmatrix}\right)e^{-ipx}\right\}$$

Here, because of the explicit expressions used for the creation and annihilation operators, we need only use 1 and 0 states in the spinors.

Now $$\psi(x) = |E', \mathbf{p}'\rangle$$

where $$a^\dagger(\mathbf{p})|0\rangle = |E, \mathbf{p}\rangle$$

$$a^\dagger(-\mathbf{p})|0\rangle = |E, -\mathbf{p}\rangle$$

$$b^\dagger(\mathbf{p})|0\rangle = |-E, \mathbf{p}\rangle$$

$$b^\dagger(-\mathbf{p})|0\rangle = |-E, -\mathbf{p}\rangle$$

So

$$\psi(x)|E', \mathbf{p}\rangle = \int d^3\mathbf{p} \left\{ \left(a(\mathbf{p})\begin{pmatrix}1\\0\end{pmatrix} + a(-\mathbf{p})\begin{pmatrix}0\\1\end{pmatrix} \right) e^{-ipx}|E', \mathbf{p}'\rangle + \left(b^\dagger(\mathbf{p})\begin{pmatrix}1\\0\end{pmatrix} + b^\dagger(-\mathbf{p})\begin{pmatrix}0\\1\end{pmatrix} \right) e^{-ipx}|E', \mathbf{p}'\rangle \right\}$$

Since $$a(\mathbf{p})|E', \mathbf{p}'\rangle = a(\mathbf{p})a^\dagger(\mathbf{p})|0\rangle$$

$$= \delta^3(\mathbf{p}_1 - \mathbf{p}_2) - a(\mathbf{p})a^\dagger(\mathbf{p})|0\rangle = |0\rangle,$$

then $$\langle 0|\psi(x)|E', \mathbf{p}'\rangle = \begin{pmatrix}1\\0\end{pmatrix} e^{-ipx}.$$

In principle, therefore, we no longer need to define an explicit process of second quantization. Our nilpotent operators are already second quantized.

6.12 The Nilpotent State

As we have seen, the nilpotent state (± ***ik****E* ± **ip** + ***j****m*) is superior to the conventional quantum state in being automatically second quantized, with in-built supersymmetry. Amplitude and phase are uniquely determined by the same operator and each is quantized in the same way. So formal second quantization is unnecessary, and quantum field integrals acting on vacuum produce the nilpotent state vector. This means that creation / annihilation operators are easily identified within the state. The fermion state (± ***ik****E* ± **ip** + ***j****m*) incorporates 4 creation (or annihilation) operators:

Fermion creation spin up	(***ik****E* + **ip** + ***j****m*)
Fermion creation spin down	(***ik****E* – **ip** + ***j****m*)
Antifermion creation spin down	(–***ik****E* + **ip** + ***j****m*)

Antifermion creation spin up	$(-ikE - \mathbf{ip} + jm)$
Antifermion annihilation spin down	$(ikE + \mathbf{ip} + jm)$
Antifermion annihilation spin up	$(ikE - \mathbf{ip} + jm)$
Fermion annihilation spin up	$(-ikE + \mathbf{ip} + im)$
Fermion annihilation spin down	$(-ikE - \mathbf{ip} + jm)$

or even two creation and two annihilation operators

Fermion creation spin up	$(ikE + \mathbf{ip} + jm)$
Fermion creation spin down	$(ikE - \mathbf{ip} + jm)$
Fermion annihilation spin up	$(-ikE + \mathbf{ip} + jm)$
Fermion annihilation spin down	$(-ikE - \mathbf{ip} + jm)$

which would maintain the zero totality between fermion and vacuum. The absolute + or – signs attributed to the spin up / down states are, of course, arbitrary, but the relative signs are significant.

We can also identify the state vectors for

Fermion	$(\pm ikE \pm \mathbf{ip} + jm)$
Fermion with reversed spin	$(\pm ikE \mp \mathbf{ip} + jm)$
Antifermion	$(\mp ikE \pm \mathbf{ip} + jm)$
Antifermion with reversed spin	$(\mp ikE \mp \mathbf{ip} + jm)$

In fact, it is unnecessary to give the full set of components to specify the state. Once the lead term is specified, the others (which we may regard as 'drone' terms) automatically follow by sign variation.

6.13 Nonlocality

The Pauli exclusion principle in the form

$$(\pm ikE \pm \mathbf{ip} + jm)(\pm ikE \pm \mathbf{ip} + jm) = 0$$

demands *nonlocality*. In effect, because of the way they are defined, nilpotent operators are specified with respect to the entire quantum field. The indirect Pauli exclusion through the antisymmetric nature of the fermion wavefunction and the product term $-8i\mathbf{p}_1 \times \mathbf{p}_2$ necessarily derives from the same source. It is probable that the relative 'phase' term for $\mathbf{p}$, which ensures that each fermionic spin state is differently oriented, is determined from the angular relations between the orthogonal projections of E, $\mathbf{p}$ and m, thus ensuring that the angular momentum operator $\mathbf{p}$, taking both amplitude and phase, contains the entire information needed to specify the fermionic state.[12] (The conventionally understood phase term, for example the exp $(-i(Et - \mathbf{p.r}))$ for a free particle, is, of course, completely determined by E, $\mathbf{p}$ and m.) In this case, all fermionic $\mathbf{p}$ vectors will intersect instantaneously at all times, leading to the construction of a Euclidean 3-D space, with a 3-dimensionality defined as that of connected objects in space, rather than of the space itself. The construction of this space, which intrinsically defines angular momentum conservation, will then be equivalent to defining the 'mechanism' for ensuring nonlocality, and the uniqueness of the 'phase angles' involved in defining unique 5-fold fermionic states will then be analogous to the uniqueness of the angles involved in defining a 5-fold tiling pattern.

We can consider the nilpotency as defining the interaction between the localized fermionic state and the unlocalized vacuum, with which it is uniquely self-dual. The phase is the mechanism through which this is accomplished. (Lorentzian 'locality', or 'relativity', with c-delayed transmission, can be defined as internal to the brackets defining the nilpotent, while the instantaneous nonlocality is external.) Defining a fermion implies simultaneous definition of vacuum as 'the rest of the universe' with which it interacts. The nilpotent structure then provides energy-momentum conservation without requiring the system to be closed, since the E and $\mathbf{p}$ terms also contain all possible interactions.

The nilpotent structure is thus naturally also *thermodynamic* (as we have seen also in connection with the antisymmetric product term $-8i\mathbf{p}_1 \times \mathbf{p}_2$), and provides a route to a mathematization of nonequilibrium thermodynamics – all systems in this formulation are open systems. Also, the formation of any new state, which is determined by the nature of all other nilpotent states, is a creation event within a unique birth-ordering. Each 'creation' event (which includes any interaction and any change in parameters, as well as entirely new fermionic creations) also necessarily changes all existing states to some degree. In this sense, a nilpotent structure uniquely allows us to conceive of the infinite while only observing the finite.

The principle, of course, applies equally to both free and bound states, as does the nilpotent formalism. In principle, this means that the full Dirac *equation* is not necessary for quantum mechanics, only the *operator* defined by $(\pm ikE \pm i\mathbf{p} + jm)$. This can be considered as a creation operator acting on vacuum. The operator, once specified, uniquely determines the phase which it must act upon to produce a nilpotent amplitude. The powerful calculation method this offers can be seen worked out for various examples in chapter 10. And, as we have seen, even the full operator is not actually necessary – only the lead term, which defines the nature of the other three terms uniquely by automatic sign variation.

Ultimately, the direct use of a nilpotent expression provides a more accurate approach to quantum mechanics than the equating to zero of a differential operator acting on a wavefunction. The latter tacitly assumes that the amplitude of the wavefunction can be treated as a constant, even when it is not, whereas the nilpotent formalism differentiates only the phase. Amplitude becomes a *consequence* of the phase factor, not a source of independent information. Strictly speaking, the wavefunction approach is only true for free particles in the relativistic case, though, in practical examples, there are steps that can be taken to overcome the difficulty. However, it may be that this is the source of problems with the more extensive calculations needed for the multiple interactions of quantum electrodynamics (QED). The direct use of the wavefunction works largely in QED because the particles are *nearly* free, but the fact that, ultimately, they are not means that the problems steadily accumulate as the calculations are extended.

Interestingly, the expression

$$(\pm ikE \pm i\mathbf{p} + jm)(\pm ikE \pm i\mathbf{p} + jm) \to 0$$

or

$$(\pm ikE \pm i\mathbf{p} + jm)(\pm ikE \pm i\mathbf{p} + jm)\phi \to 0$$

where ϕ is an arbitrary scalar factor (phase, etc.), has at least *five* distinct meanings:

classical	special relativity
operator × operator	Klein-Gordon equation
operator × wavefunction	Dirac equation
wavefunction × wavefunction	Pauli exclusion
fermion × vacuum	thermodynamics

In the last interpretation, it is evident that it is not vacuum itself which is 'nothing', but vacuum in combination with the fermion state which produces it.

6.14 BRST Quantization

The Dirac nilpotent operator, being automatically second quantized, already incorporates a full quantum field representation. More conventional approaches to field quantization, however, can be used to demonstrate the relation between charge and energy operators, which the nilpotent formalism requires. Nilpotent operators of a special kind are, in fact, already used in standard quantum field theory, and it will be instructive to make a direct link between these and terms of the form ($\pm$ $\boldsymbol{i}kE$ $\pm$ $\boldsymbol{i}\mathbf{p}$ + $\boldsymbol{j}m$), considered as both energy and charge operators. While this section requires mathematical and physical apparatus which has not yet been fully introduced, the aim, in principle, is to show that the result of these refinements reduces to the properties of the nilpotent as already defined. So it has been included here as an example of the anticipatory power of the nilpotent formalism.

In the standard theory, field quantization requires gauge fixing (or removal of gauge invariance) before propagators can be constructed. The canonical quantization of the electromagnetic field uses Coulomb gauge, but this means that Lorentz invariance must be broken. The path integral approach allows us to use any gauge, and so maintain Lorentz invariance, but the problem now is the introduction of nonphysical or 'fictitious' Fadeev-Popov ghost fields. A version used in string theory (BRST) eliminates the ghost fields by packaging all the information into a single operator, applied to the Lagrangian. (Essentially the same principle applies also in Gupta-Bleuler quantization.) Significantly, the BRST operator (δ_{BRST}) is a nilpotent. This operator can be used to construct a Noether current (J_μ), corresponding to a nilpotent BRST conserved fermionic charge (Q_{BRST}). The condition for defining a physical state then becomes

$$Q_{\mathrm{BRST}} \left|\psi\right\rangle = 0.$$

In the Dirac nilpotent formulation, ($\pm$ $\boldsymbol{i}kE$ $\pm$ $\boldsymbol{i}\mathbf{p}$ + $\boldsymbol{j}m$), which applies only to physical (mass shell) states, is already second quantized, and a nilpotent operator of the form δ_{BRST}. It is, also, a nilpotent *charge* operator of the form Q_{BRST}, but extended to incorporate weak and strong, as well as electromagnetic, charges.[13] It is, finally, in its eigenvalue form, identical to $\left|\psi\right\rangle$. So the three possible meanings for the expression ($\pm$ $\boldsymbol{i}kE$ $\pm$ $\boldsymbol{i}\mathbf{p}$ + $\boldsymbol{j}m$) apply, respectively, to: E and $\mathbf{p}$ interpreted as differential operators in time and space; E, $\mathbf{p}$ and m as coefficients determining the nature of the charges specified by $\boldsymbol{k}$, $\boldsymbol{i}$ and $\boldsymbol{j}$; and E and $\mathbf{p}$ interpreted as eigenvalues of energy and momentum. The nilpotent Dirac operator thus supplies simultaneously all the characteristics which the separate BRST terms δ_{BRST}, Q_{BRST}, and $\left|\psi\right\rangle$ require.

Chapter 7

Nonrelativistic Quantum Mechanics and the Classical Transition

BRIAN KOBERLEIN was a collaborator in section **7.9** of this chapter. Here, a derivation of the bispinor form of the Dirac equation leads to the Schrödinger and Heisenberg approximations, and to a discussion of the dualistic connection between these two main versions of nonrelativistic quantum mechanics. The nature of the quantum-classical transition is then established with some discussion of specific approaches to the classical limit, including a formulation of the Dirac nilpotent in terms of discrete differentiation. The chapter ends with a comparison of the very different ways that duality enters into the idempotent and nilpotent versions of quantum mechanics.

7.1 The Bispinor Form of the Dirac Equation

The bispinor or covariant form of the Dirac equation, with Ψ split into the component spinors ϕ and χ, can also be 'derived' from the classical relativistic conservation of energy equation in the same way as the nilpotent form (cf also 5.6). However, as the classical equation is now identified as a consequence of the original creation of the Dirac state in the nilpotent form, the derivation is ultimately a quantum one. Although the paired equations of the bispinor form do not have the fundamental significance of the single nilpotent form, they are useful in relating the nilpotent form to the more conventional ones, and in the transition to the nonrelativistic approximation. Starting with

$$E^2 - p^2 - m^2 = 0, \tag{7.1}$$

as the eigenvalue of the Dirac equation, and a plane (quantized) wave $e^{-i(Et\,-\,\mathbf{p.r})}$, where A is an arbitrary amplitude, we write

$$(E - m)\,(E + m)\;e^{-i(Et\,-\,\mathbf{p.r})} = pp\;e^{-i(Et\,-\,\mathbf{p.r})}$$

and define paired (scalar) 'wavefunctions'

$$\phi = (E + m)\, e^{-i(Et - \mathbf{p.r})}$$

$$\chi = p\, e^{-i(Et - \mathbf{p.r})}$$

as purely mathematical constructs. Then we have

$$(E - m)\, \phi = p\, \chi \tag{7.2}$$

$$(E + m)\, \chi = p\, \phi \tag{7.3}$$

These equations can also be derived in a more conventional way. We take the Dirac equation in the form

$$(i\boldsymbol{\gamma}.\mathbf{p} + m - \beta E)\, \psi = 0$$

and choose, again without loss of generality, the momentum direction $\mathbf{i}p_x = \mathbf{p}$. Here again, also, E and $\mathbf{p}$ represent the quantum differential operators, rather than their eigenvalues. This time, we make the conventional choices for β:

$$\begin{pmatrix} 1 & 0 & 0 & 0 \\ 0 & 1 & 0 & 0 \\ 0 & 0 & -1 & 0 \\ 0 & 0 & 0 & -1 \end{pmatrix}$$

and for γ^1:

$$\begin{pmatrix} 0 & 0 & 0 & -i \\ 0 & 0 & -i & 0 \\ 0 & i & 0 & 0 \\ i & 0 & 0 & 0 \end{pmatrix}$$

leading to the representation:

$$\begin{pmatrix} E-m & 0 & 0 & -p \\ 0 & E-m & -p & 0 \\ 0 & p & -E-m & 0 \\ p & 0 & 0 & -E-m \end{pmatrix} \begin{pmatrix} \psi_1 \\ \psi_2 \\ \psi_3 \\ \psi_4 \end{pmatrix} = 0 \, .$$

This again reduces to (7.2) and (7.3), with the bispinors given by

$$\phi = \begin{pmatrix} \psi_1 \\ \psi_2 \end{pmatrix}$$

and

$$\chi = \begin{pmatrix} \psi_3 \\ \psi_4 \end{pmatrix}.$$

An alternative factorisation of equation (7.1) allows us to write

$$(E - p)\,(E + p)\,e^{-i(Et - \mathbf{p.r})} = mm\,e^{-i(Et - \mathbf{p.r})}$$

with new wavefunctions:

$$\phi' = (E + p)\,e^{-i(Et - \mathbf{p.r})}$$

$$\chi' = m\,e^{-i(Et - \mathbf{p.r})}$$

and new paired equations:

$$(E - p)\,\phi' = m\,\chi' \qquad (7.4)$$

$$(E + p)\,\chi' = m\,\phi' \qquad (7.5)$$

We can, of course, using the definitions of the wavefunctions, also rewrite (7.2) and (7.3), and (7.4) and (7.5), as differential equations. Then we have:

$$(\partial/\partial t - m)\,\phi = \nabla\chi \qquad (7.6)$$

$$(\partial/\partial t + m)\,\chi = \nabla\phi \qquad (7.7)$$

and

$$(\partial/\partial t - \nabla)\,\phi' = m\,\chi' \qquad (7.8)$$

$$(\partial/\partial t + \nabla)\,\chi' = m\,\phi' \qquad (7.9)$$

So far, we have introduced no new physics, but we have, in equations (7.6) and (7.7), and (7.8) and (7.9), recognisable versions of the bispinor form of the Dirac equation. These equations, as they stand, are purely mathematical variations on the conservation of energy equation. It is only when we seek a 'physical' meaning for the mathematical wavefunctions that they will introduce any new physical content. To create a single equation, we reduce ϕ and χ, or ϕ' and χ', to a single wavefunction, using quaternionic factors, but we can also derive quaternionic versions of the paired equations. Here, we apply the plane wave solution to non-nilpotent, quaternionic versions of the Dirac equation, which explicitly allow two mass states, such as:

$$\left(-i\boldsymbol{i}\frac{\partial}{\partial t} + \boldsymbol{k}\nabla + im\right)\psi = 0$$

and

$$\left(-i\boldsymbol{i}\frac{\partial}{\partial t} + \boldsymbol{k}\nabla - im\right)\psi = 0$$

to obtain

$$(-\boldsymbol{i}\,E - i\,m)\,\phi = i\boldsymbol{k}\,p\,\chi$$

and

$$(-\boldsymbol{i}\,E + i\,m)\,\chi = i\boldsymbol{k}\,p\,\phi.$$

With χ now defined as $(\boldsymbol{k}\, p \,/\, (\boldsymbol{i}i\, E + m))\, \phi$, and ϕ as $(\boldsymbol{k}\, p \,/\, (\boldsymbol{i}i\, E - m))\, \chi$, as in the $p \,/\, (E + m)$ of the conventional treatment, the quaternion and scalar parts of the operator, $-\boldsymbol{i}i\, \partial / \partial t + \boldsymbol{k}\ \nabla$ and $\pm\, im$, are taken to apply to different wavefunctions.

When $E \gg m$ or $m \approx 0$, as in the case of neutrinos, then $E = p$, where $p = |\mathbf{p}|$, and we can write a quaternionic version of equation (7.5) in the form

$$(\boldsymbol{k}\, E + \boldsymbol{i}i\, p)\, \chi = 0,$$

or, since the sign of i is arbitrary,

$$(\boldsymbol{k}\, E - \boldsymbol{i}i\, p)\, \phi = 0,$$

where the wavefunctions χ and ϕ contain left- and right-handed spinors, u_L and u_R, of the form

$$u_L = (\boldsymbol{k}\, E - \boldsymbol{i}i\, p) = E\, (\boldsymbol{k} - \boldsymbol{i}i)$$

$$u_R = (\boldsymbol{k}\, E + \boldsymbol{i}i\, p) = E\, (\boldsymbol{k} + \boldsymbol{i}i).$$

Since $\gamma_5 = \boldsymbol{i}\boldsymbol{j}$, then

$$\gamma_5\, u_R = \boldsymbol{i}\boldsymbol{j}\, E\, (\boldsymbol{k} + \boldsymbol{i}i) = E\, (\boldsymbol{i}i + \boldsymbol{k}) = u_R,$$

and

$$\left(\frac{1-\gamma_5}{2}\right) u_R = 0.$$

In the left-handed case,

$$\gamma_5\, u_L = \boldsymbol{i}\boldsymbol{j}\, E\, (\boldsymbol{k} - \boldsymbol{i}i) = E\, (\boldsymbol{i}i - \boldsymbol{k}) = -u_L,$$

and

$$\left(\frac{1-\gamma_5}{2}\right) u_L = u_L,$$

as expected.

The factorisation of the squared terms in the conservation of energy equation allows several choices in the use of square roots of 1 and –1, but the process of adding the component wavefunctions ϕ' and χ' to produce the total wavefunction ψ would not be possible without using a noncommutative (quaternion) algebra to change the relative sign of p or m with respect to E in the paired equations (7.2) and (7.3), or (7.4) and (7.5), and this has the added bonus of applying the same algebra, on a symmetrical basis, to each of E, p and m.

7.2 The Schrödinger Approximation

The nonrelativistic Schrödinger approximation allows a relatively easy calculation of observable expectation values of quantities such as position and momentum in the 'collapse' of the wavefunction onto a particular eigenstate, which occurs, in the Copenhagen interpretation of quantum mechanics, when a quantum system confronts a classical measuring apparatus (see section 9.3). So, superposition of a large number of infinitely long waves, $Ae^{-i(Et\,-\,\mathbf{p.r})}$, with the same amplitude but different wave numbers k_x gives a maximum amplitude at the point where all the waves are in phase and thus achieves a degree of localisation, giving meaning to the space parameter (and providing an analogue to the sum of classical field energies of the form q^2 / r for all r centred at a point where a charge would be expected). Using the conjugate symmetry between coordinates and momenta, this takes the form of a Fourier transformation, as in quantum field theory, so that:

$$\psi(\mathbf{r},t) = \int_{-\infty}^{\infty} \phi(\mathbf{p}) e^{-i(Et-\mathbf{p.r})} d^3\mathbf{p} ,$$

where
$$\phi(\mathbf{p}) = \int_{-\infty}^{\infty} \psi(\mathbf{r},0) e^{-i\mathbf{p.r}} d^3\mathbf{r} .$$

Regular mathematical procedures also allow us to transform from the Schrödinger to the Heisenberg formulation, as we will see in the next section. The derivation of the Schrödinger equation requires only the coupled equations (7.2) and (7.3). Then, assuming the non-relativistic approximation $E \approx m$, for low $\mathbf{p}$, we obtain

$$\chi \approx \frac{p}{2m} \phi$$

from (7.3), and
$$(E - m)\phi = \frac{p^2}{2m} \phi , \tag{7.10}$$

by substituting for χ into (7.2). Using the same approximation, ϕ, here, also becomes ψ, and is a scalar in the same way as the wavefunctions used in (7.2) and (7.3).

Conventionally, of course, the Schrödinger equation excludes the rest mass energy m from the total energy term E, and so becomes:

$$E\psi = \frac{p^2}{2m} \psi ,$$

where $E = \mathcal{H}$ and p are, of course, the usual quantum operators, $i\hbar\,\partial/\partial t$ and $-i\hbar\nabla$. In the presence of a potential energy V, this is modified to

$$(E - V)\psi = \frac{p^2}{2m} \psi .$$

Using the operator $E = \mathcal{H}$ to specify the time-evolution of the system described by a time-dependent ket wavefunction, $|\Psi(t)\rangle$, we can write down an equation of motion in the form

$$i\hbar \frac{\partial}{\partial t}|\Psi(t)\rangle = \mathcal{H}|\Psi(t)\rangle .$$

Significantly, we may show, following earlier treatments,[1] that the Schrödinger equation can be used to derive the anomalous magnetic moment of the electron in the presence of a magnetic field **B** (see 8.8 for classical magnetic fields). Spin, of course, is purely a property of the multivariate nature of the **p** term, and has nothing to do with whether the equation used is relativistic or not. In our operator notation, the Schrödinger equation, whether field-free or in the presence of a field with vector potential **A**, can be written in the form,

$$2mE\psi = \mathbf{p}^2\psi$$

Using a multivariate, $\mathbf{p} = -i\nabla + e\mathbf{A}$, for the covariant derivative, we derive:

$$\begin{aligned}
2mE\psi &= (-i\nabla + e\mathbf{A})\,(-i\nabla + e\mathbf{A})\;\psi \\
&= (-i\nabla + e\mathbf{A})\,(-i\nabla\psi + e\mathbf{A}\,\psi) \\
&= -\nabla^2\psi - ie\,(\nabla.\psi\mathbf{A} + i\nabla\psi\times\mathbf{A} + \mathbf{A}.\nabla\psi + i\mathbf{A}\times\nabla\psi) + e^2\mathbf{A}^2\psi \\
&= -\nabla^2\psi - ie\,(\nabla.\psi\mathbf{A} + 2\mathbf{A}.\nabla\psi + i\psi\nabla\times\mathbf{A}) + e^2\mathbf{A}^2\psi \\
&= -\nabla^2\psi - ie\,(\psi\nabla.\mathbf{A} + 2\mathbf{A}.\nabla\psi) + e^2\mathbf{A}^2\psi + e\mathbf{B}\psi \\
&= (-i\nabla + e\mathbf{A}).(-i\nabla + e\mathbf{A})\;\psi + e\mathbf{B}\psi \\
&= (-i\nabla + e\mathbf{A}).(-i\nabla + e\mathbf{A})\;\psi + 2m\,\boldsymbol{\mu}.\mathbf{B}\;\psi.
\end{aligned}$$

This is the conventional form of the Schrödinger equation in a magnetic field for spin up. The wavefunction can be either scalar or nilpotent. Reversing the (relative) sign of $e\mathbf{A}$ for spin down, we obtain

$$\begin{aligned}
2mE\psi &= (-i\nabla - e\mathbf{A})\,(-i\nabla - e\mathbf{A})\;\psi \\
&= (-i\nabla - e\mathbf{A})\,(-i\nabla - e\mathbf{A})\;\psi - 2m\,\boldsymbol{\mu}.\mathbf{B}\;\psi.
\end{aligned}$$

It is significant that the standard derivation of the Schrödinger equation begins with the classical expression for kinetic energy $p^2 / 2m$, and that the factor 2 in this equation ultimately carries over into the same factor in the spin term for the electron. It is precisely because the Schrödinger equation is derived via a kinetic energy term that this factor enters into the expression for the spin, and this process is essentially the same as the process which, through the anticommuting quantities of the Dirac equation, makes $(\mathbf{L} + \hat{\boldsymbol{\sigma}} / 2)$ a constant of the motion.

Anticommuting operators also introduce the factor 2 in the Heisenberg uncertainty relation for the same reason, the Heisenberg term relating directly to the zero-point energy derived from the kinetic energy of the harmonic oscillator. In fact, as we will show in chapter 16, the origin of the factor 2, in all significant cases – classical, quantum, relativistic – is in the virial relation between kinetic and potential energies. In principle, the kinetic energy relation is used when we consider a particle as an object in itself, described by a rest mass m_0, undergoing a continuous change. The potential energy relation is used when we consider a particle within its 'environment', with 'relativistic mass', in an equilibrium state requiring a discrete transition for any change. We can consider the kinetic energy relation to be concerned with the action side of Newton's third law, while the potential energy relation concerns both action and reaction. Because of the necessary relation between them, each of these approaches is a proper and complete expression of the conservation of energy. This fundamental relation, as we saw in section 6.11, leads to the significant fact that the nilpotent wavefunctions, in principle, produce a kind of supersymmetry, with the supersymmetric partners not being so much realisable particles, as the couplings of the fermions and bosons to vacuum states.

7.3 The Heisenberg Formulation of Quantum Mechanics

The alternative version of nonrelativistic quantum mechanics due to Heisenberg can only be related to the Schrödinger formulation at the point of observation. In the Schrödinger formulation, the average or expectation value of a set of repeated observations on a system, defined by the operator $\hat{A}(x, \partial / \partial x)$, each of which observations is in an arbitrary state $\psi(x)$, is given by

$$\bar{\alpha}_\psi = \frac{\int_{-\infty}^{\infty} \psi^* \hat{A} \psi dx}{\int_{-\infty}^{\infty} \psi^* \psi dx} = \langle \psi | \hat{A} | \psi \rangle .$$

If each $\hat{A}(x, \partial / \partial x)$ is made at time t on an assembly of systems all in the same state $|\psi(t)\rangle$, then

$$\bar{\alpha}_{\psi(t)} = \langle \psi(t) | \hat{A} | \psi(t) \rangle ,$$

which can also be written

$$\bar{\alpha}_{\psi(t)} = \langle \psi | e^{iEt} \hat{A} e^{-iEt} | \psi(t) \rangle = \langle \psi | e i^{\mathcal{H}} \hat{A} \; e^{-\mathcal{H}} | \psi \rangle .$$

If we now define a time-dependent operator $\hat{A}(t) = e i^{\mathcal{H}} \hat{A} \; e^{-\mathcal{H}}$, then the average value for this operator, for a state $|\psi\rangle$ specified at $t = 0$, becomes

$$\bar{a}(t)_{\psi(t)} = \langle\psi|\hat{A}(t)|\psi\rangle = \langle\psi|\, ei^{\mathcal{H}}\,\hat{A}\; e^{-\mathcal{H}}|\psi\rangle$$

and we have converted our formulation to a Heisenberg picture.

We can also write an equation of motion by differentiating $\hat{A}(t)$ with respect to t, and writing $\hat{\mathcal{H}}(t) = ei^{\mathcal{H}}\,\hat{\mathcal{H}}e^{-\mathcal{H}}$ as $\hat{\mathcal{H}}$. We then obtain:

$$i\hbar\frac{d\hat{A}(t)}{dt} = \hat{\mathcal{H}}\hat{A}(t) + \hat{A}(t)\hat{\mathcal{H}} = [\,\hat{A}(t), \hat{\mathcal{H}}].$$

It is through this equation of motion that the Heisenberg formulation, like that of Schrödinger, allows a relatively easy transition to classical nonrelativistic physics.

7.4 Heisenberg v. Schrödinger

The duality between the two main systems of nonrelativistic quantum mechanics is of special interest. Neither Heisenberg nor Schrödinger gives a complete description of reality, and each requires completion with an *ad hoc* process of measurement. The first selects all the discrete options, while the second chooses all the continuous ones. In each case, the process of measurement is used to restore the true attributes of fundamental parameters lost in the definition of the system. Thus, while the Heisenberg theory is discrete and directly based on observables, the Schrödinger formalism is based on a continuous, and therefore unmeasurable, interpretation of space and time, in the concept of the wavefunction; particles such as electrons are delocalised and spread throughout space and time. Time, in this formalism, has no status as an observable quantity; information is derived from the wavefunction only by the application of momentum and position operators.

The two theories represent opposite extremes in both their definitions of the system and of measurement. The expressions in bold type in the diagram following represent the violations of fundamental conditions within each system, which must be corrected by the respective processes of measurement. The introduction in these measurement processes of a virtual version of what each theory excludes in the system effectively links uncertainty to duality. The Schrödinger measurement process, for example, (wavefunction collapse) allows us to restore particles in discrete space (not contained within the system), but only at the expense of knowledge of the wavelengths of the system. The price of

Heisenberg measurement, on the other hand, is the loss of the causality, which the system had retained; measurement brings in nonlocality and the vacuum.

Schrödinger's wave mechanics

	The System		Measurement	
continuous	**space**	**virtual**	restores introduces	
	charge	**particles**	discreteness	localised
	momentum		of these	particles
	angular momentum			
	time	real	not changed	
	mass	vacuum	by measurement	
	energy			

Heisenberg's quantum mechanics

	The System		Measurement	
discrete	space	real	not changed	
	charge	particles	by measurement	
	momentum			
	angular momentum			
	time	**virtual**	restores	introduces
	mass	**vacuum**	continuity	nonlocalised
	energy		of these	vacuum

Wavefunction collapse is outside of the Schrödinger *equation*, because the equation itself doesn't give the true full information, as it makes space continuous. Neither, of course, does the Heisenberg formalism, because this makes time discrete! In principle, the continuous Schrödinger wavefunction allows no direct knowledge of time or position, and so denies causality of the discrete kind required by Einstein, until a virtual causality is introduced by measurement. Thus, the interpretation suggested by Born explains the squared wavefunction (squared, of course, because it links space with time in a process of measurement) as a probability amplitude when *we* introduce a particle-like discontinuity in the *ad hoc* process of 'measurement', or collapse of the wavefunction. But the unobservable status of time in Schrödinger's theory (which remains even when position becomes observable with wavefunction collapse) does not carry over into the *alternative* Heisenberg formulation, where time is assumed to have a discrete structure, like space, and so can be brought into a meaningful uncertainty relation with energy.

A great deal of confusion has been generated by the fact that the Heisenberg and Schrödinger theories give the same basic results in application, for their physical assumptions are nonetheless incompatible, and the axioms of one cannot be used to comment on those of the other. At no time are the *correct* physical assumptions of either theory altered in the process of measurement. The Schrödinger theory, for instance, correctly assumes continuous time within the system, and, therefore, leaves time continuous in applying the process of measurement. The discrete time involved in Heisenberg uncertainty thus has no meaning in wave mechanics, and cannot be carried over from the uncertainty principle into the Schrödinger theory. Its absence from that theory is thus the cause of no philosophical difficulty whatsoever.

7.5 The Quantum-Classical Transition

The Heisenberg and Schrödinger formulations only really meet at the point of measurement, in effect at the point where quantum physics becomes classical. Although we need these formulations to effect a relatively easy transition to nonrelativistic classical physics, we need to return to the Dirac equation to fully understand what is really meant by the quantum-classical transition, and we cannot fully comprehend this subject without also referring to the laws of classical thermodynamics.

The nilpotent Dirac equation contains all the information needed for the entire structure of physics. Everything else is a derivation from it, or an approximation to it. Both the equation and the physical parameters it conjures into existence can be derived initially purely from the algebra which results from the requirement of a zero totality, without reference to physical assumptions of any kind, such as the concept of measurement, special relativity or an asymptotic approach to classical conditions (the Correspondence Principle). We simply define the Dirac state as a conserved square root of zero, requiring the application of its nonconserved (and identical) partner to return it to the zero state from which it came, and, as a result of the algebra, this state is seen as a unique member of an infinite series of such states, all entangled with each other, and incorporating charges and masses subject to universal interactions. Quantum systems, however, are defined, in the first instance, as isolated, and it becomes necessary to discover how the complexity that results affects the physics, and indeed pushes it in the direction of classicality and observability. It will be convenient, therefore, to establish both the meaning of the quantum-classical transition, and to establish the way that the more fundamental quantum approach puts forward requirements that determine the nature of classical physics.

Classical physics differs from quantum physics in being concerned with *measurement*. Measurement processes are discrete, and involve discrete sources (or charges). They rely on the $SU(3) \times SU(2)_L \times U(1)$ symmetries which apply to these sources (and whose direct expressions are 'interactions', equivalent in principle to the action of classical field terms) producing restrictions on the freedom of the individual wavefunctions to contain infinitely possible variations in space and time coordinates. As soon as a measurement is made, a system necessarily becomes at least semi-classical. In fact, we could say that, as soon as any interaction occurs to restrict the absolute freedom of space and time variation of a fermionic state, a quantum system is already on the way to becoming classical.

A hypothetically isolated system (e.g. a hydrogen atom not interacting with other hydrogen atoms) must be purely quantum. Once we have any classical element or interaction the system is no longer isolated. This is how we make a measurement. We can't make classical-type observations on an isolated system, otherwise it wouldn't be isolated. An isolated system conserves the E-$\mathbf{p}$-m state within the system, linking it with the total $\boldsymbol{k}, \boldsymbol{i}, \boldsymbol{j}$ charge values, whether 0 or unit, positive or negative. This system must remain coherent – with angular momentum operators aligned, so that addition is effectively scalar, like that of the charge units. If the system interacts with an external system (however small the interaction), then it can no longer be defined in an isolated way: the connection between the conservation laws for charge ($\boldsymbol{k}$, $\boldsymbol{i}$, $\boldsymbol{j}$) and angular momentum (E-$\mathbf{p}$-m) is broken, due to decoherence in the vector terms. Some of the 'phase space' is lost; the decoherence will also be retarded because of the unidirectionality of time.

A key result in this context is the demonstration that there are no microcanonical ensembles in nature.[2] In effect, this means that there is no such thing as a truly isolated system, in particular an isolated quantum system. Every hydrogen atom is attempting, as it were, to be part of a hydrogen molecule with some other hydrogen atom. In fact, for all the apparent similarity between the energy levels applicable to any two given hydrogen atoms, there must always be some subtle difference due to the particular interactions of each and to Pauli exclusion. Every hydrogen atom necessarily has a unique set of energy levels. It is simply an approximation to assume that there is such a thing as a set of universal energy levels applicable to hydrogen. The exactness of quantum transitions as countable units is an illusion (just as is the idea that 'rest mass' – a quantity that is never observed – actually quantizes mass). Quantization is not a process which provides exact counting units. Furthermore, the set of energy levels applicable to any given hydrogen atom must be constantly changing. Each nilpotent structure is unique but never fixed. This is what allows us to imagine an absolute order for events, a 'cosmic' or universal time irrespective of our systems of time measurement.

If the system is not isolated, then energy is not conserved *within* it, but some is lost to the 'rest of the universe' with which it interacts. Hence, we need the second law of thermodynamics, which may be seen as a necessary consequence of the entanglement of all the universe's state vectors and the universality of interactions (cf. 6.4), and, in fact, the first law, where the energy balance is only maintained globally by incorporating 'lost' energy into the equation (cf 9.6). It is also apparent why this process is our only way of apprehending the direction of time, and why the direction of time is connected with the second law of thermodynamics. Measurement is an interaction process with the rest of the universe (i.e. vacuum) – causality requires this. An interacting system is no longer canonical. Energy lost to the rest of the universe leads to decoherence. In interaction, there is decoherence of the **p** term and angular momentum loses some of its 'phase space'.

The sequence of events is irreversible precisely because it is unique. To make a measurement requires a semi-classical situation with a non-isolated system; as soon as we make a measurement, we lose energy from the system to the 'rest of the universe', so increasing the 'entropy' (or measure of decoherence).[3] The sequence of events behaves as an irreversible sequence because *continuous* time itself is itself irreversible, and a sequence of event 'measurements' must follow the same sequence; but this for any known pair of events will always require the increase in entropy that results from irreversible change.

7.6 The Classical Limit

The Dirac nilpotent gives us a number of ready-made composite parameters, in addition to space, time, mass and charge. Energy (E) and momentum (**p**) are defined automatically as fundamental quantities, while angular momentum follows on from quantum mechanics and Noether's theorem, or the symmetry of the parameter group. Quantum mechanics becomes classical physics when eigenvalues related to these quantities become observables. We can then obtain classical relativistic equations such as $E^2 - p^2 - m^2 = 0$ and classical nonrelativistic equations such as $E = p^2 / 2m$ without further derivation. Nonrelativistic classical mechanics, in its Hamiltonian form, provides a ready-made route for a quantum-classical transition from either the Schrödinger or the Heisenberg formulations of quantum mechanics. The Heisenberg equations of motion, for example, become identical with the classical ones if the Heisenberg operators are replaced by classical canonical variables.

Thus, if we apply the general expression we obtained for an operator in section 7.3 to the position and momentum operators $\hat{q}(t)$ and $\hat{p}(t)$, we obtain:

$$i\hbar \frac{d\hat{q}(t)}{dt} = \hat{\mathcal{H}}\hat{q}(t) + \hat{q}(t)\hat{\mathcal{H}} = [\hat{q}(t), \hat{\mathcal{H}}].$$

and

$$i\hbar \frac{d\hat{p}(t)}{dt} = \hat{\mathcal{H}}\hat{p}(t) + \hat{p}(t)\hat{\mathcal{H}} = [\hat{p}(t), \hat{\mathcal{H}}].$$

It is also consistent with the commutation relations (see also 7.7) to *define* $i\hbar\partial\hat{\mathcal{H}} / \partial\hat{p}$ as $= [\hat{q}, \hat{\mathcal{H}}]$ and $i\hbar\partial\hat{\mathcal{H}} / \partial\hat{q}$ as $= [\hat{p}, \hat{\mathcal{H}}]$, so we obtain equations of motion, which are of exactly the same form as Hamilton's classical equations (for which, see 8.7):

$$\frac{\partial\hat{q}}{\partial t} = \frac{\partial\hat{\mathcal{H}}}{\partial\hat{p}} \quad \text{and} \quad \frac{\partial\hat{p}}{\partial t} = -\frac{\partial\hat{\mathcal{H}}}{\partial\hat{q}}.$$

We can use either the Heisenberg or the Hamiltonian equations with $\mathbf{r}$ for q, to define a velocity operator, which, for a free particle, becomes:

$$\mathbf{v} = \dot{\mathbf{r}} = \frac{d\mathbf{r}}{dt} = \frac{1}{i\hbar}[\mathbf{r}, \mathcal{H}] = -ij\mathbf{1}c = c\boldsymbol{\alpha}.$$

With the free particle Hamiltonian given (in this notation) by $\mathcal{H} = -ijc\mathbf{1p} + ikmc^2 = c\boldsymbol{\alpha}.\mathbf{p} + \beta mc^2$, we can also write an equation of motion for the operator $-ij\mathbf{1} = \boldsymbol{\alpha}$, as a function of t:

$$\frac{d\boldsymbol{\alpha}}{dt} = \frac{1}{i\hbar}[\boldsymbol{\alpha}, \mathcal{H}] = \frac{2}{i\hbar}(c\mathbf{p} - \mathcal{H}\boldsymbol{\alpha}).$$

Since $\mathcal{H}$ is a constant, this yields the solution:

$$\boldsymbol{\alpha}(t) = \frac{\mathbf{v}(t)}{c} = \frac{\dot{\mathbf{r}}(t)}{c} = c\mathcal{H}^{-1}\mathbf{p} + [\boldsymbol{\alpha}(0) - c\mathcal{H}^{-1}\mathbf{p}]\, exp\,(2i\mathcal{H}t / h).$$

This, in turn, can be solved, to give the equation of motion for a free fermion:

$$\mathbf{r}(t) = \mathbf{r}(0) + \frac{c^2\mathbf{p}}{\mathcal{H}} t + \frac{\hbar c}{2i\mathcal{H}} [\alpha(0) - c\mathcal{H}^{-1}\mathbf{p}](exp\,(2i\mathcal{H}t / h) - 1).$$

The first term of this solution represents the initial position vector and the second term represents the displacement at time t. The third term, however, has no classical analogue, and represents a violent oscillatory motion or high-frequency vibration (*zitterbewegung*) of the particle at frequency $\approx mc^2 / \hbar$, and amplitude $\hbar / mc$, which is the Compton wavelength for the particle.

In the case of the Schrödinger equation, there is the intriguing example of the Bohmian approach, which has been notably expounded and extended by Basil Hiley.[4] Here, we express the wavefunction in polar form

$$\Psi(r, t) = R(r, t)\, e^{S(r,\, t)},$$

and separate real and imaginary parts. The imaginary part of the Schrödinger equation gives a conservation of probability equation:

$$\frac{\partial R^2}{\partial t} + \nabla . \frac{R^2 \nabla S}{m} = 0,$$

whilc the real part gives:

$$\frac{\partial S}{\partial t} + \frac{(\nabla S)^2}{2m} - \frac{\nabla^2 R}{2mR} + V = 0,$$

which, with the third term (the 'quantum potential', $Q = -\nabla^2 R / 2mR$) equated to zero, and S taken as the classical term, 'action', becomes the classical Hamilton-Jacobi equation. In fact, since, classically, by definition, $\partial S / \partial t = -E$ and $\nabla S = \mathbf{p}$, the equation becomes the classical conservation of energy equation:

$$E = \frac{p^2}{2m} + V .$$

By contrast with the conventional Schrödinger theory, Bohmian quantum mechanics, which Hiley has also derived from the Heisenberg formalism, assumes that real particle trajectories exist, and that uncertainty is introduced at the level of observation. According to our reasoning, however, it is the *concept* of observation (i.e. a hypothetical 'certainty' with respect to the observer) which is problematic in physics, and so it is not really significant whether uncertainty is considered to be introduced with the 'system' or with the observer.

7.7 The Dirac Nilpotent Using Discrete Differentiation

Another quantum mechanics formalism can be defined using a discrete version of differential calculus investigated by Lou Kauffman;[5-6] and it can be made relativistic by applying it to the nilpotent Dirac equation. Suppose we use the symbol E for $\mathcal{H}$, and define

$$F = \psi = \boldsymbol{ik}E + \boldsymbol{i}\mathbf{i}P_1 + \boldsymbol{i}\mathbf{j}P_2 + \boldsymbol{i}\mathbf{k}P_3 + \boldsymbol{j}m, \tag{7.11}$$

where, as usual, bold symbols represent multivariate vectors and bold italic symbols represent quaternions. Algebraically,

$$\begin{aligned} -\boldsymbol{ik}\ \psi\, \boldsymbol{ik} &= \psi - 2\,\boldsymbol{ik}E \\ \boldsymbol{i}\mathbf{i}\ \psi\, \boldsymbol{i}\mathbf{i} &= \psi - 2\,\boldsymbol{i}\mathbf{i}P_1 \\ \boldsymbol{i}\mathbf{j}\ \psi\, \boldsymbol{i}\mathbf{j} &= \psi - 2\,\boldsymbol{i}\mathbf{j}P_2 \\ \boldsymbol{i}\mathbf{k}\ \psi\, \boldsymbol{i}\mathbf{k} &= \psi - 2\,\boldsymbol{i}\mathbf{k}P_3 \\ \boldsymbol{j}\ \psi\, \boldsymbol{j} &= \psi - 2\boldsymbol{j}m. \end{aligned}$$

We now define the operator

$$\mathcal{D} = -\boldsymbol{k}\frac{\partial}{\partial t} - \boldsymbol{i}i\mathbf{i}\frac{\partial}{\partial X_1} - \boldsymbol{i}i\mathbf{j}\frac{\partial}{\partial X_2} - \boldsymbol{i}i\mathbf{k}\frac{\partial}{\partial X_3},$$

with $$i\frac{\partial F}{\partial t} = [F, H] = [F, E] \quad \text{and} \quad -i\frac{\partial F}{\partial X_i} = [F, P_i],$$

in the discrete calculus. (We can use $\partial F / \partial t$ rather than dF / dt, here, because we are making no explicit use of a velocity variable.) The negative operators are those which produce amplitude (7.11) in the continuum version. (The negative signs can be removed if we use a complexified version of (7.11).) This means that

$$\begin{aligned} -\boldsymbol{k}\frac{\partial \psi}{\partial t} &= \boldsymbol{ik}[\psi, E] = \boldsymbol{ik}\,\psi E - \boldsymbol{ik}E\psi \\ &= \boldsymbol{ik}\,\psi \boldsymbol{ikik}E - \boldsymbol{ik}E\psi \\ &= -\psi \boldsymbol{ik}E - \boldsymbol{ik}E\psi + 2\boldsymbol{ik}E\boldsymbol{ik}E \\ &= -\psi \boldsymbol{ik}E - \boldsymbol{ik}E\psi + 2E^2. \end{aligned} \tag{7.12}$$

$$\begin{aligned} -\boldsymbol{i}i\mathbf{i}\frac{\partial \psi}{\partial X_1} &= \boldsymbol{i}\mathbf{i}[\psi, P_1] = \boldsymbol{i}\mathbf{i}\,\psi P_1 - \boldsymbol{i}\mathbf{i}P_1\psi = -\boldsymbol{i}\mathbf{i}\,\psi \boldsymbol{i}\mathbf{i}\boldsymbol{i}\mathbf{i}P_1 - \boldsymbol{i}\mathbf{i}P_1\psi \\ &= -\psi \boldsymbol{i}\mathbf{i}P_1 - \boldsymbol{i}\mathbf{i}P_1\psi + 2\boldsymbol{i}\mathbf{i}P_1\boldsymbol{i}\mathbf{i}P_1 = -\psi \boldsymbol{i}\mathbf{i}P_1 - \boldsymbol{i}\mathbf{i}P_1\psi - 2P_1^2 \end{aligned} \tag{7.13}$$

$$-i\boldsymbol{i}\frac{\partial\psi}{\partial X_2} = \boldsymbol{i}\mathbf{i}[\psi, P_2] = \boldsymbol{i}\mathbf{i}\,\psi P_2 - \boldsymbol{i}\mathbf{i}P_2\psi = -\boldsymbol{i}\mathbf{i}\,\psi\boldsymbol{i}\mathbf{i}\boldsymbol{i}\mathbf{i}P_2 - \boldsymbol{i}\mathbf{i}P_2\psi$$

$$= -\psi\boldsymbol{i}\mathbf{i}P_2 - \boldsymbol{i}\mathbf{i}P_2\psi + 2\boldsymbol{i}\mathbf{i}P_2\boldsymbol{i}\mathbf{i}P_2 = -\psi\boldsymbol{i}\mathbf{i}P_2 - \boldsymbol{i}\mathbf{i}P_2\psi - 2P_2^2 \tag{7.14}$$

$$-i\boldsymbol{i}\frac{\partial\psi}{\partial X_3} = \boldsymbol{i}\mathbf{i}[\psi, P_3] = \boldsymbol{i}\mathbf{i}\,\psi P_3 - \boldsymbol{i}\mathbf{i}P_3\psi = -\boldsymbol{i}\mathbf{i}\,\psi\boldsymbol{i}\mathbf{i}\boldsymbol{i}\mathbf{i}P_3 - \boldsymbol{i}\mathbf{i}P_3\psi$$

$$= -\psi\boldsymbol{i}\mathbf{i}P_3 - \boldsymbol{i}\mathbf{i}P_3\psi + 2\boldsymbol{i}\mathbf{i}P_3\boldsymbol{i}\mathbf{i}P_3 = -\psi\boldsymbol{i}\mathbf{i}P_3 - \boldsymbol{i}\mathbf{i}P_3\psi - 2P_3^2 \tag{7.15}$$

Now, if m is a scalar, we may use the identity

$$0 \equiv \boldsymbol{j}\,\psi m - \boldsymbol{j}m\,\psi = -\boldsymbol{j}\,\psi\boldsymbol{j}\boldsymbol{j}m - \boldsymbol{j}m\,\psi$$

$$= -\psi\boldsymbol{j}m - \boldsymbol{j}m\,\psi - 2\boldsymbol{j}m\boldsymbol{j}m = -\psi\boldsymbol{j}m - \boldsymbol{j}m\,\psi - 2m^2 . \tag{7.16}$$

Combining equations (7.12)-(7.16), term by term, we obtain

$$\not{D}\,\psi = -\psi(i\boldsymbol{k}E + \boldsymbol{i}\mathbf{i}P_1 + \boldsymbol{i}\mathbf{j}P_2 + \boldsymbol{i}\mathbf{k}P_3 + \boldsymbol{j}m)$$

$$-(i\boldsymbol{k}E + \boldsymbol{i}\mathbf{i}P_1 + \boldsymbol{i}\mathbf{j}P_2 + \boldsymbol{i}\mathbf{k}P_3 + \boldsymbol{j}m)\psi + 2(E^2 - P_1^2 - P_2^2 - P_3^2 - m^2).$$

When ψ is nilpotent, then

$$\not{D}\,\psi = 0.$$

This can also be written

$$\left(\boldsymbol{k}\frac{\partial}{\partial t} + i\boldsymbol{i}\nabla\right)\psi = 0 \tag{7.17}$$

or, in fuller form:

$$\left(\pm\boldsymbol{k}\frac{\partial}{\partial t} \pm i\boldsymbol{i}\nabla\right)\psi = 0,$$

where

$$\psi = \pm\, i\boldsymbol{k}E \pm \boldsymbol{i}\mathbf{i}P_1 \pm \boldsymbol{i}\mathbf{j}P_2 \pm \boldsymbol{i}\mathbf{k}P_3 + \boldsymbol{j}m.$$

This is the discrete form of the nilpotent Dirac equation. Significantly, the operator does not require a mass term (which is redundant when we have exact knowledge of E and P), although, for reasons of symmetry, we could define a mass operator $\boldsymbol{j}[\psi, m] \equiv 0$. Also not specifically required is the use of i (or $\hbar$) in defining $\partial F / \partial t$ and $\partial F / \partial X_i$, which means that a smooth transition from classical to quantum conditions may be effected. These scalar coefficients emerge only in the relativistic version of quantum mechanics, where the scaling relations between E, $\mathbf{p}$ and m, and the pseudoscalar nature of the energy operator, are necessarily fixed at the same time.

Kauffman's direct derivations from the discrete calculus include the Schrödinger equation, the diffusion equation, Hamilton's equations as 'a mathematical pattern independent of physics', and gauge field curvature as an immediate 'tautology of the mathematical formalism'; in particular, the Levi-Civita affine connection (the 'covariant derivative' which preserves the Riemannian metric in general relativity, and removes the need for inertia) becomes a consequence of the Jacobi identity. The fact that Hamilton's equations seemingly require no physical input might be taken to be a consequence of the fact that, in the discrete differentiation process (in the language of chapter 2), nonconservation (differentiation) is directly defined in relation to an abstract idea of conservation (represented by $\mathcal{H}$). Also, the fact that the creation of a discrete calculus (or a discrete sequence of events in the process of change) requires anticommutativity seems to connect with the idea of anticommutativity as the origin of discreteness in chapter 1.

In relation to the relativistic calculation in this section, with only one well-defined direction for spin, the minimalist representation of $\mathcal{D}$ in equation (7.17) would seem to be in accordance with the holographic principle, where a bounding 'area' contains all the information relevant to a canonically defined 'system' (see chapter 21). It is significant also that phase plays no part in this discrete form of relativistic quantum mechanics, the 'wavefunction' ψ being specified by the amplitude alone. In fact, equation (7.17) suggests that $\partial F / \partial t$ and ∇ can be regarded as *equivalent* pieces of information as, say, respectively, phase and amplitude, rather than independent ones, which means that the (angular) momentum operator ∇ (in the nilpotent formalism) can be structured to carry all the available information about a fermionic state.

Of particular interest is the fact that the discrete form of a creation operator, defined by $\mathcal{D}$, has a corresponding annihilation operator which, in the absence of a mass term, becomes its *exact negative* ($-\mathcal{D}$). In this minimalist interpretation, then, the zero totality of the 4-spinor applies algebraically as well as physically. We can also extend the definition of $\mathcal{D}$, following Kauffman, to include covariant terms, such as A_i, so that $\mathcal{D}$ becomes $\mathcal{D} - A_i$; and even to combine these ideas. Thus the covariant terms A_i can be seen as representing either a field source or an expression of the distortion of the Euclidean space-time structure – for example, that produced by the presence of mass in general relativity. This means that, if we *choose* to use structures of this kind to replace the direct use of mass, then a massless covariant $\mathcal{D}$ operator provides us with a convenient route to achieving this. For example, using Finsler geometry (a quartic generalisation of the Riemannian geometry used in GR for an anisotropic metric), Bogoslovsky[7] considers the field of a fermion-antifermion condensate as a source of space-time anisotropy, with a phase transition in which the particles acquire masses from the space-time, the mass shell taking the form of two hyperboloid inscribed cones. The connection with a double nilpotent representation of bosonic states using the massless covariant $\partial\!\!\!/$ operator is immediately apparent.

7.8 Idempotent and Nilpotent Versions of Quantum Mechanics

An idempotent aspect of the nilpotent quantum mechanics is obvious from the vacuum operators. As we have seen, a nilpotent amplitude like $(\boldsymbol{k}E + \mathbf{ip} + \boldsymbol{j}m)$ is effectively unchanged by postmultiplication by, say, $\boldsymbol{k}\,(\boldsymbol{ik}E + \mathbf{ip} + \boldsymbol{j}m)$ (which is a vacuum operator). In other words, we can write $(\boldsymbol{ik}E + \mathbf{ip} + \boldsymbol{j}m)$ as

$$(\boldsymbol{ik}E + \mathbf{ip} + \boldsymbol{j}m)\,\boldsymbol{k}(\boldsymbol{ik}E + \mathbf{ip} + \boldsymbol{j}m)\,\boldsymbol{k}(\boldsymbol{ik}E + \mathbf{ip} + \boldsymbol{j}m)\,\boldsymbol{k}(\boldsymbol{ik}E + \mathbf{ip} + \boldsymbol{j}m)\,\ldots \quad (7.18)$$

The only change is a scalar multiple which can be normalised away. Exactly the same is true for

$$(\boldsymbol{ik}E + \mathbf{ip} + \boldsymbol{j}m)\,\boldsymbol{i}(\boldsymbol{ik}E + \mathbf{ip} + \boldsymbol{j}m)\,\boldsymbol{i}(\boldsymbol{ik}E + \mathbf{ip} + \boldsymbol{j}m)\,\boldsymbol{i}(\boldsymbol{ik}E + \mathbf{ip} + \boldsymbol{j}m)\,\ldots \quad (7.19)$$

and

$$(\boldsymbol{ik}E + \mathbf{ip} + \boldsymbol{j}m)\,\boldsymbol{j}(\boldsymbol{ik}E + \mathbf{ip} + \boldsymbol{j}m)\,\boldsymbol{j}(\boldsymbol{ik}E + \mathbf{ip} + \boldsymbol{j}m)\,\boldsymbol{j}(\boldsymbol{ik}E + \mathbf{ip} + \boldsymbol{j}m)\,\ldots \quad (7.20)$$

Multiplying (7.18) from the left by $\boldsymbol{j}$, for example, now creates an idempotent:

$$\boldsymbol{j}(\boldsymbol{ik}E + \mathbf{ip} + \boldsymbol{j}m)\,\boldsymbol{j}(\boldsymbol{ik}E + \mathbf{ip} + \boldsymbol{j}m)\,\boldsymbol{j}(\boldsymbol{ik}E + \mathbf{ip} + \boldsymbol{j}m)\,\boldsymbol{j}(\boldsymbol{ik}E + \mathbf{ip} + \boldsymbol{j}m)\,\ldots$$

and similar procedures could be applied to (7.19) and (7.20).

Now, we can write the Dirac equation for a free particle in the nilpotent form

$$(\boldsymbol{ik}\partial / \partial t + \boldsymbol{i}\nabla + \boldsymbol{j}m)\,(\boldsymbol{ik}E + \mathbf{ip} + \boldsymbol{j}m)\,e^{-i(Et - \mathbf{p.r})} = 0. \quad (7.21)$$

But, since $\boldsymbol{j}\boldsymbol{j} = -1$, we could equally well write (7.21) in a form such as

$$(\boldsymbol{ik}\partial / \partial t + \boldsymbol{i}\nabla + \boldsymbol{j}m)\,\boldsymbol{j}\boldsymbol{j}\,(\boldsymbol{ik}E + \mathbf{ip} + \boldsymbol{j}m)\,e^{-i(Et - \mathbf{p.r})} = 0. \quad (7.22)$$

Here, we can see that the nilpotent equation actually incorporates an idempotent equation. The equations are precisely the same – the difference is purely one of interpretation. There isn't even a transformation required, just a redistribution of algebraic operators between differential operator and amplitude. So the alternative interpretations are:

IDEMPOTENT

$$[(\boldsymbol{ik}\partial / \partial t + \boldsymbol{i}\nabla + \boldsymbol{j}m)\,\boldsymbol{j}]\,[\boldsymbol{j}\,(\boldsymbol{ik}E + \mathbf{ip} + \boldsymbol{j}m)\,e^{-i(Et - \mathbf{p.r})}] = 0.$$

operator *wavefunction*

NILPOTENT

$$[(\boldsymbol{ik}\partial / \partial t + \boldsymbol{i}\nabla + \boldsymbol{j}m)\,\boldsymbol{j}\boldsymbol{j}]\,[(\boldsymbol{ik}E + \mathbf{ip} + \boldsymbol{j}m)\,e^{-i(Et - \mathbf{p.r})}] = 0.$$

operator *wavefunction*

With these two possibilities, we seem to be close to a quantum realisation of Spencer-Brown's Laws of Form,[8] in which the mark $\urcorner$, which makes a distinction between *inside* and *outside*, produces two laws of transformation, suggestive of alternative idempotent and nilpotent properties. In the law of calling, two marks, neither of which is inside the other, produce a single mark

$$\urcorner\,\urcorner = \urcorner$$

while the law of crossing ensures that one mark inside the other produces an unmarked state, equivalent to nothing

$$\overline{\urcorner|} =$$

The laws then extend into the logical functions AND, OR, NOT, NAND, NOR, etc., and other aspects of mathematics.

It should be noted, however, that, for all their mathematical equivalence, there is, of course, a considerable difference in significance between idempotent and nilpotent interpretations of equation (7.22), with only the nilpotent interpretation including the constraints which allow the massive reduction in information required to specify the fermionic state.[9]

7.9 A Fundamental Quantum Mechanical Duality

An intriguing duality that arises within the mathematical structure of quantum mechanics may be used to make an indirect comment on the significance of the Bohmian interpretation, with parallels between different systems which reflect and relate to the differences between classical Lagrangian and Hamiltonian dynamics. This, again, is based on the work of Hiley, who has produced a double system of nonrelativistic quantum mechanics – conventional and Bohmian – by using an algebra of process to generate Clifford algebras.[10,11] Using mathematical expressions for the strength of process, directed process, the order of succession, and the order of existence, he produces symplectic spinors and orthogonal Clifford algebras. The methodology requires the definition of an instant in time, and this leads to a doubling of the Clifford algebras with two expressions for the wavefunction. A left ideal represents ideas coming from the past while a right ideal represents information coming from the future. Together, the left and right spinors produce *zitterbewegung*. Heisenberg symplectic spinors represent the left and right ideals, while a Dirac idempotent is introduced via the Heisenberg algebra.

In a physical system, however, energy is unique as well as time, and the nilpotent Dirac formulation uses a fixed or unique element of energy, rather than a fixed moment in time. The nilpotent formulation is thus not about time directly. This comes from the fact that, in the fundamental parameter group, energy is conserved while time is not. Reversing their roles will produce a reversed algebra. Another fundamental difference is that energy really is unique. It only has one sign; mass is real and positive, and we have *CPT* rather than *MCPT* symmetry. Time, on the other hand, while physically unique, has two signs. The nilpotent process thus fixes primarily on conservation, rather than nonconservation. It says that mass-energy is a conserved quantity, and that it can, therefore, be described uniquely in mathematical terms. Thus, in principle, it ought to be possible to fix a unique-energy wavefunction as a single object, rather than as the two separate objects required by fixing a moment in time. So, in this formalism, ikE and $-ikE$ are contained in a single wavefunction.

In principle, if you choose unique energy rather than unique time, it becomes a *single* representation because energy has only one real physical solution, but unique time requires a *double* representation because time has a built-in $\pm\, i$ because of its complex nature. So rather than doubling the algebra, the nilpotent formulation doubles the physical representation. The particular value of this formulation with relation to particle structures arises *because* it uses a single mathematical representation with everything compactified within it. The Dirac nilpotent is self-dual, which means that it is already second quantized, and immediately creates boson states from fermions and vice versa. As a square root of zero, the nilpotent produces both solutions at the same time. In more general terms, it uses the method of getting discreteness from continuity which occurs in all aspects of physics. The change from 2π rotation to 4π rotation (≡ spin ½) for fermion states occurs, in both idempotent and nilpotent systems, by doubling the algebra. This is explicit in the idempotent algebra, but compactified in the nilpotent formulation, where we can *privilege* positive mass-energy. The nilpotent algebra is thus already a double algebra, and a duality stemming from a zero totality explains why it is nilpotent.

The nilpotent Dirac wavefunction, however, is *not* the same as the nilpotent projection operator used in the idempotent system, and it is a quantum field operator rather than a conventional wavefunction. The Dirac nilpotent uses a 4-spinor not a 2-spinor, and all the terms in the free-particle state have the same exponential phase term. Many representations of the Dirac wavefunction use two 2-spinors (as in section 6.1), which are split in conventional calculations, but this one keeps the wavefunction intact, and we have no need to regard the spinor as a bispinor at all. As we will see in chapter 11, when we write down propagators we have only one summation, whereas conventional Dirac algebra uses two summations, one for each spin state, and, when we create the energy states for the hydrogen atom, we use only one series of simultaneous equations, derived from coefficients of r^2, r^1, r^0, r^{-1}, r^{-2}, etc., not two, as in the conventional method (see 10.1). The strong interaction solution, in chapter 10, will show how we can extend this to a problem where the solution is not well worked out.

The method seems to be comparable with using Hamiltonian rather than Lagrangian dynamics (cf 8.7) – one equation is needed rather than two. This is because $\pm E$ can be grouped as well as $\pm \mathbf{p}$, and it makes no sense there to group just one of them. E, $\mathbf{p}$ together is effectively the same as including space and time together as the Hamiltonian method does, not as separate variations as in the Lagrangian case. Doubling the algebra is the same as the general symmetry-doubling, as in the discrete-continuous symmetry. The doubling occurs because we are making something discrete (space, charge, the Dirac state) from something continuous (time, the vacuum). It is also an expression of the duality of conserved / nonconserved properties; we are creating a conserved quantity against a 'background' of nonconservation (space, time).

Doubling is thus present in both idempotent and nilpotent formulations; it is the *method* of doubling that distinguishes them. Fixing space-time is a Heisenberg option and is used in the idempotent method, leading to two expressions; fixing E, $\mathbf{p}$, is a Schrödinger option, which is used in the nilpotent method, but here they are *combined* in a single expression. The idempotent has a duality between space and momentum; but the nilpotent has a duality between the combined space-time and the combined energy-momentum. This is precisely because the spinor is not split into two, so the E and $\mathbf{p}$ and the + and – values of each are all on the same footing, producing 4 equal states. While the conventional Dirac formalism splits either the space ($\mathbf{p}$) and the time (E) into two (with different exponentials), the nilpotent structure incorporates all 4 by using the same phase, and transferring the sign variation onto the differential operator. There is then no split between space and time, and the nilpotent becomes dual.

The duality of the idempotent is that of x and $\mathbf{p}$, with each expression incorporating half of the total duality available at any time. The duality comes in separate expressions because there are left and right ideals (corresponding to x and $\mathbf{p}$); x corresponds to Bohmian and $\mathbf{p}$ to conventional quantum mechanics, showing the duality between them, but the nilpotent collapses this to one because left and right ideals become identical, and the duality is incorporated directly, leading to a quantum field representation, and automatic second quantization, which is more directly relevant to describing particle states.

To obtain such second quantization requires the combination of the Heisenberg and Schrödinger approaches. Heisenberg, the basis for the idempotent, puts the **p** operator on space and the E operator on time, which achieves first quantization; second quantization requires their combination. The nilpotent, which is founded on Schrödinger, combines the Heisenberg-like vector term $(\boldsymbol{ik}E + \mathbf{ip} + \boldsymbol{j}m)$ with the Schrödinger-like phase term. (Because of the variability of time, it is probably easier to incorporate Heisenberg using the Schrödinger approach, as here, than it would be to incorporate Schrödinger using the Heisenberg approach.) The idempotent algebra uses the symplectic group for the Heisenberg-related spinors, and 2-equation duality, but it seems to be the metaplectic group, as the double cover of the symplectic group (with twice as many generators), which leads to the integrated duality of the nilpotent.[12]

In more mathematical terms, the idempotent method derives the quantum equation from an algebraic version of the Heisenburg formalism, that is

$$[\mathcal{H}, e] = Ee,$$

where $\mathcal{H}$ is the Hamiltonian, E is the energy operator, and e is a left / right dual operator. This is then defined formally, in terms of two vector states A and B, such that

$$[\mathcal{H}, e] = \mathcal{H}e - e\mathcal{H} = (\mathcal{H}A)B - A(B\mathcal{H}) = i\frac{\partial A}{\partial t}B + iA\frac{\partial B}{\partial t},$$

before being separated into the two conjugate Schrödinger equations,

$$i\frac{\partial A}{\partial t} = \mathcal{H}A$$

$$-i\frac{\partial B}{\partial t} = B\mathcal{H},$$

in which A and B are operators in the algebra (not the usual states). They are related to the usual states by saying that $A^* = B$, forming the position / momentum duals. However, if that is the case, then AB is time independent, and their total energy is zero, giving

$$[\mathcal{H}, e] = \mathcal{H}e - e\mathcal{H} = 0.$$

We can, in fact, propose a different form of this algebraic equation, by defining $e = AB$. The equation for quantum mechanics then becomes

$$\mathcal{H}e = Ee,$$

where E is the net energy and $\mathcal{H}$ is a one-directional operator:

$$\mathcal{H}e = \mathcal{H}(AB) = (\mathcal{H}A)B + A(\mathcal{H}B) = (\mathcal{H}A)B - A(B\mathcal{H}) = [\mathcal{H}, e].$$

Thus, it is equivalent to the idempotent form. If one then imposes the condition that this state is the product of the position and momentum values of the same state, then $E = 0$ and $\mathcal{H}e = 0$. Thus, they form a nilpotent identity. Here both $\mathcal{H}$ and e are 'operators', even though $\mathcal{H}$ is the Hamiltonian operator, and e is the 'state' operator. Thus the operator and state are on the same footing. One could then take this further by transforming the operators to be equivalent. This would then give the general equation

$$\mathcal{H}\mathcal{H} = E\mathcal{H}$$

or, in the dual case, $$\mathcal{H}\mathcal{H} = 0.$$

If we normalize things so that $E = 1$, then the state operator is either nilpotent (and it is then a pure dual state), or it is idempotent (and it is a product state). (The idempotent is the product state of the nilpotent, but it can also be turned around; it seems to be linked to the boson / fermion duality, and supersymmetry operators, in that one can be seen as the product of the other.) The nilpotent algebra thus becomes equivalent to the two separate algebras of the idempotent method combined. Of course, separate algebras serve to show the equivalence between Bohmian and conventional nonrelativistic quantum mechanics, but a combined algebra is more appropriate to the requirement for a self-dual quantum-field equivalent, which is also relativistic. A fully combined system is also unlikely to be achieved within any system which is nonrelativistic.

It is interesting that this combination can be obtained by *incorporating* Bohmian quantum mechanics into the conventional version, and that this becomes equivalent to incorporating the vacuum state with the fermion, or the fermion with the vacuum state, or even to incorporating Heisenberg (fermion) within Schrödinger (vacuum).[13] This leads us on to the consideration of that whole class of dualities which make physics workable, and which are characterized by the appearance of alternative representations related by the numerical factor 2 (see chapter 16).

Chapter 8

The Classical and Special Relativistic Approximations

If quantum mechanics, as introduced in chapters 3-7, describes the system, then classical physics introduces the idea of measurement and observation. Though classical results can be derived by approximation from quantum mechanical equations, once the mechanism is established, they can also be derived more directly from the scaling relations needed in the original definition of the Dirac state. This, in effect, pinpoints the significant aspect of the Dirac nilpotent which is carried over into the classical context. Special relativity is first derived, as a purely mathematical structure, directly from the Dirac nilpotent (a reversal of the usual procedure); its physical interpretation is then seen as a result of the concept of 'observation' used in the quantum-classical transition. Classical mechanics and electromagnetic theory are then derived formally (in several different forms).

8.1 Linear Versus Orbital Dynamics

Despite the various formal similarities, there is a significant change in methodology between quantum and classical domains, and a more 'fundamentalist' approach is needed to fully incorporate the idea of 'interaction' which emerges from the relationships within the parameter group. The change in methodology seems to be related to greater emphasis on linear, as opposed to orbital, motion in classical physics. Newtonian mechanics, for example, is deliberately structured on the privileging of straight line over orbital motion as the inertial system, or the system operating without the action of force. It is certainly possible to choose to privilege orbital motion classically, as angular momentum is a conserved quantity in classical physics, and Pope and Osborne have produced a quantum-classical theory of this kind on the basis that orbital motion is more 'natural' than linear.[1] In *pure* quantum theory, this is true because angular momentum carries the information relevant to charge, and the fundamental interactions between particles of matter, but quantum mechanics, unlike classical physics, is not a system of *measurement*, and angular momentum is not strictly an observable. Measurement (and the scaling process with respect to space on which it necessarily depends) is inconceivable except in straight lines, and a theory of measurement, and the interaction between measurement and quantum systems, must comprehend linear dynamical quantities as well as orbital ones.

There is, of course, no such thing as a perfect straight-line motion, but there is equally no such thing as perfect orbital motion either. Newton found in his *Principia* that no planet in the solar system ever traversed the same orbit twice – nor does any hydrogen atom in a quantum context. In either case, it is necessary to use the concept of *force* to show how the defined inertial condition is altered. Orbital behaviour may be natural for systems on the quantum or even classical scale, but the classical concept of measurement – which is dually distinct from systemic behaviour – leads naturally to the conventional definitions of force and linear momentum. An angular momentum approach will always favour systems which can be described as 'microcanonical', or even 'macrocanonical'. Pope and Osborne, for example, have shown that it can be used to derive the effect of the spin of a body on its gravitational orbit in the form of an effectively altered value of the gravitational constant. A Newtonian result is not easily accessible in such cases, because a spinning body defines a noninertial frame, and so defines a problem outside the boundaries of Newtonian physics. However, the concept is not *contradictory* to Newtonian physics, as the Newtonian methodology handles such frames by incorporating fictitious inertial forces, which create related effects, and, in the 'nonequilibrium' conditions applicable to complex systems on a macroscopic scale, a linear dynamics, closer to the process of measurement, will be required. The same will be true for physics close to the quantum-classical boundary.

8.2 Scaling Relations

The creation of the Dirac nilpotent has many significant consequences outside of pure quantum mechanics. One is that the units of the individual components must be related *numerically*. Through the Dirac equation, E and t, and $\mathbf{p}$ and $\mathbf{r}$, become conjugate variables, that is, ones which exchange statements about conservation into equivalent statements about nonconservation, and vice versa. This requires both direct and inverse numerical relations. Numerical relationships must also be established between the units of E and **p**, and between those of t and **r**. We are thus required to introduce the constants $\hbar$ and c, which are here equated, formally, to 1. A third constant, G, is required to relate charge to mass units (or direct relationships to inverse ones). Another way of interpreting these constants (one which amounts in principle to the same thing) is to say that they are the natural result of binary operations within the parameter group.

The canonical D_2 group elements are required to be their own inverses, and to be each group identities. In addition, the group multiplication rule (when all possible arrangements are taken into consideration) requires such operations as

charge* time = space* mass.

In principle, the units of space (r), time (t), mass (m) and charge (q), must be related by a set of equations of the form:

$$r \propto t \propto m \propto q \propto \frac{1}{r} \propto \frac{1}{t} \propto \frac{1}{m} \propto \frac{1}{q}$$

where the charge component represents an idealised value assumed to exist under Grand Unification of the three nongravitational forces. The squared multiplication of the units, which automatically exists once we have defined the nature of the Dirac state, is a binary operation which makes this possible. The existence of this particular binary operation within the parameter group is, of course, linked to the same operation being responsible for the 4-dimensionality of space-time and mass-charge.

We use the scaling constants (or rather scaling parameters, since they need not be actually constant if they are known to vary according to some fixed rule) to create the necessary number of independent fundamental relationships. And since the system has inherent duality in making each quantity its own inverse, then we must define a relation between each quantity and the inverse of every other, for which, after all the direct relationships have been established, one further scaling constant (or parameter) will suffice. So, the group relationship predicts that such fundamental constants must exist, while effectively ensuring that their individual values have no independent meaning. To relate these to familiar scales of measurement, we create them from combinations of the four historically-generated fundamental constants G, c, $\hbar$ $(= h / 2\pi)$ and $4\pi\varepsilon_0$. Here, for convenience, we assume that 'charge' has the electromagnetic value, though this is not a necessary assumption, and a grand unified value could be used instead (a more fundamental unit of charge is used in chapter 16); the actual 'values' of the constants are not particularly significant – only the fact that some such scaling must exist. However, since the electric charge is the one in the Dirac state that is associated only with a scalar, it is a reasonable assumption that it is the one whose interaction is the closest in form to the ideal state, a pure scalar phase.

We can now express the scaling relations between the units of space (r), time (t), mass (m), and charge (q) as follows (with the equality sign being interpreted as meaning 'equivalence'):

$$r = ict \tag{8.1}$$

$$r = \frac{G}{c^2} m \tag{8.2}$$

$$iq = (4\pi\varepsilon_0 G)^{1/2}\, m \tag{8.3}$$

The respective imaginary and quaternion operators required by t and q are significant in determining the signs of their squared units. These operators are normally subsumed within the symbols t and q, but here they are added for emphasis.

The further relations between any parameter and the inverse of any other can all be derived from:

$$m = \frac{\hbar}{c^2} \frac{1}{it} \tag{8.4}$$

This last result (the inverse one) is the one that we recognise as being responsible for quantization of energy and other physical properties. Relationships such as $E = \hbar\,\omega = h\nu$ and $p = \hbar k = h / \lambda$ then relate energy and momentum to such time- and space-related observables as angular frequency (ω) or frequency (ν), and wave number (k) or wavelength (λ). Quantization could thus be said to be a result of the fact that each parameter is its own inverse (which is the same thing as saying that a squaring operation is needed to construct the relationships between parameters incorporated into the Dirac nilpotent). Quantization and duality of scale are aspects of the same phenomenon.

The quantities G, c, $\hbar$ and $4\pi\varepsilon_0$ are, of course, merely consequences of the historical choice of units and so have no ultimate significance; their accidental natures are shown by the fact that the true constants relating the four parameters to each other must be derived from their combinations. So the four independent scaling constants become c, (G / c^2), $(4\pi\varepsilon_0 G)^{1/2}$, and $(\hbar / c^2)$. The constants which do have significance are those related, as we have stated, to the size and mass, etc., of individual particles, such as electrons, and these must ultimately be derived in any fundamental theory.

Like all classical physics equations, the scaling relations are not statements of identity, but only of proportionality relationships between differently-measured units. They also lead to the existence of the fundamental (Planck-type) absolute units for space, time, mass and charge, which, again, have long been known. (The Planck length, time and mass are, respectively, $(G\hbar / c^3)^{1/2}$, $(G\hbar / c^5)^{1/2}$ and $(\hbar c / G)^{1/2}$, while $(\hbar c)^{1/2}$ defines a fundamental unit of charge.) Since they are based on *scaling*, in effect against the value of a length between spatial positions – the only thing that can, in fact, be measured, even approximately – we can use them in constructing classical physics, specifically as a process of *measurement*, together with the principle that physics, whether quantum or classical, structures itself by defining systems in which conserved quantities remain fixed while nonconserved quantities vary absolutely. In principle, also, we can assume that any term related to another by a scaling relation in a meaningful physical equation can be replaced by the alternative term to produce another meaningful equation. In combination with the principles of conservation of mass and charge and of nonconservation of space and time, the scaling relations lead to derivations of the laws of classical mechanics and electromagnetic theory. The presence of c^2 and $\hbar$, here, informs us that these quantities are fundamental to physics, whether classical, relativistic or quantum.

8.3 Special Relativity

A process regularly used in theoretical physics is to derive a nonrelativistic formulation first, which is then made relativistic by incorporating a 4-vector relationship between space and time, or energy and momentum. This is used in the standard derivation of the relativistic Dirac equation from the non-relativistic Schrödinger equation (via the Klein-Gordon equation). As is well known, of course, the derivation cannot be done deductively because the relativistic connection between space and time can only be accomplished by incorporating matrix terms of unknown origin into the final equation. However, in the derivation of the Dirac equation from the zero-totality algebra, relativity isn't even mentioned. There is absolutely no discussion about measurement, observers or light signals – or even the existence or non-existence of the aether, as such. The various terms take their place and status in the equation for purely fundamental reasons and the physical consequences are what follow from this. The consequences of this are profound, especially in the area of general relativity and the possibility of quantum gravity.

The conjugate nature of E, $\mathbf{p}$ and t, $\mathbf{r}$ means that we can also establish a nilpotent structure connecting t and $\mathbf{r}$, with another term τ (described as 'proper time') in the position occupied by m. This is, strictly, a rest mass-related, rather than time-related concept, though the overall structure may be thought of as a 3-D 'time' comparable to the 3-D 'mass' represented by ($\boldsymbol{ik}E$ + $\boldsymbol{i}\mathbf{p}$ + $\boldsymbol{j}m$), and opposed to the 1-D 'charge' and 'space' created by the single well-defined direction for quantized angular momentum. Defining the nilpotent ($\pm\, \boldsymbol{ik}t \pm \boldsymbol{i}\mathbf{r} + \boldsymbol{j}\tau$) or ($\pm kt \pm i\boldsymbol{i}\mathbf{r} + i\boldsymbol{j}\tau$) means that we have essentially established special relativity purely in terms of algebra and a relationship between the units. Without any physical assumptions, we derive the equation for 'Lorentz invariance':

$$(\pm\, \boldsymbol{ik}t \pm \boldsymbol{i}\mathbf{r} + \boldsymbol{j}\tau)\,(\pm\, \boldsymbol{ik}t \pm \boldsymbol{i}\mathbf{r} + \boldsymbol{j}\tau) = -\,t^2 + r^2 + \tau^2 = 0,$$

or, in differential terms,

$$-\,dt^2 + dr^2 + d\tau^2 = 0.$$

The 'physical' content of 'relativity' then becomes whatever interpretation we choose to place upon the mathematical consequences. It depends on whether we decide, in the case of a non-free state, to retain the constancy of c as a convention, or allow it to vary in an appropriate way. The choice only matters in the classical domain, where we have the concept of measurement. The so-called 'principle of relativity' is, in fact, common to all dynamical theories. The laws of physics cannot distinguish between states of rest and states of relative uniform motion. That is, frames of reference in uniform relative motion are inertial, like states of rest, because the laws of physics must be framed in terms of squared time, or the rate of change of uniform motion. The real differences between special relativity and other possible theories lies in their differing interpretations of the effect of the observer's motion on the one-way 'speed of light' (c) received from a distant source, and the consequences this has for the idea of 'simultaneity'. (No theory assumes that motion of the source has any effect.)

If we choose to use Einstein's definition of 'simultaneity' (events at two points are simultaneous if they occur when they receive light signals sent simultaneously in both directions from a source midway between them) and assumed *constancy* of the speed of light, then we obtain the familiar light-cone, the Lorentz transformations, the velocity addition law, and the necessity of c as a maximum physical speed to preserve causality or the assumed absolute order of physical events (effectively, absolute time). The choice, however, is arbitrary, like choosing a 'gauge' condition, and no choice at all is necessary at the quantum level where we are not thinking in terms of physical measurement, and where simultaneity, as a measurable concept in the Einsteinian sense, has no meaning. This, of course, is why quantum experiments, like that of Aspect, of 1982,[2] violate the Bell inequality, and require explanation in terms of nonlocality.

In general, the vector nature of space means that we can choose components along any mutually perpendicular axes x, y, z to specify vector **r**, as long as the component magnitudes are such that

$$x^2 + y^2 + z^2 = r^2.$$

We say that vector **r** is invariant to choice of x, y, z; and this is an expression of the translation-rotation symmetry of space. Where an observational definition of position is possible in a relativistic context, all vectors with 3 components must be replaced with vectors with 4 components, one of which is imaginary. Space and time are linked in a single coordinate system such that

$$x^2 + y^2 + z^2 + (ict)^2 = r^2$$

is the invariant quantity under arbitrary transformation of axes. According to the standard application of 4-vector theory, we find that, because space and time are now linked in an algebraic relationship, the coordinates measured for systems in relative motion with respect to ourselves are observed to depend on the relative velocity of the motion.

Time dilation is, of course, an automatic consequence of the nilpotent formalism, once we accept invariant c, since we can write

$$(\pm\, \boldsymbol{i}kct \pm \mathbf{i}\mathbf{r} + \boldsymbol{j}\tau)\,(\pm\, \boldsymbol{i}kct \pm \mathbf{i}\mathbf{r} + \boldsymbol{j}\tau) = -\,c^2t^2 + r^2 + c^2\tau^2 = 0,$$

making explicit use of c, or

$$c^2\tau^2 = c^2t^2 - r^2,$$

in the form

$$c^2{t_0}^2 = c^2t^2 - r^2.$$

From this equation, we obtain

$$t_0^2 = t^2\left(1 - \frac{r^2}{c^2t^2}\right) = t^2\left(1 - \frac{v^2}{c^2}\right)$$

and

$$t = \frac{t_0}{\sqrt{\left(1 - \dfrac{v^2}{c^2}\right)}} = \gamma t_0\,.$$

Length contraction by the same factor (of the coordinate $r = l$ in the direction of motion) is then a necessary consequence of maintaining the constancy of c (or the principle of relativity):

$$l = l_0 \sqrt{\left(1 - \frac{v^2}{c^2}\right)} = \frac{l_0}{\gamma}.$$

The formulae for t and l are usually taken to mean that times, as measured by moving clocks are dilated by the factor $\gamma = (1 - v^2 / c^2)^{1/2}$, while lengths of moving measuring rods are contracted by the same amount. It can also be shown that masses of moving objects are proportionately increased.

Using the Einstein definition of synchronization, the Lorentz transformations, which are a straightforward derivation from these contracted lengths and dilated times, show the mathematical form under which each physical quantity in a relatively moving system will remain invariant under arbitrary transformations of the space-time axes:

$$x' = \frac{x - vt}{\left(1 - v^2 / c^2\right)^{1/2}}$$

$$y' = y$$

$$z' = z$$

$$t' = \frac{t - vx / c^2}{\left(1 - v^2 / c^2\right)^{1/2}}.$$

The quantities to be used in relativistic equations are thus *Lorentz-invariant* quantities, rather than quantities which are invariant only under ordinary vector transformations (although in a few cases these are identical). However, if we *assume* that the quantities used are Lorentz-invariant, the *form* of the equations involving them remains the same as those derived from using ordinary vector theory. Since the Lorentz transformations are a completely deductive consequence of the assumption of 4-vector space-time, no detailed account need be given of the standard methods of derivation.

The classical Galilean transformations, which are the same equations for the special case when $v^2 / c^2 \rightarrow 0$, allow the same invariance of physical quantities between systems in relative motion when conditions are non-relativistic (or when $c >> v$).

$$x' = x - vt$$

$$y' = y$$

$$z' = z$$

$$t' = t \, .$$

The first equation alone is sufficient to show that, even under non-relativistic conditions, events which are observed as simultaneous in one frame need not be observed as simultaneous in another, or that the apparent order of events in one frame may be seen as reversed in another. (For example, a faster signal sent out later from the same place as an earlier, slower one, may arrive earlier at the same destination.) This is a purely kinematical, and most often purely visual, effect, based on the relative positions of objects seen at different times, which is clearly not specific to special relativity, and, despite a wealth of popular writing on the subject, has no fundamental philosophical implications for that theory. Even though the relativistic and non-relativistic equations suggest different conditions under which the effect would occur, the *effect itself* has no intrinsic significance, say, for understanding the nature of time. In addition, as the observed motions of quasars show, kinematic-visual observations may even appear to show objects moving at speeds much greater than c without in any way fundamentally affecting our knowledge that c is a limiting speed for all known physical matter. If we rightly deny philosophical implications to these purely visual observations (for example, that c may be exceeded or time reversed) by invoking special relativity and the Lorentz transformations, then we must consider it illogical or inconsistent to use the same theory to derive philosophical implications about the 'nature of time' from kinematical observations in other contexts. In principle, the parameter t used in the Lorentz transformations does not tell us anything about the nature of time as a fundamental parameter and its relation to the causal order of events.

8.4 The Significance of the Proper Time

In the Einsteinian version of relativity, the velocity v in the transformation equations represents only the value relative to the observer, and has no absolute significance; the space-time coordinates ensure that the one-way speed of light remains constant. As this speed remains strictly non-measurable, this assumption is effectively equivalent to a gauge condition; an alternative hypothesis would allow the one-way speed to vary, as long as the (measurable) two-way speed for a round trip remained constant. The velocities in this theory, which is associated particularly with Poincaré and Lorentz, are those measured relative to an absolute reference or rest frame, provided by an all-pervading aether or vacuum. The two-way speed of light is preserved as a result of the same 'relativistic' effects of time dilation, length contraction and mass increase, though these are now assumed to be due to the direct action of the aether through which the affected object moves.

For all measurable consequences based on time-delayed interactions, these are equivalent theories (cf 9.5 for further discussion); no experiment so far conceived will be able to measure the one-way speed of light or use information from a mechanism using light-speed transfer to detect a 'drift' through the aether. And neither theory, in our terms is fundamental, as both are constructed in terms of the measurement approximation, rather than in the quantum relationship between space and time which emerges from the fundamental algebraic structure of the Dirac nilpotent. However, the discovery, in 1977, of an anisotropy in the cosmic microwave background radiation due to the Earth's motion through space led to headlines such as 'Aether drift detected at last' in *Nature*[3] and 'The Cosmic Background Radiation and the New Aether Drift' in *Scientific American*.[4]

This experiment seemingly succeeded in measuring motion with respect to an absolute standard of rest, where all others had failed, because it was measuring the *background* and not the signal. The fundamental creation of a 4-vector relationship between space and time in the Dirac nilpotent makes it impossible to detect absolute motion using any signal also structured as a 4-vector, that is by a Lorentzian measurement, but an isotropic temperature background is unstructured or random in this sense, and everything else, such as the Earth's motion, becomes structured with respect to it. *If* the microwave background can be shown to be a vacuum effect, then it provides the absolute rest frame or absolute 'space' suggested by the continuous nature of mass and the proof of quantum nonlocality. However, since both the Einstein and Lorentz-Poincaré theories of special relativity are concerned with the classical process of *measurement* and not with the quantum relation between space and time, or momentum and energy, then it is immaterial whether or not we choose a system in which an undetectable vacuum or 'aether' is directly employed.[5]

A significant aspect of the Einstein version of special relativity is that it is in every sense a *classical approximation* to a theory that, fundamentally, requires a quantum explanation. This has, ever since, been a source of philosophical confusion. Paradoxes (in particular, the twin paradox) have arisen because the kinematic approach to relativity has been privileged as fundamental, rather than as an almost opportunistic approach to a convenient approximation which results from the elimination of a specific term in the equations under classical conditions. Einstein uses the quantum process of light-signalling (almost certainly based on his own discovery in the same year of the light photon) as though it were classical. He creates concepts of simultaneity, light-signalling, and 'measurement' of a classical one-way 'speed of light', as though they actually have intrinsic meaning. Of course, as quantum processes, the absorption and emission of light photons are irreconcileable with *any* concept of simultaneity. Any two quantum events must have an unchangeable temporal order, which denies absolute meaning to any kinematics based on the idea that they can be simultaneous.

The term which is missing in Einsteinian relativity is the proper time; we are told that the space-time combination is an invariant, but not what this invariant is, or why it's an invariant. Proper time and causality are added in as a 'common-sense' extra. But there is nothing common-sense about it at all. Proper time – the causality term – occupies the position in the space-time relation that rest mass does in the momentum-energy relation. It has a specific mathematical origin, and is linked with a specific mathematical operator (a quaternion unit) in the nilpotent formalism. It is also directly associated with causality. Causality also has a very specific physical origin in quantum mechanics, which is intimately connected with the idea of the vacuum and nonlocality. And even the rest mass has a vacuum origin in the Higgs mechanism. Einsteinian relativity is able to dispense with the 'aether' (vacuum) because it leaves the 'aether' or causality term out of the kinematically-derived equations, which treat space and time as a pure 4-vector. Of course, this is entirely reasonable and even desirable within the classical domain, but the point at which this domain fails is where quantum causality determines that two events cannot be exactly simultaneous.

In fact, though momentum-energy, and space-time, are regarded as pure 4-vector combinations in special relativity, in a quantum system, this is not strictly true, because the Dirac energy and momentum terms are only fully represented mathematically when each has a different quaternion operator applied to its respective pseudoscalar or vector. When taking the invariant scalar product, of course, these operators disappear, but their significance becomes apparent when we introduce a third term, with yet another quaternion operator (rest mass) to convert them into a nilpotent. A pure 4-vector could not be made into a nilpotent in this way. The same must apply, in general, to relativistic time and space, whether the system is quantum or classical, and to the components of the Dirac differential operator (∂^μ). And, of course, it is the special relativistic assumption that space and time constitute a true 4-vector that is responsible ultimately for the additional paradox of wave-particle duality, where we force a physical union between quantities which are intrinsically unlike. The proper time, the additional term which completes the nilpotent with space and time, is a real scalar, like rest mass, and its squared value must exactly cancel the scalar product of the time and space components. The fact that it can never be negative, by analogy with rest mass, means that only retarded solutions are possible for the space-time combination. The validity of proper time in both classical and quantum contexts is an indication that the link between these two domains is essentially through the scalar additive nature of the rest mass or 'inertia' of the component systems.

It is apparent from this analysis that the space, time, and proper time, or momentum, energy, and rest mass, have a 'dimensional' character, which results from the process by which they are quantized. Interestingly, this emphasizes the fact that time and proper time are fundamentally different physical concepts, although the proper time is *numerically* equivalent to the time value in the context of a zero space component (that is, the time value measured by a clock carried with a moving object). The 'relativistic' properties are purely a result of this dimensional property, as are the constants $\hbar$ and c. (In this sense, quantization and relativity are essentially the same process, and the discreteness of quantization is essentially the same as the discreteness produced by time-delayed transfer of information, as is effectively the case in quantum field theory, where interactions are mediated by gauge bosons.[6]) We may relate this to the 3-dimensional quantized 'orthogonality' in the synthesis proposed by Pope and Osborne, which also defines 'relativity' in an algebraic manner.

The quantization involved in creating the Dirac state may be thought of as a localization of energy-related processes, or an expression of time-delayed action between discrete 'sources' (charges), with a maximum transmission rate of c for a zero-rest mass system, and it is significant that this transmission rate only has meaning for discrete sources.[7] However, the entire nilpotent structure is founded on the principle of instant correlation for all fermionic state vectors, and is nonlocal in this sense, because a structure of this kind makes the existence of two identical wavefunctions instantaneously impossible. The specification of uniqueness requires instant correlation, at the same time as the nilpotent nature of the operator and its built-in 4-vector component require time-delayed action between the discrete charges. The fundamentally continuous nature of mass-energy, which is related to the absolute unchanging order of events – the principle of causality required by both special relativity and Newtonian physics – suggests that the carrier of the nonlocal correlation between the Dirac states is the interaction we describe as gravity, which does not rely, as the others do, on discrete sources.

8.5 The Nature of Classical Physics

Classical, unlike quantum, physics is concerned with measurement. The idea is, of course, an approximation. Since classical physics incorporates the effects of many micro-quantum-systems, the absolute variability of the space parameter, in particular, becomes increasingly restricted by interactions, allowing an effective degree of localisation, which we calculate as an 'expectation' value. So, the purpose of classical physics is to regulate the effect of such localisation on the fundamental parameters of complex systems, and to determine the degree to which they become redefinable as observable quantities, and which approximations make this possible. The key quantities, however, and many of the fundamental principles, remain largely the same, and some (such as the behaviour of rest mass) are carried over unchanged.

The most fundamental laws of physics are essentially definitions and conservation laws. Classical mechanics, for example, is structured on only two fundamental requirements: the construction of a quantity involving conserved and nonconserved parameters (force, energy, momentum, action, Lagrangian or Hamiltonian) and the definition of its behaviour under variation of the variable components, that is, whether it is defined to be zero, invariant, or an extremum. Essentially, the laws of classical mechanics are set up to define what is meant by a conservative system, or to isolate for observation a particular section of the unchanging mass of the universe; they become directly useful as physical conditions allow a near approximation to this observation to be made. Quantum physics does not violate the notion of a conservative system; it merely tells us that physical observation is incompatible with its exact definition. This is why classical physics based on observation is only an approximation.

As already stated, the nonconservation of space and time means that, in a system, these quantities act as *variables*, while mass, charge, energy, momentum and angular momentum remain constant. Absolute variability requires the possibility of infinitesimal changes, which we express mathematically in terms of the infinitesimal space and time elements dr and dt. By simple differentiation, we may show that proportionality relations between the units of space and time may also apply to their infinitesimal elements. The key to further development now lies in establishing precisely the way in which variable and constant quantities impose exacting conditions on the symmetry, or scaling, relations. The symmetry relations alone merely indicate the existence of quantities which are of equivalent status in physical equations, but they become of precise *physical* significance when they relate the conservation of mass and charge to the variation of space and time.

The defined quantities of classical physics invariably combine the universal conserved quantity mass with differentials expressing the variation of the nonconserved quantities space and time; the defined quantities, which serve to link the conserved and nonconserved parts of the system, are then observed to be conserved, or zero, or a maximum or a minimum, as the nonconserved quantities are subjected to continuous variation. In principle, the behaviour of the defined quantities expresses the conservation of mass, a quantity which will uniquely define a system, and sometimes also of charge, as the space and time coordinates continuously change.

In connection with the laws of physics, nonconservation of space and time is manifested in the property of *gauge invariance*: states distinguishable only by arbitrary changes in space and time coordinates have equal probability of occurring. Elements of space and time have no unique identity, and so must be allowed equally to take all possible values which preserve the values of the conserved quantities, whether primary, like mass and charge, or composite, like energy and momentum. In classical electromagnetic theory, electric and magnetic field terms remain invariant under arbitrary changes of scalar and vector potentials brought about essentially by translations or rotations in the space and time coordinates. Scalar and vector potentials, of course, are the products of charge terms and functions of the space and time coordinates (cf 8.6 and 8.7); they do not contribute directly to the energy or any other conserved quantity. In this case, the principle of gauge invariance tells us that a system will remain conservative under changes which affect only the space and time coordinates, and which do not involve changes to its energy, momentum or angular momentum.

The 'arbitrary' changes which are thus allowed are changes in space or time coordinates which do not change the mass or charge of the system, or the value of any of the conserved quantities related to mass. From standard electromagnetic theory, it is clear that such changes represent changes of *phase* in the electromagnetic wave equations and so 'gauge' invariance is actually a local phase invariance. In more formal language, conservation of electric charge is equivalent to invariance of the Lagrangian (as defined in 8.7) under arbitrary phase changes of the charged particle wavefunctions to which it applies.

Essentially, charge and energy conservation become equated to an invariance under the transformation of electrostatic potential by a constant which represents such changes of phase, and the phase changes are produced by changes in the coordinate system; once again, the conservation of mass and charge implies, at the same time, the nonconservation of space and time. It is significant that the nonconservation of space and time, expressed in this way, is as local as the conservation of mass and charge, for the arbitrary phase transformations which, in terms of the Yang-Mills principle, are required in all successful gauge theories, are invariably local rather than global.

It is gauge invariance which allows us to see how our predicted direct connection between conservation of charge and conservation of momentum (see chapter 2) may be realised in a physical system, for charge conservation is a direct consequence of the fact that arbitrary changes in electrostatic potential are allowed. Now, in a conservative system, electrostatic potential varies only with the values of the space coordinates. The principle of charge conservation under gauge invariance is thus a particular case of the general principle which we have derived from symmetry alone: electric charge is conserved *precisely because space is not*. With our previous association of conservation of momentum with nonconservation of space, we now have a direct connection, as predicted, between momentum conservation and the conservation of charge.

We have, of course, also predicted a second relation, this time between the conservation of angular momentum and the *type* of charge. However, this appears to be a relationship which is valid for quantum, rather than classical systems. An instance of this is almost certainly to be found in the connection observed in fundamental particles between spin and statistics. Fermions, as we have already stated, appear to have weak units of charge, where bosons have none, and these two classes of particles are associated, respectively, with half-integral and integral values of spin. The respective antisymmetric and symmetric wavefunctions which are responsible for these spin values are undoubtedly connected with the presence or absence of the weak unit of charge and its parity-violating properties; they are certainly not related to the presence or absence of electromagnetic or strong charge units. In addition, the conservation of the left- or right-handedness of the angular momentum term may be seen as related to the conservation of the sign of weak charge as determining fermionic / antifermionic status. It is, therefore, the presence or absence of a *particular type* of charge which determines the spin component of the angular momentum of the particle. Now, the parity or space-reflection operator reverses momentum, but not angular momentum; it is significant, therefore, that the weak interaction, in violating parity, shows itself indifferent to the sign of weak charge, but not to its type. Charges, we may say, which are subject to parity operations which preserve angular momentum, may change their signs, but not their type.

8.6 Constructed Quantities

The key concepts in classical mechanics, as in other aspects of physics, are those which combine the minimum information necessary to distinguish the conserved and nonconserved parameters. Of the conserved quantities, mass is universal and never zero, and therefore must be present; charge, however, is local, and can take zero values. Charges are nonreal, so a system cannot be defined for a single charge. There must be at least two charges with a finite space r that defines their discreteness, and there must be an interaction between the charges, which we may write as $\boldsymbol{i}q_1\boldsymbol{i}q_2 = -\, q_1q_2$. Though individual charges cannot produce mass, mass can be produced as a result of interactions. In the quantum state, interactions can occur with vacuum fields, but classical physics is concerned mainly with observables.

To specify the conservation or invariability of mass, we also need to specify the nonconservation or variability of space and time; hence, these parameters are included in differential form. A convenient way to define a system, therefore, would be the construction of a quantity containing mass and the differentials of space and time. The most immediately useful constructs then include $p = m\ dr\ /\ dit$ and $F = dp\ /\ dit$ (with time, most conveniently specified as the independent variable). The second quantity, as has been previously explained, has the advantage of producing a real rather than an imaginary construct. Now, space, of course, is really a vector (neglecting, for the moment, any 4-vector aspects); to incorporate this aspect, we may multiply both terms by the unit vector $\mathbf{r}\ /\ r$, to yield the familiar quantities, *momentum*,

$$\mathbf{p} = m\frac{d\mathbf{r}}{dit}$$

and *force*,

$$\mathbf{F} = \frac{d\mathbf{p}}{dit}.$$

(Imaginary and quaternion labels are retained here for emphasis but would not, of course, normally be used.)

The definitions of such quantities are, as yet, purely mathematical and convey no additional physical information. This can now be supplied, however, by using the scaling rclations to find other quantities to which these defined ones can be related, while at the same time applying the conditions for conservation and nonconservation. This enables us to set up a system of equations for classical mechanics and its direct corollary, electromagnetic theory. It is notable that classical mechanics, derived in this sense, already incorporates the quantum notion and the interrelation of mass and energy, and so is not a system entirely separated from these concepts.

8.7 Classical Mechanics

The equations so far derived are merely general relations between units and have not yet been applied to physical systems. We may define a system in terms of a fixed number of elements of mass and charge, the spaces and times associated with them being variable. To make the system physically meaningful, we have to find out how these various quantities are associated. The numerical relations established between the parameters through the group, can be combined with conservation and nonconservation conditions to provide mathematical derivations of the laws of both classical mechanics and electromagnetic theory. A system which is not redefined after changes in its coordinates is described as *conservative*.

In classical physics, it is the 'interactions' or squaring of mass and charge values which create the equivalent of the creation of the Dirac state in quantum mechanics. Such interactions, which are required by the binary operation between group members and which overcome the problem of having imaginary charge as a component, are directly incorporated in some forms of the scaling relationships. From scaling relations (8.2) and (8.4), remembering that each element of mass is unique, we may derive the expression

$$Gm_1m_2 = \hbar\frac{r}{it}.$$

In differential form, under the specific conservation of mass elements,

$$Gm_1m_2 = \hbar\frac{dr}{dit},$$

from which

$$\frac{Gm_1m_2}{c^2it} = m\frac{dr}{dit} = p.$$

The mass term on the right hand side, of course, is a new mass unit, distinguishable from m_1 and m_2.

By differentiation, and a further substitution,

$$-\frac{Gm_1m_2}{c^2i^2t^2} = -\frac{Gm_1m_2}{r^2} = \frac{dp}{dit}.$$

Applying the unit vector, $\mathbf{r} / r$, this becomes

$$-\frac{Gm_1m_2}{r^3}\mathbf{r} = \frac{d\mathbf{p}}{dit},$$

which is a combination of Newton's law of gravitation and second law of motion, with the left hand side a new equivalent quantity for force, conventionally described as *gravitational force*. For constant mass m,

$$-\frac{Gm_1m_2}{r^3}\mathbf{r} = m\frac{d^2\mathbf{r}}{dt^2}, \qquad (8.5)$$

Newton's first law of motion defines the case where $d\mathbf{p} / dt$ or $d^2\mathbf{r} / dt^2 = 0$.

Neither m_1 nor m_2 is, of course, privileged, and so interpreting the vectors $\mathbf{r}$ and $\mathbf{p}$ as directed from m_1 to m_2 means that reversing the mass terms produces reversed vectors, from m_2 to m_1, as required by Newton's third law of motion, which makes the vector sum of forces in a conservative system equal to zero. The mutuality of the interaction means that m in equation (8.5) can be either m_1 or m_2. Writing the equation in either form introduces what is called the 'principle of equivalence' – that the gravitational mass, m_1 or m_2, and inertial mass, m, are indistinguishable. Though this plays a major part in the general theory of relativity, it is a fundamental principle of classical theory also.

The equivalent case for charges defines Coulomb's law of electrostatics and introduces *electrostatic force* (with the opposite sign, and hence reversed vector, for identically valued charges):

$$\frac{q_1q_2}{4\pi\varepsilon_0 r^3}\mathbf{r} = \frac{d\mathbf{p}}{dit}.$$

Force, in any form, becomes a particularly significant quantity in physics in relating the squared modulus of the conserved parameters to the squared modulus of the nonconserved parameters.

A conservative system must acquire or lose energy through the infinitesimal changes in the coordinates of its components in order to balance the changes in its total *potential energy* (V), derived from the integral of force over distance.

Symmetry relations between mc^2 and terms such as $m\ (dr\ /\ dt)^2$ and $m\ (r\ /\ t)^2$ show that energy may be gained or lost in this form. When a component mass m acquires velocity $\mathbf{v} = d\mathbf{r}\ /\ dt$ by infinitesimal change over some time interval t, it can be shown, by integration of the force equation,

$$-\frac{Gm_1m_2}{r^3}\mathbf{r} = m\frac{d^2\mathbf{r}}{dt^2}$$

or

$$\frac{q_1q_2}{4\pi\varepsilon_0 r^3}\mathbf{r} = m\frac{d^2\mathbf{r}}{dt^2}$$

that the energy acquired, $Gm_1m_2\ /\ r$ or $-\ q_1q_2\ /\ 4\pi\varepsilon_0 r$ is equivalent to $m\ (dr\ /\ dt)^2\ /\ 2$ or $mv^2\ /\ 2$. Energy of this type is described as *kinetic energy* (T).

The normal symmetry relation

$$\frac{Gm_1m_2}{r} = m\left(\frac{dr}{dt}\right)^2$$

which applies when the system is conservative and there is no overall change in coordinate r is known as the *virial relation*, and the expression

$$\overline{V} = -2\overline{T}$$

applies for the time-averaged values in a steady-state conservative system subject to inverse-square law forces. Many large-scale systems, with only small-scale dissipative losses, approach very closely to this ideal relationship when treated as a single unit.

All the other significant relations of classical mechanics, in any of its forms, can be derived now by purely mathematical manipulation. For example, interpreting a 'system' to mean any combination of unit masses, the conservation of (linear) momentum follows by integration of the total force over time, and the conservation of angular momentum (defined at a fixed spatial displacement $\mathbf{r}$ from a given axis of rotation as $\mathbf{L} = \mathbf{r} \times \mathbf{p}$) from the fact that $d\mathbf{p}\ /\ dt$ in a conservative system is zero. That is

$$\frac{d\mathbf{L}}{dt} = \mathbf{r} \times \frac{d\mathbf{p}}{dt} = 0$$

in a conservative system, and the angular momentum remains unchanged. The principle of conservation of momentum requires that, for a conservative system, the total rate of change of momentum or force in the system is zero, which, in the simplest case, a system defined by two components, reverts to Newton's third law of motion: to every action there is an equal and opposite reaction.[8]

Direct manipulation of the scaling relations reveals that momentum terms are equivalent to $m c \mathbf{r} / r$, and that scalar terms of the form $G m_1 m_2 / r$ and $q_1 q_2 / 4\pi\varepsilon_0 r$, which we may describe as *gravitational and electrostatic potential energies*, are equivalent to those of the form mc^2; in each of these cases, m may be described as an 'equivalent mass'. Though these results normally emerge only from relativity theory, they are actually inherent in the structure from which classical mechanics must be derived.

Further results follow on immediately from the mathematical definition of new concepts. Thus, defining velocity as $\mathbf{v} = d\mathbf{r} / dit$ and acceleration as $\mathbf{a} = d\mathbf{v} / dit = (-)\, d^2 r / dt^2$, and field intensity as $\mathbf{F} / m$, we have, in the case of constant mass, $\mathbf{F} = m\mathbf{a}$, and can define *gravitational field intensity* as

$$\mathbf{g} = -\frac{Gm}{r^3}\mathbf{r}$$

and *electrostatic field intensity* as

$$\mathbf{E} = -\frac{iq}{4\pi\varepsilon_0 r^3}\mathbf{r}\,.$$

These gravitational and electrostatic field terms are directly additive as vectors without regard to the masses or charges with which they are associated. Since the interactions $m_1 m_2$ and $q_1 q_2$ apply to all masses and charges, the field terms extend the gravitational and electrostatic interactions into universal interactions.

The imaginary nature of time ensures that time measurement takes place only in terms of t^2 and not in terms of t. Hence time measurement requires the agency of force and acceleration and not that of unaccelerated motion. Imaginary time also leads to a force vector defined in terms of $m d^2\mathbf{r} / dt^2$ or equivalent being negative where the space vector is positive. It is because the ultimately important quantities in observational physics are real numbers that the results of interactions cannot be described in terms of imaginary quantities such as unaccelerated motion ($d\mathbf{r} / dit$) or unchanging momentum ($m d\mathbf{r} / dit$).

From vector theory, we can show that, for the related *scalar potentials*, $\phi = -Gm / r$ and $\phi = -iq / 4\pi\varepsilon_0 r$,

$$\mathbf{g} = -\nabla\phi$$

and

$$\mathbf{E} = -\nabla\phi,$$

and, also, that the respective force laws are equivalent to the Laplace equations

$$-\nabla^2\phi = \nabla.\mathbf{g} = 0$$

and $$-\nabla^2\phi = \nabla.\mathbf{E} = 0,$$

in a space without sources,[9] and to the Poisson equations,

$$-\nabla^2\phi = \nabla.\mathbf{g} = 4\pi\rho G$$

and $$-\nabla^2\phi = \nabla.\mathbf{E} = \rho / \varepsilon_0,$$

in a space with them. None of this requires any new physical argument. It can also be shown that the absolute value of potential terms is arbitrary to the point where it satisfies these equations and that an infinite number of possible values of potential may produce the same conditions of field intensity and force. This is an example of the general principle of gauge invariance.

Some particularly important results are expressed in terms of the *Hamiltonian function* $\mathcal{H}$, which is defined as $T + V$, in classical, as well as quantum contexts, and others in terms of the *Lagrangian* $\mathcal{L}$, defined as $T - V$. One of these is *Hamilton's principle*, which states that the Lagrangian function for a conservative system is either a maximum or a minimum. This follows directly from

$$\mathcal{L} = T - V = \frac{1}{2}m\left(\frac{dr}{dt}\right)^2 - \frac{Gm_1m_2}{r}.$$

Hence $$\frac{d\mathcal{L}}{dt} = \frac{dr}{dt}\left(m\frac{d^2r}{dt^2} + \frac{Gm_1m_2}{r^2}\right),$$

which, by Newton's second law, is 0. In the case of quantum fields, of course, which extend throughout space, the term actually used is the Lagrangian *density*, which may be integrated over all space to find the total Lagrangian. The Lagrangian density of a particle may be expressed as a function of the particle field defined by ψ, and Hamilton's principle provides the equations of motion from $\mathcal{L}$ in terms of particle wavefunctions.

If we write the Hamiltonian for this system in the form,

$$\mathcal{H} = T + V = \frac{p^2}{2m} + \frac{Gm_1m_2}{r},$$

then we also can see immediately that Hamilton's equations,

$$\frac{\partial q}{\partial t} = \frac{\partial \mathcal{H}}{\partial p} \quad \text{and} \quad \frac{\partial p}{\partial t} = -\frac{\partial \mathcal{H}}{\partial q},$$

apply for position and momentum coordinates q and p, for the first gives p / m and the second $-\, Gm_1m_2 / r^2 = m\, dv / dt$.

At the same time, Lagrange's equations of motion

$$\frac{d}{dt}\frac{\partial \mathcal{L}}{\partial \dot{q}_i} - \frac{\partial \mathcal{L}}{\partial q_i} = 0$$

apply to

$$\mathcal{L} = T - V = \frac{1}{2}mv^2 - \frac{Gm_1m_2}{r}$$

if we write Newton's second law in the form

$$\frac{d(mv)}{dt} + \frac{Gm_1m_2}{r^2} = 0 .$$

Newton's laws of motion, or the conservation laws of energy, momentum and angular momentum are the only principles needed for the complete deductive development of classical mechanics, which describes the behaviour (i.e. space-time variation) of systems under the action of specific force laws such as Newton's law of gravitation and Coulomb's law of electrostatics, and this is true whether we describe the system mathematically in Newtonian, Lagrangian or Hamiltonian forms.

8.8 Classical Electromagnetic Theory

Electromagnetic theory follows on from classical mechanics on the basis of three fundamental principles: the automatic transformation of vector quantities into 4-vectors; the local conservation of charge; and Coulomb's law of electrostatics. To replace nonrelativistic equations with relativistic ones, we simply replace all vector terms with 4-vectors, **r**, for example, being replaced by (**r**, *ict*). This procedure can be done, of course, with purely mechanical equations, to generate the standard results of special relativistic mechanics, but it is particularly significant in classical electromagnetic theory, which follows on immediately from applying 4-vector terms to the definition of electrostatic force.

Charge conservation carries over unchanged from quantum to classical physics. The significant fact here is that charge is *locally* conserved, and, hence, by a standard argument, the continuity equation,

$$\frac{\partial \rho}{\partial t} + \nabla . \mathbf{j} = 0 ,$$

must apply, with ρ defined as the charge density and $\mathbf{j} = \rho\, \mathbf{v}$ as the current density – for this equation relates the rate of decrease of charge in any volume to the flux of current out of its surface. The differential operator in this equation is a

4-vector, and so, recognisably, is the quantity with scalar and vector parts, ρ and $\mathbf{j} / c$.

Now, the scalar part of this latter quantity appears in Poisson's equation,

$$-\nabla^2\phi = \rho / \varepsilon_0,$$

which is the differential form of Coulomb's law, and so we should expect to find an equivalent vector part ($\mathbf{A} / c$) for ϕ, and an equivalent scalar part ($-(1 / c^2)\, \partial^2 / \partial t^2$) for ∇^2. Since the new (Dalembertian) differential operator $\Box = ((1 / c^2)\, \partial^2 / \partial t^2 - \nabla^2)$ is itself a universal scalar, we may separate out the scalar and vector parts of the total equation to give the *wave equations*:

$$\Box\, \phi = \rho / \varepsilon_0$$

and

$$\Box\, \mathbf{A} = \mathbf{j} / \varepsilon_0.$$

It is significant that ϕ and $\mathbf{A}$ are arbitrary to the point where they satisfy these equations (the condition of gauge invariance, as previously discussed). For convenience, we can arbitrarily restrict the values using a *gauge condition*. If we choose the so-called Lorentz gauge, in which

$$\frac{\partial \phi}{\partial t} + \nabla.\mathbf{A} = 0,$$

we can define new vectors **E** and **B**, without reference to physical characteristics, such that

$$\mathbf{E} = -\nabla\phi + \frac{\partial \mathbf{A}}{\partial t}$$

and

$$\mathbf{B} = \nabla \times \mathbf{A}.$$

Because of its origin in the time component, the scalar part of a 4-vector is an imaginary quantity while the vector part is a real quantity; thus ϕ and ρ are imaginary while **A** and **j** are real. Also, **E** is imaginary while **B** is real. (The complex nature of 4-vectors may be assumed in all relevant equations by taking c to mean ic where this is not specifically stated.)

We can now obtain the four (inhomogeneous) Maxwell equations in their standard form, and identify **E** and **B** as the electrostatic and magnetic field vectors.

$$\nabla.\mathbf{B} = 0 \tag{8.6}$$

follows automatically from the vector identity $\nabla.(\nabla \times \mathbf{A}) = 0$.

$$\nabla.\mathbf{E} = -\nabla^2\phi - \nabla.\frac{1}{c}\frac{\partial \mathbf{A}}{\partial t}$$

$$= \frac{\rho}{\varepsilon_0} - \frac{1}{c^2}\frac{\partial^2\phi}{\partial t^2} - \frac{1}{c}\nabla.\frac{\partial \mathbf{A}}{\partial t}$$

$$= \frac{\rho}{\varepsilon_0} - \frac{1}{c}\frac{\partial}{\partial t}\left(\frac{1}{c}\frac{\partial\phi}{\partial t} + \nabla.\mathbf{A}\right). \tag{8.7}$$

Applying the Lorentz condition,

$$\nabla.\mathbf{E} = \frac{\phi}{\varepsilon_0}.$$

This identifies **E** with the electrostatic field intensity of Coulomb's law.

$$\nabla\times\mathbf{E} = -\nabla\times(\nabla\phi) - \frac{1}{c}\frac{\partial}{\partial t}(\nabla\times\mathbf{A})$$

$$= 0 - \frac{1}{c}\frac{\partial \mathbf{B}}{\partial t}. \tag{8.8}$$

$$\nabla\times\mathbf{B} = \nabla\times\nabla\times\mathbf{A} = \nabla(\nabla.\mathbf{A}) - \nabla^2\mathbf{A}$$

$$= -\frac{1}{c}\frac{\partial}{\partial t}(\nabla\phi) - \nabla^2\mathbf{A}$$

$$= \frac{1}{c}\frac{\partial \mathbf{E}}{\partial t} + \frac{1}{c}\frac{\partial^2\mathbf{A}}{\partial t^2} - \nabla^2\mathbf{A}$$

$$= \frac{\mathbf{j}}{\varepsilon_0} + \frac{1}{c}\frac{\partial \mathbf{E}}{\partial t}. \tag{8.9}$$

Equations (8.8) and (8.9) represent Faraday's law of electromagnetic induction and Ampère's law. The second term in Ampère's law represents the so-called aethereal displacement current, introduced by Maxwell to make the equations symmetrical.

The homogeneous equations follow from the assumption of empty space (ρ = 0, **∇.E** = 0, **j** = 0). From (8.6) and (8.9), **B** becomes identical with the physical quantity which we describe as *magnetic field intensity*, and which provides the 4-vector correction ($- q\mathbf{E}v^2 / c^2$) to the electrostatic force $q\mathbf{E}$. The wave equations for **E** and **B** in empty space,

$$\Box\,\mathbf{E} = 0$$

and
$$\Box\,\mathbf{B} = 0,$$

follow directly from consideration of $\nabla \times \nabla \times \mathbf{E}$ and $\nabla \times \nabla \times \mathbf{B}$. Thus

$$\nabla \times \nabla \times \mathbf{E} \equiv \nabla(\nabla.\mathbf{E}) - \nabla^2\mathbf{E} = -\nabla^2\mathbf{E}$$

$$= \nabla \times \left(-\frac{1}{c}\frac{\partial \mathbf{B}}{\partial t} \right) = -\frac{1}{c}\frac{\partial(\nabla \times \mathbf{B})}{\partial t} = -\frac{1}{c^2}\frac{\partial^2 \mathbf{E}}{\partial t^2}$$

and
$$\nabla \times \nabla \times \mathbf{B} \equiv \nabla(\nabla.\mathbf{B}) - \nabla^2\mathbf{B} = -\nabla^2\mathbf{B}$$

$$= \nabla \times \left(\frac{1}{c}\frac{\partial \mathbf{E}}{\partial t} \right) = \frac{1}{c}\frac{\partial(\nabla \times \mathbf{E})}{\partial t} = -\frac{1}{c^2}\frac{\partial^2 \mathbf{E}}{\partial t^2}.$$

Gauge invariance in classical electromagnetic theory is expressed, as we have seen, in the fact that the potentials $\mathbf{A}$ and ϕ are not unique for given $\mathbf{E}$ and $\mathbf{B}$. Maxwell's equations are invariant under the transformation (most conveniently expressed in tensor notation)

$$A^\mu \rightarrow A'^\mu = A^\mu - \partial^\mu F$$

where F is an arbitrary function. That is, they are invariant under the transformations

$$\mathbf{A} \rightarrow \mathbf{A}' = \mathbf{A} + \nabla F$$

and
$$\phi \rightarrow \phi' = \phi - \partial F / \partial t.$$

Since ϕ and $\mathbf{A}$ transform in the same way as ct and x, y, z, this may be taken as a direct result of the translation-rotation symmetry of space-time introduced with the 4-vector system. In quantum electrodynamics, or quantum field theory, as we have seen in chapter 6, the $U(1)$ symmetry represented by the arbitrary function F, is interpreted as a process of absorption and emission by charged particles of virtual quanta of the electromagnetic field (photons), with the quantity of charge or coupling constant taken as a measure of the probability of absorption or emission. (This will be discussed further in chapter 11.)

The full development of classical electromagnetic theory requires the Lorentz force term in addition to Maxwell's equations to link the theory to dynamical equations of motion. (Some other results – such as the Poynting theorem, the Larmor radiation formula and the Larmor precession in a magnetic field – are deductive consequences, which will be used in later chapters.) The derivation of the Lorentz force formula is a standard development from classical mechanics and is most conveniently done using the Lagrangian formalism. Here, we make use of the fact that the electrostatic field may be defined in terms of a generalised scalar potential θ such that $\mathbf{E} = -\nabla\theta$. This potential describes the action of a charged particle in both electric and magnetic fields. Since $\nabla(\mathbf{A.v})$ is identical to $\partial\mathbf{A} / \partial t$, then a charged particle in an electromagnetic field may be supposed to have potential energy

$$U = q\phi - \frac{q}{c}\mathbf{A.v}$$

such that $$q\mathbf{E} = -\nabla U.$$

Applying the Lagrangian force equation

$$Q_j = \frac{d}{dt}\left(\frac{\partial U}{\partial q_j}\right) - \frac{\partial U}{\partial q_j}$$

for generalised coordinates q_j, we find that the component of force in the x direction is given by

$$F_x = \frac{d}{dt}\left(\frac{\partial U}{\partial v_x}\right) - \frac{\partial U}{\partial q_x},$$

$$= q\left(-\frac{\partial}{\partial x}\left(\phi - \frac{1}{c}\mathbf{v.A}\right)\right) - \frac{q}{c}\frac{d}{dt}\left(\frac{\partial}{\partial v_x}\mathbf{A.v}\right)$$

$$= q\left(-\frac{\partial\phi}{\partial x} - \frac{1}{c}\frac{\partial A_x}{\partial t}\right) + \frac{q}{c}\left(\frac{\partial \mathbf{v.A}}{\partial t} - \frac{dA_x}{dt} + \frac{\partial A_x}{\partial t}\right).$$

Since $$\left(\nabla\times\nabla\times\mathbf{A}\right)_x = \frac{\partial}{\partial x}(\mathbf{v.A}) - \frac{dA_x}{dt} + \frac{\partial A_x}{\partial t},$$

this becomes $$F_x = qE_x + \frac{q}{c}\mathbf{v}\times\mathbf{B}.$$

Generalising to three dimensions, we obtain the Lorentz force law

$$\mathbf{F} = q\mathbf{E} + \frac{q}{c}\mathbf{v}\times\mathbf{B}.$$

Slightly different equations would, of course, apply if we were to adopt a convention in which, for reasons of formal symmetry, the electric charge had a non-zero magnetic, as well as static, component (a classical magnetic monopole). Then $\nabla.\mathbf{B}$ would become $\mu_o\rho_m$ and the term $-\mu_o\mathbf{j}_m$ would be added to the expression for $\nabla \times \mathbf{E}$, while the Lorentz force would acquire an extra $g(\mathbf{B} - (\mathbf{v} \times \mathbf{E}) / c^2)$.

The fundamental equations for electromagnetic theory have now been derived, along with those for classical mechanics, in an unusual way, from the symmetry relations of the parameter group, as structured in thc nilpotent formalism. This derivation has had two significant effects. One is that it has demonstrated that there is no fundamental divide between classical and quantum physics, as the quantum condition is intrinsic to the derivation, and the other is that it has established that the quantities E and $\mathbf{p}$, which were introduced essentially as labels in the original nilpotent structure, really do have the properties that in classical physics are associated with energy and momentum.

The equations derived for the electromagnetic interaction should apply in principle to the strong and weak interactions also. That is, all three interactions would be of the same type and of equal strength if measured under the same conditions. Any differences should be applicable to the fact that the strong and weak charges cannot act as though free of the structures which contain them. This is the basis of Grand Unified Theories. The electromagnetic charge, though acting with freedom with regard to such structures (because of its association with a scalar quantity in the Dirac equation), is restricted in value by them. In chapter 15, we will show that Grand Unification can be accomplished in such a way as suggests that all three nongravitational interactions acquire a scalar phase invariance of exactly the same type as the electromagnetic interaction (and so an identical type of physical manifestation) at the energy value equivalent to the Planck mass. This is the energy value associated with quantum gravity, and suggests that gravity may also be associated in the unification in the same manner, with a linear set of Maxwell-type equations. Chapter 17 shows how this may be achieved, within the mathematical formalism of general relativity, by reinterpreting that theory as a combination of classical gravity with a Machian understanding of inertia.

Chapter 9

The Resolution of Paradoxes

This short chapter analyses paradoxes that arise at the quantum / classical boundary, concerning conservation and nonconservation, continuity and discontinuity, and irreversibility and causality, before going on to explain the mass frame and zero-point energy in the context of the fundamental continuity of mass, and the result this has on creating the laws of thermodynamics and the observed arrow of time.

9.1 Paradoxes Relating to Conservation and Nonconservation

The relationship between quantum and classical physics, which is significant in understanding both, is ultimately dualistic in origin. The duality between 'measurement' and the 'system' (equivalent to 'probe' and 'response') is a fundamental consequence of the symmetry of the parameter group, and the zero conceptual totality which makes physics possible. It leads, however, to a number of apparent paradoxes at the quantum-classical boundary, though the origin of the boundary itself is really the decreasing degree of isolation that is possible in complex systems.

It is proposed here that the source of the apparent paradoxes of quantum uncertainty and the origin of wave-particle duality lies in the fundamental symmetries between space, time, mass and charge, which determine that conservation and nonconservation, and continuity and discontinuity, remain *exactly opposite* properties. These built-in physical oppositions ensure that indeterminacy is inherent within the formal structure that underlies the whole of physics and that no fundamental choice can be made, on any physical grounds, between wave and particle theories, between quantum mechanics and stochastic electrodynamics (which is a theory based on random fluctuations of a vacuum filled with virtual zero point energy modes corresponding to $\hbar \omega / 2$ at each wavelength), and between competing physical interpretations of the Minkowski formalism for space-time invariance.

We begin with the concepts of conservation and nonconservation. These concepts have long been used in physics, but the full range of their possibilities has never been completely exploited. For this to be accomplished, it needs to be established that they apply specifically to the fundamental parameters, mass, charge, space and time, and that in this context they are exactly opposite properties, many of whose aspects may be discovered by applying precise rules of symmetry:

conserved quantities	**nonconserved quantities**
translation asymmetry	translation symmetry
rotation asymmetry	rotation symmetry
elements unique	elements nonunique
local identity	local nonidentity
noninvariance	gauge invariance
fixed in system	differentiable
fixed units	variable grain size

The mysteries of quantum mechanics become significantly less mysterious if we accept the principle that nonconservation is a fundamental and exact property, for, then, in addition to being 'quantum' (in the sense of incorporating the quantity $\hbar$), *all* physics becomes probabilistic. This probabilistic nature of physics stems from the fact that space and time are nonconserved quantities, while mass and charge are conserved. In principle, these are absolute requirements, so space and time should vary infinitely while mass and charge vary not at all. It is then *inherent* in the nature of space and time that they do not possess fixed values and that they should be arbitrary over a range limited only by the restrictions imposed by conservation laws related to mass and charge. This is the fact reflected in the phenomenon of gauge invariance, which applies alike to both classical and quantum systems, and it is also the ultimate source of quantum mechanical uncertainty and of its macroscopic manifestations in areas like molecular theory. States distinguishable *only* by arbitrary changes in space and time coordinates – in effect, changes of phase – have equal probability of occurring, and the laws of physics must be constructed so as not to make physical distinctions between them.

Quantum mechanics is therefore about describing the absolute variation of the nonconserved parameters while conservation laws are preserved. With space and time as inherently nonconserved quantities, the absolute measurement of particle coordinates becomes an intrinsic impossibility, which needs to be reflected in the structure of fundamental physical laws. Heisenberg's uncertainty principle (which will be derived formally in 16.6) tells us, in effect, that a *physical* conservative system cannot be realised in practice because a 'measurement' fixes the values of space and time, quantities that, in principle, ought not to be fixed. In other words, an ideal system requires that $\Delta p = 0$, $\Delta E = 0$, $\Delta x = \infty$, $\Delta t = \infty$; changes of energy and momentum should never happen, changes of space and time should always happen. The consequence is that the system, so fixed, ceases to be conservative.

Ideally, in a conservative system, variations in momentum and energy should be zero and variations in space and time infinite, but, if we elect to make the system nonconservative, we can reduce the relationship to finite variations in each; the extremely small value, in classical terms, of the constant relating the units of energy and inverse time, or momentum and inverse displacement, means that the nonconservative aspects of fixing particle coordinates become relatively insignificant on the large scale, and close approximations to conservative systems may be found.

Space and time, however, cannot really be fixed, even by making a 'measurement'. If, at some instant, we assume a classically precise state in space and time for a system with an infinite number of equally probable alternative states, the effect will be a degree of purely random variation from this state – that is, a random variation of the space and time coordinates – to allow an equal representation of the alternative states, although always in such a way as to preserve energy and momentum within a conservative system. The random variation of space with time will take place with respect to the three symmetries of space which yield equally probable states (translation, rotation and reflection) and the single corresponding symmetry of time (translation), to produce the translational, rotational and vibrational modes of motion which are characteristically associated with the internal energy function of classical thermodynamics. In this sense, there is no real quantum-to-classical transition, no real 'collapse' of the wavefunction, no need to invoke a classical 'measuring apparatus': 'classical' conditions are created when the field terms incorporated within the E and $\mathbf{p}$ of the quantum system are such as to restrict the random variation to a level below observation (ideally, to the Heisenberg uncertainty limit).

In the case of thermodynamics, classical and quantum approaches merge almost imperceptibly; classical thermodynamics is not really 'classical' because the fundamental thermodynamic quantity, 'heat', is not classical. Although it is often stated that the indeterminacy of classical thermodynamics is different in kind from the indeterminacy of quantum mechanics because in classical systems indeterminacy is not intrinsic while in quantum systems it is, heat is undoubtedly quantum mechanical in origin and cannot be treated as though it is not. The concepts of indeterminacy in classical electrodynamics, and probability in quantum mechanics stem from precisely the same source: gauge invariance, or the translation-rotation symmetry of space-time, and, ultimately, from space-time nonconservation. The random modes of motion of, say, molecules in gases with respect to an assumed 'fixed' position, have precisely the same meaning as the probability densities of quantum mechanics with respect to a 'measured' position determined by the collapse of a wavefunction onto a particular eigenstate. They can also be seen as harmonic oscillator solutions of the Dirac equation.

Gauge invariance in both quantum and classical physics requires that interactions have no absolute phase and the mass, energy and momentum associated with the exchange particles in these interactions have no specified positions at any given time instant. The same is likely to be true even of the 'fixed' masses associated with particles of 'real' matter, for these undoubtedly stem in some way from the action of gauge-invariant 4-vector fields. As a result of gauge invariance, all equally probable states occur, as it were, at once. To make a measurement, one state must be selected arbitrarily at a particular instant and given classically 'precise' coordinates. The classical or quantum nature of this process is purely a matter of the degree to which an expectation value can be regarded as a 'fixed' position. The result is a random variation from this state to allow equal representation of all the alternative states. In a measurement we choose one phase in an infinite series of equally probable states and then 'observe' the system progress through the rest of the series. That is, as time varies monotonically, a variation occurs in all the states of space which are symmetric by translation, rotation or reflection, but which retain fixed values of mass, energy and momentum.

It is for this very reason that it has been possible to develop a 'stochastic electrodynamics', based on a combination of classical physics and a fluctuating electromagnetic zero-point field, which is equivalent in nearly all significant respects to the standard versions of quantum mechanics, and which can account for the black body spectrum, *zitterbewegung*, spontaneous emission, and other basic quantum phenomena, in addition to leading to a relatively straightforward derivation of the Schrödinger and Dirac equations.[1-6] Stochastic electrodynamics is in no way a 'hidden variables' theory; the random fluctuations of the zero-point field are just as intrinsic as those of quantum mechanics, and this is because they stem from exactly the same source. The two theories are merely alternative expressions for the same physical information, and both derive from the fact that random variation is an inherent component of the nonconserved parameters, space and time.[7]

A true understanding of the idea of nonconservation reveals that there is nothing whatever mysterious about the intrinsically random nature of quantum mechanics or 'God playing dice' with the universe. It is simply a result of the absolutely symmetrical nature of the opposition of nonconservation, as manifested by space and time, to conservation, as manifested by charge and mass. Symmetry alone constrains space and time to be translation and rotation symmetric, for their units to have no fixed identity, and for their values to be indeterminate within limits set only by the necessity of conserving mass and charge. Quantum mechanics is, therefore, not something of a different, probabilistic, nature imposed on a 'deterministic' system of classical physics; it is the logical culmination of a system in which the parameters space and time are *intrinsically* indeterminate.[8]

9.2 Paradoxes Relating to Continuity and Discontinuity

Heisenberg uncertainty or indeterminacy is not the explanation of wave-particle duality, although it is frequently advanced as an *application*. In conventional accounts of quantum mechanics, the fundamental duality condition, $p = h / \lambda$, is applied to the Heisenberg uncertainty relation between momentum and displacement in order to investigate indeterminacy within a dualistic context, but the use of the Heisenberg relation does not contribute to the understanding of the origin of the duality condition itself. Indeterminacy results from the distinction between conserved and nonconserved quantities, but duality comes from a separate distinction: that between quantities which are continuous or discontinuous, or, as we may also call them, indivisible or divisible.

According to the Copenhagen interpretation of quantum mechanics, measurement takes place at the interface between the 'world' and the 'measuring apparatus'. We have *measurement*, which requires fixed space and time; and a *system*, which requires nonfixed space and time. Now, physical laws are constructed in terms of conservative systems; physical measurements are not. Fixing space and time breaks continuity and variability; irreversibility of time shows up when we break its continuity. Coarse-graining processes, for example, produce irreversibility because every process that interrupts smoothness does, not because they are its intrinsic source. Irreversibility cannot be manifested until we interrupt time, but becomes immediately apparent as soon as we do so. Measurement 'breaks the rules': continuous or indivisible time becomes countable, so its direction becomes manifest.

Now, the whole of 'measurement' arises from the quaternion, and ultimately algebraic, link-up between mass and charge and the corresponding 4-vector link-up between space and time, which are incorporated into what we describe as 'Lorentz-invariance'. (It is the *further* link-up, between the conserved quaternion mass-charge and the nonconserved 4-vector space-time, in the Dirac nilpotent, which is responsible, as we have seen, for 'uncertainty'.) Lorentz invariance is at the interface between the continuous and discontinuous, and the real and imaginary. With space and time we have two parameters, one of which is discontinuous and the other continuous; and, because of the imaginary system of numbers which we have used, we must combine them in a single mathematical structure. Lorentz invariance makes space continuous or it makes time discontinuous. Either way is possible; the choice provides the origin of wave-particle duality. The combination introduces measurement, which manifests itself clearly in the discontinuous option; but the fundamental sharing of properties occurs in 'measurement' only, and not in ultimate 'reality'. The same options occur in the case of mass and charge.

It is, therefore, the Lorentz invariance between space and time, and the parallel connection between mass and charge – the ultimate source of the process and units of 'measurement' as independent of the laws of physics – which forces us to link one quantity which is continuous (that is, time or mass) with one which is discrete (that is, space or charge). Lorentz invariance arises solely from the mathematical description of the parameters in terms of real or imaginary numbers (quaternions requiring an added real part, and 4-vectors an added imaginary) but the result of this forced union is that, *for the purpose of measurement*, either a quantity which is continuous must become discrete, or a quantity which is discrete must become continuous. The first is the particle option (time and mass become discrete), while the second leaves us with waves (space and charge become continuous). Neither solution is characteristic of what may be called ultimate 'reality', the idealised state in which no parameters exchange properties;

each is simply introduced as an artefact of measurement. Lorentz invariance, of course, leads us to what is usually called the 'wave' equation, but this is just as much a 'particle' equation, as quantum theory has shown (and Hamilton did earlier in a classical context), and the formal expression cannot predict the physical nature of its individual solutions.

Now, the Heisenberg relations are directly connected with conservation laws involving the fundamental parameters mass and charge. Thus

$$\Delta E \, \Delta t \geq \hbar \, / \, 2$$

is associated with conservation of mass,

$$\Delta p \, \Delta x \geq \hbar \, / \, 2$$

with conservation of charge magnitude, and

$$\Delta J \, \Delta \theta \geq \hbar \, / \, 2$$

with conservation of charge type. The spin term $\hbar \, / \, 2$ suggests a minimum discrete value for angular momentum. The fact that electron spin has constant magnitude in any given direction is an extreme example of the uncertainty relations; zero variation of spin is associated with infinite variation of angle.

To make physical *measurements*, as we have seen, it is necessary to fix, to some degree, the nonconserved quantities, space and time, at the expense of unfixing the conserved quantities, mass and charge. Of course, as we have also observed, this does not mean that classical conservation laws are violated, only that the physical measurements are not made within systems defined as conservative. This is why pairs of variables exist that, in the Heisenberg formulation, do not commute, and their noncommutation is expressed in terms of the constant which relates their *reciprocal* units. Energy (which corresponds with mass as a conserved quantity) does not commute with time; and momentum (which corresponds with charge) does not commute with space. The reason for these particular pairings is that energy, mass and time are nondimensional; while momentum, charge and space are (3-)dimensional. The final anticommuting relation also exists to cover the rotational aspects of the latter. Angular momentum (corresponding with type of charge) does not commute with angle (i.e. the angle of the phase).

In addition, the parameters in the first pairing are fundamentally continuous, while those in the second pairing are fundamentally discrete; and, in their combination, we have the ultimate reason for wave-particle duality. The quantity $\hbar$ thus links pairings of continuous *and* of discontinuous quantities, which are separately linked by Lorentz-invariance, or its quaternion equivalent: space-time, momentum-energy, charge-mass. It is because discrete charge is linked with the microscopic aspects of matter that we have hitherto assumed that 'quantization' (meaning discontinuity) is particularly characteristic of this aspect of physics. But the continuity of the parameter mass makes it possible to reconstruct physics at the most fundamental level on the basis of a continuous vacuum field (or 'aether').

9.3 Irreversibility and Causality

Time is essentially irreversible because it is absolutely continuous, in the same way as mass is unipolar because it is an absolutely continuous scalar field. The continuity of time is opposed to the essential discontinuity of space used in the process of measurement. Measurement is not an absolute process because we change some of the conditions required for absolute truth. According to the laws of physics, whether relativistic, quantum or classical, time has two indistinguishable directions of mathematical symmetry; this is characteristic of quantities determined by imaginary numbers. Physical irreversibility allows only one time direction, but, because of the mathematical indistinguishability of imaginary numbers, this can never be known in absolute terms. According to standard mathematical arguments, imaginary numbers of one sign cannot be privileged in any way with respect to those of the other; so, the laws of physics, in being constructed always for quantities involving the second power of time, prevent a mathematical realisation of time's direction.

Now, 'relativity' is a term with a complex meaning. The expression 'relativity' is used variously to cover more than one aspect of the theories of that name, and to locate the ultimate sources for the theories, it is necessary to separate out these aspects according to their particular origins. Thus, space and time are 'relative', in one sense, because they are nonconserved and have no absolute meaning; but uniform motion is, additionally, 'relative' because it is dependent on the first power of time, and is, therefore, an imaginary quantity. Accelerated motion is privileged, unlike uniform motion, because it is no longer concerned with space and time only, but, by relation, with conserved quantities, such as mass, and this is because it is also concerned with a real quantity of motion, with time taken to the second power.

This aspect of 'relativity', of course, is also apparent in nonrelativistic Newtonian physics, but, in *truly relativistic* theories, there is the additional element of the mathematical combination of space and time, which extends the aspect of nonconservation or 'relativity' known as space rotation invariance to a combined space-time rotation, or Lorentz, invariance. The theories which incorporate Lorentz invariance as a fundamental component necessarily involve a mathematical combination of unlike physical quantities – space and time are only similar in their shared property of nonconservation. The combination is not, in fact, a true physical process; the consequence is that the mathematical procedure of Lorentz-invariance has no completely recoverable physical meaning.

Lorentz invariance involves squaring of quantities; space and time are only linked when they are squared for Pythagorean addition. Squaring in the parallel case of mass and charge represents an 'interaction', and it is interaction which produces irreversibility. It is, indeed, the process of interaction, which is unique for any set of elements of mass and charge, which leads to irreversibility in physical terms. Interactions are responsible for wavefunction collapse in quantum mechanics; variously 'charged' particles, or sources of 'interactions', provide the so-called 'apparatus' required for quantum mechanical 'measurements'; there is no need for 'conscious' observers. A so-called wavefunction 'collapse' is simply the use of interactions or fields to restrict the absolute variability of a quantum system to some minimal level, possibly determined by Heisenberg uncertainty. All interactions involve some degree of collapse – there is no dramatic point at which an uncollapsed wavefunction becomes a collapsed one – and all wavefunctions involve some degree of collapse. The Lorentz invariance is not therefore directly responsible for interaction and for irreversibility, but it is responsible for them, indirectly, by symmetry.

The significance of the Lorentzian relation for *measurement* is that this operation requires the fixing of space and time, as well as their relationship to each other. That fixing has two aspects: it changes the space-time properties of nonconservation, and it violates time's additional property of continuity. In the former case, we give space-time the characteristics of mass-charge. In the latter case, we give time the characteristics of space. Stop time at any point to make a measurement and it is no longer continuous; so irreversibility manifests itself. Measurement therefore requires us to determine the arrow of time, to specify the extra thing we need to know, in addition to the fixed mass, charge, etc., within the system. The identical mass-charge link-up gives us fixed values of mass.

Time is, in fact, absolute in one sense, relative in another. Classical and relativistic theories, contrary to popular opinion, use *both* absolute and relative time. The absolute order of causally-connected events cannot be reversed. This is what Newton meant by absolute time: events cannot precede their causes. In what is probably one of the most often quoted, and most misunderstood, passages in the *Principia*, he writes: 'Absolute, true, and mathematical time, of itself, and from its own nature, flows equably without relation to anything external, and by another name is called duration: relative, apparent, and common time, is some sensible and external (whether accurate or unequable) measure of duration by the means of motion, which is commonly used instead of true time ... .'[9] But, as he also writes: 'It may be, that there is no such thing as an equable motion, whereby time may be accurately measured. All motions may be accelerated and retarded, but the flowing of absolute time is not liable to any change. The duration of perseverance of the existence of things remains the same, whether the motions are swift or slow, or none at all: and therefore this duration ought to be distinguished from what are only sensible measures thereof ... .' Absolute time, to Newton, is an order of events, not a 'sensible' measure: 'As the order of the parts of time is immutable, so also is the order of the parts of space. ... All things are placed in time as to order of succession; and in place as to order of situation. ... in philosophical disquisitions, we ought to abstract from our senses, and consider things themselves, distinct from what are only sensible measures of them.' In principle, this is no different from the Einsteinian order determined by causality.

Of course, a complete causal sequence can never be established, because we never have complete information about a physical situation through measurement, but such would certainly exist on a cosmic scale, and Einstein's general theory also presupposes it. Order, as Mogens Wegener has put it, is absolute, while duration (taken as a 'measured' quantity) is relative,[10] and this fact seems to be the basis of both Newtonian and Einsteinian positions. Absolute order, of course, stems from the irreversibility or continuity of time, while the 'relativity' of the measurement of duration is a result of its nonconservation. But causality is only known relatively, that is when we interrupt the flow of time. The scale of time is also arbitrary, and therefore relative, like that of space, but the unidirectionality of time makes it unlike space in this respect. Hence, 'relativity' of time is not the same as 'relativity' of space – which includes nonconservation of magnitude, direction, components, grain size, element identity, zero position and discontinuity.

Causality requires discontinuity, as in the Einstein approach to relativity. As soon as we make time discontinuous at all, we directly introduce causality or time's specific direction. Before this, we don't know what its direction is, and we have no direct knowledge of causality. In quantum mechanics, we don't need causality before we make a measurement. This is why quantum mechanical 'information' appears to be transferred faster than light, when concerned with virtual processes, as in the experimentally-observed correlations between the angular momentum states of once-connected photons. 'Interactions', in this sense, are immediate. As a result of gauge invariance, we have no absolute knowledge of the phase of an interaction, and so we cannot specify it as beginning at one point in time and ending at another. It is only when there is an actual so-called 'transfer of energy', a localisation or 'collapse of the wavefunction' in an irreversible event, a 'measurement' or interaction with a 'measuring apparatus', that we can specify the velocity of light as a limiting speed. (The 'transfer of energy' is only 'so-called' because the actual transfer involved is of structure or discreteness, not energy.) Causality also requires mass to be discontinuous, as in Einstein's theory, where relativity is linked with the lightquantum and signalling at the speed of light.[11] The alternative, wave, picture (with its built-in gauge invariance) is not based on an identifiable sequence of causally related discrete events.

The second law of thermodynamics is the physical expression of irreversibility and causality. It is only half a conservation law[12] – entropy can be created but not destroyed – because time is irreversible and we choose only one of two options when we make a measurement. The actual direction of time is that which corresponds to the single allowed sign of mass-energy (arbitrarily described as 'positive') to which it is symmetrical. Roger Penrose has speculated that gravity will solve the irreversibility paradox;[13] this is true in the sense that positive mass means positive time. If we like, we can say that the negative form of mass is virtual; so is the negative form of time.

9.4 The Mass Frame and Zero-Point Energy

What is known as absolute space, or aether, or vacuum, is the mass frame. This frame is unobservable by definition, although we can conceive of it theoretically, for, just as we can transfer the discontinuous properties of space to mass, so we can transfer the continuous properties of mass to space. Observation, of course, requires discontinuity. The frame is also not subject to variation, and is non-Lorentzian, but it appears to be the frame in which gravity operates, and it is only in such a frame that fields can be described as continuous. It is only in the *Lorentzian* frame, however, that we have variation, and that we have concentrations of mass brought about by charged particles, etc. In the absolute frame, mass would be unchanged from place to place.[14] The rest frame of the photon is of this kind, photons experiencing zero time and zero space, and so transferring their energy immediately and without change of location.

Vacuum, as we know, is not empty space, but contains an apparently infinite amount of zero-point energy. Zero-point energy, according to exponents such as T. H. Boyer, can be interpreted as a classical phenomenon, arising from electromagnetic radiation, with an energy spectrum of $\hbar \omega / 2$ per normal mode of vibration, derived from Lorentz invariance, and leading to classical explanations of the Planck black body spectrum and Bose-Einstein statistics. (The zero-point energy is, of course, seen from within a Lorentzian frame in the context of measurement.) The Schrödinger equation has been reduced to Newtonian mechanics combined with this extra component, while the third law of thermodynamics has been shown to be derivable classically if the zero-point energy component is taken into account. Classical thermodynamics, including the theory of specific heats, certainly becomes completely consistent when it is included, and equipartition leads directly to the Planck law for black body radiation and not to the Rayleigh-Jeans version. At the same time, quantum field theory becomes, in effect, classical field theory with zero-point fluctuations. The whole development of stochastic electrodynamics, as it is called, has shown, more or less conclusively, that quantum results can be duplicated by classical theories which take full account of the existence of zero-point radiation.[1-6] In principle, continuous 'aether' theories work if the effects of aether are taken as equivalent to discontinuous quanta.

It is interesting to note that the delocalised energy assumed in the continuous theories has the characteristics of both the classical aether and the quantum mechanical vacuum, and need not be finite in value. Rather than finding the infinite energy density of the vacuum a problem to be avoided by some arbitrary cut-off, we should rather expect it to be true, and suggestive of a medium along the lines of Dirac's infinite number of filled negative energy states, and related to the exclusion of the negative direction of time. Vacuum energy in this form would certainly remain undetectable by direct means.

The term $\hbar / 2$ which appears in the zero-point energy expression is significant because it is identical to the minimum value of action in the Heisenberg uncertainty relations, and is also the spin angular momentum unit for a fermion. Spin is not, as is often thought, a relativistic or a quantum concept, though it found its first explanation in Dirac's relativistic quantum equation for the electron. It is actually an expression of pure space rotation invariance, and can be found in classical, as well as in quantum, contexts.[15] It can also be derived, as we have seen (in 7.2), from the nonrelativistic Schrödinger equation, as well as the relativistic one of Dirac, given a more extensive understanding of the vector properties of space.

The factor ½ in the spin term $\hbar\ /\ 2$ derived from the Schrödinger equation comes from the kinetic energy expression $p^2\ /\ 2m$. From the same origin is derived the zero-point energy ($\hbar\omega\ /\ 2$) in the Schrödinger expression for the linear harmonic oscillator (cf 16.5). But the total energy of an electromagnetic oscillator also contains an equal component from electromagnetic field modes (which is why the spontaneous emission coefficient is twice that for stimulated emission, cf 16.8). The $\hbar / 2$ term in the Heisenberg uncertainty relations, being identical to that derived from the electromagnetic zero-point harmonic oscillator, also has the same ultimate origin in the classical kinetic energy term. The relativistic spin $\hbar\ /\ 2$ equation predicts, in addition, as we have seen in 7.6, a *zitterbewegung* or additional random motion, with frequency $2mc^2\ /\ \hbar$, superimposed on the normal directed motion, which represents an interference between the positive and negative virtual energy components (which both exist where, as here, causality and a unique time-direction are not specified).

Now, the Klein-Gordon equation for the photon is based on $E = mc^2$, with twice the mass applied compared to that in the Dirac equation, and this is, essentially, because it is derived from a potential energy term (mc^2), rather than from the classical kinetic energy, $mv^2\ /\ 2$. Hence, the spin in this case is derived by dividing the momentum operator by m, rather than by $2m$. Consequently, the spin is a unit of $\hbar$, and the energy of emission or absorption of photons is $\hbar\,\omega$, rather than $\hbar\,\omega\ /\ 2$. This is why transformations between states in the Heisenberg formalism are in integer units of $\hbar\,\omega$. In principle, the term $\hbar\,\omega$ is connected with an exchange particle and hence represents potential energy (the mass of the particle); while $\hbar\,\omega / 2$ is associated with intrinsic rotational energy (and wave-type motions) and hence represents kinetic energy. The distinction is essentially that of the classical virial theorem from which the relationship between kinetic and potential energies ultimately derives, though the ultimate origin of that theorem is also, of course, quantum mechanical. (See chapter 16.)

So, it would seem that the two major representations of quantum mechanics, like the two major physical interpretations of relativity theory offer choices between discontinuous and continuous formalisms, and the same choice is offered between quantum mechanics itself and a stochastic theory based on a classical vacuum field. Heisenberg's representation of quantum mechanics is a discontinuous theory, with integral spin $\hbar$ boson exchange (equivalent to fixed potential energy) as the mechanism for interaction, and unit values of $\hbar$ in the commutators; while Schrödinger's formulation is continuous, with gradualistic energy exchange, just as the classical kinetic energy $mv^2 / 2$ or $p^2 / 2m$ represents integration over a continuous range of energy values. In the same way, Einstein's special relativity is a discontinuous theory, with events caused by a localised exchange of particles with unit $\hbar$; while the Lorentz-Poincaré aether provides an alternative model in which the emphasis is on continuity provided by the delocalised energy provided by a continuous vacuum field.

In principle, the choice seems to be between potential or kinetic energy mechanisms, via spin $\hbar$ or spin $\hbar / 2$ exchanges, boson or fermion carriers, abrupt or gradualistic transitions, using localised or nonlocalised energies. The photon will never be absolutely necessary to the construction of a theory of radiation, though it has an undoubted significance in fundamental particle theories, and it is not strictly true to regard 'classical' theory as requiring continuous waves and 'quantum' theory as introducing discontinuity at a more microscopic level, for quantization is really only a quasi-discrete theory. The charges that produce quantization in the Dirac nilpotent structure are certainly discrete quantities, but the composite quantities that they produce (energy and momentum) are only discrete in certain aspects. Nor is classical theory essentially continuous, for the idea of measurement which it introduces is impossible except in discrete terms. In addition, real physical systems, because of the fundamental quantum connections which make up the universe we observe, can never be described as either fully quantum or fully classical.

The existence of the constant $\hbar$, and its undoubted universal validity, does not by itself make any comment on the discontinuity of the radiation field; just as the discontinuous theory uses integral multiples of this constant to define its energies, so the continuous theory uses half-integral values. It is almost certainly always possible to use either method, in each physical instance where the question arises. Duality is very probably completely absolute across the whole range of physics. According to the analysis in this chapter, it stems from fundamental symmetries involving the irreducible parameters space, time, mass and charge, in particular, the existence of both direct and inverse relationships between the basic units of each, and the pairing off of parameters as real or imaginary, conserved or nonconserved, discrete or continuous, in addition to the fact that the act of measurement effectively requires a violation of the fundamental conditions determining these relationships. There is no fundamental dividing line between classical and quantum physics – all physics is 'quantum' in requiring the constant $\hbar$. As we have seen, fundamental derivations of the equations of classical mechanics and electromagnetic theory require it, just like those of quantum mechanics and quantum electrodynamics. But we can use $\hbar$ either in a continuous sense ($\hbar\omega / 2$, zero-point energy) or in a discontinuous sense ($\hbar\omega$, quantum energy), just as we can use kinetic or potential energy, related by the same factor, in understanding the behaviour of material gases (see chapter 16).

9.5 Two Versions of Relativity

The duality between fundamentally continuous and fundamentally discontinuous approaches to physics is also relevant to the relationship and relative status of the Einstein-Minkowski version of special relativity with respect to the aether-based theories associated with Poincaré and Lorentz. This problem has been one of the most contentious in recent scientific history, but we can now see that the relationship between the two relativity theories is very clearly related to the problem of duality in the physical nature of radiation. And, in fact, there are many sets of alternatives in fundamental physics which are merely different ways of making the same basic choice. The choice, as we have seen, also extends to areas of pure mathematics; for example, to that between Leibnizian differentials (based on space) and Newtonian fluxions (based on time), or to that between real numbers, representing space, which are ultimately based on a countable number of algorithmic processes, and the uncountable set of real numbers defined by Cantor, representing time. So, we have:

particles	waves
relativity	aether
quantum mechanics	wave mechanics
QED	SED
$\hbar$	$\hbar / 2$
potential energy	kinetic energy
charge-like	mass-like
space-like	time-like
momentum-related	energy-related
spin 1 exchange	spin ½ exchange
boson exchange	fermion exchange

Einsteinian special relativity is at one extreme of a range of options open to us as a result of Lorentz invariance. The derivation of expressions like that for the relativistic Doppler shift of light in vacuum does not require the explicit exclusion of a medium; *any* inertial frame can be used in the derivation, and a privileged one is by no means excluded. The possibility of such a privileged frame is even suggested by the existence, in addition to the zero-point energy, of an isotropic microwave background radiation relative to which the Earth's 'absolute' motion can be detected. Again, tests of Lorentz invariance are not identical to absolute tests of the assumptions of special relativity; a perfectly uniform aether would show no deviations either. Because the zero-point energy is virtual in the Heisenberg formulation, based on discrete energy exchange – a result of Heisenberg uncertainty – there is a degree of option in our employment of the concept, even where we are using the related discrete energy exchange mechanism of Einsteinian relativity. The fact that we have the option of excluding it in kinematic considerations does not mean that we must necessarily do so.

Of course, as Heisenberg's quantum mechanics is based on a discrete model of radiation, it is an equally valid (though non-Heisenberg) option to treat the zero-point energy as real. This is the basis of stochastic electrodynamics, the Lorentz-Poincaré version of relativity, and even the Schrödinger formulation of quantum mechanics. Neither option gives an entirely true picture of reality, as each is limited by the processes of measurement. There is undoubtedly a truly continuous real distribution of energy or mass in the vacuum, but matter, on the other hand (representing 'charged' particles), is discrete. Einstein, Minkowski, Heisenberg and QED, taken to their logical conclusions, would deny the existence of real continuous mass; Lorentz, Poincaré, Schrödinger and SED, taken to their logical conclusions, would deny the existence of real discontinuous charged matter; each, of course, has to accommodate the alternative possibilities in a virtual form. Essentially, to maintain Lorentz-invariance for the purpose of measurement, we have to assume, either that continuous mass is discrete, or that discontinuous charge is continuous; either way, the choice represents a deviation from fundamental 'reality'.

In the same sense, as we have seen in chapter 7, Heisenberg's quantum mechanics and Schrödinger's wave mechanics are quite different theories physically, though their mathematical formulations are interchangeable, as Schrödinger and others have demonstrated; but there is no information on which we can decide that one is superior or closer to physical 'reality' than the other. It is like the Heisenberg formulation itself with its arbitrary choices between momentum and position, or energy and time. It is not Lorentz-invariance itself that introduces the wave aspect into physics. The so-called 'wave equation' has both wave and particle solutions. Einstein's version of relativity is a discontinuous particle theory, and we have both continuous (Schrödinger) and discontinuous (Heisenberg) versions of wave or quantum mechanics.

And, indeed, it is a crucial point in setting up such alternative theories that the differences between them are *beyond the scope of measurement*, because measurement violates the conditions required for absolute knowledge. Thus, while the Lorentz-Poincaré theory requires an aether which cannot be detected (by electromagnetic signals), a velocity of light change that cannot be observed, the Einstein-Minkowski version requires a *relativity* which cannot be observed either, for the lesson of the whole clock paradox argument as is that, under no conceivable physical circumstances, can we conceive of a situation in which the two clocks are perfectly symmetric. In other words, there is no such thing in the physical world as *relativity*, or, very probably, simultaneity, either (that is, at least, in measurable terms, any measurement requiring a causal sequencing of the events involved); we can use the idea of 'relativity' precisely because it is an absolute or extreme position which can never be realised physically. The special theory of relativity, as we have seen – contrary to the views expressed by many of its exponents – uses absolute time in the true Newtonian sense, that is, in the sense of an irreversible absolute order of events. The impossibility of transmitting energy faster than c means that there can be no reversal of causality; events cannot precede their causes.

Again, the whole point of quantum mechanics, and the Heisenberg uncertainty principle in particular, is that the act of measurement violates the principles on which a 'system' is built, namely, the conservation principles of mass, energy, momentum, angular momentum and charge, and the *nonconservation* principles of space and time (translation and translation-rotation symmetries, gauge invariance, and so forth). To the extent where we allow fixing of space and time, we must have simultaneous 'nonfixing' of the conserved quantities; it is the extent to which we can reduce this latter to a minimum which determines our success in using a classical formalism. It is significant, of course, that the 'aether' of quantum mechanics is a *virtual* aether, unattainable in the normal sense, and a product of the uncertainty of fixing absolute position in space.

In no sense does this lead to a position of complete relativity of knowledge;

overall absolute positions are certainly still possible, and even required. The relationship between the components of the overall theory is a precise one – varying assumptions are made within *fixed* areas of indeterminacy. The fundamental axioms may be contradictory, but this is always allowed within the general theory. Special relativity and the alternative aether theories are such options; and these in effect presuppose the quantum and classical theories of radiation. We can have an idealised absolute motion if we like, or we can specify an idealised absence of absolute motion. It is a choice available to us, like a gauge condition in electrodynamics.

It so happens that the fundamental structure of physics requires the incorporation of the opposing concepts of continuity and discontinuity on an equal basis; in principle, one basic parameter, mass, is continuous while another, charge, is discrete; these two fundamental quantities correspond, in the older style of physics, to the categories of aether and matter. Radiation is the connecting link between mass and charge, or as late nineteenth century physicists described it, aether and matter, and we can look on it either as discrete, like charge, or continuous, like mass; the position is indeterminate, and indeterminacy of this kind is fundamental within physics.

9.6 Thermodynamics and the Arrow of Time

Some account of thermodynamics has already been given, together with its connection with irreversibility and the direction of time. Each of the laws of thermodynamics, in fact, is, in its way, an expression of the character of the fundamental properties of time. These arise, as we have seen, from symmetry with space, mass and charge. From symmetry with space, we learn that it has no fixed units and is translation asymmetric; from symmetry with charge, that it is imaginary – alternative mathematically symmetric equations must exist for equations with the opposite sign of time; and from symmetry with mass, that it is a continuum. None of these properties can be explained in anything but abstract terms; they are purely a result of mathematical symmetry.

For thermodynamics, the last is the most important property. Time is uniform, without branching or dimensions, jumps, gaps, or any interruption to smooth flow. (This is true, as we have seen, even in special relativity, which rejects the reversal of causally related events.) It cannot become zero or change sign. This is the fundamental *arrow of time*. The second law of thermodynamics is a *manifestation* of this arrow, but it is certainly not the *cause*. Time is a *fundamental*, and intrinsically abstract, concept and is not in any way to be explained on the *a priori* assumption of a special structure within complex physical systems. It is the *arrow* which is fundamental, not the second law.[16]

The arrow is an expression of the *retardation* involved in measurement related to discrete sources, which we saw was a necessary result, due to compactification, of the Dirac nilpotent structure ($\pm\, ikt \pm \mathbf{i}\mathbf{r} + j\tau$), with τ occupying the same position as the unipolar term m in ($\pm\, ikE \pm \mathbf{i}\mathbf{p} + jm$). The combination also produces quantization, and there is a simple link which connects c-retarded (4-vector) interactions between such discrete sources as electric charges and quantization through energy proportional to inverse distance (or discrete separation), which, being c-retarded, becomes proportional to inverse time of interaction, or frequency of 'signalling'. In this sense, quantization and Lorentz invariance are different formulations of the same condition: one presupposes the other. The retardation, which arises from the unidirectionality of time, is also expressed in both classical and quantum terms through the concept of 'uncertainty' (another measure of nonconservation). The nilpotent packages all the effects into one. It is, also, as noted in 6.4, *naturally* thermodynamic in requiring a specification of the rest of the universe for its own definition.

Now, an 'interacting' system – or one in which the retardation becomes explicit – is no longer an isolated or microcanonical one, so the conservation principles will not be satisfied internally, and an external energy exchange will take place, causing the system to be redefined.[17] The measure of whether such an exchange is taking place (in the absence of a phase transition) is described as a difference in *temperature*; the system which loses energy during the exchange is described as being at the higher temperature, and the point at which no further exchange (or transfer of 'heat', dQ) is needed is called *thermal equilibrium*. An exchange also decoheres the angular momentum states in a microcanonical system, decoupling the angular momentum conservation from that of charge, and extending the element of uncertainty and randomness with respect to the classical *measurement* that is taking place (cf the exchange between $\mathbf{p}_1$ and $\mathbf{p}_2$ in 6.4). This is described as an increase in *internal energy* (dU). The fact that two systems cannot be in thermal equilibrium if *one* is acting with a third (and therefore involved in an energy exchange) constitutes the zeroth law of thermodynamics, which is effectively a definition of temperature.

Interactions between discrete systems are, of course, retarded, so thermal equilibrium is time-delayed, and may never be achieved before an exchange takes place with another system.[18] (To attain *complete* thermal equilibrium would, in fact, require an infinite time, and, as we also have seen, in 6.5, the natural condition for nilpotent fermions is nonequilibrium – which is precisely what drives time relentlessly in a single direction.) In addition, energy exchange between systems and decoherence obviously cannot occur in such a way as to diminish the effect of retardation (cf 7.5). So classical measurement describes the concept of 'work done' ($\bar{d}W$), in the first law of thermodynamics, as the difference between the heat supplied and the increase in internal energy ($dQ - dU$) – the inexact differential, $\bar{d}W$, reflecting the fact that this is a macroscopic or classical and statistical concept. The increasing effect of retardation and decoherence are then reflected in the second law, which is also necessarily statistical, but, as Carathéodory showed by pure mathematical reasoning, defines a quantity, entropy (S), involving both heat (Q) and temperature (T), to measure this increasing state of 'disorder', such that $dQ = TdS$. If temperature is defined in energy units – by equating the historically-defined Boltzmann constant k to 1 – then entropy becomes a pure number, an index of the number of equally probable states (Γ). Textbooks on thermodynamics, following Boltzmann, show that the mathematical function $ln\ \Gamma$ has the properties required of entropy in a microcanonical ensemble.

As all natural processes involve an exchange, and therefore a loss of microcanonical status, all involve an increase in entropy, and all are irreversible. Because all energy exchanges are retarded in the way described, the irreversible increase in entropy due to natural processes becomes associated with the unidirectionality of time, and the direction of time becomes associated with the progression of a sequence of irreversible dynamic events. Though entropy is associated generally with the concept of disorder, order is created within systems as a result of the separate conservation processes which apply to the different types of charge.

Some of these are relevant in the context of phase transitions. The creation and annihilation of fermions is, of course, a phase transition, and involves the creation and annihilation of units of weak, and other, charges. The weak charge is exclusive to fermionic matter, and its properties are involved in most kinds of 'phase' behaviour. It will be shown in 10.11 that the dipolar weak interaction produces a harmonic oscillator solution for the Dirac equation, and most of the properties of gaseous and condensed matter relate to the harmonic oscillator behaviour of its components, while the dipolar Van der Waals force, which expresses in its most fundamental aspect the nature of the weak vacuum (cf 10.11), plays a significant role in all material phases. In addition, the properties of the solid state are determined by the Pauli exclusion principle that invariably accompanies the presence of weak charge, while Bose-Einstein condensation is effectively the elimination of this charge and its dipolarity. Another phase transition of the Van der Waals-type occurs with the creation of interbaryonic, or nuclear, matter through a remnant of the strong forces between quarks, and this can be seen, in at least a partial sense, as a Bose-Einstein condensation. (Here, in some sense, the weak force dominates over the strong.)

Nonconservation of space and time is the ultimate cause of change in physical systems, and change occurs as a result of the natural tendency to create a state of thermal equilibrium as a result of interaction; this, in turn, arises from gauge invariance and translation-rotation symmetry, the direct manifestations of the principle of nonconservation. By being irreversible and occurring in all natural processes, the increase of entropy, which is the direct result of the attempted attainment of thermal equilibrium, becomes identical in principle to the inevitable change which is the result of the nonconservation of space and time.

Some kinetic energy is always present in physical systems, and, as zero-point energy, even in a space devoid of charged particles, simply as a result of the requirements of gauge invariance, and translational-rotational symmetry. Vacuum is thus merely the state of lowest energy. According to Heisenberg's uncertainty principle, oscillations of the radiation field can only be in the rest position if they have infinite momentum, and, even in classical theory, there is Maxwell's displacement current in 'empty' space. Clearly, the kinetic energy of a natural system and the temperature, or relative disorder of its components, can never be reduced to zero. However, we should expect those materials with the most coherent structures, such as crystals, to experience a lesser degree of disorder than the less coherent ones, and to tend towards a state of minimum entropy as the absolute zero of temperature is approached. This is the experimentally-established version of the third law of thermodynamics – though the unattainability of absolute zero is really a more fundamental condition. However, it is really with the first and second laws that thermodynamics is established as a subject, and the nature of the quantum to classical transition fully understood.

Chapter 10

Electric, Strong and Weak Forces

Concerning chapters 10-13, see Preface, p. xvi. With both quantum and classical physics now established, in principle, this chapter returns to the detailed formal discussion of the quantum processes involved in the three nongravitational interactions, with initial emphasis on the electromagnetic interaction as the simplest case. Though largely using established techniques, it shows the special power of the nilpotent formalism. The Dirac equation is first solved for a fermion in a spherically symmetric Coulomb (or static electric) field with an inverse-distance potential (the 'hydrogen atom' solution), and immediately we noticc that only one set of equations is required for the solution, as opposed to the two sets needed conventionally. This is because the nilpotent, unlike any other formalism, works as a complete 'package' of all the significant information. It is particularly significant that a point source, with spherical symmetry, defining angular momentum conservation, requires a minimal condition of a Coulomb field to a provide nilpotent solution. It is then shown how the actual *structure* of the Dirac nilpotent leads to the fundamental differences between the electromagnetic, strong and weak interactions. $SU(3)$ for the strong interaction is shown to be a consequence of the behaviour of the **p** term, and the baryon wavefunction is used to show how it operates. The nature of the potential involved is established as varying linearly with distance, and this potential is used to obtain a completely analytic solution of the Dirac equation, in nilpotent form, for quark-antiquark and three-quark interactions. The angular momentum conservation process involved is shown to extend also to weak and electric interactions, in illustration of the theorem already derived by symmetry, connecting this with conservation of *type of* charge. The $SU(2)_L \times U(1)$ structure connecting weak and electric interactions is then seen to arise from the pseudoscalar-scalar combination of the iE and m terms in the Dirac nilpotent. A discussion of the vacuum shows how a filled weak vacuum, corresponding to a continuous mass-energy distribution, leads to the Higgs mechanism for generating rest mass, and can be related to the conventional treatment, while the *weak isospin* $SU(2)$ results from the option of filling or emptying the electromagnetic vacuum. The solution of the Dirac equation for a spherically-symmetric distance-related potential which is not linear (like the strong interaction potential) or inverse linear (like the electromagnetic interaction potential) is seen to lead invariably to a harmonic oscillator, exactly like a dipolar field; this can be equated with the behaviour of the weak interaction, whose dipole nature comes from the dual solution of its complex nilpotent source term (iE). The $SU(2)_L \times U(1)$ combination thus emerges ultimately from the use of a complex scalar combination for the electroweak field, exactly as required to generate mass by the Higgs mechanism, while the three solutions of the Dirac equation for a *spherically-symmetric* field will necessarily be those which are concerned with the conservation of angular momentum. A key result is that the idealised electroweak mixing parameter, $\sin^2\theta_W$, must be 0.25.

10.1 The Dirac Equation in the Coulomb Field

Using the nilpotent quaternion state vector, we can reduce the procedure of describing the behaviour of single fermions under the actions of different potentials to relatively simple algebra applied to a single equation. In principle, we begin with the nilpotent equation

$$(\pm \boldsymbol{k}E \pm \boldsymbol{i}i\,\mathbf{p} + \boldsymbol{i}j\,m)\,(\pm \boldsymbol{k}E \pm \boldsymbol{i}i\,\mathbf{p} + \boldsymbol{i}j\,m) = 0.$$

As a product of row and column vectors, this can be written:

$$((\boldsymbol{k}E + i\boldsymbol{i}\mathbf{p} + i\boldsymbol{j}m)\,(\boldsymbol{k}E - i\boldsymbol{i}\mathbf{p} + i\boldsymbol{j}m)\,(-\boldsymbol{k}E + i\boldsymbol{i}\mathbf{p} + i\boldsymbol{j}m)\,(-\boldsymbol{k}E - i\boldsymbol{i}\mathbf{p} + i\boldsymbol{j}m))$$

$$\times \begin{pmatrix} \boldsymbol{k}E + i\boldsymbol{i}\mathbf{p} + i\boldsymbol{j}m \\ \boldsymbol{k}E - i\boldsymbol{i}\mathbf{p} + i\boldsymbol{j}m \\ -\boldsymbol{k}E + i\boldsymbol{i}\mathbf{p} + i\boldsymbol{j}m \\ -\boldsymbol{k}E - i\boldsymbol{i}\mathbf{p} + i\boldsymbol{j}m \end{pmatrix} = 0 \,. \qquad (10.1)$$

We can consider $(\pm \boldsymbol{k}E \pm \boldsymbol{i}i\,\mathbf{p} + \boldsymbol{i}j\,m)$, taken as an amplitude, to be the result of the action of a differential operator on a vacuum phase. For a free fermion, this leads to the Dirac equation:

$$\left(\pm i\boldsymbol{k}\frac{\partial}{\partial t} \pm i\boldsymbol{i}\boldsymbol{\sigma}.\nabla + i\boldsymbol{j}m\right)\left(\pm \boldsymbol{k}E \pm i\boldsymbol{i}\mathbf{p} + i\boldsymbol{j}m\right)e^{-i(Et-\mathbf{p.r})} = 0\,,$$

where the differential operator is understood to be a row vector, and the wavefunction a column vector, with the same four components as in eq. (10.1).

An important result emerges as soon as we set about applying spherical symmetry to the Dirac equation. Suppose we consider the behaviour of a fermion within the range of a point source. This is a situation which is intrinsically spherically symmetric, so it is convenient to write the Dirac operator in polar coordinates. To use the standard conversion of the ∇ or $\boldsymbol{\sigma}.\nabla$ term into polar coordinates, we need to use ordinary vectors, rather than multivariate vectors, which means that we require an explicit introduction of fermionic spin or total angular momentum. So we write:

$$\boldsymbol{\sigma}.\nabla = \left(\frac{\partial}{\partial r} + \frac{1}{r}\right) \pm i\,\frac{j+\frac{1}{2}}{r}.$$

We can now set up an operator of the form:

$$\left(\boldsymbol{k}\big(E + V(r)\big) + \boldsymbol{i}\left(\frac{\partial}{\partial r} + \frac{1}{r} \pm i\,\frac{j+\frac{1}{2}}{r}\right) + \boldsymbol{ij}m\right),$$

where $V(r)$ is the radially-dependent potential energy term. It will quickly become apparent that, unless $V(r)$ contains an expression of the form $A\ /\ r$, or $-A\ /\ r$, to compensate for those in the term beginning with $\boldsymbol{i}$, then no nilpotent solution can be found. We can regard this *Coulomb term* as the minimum requirement for spherical symmetry. It is, in fact, an expression of the magnitude of the charge, or the coupling constant.

It is remarkably easy to find spherically-symmetric nilpotent solutions for the field-dependent defragmented Dirac operator constructed in this way, as the energy, momentum and mass operators maintain their separate identities through all operations. This means that we don't need to break up the wavefunction into separate energy and momentum eigenfunctions to do calculations. The method is also completely general, and can be applied to any type of potential. In principle, all we need to do is to set up a differential operator with the appropriate field terms, and then find the function which will make its eigenvalue (taken over the four solutions equivalent to $\pm E$, $\pm \mathbf{p}$) a nilpotent. The method is exact and analytic and provides the full 'hydrogen atom' solution (involving hyperfine levels) in the case of the pure Coulomb interaction in just six steps, though a few more will be added here for clarity. It could even be argued that it is more strictly correct than the conventional method, which assumes that amplitudes that vary with position can be treated as constant.

Conventionally, applying a potential ϕ, to a fermion of charge $-e$, we use the covariant derivative

$$i\frac{\partial}{\partial t} \rightarrow i\frac{\partial}{\partial t} + e\phi ,$$

or, for eigenvalue E,

$$E \rightarrow E + e\phi.$$

This produces a Dirac equation of the form

$$(\pm \boldsymbol{k}(E + e\phi) \pm \boldsymbol{i\sigma}.\nabla + \boldsymbol{ij}m)\ \psi = 0,$$

and, for the hydrogen atom, or any system with a spherically symmetric Coulomb field, we assume a potential energy $e\phi$ of the form $A / r = -Ze^2 / r$. So we obtain as the Dirac equation for a fermion in a Coulomb potential:

$$\left(\boldsymbol{k}\left(E + \frac{A}{r} \right) + \boldsymbol{i}\left(\frac{\partial}{\partial r} + \frac{1}{r} \pm i\frac{j + ½}{r} \right) + \boldsymbol{ij}m \right)\psi = 0 .$$

The $\pm$ values of $\boldsymbol{k}$ and $\boldsymbol{i}$ still lead to four solutions within ψ. In the case of stationary states, this implies that ψ contains a nilpotent column vector of the form ($\pm\ \boldsymbol{k}E' \pm\ \boldsymbol{i}i\mathbf{p}' + \boldsymbol{ij}m$), where E' and $\mathbf{p}'$ are terms with the respective dimensions of energy and momentum.

In our methodology, however, the wavefunction is superfluous. We need only specify the operator required for spherical symmetry:

$$\left(\boldsymbol{k}\left(E + \frac{A}{r} \right) + \boldsymbol{i}\left(\frac{\partial}{\partial r} + \frac{1}{r} \pm i\frac{j + ½}{r} \right) + \boldsymbol{ij}m \right)$$

and then find the phase which will make the amplitude (or eigenvalue) nilpotent. So, we try the standard solution, and suppose that the phase factor has the form:

$$F = e^{-ar} r^{\gamma} \sum_{\nu=0} a_{\nu} r^{\nu} .$$

Then
$$\frac{\partial F}{\partial r} = \left(-a + \frac{\gamma}{r} + \frac{\nu}{r} + \ldots \right) F ,$$

where, for a bound state, a is real and positive. The amplitude produced by the differential operator then becomes

$$\left(\pm \boldsymbol{k}\left(E + \frac{A}{r} \right) \pm \boldsymbol{i}\left(-a + \frac{\gamma}{r} + \frac{\nu}{r} + \ldots \frac{1}{r} \pm i\frac{j + ½}{r} \right) + \boldsymbol{ij}m \right).$$

Written out in full column vector form, this would be:

$$\begin{pmatrix} \left(\mathbf{k}\left(E+\frac{A}{r}\right)+\mathbf{i}\left(-a+\frac{\gamma}{r}+\frac{\nu}{r}+\ldots\frac{1}{r}+i\frac{j+½}{r}\right)+\mathbf{ij}m \right) \\ \left(\mathbf{k}\left(E+\frac{A}{r}\right)-\mathbf{i}\left(-a+\frac{\gamma}{r}+\frac{\nu}{r}+\ldots\frac{1}{r}-i\frac{j+½}{\mathrm{r}}\right)+\mathbf{ij}m \right) \\ \left(-\mathbf{k}\left(E+\frac{A}{r}\right)+\mathbf{i}\left(-a+\frac{\gamma}{r}+\frac{\nu}{r}+\ldots\frac{1}{r}+i\frac{j+½}{r}\right)+\mathbf{ij}m \right) \\ \left(-\mathbf{k}\left(E+\frac{A}{r}\right)-\mathbf{i}\left(-a+\frac{\gamma}{r}+\frac{\nu}{r}+\ldots\frac{1}{r}-i\frac{j+½}{r}\right)+\mathbf{ij}m \right) \end{pmatrix}.$$

Squaring, and applying the nilpotency condition, we obtain:

$$4\left(E+\frac{A}{r}\right)^2 = -2\left(-a+\frac{\gamma}{r}+\frac{\nu}{r}+\ldots\frac{1}{r}+i\frac{j+½}{r}\right)^2 - 2\left(-a+\frac{\gamma}{r}+\frac{\nu}{r}+\ldots\frac{1}{r}-i\frac{j+½}{r}\right)^2 + 4m^2$$

In conventional terms, we would say that the amplitude of the wavefunction (for stationary states) must be of the exact form to zero the eigenvalues produced by the differential operator; and so we take the eigenvalues of the differential operator as a row vector of exactly the same form, and equate the product of the row and column vectors to zero.

Equating constant terms now leads to

$$E^2 = -a^2 + m^2$$

$$a = \sqrt{m^2 - E^2}\,.$$

Equating terms in $1/r^2$, with $\nu = 0$, we obtain:

$$\left(\frac{A}{r}\right)^2 = -\left(\frac{\gamma+1}{r}\right)^2 + \left(\frac{j+½}{r}\right)^2,$$

from which, excluding the negative root (as usual),

$$\gamma = -1 + \sqrt{(j+½)^2 - A^2}\,.$$

Assuming the power series terminates at n', and equating coefficients of $1/r$ for $\nu = n'$,

$$2EA = 2\sqrt{m^2 - E^2}\,(\gamma + 1 + n'),$$

the terms in $(j + ½)$ cancelling over the summation of the four multiplications, with two positive and two negative. From this we may derive

$$\frac{E}{m} = \frac{1}{\sqrt{1 + \dfrac{A^2}{(\gamma + 1 + n')^2}}} = \frac{1}{\sqrt{1 + \dfrac{A^2}{\left(\sqrt{(j + ½)^2 - A^2} + n'\right)^2}}},$$

which is applicable to any system with a spherically symmetric potential A / r or $- A / r$ (including classical Newtonian gravity) acting on a quantum fermionic state. With $A = Ze^2$, we obtain the hyperfine or fine structure formula for a one-electron nuclear atom or ion:

$$\frac{E}{m} = \frac{1}{\sqrt{1 + \dfrac{(Ze^2)^2}{(\gamma + 1 + n')^2}}} = \frac{1}{\sqrt{1 + \dfrac{(Ze^2)^2}{\left(\sqrt{(j + ½)^2 - (Ze^2)^2} + n'\right)^2}}} \tag{10.2}$$

and, with $Z = 1$, it becomes applicable to hydrogen. In fact, the formula we have derived is well-known as the standard derivation of the hyperfine structure for the hydrogen atom, and its derivation (six lines in the most abbreviated version) may be taken as a useful test of the power of any Dirac formalism.

Writing the total quantum number n as

$$n = \gamma + 1 + n' = \sqrt{(j + ½)^2 - (Ze^2)^2} + n',$$

equation (10.2) can be written in the form

$$\frac{E}{m} = \frac{1}{\sqrt{1 + \dfrac{(Ze^2)^2}{n^2}}}. \tag{10.3}$$

Expanding (10.3), we obtain

$$E = m\left(1 - \frac{(Ze^2)^2}{2n^2} \ldots\right),$$

giving us, to a first approximation, the energy levels of the nonrelativistic theory:

$$E_n = m\frac{(Ze^2)^2}{2n^2} = \frac{Z^2e^2}{2a_0n^2},$$

where a_0 is the Bohr radius, $1 / me^2$, or $4\pi\varepsilon_0\hbar^2 / me^2$ in conventional units (which is derived, classically, from the combination of the equations $mv^2 = e^2 / 4\pi\varepsilon_0 a_0$

and $mva_0 = \hbar$). This, of course, explains the structure of the formulae for the energy level transitions in the series for hydrogen, described by Balmer and others, i.e.:

$$\frac{\Delta E}{hc} = \frac{1}{\lambda} = R\left(\frac{1}{m^2} - \frac{1}{n^2}\right),$$

where m and n are integers ($n > m$) and the Rydberg constant R becomes, in conventional units, $m_e e^4 / (4\pi e_0)^2 \hbar^3 4\pi c$.

We can also use (10.3) to find a value of the term a in the expression for phase. Here, we write

$$\frac{a^2}{m^2} = \frac{m^2 - E^2}{m^2} = 1 - \frac{1}{1 + \frac{(Ze^2)^2}{n^2}},$$

from which, to a first approximation, we may derive

$$a = \frac{Ze^2 m}{n} = \frac{Z}{na_0},$$

which means that the phases, in the wavefunctions, associated with the first two energy levels, and approximating to exp ($-ar$), become exp ($-Zr / a_0$) and exp ($-Zr / 2a_0$), as in the standard nonrelativistic theory.

We can now supply standard techniques to find the normalization constants associated with these wavefunctions. For example, assuming that

$$\int \psi\psi * dV = 1,$$

in all cases, where dV is an element of volume, if we suppose that the wavefunction for the first orbital approximates to C *exp* ($-Zr / a_0$), where C is the normalization constant, then

$$\int_0^\infty \int_0^\pi \int_0^{2\pi} C^2 e^{-2Zr/a_0} r^2 \sin\theta dr d\theta d\phi = 1.$$

From this, we obtain

$$\pi\left(\frac{a_0}{Z}\right)^3 C^2 = 1$$

or

$$C = \left(\frac{Z^3}{\pi a_0^3}\right)^{1/2}$$

and $$\psi = \left(\frac{Z^3}{\pi a_0^3} \right)^{1/2} e^{-2Zr/a_0} .$$

The Coulomb field provides the simplest solution of the Dirac equation for a spherically-symmetric potential. In classical theory it is the only potential which provides the conditions for a stable orbit, other than the harmonic oscillator with force $\propto r$ (Bertrand's theorem). (A constant potential, or zero force, would, of course, merely change the origin fixed for the E term.) From the derivation in this section, we see that have no need to assume that a Coulomb potential is applied to the fermion, as any point source (which has spherical symmetry by definition) automatically requires a Coulomb term to be included in $\boldsymbol{k}E$, because, once polar coordinates are applied to $\boldsymbol{i}\nabla$, with $1 / r$ terms added, we can only obtain nilpotent solutions when there is a compensating $1 / r$ term in $\boldsymbol{k}E$. Thus, spherical symmetry alone – which is, of course, an expression of the conservation of angular momentum, and hence of charge – demands the existence of a Coulomb term in the potential, and this will be found necessary also to strong, weak, and inertial, as well as electric, interactions. So, for any field derived from specified single particle states, the Coulomb potential will be automatic.

10.2 Condensed Matter: The Kronig-Penney Model

Two big areas of application for the quantum Coulomb interaction are condensed matter theory and quantum chemistry. These are, of course, already extensive areas of theoretical development, with well-established techniques. No new results in these areas will be shown here, though they can be expected from nilpotent theory at a future time. Preliminary investigations in both areas, however, show that existing methods are certainly compatible with nilpotent theory.

In condensed matter theory, for example, the Kronig-Penney model for a wavefunction in a periodic potential takes the same form in the nilpotent theory as it does in the conventional derivation from the Schrödinger equation. Only the form of the energy eigenvalue is different. We assume that the Bloch theorem holds; so that, in the case of a periodic potential, the solution has the form $u_{\mathbf{k}}(\mathbf{r}) \exp(i\mathbf{k}.\mathbf{r})$, where $u_{\mathbf{k}}(\mathbf{r}) = u_{\mathbf{k}}(\mathbf{r} + \mathbf{T})$ has the period of the crystal lattice, and $\mathbf{k}$ is the wavevector. Once this has been established, then no *new* recourse to the wave equation is required. If we now imagine a square-well periodic potential in the x direction, then we may write down a nilpotent operator of the form:

$$\left(k\left(\varepsilon-U(x)\right)+i\left(\mathbf{i}\frac{\partial}{\partial x}+\mathbf{j}p_y+\mathbf{k}p_z\right)+ijm\right)$$

where p_x and p_y are constants, taken to be 0, so $p_x = p$. In the region $0 < x < a$, $U = 0$, the eigenfunction ψ will be the linear combination

$$Ae^{ipx} + Be^{-ipx}$$

of plane waves travelling to the right and to the left, with energy eigenvalue

$$\varepsilon = \sqrt{p^2 + m^2}$$

as opposed to the nonrelativistic $p^2 / 2m$. In the region $a < x < a + b$, $U = 0$, the eigenfunction (including the phase term) is the linear combination

$$Ce^{Qx} + De^{-Qx}$$

with $$U_0 - \varepsilon = \sqrt{Q^2 + m^2}\,.$$

We now determine that ψ and $d\psi / dx$ must be continuous at $x = 0$. So

$$A + B = C + D$$

$$ipA - ipB = QC - QD.$$

Continuity is also required at $x = a$. So

$$Ae^{ipa} + Be^{-ipa} = Ce^{Qa} + De^{-Qa}; \tag{10.4}$$

$$ipA^{ipa} - ipBe^{-ipa} = CQe^{Qa} - Q\,De^{-Qa}. \tag{10.5}$$

In addition, as a result of the form of the Bloch function, $u_{\mathbf{k}}(\mathbf{r})\ exp\ (i\mathbf{k.r})$, the values of ψ and $d\psi / dx$ at $x = a$ will be required to be equal to those at $x = -b$, but advanced by a phase factor $exp\ [ik(a + b)]$, which means we now have:

$$A\ exp\ [ipa - ik(a + b)] + B\ exp\ [-ipa - ik(a + b)] = Ce^{Qa} + De^{-Qa}; \tag{10.6}$$

$$ipA[ipa - ik(a + b)] - ipB[ipa - ik(a + b)] = CQe^{Qa} - Q\,De^{-Qa}. \tag{10.7}$$

To solve for A, B, C, D in equations (10.4)-(10.7), we zero the determinant of their coefficients, which means that

$$[(Q^2 - p^2) / 2\,Qp] \sinh Qb \sin pa + \cosh Qb \cos pa = \cos k\,(a + b). \tag{10.8}$$

A simplification of this result arises if we approach the periodic delta function limit, $b = 0$ and $U_0 = \infty$, in such a way that $Q^2ba / 2$ becomes a finite quantity P. Then (10.8) becomes

$$(P / pa) \sin pa + \cos pa = \cos ka. \tag{10.9}$$

The allowed values for p and $\varepsilon = \sqrt{p^2 + m^2}$ become those for which the function $(P / pa) \sin pa + \cos pa$ lies between -1 and $+1$, for given values of P.

10.3 The Helium Atom

The theory of the helium atom is not, of course, quantum chemistry, though it sets us on the way to it. An approximate solution for the helium atom requires only a solution for the hydrogen atom. Once again, no new recourse is required to the wave equation originally used, and we can use entirely conventional methods to obtain the solution once we have used the nilpotency to obtain the phase factor and amplitude for the hydrogen atom. We can express the nilpotent operator for a helium atom in the form:

$$\left(\mathbf{k}\left(E - \frac{2e^2}{r_1} - \frac{2e^2}{r_2} + \frac{e^2}{r_{12}} \right) + i\left(\frac{\partial}{\partial r_1} + \frac{1}{r_1} \pm i\frac{j_1 + ½}{r_1} + \frac{\partial}{\partial r_2} + \frac{1}{r_2} \pm i\frac{j_2 + ½}{r_2} \right) + ijM \right)$$

If we exclude the interaction term e^2 / r_{12}, where $r_{12} = |\mathbf{r}_1 - \mathbf{r}_2|$ is the distance between the two electrons, the ground-state eigenfunction for this operator can be taken as the product of two normalised hydrogenic eigenfunctions:

$$\psi(r_1, r_2) = \left(\frac{Z^3}{\pi a_0^3} \right) \exp((-Z / a_0)(r_1 + r_2))$$

with Bohr radius a_0. The expectation value for the kinetic energy for the ground state of a hydrogen atom is $e^2 / 2a_0$, while the potential energy is $-e^2 / a_0$. So, for the case of an atom with two electrons and proton number Z (= 2 for helium) the total energy, excluding the interaction energy between the electrons, is $(e^2Z^2 / a_0) - (4e^2Z^2 / a_0)$. The expectation value for the interaction energy between the electrons is

$$\iint \bar{\psi}(r_1, r_2) \frac{e^2}{r_{12}} \psi(r_1, r_2) dr_1 dr_2 = e^2 \left(\frac{Z^3}{\pi a_0^3} \right) \iint \frac{1}{r_{12}} \exp((-2Z / a_0)(r_1 + r_2)) dr_1 dr_2 \tag{10.10}$$

This integral can be evaluated using the approximation that it arises from the overlap of two spherically symmetric charge distributions. We can now expand $1 / r_{12}$ in spherical harmonics, using the generating function for the Legendre polynomials:

$$\frac{1}{r_{12}} = \left(\frac{1}{r_1} \right) \sum_{l=0}^{\infty} \left(\frac{r_2}{r_1} \right)^l P_l(\cos\theta) \qquad \text{for } r_1 > r_2$$

$$\frac{1}{r_{12}}=\left(\frac{1}{r_2}\right)\sum_{l=0}^{\infty}\left(\frac{r_1}{r_2}\right)^l P_l(\cos\theta) \qquad \text{for } r_1 < r_2 \tag{10.11}$$

where θ is the angle between $\mathbf{r}_1$ and $\mathbf{r}_2$. With θ_1,ϕ_1 and θ_2,ϕ_2 as the respective polar angles of the vectors $\mathbf{r}_1$, $\mathbf{r}_2$. From the scalar product of these vectors in rectangular coordinates, we obtain $\cos\theta = \cos\theta_1 \cos\theta_2 + \sin\theta_1.\sin\theta_2 \cos(\phi_1 - \phi_2)$. It can be shown, mathematically, that

$$P_l(\cos\theta)=P_l(\cos\theta_1)P_l(\cos\theta_2)+2\sum_{m=1}^{l}\frac{(l-m)!}{(l+m)!}P_l^m(\cos\theta_1)P_l^m(\cos\theta_2)(\phi_1-\phi_2) \tag{10.12}$$

Substituting (10.11) and (10.12) into (10.10), and making use of the orthogonality of spherical harmonics, integration over the polar angles of $\mathbf{r}_1$ makes all terms vanish except for that for which l and m are zero. The expectation value for the interaction energy then becomes

$$(4\pi)^2\int_0^\infty\left[\int_0^r\frac{1}{r_1}\exp((-2Z/a_0)(r_1+r_2))r_2^2dr_2+\int_0^r\frac{1}{r_2}\exp((-2Z/a_0)(r_1+r_2))r_2^2dr_2\right]$$

$$\times\left(\frac{Z^3}{\pi a_0^3}\right)r_1^2dr_1$$

which works out at

$$\frac{5\pi^2a_0^5}{8Z^5}=\frac{5e^2Z}{8a_0}.$$

The expectation value for the Hamiltonian or total energy now becomes:

$$<\mathcal{H}> = \frac{e^2Z^2}{a_0}-\frac{4e^2Z}{a_0}+\frac{5e^2Z}{8a_0}=\frac{e^2}{a_0}\left(Z^2-\frac{27}{8}Z\right).$$

Differentiating with respect to Z and equating to zero, we obtain a minimum when $Z_{eff} = 27/16 = 1.69$ (reduced from 2 by each electron screening the other from the nucleus). The lowest value for the upper limit of the ground state energy for the helium atom then becomes

$$-Z_{eff}^2\frac{e^2}{a_0}=-\left(\frac{27}{16}\right)^2\frac{e^2}{a_0}=-2.85\frac{e^2}{a_0},$$

in comparison with the experimental value of $-2.904\ e^2/a_0$.

10.4 *SU*(3)

A more fundamental analysis (see 6.6) shows that the vector nature of the **p** term in the Dirac nilpotent state vector produces a natural *SU*(3) symmetry for the strong interaction, as is evident from the possible phases of the baryon state:

$$(\boldsymbol{k}E \pm \boldsymbol{ii}\, p_x + \boldsymbol{ij}\, m)\, (\boldsymbol{k}E \pm \boldsymbol{ii}\, p_y + \boldsymbol{ij}\, m)\, (\boldsymbol{k}E \pm \boldsymbol{ii}\, p_z + \boldsymbol{ij}\, m).$$

The *SU*(3) symmetry then becomes simply a straightforward expression of perfect gauge invariance between all the possible phases. Gauge invariance is really an expression of the nonconservation codified within a differential operator, and the conventional way of defining this is (as in QED) via a covariant derivative, which, for an *SU*(3) symmetry, takes the form:

$$\partial_\mu \rightarrow \partial_\mu + ig_s \frac{\lambda^a}{2} A^{a\mu}(x),$$

or, in terms of the component coordinates:

$$ip_1 = \partial_1 \rightarrow \partial_1 + ig_s \frac{\lambda^a}{2} A^{a1}(x)$$

$$ip_2 = \partial_2 \rightarrow \partial_2 + ig_s \frac{\lambda^a}{2} A^{a2}(x)$$

$$ip_3 = \partial_3 \rightarrow \partial_3 + ig_s \frac{\lambda^a}{2} A^{a3}(x)$$

$$E = i\partial_0 \rightarrow i\partial_0 - g_s \frac{\lambda^a}{2} A^{a0}(x).$$

For an *SU*(3) structure, we require a field with eight generators.

If we insert the coordinate expressions into the differential form of the baryon state vector, we obtain:

$$\left(\boldsymbol{k}\left(E - g_s \frac{\lambda^a}{2} A^{a0}\right) \pm \boldsymbol{i}\left(\partial_1 + ig_s \frac{\lambda^a}{2} A^{a1}\right) + \boldsymbol{ij}m\right)$$

$$\times \left(\boldsymbol{k}\left(E - g_s \frac{\lambda^a}{2} A^{a0}\right) \pm \boldsymbol{i}\left(\partial_2 + ig_s \frac{\lambda^a}{2} A^{a2}\right) + \boldsymbol{ij}m\right)$$

$$\times \left(\boldsymbol{k}\left(E - g_s \frac{\lambda^a}{2} A^{a0}\right) \pm \boldsymbol{i}\left(\partial_3 + ig_s \frac{\lambda^a}{2} A^{a3}\right) + \boldsymbol{ij}m\right)$$

If we write $\left(E - g_s \frac{\lambda^a}{2} A^{a0}\right)$ in the form E', the possible phases then become:

$$\left(kE' \pm i\left(\partial_1 + g_s \frac{\lambda^\alpha}{2} \mathbf{A}^\alpha\right) + ijm\right)(kE' + ijm)(kE' + ijm)$$

p_x active

$$(kE' + ijm)\left(kE' \pm i\left(\partial_2 + g_s \frac{\lambda^\alpha}{2} \mathbf{A}^\alpha\right) + ijm\right)(kE' + ijm)$$

p_y active

$$(kE' + ijm)(kE' + ijm)\left(kE' \pm i\left(\partial_3 + g_s \frac{\lambda^\alpha}{2} \mathbf{A}^\alpha\right) + ijm\right)$$

p_z active

in parallel to the six forms incorporated in

$$\psi \sim (BGR - BRG + GRB - GBR + RBG - RGB).$$

The state also exhibits *hyperentanglement*, with all components sharing the same values of E, $\mathbf{p}$ and m.

Conventionally, we describe three quark 'colours' (**R**, **G**, **B**), which are as inseparable as the three dimensions of space. Though all 'phases' of the interaction are, of course, equally probable, and present at the same time, we can imagine arbitrarily isolating one phase as the carrier of the 'colour' or 'active' component of the interaction (i.e. the vector term, $ig_s \lambda^\alpha \mathbf{A}^\alpha / 2$), or, alternatively, the strong charge (s); and then picture this as being 'transferred', at a constant rate, to create the next phase, along with the momentum and spin-carrying term (**p**). The 'current' effecting the 'transfer' of strong charge or 'colour' field will then be carried by the eight generators of the strong field, or 'gluons'; and the 'transfer' will be, simultaneously, an expression of the conservation of the directional aspect of angular momentum. Deriving entirely from the nilpotent structure of the baryon state vector, the interaction will necessarily be nonlocal, and the constant rate of momentum 'transfer', will be equivalent to a force which does not depend on the physical separation of the components. Such a force, requires, in mathematical terms, a potential which is linear with distance, though, as has been shown in section 10.1, an additional Coulomb component ($-g_s \lambda^\alpha A^{0\alpha} / 2$) is needed for spherical symmetry. This is the scalar or 'passive' component, which is responsible for the coupling constant. The total potential then consists of a term linearly proportional to r combined with an inverse linear term. In addition, exactly the same structure of 'colour' phases and interaction should apply even when the bound state is a bosonic state composed of quark and antiquark, that is, a meson, rather than a three-quark baryon.[1]

10.5 The Quark-Antiquark and Three-Quark Interactions

With the assumption of a linear potential for the strong interaction, we can immediately use the nilpotent form of the Dirac equation to obtain an analytical solution for both the quark-antiquark and three-quark potentials, which predicts both infrared slavery and asymptotic freedom. Beginning with the idea of a linear potential for the strong interaction, let us suppose that the quark-antiquark potential in the bound meson state is of the form:

$$V = \sigma r + D,$$

where D is a function of r, yet to be determined. With a strong or colour charge for the quark of strength q $(= \sqrt{\alpha_s})$, and with $B = q\sigma$, this is equivalent to a potential energy

$$W = Br + qD$$

for the quark-antiquark interaction. Let us assume that, if D contains any constant term (C), its effect will be merely to shift the value of E to $E' = E + qC$. It will be convenient to refer to this, simply, as E; and it becomes clear, from 10.1, that qD must be a scalar phase or Coulomb term of the form A / r, as would be expected from spherical symmetry or equality in all directions. (A potential of this form is also required in standard QCD theory for the one gluon exchange.)

To determine the effect of this potential, we now construct the appropriate form of the Dirac equation. Assuming spherical symmetry, it is convenient, as in 10.1, to choose a form of the $\mathbf{p}$ operator in which helicity ($\boldsymbol{\sigma}.\mathbf{p}$) is explicit and ∇ becomes an ordinary vector; $\boldsymbol{\sigma}.\nabla$ can then be expressed as a function of r in polar coordinates, with the explicit addition of the angular momentum term which would be required using a multivariate form of ∇. Then, asssuming constant total energy, we can use this expression to construct a nilpotent differential operator of the form:

$$\left(\boldsymbol{k}\left(E + \frac{A}{r} + Br\right) + \boldsymbol{i}\left(\frac{\partial}{\partial r} + \frac{1}{r} \pm i\frac{j+½}{\mathrm{r}}\right) + \boldsymbol{ij}m\right).$$

We now need to identify the phase factor to which this operator applies. We suppose, on the basis of the parallel calculations for the Coulomb potential, that it is of the form:

$$\psi = \exp\left(-ar - br^2\right) r^{\gamma} \sum_{\nu=0} a_\nu r^\nu ,$$

and consider the ground state (with $\nu = 0$) over the four Dirac solutions. The four-part nilpotent state vector defines the condition:

$$E^2 + 2AB + \frac{A^2}{r^2} + B^2 r^2 + \frac{2AE}{r} + 2BEr = m^2$$

$$-\left(a^2 + \frac{(\gamma+\nu+...+1)^2}{r^2} - \frac{(j+½)^2}{r^2} + 4b^2r^2 + 4abr - 4b(\gamma+\nu+...+1) - \frac{2a}{r}(\gamma+\nu+...+1) \right)$$

with the positive and negative $i(j + ½)$ terms cancelling out over the four solutions. Then, assuming a termination in the power series, we can equate:

(1) coefficients of r^2: $\quad B^2 = -4b^2$

(2) coefficients of r: $\quad 2BE = -4ab$

(3) coefficients of $1/r$: $\quad 2AE = 2a(\gamma+\nu+1)$

(4) coefficients of $1/r^2$: $\quad A^2 = -(\gamma+\nu+1)^2 + (j+½)^2$

(5) constant terms: $\quad E^2 + 2AB = -a^2 + 4b(\gamma+\nu+1) + m^2$

The first three equations immediately lead to:

$$b = \pm\frac{iB}{2}$$

$$a = \mp iE$$

$$\gamma + \nu + 1 = \mp iA .$$

The case where $\nu = 0$ then requires a state vector with phase factor

$$\psi = \exp\left(\pm iEr \mp iBr^2/2\right) r^{\mp iqA-1} .$$

The imaginary exponential terms in ψ can be seen as representing asymptotic freedom, the exp ($\mp iEr$) being typical for a free fermion. The complex $r^{\gamma-1}$ term can be written as a component phase, $\phi(r) = \exp(\pm iqA \ln(r))$, which varies less rapidly with r than the rest of ψ. We can therefore write ψ as

$$\psi = \frac{\exp(kr + \phi(r))}{r},$$

where $$k = \pm iE \mp iBr/2 .$$

At high energies, where r is small, the first term dominates, approximating to a free fermion solution, which can be interpreted as asymptotic freedom. At low energies, when r is large, the second term dominates, with its confining potential Br, and this can be interpreted as infrared slavery. Significantly, the Coulomb term, which is required to maintain spherical symmetry, is the component which here defines the strong interaction phase, $\phi(r)$, and this can be related to the directional status of **p** in the state vector.

Reducing the quark-quark potential to the Coulomb term, which is what we suppose might happen effectively at short distances, produces a hydrogen-like spectral series. Here, again, we have, as in 10.1,

$$4\left(E+\frac{A}{r}\right)^2 = -2\left(-a+\frac{\gamma}{r}+\frac{\nu}{r}+\ldots\frac{1}{r}+i\frac{j+½}{r}\right)^2 - 2\left(-a+\frac{\gamma}{r}+\frac{\nu}{r}+\ldots\frac{1}{r}-i\frac{j+½}{r}\right)^2 + 4m^2$$

where the phase factor has the form

$$\psi = e^{-ar} r^{\gamma} \sum_{\nu=0} a_{\nu} r^{\nu} .$$

Applying this over the four Dirac solutions, and expanding (for the ground state), we obtain:

$$4\left(E+\frac{A}{r}\right)^2 = -2\left(-a+\frac{\gamma}{r}+\frac{\nu}{r}+\ldots\frac{1}{r}+i\frac{j+½}{r}\right)^2 - 2\left(-a+\frac{\gamma}{r}+\frac{\nu}{r}+\ldots\frac{1}{r}-i\frac{j+½}{r}\right)^2 + 4m^2$$

as before, from which, equating coefficients of $1/r$, coefficients of $1/r^2$, and constant terms, we obtain the same result as in 10.1, with

$$a = \sqrt{m^2 - E^2}$$

and $$m^2 = E^2\left(1+\frac{A^2}{(\gamma+\nu+1)^2}\right).$$

According to the last equation, there will be a certain value of E, below which a is real, suggesting a confined solution, with equations which are identical in form to those for the Coulomb potential defined for atomic states, but with A replacing Ze^2. We assume a state vector, with phase factor:

$$\psi = \exp\left(-\sqrt{m^2 - E^2}\,r\right) r^{\gamma} \sum_{\nu=0} a_{\nu} r^{\nu} ,$$

and, allowing the power series to terminate at $\nu = n'$, we obtain the characteristic Coulomb-type solution:

$$\frac{E}{m} = \left(1 + \frac{A^2}{(\gamma + 1 + n')^2}\right)^{-1/2},$$

or

$$\frac{E}{m} = \left(1 + \frac{A^2}{\left(\sqrt{(j + ½)^2 - A^2} + n'\right)^2}\right)^{-1/2}.$$

The condition resulting from $E^2 > m^2$ is that of *asymptotic* freedom, rather than escape, because of the continued presence (though reduced effect) of the confining linear potential. We can use the full and Coulomb-like solutions to investigate the transition point at which infrared slavery becomes effective. From the full solution, let

$$k = \pm iE \mp i\frac{Br}{2} = \frac{2\pi}{\lambda} = 0$$

at zero effective energy (or infrared slavery, with $\lambda = \infty$). Then

$$r = \frac{2E}{B}.$$

If, from the Coulomb-like solution, we take the 'free-particle' transition energy as the mass of the state m, and assume that this mass is mostly dynamic (gluonic) in origin, then we find $Br = 2E$, suggesting a virial relationship, as would be expected with a linear potential. From the Coulomb-like solution, we may also take E as the mass or reduced mass of the c quark, in the case of charmonium ($c\bar{c}$) ($\approx$ 1.5 GeV). Taking $\sigma \approx 1$ GeV fm^{-1} and $q \approx 0.4$, we find $r \approx 4$ fm.

Virtually identical arguments apply to the three-quark or baryon system. Here, the potential is of the form[2]:

$$V_{3Q} = -A_{3Q} \sum_{i<j} \frac{1}{|\mathbf{r}_i - \mathbf{r}_j|} + \sigma_{3Q} L_{min} + C_{3Q}.$$

where L_{min}, the minimal total length of the colour flux tubes linking three quarks, arranged in a triangle with sides, a, b, c, is given by

$$L_{min} = \left[\frac{1}{2}\left(a^2 + b^2 + c^2\right) + \frac{\sqrt{3}}{2}\sqrt{(a+b+c)(-a+b+c)(a-b+c)(a+b-c)}\right]^{1/2}.$$

For perfect spherical symmetry, when $a = b = c$, L_{min} becomes a multiple of the distance r of any quark from the centre of the flux tubes, and

$$\sum_{i<j} \frac{1}{|\mathbf{r}_i - \mathbf{r}_j|}$$

becomes a multiple of $1 / r$. The potential V_{3Q} then has exactly the same form as $V_{Q\bar{Q}}$, and the same solutions will apply, with variations in the values of the constants A, σ and C. The model of Takahashi *et al*[2] suggests that $\sigma_{3Q} \approx \sigma_{Q\bar{Q}}$ and $A_{3Q} \approx A_{Q\bar{Q}} / 2$, which accords with the theoretically-assumed value of $2\alpha_s / 3$ for qA. It is highly likely that the relationship $A_{3Q} \approx A_{Q\bar{Q}} / 2$ is virial in origin.

It is possible that the results may imply that quark mass (m) and intrinsic angular momentum (j + ½) are both zero for quarks in the asymptotically free state, perhaps indicating that quark 'masses' might run to zero at that state, and to lepton masses at grand unification, and that the spins and masses of baryon states do not come from the individual valence quarks, but from the system as a whole, as both the baryon wavefunctions and the rules on particle structures imply (cf chapters 13 and 15), in addition to the results following on from the EMC experiment of 1987.[3] The phase term in the full solution is interestingly proportional to α_s^2, and is the only place where A appears in the expression. Thus the Coulomb part of the potential – which is the component we believe to be significant in grand unification – results in a scalar phase term (as does the $U(1)$ term for the electromagnetic interaction). It may be that we can regard this phase term as the one representing the gauge invariant 'transfer' of strong charge, or angular momentum, or vector part of the $SU(3)$ covariant derivative, between the 'coloured' components of baryons and mesons. It is, finally, this process of gauge invariant 'transfer' (as outlined in the previous section) which allows us to suggest the derivation of the form of the confining potential from first principles, for the 'carrier' of the strong charge (or vector part of the covariant $SU(3)$ derivative) is the angular momentum, and constant force of equal magnitude in all directions is equivalent to a constant rate of change of momentum $\mathbf{p}$ (and hence of angular momentum defined through $\boldsymbol{\sigma}.\mathbf{p}$). The exact equivalence of all possible phases is identical to a constant rate of imagined rotational transfer of the strong charge via gluons in the same way as c represents the constant rate of transfer of the electromagnetic force via virtual photons (and makes the relationship $\sigma_{3Q} \approx \sigma_{Q\bar{Q}}$ highly probable). From this fact alone, we can derive the necessity for a confining potential $\propto r$, which the nilpotent Dirac algebra requires to be supplemented by a Coulomb term representing the phase.

The nilpotent structures may be used to suggest explanations of some aspects of the interaction of protons, as well as quarks. At intermediate distance, this interaction looks principally like the exchange of a single gluon, with minor adjustments needed to maintain the colour singlet state. For any given phase, each proton will always be effectively represented as though by a single fermionic state, say ($\boldsymbol{k}E_1 \pm \boldsymbol{ii}\,\mathbf{p}_1 + \boldsymbol{ij}\,m_1$) and ($\boldsymbol{k}E_2 \pm \boldsymbol{ii}\,\mathbf{p}_2 + \boldsymbol{ij}\,m_2$). Now, the strong interaction between two such states, each in a single (though unspecified) phase, can be represented by a momentum transfer, which is exactly of the form involved in gluon exchange between the component parts of the proton. This will also be still beyond the threshold at which the Coulombic or scalar part of the interaction dominates over the vector part.

However, the interaction between protons within a nuclear-type structure (separation 1 fm) is a saturated potential, of the form r^n, where $n \leq -2$, or polynomial combinations of that form. This is characteristic of a dipolar or multipolar force, and, of course, nuclear matter exists in an energy regime at which the interacting particle is a strong-dipolar massive pion involved in the Coulombic or scalar part of the strong force, rather than the massless gluon exchange involved in the vector part. In effect, the pion is created because nuclear matter has undergone a phase transition, in which the weak force also plays a part, for the pion is a weak, as well as a strong dipole. The weak force appears to be involved generally within phase transitions because (as we will see in 10.13) it is *fundamentally* dipolar, in response to its origin within the pseudoscalar, or energy, term in the Dirac nilpotent state. By this, we mean that the weak interaction creates a dipole between the fermion state and the vacuum (if the fermion states has no real partner), leading to spin ½ and *zitterbewegung*, and the weak interaction, as a uniquely one-handed force, also uniquely has a dipole moment, which is manifested through the spin. It is this dipole moment which is responsible for the tendency of fermions to structure themselves as aggregated matter, finding real, rather than vacuum, partners among other fermions or fermion-like structures, and creating dipolarity and multipolarity within the other forces. A weak interaction between two fermionic sources always includes the vacuum partner as the other dipole component. Dipolar / multipolar states are associated with harmonic oscillator-type regimes, or creation and annihilation processes, and it is precisely such processes which are involved in the concept of phase transition (see 10.12-10.13). The pionic state is essentially a colour or strong singlet because of the necessity of making a bosonic state a weak singlet.

10.6 Angular Momentum

According to the mechanism outlined in the previous sections, angular momentum carries the information relevant to the process of strong charge 'transfer', or gauge invariance, which defines the meaning of the strong interaction. The three-dimensionality of the (angular) momentum operator not only allows for the creation of a three-part (i.e. three-quark) fermionic nilpotent for the baryon, but also generates an *SU*(3) structure for the strong interaction. Spin, as a consequence, becomes a property of the baryon as a whole, not of the component quarks. A theory, by Brodsky, Ellis and Karliner, equating baryon spin to the orbital angular momentum of the quarks, is possible in this context.[4]

Angular momentum is also important, however, to the descriptions of weak and electric interactions. The reasons for this lie deep in the foundations of physics. Here, we require our extended version of Noether's theorem, which relates conserved quantities to symmetries or invariance under particular transformations. The theorem is, in fact, an example of the principle of duality in action. Invariance under transformation is really only another way of describing nonconservation, and duality requires every nonconserved quantity to be paralleled by an equivalent conserved one. The conjugate variables provide the most obvious examples. So, the translation symmetry of space (or non-identifiability of its elements) becomes equivalent to the conservation of the conjugate linear momentum, while the translation symmetry of time (or non-identitifiability of its elements) is equivalent to the conservation of the conjugate energy. At the same time, the rotation symmetry of space (or non-identity of directions in space) requires the conservation of a new conjugate parameter, angular momentum.

Charge, as a conserved quantity, has units which should be conserved in type as well as number, So, if we consider charge units to be arranged along axes, separately representing the electric, strong and weak charges, then duality suggests that these axes, unlike those of space, should be fundamentally irrotational, so that one type of charge (say, electric) can never be converted into another (say, weak or strong). This is, of course, the basis of the laws of lepton and baryon conservation. It is also the reason why baryon decay has never been detected, and it can be seen as the fundamental basis for defining interactions in which each type of charge acts in such a way that it is oblivious to the presence or absence of charge of a different type. If the property of conservation of charge type is fundamental, it ought to be linked directly to the conservation of angular momentum, according to the scheme proposed for an extended interpretation of Noether's theorem:

symmetry	**conserved quantity**	**linked conservation**
space translation	linear momentum	value of charge
time translation	energy	value of mass
space rotation	angular momentum	type of charge

A special case of the relationship is evident, as we have seen, in the connection between spin and statistics, which requires fermions, with nonzero weak charge, and bosons, with zero weak charge, to have different values of spin angular momentum. The actual spins are easily calculated using formal procedures, but the relationship between the spin ½ of the fermion and its status as ½ of a dual state will also be immediately evident. Another aspect is evident in the weak interaction itself where a change of parity state may produce a change in the sign of weak charge. Parity reverses momentum but not angular momentum. Hence the weak charge changes its sign but does not become a charge of a different kind.

However, it is also possible to show that the conservation of angular momentum requires the separate conservation of weak, strong and electric charges, in a much more fundamental way, through the conservation of the separate properties of orientation (with respect to the linear momentum) (i.e. handedness), direction, and magnitude. Essentially, charge is defined for a point source, with spherical symmetry. Spherical symmetry effectively determines angular momentum conservation, and has three fundamental aspects, namely the facts that the symmetry is independent of the length of the radius vector, of its direction (or choice of axes), and of the handedness of the rotation (left or right). We can see immediately that these correspond to the respective $U(1)$, $SU(3)$ and $SU(2)$ or $O(3)$ symmetries. We can also show, mathematically, that only three conditions exist in which a nilpotent state vector can maintain spherical symmetry. All require the actions of scalar potentials: and these are inverse linear with distance, which corresponds to $U(1)$; direct linear with distance plus inverse linear, which corresponds to $SU(3)$; and any other spherically symmetrical relation with distance plus inverse linear. The first two of these have been covered in 10.1 and 10.5. The last, which will be discussed in 10.12, incorporates the special case of inverse third power plus inverse linear, which is characteristic of a dipole-dipole force, such as the pure weak interaction requires, and which provides a harmonic oscillator solution, exactly corresponding to the characteristic behaviour of the weak interaction as a creator and destroyer of fermion-antifermion pairs, or, in effect, weak dipoles.

It is precisely because of the connection between the conservation of the different types of charge and the three separate properties associated with the conservation of angular momentum that the electric and weak charges are found to be directly linked in the $SU(2)_L \times U(1)$ symmetry. With the strong charge being the one associated with the **p** operator, it is the *combination* of the other two charges that affects this, rather than either taken separately.

10.7 The Weak Filled Vacuum

The conservation properties of the weak and electromagnetic charges are certainly determined by those of the angular momentum operator, and, in the case of a quark-type arrangement, might be expected to operate the same system of 'privileging' one charge in three during the complete phase cycle (with only one component of angular momentum well-defined). This is because the 'quarks' are effectively only a way of identifying separate phases for a particular interaction with vector properties. The weak and electric charges, however, are not directly attached to the **p** operator, like the strong charge, and so their 'privileged' phases will not necessarily coincide with that of the strong charge or with each other. The weak charge (w) is, in fact, attached to E and the electric charge (e) to m, in the Dirac state, and it is their *combination* which affects **p**. It is because of this that we tend to think of the electric and weak forces as being in some way combined, but the two charges are actually governed by quite separate symmetries.

Just as the character of the strong force and its relationship with the conservation of angular momentum is determined by its association with a vector operator, so we can expect that the pseudoscalar nature of iE and the scalar nature of m will determine the respective character and angular momentum relation of the weak and electric forces. The weak charge (w), which is the one associated with the quaternion label $\boldsymbol{k}$, produces two sign options for iE, because the algebra demands complexification of E, as it does of the parent-parameter time, and, consequently, two mathematical solutions. The sign option, in effect, determines the helicity state, or handedness with respect to the direction of motion, and it is this aspect of angular momentum conservation which is linked to the weak interaction.

One of the special properties of the weak interaction is its confinement to a single helicity state for fermions, with the opposite state reserved for antifermions. This is entirely a result of the fundamental group duality requiring mass-energy to be a continuum, and the consequent generation of a filled vacuum state. Essentially, there is no *physical* state corresponding to $-E$, although the use of a complex operator requires that $-iE$ has the same *mathematical* status as iE. Charge conjugation, however, or reversal of the signs of quaternion labels, *is* permitted physically. So $-\boldsymbol{k}E$ states are interpreted as antifermion or charge-conjugated states; and the mass-energy continuum becomes a filled vacuum for the ground state of the universe, in which such states would not exist.

A filled vacuum of this type was invoked by Dirac in the process of deriving the antiparticle concept, but the filled vacuum is now specifically a $\boldsymbol{k}$ or *weak* vacuum. Its manifestation is a violation of charge conjugation symmetry for the weak interaction, with consequent violation of either time reversal symmetry or parity to maintain the invariance of *CPT*. In principle, though the weak interaction can tell the difference between particle and antiparticle, it cannot distinguish between + and – signs of weak charge, and making the transition in the sign of the $\boldsymbol{k}$ operator (equivalent to *T*), because it is now interpreted as a charge conjugation (*C*), comes at the price of switching the sign of the $\boldsymbol{i}$ operator (*P*) as well.

This is also connected with the existence of just four solutions to the Dirac equation. Antifermions represent conjugate charge states to fermions, and we should, ideally, have eight independent combinations of $\pm \boldsymbol{k}w \pm \boldsymbol{i}s \pm \boldsymbol{j}e$. However, the Dirac equation allows only four solutions, and the charge parameters (*w*-*s*-*e*), like the E-$\mathbf{p}$-m terms, require mapping onto a quaternion or 4-vector 4-space. (Ultimately, we lose a degree of freedom when we compactify to the nilpotent state, because the fermion and vacuum do not then represent independent sources of information.) As it happens, the E-$\mathbf{p}$-m terms and charge parameters present us with alternative problems, for, while E-$\mathbf{p}$-m offers too few solutions, *w*-*s*-*e* requires too many. For E-$\mathbf{p}$-m, we have only the + and – states of $\mathbf{p}$ (which, as a vector, automatically provides these alternatives), while E and m are both, strictly speaking, confined to positive values. In the case of the charges, the eight possible combinations of $\pm \boldsymbol{k}w \pm \boldsymbol{i}s \pm \boldsymbol{j}e$ have to be reduced to four. To overcome the problem, each effectively 'borrows' aspects of the other: E-$\mathbf{p}$-m uses charge; *w*-*s*-*e* uses mass.

Unphysical $-E$ states appear in the Dirac equation to create the extra alternatives for E-$\mathbf{p}$-m, which are explained physically by Dirac's assumption of a filled vacuum for antifermions in the ground state, with a natural preponderance of matter over antimatter. In the case of *w*-s-*e*, we apply the category of antifermions using the negative sign of *s*, retaining the positive sign for fermions. We then require one other charge (here, assumed to be *e*) to adopt + and – signs within matter, producing what is called (weak) 'isospin', as the equivalent of the spin variation produced by the two signs of **p**. The filled vacuum then gives us the opportunity to remove the unwanted degree of freedom in the sign of *w*, by making the *effective* signs of *w* for matter and antimatter linked to those of *s*. Where *s* charges are present, the effective sign of *w* is determined by that of *s*, reducing the degrees of freedom in the charge structures from the eight of $\pm w \pm s \pm e$ to the four of $\pm (w + s) \pm e$, because of the linking of the signs of two of the quaternion operators. If fermions take up an *effective* + sign for *w*, then antifermions take up an *effective* – sign, regardless of the actual sign. The extra $-w$ for fermions or $+w$ for antifermions is suppressed by violating charge conjugation symmetry and consequently either parity or time reversal invariance as well.

It is, therefore, *w*, in effect, that determines the status of matter and antimatter, rather than *s*, though the sign of *s* is linked with the effective sign of *w*. As a result of this, both quarks and free fermions become mixed states, containing $+w$, and suppressed $-w$, states, and involving alternative violations of parity and time reversal symmetry. The removal of a degree of freedom from the charges $\pm w \pm s \pm e$ thus coincides exactly with the acquisition of a degree of freedom by $E \pm \mathbf{p}$, when it increases from the physical two to the mathematical four of $\pm E \pm \mathbf{p}$; in each case the sign of the $\boldsymbol{k}$ operator determines that the Dirac state has the four solutions which result from its quaternionic structure and its 4-D space-time. In principle, the unique $2^{n/2} \times 2^{n/2}$ matrix representation of the Clifford algebra, where $n = 2$, permits an exact quaternionic structure only because it is situated within a universe which has a filled $\boldsymbol{k}$ vacuum, and a single sign for the term $\boldsymbol{j}m$ in either fermion or antifermion states; and, ultimately, this is possible only because of the 4-dimensionality of the space-time signature which we have applied to the equation. (The 'coincidence' which makes $2^{n/2} \times 2^{n/2}$ into 4 because $2n = 4$, is, in fact, an expression of the fact that dimensionalization, to create a 3-space, and complexification to convert this to a space-time, are 'dual' processes of exactly equivalent status.) The process is outlined in Figure 10.1.

Figure 10.1

Exactly four Dirac solutions required:

	E	p	m
	+	+	+
	– *	–	
	*only with filled fermion vacuum	spin up spin down	
→	– = antifermions		
→	ground state has fermions only		

Exactly four charge allocation solutions required:

w	s	e
+ *	+	+
– *	–	–
+* fermions –*antifermions	fermions antifermions	isospin up isospin down
*effective because of filled weak (fermion) vacuum		

10.8 The Origin of the Higgs Mechanism

Just as the problem with E-$\mathbf{p}$-m was solved by invoking charge, so the problem with w-s-e is solved by invoking mass. Physically, the loss of a degree of freedom for w means that both quarks and free fermions become mixed states, containing both $+w$, and suppressed $-w$, states, and involving respective violations of parity and time reversal symmetry for the latter. A violation of parity or time reversal symmetry, consequent upon the violation of charge conjugation, as we have said, also means that only one state of helicity or $\boldsymbol{\sigma}.\mathbf{p}$ exists for the pure weak interaction for fermions, with the opposite helicity applying to antifermions. Because (according to the Dirac equation) $\boldsymbol{\sigma} = -\mathbf{1}$, the fermionic state acquires negative helicity or left-handedness. The Dirac formalism, however, requires the creation of alternative states of positive helicity or right-handedness, through the existence of $-\mathbf{p}$. If we wish to create these states, then the only remaining mechanism is through the introduction of rest mass in the term jm. The nilpotent version of the Dirac state thus associates the mass with the ***j*** quaternion label, which defines what we call the electric charge; and the presence of m simultaneously mixes E and **p** terms, right-handed and left-handed components, and the effects of e and w charges.

In fact, for a particle with any other kind of charge as well as weak, the charge conjugation violation is not absolute and the alternative state of helicity is allowed. However, a finite probability of alternative helicity requires a nonzero rest mass (because the asssociated particle speed must be $< c$), the amount being determined by the probability of the state and the strength of the interaction involved (e, s). Since it is the 'filled' weak vacuum (that is, one with a nonzero expectation value, or 'Higgs field'), that gives rise to the nonzero rest mass of the fermions involved, then the mass of a particle must be determined by the strength of its coupling to this field; and the strength of the coupling will depend ultimately on the degree of symmetry-breaking which the creation of that particle requires. The process is shown in Figure 10.2.

Figure 10.2

Filled weak vacuum

→ Violation of weak charge conjugation symmetry (+ P / T)

→ Fermion with only weak charge in a single state of helicity

→ For fermion with other charge, violation not absolute

→ Finite probability of alternative helicity

→ Nonzero rest mass (speed $< c$)

→ Amount of mass depends on probability of state (e.g. number of zero charges) and strength of the interaction involved (e, s)

Filled weak vacuum (with nonzero expectation value) = 'Higgs field'

→ nonzero rest mass of fermions involved

→ mass of fermion determined by the strength of coupling to this field

→ strength of the coupling depends on degree of symmetry-breaking in creation of fermion (e.g. number of missing or zero charges)

10.9 $SU(2)_L \times U(1)$

Just as the character of the strong force and its relationship with the conservation of angular momentum is determined by its association with a vector operator, so we can expect that the pseudoscalar nature of iE and the scalar nature of m will determine the respective character and angular momentum relations of the weak and electric forces. In the case of the weak force, we have two sign options for iE, because we are using complex algebra, and there are necessarily two mathematical solutions. The sign option, in effect, determines the helicity state, and it is this aspect of angular momentum conservation which is linked to the weak interaction.

The separate conservation laws for w, s, and e charges, which are axiomatic in this theory, determine that each type of charge must be independent of the others. It is particularly essential to the *characterization* of the weak interaction to express its independence from the presence or absence of electric charges, for it is precisely this independence that creates the characteristic $SU(2)_L$ 'isospin' pattern associated with the interaction. If the mixing of E and **p** terms, or right-handed and left-handed components, is also equivalent to the mixing of e and w charges, then it is important to establish that this mixing *does not affect the weak interaction* as such. Otherwise, the whole idea of defining the weak interaction through charge-conjugation violation would be compromised. The weak interaction must be simultaneously left-handed for fermion states and indifferent to the presence or absence of the electric charge, which introduces the right-handed element.

There are two possible $SU(2)_L$ states: with electric charge or without electric charge. These are the two states of weak isospin, and the weak interaction must behave in such a way that they are indistinguishable. Mathematically, these two $SU(2)_L$ states are described by a quantum number t_3 (the third component of weak isospin, by analogy with the $SU(2)$ of spin), whose value is such that $(t_3)^2 = (½)^2$ in half the total number of possible states, that is, in the left-handed ones. For the electric force, in the case of free fermions, the relevant quantum number (Q) is determined by the absence or presence of the electric charge, and takes the values 0 and −1, equivalent to the charges 0 and $-e$, the negative sign being purely historical in origin, with the positive sign reserved for antistates. So $Q^2 = 1$ again in half the total number of possible states (though it is a different half, including the right-handed ones), and 0 in the others. By a standard argument,[5,6] it can be shown that, if the weak and electric interactions are described by some grand unifying gauge group, irrespective of its particular structure, then, to satisfy orthogonality and normalisation conditions, the parameter which describes the mixing ratio, $\sin^2\theta_W$, is precisely determined by the ratio of the traces of the quantum numbers, $Tr\,(t_3^2)\ /\ Tr\,(Q^2)$, which in this case must be 0.25 (cf 15.6 for the actual calculation).

However, the ratio cannot apply only to free fermions. The weak interaction is also required to be indifferent to the presence or absence of the strong charge, that is, to the directional state of the angular momentum operator, and so the same mixing proportion, as observed in free fermion states, should exist also for quark states, and separately for each 'colour' phase, so that none is preferred, and colour is not directly detected through w. Applying this to quarks, we can create the same weak isospin states for one lepton-like 'colour', that is, we have one quark state with alternative Q values of –1 and 0, or charge values of $-e$ and 0. We now find that the only corresponding isospin states for the other colours that retain both the accepted value of $\sin^2\theta_W$ and the variation of only one 'privileged' quark phase 'instantaneously' in three, are 1 and 0 (or e and 0). In effect, the variation 0 0 $-e$ must be taken against either an empty background or 'vacuum' (0 0 0) or a full background (e e e), so that the two states of weak isospin in the three colours become:

$$\begin{array}{ccc} e & e & 0 \\ 0 & 0 & -e. \end{array}$$

A filled 'electromagnetic' vacuum might be considered to generate an antifermion 'image' of the form $\boldsymbol{j}$ ($\pm$ $\boldsymbol{k}E$ $\pm$ $\boldsymbol{i}i\mathbf{p}$ + $\boldsymbol{i}jm$) for a fermion with state vector ($\pm$ $\boldsymbol{k}E$ $\pm$ $\boldsymbol{i}i\mathbf{p}$ + $\boldsymbol{i}jm$), and, significantly, the bosonic form generated would have spin 0.

If the weak interaction is characterized by $SU(2)$, the electromagnetic interaction takes on the required $U(1)$ structure for a pure scalar magnitude by introducing a required phase. Conventionally, if $SU(2)$ breaks parity, the only way of maintaining a group structure, and the only way of ensuring that $SU(2)$ remains renormalizable, is to incorporate $U(1)$. This becomes significant in defining a Higgs ground state which is nonsymmetric and parity violating through finding the one such state that $SU(2)$ and $U(1)$ have in common.

10.10 The Weak Interaction and the Dirac Formalism

The argument here suggests that the pattern of $SU(3) \times SU(2)_L \times U(1)$ for the strong, weak and electric interactions between fermions can be established from first principles, and that the reasoning applied to the state vectors for $SU(3)$ can also be applied to those used for $SU(2)_L \times U(1)$, together with the formalisms relating to these symmetries for the derivation of Lagrangians, generators, and covariant derivatives. To apply the Dirac formalism to the weak interaction, we observe, first, that, experimentally, weak interactions all follow a pattern, which is determined by the $SU(2)_L$ symmetry. In the case of leptons, it is

$$e + \nu \rightarrow e + \nu.$$

For quarks, it is

$$u + d \rightarrow u + d,$$

and, for weak interactions involving both leptons and quarks (for example, β decay):

$$d + \nu \rightarrow e + u.$$

These can all be seen to involve the same two-isospin state structure, as should apply irrespective of the presence or absence of strong charges.

Considering the lepton case as exemplar, we find that there are four possible vertices (assuming left-handed components only) (Figure 10.3).

Figure 10.3

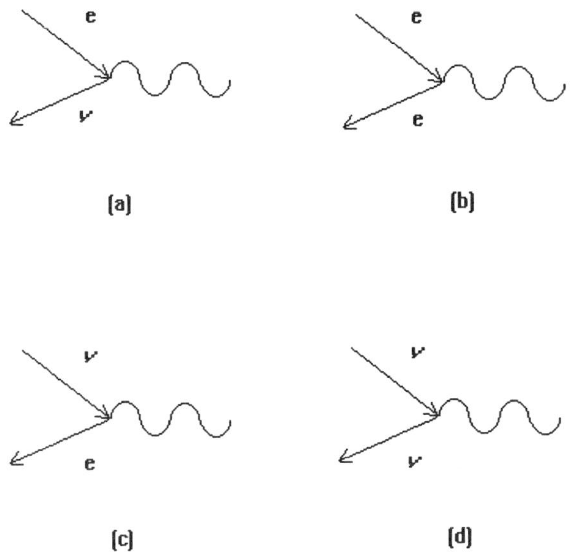

All the vertices occur at once, and so the interaction can be described as a mixing or superposition of the four possibilities. However, vertex (b), and this one alone, also represents a possible electromagnetic interaction, giving us a 1 to 4 ratio for the occurrence of the electromagnetic to weak interaction at the energy which the vertices characteristically represent (phenomenologically, that of the *W* / *Z* bosons). This results from the fact that particle charge structures at this energy are such that the electroweak mixing ratio becomes

$$\sin^2\theta_W = \frac{e^2}{w^2} = \frac{Tr(t_3^2)}{Tr(Q^2)} = 0.25\,.$$

The spin 1 interaction vertex in these cases can be considered to be of the general form:

$$(\pm \boldsymbol{k}E \pm i\boldsymbol{i}\mathbf{p} + i\boldsymbol{j}m)\,(\mp \boldsymbol{k}E \pm i\boldsymbol{i}\mathbf{p} + i\boldsymbol{j}m).$$

The *m* term arises, as we have seen, from the fact that **p** is not purely composed of left-handed helicity states (with –**p** right-handed), but incorporates a right-handed component, which itself cannot contribute to the weak interaction because of charge-conjugation violation and the presence of a weak filled vacuum. In the Standard Model, the right-handed component can only arise from the presence of the electromagnetic interaction. The weak interaction, therefore, cannot exist as a pure left-handed interaction, without a mixing with the electromagnetic interaction to produce the necessary non-zero mass through the introduction of right-handed states. Beyond the Standard Model, other possibilities exist for the introduction of the right-handed element, such as neutrinos being Majorana particles, mixing with their antistates.

If we now put into the *E* and **p** terms of the state vector the covariant derivatives for an $SU(2)_L \times U(1)$ electroweak interaction, the (pseudo)scalar part goes with *E* and the vector part with **p**. Mass is produced by the mixing of *E* with **p** via the relativistic connection between these terms. It is, similarly, produced by the mixing of one gauge field B^0 with the other three: W^+, W^0, and W^-, where the four are now identified with the respective vertices (d), (a), (b), and (c). Choosing the single, well-defined direction of spin or angular momentum (**p**) to be, in principle, the one where the total value for the interacting fermion-antifermion combination is 0, we can ensure that the mixing is specifically between the neutral components, B^0 and W^0, and create one massless *combination* to represent the carrier of the pure electromagnetic interaction (γ), with the other being the massive neutral weak carrier Z^0. If the mixing must be such as to define the ratio of the two interactions, $\sin^2\theta_W$, at 0.25, the other two vertices, W^+ and W^-, then fulfil the requirements for the existence of states corresponding to total spin values of +1 and –1.

For left-handed leptons, we have the covariant derivatives:

$$\partial_\mu \rightarrow \partial_\mu + ig\frac{\boldsymbol{\tau}.\mathbf{W}^\mu}{2} - ig'\frac{B^\mu}{2},$$

and, for right-handed:

$$\partial_\mu \rightarrow \partial_\mu - ig'\frac{B^\mu}{2}.$$

The energy operator and the single well-defined component of spin angular momentum give us:

$$E = i\partial_0 \rightarrow i\partial_0 + ig'\frac{B^0}{2} + ig'\frac{B^3}{2},$$

and

$$ip_3 = \partial_3 \rightarrow \partial_3 + ig\frac{\boldsymbol{\tau}.\mathbf{W}^3}{2} + ig\frac{\boldsymbol{\tau}.\mathbf{W}^0}{2}.$$

Applying these covariant derivatives to the nilpotent vertex which describes the weak interaction, we find that we can represent three components (the vector ones) as 'active', and one (the scalar) as 'passive'.

A vertex for a standard electroweak transition can be written in the form:

$$(\pm \boldsymbol{k}E \pm i\boldsymbol{i}\mathbf{p} + i\boldsymbol{j}m)\,(\mp \boldsymbol{k}E \pm i\boldsymbol{i}\mathbf{p} + i\boldsymbol{j}m) =$$

$$\left(\pm\boldsymbol{k}\left(\partial_0 + g'\frac{B^0}{2} + g'\frac{B^3}{2}\right) \pm \boldsymbol{i}\left(\partial_3 + ig\frac{\boldsymbol{\tau}.\mathbf{W}^3}{2} + ig\frac{\boldsymbol{\tau}.\mathbf{W}^0}{2}\right) + i\boldsymbol{j}m\right) \times$$

$$\left(\mp\boldsymbol{k}\left(\partial_0 + g'\frac{B^0}{2} + g'\frac{B^3}{2}\right) \pm \boldsymbol{i}\left(\partial_3 + ig\frac{\boldsymbol{\tau}.\mathbf{W}^3}{2} + ig\frac{\boldsymbol{\tau}.\mathbf{W}^0}{2}\right) + i\boldsymbol{j}m\right).$$

With m determined from the combination of E and $\mathbf{p}$, we can, by appropriate choice of the value of m, make these compatible by additionally defining a combination of the coupling constants related to the $SU(2)_L$ and $U(1)$ symmetries, g' and g, which removes B^3 from E and W^0 from $\mathbf{p}$. It is, of course, significant here that it is B^μ which is characteristic of right-handed lepton states, and therefore associated with the production of mass. Writing these combinations as γ^0 and Z^3, and those of g' and g, as e and w (= g), we obtain:

$$(\pm \boldsymbol{k}E \pm i\boldsymbol{i}\mathbf{p} + i\boldsymbol{j}m)\,(\mp \boldsymbol{k}E \pm i\boldsymbol{i}\mathbf{p} + i\boldsymbol{j}m) =$$

$$\left(\pm\boldsymbol{k}\left(\partial_0 + e\frac{\gamma^0}{2}\right) \pm \boldsymbol{i}\left(\partial_3 + ig\frac{\boldsymbol{\tau}.\mathbf{Z}^3}{2}\right) + i\boldsymbol{j}m\right)\left(\mp\boldsymbol{k}\left(\partial_0 + e\frac{\gamma^0}{2}\right) \pm \boldsymbol{i}\left(\partial_3 + ig\frac{\boldsymbol{\tau}.\mathbf{Z}^3}{2}\right) + i\boldsymbol{j}m\right).$$

Here, $\gamma^0 / 2$ becomes the same as the electrostatic potential ϕ, the purely 'passive' component of the interaction So, we can write this in the form:

$$(\pm \boldsymbol{k}E \pm \boldsymbol{i}i\mathbf{p} + \boldsymbol{i}jm)\,(\mp \boldsymbol{k}E \pm \boldsymbol{i}i\mathbf{p} + \boldsymbol{i}jm) =$$

$$\left(\pm \boldsymbol{k}(\partial_0 + e\phi) \pm \boldsymbol{i}\left(\partial_3 + iw\frac{\boldsymbol{\tau}.\mathbf{Z}^3}{2}\right) + \boldsymbol{i}jm\right)\left(\mp \boldsymbol{k}(\partial_0 + e\phi) \pm \boldsymbol{i}\left(\partial_3 + iw\frac{\boldsymbol{\tau}.\mathbf{Z}^3}{2}\right) + \boldsymbol{i}jm\right).$$

With e and w, the combinations of g' and g, now representing the pure electromagnetic and weak coupling constants, we must necessarily obtain the ratio $e^2 / w^2 = 0.25$, and both quarks and leptons must be structured to observe this.

Significantly, the *exchange* of electromagnetic charge, through W^+ or W^-, is not itself an electromagnetic interaction, but rather an indication of the weak interaction's indifference to the presence of the electromagnetic charge. A 'weak interaction', in principle, is a statement that all states of a particle with the same weak charge are equally probable, given the appropriate energy conditions, and that gauge invariance is maintained with respect to them. Weak bosons are massive because they act as carriers of the electromagnetic charge, whereas electromagnetic bosons (or photons) are massless because they do not. The quantitative value of the mass must be determined from the coupling of the weak charge to the asymmetric vacuum state which produces the violation of charge conjugation in the weak interaction. The weak interaction is also indifferent to the presence of the strong charge, and so cannot distinguish between quarks and leptons (hence, the intrinsic identity of purely lepton weak interactions with quark-lepton or quark-quark ones) and, in the case of quarks, it cannot tell the difference between a filled 'electromagnetic vacuum' (up quark) and an empty one (down quark). The weak interaction, in addition, is also indifferent to the sign of the weak charge, and responds (via the vacuum) only to the status of fermion or antifermion – hence, the Cabibbo-Kobayashi-Maskawa mixing between the three generations of quarks and leptons (cf 13.11).

10.11 The Higgs Mechanism for $U(1)$ and $SU(2)_L$

It is now possible to relate the work in the preceding sections to the conventional treatment of the Higgs mechanism.[7] It is usual to first illustrate the procedure using a $U(1)$ symmetry group. We take the Lagrangian for a complex scalar field $\phi = (\phi_1 + i\phi_2) / \sqrt{2}$,

$$\mathcal{L} = (\partial_\mu \phi)^* (\partial^\mu \phi) - V(\phi^*, \phi) = (\partial_\mu \phi)^* (\partial^\mu \phi) - \mu^2 \phi^* \phi - \lambda(\phi^* \phi)^2,$$

where $\lambda(\phi^* \phi)^2$ is a self-interaction, and make this invariant under a $U(1)$ local gauge transformation,

$$\phi \to e^{i\alpha(x)} \phi,$$

by replacing ∂_μ with the covariant derivative,

$$D_\mu = \partial_\mu - ieA_\mu,$$

with the gauge field transforming as

$$A_\mu \to A_\mu + \frac{1}{e} \partial_\mu \alpha .$$

The gauge invariant Lagrangian then becomes of the same form as the QED Lagrangian for a charged scalar particle of mass μ without the self-interaction term:

$$\mathcal{L} = \left(\partial^\mu + ieA_\mu\right)\phi * \left(\partial^\mu - ieA_\mu\right)\phi - \mu^2 \phi * \phi - \lambda(\phi * \phi)^2 - \frac{1}{4} F_{\mu\nu} F^{\mu\nu} .$$

We now need to introduce the specific changes that will produce a filled vacuum state or spontaneous symmetry-breaking. To spontaneously break the symmetry, we take $\mu^2 < 0$, $\lambda > 0$. The potential V now has a local maximum at $\phi = 0$, and a minimum at

$$\phi_1^2 + \phi_2^2 = v^2 = -\frac{\mu^2}{2\lambda} .$$

Without loss of generality, we are free to choose one of the degenerate vacua represented by this equation as the physical one. So we choose $\phi_1 = v$ and $\phi_2 = 0$, the so-called physical or unitary gauge. Now, expanding the Lagrangian about this vacuum by defining fields, $\eta(x)$ and $\xi(x)$, so that

$$\phi(x) = \sqrt{\frac{1}{2}} \left(v + \eta(x) + i\xi(x)\right).$$

and substituting $\phi(x)$ into the Lagrangian, we obtain

$$\mathcal{L} = \frac{1}{2}\left(\partial_\mu \xi\right)^2 + \frac{1}{2}\left(\partial_\mu \eta\right)^2 - v^2 \lambda \eta^2 + \frac{1}{2} e^2 v^2 A_\mu A^\mu - evA_\mu \partial^\mu \xi - \frac{1}{4} F_{\mu\nu} F^{\mu\nu}$$
$$+ \textit{interaction terms.}$$

The η-field mass now becomes $\sqrt{2\lambda v^2}$, and the A-field mass is ev, but the ξ-field has only a kinetic energy term and no mass. This Goldstone boson is a massless spin 0 scalar, which is denied physical existence by the nilpotent algebra, but we can choose a gauge to eliminate it, in which, using polar coordinates,

$$\phi(x) \to \sqrt{\frac{1}{2}}(v + h(x))e^{i\theta(x)/v},$$

and

$$A_\mu \to A_\mu + \frac{1}{ev}\partial_\mu \theta.$$

With h real, we now obtain

$$\mathcal{L} = \frac{1}{2}(\partial_\mu h)^2 - v^2 \lambda h^2 + \frac{1}{2}e^2 v^2 A_\mu^2 - \lambda v h^3 - \frac{1}{4}\lambda h^4 + \frac{1}{2}e^2 A_\mu^2 h^2 - ve^2 A_\mu^2 h - \frac{1}{4}F_{\mu\nu}F^{\mu\nu}$$

The Lagrangian now includes only two massive particles, the vector gauge boson A_μ and the massive spin 0 scalar (Higgs boson) h.

The application to $U(1)$ was simply an illustration of the mechanism, but, for $SU(2)$, where we believe the symmetry is truly broken, we will need to justify our assumptions on a fundamental basis. In the case of an $SU(2)$ local gauge symmetry, we apply an $SU(2)$ doublet of complex scalar fields,

$$\phi = \sqrt{\frac{1}{2}}\begin{pmatrix} \phi_1 + i\phi_2 \\ \phi_3 + i\phi_4 \end{pmatrix},$$

to what is essentially the same Lagrangian:

$$\mathcal{L} = (\partial_\mu \phi)^\dagger (\partial^\mu \phi) - V(\phi^\dagger, \phi) = (\partial_\mu \phi)^\dagger (\partial^\mu \phi) - \mu^2 \phi^\dagger \phi - \lambda(\phi^\dagger \phi)^2.$$

The complex doublet is chosen here because it is the simplest that will produce the gauge fields that we already know must exist for a spontaneously-broken $SU(2)$.

Replacing ∂_μ with the covariant derivative,

$$D_\mu = \partial_\mu - ig\frac{1}{2}\tau_a W_\mu^a,$$

with W_μ^a representing the three new gauge fields where $a = 1, 2, 3$, and $\tau_a W_\mu^a$ can be written as $\boldsymbol{\tau}.\mathbf{W}_\mu$, the gauge invariant Lagrangian then becomes

$$\mathcal{L} = \left(\partial_\mu \phi + ig\frac{1}{2}\boldsymbol{\tau}.\mathbf{W}_\mu\right)^\dagger \left(\partial^\mu \phi + ig\frac{1}{2}\boldsymbol{\tau}.\mathbf{W}_\mu\right)^\dagger \frac{1}{2}(\partial_\mu h)^2 - \mu^2 \phi^\dagger \phi - \lambda(\phi^\dagger \phi)^2 - \frac{1}{4}\mathbf{W}_{\mu\nu}\mathbf{W}^{\mu\nu}$$

with the last term representing the additional kinetic energy of the gauge fields. Again assuming $\mu^2 < 0$, $\lambda > 0$, to spontaneously break the symmetry, and produce the filled vacuum state that we already know must exist for the weak interaction, we find the minimum potential occurs when

$$\phi^\dagger \phi = \frac{1}{2}\left(\phi_1^2 + \phi_2^2 + \phi_3^2\right) = -\frac{\mu^2}{2\lambda} .$$

Choosing $\phi_1^2 = \phi_2^2 = \phi_4^2 = 0$, and $\phi_3^2 = v^2 = -\mu^2 / 2\lambda$, as our gauge (as we are free to do), and expanding $\phi(x)$ about the vacuum,

$$\phi_0 = \sqrt{\frac{1}{2}} \begin{pmatrix} 0 \\ v \end{pmatrix},$$

we substitute

$$\phi(x) = \sqrt{\frac{1}{2}} \begin{pmatrix} 0 \\ v + h(x) \end{pmatrix}$$

into the Lagrangian to gauge away the unphysical Goldstone bosons, which, according to our nilpotent formalism, simply cannot exist. The fluctuations from the vacuum can be parameterized in terms of the four real fields $\boldsymbol{\theta}$ and h, using

$$\phi(x) \rightarrow \sqrt{\frac{1}{2}} \begin{pmatrix} 0 \\ (v + h(x)) \end{pmatrix} e^{i\boldsymbol{\tau}.\boldsymbol{\theta}(x)/v} .$$

Substituting ϕ_0 into the Lagrangian, we obtain an expression containing the term

$$\left(ig\frac{1}{2}\boldsymbol{\tau}.\mathbf{W}_\mu\right)^\dagger \left(ig\frac{1}{2}\boldsymbol{\tau}.\mathbf{W}_\mu\right)$$

$$= \frac{g^2}{8}\left(\begin{pmatrix} W_\mu^3 & W_\mu^1 - iW_\mu^2 \\ W_\mu^1 + iW_\mu^2 & W_\mu^3 \end{pmatrix}\begin{pmatrix} 0 \\ v \end{pmatrix}\right)^\dagger \left(\begin{pmatrix} W_\mu^3 & W_\mu^1 - iW_\mu^2 \\ W_\mu^1 + iW_\mu^2 & W_\mu^3 \end{pmatrix}\begin{pmatrix} 0 \\ v \end{pmatrix}\right)$$

$$= \frac{g^2}{8}\left[\left(W_\mu^1\right)^2 + \left(W_\mu^2\right)^2 + \left(W_\mu^3\right)^2\right].$$

The Lagrangian now describes three massive vector gauge fields and one massive scalar h. The first three are required to explain the weak interaction, as we already understand it physically; the last is the Higgs boson, a spin 0 particle, which must necessarily be massive in the nilpotent formalism.

10.12 The Spherical Harmonic Oscillator

To gain a deeper understanding of the weak interaction, we need to explore further the solutions of the Dirac equation for spherically symmetric potentials.

The spherical harmonic oscillator provides a particularly significant case. Here, we have, say, a potential energy of the form ½ Cr^2. We may also suppose that spherical symmetry requires a Coulomb 'phase' term, say A / r, the exact form and significance of which will become apparent in the calculation. (The nilpotent method, as we have seen, shows that spherically symmetric solutions are impossible without such a term.) The covariant form of the differential operator will then become

$$\left(\boldsymbol{k}\left(E+\frac{A}{r}+\frac{1}{2}Cr^2\right)+\boldsymbol{i}\left(\frac{\partial}{\partial r}+\frac{1}{r}\pm i\frac{j+½}{\mathrm{r}}\right)+\boldsymbol{ij}m\right).$$

As usual, the solution will require the phase factor ϕ which will make the eigenvalue nilpotent. Polynomial potential terms which are multiples of r^n require the incorporation into the exponential of terms which are multiples of r^{n+1}. So, extending our work on the strong interaction and the Coulomb field, we may suppose that the solution is of the form:

$$\phi = \exp\left(-ar+cr^3\right)r^\gamma \sum_{\nu=0} a_\nu r^\nu .$$

So
$$\frac{\partial\phi}{\partial r}=\left(-a+3cr^2+\frac{\gamma}{r}+\frac{\nu}{r}+...\right)\phi,$$

and the amplitude then becomes:

$$\left(\pm\boldsymbol{k}\left(E+\frac{A}{r}+\frac{1}{2}cr^2\right)\pm\boldsymbol{i}\left(-a+3cr^2+\frac{\gamma}{r}+\frac{\nu}{r}+...\frac{1}{r}\pm i\frac{j+½}{r}\right)+\boldsymbol{ij}m\right).$$

Assuming that this is nilpotent, and that the power series terminates, we obtain:

$$4\left(E+\frac{A}{r}+\frac{1}{2}Cr^2\right)^2=-2\left(-a+3cr^2+\frac{\gamma}{r}+\frac{\nu}{r}+...\frac{1}{r}+i\frac{j+½}{r}\right)^2$$
$$-2\left(-a+3cr^2+\frac{\gamma}{r}+\frac{\nu}{r}+...\frac{1}{r}-i\frac{j+½}{r}\right)^2+4m^2.$$

Equating constant terms, we find

$$E^2 = -a^2+m^2$$

$$a=\sqrt{m^2-E^2} \qquad (10.13)$$

Equating terms in r^4, with $\nu = 0$, we obtain:

$$\frac{1}{4}C^2 = -9c^2$$

from which
$$c^2 = \pm\frac{iC}{6}.$$

When C is real, c must be imaginary.

Equating coefficients of r, where $\nu = 0$, we find

$$AC = -\, 6c\,(1 + \gamma)$$

and
$$(1 + \gamma) = \pm\, iA\,.$$

This means that, if $(1 + \gamma)$ is real, then A must be imaginary – though the opposite could, of course, also be true.

Equating coefficients of $1 \,/\, r^2$ and coefficients of $1 \,/\, r$, and assuming the power series terminates in $\nu = n'$, we obtain

$$A^2 = -\,(1 + \gamma + n')^2 + (j + ½)^2 \qquad (10.14)$$

and
$$EA = a\,(1 + \gamma + n'). \qquad (10.15)$$

Using (10.13), (10.14) and (10.15), we obtain

$$\left(\frac{m^2 - E^2}{E^2}\right)(1 + \gamma + n')^2 = -(1 + \gamma + n')^2 + (j + ½)^2$$

or
$$E = -\frac{m}{(j + ½)}(\pm\, iA + n').$$

If we now take A to have a half-unit value ($\pm$ ½ i), in line with the value of spin, we obtain a set of energy levels of the form expected in the simple harmonic oscillator:

$$E = -\frac{m}{(j + ½)}(½ + n').$$

We can now associate the phase term required for spherical symmetry ($A = \pm$ ½ i), directly with the random directionality of the spin of the fermion. In the case of the harmonic oscillator, the term, with its unit value and imaginary coefficient, is effectively brought in when the spin component is added to the $\boldsymbol{\sigma}.\nabla$ term in transforming from rectilinear to polar coordinates (although it is, of course, present implicitly where $\boldsymbol{\sigma}$ and ∇ are taken as multivariate vectors). (It can, of course, be converted to a real term by making the other term in the potential function imaginary instead – only the *relative* status of the terms can be determined (cf 10.12).) It is also significant that, although the nilpotent produces the same three equations as were required to generate the energy level series in the case of the pure Coulomb interaction, this solution cannot be applied to the case of the harmonic oscillator as it would result in a series of imaginary or complex energy levels.

The same method can be applied to any case where the potential can be expressed as a polynomial function of the radial distance. The Lennard-Jones potential,

$$V = \frac{C}{r^6} - \frac{D}{r^{12}},$$

which is a key component of the theory of molecular forces, provides a characteristic instance. Again incorporating a Coulomb or phase term for spherical symmetry, the Dirac operator becomes:

$$\left(\mathbf{k}\left(E + \frac{A}{r} + \frac{C}{r^6} - \frac{D}{r^{12}} \right) + \mathbf{i}\left(\frac{\partial}{\partial r} + \frac{1}{r} \pm i\frac{j+½}{\mathrm{r}} \right) + \mathbf{ij}m \right)$$

suggesting a phase factor of the form:

$$\phi = exp\left(-ar - cr^5/5 - dr^{11}/11\right) r^\gamma \sum_{\nu=0} a_\nu r^\nu .$$

The squared nilpotent amplitude now becomes:

$$\left(\pm\mathbf{k}\left(E + \frac{A}{r} + \frac{C}{r^6} - \frac{D}{r^{12}} \right) \pm \mathbf{i}\left(-a + \frac{c}{r^6} + \frac{d}{r^{12}} + \frac{\gamma}{r} + \frac{\nu}{r} + \ldots \frac{1}{r} \pm i\frac{j+½}{r} \right) + \mathbf{ij}m \right)^2 = 0 .$$

Equating constant terms, as usual,

$$E^2 = -a^2 + m^2$$

$$a = \sqrt{m^2 - E^2} .$$

Equating respective coefficients of r^{-24} and r^{-18}, we find that $D^2 = -d^2$ and $CD = cd$, from which we obtain $d = \pm\, iD$ and $c = \mp\, iC$, with the two coefficients having opposite signs. Equating coefficients of r^{-7}, for the case when $\nu = 0$, leads to

$$AB = -b\,(1 + \gamma),$$

with $$(1 + \gamma) = \pm\, iA,$$

which is the same result as for the harmonic oscillator, with potential energy $A/r + ½Cr^2$. Finally, equating respective coefficients of r^{-2} and r^{-1}, for a series terminating in $\nu = n'$, produces the identical relations:

$$A^2 = -(1 + \gamma + n')^2 + (j + ½)^2,$$

$$EA = a\,(1 + \gamma + n'),$$

and
$$E = -\frac{m}{\left(j + \frac{1}{2}\right)}\left(\tfrac{1}{2} + n'\right),$$

which demonstrate that the energy levels are again those of the harmonic oscillator.

This result is not sensitive to the particular terms used in the polynomial form of the potential, nor to the number of such terms. For any spherically symmetric polynomial potential with terms of the form Ar^n, where $|n| \geq 2$, the solution will be that of a harmonic oscillator. So, if we take

$$V = \frac{A}{r} + \frac{C}{r^m} + \frac{D}{r^n} \ldots ,$$

again incorporating a Coulomb or phase term for spherical symmetry, the operator takes the form:

$$\left(\boldsymbol{k}\left(E + \frac{A}{r} + \frac{C}{r^m} + \frac{D}{r^n} \ldots\right) + \boldsymbol{i}\left(\frac{\partial}{\partial r} + \frac{1}{r} \pm i\frac{j + \frac{1}{2}}{\mathrm{r}}\right) + \boldsymbol{i}\boldsymbol{j}m\right)$$

leading to a phase factor such as:

$$\phi = \exp\left(-ar - cr^{m-1}/(m-1) - dr^{n-1}/(n-1) - \ldots\right)r^{\gamma} \sum_{\nu=0} a_{\nu} r^{\nu} .$$

Particularly important examples are the dipolar case, where $V \propto r^{-3}$, and the multipolar case, with $V \propto r^{-n}$, where $n > 3$. Only in the special cases of $n = 1$ (the strong interaction potential) and $n = -1$ (the pure electrostatic or gravitational Coulomb potential), do we expect particular bound solutions, and the special nature of these solutions is a result of the symmetry of 3-dimensional space, just as it is in the analogous case of classical physics, where only constant and inverse-square forces produce a virial relation between the potential and kinetic energies of exactly 2.

10.13 The Weak Interaction as a Harmonic Oscillator

It would appear that there are three solutions of the Dirac equation for spherically-symmetric distance-dependent potentials. This becomes particularly significant in that specifying the spherical symmetry of space is an equivalent way of expressing the conservation of angular momentum. So, the Dirac equation effectively specifies three types of interacting potential under which angular momentum conservation is preserved, and we may imagine that these are also equivalent to specifications of the types of *charge* states that can be conserved.

Now, V proportional to $1 / r$ gives the Coulomb solution for the electric force. V proportional to r gives the quark-confinement solution for the strong force. Any other polynomial-type r dependence gives a harmonic oscillator. Let us suppose that this is *the solution for the weak force*. It doesn't actually matter what the shape of the function is, the solution will be a harmonic oscillator as long as it is not proportional to r or $1 / r$. In fact, there is a good reason why it will not be either of these for the weak interaction. Essentially the weak interaction is *always dipolar*. It always involves one fermion-antifermion combination becoming another. This is not true of the electric interaction where the interacting fermions remain unchanged. (Even in e^+e^- pair production / annihilation, there is always an amplitude for the weak interaction.) We can therefore think of the weak charge as only manifesting itself when it is part of a dipole-dipole interaction. If this is true of the weak charge, then the weak *field* will behave in essentially the same way.

In addition, the weak charge should have a dipole moment because of its left-handedness for fermion and right-handedness for antifermion states. Searches for electric and strong dipole moments have produced negative results to many orders of magnitude, implying the indistinguishability of left- and right-handed states, but weak dipole moments have not even been conceived. Yet the single-handedness of the weak charge necessarily implies that a weak dipole moment should exist; and if weak interactions always involve a weak dipole, then a weak dipole moment, as the direct expression of single-handedness, should, in some sense, be the very *manifestation* of the weak interaction.

From this it would seem that any interaction between a weak dipole and a weak field will be manifested as a dipole-dipole or dipole-multipole interaction. In this case, the weak potential V will be proportional to r^{-n}, where n is 3 or greater, or to a polynomial incorporating terms of this kind; and, for a single weak charge (fermion or antifermion), taken separately, the action of the field will require a similar potential with $n = 2$ or greater. In either case, the solution of the Dirac equation will be a harmonic oscillator.

Now, the harmonic oscillator is a classic way of representing the production of spin ½ fermion (and antifermion) states emerging from, or disappearing into, the vacuum using the appropriate creation or annihilation operators; and these operators are, of course, essentially the same in principle as those used in Quantum Field Theory. So we can represent the action of the weak dipole moment as the *cause* of the production of fermion-antifermion combinations from the vacuum (or real bosonic states) or of the reverse process of mutual annihilation. It is, of course, the presence of a single unit of positive or negative weak charge which distinguishes fermions from bosons or characterizes spin ½ states (see chapter 13). In the case of neutrinos and antineutrinos, it is their *only* charge-related characteristic. It is the weak interaction, therefore, that is particularly significant in the production of fermion and antifermion states; and the fact that it must be single-handed produces, via the weak dipole moment, the driving mechanism for the creation (or annihilation) of spin ½ states.

The handedness is, of course, ultimately a vacuum property because it stems from the existence of a filled weak vacuum, with a zero point energy, with ½ quantum values, corresponding to the ½ quantum value for intrinsic fermion spin, that could be taken to be the physical manifestation of the weak dipole moment.[8] What this means in effect is that the weak vacuum is in a continual state of proclaiming its filled state, by creating weak dipoles which have a dipole moment or specific handedness. So it is not surprising that 'fluctuations' in this vacuum are the same thing as the production or annihilation of a weak dipolar fermion-antifermion pair, each of spin ½, via a harmonic oscillator creation-annihilation mechanism. The same fluctuations are also responsible for the Casimir or Van der Waals force, which in the simplest case produces a V proportional to r^{-3} from the zero point energy, corresponding to the potential for a fluctuating dipole-dipole interaction.[9] Their existence is further proclaimed by the three 'vacuum' terms in the Dirac 4-spinor, one corresponding to each charge type (through the connection with the hidden $\mathbf{i}, \mathbf{j}, \mathbf{k}$ column operators), in addition to each Dirac solution, and, physically, by the *zitterbewegung*.

It is the dipole moment that privileges the real fermion (or antifermion) over its accompanying vacuum states; and it is the necessity for the *zitterbewegung* (see 7.7) that demonstrates that the weak interaction, like the electromagnetic and strong interactions, is, intrinsically, *a property of the nilpotent structure alone*, for the exchange of one fermion state for another requires weak processes of exactly the same kind as those described in section 10.10. (This will be further discussed in section 12.8.)

Significantly, all three interactions require a Coulombic phase term (V proportional to $1 / r$) in a nilpotent version of the Dirac equation. In the case of the electric field, of course, this produces the entire interaction, in the case of the strong interaction it is the one gluon exchange, and in the case of the weak interaction it is associated with the spin and the weak hypercharge; but in all cases it represents spherical symmetry – or arbitrary direction of spin in the absence of a field external to the system. The harmonic oscillator case, however, uses a complex phase term, presumably relating to the complex nature of E in the nilpotent state vector on the site corresponding to the position of the weak charge. It is the complex nature of this term which associates the two states of handedness with the respective concepts of fermion and antifermion, and which allows the possibility of pair creation or annihilation.

The complexity of the term, in this sense, drives the creation of a dipolar field, the algebra requiring simultaneous positive and negative solutions and privileging neither. The nilpotent nature of the Dirac operator also demands that one of the three charges or E-$\mathbf{p}$-m terms is complex; so a spin ½ or nilpotent state is, in principle, impossible without a complex aspect. It is significant that the three terms involved in the Dirac nilpotent operator – iE, $\mathbf{p}$ and m – which we have associated with the respective weak, strong and electric charge operators $\boldsymbol{k}$, $\boldsymbol{i}$, $\boldsymbol{j}$, are different algebraic objects, being, respectively, pseudoscalar, vector and real scalar. While the pseudoscalar suggests a dipolar mechanism (with $V \alpha r^{-n}$), one can imagine a gauge invariant vector term leading to a mechanism of constant rate of change of vector momentum with respect to distance (i.e. constant force, or $V \alpha r$), while a gauge invariant real scalar structure produces a simple $U(1)$ symmetry (requiring $V \alpha 1 / r$). These will be the appropriate variations with respect to r applied to find the covariant form of the time differential.

The dipolar nature of the weak force is thus intrinsically connected with the complex associations of the weak charge. Complex numbers are not privileged as to sign, and the weak charge tends to behave in such a way as to make its sign irrelevant; fermion and antifermion are distinguishable, but $+w$ and $-w$ are not. Complex equations necessarily have dual solutions, and we can consider the weak charge as carrying with it its alternative sign as a vacuum image, the dipole moment thus created setting up the handedness which ensures chirality.[10] In addition, complexity is associated with CP violation, through complex coupling constants or vacuum states, and CP violation is uniquely associated with the weak interaction.

10.14 A Strong-Electroweak Solution of the Dirac Equation

Here, we take a potential energy with a total Coulomb (scalar phase) term A / r (for the electromagnetic and strong interactions), combined with harmonic oscillator term Cr^s, for the weak interaction, a linear term Br for the strong interaction, and, possibly, a pseudoscalar phase term $\pm iD / r$. The covariant form of the differential operator then becomes:

$$\left(\mathbf{k}\left(E + \frac{A}{r} \pm i\frac{D}{r} + Br + Cr^s \right) + \mathbf{i}\left(\frac{\partial}{\partial r} + \frac{1}{r} \pm i\frac{j+\frac{1}{2}}{\mathrm{r}} \right) + \mathbf{i}\mathbf{j}m \right). \qquad (10.16)$$

Again, we need to choose a phase factor which will make the amplitude nilpotent. This time, we have:

$$\phi = exp\left(-ar + br^{s+1} + \tfrac{1}{2}cr^2\right) r^{\gamma} \sum_{\nu=0} a_\nu r^\nu ,$$

so that
$$\frac{\partial \phi}{\partial r} = \left(-a + (s+1)br^s + cr + \frac{\gamma}{r} + \frac{\nu}{r} + \ldots \right)\phi .$$

With a nilpotent amplitude, we obtain:

$$2\left(E + \frac{A}{r} + i\frac{D}{r} + Br + Cr^s \right)^2 + 2\left(E + \frac{A}{r} - i\frac{D}{r} + Br + Cr^s \right)^2$$

$$= -2\left(-a + (s+1)br^s + cr + \frac{\gamma}{r} + \frac{\nu}{r} + \ldots \frac{1}{r} + i\frac{j+\frac{1}{2}}{r} \right)^2$$

$$-2\left(-a + (s+1)br^s + cr + \frac{\gamma}{r} + \frac{\nu}{r} + \ldots \frac{1}{r} - i\frac{j+\frac{1}{2}}{r} \right)^2 + 4m^2 .$$

Then, assuming that the power series terminates at n', we equate

constant terms:	$E^2 + 2AB = -a^2 + m^2$
coefficients of r^{-2}:	$A^2 - D^2 = -(\gamma + n' + 1)^2 + (j + \frac{1}{2})^2$
coefficients of r^{-1}:	$2EA = 2a\,(\gamma + n' + 1)$
coefficients of r:	$2EB = 2ac$
coefficients of r^2:	$B^2 = -c^2$
coefficients of r^{s-1}:	$2AC = -2(\gamma + n' + 1)\,(s + 1)\,b$
coefficients of r^s:	$2EC = 2a\,(s + 1)\,b$
coefficients of r^{s+1}:	$2BC = -2\,(s + 1)\,bc$
coefficients of r^{2s}:	$C^2 = -(s + 1)^2\,b^2$

These equations lead to the following results:

$$E = ia$$

$$B = -ic$$

$$C = -i(s+1)b$$

$$A = -i(\gamma + n' + 1)$$

$$2(\gamma + n' + 1)c = m^2$$

$$D = i(j + ½)$$

So, the phase factor takes the form

$$\phi = \exp\left(iEr + iCr^{s+1}/(s+1) + ½iB\text{r}^2\right)r^\gamma \sum_{\nu=0} a_\nu r^\nu ,$$

where $$\gamma = -(im^2/B) - n' - 1.$$

If $(\gamma + n' + 1)$ is taken as real, then the electromagnetic and strong coefficients, A and B, will become imaginary, reversing their usual status. However, if $(\gamma + n' + 1)$ is taken as imaginary, and γ complex, A and B will become real, as in the usual representation; iD will also become real, allowing a real $A + iD$ to link all three scalar phase terms (and also that for the classical gravitational interaction, which, like a real iD, may be subsumed within A), while the weak coefficient C remains imaginary, retaining the status with respect to the scalar phase term that it had in the pure weak interaction.

An interesting case which is almost exactly equivalent in form to (10.16) (with real $A + iD$) occurs in the case of the Yukawa potential, which first arose in the case of massive mesonic exchange between baryons (originally pions between nucleons). In this case, the Laplace equation, which, for a Coulomb interaction (potential ϕ) involving an exchange of massless bosons (photons), is given by

$$\nabla^2\phi = 0,$$

becomes modified relativistically to

$$\left(\nabla^2 - m^2\right)\phi = 0,$$

when the exchange bosons acquire mass m. Now, for a point source with spherically symmetry, $\nabla^2\phi$ may be written as

$$\nabla^2\phi = \frac{1}{r}\frac{\partial^2}{\partial r^2}(r\phi),$$

which means that

$$\frac{1}{r}\frac{\partial^2}{\partial r^2}(r\phi) = m^2 r\phi .$$

The solution to this equation takes the form

$$\phi = \frac{g^2}{r}\exp(-mr)$$

where g is the coupling constant for the interaction ($g^2 = e^2 / 4\pi\varepsilon_0$ in the pure electrostatic case). With explicit use of the constants $\hbar$ and c, $\exp(-mr)$ becomes $\exp(-mcr/\hbar)$.

The spherically symmetric differential operator with Yukawa potential,

$$\left(\mathbf{k}\left(E + \frac{g^2}{r}exp(-mr) + \ldots\right) + \mathbf{i}\left(\frac{\partial}{\partial r} + \frac{1}{r} \pm i\frac{j+½}{\mathrm{r}}\right) + \mathbf{ij}m\right),$$

yields a nilpotent solution if we expand the exponential to give

$$\left(\mathbf{k}\left(E + \frac{g^2}{r} - g^2 m + \frac{g^2 m^2 r}{2!} - \frac{g^2 m^3 r^2}{3!}\ldots\right) + \mathbf{i}\left(\frac{\partial}{\partial r} + \frac{1}{r} \pm i\frac{j+½}{\mathrm{r}}\right) + \mathbf{ij}m\right).$$

This is of the same form as (10.16) with a constant term ($E - g^2m$), a Coulomb term (g^2 / r), a confining term ($g^2m^2\mathrm{r} / 2!$), and a polynomial series of harmonic oscillator terms ($- g^2m^3r^2 / 3! + \ldots$). So the solution must, in principle, be the same, using a phase factor

$$\phi = exp\left(i\left(E - g^2 m\right)r + ig^2 m^2 r^2 / 4 + ig^2 m^3 r^3 / 3! + \ldots\right) r^\gamma \sum_{\nu=0} a_\nu r^\nu .$$

In principle, the space variations for the strong, electric and weak potentials in the covariant time derivative are definitions of all those states that are equivalent under the conditions of conservation of charge and energy. We avoid using the space derivative (with vector potential terms and time variation) by making our frame of reference the 'static' one, while the mass term in the differential operator is necessarily a constant. This allows us to use spherical spatial symmetry (or no variation with respect to spatial direction) to convert the problem to one conceived in terms of angular momentum, and by implication pure charge, conservation. In principle, then, the three interaction potentials are describing, as we have said, three separate requirements for conserving angular momentum – its behaviour with respect to space (the vector or directional term), its behaviour with respect to time (the pseudoscalar or helicity term), and its behaviour with respect to mass (the scalar or pure magnitude term), although all the terms necessarily have a scalar or magnitude component. While, ultimately, this may suggest that it is not necessary, in principle, for a fermion state to have mass, the presence of electric charges in quark and lepton states, and the interchangeability of $+w$ an $-w$ in neutrino states, means that a nonzero mass term is required. (See also 14.11.)

Converting from the differential operator formalism to the state vector eigenvalue links the respective directional, helicity and magnitude terms with **p**, iE and m; while associating the potentials via their r-dependence with the respective ones produced by strong, weak and electric charges enables us to specify the aspects of angular momentum conservation which are preserved by the separate conservation properties of each charge type. In view of the separate conservation laws applying to w, s and e, it is significant that each charge type requires a separate statement of angular momentum conservation in the Dirac equation; so the presence or absence of each charge can be defined by a separate angular momentum operator. As we will see chapter 13, to define a unified fermion state it is convenient to assign angular momentum operators which are arbitrarily variable, but orthogonal, components of a single angular momentum pseudovector. Different fermionic states arise according to whether these are or are not aligned to each other, or to a defined projection of the actual angular momentum state of the fermion. All fermion states have weak charges because all necessarily have a helicity state, but, where the arbitrarily defined components are aligned with each other, separate information on charge conservation becomes unnecessary and the charge structure of the fermion will reflect this. Investigation of the broken symmetry between electromagnetic, strong and weak interactions leads naturally, it would seem, to a consideration of particle structures.

Chapter 11

QED and its Analogues

Quantum electrodynamics (QED), involving multiple fermion-boson processes, is dealt with extensively in nilpotent form to show that the packaged nilpotent automatically removes the infrared divergence in the fermion propagator, and that renormalization, as such, is not needed in this formalism, although *rescaling* of charge values will occur during interactions. The process is then extended to QCD (for the strong interaction) and QFD (for the weak), with the latter providing a particularly neat nilpotent explanation. This chapter is necessarily the one most heavily dependent on conventional theoretical techniques (Green's functions, Feynman diagrams, propagators, perturbation theory, renormalization, etc.) though all are given at least an outline derivation in terms of the mathematical structures created in earlier chapters. Some of the results are entirely new, but, even where they are not, it is evident that the nilpotent method produces a considerable advance in both convenience of calculation and explanatory power.

11.1 A Perturbation Expansion of the Dirac Equation for QED

The behaviour of a fermion in a spherically-symmetric static potential provides one of the simplest cases of the electromagnetic interaction at the fundamental level. When we extend the formalism to more general considerations, we are confronted by the apparent infinities that emerge in QED and their removal by the method of renormalization. Certain aspects of the nilpotent algebra, however, suggest that it might be possible to develop a more fundamental understanding of this process. For example, a formalism which doesn't appear to require normalization, might not require renormalization either. Also certain aspects of the algebra indicate that it has a built-in second quantization and a natural supersymmetry without requiring extra supersymmetric particles, which suggests that infinite sums must add up to finite values and that any necessary cancellations might be automatic. It appears to be significant, for example, that the infinite series

$$(kE + ii\mathbf{p} + ijm)\, k\, (kE + ii\mathbf{p} + ijm)\, k\, (kE + ii\mathbf{p} + ijm)\, k\, (kE + ii\mathbf{p} + ijm) \ldots,$$

which may be taken as representative as the individual terms in a fermion state vector acting on vacuum, is identical to

$$(kE + ii\mathbf{p} + ijm)\, (-kE + ii\mathbf{p} + ijm)\, (kE + ii\mathbf{p} + ijm)\, (-kE + ii\mathbf{p} + ijm) \ldots,$$

which may be taken as the creation by a fermion of an infinite series of alternate boson and fermion structures, presumably in the vacuum in a kind of supersymmetric series, with each particle or particle plus vacuum being its own supersymmetric partner. This suggests the existence of a kind of supersymmetry without a new set of particles. The aim, therefore, of the next few sections will be to develop the formalism to incorporate the existing techniques of QED, and to discover if there are any aspects of it which help either to explain or to remove apparent anomalies in the existing structures.

We will assume here that, if fermions can be taken as having the four terms in the state vector in a particular order, say,

$$(kE + ii\,\mathbf{p} + ij\,m)\,;\, (kE - ii\,\mathbf{p} + ij\,m);\, (-kE + ii\,\mathbf{p} + ij\,m);\, (-kE - ii\,\mathbf{p} + ij\,m),$$

then the order of terms in the equivalent antifermion (with the same spin) will be automatically determined as:

$$(-kE + ii\,\mathbf{p} + ij\,m)\,;\, (-kE - ii\,\mathbf{p} + ij\,m);\, (kE + ii\,\mathbf{p} + ij\,m);\, (kE - ii\,\mathbf{p} + ij\,m),$$

so that to specify either a fermion or antifermion, it will often be convenient just to specify the first of the four terms, with the presence of the other three automatically understood. We will, for example, refer on occasions to

$$(kE + ii\,\mathbf{p} + ij\,m)\, (-kE + ii\,\mathbf{p} + ij\,m)$$

being a scalar on the basis of this understanding.

We begin with the Dirac equation for a fermion in the presence of the electromagnetic potentials ϕ, $\mathbf{A}$.

$$\left(k\frac{\partial}{\partial t} + ii\boldsymbol{\sigma}.\nabla + ijm\right)\psi = -e(ik\phi - i\boldsymbol{\sigma}.\mathbf{A})\psi$$

and apply a perturbation expansion to ψ, so that

$$\psi = \psi_0 + \psi_1 + \psi_2 + \ldots,$$

where $\psi_0 = (kE + ii\boldsymbol{\sigma}.\mathbf{p} + ijm)\, e^{-i(Et - \mathbf{p.r})}$ is the solution of the unperturbed equation:

$$\left(k\frac{\partial}{\partial t}+ii\boldsymbol{\sigma}.\nabla+ijm\right)\psi_0=0$$

and represents zeroth-order coupling, or a free electron of momentum **p**. Here, of course, as always, we take ($\boldsymbol{k}E$ + $\boldsymbol{i}i\boldsymbol{\sigma}.\mathbf{p}$ + $\boldsymbol{j}m$), and the equivalent differential operator, as a 4-component vector incorporating the four solutions ($\pm \boldsymbol{k}E \pm \boldsymbol{i}i\boldsymbol{\sigma}.\mathbf{p}$ + $\boldsymbol{j}m$) in the kind of fixed order already specified, say ($\boldsymbol{k}E$ + $\boldsymbol{i}i\boldsymbol{\sigma}.\mathbf{p}$ + $\boldsymbol{j}m$), ($-\boldsymbol{k}E$ + $\boldsymbol{i}i\boldsymbol{\sigma}.\mathbf{p}$ + $\boldsymbol{j}m$), ($\boldsymbol{k}E - \boldsymbol{i}i\boldsymbol{\sigma}.\mathbf{p}$ + $\boldsymbol{j}m$), ($-\boldsymbol{k}E - \boldsymbol{i}i\boldsymbol{\sigma}.\mathbf{p}$ + $\boldsymbol{j}m$). This means that ($\boldsymbol{k}E$ + $\boldsymbol{i}i\boldsymbol{\sigma}.\mathbf{p}$ + $\boldsymbol{j}m$) ($-\boldsymbol{k}E$ + $\boldsymbol{i}i\boldsymbol{\sigma}.\mathbf{p}$ + $\boldsymbol{j}m$) becomes a scalar quantity, while ($\boldsymbol{k}E$ + $\boldsymbol{i}i\boldsymbol{\sigma}.\mathbf{p}$ + $\boldsymbol{j}m$) ($\boldsymbol{k}E$ + $\boldsymbol{i}i\boldsymbol{\sigma}.\mathbf{p}$ + $\boldsymbol{j}m$) = 0.

Using the perturbation expansion, we can write

$$\left(k\frac{\partial}{\partial t}+ii\boldsymbol{\sigma}.\nabla+ijm\right)(\psi_0+\psi_1+\psi_2+\ldots)=-e(ik\phi-i\boldsymbol{\sigma}.\mathbf{A})(\psi_0+\psi_1+\psi_2+\ldots),$$

leading to the series

$$\left(k\frac{\partial}{\partial t}+ii\boldsymbol{\sigma}.\nabla+ijm\right)\psi_0=0$$

$$\left(k\frac{\partial}{\partial t}+ii\boldsymbol{\sigma}.\nabla+ijm\right)\psi_1=-e(ik\phi-i\boldsymbol{\sigma}.\mathbf{A})\psi_0$$

$$\left(k\frac{\partial}{\partial t}+ii\boldsymbol{\sigma}.\nabla+ijm\right)\psi_2=-e(ik\phi-i\boldsymbol{\sigma}.\mathbf{A})\psi_1$$

Expanding ($i\,\boldsymbol{k}\phi - \boldsymbol{i}\,\boldsymbol{\sigma}.\mathbf{A}$) as a Fourier series, and summing over **k**, we obtain

$$(ik\phi-ii\boldsymbol{\sigma}.\mathbf{A})=\sum(ik\phi(\mathbf{k})-ii\boldsymbol{\sigma}.\mathbf{A}(\mathbf{k}))e^{i\mathbf{k}.\mathbf{r}},$$

so that $$\left(k\frac{\partial}{\partial t}+ii\boldsymbol{\sigma}.\nabla+ijm\right)\psi_1=-e\sum(ik\phi(\mathbf{k})-i\boldsymbol{\sigma}.\mathbf{A}(\mathbf{k}))e^{i\mathbf{k}.\mathbf{r}}\psi_0$$

$$=-e\sum(ik\phi(\mathbf{k})-i\boldsymbol{\sigma}.\mathbf{A}(\mathbf{k}))e^{i\mathbf{k}.\mathbf{r}}(kE+ii\boldsymbol{\sigma}.\mathbf{p}+ijm)e^{-i(Et-\mathbf{p}.\mathbf{r})}$$

$$=-e\sum(ik\phi(\mathbf{k})-i\boldsymbol{\sigma}.\mathbf{A}(\mathbf{k}))(kE+ii\boldsymbol{\sigma}.\mathbf{p}+ijm)e^{-i(Et-(\mathbf{p}+\mathbf{k}).\mathbf{r})}.$$

Suppose we expand ψ_1 as

$$\psi_1=\sum\nu_1(E,\mathbf{p}+\mathbf{k})e^{i(Et-(\mathbf{p}+\mathbf{k})).\mathbf{r}}.$$

Then $$\sum\left(k\frac{\partial}{\partial t}+ii\boldsymbol{\sigma}.\nabla+ijm\right)\nu_1(E,\mathbf{p}+\mathbf{k})e^{-i(Et-(\mathbf{p}+\mathbf{k}).\mathbf{r})}$$

$$=-e\sum(ik\phi(\mathbf{k})-i\boldsymbol{\sigma}.\mathbf{A}(\mathbf{k}))(kE+ii\boldsymbol{\sigma}.\mathbf{p}+ijm)e^{-i(Et-(\mathbf{p}+\mathbf{k}).\mathbf{r})}.$$

$$\sum \left(kE + ii\boldsymbol{\sigma}.(\mathbf{p}+\mathbf{k}) + ijm\right)v_1(E, \mathbf{p}+\mathbf{k})e^{-i(Et-(\mathbf{p}+\mathbf{k}).\mathbf{r})}$$
$$= -e\sum \left(ik\phi(\mathbf{k}) - i\boldsymbol{\sigma}.\mathbf{A}(\mathbf{k})\right)\left(kE + ii\boldsymbol{\sigma}.\mathbf{p} + ijm\right)e^{-i(Et-(\mathbf{p}+\mathbf{k}).\mathbf{r})},$$

and, equating individual terms,

$(kE + ii\boldsymbol{\sigma}.(\mathbf{p} + \mathbf{k}) + ijm)v_1(E, \mathbf{p} + \mathbf{k}) = -e\,(i\,k\phi\,(\mathbf{k}) - i\,\boldsymbol{\sigma}.\mathbf{A}\,(\mathbf{k}))\,(kE + ii\boldsymbol{\sigma}.\mathbf{p} + ijm)$.

We can write this in the form

$$v_1(E, \mathbf{p} + \mathbf{k}) = -e\,[kE + ii\,\boldsymbol{\sigma}.(\mathbf{p} + \mathbf{k}) + ijm]^{-1}\,(i\,k\phi\,(\mathbf{k}) - i\,\boldsymbol{\sigma}.\mathbf{A}\,(\mathbf{k}))$$
$$\times\,(kE + ii\boldsymbol{\sigma}.\mathbf{p} + ijm),$$

which means that

$$\psi_1 = -e\sum\left[kE + ii\boldsymbol{\sigma}.(\mathbf{p}+\mathbf{k}) + ijm\right]^{-1}\left(ik\phi(\mathbf{k}) - i\boldsymbol{\sigma}.\mathbf{A}(\mathbf{k})\right)\left(kE + ii\boldsymbol{\sigma}.\mathbf{p} + ijm\right)e^{-i(Et-(\mathbf{p}+\mathbf{k}).\mathbf{r})}$$

This is the wavefunction for first-order coupling, with an electron (for example) absorbing or emitting a photon of momentum **k**. Applying this to the second term in the perturbation series, we obtain:

$$\left(k\frac{\partial}{\partial t} + ii\boldsymbol{\sigma}.\nabla + ijm\right)\psi_2 = -e\left(ik\phi(\mathbf{k}) - i\boldsymbol{\sigma}.\mathbf{A}(\mathbf{k})\right)e^{-i\mathbf{k}.\mathbf{r}}\psi_0$$
$$\times\sum(-e)\left[kE + ii\boldsymbol{\sigma}.(\mathbf{p}+\mathbf{k}) + ijm\right]^{-1}\left[\left(ik\phi(\mathbf{k}) - i\boldsymbol{\sigma}.\mathbf{A}(\mathbf{k})\right)\left(kE + ii\boldsymbol{\sigma}.\mathbf{p} + ijm\right)e^{-i(Et-(\mathbf{p}+\mathbf{k}).\mathbf{r})}\right.$$

We now introduce $\sum\left(ik\phi(\mathbf{k}') - i\boldsymbol{\sigma}.\mathbf{A}(\mathbf{k}')\right)e^{i\mathbf{k}'.\mathbf{r}}$ for the 4-potential involved with this term, and summing over **k** and **k'**,

$$\psi_2 = \sum\sum v_2(E, \mathbf{p}+\mathbf{k}+\mathbf{k}')e^{-i(Et-(\mathbf{p}+\mathbf{k}+\mathbf{k}').\mathbf{r})}.$$

We thus obtain

$$\sum\left(k\frac{\partial}{\partial t} + ii\boldsymbol{\sigma}.\nabla + ijm\right)v_2(E, \mathbf{p}+\mathbf{k}+\mathbf{k}')e^{-i(Et-(\mathbf{p}+\mathbf{k}+\mathbf{k}').\mathbf{r})} =$$
$$-e\sum\sum\left(ik\phi(\mathbf{k}') - i\boldsymbol{\sigma}.\mathbf{A}(\mathbf{k}')\right)\left[kE + ii\boldsymbol{\sigma}.(\mathbf{p}+\mathbf{k}) + ijm\right]^{-1}$$
$$\times(-e)\left(ik\phi(\mathbf{k}) - i\boldsymbol{\sigma}.\mathbf{A}(\mathbf{k})\right)\left(kE + ii\boldsymbol{\sigma}.\mathbf{p} + ijm\right)e^{-i(Et-(\mathbf{p}+\mathbf{k}+\mathbf{k}').\mathbf{r})}.$$

Comparing coefficients, we find that

$$(kE + ii\,\boldsymbol{\sigma}.(\mathbf{p} + \mathbf{k} + \mathbf{k}') + ijm)\,v_2(E, \mathbf{p} + \mathbf{k} + \mathbf{k}') = -e\,(i\,k\phi\,(\mathbf{k}') - i\,\boldsymbol{\sigma}.\mathbf{A}\,(\mathbf{k}'))$$
$$\times\,[kE + ii\,\boldsymbol{\sigma}.(\mathbf{p} + \mathbf{k}) + ijm]^{-1}\,(-e)\,(i\,k\phi\,(\mathbf{k}) - i\,\boldsymbol{\sigma}.\mathbf{A}\,(\mathbf{k}))\,(kE + ii\boldsymbol{\sigma}.\mathbf{p} + ijm).$$

Hence

$$v_2(E, \mathbf{p} + \mathbf{k} + \mathbf{k}') = [kE + ii\,\boldsymbol{\sigma}.(\mathbf{p} + \mathbf{k} + \mathbf{k}') + ijm]^{-1}(-e)(i\,k\phi\,(\mathbf{k}') - i\,\boldsymbol{\sigma}.\mathbf{A}\,(\mathbf{k}'))\times$$
$$[kE + ii\,\boldsymbol{\sigma}.(\mathbf{p} + \mathbf{k}) + ijm]^{-1}\,(-e)\,(i\,k\phi\,(\mathbf{k}) - i\,\boldsymbol{\sigma}.\mathbf{A}\,(\mathbf{k}))\,(kE + ii\boldsymbol{\sigma}.\mathbf{p} + ijm),$$

and

$$\psi_2 = \Sigma\Sigma\left[kE + ii\boldsymbol{\sigma}.(\mathbf{p}+\mathbf{k}+\mathbf{k}') + ijm\right]^{-1}(-e)\left(ik\phi(\mathbf{k}') - i\boldsymbol{\sigma}.\mathbf{A}(\mathbf{k}')\right) \times$$
$$\left[kE + ii\boldsymbol{\sigma}.(\mathbf{p}+\mathbf{k}) + ijm\right]^{-1}(-e)\left(ik\phi(\mathbf{k}) - i\boldsymbol{\sigma}.\mathbf{A}(\mathbf{k})\right)\left(kE + ii\boldsymbol{\sigma}.\mathbf{p} + ijm\right)e^{-i(Et-(\mathbf{p}+\mathbf{k}+\mathbf{k}').\mathbf{r})}$$

This is the wavefunction for second-order coupling, with an electron (for example) absorbing and / or emitting two photons of momenta **k** and **k'**.

11.2 Integral Solutions of the Dirac Equation

We may suppose that the equation

$$\left(k\frac{\partial}{\partial t} + ii\boldsymbol{\sigma}.\nabla + ijm\right)\psi_1 = -e\left(ik\phi - i\boldsymbol{\sigma}.\mathbf{A}\right)\psi_0$$

has an integral solution of the form

$$\psi_1(x) = \int_{-\infty}^{\infty} G_1(x, x')\ \psi_0(x')\ dx\ ,$$

where x and x' are 4-vectors, and the *Green's function*,

$$G_1(x, x') = \frac{1}{(2\pi)^4}\int_{-\infty}^{\infty} d^4p\ [kE + ii\ \boldsymbol{\sigma}.\mathbf{p} + ijm]^{-1}\ (-e)\ (i\ k\phi\,(x') - i\ \boldsymbol{\sigma}.\mathbf{A}\ (x'))$$
$$\times\ exp\ -ip(x - x'),$$

where p is also a 4-vector. Defining the Fourier transform of $(ik\phi\,(x') - i\ \boldsymbol{\sigma}.\mathbf{A}\ (x'))$ as

$$\frac{1}{(2\pi)^4}\int_{-\infty}^{\infty} d^4p'\ (i\ k\phi\,(p') - i\ \boldsymbol{\sigma}.\mathbf{A}\ (p'))\ \exp -ip'\,(x - x'),$$

we obtain

$$G_1(x, x') = \frac{1}{(2\pi)^8}\int\int d^4p\ d^4p'\ [kE + ii\ \boldsymbol{\sigma}.\mathbf{p} + ijm]^{-1}$$
$$\times\,(-e)\ (i\ k\phi\,(p') - i\ \boldsymbol{\sigma}.\mathbf{A}\ (p'))\ \exp -i(p + p')\ (x - x').$$

The procedure can be extended to $\psi_2(x)$ and

$$G_2(x'', x') = \frac{1}{(2\pi)^8}\int\int d^4p''\ d^4p'''\ [kE'' + ii\ \boldsymbol{\sigma}.\mathbf{p''} + ijm]^{-1}$$
$$\times\,(-e)\ (i\ k\phi\,(p'') - i\ \boldsymbol{\sigma}.\mathbf{A}\ (p''))\ \exp -i(p'' + p''')\ (x'' - x').$$

Reversing the order of x and x' in the first term, the product of G_2G_1 becomes

$$G_2G_1 = \frac{1}{(2\pi)^{16}} \int\int\int\int d^4p\, d^4p'\, d^4p''\, d^4p''' \, [kE'' + ii\, \boldsymbol{\sigma}.\mathbf{p}'' + ijm]^{-1}$$

$$\times (-e)\, (i\, \boldsymbol{k}\phi(p''') - \boldsymbol{i}\, \boldsymbol{\sigma}.\mathbf{A}\, (p'''))\, [kE + ii\, \boldsymbol{\sigma}.\mathbf{p} + ijm]^{-1}$$

$$\times (-e)\, (i\, \boldsymbol{k}\phi(p') - \boldsymbol{i}\, \boldsymbol{\sigma}.\mathbf{A}\, (p'))\, \exp -i(p'' + p''')\, (x'' - x')\, \exp -i(p + p')\, (x' - x).$$

Here, $[kE'' + ii\, \boldsymbol{\sigma}.\mathbf{p}'' + ijm]^{-1}\, (-e)\, (i\, \boldsymbol{k}\phi(p''') - \boldsymbol{i}\, \boldsymbol{\sigma}.\mathbf{A}\, (p'''))\, [kE + ii\, \boldsymbol{\sigma}.\mathbf{p} + ijm]^{-1}\, (-e)\, (i\, \boldsymbol{k}\phi(p') - \boldsymbol{i}\, \boldsymbol{\sigma}.\mathbf{A}\, (p'))$ has the same form as $v_2(E, \mathbf{p} + \mathbf{k} + \mathbf{k}')$.

11.3 Renormalization

In standard QED, renormalization is not needed at the tree level, where there are no loops or closed paths in the diagrams, but is needed when loops are required in the calculation. There are essentially three types of graph which remain divergent after the application of gauge invariance and Furry's theorem (that a Feynman loop diagram vanishes if it has an odd number of photon lines). These are the electron self-energy graph (with divergence $D = 1$), the electron-photon vertex graph (with $D = 0$), and the photon vacuum polarization graph (with $D = 2$, which can be reduced to $D = 0$, by gauge invariance). Let us take the electron self-energy graph, for a free electron not interacting with any external field. In effect, this is a representation of an electron emitting a virtual photon and then reabsorbing it. The perturbation expansion for a first-order coupling of this kind (for either emission or absorption) produces a wavefunction of the form

$$\Psi_1 = -e\, \Sigma\, [kE + ii\, \boldsymbol{\sigma}.(\mathbf{p} + \mathbf{k}) + ijm]^{-1}\, (ik\phi - \boldsymbol{i}\, \boldsymbol{\sigma}.\mathbf{A})\, (kE + ii\boldsymbol{\sigma}.\mathbf{p} + ijm) \times e^{-i(Et - (\mathbf{p} + \mathbf{k}).\mathbf{r})},$$

where $\mathbf{k}$ is the extra momentum acquired by the electron from the photon. If we observe the process in the rest frame of the electron and eliminate any external source of potential, then $\mathbf{k} = 0$, and $(ik\phi - \boldsymbol{i}\, \boldsymbol{\sigma}.\mathbf{A})$ reduces to the static value, $\boldsymbol{ik}\phi$. In this case, Ψ_1 becomes

$$\Psi_1 = -e\, \Sigma\, [kE + ii\, \boldsymbol{\sigma}.\mathbf{p} + ijm]^{-1}\, ik\phi\, (kE + ii\boldsymbol{\sigma}.\mathbf{p} + ijm)\, e^{-i(Et - (\mathbf{p} + \mathbf{k}).\mathbf{r})}.$$

Writing this as

$$\Psi_1 = -e\, \Sigma\, [kE + ii\, \boldsymbol{\sigma}.\mathbf{p} + ijm]^{-1}[-kE + ii\, \boldsymbol{\sigma}.\mathbf{p} + ijm]^{-1}(-kE + ii\, \boldsymbol{\sigma}.\mathbf{p} + ijm)\, ik\phi \times (kE + ii\boldsymbol{\sigma}.\mathbf{p} + ijm)\, e^{-i(Et - (\mathbf{p} + \mathbf{k}).\mathbf{r})},$$

we obtain $\Psi_1 = 0$, for any fixed value of ϕ. In other words, a *non-interacting* electron requires no renormalization as a result of its self-energy.[1] We can extend this idea to the photon vacuum polarization, which creates and then annihilates an electron-positron pair. These processes are equivalent, in effect to producing a diagram in which an electron and positron travel in mutual opposite directions from or towards a vertex with an incoming or outgoing photon, which, in the absence of any external potential, is, in principle, the same as an electron emitting and absorbing a photon and being undeviated by any momentum change ($\mathbf{k}$ = 0), as in the self-energy diagram. The calculation of $\Psi_1 = 0$ for the self-energy should, therefore, also apply in this case. The electron-photon vertex graph diverges only beyond the tree level, when a photon loop links the electron on each side of a vertex formed with an incoming photon. In the context of $\mathbf{k} = 0$, this reduces to another equivalent of electron self-energy or vacuum polarization.

It appears, from this calculation that $[\boldsymbol{k}E + \boldsymbol{i}i\ \boldsymbol{\sigma}.\mathbf{p} + \boldsymbol{i}jm]^{-1}$ is equivalent to the vacuum 'image' of $(\boldsymbol{k}E + \boldsymbol{i}i\ \boldsymbol{\sigma}.\mathbf{p} + \boldsymbol{i}jm)$, that is, $(-\boldsymbol{k}E + \boldsymbol{i}i\ \boldsymbol{\sigma}.\mathbf{p} + \boldsymbol{i}jm)$, and this seems to be its physical meaning. In the nilpotent formulation, a fermion with quaternion state vector $(\boldsymbol{k}E + \boldsymbol{i}i\ \boldsymbol{\sigma}.\mathbf{p} + \boldsymbol{i}jm)$ generates a vacuum image of the form $(-\boldsymbol{k}E + \boldsymbol{i}i\ \boldsymbol{\sigma}.\mathbf{p} + \boldsymbol{i}jm)$ by acting on the vacuum operator $\boldsymbol{k}(-\boldsymbol{k}E + \boldsymbol{i}i\ \boldsymbol{\sigma}.\mathbf{p} + \boldsymbol{i}jm)$. However, $(\boldsymbol{k}E + \boldsymbol{i}i\ \boldsymbol{\sigma}.\mathbf{p} + \boldsymbol{i}jm)\ \boldsymbol{k}\ (\boldsymbol{k}E + \boldsymbol{i}i\ \boldsymbol{\sigma}.\mathbf{p} + \boldsymbol{i}jm)$ is in principle identical to $(\boldsymbol{k}E + \boldsymbol{i}i\ \boldsymbol{\sigma}.\mathbf{p} + \boldsymbol{i}jm)$, as is $(\boldsymbol{k}E + \boldsymbol{i}i\ \boldsymbol{\sigma}.\mathbf{p} + \boldsymbol{i}jm)\ \boldsymbol{k}\ (\boldsymbol{k}E + \boldsymbol{i}i\ \boldsymbol{\sigma}.\mathbf{p} + \boldsymbol{i}jm)\ \boldsymbol{k}\ (\boldsymbol{k}E + \boldsymbol{i}i\ \boldsymbol{\sigma}.\mathbf{p} + \boldsymbol{i}jm)\ \boldsymbol{k}\ (\boldsymbol{k}E + \boldsymbol{i}i\ \boldsymbol{\sigma}.\mathbf{p} + \boldsymbol{i}jm) \ldots$, in an infinite series of actions upon the vacuum; and this is mathematically and physically equivalent to $(\boldsymbol{k}E + \boldsymbol{i}i\ \boldsymbol{\sigma}.\mathbf{p} + \boldsymbol{i}jm)\ (-\boldsymbol{k}E + \boldsymbol{i}i\ \boldsymbol{\sigma}.\mathbf{p} + \boldsymbol{i}jm)\ (\boldsymbol{k}E + \boldsymbol{i}i\ \boldsymbol{\sigma}.\mathbf{p} + \boldsymbol{i}jm)\ (-\boldsymbol{k}E + \boldsymbol{i}i\ \boldsymbol{\sigma}.\mathbf{p} + \boldsymbol{i}jm) \ldots$, which generates an infinite series of alternate fermion and boson states. We can, therefore, conceive of this in terms of a fermion generating an alternate series of (positive) boson and (negative) fermion loops, which sum up to return to the pure fermion state itself. It is effectively a form of supersymmetry in which the supersymmetric partners are not new particles but merely vacuum images or couplings of the original state.

In addition to this, as we have shown in chapter 6, the nilpotent representation of the Dirac equation is already second quantized because the operator and state vector are necessarily identical, and each is quantized. This means that the quantum field integrals reduce in principle to the form of a single quaternion state vector. This representation says nothing directly about the energy value associated with the coupling of a charge to an electromagnetic field, but it does define the fact that such a single representation must be possible, and, that the infinite number of possible fluctuations must be constrained in such a way that they sum up to a single finite value for the coupling in any given state. From this perspective, the existence of a method for eliminating apparent infinities is no more surprising than the fact that complicated physical systems act in such a way that they conserve energy or that the complicated mess of quarks and gluons inside a proton somehow 'conspires' to deliver a total spin value of ½ to the composite particle. It is also not surprising that there is a maximum value of energy involved which ensures that the summation is finite. The existence of the quaternion state vector as a single representation of the entire quantum field is a statement that a finite summation exists. The calculation at the beginning of this section is an illustration of the fact. There is even a mechanism for ensuring that it happens in the interactions of quaternion state vectors with the vacuum.

When we wish to determine the relative strength of the coupling involved in an electron interacting with an external field, for example in the process of electron-muon scattering or the coupling of the electron to a vector potential to determine its magnetic moment, we *then* need to sum a perturbative series of real interacting terms. (The calculation is essentially an expression of the fact that the interacting nilpotent is an open system.) The process, however, will not be divergent, and it will naturally cut off at the value of energy described as the Planck mass (M_P), the energy at which the effects of what is usually referred to as quantum gravity take place.[2] This is, very probably, because gravity is the carrier of the wavefunction correlations involved in nonlocality, and the Planck mass is the quantum of the inertial interactions which result from the effect of gravity on the time-delayed nature of the nongravitational interactions, which, in turn, produces the inertial mass associated with 'charged' particles (cf chapter 17). It is, of course, the fact that such particles have inertial mass which creates the time delay in the transmission of energy; and the inertial mass may also be seen as the result of a coupling to the Higgs field, which produces the filled vacuum which allows the instantaneous transmission of gravity and wavefunction correlations. Cutting off at the Planck mass ($(\hbar c / G)^{1/2}$), which is c-dependent, is the same thing, in effect, as saying that the real interactions (and inertial reactions) are time-delayed or that the fermions involved have inertial mass, and it is significant that, in QED, the process actually determines the mass-value of the electron. The process is, in effect, creating a c-related 'event horizon' such as we would expect for an inertial process as described.

The significant result from current QED calculations is that the first-order correction term for the electric fine structure constant, for purely leptonic interactions, at energy of interaction μ, is $(1 / 3\pi) \ln (M_P^2 / \mu^2)$, which has the same form as the term which we will derive for the quark theory in chapter 15, using the idea of lepton-type quarks, and also the correction term which is needed to achieve Grand Unification at the Planck mass.

11.4 Green's Function Solution

To apply the method of Green's functions (as in 11.2), we solve for a unit source and then sum over the whole distribution. Beginning with

$$\left(k\frac{\partial}{\partial t}+ii\boldsymbol{\sigma}.\nabla+ijm\right)\psi=-e(ik\phi-i\boldsymbol{\sigma}.\mathbf{A})\psi$$

or

$$\left(k\frac{\partial}{\partial t}+ii\boldsymbol{\sigma}.\nabla+ijm\right)(kE+ii\boldsymbol{\sigma}.\mathbf{p}+ijm)e^{ipx}=-e(ik\phi-i\boldsymbol{\sigma}.\mathbf{A})(kE+ii\boldsymbol{\sigma}.\mathbf{p}+ijm)e^{ipx}$$

where p and x are 4-vectors, we solve for a unit source using the δ-function

$$\left(k\frac{\partial}{\partial t}+ii\boldsymbol{\sigma}.\nabla+ijm\right)G_F=\delta^{(4)}(x-x'),$$

where G_F is a wave produced at x by a unit source at x'. Translational invariance shows that G_F is a function of $(x - x')$. Then the solution for the whole distribution becomes

$$\psi(x)=-e\int d^4x'\ G_F(x-x')\,(i\,k\phi-i\,\boldsymbol{\sigma}.\mathbf{A})\ \psi(x')$$

or

$$(kE+ii\boldsymbol{\sigma}.\mathbf{p}+ijm)e^{ip.x}=\ e\int d^4x'\ G_F(x-x')\,(i\,k\phi-i\,\boldsymbol{\sigma}.\mathbf{A})\,(kE+ii\boldsymbol{\sigma}.\mathbf{p}+ijm)e^{ip.x'}.$$

Taking the Fourier transform into momentum space, we obtain

$$G_F(x-x')=\frac{1}{(2\pi)^4}\int d^4p\ S_F(p)\ e^{-ip.(x-x')}.$$

Then, using the Fourier representation of the δ-function, we find

$$\frac{1}{(2\pi)^4}\int d^4p\left(k\frac{\partial}{\partial t}+ii\boldsymbol{\sigma}.\nabla+ijm\right)S_F(p)\ e^{-ip.(x-x')}=\frac{1}{(2\pi)^4}\int d^4p\ e^{-ip.(x-x')},$$

from which

$$(kE+ii\boldsymbol{\sigma}.\mathbf{p}+ijm)\ S_F(p)=1$$

and

$$S_F(p)=\frac{1}{(kE+ii\boldsymbol{\sigma}.\mathbf{p}+ijm)}$$

which is the *electron propagator* in the Feynman formalism.

Conventional theory assumes that

$$S_F(p) = \frac{1}{\not{p} - m} = \frac{\not{p} + m}{p^2 - m^2},$$

where $\not{p}$ represents $\gamma^\mu \partial_\mu$, or its eigenvalue, and that there is a singularity or 'pole' (p_0) where $p^2 - m^2 = 0$, the 'pole' being the origin of positron states. Effectively, on either side of the pole we have positive energy states moving forwards in time, and negative energy states moving backwards in time, the terms ($\not{p} + m$) and ($- \not{p} + m$) being used to project out, respectively, the positive and negative energy states. We then add the infinitesimal term $i\varepsilon$ to $p^2 - m^2$, so that $iS_F(p)$ becomes

$$iS_F(p) = \frac{i(\not{p} + m)}{p^2 - m^2 + i\varepsilon} = \frac{(\not{p} + m)}{2p_0}\left(\frac{1}{p_0 - \sqrt{p^2 + m^2} + i\varepsilon} + \frac{1}{p_0 + \sqrt{p^2 + m^2} - i\varepsilon}\right)$$

and take a contour integral over the complex variable to give the solution

$$S_F(x - x') = \int d^3p \frac{1}{(2\pi)^3} \frac{m}{2E}\left[-i\theta(t - t')\sum_{r=1}^{2}\Psi(x)\overline{\Psi}(x') + i\theta(t' - t)\sum_{r=3}^{4}\Psi(x)\overline{\Psi}(x')\right]$$

with summations over the up and down spin states.

In the nilpotent formalism, however, the $i\varepsilon$ term is unnecessary and there is no infrared divergence at the pole, because the denominator of the propagator term is a positive nonzero scalar. We write

$$S_F(p) = \frac{1}{(\boldsymbol{k}E + \boldsymbol{i}i\boldsymbol{\sigma}.\mathbf{p} + \boldsymbol{i}\boldsymbol{j}m)},$$

and are free to choose our usual interpretation of the reciprocal of a nilpotent:

$$\frac{1}{(\boldsymbol{k}E + \boldsymbol{i}i\boldsymbol{\sigma}.\mathbf{p} + \boldsymbol{i}\boldsymbol{j}m)} = \frac{(-\boldsymbol{k}E + \boldsymbol{i}i\boldsymbol{\sigma}.\mathbf{p} + \boldsymbol{i}\boldsymbol{j}m)}{(\boldsymbol{k}E + \boldsymbol{i}i\boldsymbol{\sigma}.\mathbf{p} + \boldsymbol{i}\boldsymbol{j}m)(-\boldsymbol{k}E + \boldsymbol{i}i\boldsymbol{\sigma}.\mathbf{p} + \boldsymbol{i}\boldsymbol{j}m)} = \frac{(-\boldsymbol{k}E + \boldsymbol{i}i\boldsymbol{\sigma}.\mathbf{p} + \boldsymbol{i}\boldsymbol{j}m)}{(E^2 + p^2 + m^2)}$$

which is finite at all values. We can also, for example, write

$$\frac{1}{(\boldsymbol{k}E + \boldsymbol{i}i\boldsymbol{\sigma}.\mathbf{p} + \boldsymbol{i}\boldsymbol{j}m)} = \frac{(-\boldsymbol{k}E + \boldsymbol{i}i\boldsymbol{\sigma}.\mathbf{p} + \boldsymbol{i}\boldsymbol{j}m)}{-2\boldsymbol{k}E}\left(\frac{1}{(\boldsymbol{k}E + \boldsymbol{i}i\boldsymbol{\sigma}.\mathbf{p} + \boldsymbol{i}\boldsymbol{j}m)} - \frac{1}{(-\boldsymbol{k}E + \boldsymbol{i}i\boldsymbol{\sigma}.\mathbf{p} + \boldsymbol{i}\boldsymbol{j}m)}\right)$$

and other, similar, forms. Our integral is now simply

$$S_F(x - x') = \int d^3p \frac{1}{(2\pi)^3} \frac{m}{2E}\, \theta(t - t')\, \Psi(x)\, \overline{\Psi}(x'),$$

in which $\Psi(x) = ((\boldsymbol{k}E + \boldsymbol{i}i\,\mathbf{p} + \boldsymbol{i}\boldsymbol{j}\,m) \ldots) \exp(ipx)$,

where $((\boldsymbol{k}E + \boldsymbol{i}i\,\mathbf{p} + \boldsymbol{i}\boldsymbol{j}\,m) \ldots)$ is the bra matrix with the terms:

$$(\boldsymbol{k}E + \boldsymbol{ii}\ \mathbf{p} + \boldsymbol{ij}\ m);\ (\boldsymbol{k}E - \boldsymbol{ii}\ \mathbf{p} + \boldsymbol{ij}\ m);\ (-\boldsymbol{k}E + \boldsymbol{ii}\ \mathbf{p} + \boldsymbol{ij}\ m);\ (-\boldsymbol{k}E - \boldsymbol{ii}\ \mathbf{p} + \boldsymbol{ij}\ m)$$

and the adjoint term becomes

$$\bar{\Psi}(x') = ((\boldsymbol{k}E - \boldsymbol{ii}\ \mathbf{p} - \boldsymbol{ij}\ m)\ \dots)\ (\boldsymbol{ik})\ exp\ (-ipx'),$$

with $((\boldsymbol{k}E - \boldsymbol{ii}\ \mathbf{p} - \boldsymbol{ij}\ m)\ \dots)\ (\boldsymbol{ik})$ a ket. The reason for this success is apparent. The nilpotent formulation is automatically second quantized and the negative energy states appear as components of the nilpotent wavefunction on the same basis as the positive energy states. No averaging over spin states or 'interpreting' $-E$ as a reversed time state is necessary; the 'reversed time' state occurs with the t in the operator $\partial / \partial t$, and there is no need to separate out the states on opposite sides of the pole.

We can use the electron propagator to define the photon propagator. Conventionally, we derive the photon propagator, $\Delta_F(x - x')$, directly from the Klein-Gordon equation, while recognizing that its mathematical form depends on the choice of gauge. In the presence of a source field, represented by current $j(x)$, we can write the Klein-Gordon equation in the form

$$\left(\frac{\partial}{\partial t} - \nabla^2 - m^2\right)\phi(x) = j(x),$$

and, with $\Delta_F(x - x')$ as Green's function, we have

$$\left(\frac{\partial}{\partial t} - \nabla^2 - m^2\right)\Delta_F(x - x') = -\delta^{(4)}(x - x'),$$

for which the solution is

$$\phi(x) = \phi_0(x) - \int d^4x'\ \Delta_F\ (x - x')\ j(x'),$$

where $\phi_0(x)$ is a solution of the equation in the absence of sources:

$$\left(\frac{\partial}{\partial t} - \nabla^2 - m^2\right)\phi(x) = 0\,.$$

Using the Fourier transform

$$\Delta_F\ (x - x') = \frac{1}{(2\pi)^4}\int d^4p\ \exp -ip(x - x')\ \Delta_F\ (p),$$

and applying the operator $\left(\frac{\partial}{\partial t} - \nabla^2 - m^2\right)$ to both sides, we obtain:

$$\Delta_F(x - x') = \frac{\not{p} + m}{p^2 - m^2}\,.$$

In the nilpotent formalism, we do not need to use the Klein-Gordon equation, which is not specific to boson states or an identifier of them, to define a boson propagator. (In conventional theory, the Klein-Gordon operator is the only scalar product which can emerge from a differential operator defined as an ideal, but it does not correspond to any of the known bosonic states – it merely defines a universal zero condition which is true for all states, whether bosonic or fermionic.) Instead, we have *three* boson propagators.

Spin 1: $$\Delta_F(x-x') = \frac{1}{(\pm \mathbf{k}E \pm \mathbf{i}i\boldsymbol{\sigma}.\mathbf{p} + \mathbf{i}\mathbf{j}m)(\mp \mathbf{k}E \pm \mathbf{i}i\boldsymbol{\sigma}.\mathbf{p} + \mathbf{i}\mathbf{j}m)},$$

Spin 0: $$\Delta_F(x-x') = \frac{1}{(\pm \mathbf{k}E \pm \mathbf{i}i\boldsymbol{\sigma}.\mathbf{p} + \mathbf{i}\mathbf{j}m)(\mp \mathbf{k}E \mp \mathbf{i}i\boldsymbol{\sigma}.\mathbf{p} + \mathbf{i}\mathbf{j}m)}$$

and Bose-Einstein condensate / nonzero Berry phase:

$$\Delta_F(x-x') = \frac{1}{(\pm \mathbf{k}E \pm \mathbf{i}i\boldsymbol{\sigma}.\mathbf{p} + \mathbf{i}\mathbf{j}m)(\pm \mathbf{k}E \mp \mathbf{i}i\boldsymbol{\sigma}.\mathbf{p} + \mathbf{i}\mathbf{j}m)}.$$

Where the spin 1 bosons are massless (as in QED), we will have expressions like:

$$\Delta_F(x-x') = \frac{1}{(\pm \mathbf{k}E \pm \mathbf{i}i\boldsymbol{\sigma}.\mathbf{p})(\mp \mathbf{k}E \pm \mathbf{i}i\boldsymbol{\sigma}.\mathbf{p})}.$$

Clearly, the relationship of the electron and photon propagators is of the form

$$S_F(x-x') = (i\gamma^\mu \partial_\mu + m)\,\Delta_F(x-x'),$$

or, in our notation,

$$S_F(x-x') = ((\mathbf{k}E + \mathbf{i}i\,\mathbf{p} + \mathbf{i}\mathbf{j}\,m)\ \ldots)\,\Delta_F(x-x').$$

This is exactly what we would expect in transferring from boson (Klein-Gordon field) to fermion (Dirac field), using our single vector operator.

Now, using

$$iS_F(p) = \frac{1}{2p_0}\left(\frac{1}{p_0 - \sqrt{p^2+m^2} + i\varepsilon} + \frac{1}{p_0 + \sqrt{p^2+m^2} - i\varepsilon}\right),$$

which is the same as the conventional electron propagator up to a factor ($\not{p} + m$), we can perform virtually the same contour integral as in the case of the electron to produce

$$i\Delta_F(x-x') = \int d^3p \frac{1}{(2\pi)^3}\frac{1}{2\omega}\,\theta(t-t')\,\phi(x)\phi^*(x'),$$

where ω takes the place of E / m. This time, of course, $\phi(x)$ and $\phi(x')$ are scalar wavefunctions. In our notation, they are each scalar products of the 4-component bra term $((\boldsymbol{k}E + \boldsymbol{ii}\,\mathbf{p} + \boldsymbol{ij}\,m)\ \ldots)$ and the 4-component ket term $((-\boldsymbol{k}E + \boldsymbol{ii}\,\mathbf{p} + \boldsymbol{ij}\,m)\ \ldots)$, multiplied respectively by exponentials exp (ipx) and exp (ipx'), expressed in terms of the 4-vectors p, x and x'. In the nilpotent formulation, $\phi(x)\phi^*(x')$ reduces to a product of a scalar term (which can be removed by normalization) and exp $ip(x - x')$.

In general, in off-mass-shell conditions (where $E^2 \neq p^2 + m^2$), poles in the propagator are a mathematical, rather than physical problem; but, in the specific case of massless bosons, conventional theory states that 'infared' divergencies occur when such bosons are emitted from an initial or final stage which is on the mass shell. Such divergencies, however, will not occur where there is no pole.[3]

Now, the product $\Psi(x)\ \bar{\Psi}(x')$, for the electron propagator, can be interpreted as the product of $((\boldsymbol{k}E + \boldsymbol{ii}\,\mathbf{p} + \boldsymbol{ij}\,m)\ \ldots)$ exp (ipx) and the 'vacuum operator', $\boldsymbol{ik}$ $((\boldsymbol{k}E + \boldsymbol{ii}\,\mathbf{p} + \boldsymbol{ij}\,m)\ \ldots)$ exp $(-ipx')$, which reduces to the product of a scalar term (again removed by normalization), $((\boldsymbol{k}E + \boldsymbol{ii}\,\mathbf{p} + \boldsymbol{ij}\,m)\ \ldots)$ and exp $ip(x - x')$. The integrals, like the propagators, are related by the factor $((\boldsymbol{k}E + \boldsymbol{ii}\,\mathbf{p} + \boldsymbol{ij}\,m)\ \ldots)$, which suggests that we should *define* the photon propagator, in terms of the electron propagator, for which, in the nilpotent formulation, we can do a completely non-divergent integral without additional infinitesimal terms, rather than as a result of the Klein-Gordon equation, which is not, strictly speaking, confined to purely bosonic states. In principle, this must reflect the fact that photon processes cannot ultimately be described independently of processes involving paired fermions or fermions and antifermions, one of which may be manifested indirectly in terms of a potential, rather than directly as a charge. We can then write

$$(\boldsymbol{k}E + \boldsymbol{ii}\,\mathbf{p} + \boldsymbol{ij}\,m)\ \Delta_F\,(x - x') = S_F\,(x - x') = (-\boldsymbol{k}E + \boldsymbol{ii}\,\mathbf{p} + \boldsymbol{ij}\,m),$$

which means that (after normalization) $S_F\,(x - x')$ takes the expected form of $q^{-2} = 1 / (E^2 - p^2)$, with $m^2 = 0$. The apparently divergent nature of the photon propagator integral, in conventional theory, is simply a reflection of the fact that photons, unlike electrons, have no independent existence and are not conserved objects. However, if we define the photon propagator in terms of the nilpotent fermion propagator integral, we will be using the automatic cancellation process which the nilpotent representation of the fermion introduces.

The significant aspect of the analysis is that it shows that one of the principal divergences in quantum electrodynamics is, as the procedure used to remove it would suggest, merely an artefact of the mathematical structure we have imposed, and not of a fundamentally physical nature. As with the 'infinite' self-energy of the non-interacting fermion, it is a classic case of the action of a 'redundancy barrier'. Its automatic removal in the nilpotent formalism is another indication of the method's power and general applicability.

11.5 The Propagator Method in Lowest Order

The Dirac equation in the presence of external Coulomb potentials,

$$\left(\boldsymbol{k} \frac{\partial}{\partial t} + \boldsymbol{i}i\boldsymbol{\sigma}.\nabla + \boldsymbol{i}\boldsymbol{j}m \right) \psi = -e(i\boldsymbol{k}\phi - \boldsymbol{i}\boldsymbol{\sigma}.\mathbf{A})\psi$$

may be solved by introducing the Dirac propagator, so that

$$\left(\boldsymbol{k} \frac{\partial}{\partial t} + \boldsymbol{i}i\boldsymbol{\sigma}.\nabla + \boldsymbol{i}\boldsymbol{j}m \right) S_F(x - x') = -e(i\boldsymbol{k}\phi - \boldsymbol{i}\boldsymbol{\sigma}.\mathbf{A})\delta^{(4)}(x - x').$$

Then $\psi(x) = \psi_0(x) + \int d^4x'\, S_F(x - x')\,(-e)\,(i\,\boldsymbol{k}\phi - \boldsymbol{i}\,\boldsymbol{\sigma}.\mathbf{A})\,\psi,$

where $\psi_0(x)$ is the solution of

$$\left(\boldsymbol{k} \frac{\partial}{\partial t} + \boldsymbol{i}i\boldsymbol{\sigma}.\nabla + \boldsymbol{i}\boldsymbol{j}m \right) \psi_0 = 0.$$

The Dirac equation with potentials may also be written in terms of the Hamiltonian,

$$\mathcal{H} = i\,\boldsymbol{k}\,e\phi + i\,\boldsymbol{i}\,\boldsymbol{\sigma}.(\nabla + i\,e\mathbf{A}) + \boldsymbol{i}\boldsymbol{j}m = \mathcal{H}_0 + \mathcal{H}_1,$$

where $\mathcal{H}_0 = i\,\boldsymbol{i}\,\boldsymbol{\sigma}.\nabla + \boldsymbol{i}\boldsymbol{j}m,$

and $\mathcal{H}_1$ is the interaction term,

$$\mathcal{H}_1 = i\,\boldsymbol{k}\,e\phi - \boldsymbol{i}\,e\,\boldsymbol{\sigma}.\mathbf{A}.$$

Then $$\left(\boldsymbol{k} \frac{\partial}{\partial t} + i\mathcal{H} \right) \psi = 0$$

and $$\left(\boldsymbol{k} \frac{\partial}{\partial t} + i\mathcal{H}_0 + i\mathcal{H}_1 \right) G(x - x') = \delta^{(4)}(x - x').$$

Solving the Green's function, $G(x - x')$, for the general case involving interactions, allows us to use *Huygens' principle* to find the time evolution of the wave, by summing all the secondary wavelets produced by infinitesimal point sources on the wavefront. In mathematical terms (changing over the variables from 4-vector notation),

$$\psi(\mathbf{x}, t) = \int d^3x' \, G_F(\mathbf{x} - \mathbf{x}', t - t') \, \psi(\mathbf{x}', t').$$

Using the symbolic identity

$$G = G_0 + G_0\mathcal{H}_1G_0 + G_0\mathcal{H}_1G_0\mathcal{H}_1G_0 + \ldots,$$

and a power expansion, and applying Huygens' principle, the time evolution of the wavefunction becomes described by

$$\psi(\mathbf{x}, t) = \psi_0(\mathbf{x}, t) + \int d^4x_1 G_0(\mathbf{x} - \mathbf{x}_1, t - t_1) \, \mathcal{H}_1(\mathbf{x}_1, t_1) \, \psi_0(\mathbf{x}_1, t_1)$$

$$+ \int d^4x_1 \, d^4x_2 \, G_0(\mathbf{x} - \mathbf{x}_1, t - t_1) \, \mathcal{H}_1(\mathbf{x}_1, t_1) \, G_0(\mathbf{x}_1 - \mathbf{x}_2, t_1 - t_2) \, \psi_0(\mathbf{x}_2, t_2) + \ldots .$$

To calculate the transition probability, in lowest order, that a free plane wave in initial state i emerges in final state f after scattering off a potential, we multiply the equation

$$\psi(\mathbf{x}, t) = \phi_i(\mathbf{x}, t) + \int d^4x' \, G_0(\mathbf{x} - \mathbf{x}', t - t') \, \mathcal{H}_1(\mathbf{x}', t') \, \phi_i(\mathbf{x}', t')$$

on the left by ϕ_j^* and integrate, obtaining the scattering matrix,

$$S_{fi} = \delta_{fi} + i \int d^4x \, \phi_f^*(x') \, \mathcal{H}_1(x') \, \phi_i(x') + \ldots$$

Considering, now, an electron interacting with external Coulomb potentials, we introduce

$$\psi(x) = \psi_0(x) + \int d^4x' \, S_F\,(x - x') \, (-e) \, (i\,\boldsymbol{k}\phi - \boldsymbol{i}\,\boldsymbol{\sigma}.\mathbf{A}) \, \psi,$$

and obtain $$\psi(x) = \psi_i(x) + (-e) \int d^4x' \, \theta\,(t - t') \int d^3p \frac{1}{(2\pi)^3} \frac{m}{2E} \, \psi(x) \, \bar{\psi}(x')$$

$$\times \, i\,\boldsymbol{k}\phi(x') - \boldsymbol{i}\,\boldsymbol{\sigma}.\mathbf{A}(x')) \, \psi(x').$$

Multiplying both sides of the equation from the left by $\bar{\psi}_f$ and integrating over all space and time, we obtain the scattering matrix, to lowest order:

$$S_{f1} = \delta_{f1} - ie\int \bar{\psi}_f \gamma_\mu A^\mu \psi_\iota d^4x .$$

11.6 Electron Scattering

From the previous section, we find that the amplitude for the scattering of an electron from initial state ψ_i to final state ψ_f is given by

$$T_{fi} = ie \int \bar{\psi}_f \ \gamma_\mu A_\mu \ \psi_i \ d^4x$$

$$= ie \int ik\,(\boldsymbol{k}E_f + \boldsymbol{i}i\boldsymbol{\sigma}.\mathbf{p}_f + \boldsymbol{i}\boldsymbol{j}m_f)\ \exp\,(i\,(E_f t - \mathbf{p}_f.\mathbf{r}))$$

$$\times\,(\boldsymbol{ik}\phi\ - \boldsymbol{i}\ \boldsymbol{\sigma}.\mathbf{A})\ (\boldsymbol{k}E_i + \boldsymbol{i}i\boldsymbol{\sigma}.\mathbf{p}_i + \boldsymbol{i}\boldsymbol{j}m_i)\ exp\,(i\,(E_i t - \mathbf{p}_i.\mathbf{r}))\ d^3r\ dt$$

$$= ie \int ik\,(\boldsymbol{k}E_f + \boldsymbol{i}i\boldsymbol{\sigma}.\mathbf{p}_f + \boldsymbol{i}\boldsymbol{j}m_f)\ (\boldsymbol{ik}\phi\ - \boldsymbol{i}\ \boldsymbol{\sigma}.\mathbf{A})$$

$$\times\,(\boldsymbol{k}E_i + \boldsymbol{i}i\boldsymbol{\sigma}.\mathbf{p}_i + \boldsymbol{i}\boldsymbol{j}m_i)\ exp\,(-\,i\,((E_i - E_f)t - (\mathbf{p}_i - \mathbf{p}_f).\mathbf{r}))\ \ d^3r\ dt.$$

It is convenient to write this in terms of the electromagnetic transition current between the initial and final states,

$$j^\mu_{f1} = -e\overline{\psi}_f \gamma_\mu \psi_\iota\,,$$

but this is only possible if we redefine the scalar product $\gamma_\mu A^\mu$ as a product of diagonal matrix (γ_μ) and column vector (A^μ), to separate out the terms $\boldsymbol{ik}$ and $-\boldsymbol{i}\boldsymbol{\sigma}$, within γ_μ, which have quite different actions on ($\boldsymbol{k}E_i + \boldsymbol{i}i\boldsymbol{\sigma}.\mathbf{p}_i + \boldsymbol{i}\boldsymbol{j}m_i$). In effect, $\boldsymbol{k}$ reverses the sign of E (time-reversal) and $\boldsymbol{i}$ reverses the sign of $\mathbf{p}$ (parity), so transforming γ_μ into a matrix is like applying the metric tensor, $g^{\mu\nu}$. Thus, the row vector γ_μ now becomes a diagonal matrix of the form:

$$\begin{pmatrix} \boldsymbol{ik} & 0 & 0 & 0 \\ 0 & -\boldsymbol{i}\sigma_1 & 0 & 0 \\ 0 & 0 & -\boldsymbol{i}\sigma_2 & 0 \\ 0 & 0 & 0 & -\boldsymbol{i}\sigma_3 \end{pmatrix}$$

It is convenient, also, to write $\exp\,(-\,i\,((E_i - E_f)t - (\mathbf{p}_i - \mathbf{p}_f).\mathbf{r})) = \exp\,(-i\ q.x)$, in terms of a 4-vector difference x and a 4-momentum difference q, of which the vector part is $\mathbf{q} = \mathbf{p}_i - \mathbf{p}_f$. So, the electron current becomes

$$j^\mu_{f1} = -e\overline{\psi}_f \gamma_\mu \psi_\iota = -e\boldsymbol{ik}\left(\boldsymbol{k}E_f + \boldsymbol{i}i\boldsymbol{\sigma}.\mathbf{p}_f + \boldsymbol{i}\boldsymbol{j}m_f\right)\gamma_\mu\left(\boldsymbol{k}E_i + \boldsymbol{i}i\boldsymbol{\sigma}.\mathbf{p}_i + \boldsymbol{i}\boldsymbol{j}m_i\right)e^{iq.\mathbf{x}}.$$

Defining $A^\mu(x)$ as the 4-vector potential associated with the static charge, and $A^\mu(q)$ as its Fourier transform, with vector component $A^\mu(\mathbf{q})$,

$$A^\mu(q) = \int x\ e^{-\,iq.x}\ A^\mu(x)\ d^4x,$$

the amplitude for the process becomes

$$T_{fi} = ie \int \overline{\psi}_f \; \gamma_\mu \; \psi_i \, A^\mu(x) \, d^4x = -\, i \int j^\mu_{fi} \, A^\mu(x) \, d^4x$$

$$= -\, e \, \boldsymbol{k} \, (\boldsymbol{k}E_f + \boldsymbol{ii\sigma}.\mathbf{p}_f + \boldsymbol{ij}m_f) \, \gamma_\mu \, (\boldsymbol{k}E_i + \boldsymbol{ii\sigma}.\mathbf{p}_i + \boldsymbol{ij}m_i) \, A^\mu(q).$$

For a static source, with time-independent $A^\mu(x)$,

$$A^\mu(q) = \int \exp\left(-\, i \, (E_i - E_f) \, t\right) dt \int e^{i\mathbf{q}.\mathbf{x}} \, A^\mu(\mathbf{x}) \, d^3x$$

$$= 2\pi \, \delta\,(E_f - E_i) \, A^\mu(\mathbf{q}).$$

Applying Maxwell's equations for time-independent $A^\mu(x)$, we have

$$\nabla^2 A^\mu(\mathbf{x}) = -\, j^\mu(\mathbf{x}),$$

and so

$$\int e^{i\mathbf{q}.\mathbf{x}} \, A^\mu(\mathbf{x}) \, d^3x = \int (\nabla^2 A^\mu(\mathbf{x})) \, e^{i\mathbf{q}.\mathbf{x}} \, d^3x$$

$$= -\, |\mathbf{q}|^2 A^\mu(\mathbf{q}) = -\, j^\mu(\mathbf{q}).$$

Hence

$$A^\mu(\mathbf{q}) = \frac{1}{|\mathbf{q}|^2} \, j^\mu(\mathbf{q}),$$

and

$$T_{fi} = -2i\pi\delta\left(E_f - E_i\right)e\boldsymbol{k}\left(\boldsymbol{k}E_f + \boldsymbol{ii\sigma}.\mathbf{p}_f + \boldsymbol{ij}m_f\right)\gamma_\mu\left(\boldsymbol{k}E_i + \boldsymbol{ii\sigma}.\mathbf{p}_i + \boldsymbol{ij}m_i\right)\frac{1}{|\mathbf{q}|^2}\, j^\mu(\mathbf{q}).$$

Removing the δ-function, we obtain the invariant amplitude:

$$-\, i\mathcal{M} = -e\boldsymbol{k}\left(\boldsymbol{k}E_f + \boldsymbol{ii\sigma}.\mathbf{p}_f + \boldsymbol{ij}m_f\right)\gamma_\mu\left(\boldsymbol{k}E_i + \boldsymbol{ii\sigma}.\mathbf{p}_i + \boldsymbol{ij}m_i\right)\frac{1}{|\mathbf{q}|^2}\, j^\mu(\mathbf{q}).$$

Applying energy conservation, $E_f = E_i$, and $q^2 = -\, |\mathbf{q}|^2$, which means that the invariant amplitude may be written:

$$-\, i\mathcal{M} = -e\boldsymbol{k}\left(\boldsymbol{k}E_f + \boldsymbol{ii\sigma}.\mathbf{p}_f + \boldsymbol{ij}m_f\right)\gamma^\mu\left(\boldsymbol{k}E_i + \boldsymbol{ii\sigma}.\mathbf{p}_i + \boldsymbol{ij}m_i\right)\frac{-ig_{\mu\nu}}{q^2}\left(-\, ij^\nu(\mathbf{q})\right).$$

We can, of course, use this equation, for example, with $\gamma_\mu = \gamma_0$ and $g_{\mu\nu} = 1$, and iZe replacing $ij^\nu(\mathbf{q})$, to find a value for the angular distribution ($|M|^2$) in the case of Rutherford scattering off a static nuclear charge Ze; and we can also assume that more complicated calculations can be done using the Feynman rules which codify the perturbative method. However, our main concern is with developing an approach to renormalization. Let us suppose, therefore, that the exchanged photon fluctuates into an electron-positron pair. Application of the Feynman rules requires a factor $(-1)^n$ for a diagram containing n fermion loops (in this case, $n = 1$). It also requires an integral over $d^4p / (2\pi)^4$ to sum over all possible values of the unobservable p, in the part of the invariant amplitude referring to the loop. The invariant amplitude is, therefore, now:

$$-i\mathcal{M} = -(-1)^1 ek(kE_f + ii\boldsymbol{\sigma}.\mathbf{p}_f + ijm_f)\gamma^\mu(kE_i + ii\boldsymbol{\sigma}.\mathbf{p}_i + ijm_i)\frac{-ig_{\mu\mu'}}{q^2}$$

$$\times\frac{1}{(2\pi)^4}\int(ie\gamma^\mu)_{\alpha\beta}\frac{i(kE + ii\boldsymbol{\sigma}.\mathbf{p} + ijm)_{\beta\lambda}}{(E^2 - p^2 - m^2)}$$

$$\times\left((ie\gamma^{\nu'})_{\lambda\tau}\frac{i(ii\boldsymbol{\sigma}.(\mathbf{p}-\mathbf{q}) + ijm)_{\beta\lambda}}{((p-q)^2 - m^2)}d^4p\times\frac{-ig_{\nu'\nu}}{q^2}\left(-ij^\nu(\mathbf{q})\right)\right).$$

In principle, the addition of this term for the photon loop to the lowest-order invariant amplitude for electron scattering can be achieved by the addition of a modifying term to the lowest-order propagator, so that

$$\frac{-ig_{\mu\nu}}{q^2} \to \frac{-ig_{\mu\mu'}}{q^2} + \frac{-ig_{\nu'\nu}}{q^2}I^{\mu'\nu'}\frac{-ig_{\mu\nu}}{q^2} \to \frac{-ig_{\mu\nu}}{q^2} + \frac{-i}{q^2}I_{\mu\nu}\frac{-i}{q^2}$$

with

$$I_{\mu\nu}(q^2) = (-1)^1\frac{1}{(2\pi)^4}\int Tr\left((ie\gamma_\mu)\frac{i(kE + ii\boldsymbol{\sigma}.\mathbf{p} + ijm)}{(E^2 - p^2 - m^2)}(ie\gamma_\nu)\frac{i(ii\boldsymbol{\sigma}.(\mathbf{p}-\mathbf{q}) + ijm)}{((p-q)^2 - m^2)}\right)d^4p$$

Bjorken and Drell[4] and others show that, with the omission of terms which disappear when the propagator is coupled to external charges or currents, $I_{\mu\nu}$ can be written as

$$I_{\mu\nu} = -g_{\mu\nu}q^2 I(q^2)$$

with $$I(q^2) = \frac{\alpha}{3\pi}\int_{m^2}^{\infty}\frac{dp^2}{p^2} - \frac{2\alpha}{\pi}\int_0^1 z(1-z^2)\,ln\left(1 - \frac{q^2z(1-z^2)}{m^2}\right)dz,$$

where m is the electron mass. In principle, the first term of this integral is divergent, but with M_P representing a natural cut-off value for the mass, $I(q^2)$ becomes the convergent:

$$I(q^2) = \frac{\alpha}{3\pi}\int_{m^2}^{M_P^2}\frac{dp}{p^2} - \frac{2\alpha}{\pi}\int_0^1 z(1-z^2)\,ln\left(1-\frac{q^2z\left(1-z^2\right)}{m^2}\right)dz.$$

For small values of $(-q^2)$, we have

$$I\left(q^2\right) \approx \frac{\alpha}{3\pi}ln\frac{M_P^2}{m^2} + \frac{\alpha}{15\pi}\frac{q^2}{m^2},$$

while large values of $(-q^2)$ lead to

$$I\left(q^2\right) \approx \frac{\alpha}{3\pi}ln\frac{M_P^2}{m^2} - \frac{\alpha}{3\pi}\frac{-q^2}{m^2} = \frac{\alpha}{3\pi}ln\frac{M_P^2}{-q^2}.$$

The first term in each case may be considered as a modification to the fine structure constant or the electric charge value as produced by the lowest-order Feynman diagram. The modified (or measured) charge (e') then becomes:

$$e' = e\left(1-\frac{e^2}{3\pi}ln\frac{M_P^2}{m^2}\right)^{1/2}.$$

In calculations, it is convenient to invert this to provide an expression for the bare charge (e), of the form:

$$e = e'(1 + \tfrac{1}{2}I(q^2) + \ldots \textit{higher order terms})$$

Then, using the value of $I(q^2)$ for large values of $(-q^2)$, say Q^2 (using a positive expression for convenience), we see that the invariant amplitude, for a process such as ee or $e\mu$ scattering can be calculated using e, rather than e', at the vertices, to give results in terms of the finite e'. Apart from the $O(e^2)$ term, the tree diagram for $O(e'^2)$ produces a ½-loop contribution for $O(e^4)$, at the photon momentum μ, or 'renormalization mass', that is added in for the replacement of e by e' at each of the two loop vertices. This is in addition to the usual $O(e^4)$ that is subtracted for the first loop diagram at Q^2. This means that the total term subtracted at $O(e^4)$, or for one loop, is

$$-\frac{\alpha}{3\pi}ln\frac{M^2}{Q^2} + \frac{\alpha}{3\pi}ln\frac{M^2}{\mu^2} = \frac{\alpha}{3\pi}ln\frac{\mu^2}{Q^2},$$

which is finite, meaning that the arbitrary cut-off can be eliminated for $O(e^4)$ irrespective of whether M is or is not M_P, and, of course, similar procedures can be employed for $O(e^6)$ and higher orders, for there will always be a subtraction between the $O(e^{2n})$ terms produced by a new loop diagram and the $O(e^{2n})$ terms produced by the correction from the previous diagram.

Clearly, the invariant amplitude should not be dependent on the value chosen for the point of subtraction, μ, and this requirement is expressed in the *renormalization group equation*:

$$\frac{\partial \mathcal{M}}{\partial \mu} = \left(\mu \frac{\partial}{\partial \mu}\bigg|_e + \mu \frac{\partial e}{\partial \mu} \frac{\partial}{\partial e} \right) \mathcal{M} = 0 \,,$$

in which there is a mutual cancellation between the dependence of $\mathcal{M}$ and the dependence of $e(\mu^2)$ on μ.

In principle, of course, 'charge' really represents an existence condition rather than a value. The 'value' of charge is a measure of the coupling to the energy of the field. Charge, as a fundamentally imaginary quantity, has no value. Its 'value' can only be determined by its coupling to another charge, and then is only determined in terms of the energy of the coupling. In this respect, charge is unlike mass, which, as a real quantity, has a value independent of its coupling with other masses. We can only know that a charge exists by its interaction with other charges, but we can know of the existence of a mass through its inertia, irrespective of the existence of other masses. The scaling of the interaction value associated with an electromagnetic charge at different energies is, therefore, an expected process, which is related to a similar scaling of the electron mass (calculated from the self-energy diagram), and stems from the same vacuum origin by which this is created. In the absence of an ultraviolet divergence in the integral, it produces the expected result, demanded by the nilpotent algebra, of a finite (but not independently determined) value of charge at all energies.

It is significant that the alternative (BPHZ) method of renormalization, via the adding of counterterms to the action, results in the same scaling of mass and charge as the direct multiplicative method described above. We can consider this as equivalent to the automatic addition of inherent supersymmetric partners in the nilpotent algebra, which has the advantage of being a required property at each successive order. A similar process of cancelling the contributions from electrons and those from electrons plus very soft (undetectable) photons has been used to remove the infrared divergences in bremsstrahlung (or radiation from accelerating charges) and pair production. Such procedures can be related to our analyses of divergent diagrams in section 11.3, and our discussion of the creation of alternate fermionic and bosonic vacuum 'images' in section 11.1, where the additional photon energy is actually zero, and so necessarily undetectable.

11.7 Strong and Weak Analogues

Essentially all renormalizable gauge theories follow the pattern provided by QED, though with differences due to the symmetry-breaking between the three interactions, which are discussed in chapter 10. Weak boson exchange, for example, in quantum flavour dynamics (or QFD), the weak interaction theory, follows the Feynman rules for electrically charged particle-photon interactions in QED, though with the weak charge ($s = \sqrt{a_2}$) replacing the electric charge ($e = \sqrt{a}$) at each vertex; because the weak bosons are massive; however, renormalization is only possible in the context of a combined electroweak theory. The quark-gluon interactions of quantum chromodynamics (or QCD), the strong interaction theory, also follow the same rules; this time it is strong charge ($s = \sqrt{a_3}$) that replaces the electric charge ($e = \sqrt{a}$) at the vertices, and there is, in addition, a numerical colour factor (C_F) representing the fact that there are three 'colours' involved in the interaction. For the exchange of a gluon between two quarks in a baryon, the colour factor is 2/3. For the same exchange between quark and antiquark in a meson it is 4/3. Local gauge invariance ensures that the gluons remain massless like photons, and that QCD is renormalizable like QED. The difference, this time, is that the gauge theory is non-Abelian: gluons themselves must carry 'colour', so there is an additional term representing gluon-gluon interactions or the three gluon vertex.

Both strong and weak interactions have relationships between fermion and boson propagators which follow the overall QED template. The quark propagator in the strong interaction is simply that for any fermion, and the fermion propagator, $S_F(p)$ or $iS_F(p)$, applies to the weak, as well as to the electromagnetic interaction. The conventional

$$iS_F(p) = \frac{i}{\not p - m} = \frac{i(\not p + m)}{p^2 - m^2}$$

becomes

$$iS_F(p) = \frac{i}{(k E + ii\boldsymbol{\sigma}.\mathbf{p} + ij m)} = \frac{i(-k E + ii\boldsymbol{\sigma}.\mathbf{p} + ij m)}{(E^2 + p^2 + m^2)}$$

in the nilpotent formulation, and its general applicability means that we can eliminate the infrared divergence in the fermion propagators for all of the interactions.

Of course, as is well known, the weak interaction theory is not renormalizable taken on its own, but only when combined with the $U(1)$ electromagnetic theory, which provides a necessary scalar phase term. It is interesting that the renormalizability of the combined electroweak interaction is related to the very mechanism which gives masses to the fermions and gauge bosons. For obvious reasons, a quantum field integral taken over all values of p will only be finite, as the nilpotent algebra demands, if the index of p in the integrand (or divergence D) is less than 0. Now, the boson propagator for the combined electroweak gauge bosons is

$$\Delta_{\mu\nu} = \frac{1}{p^2 - m^2}\left(- g_{\mu\nu} + (1-\xi)\left(\frac{p_\mu p_\nu}{p^2 - \xi m^2} \right)\right),$$

where ξ is the 't Hooft gauge term, which appears in the gauge fixing term in the Lagrangian for the interaction:

$$-\frac{1}{2\xi}\left(\partial_\mu A^\mu + \xi m \phi_2\right)^2 .$$

This term removes the unphysical (massless scalar) Goldstone boson ϕ_2, which arises from the spontaneous symmetry breaking produced by the filled weak vacuum used to eliminate negative energy states. If ξ is finite, then, as $p_\mu \to \infty$, $\Delta_{\mu\nu} \to p^{-2}$, like the pure photon propagator, which, in the absence of any gauge choice, becomes:

$$\Delta_{\mu\nu} = \frac{1}{q^2}\left(- g_{\mu\nu} + (1-\xi)\left(\frac{q^\mu q^\nu}{q^2} \right)\right).$$

However, for $\xi \to \infty$, we have the propagator for a massive vector boson theory without massless component,

$$\Delta_{\mu\nu} = \frac{1}{p^2 - m^2}\left(- g_{\mu\nu} + \left(\frac{p_\mu p_\nu}{m^2} \right)\right),$$

which becomes a constant when $p_\mu \to \infty$, leading to infinite sums in the diagrams equivalent to those in QED.

One of the most convenient choices of gauge is ξ = 1 (Feynman gauge), which leads to an electroweak boson propagator,

$$\Delta_{\mu\nu} = \frac{- g_{\mu\nu}}{p^2 - m^2},$$

entirely analogous to that for the photon in the same gauge, and similarly linked by the factor ($\boldsymbol{k}E$ + $\boldsymbol{ii}\mathbf{p}$ + $\boldsymbol{ij}m$) to the fermion propagator, as in QED. (It may be possible, here, to link the existence of massive weak bosons to the creation, in the nilpotent representation, of massive bosonic states via the interactions of fermions with the vacuum.)

The gluon propagator in QCD is again of the same form as that of the other gauge bosons:

$$\Delta_{\mu\nu} = \frac{1}{q^2}\left(-g_{\mu\nu} + (1-\xi)\left(\frac{q^\mu q^\nu}{q^2}\right)\right)\delta^{\alpha\beta},$$

which reduces, in Feynman gauge, to

$$\Delta_{\mu\nu} = \frac{1}{q^2}\left(-g_{\mu\nu}\right)\delta^{\alpha\beta},$$

and which only differs from the electroweak boson propagator by a factor, $\delta^{\alpha\beta}$, arising from the fact that freely propagating gluons have fixed colours.

11.8 QFD Using Nilpotents

In addition to its contribution to the propagator formalism, nilpotent algebra can be used in a direct way in the usual four-point Fermi interaction approximation to QFD. Conventionally, in describing a weak interaction, such as muon decay, we calculate the traces of the tensors using the trace theorem:

$$Tr\,[\gamma^\mu (1-\gamma^5)\,\not{p}_1\gamma^\nu (1-\gamma^5)\,\not{p}_2]\;Tr\,[\gamma_\mu(1-\gamma^5)\,\not{p}_3\gamma_\nu (1-\gamma^5)\,\not{p}_4]$$
$$= 256\,(p_1 \cdot p_2)\,(p_3 \cdot p_4).$$

This is because, for an invariant amplitude $\mathcal{M}$, for muon decay,

$$|\mathcal{M}|^2 = \frac{G^2}{2} Tr\left[\gamma_\mu(1-\gamma^5)\bar{\nu}_\mu(p_1)\gamma^\nu(1-\gamma^5)\mu(p_2)\right] Tr\left[\bar{\nu}_e(1-\gamma^5)(p_3)e(1-\gamma^5)(p_4)\right]$$

and the spin-averaged probability, $|\mathcal{M}|^2$, is given by

$$\tfrac{1}{2}\sum_{spins} Tr[\gamma^\mu(1-\gamma^5)\bar{\nu}_\mu(p_1)\gamma^\nu(1-\gamma^5)\mu(p_2)]Tr[\gamma_\mu(1-\gamma^5)\bar{\nu}_e(p_3)\gamma_\nu(1-\gamma^5)e(p_4)]$$
$$= 64G^2(p_1 \cdot p_2)(p_3 \cdot p_4)$$

Figure 11.1

Using nilpotents we can take a different approach, by directly investigating the 'bosonic' states at the two vertices (Fig. 11.1). First we take:

$$\gamma^\mu = \gamma^0 + \gamma^1 + \gamma^2 + \gamma^3 = \boldsymbol{ik} + \mathbf{i}\boldsymbol{i} + \mathbf{j}\boldsymbol{i} + \mathbf{k}\boldsymbol{i} = \boldsymbol{ik} + \mathbf{1}\boldsymbol{i},$$

combining the vectors **i**, **j**, **k**, for convenience, into the single unit **1**. Then:

$$(1 - \gamma^5) = (1 - \boldsymbol{ij})$$

$$\gamma^\mu (1 - \gamma^5) = (\boldsymbol{ik} + \mathbf{1}\boldsymbol{i})\,(1 - \boldsymbol{ij}) = \boldsymbol{ik} + \boldsymbol{i} + \mathbf{1}\boldsymbol{i} - \boldsymbol{ik}\mathbf{1}$$

Here, we assume:

$$\not{p}_1 = (\pm\,\boldsymbol{k}E_1 \pm \boldsymbol{i}\mathbf{i}\,\mathbf{p}_1),$$

but, by directly incorporating a mass term, we could use:

$$(\pm\,\boldsymbol{k}E_1 \pm \boldsymbol{i}\mathbf{i}\,\mathbf{p}_1 + \boldsymbol{i}\mathbf{i}\,m_1).$$

So
$$\gamma^\mu (1 - \gamma^5)\,\not{p}_1 \gamma_\nu (1 - \gamma^5)\,\not{p}_3$$
$$= (\boldsymbol{ik} + \mathbf{1}\boldsymbol{i})(1 - \boldsymbol{ij})(\pm\,\boldsymbol{k}E_1 \pm \boldsymbol{i}\mathbf{i}\,\mathbf{p}_1)\,(\boldsymbol{ik} + \mathbf{1}\boldsymbol{i})(1 - \boldsymbol{ij})(\pm\,\boldsymbol{k}E_2 \pm \boldsymbol{i}\mathbf{i}\,\mathbf{p}_2)$$
$$= (\boldsymbol{ik} + \mathbf{1}\boldsymbol{i} - \boldsymbol{i} - \boldsymbol{i}\mathbf{1}\boldsymbol{k})(\pm\,\boldsymbol{k}E_1 \pm \boldsymbol{i}\mathbf{i}\,\mathbf{p}_1)\,(\boldsymbol{ik} + \mathbf{1}\boldsymbol{i} - \boldsymbol{i} - \boldsymbol{i}\mathbf{1}\boldsymbol{k})(\pm\,\boldsymbol{k}E_2 \pm \boldsymbol{i}\mathbf{i}\,\mathbf{p}_2).$$

Using

$$(\boldsymbol{ik})(\pm\,\boldsymbol{k}E_1 \pm \boldsymbol{i}\mathbf{i}\,\mathbf{p}_1)\,(\boldsymbol{ik}) = (\pm\,\boldsymbol{k}E_1 \mp \boldsymbol{i}\mathbf{i}\,\mathbf{p}_1)$$

$$(\mathbf{1}\boldsymbol{i})(\pm\,\boldsymbol{k}E_1 \pm \boldsymbol{i}\mathbf{i}\,\mathbf{p}_1)\,(\mathbf{1}\boldsymbol{i}) = (\pm\,\boldsymbol{k}E_1 \mp \boldsymbol{i}\mathbf{i}\,\mathbf{p}_1)$$

$$(-\,\boldsymbol{i})(\pm\,\boldsymbol{k}E_1 \pm \boldsymbol{i}\mathbf{i}\,\mathbf{p}_1)\,(-\,\boldsymbol{i}) = (\pm\,\boldsymbol{k}E_1 \mp \boldsymbol{i}\mathbf{i}\,\mathbf{p}_1)$$

$$(\boldsymbol{i}\mathbf{1}\boldsymbol{k})(\pm\,\boldsymbol{k}E_1 \pm \boldsymbol{i}\mathbf{i}\,\mathbf{p}_1)\,(\boldsymbol{i}\mathbf{1}\boldsymbol{k}) = (\pm\,\boldsymbol{k}E_1 \mp \boldsymbol{i}\mathbf{i}\,\mathbf{p}_1)$$

we obtain a total of $4\,(\pm\,\boldsymbol{k}E_1 \mp \boldsymbol{i}\mathbf{i}\,\mathbf{p}_1)$ for this scalar product, or, for a state vector representing an antifermion (where $E_1 \rightarrow -\,E_1$), this would become $4\,(\mp\,\boldsymbol{k}E_1 \mp \boldsymbol{i}\mathbf{i}\,\mathbf{p}_1)$.

For a vertex involving a fermion, with state vector $(\pm\,\boldsymbol{k}E_3 \pm \boldsymbol{i}\mathbf{i}\,\mathbf{p}_3)$, taking over all four terms in the Dirac spinor, each

$$4\,(\mp\,\boldsymbol{k}E_1 \mp \boldsymbol{i}\mathbf{i}\,\mathbf{p}_1)\,(\pm\,\boldsymbol{k}E_3 \pm \boldsymbol{i}\mathbf{i}\,\mathbf{p}_3) = 4 \times 4\,(E_1E_3 - \mathbf{p}_1\mathbf{p}_3) = -\,16\,(p_1 \cdot p_3),$$

and the equivalent of $Tr\,[\gamma^\mu (1 - \gamma^5)\,\not{p}_1 \gamma^\nu (1 - \gamma^5)\,\not{p}_2]\;Tr\,[\gamma_\mu (1 - \gamma^5)\,\not{p}_3 \gamma_\nu (1 - \gamma^5)\,\not{p}_4]$ becomes $256\,(p_1 \cdot p_3)\,(p_2 \cdot p_4)$, leading once again to a spin-averaged probability:

$$|\mathcal{M}|^2 = 64G^2(p_1 \cdot p_2)(p_3 \cdot p_4).$$

This approach is only valid: (a) for antifermion-fermion vertices; and (b) where the $V - A$ term $\gamma^\mu (1 - \gamma^5)$ is included – that is, where the interaction is dipolar and single-handed. (Otherwise, the product of the two scalar products does not correspond with the product of the two traces.) In this method, however, the terms $p_1 \cdot p_3$ and $p_2 \cdot p_4$ can be easily extended to become scalar products of nilpotent operators where mass is to be taken into account.

11.9 The Success of the Nilpotent Method

It is not necessary to follow in any further detail the procedures of QED and its analogues, or to derive further results by examining specific Feynman diagrams, as the procedures are well-established using conventional techniques. The outline derivations in this chapter show that the nilpotent method is not only amenable to QED, QCD and QFD calculations, but also has something additional to offer. It removes the infrared divergence from the integral for the fermion propagator by, in effect, combining the fermion and antfermion solutions in a single expression. By being already second quantized it requires the existence of finite values for fundamental quantities such as mass and charge at all energies, whatever integrals are performed to incorporate vacuum fluctuations, and the algebra itself provides successful methods for the automatic removal of all possible divergences. Since, the nilpotent formulation does not incorporate direct information about the *measure* of the coupling of charge to the field energy, then there is no fixed 'value' of charge to be renormalized, and perturbation expansions merely correlate the different measures of coupling which occur at different external field energies. Even when perturbation expansions need to be used, the theory removes divergences by *demanding* that there must be a finite upper limit to the energy used, which other work on the algebra tells us must be the Planck mass (cf chapter 15). Because the process involved (via a filled vacuum) is the one which actually *creates* the masses of fermions, then our measure of the interaction of a charge with the field energy becomes simply an expected scaling mechanism, as happens also in classical contexts, rather than a desperate means of avoiding otherwise catastrophic results. It would appear that both the concept of 'renormalization' and the problem of divergences may be eliminated by using the nilpotent algebra, though the successful mathematical structure of QED and the other renormalizable theories remain intact.

Chapter 12

Vacuum

Vacuum is an essential component of a physics which includes the second law of thermodynamics. Its origins can be seen in the Dirac nilpotent structure which arises from the concept of zero totality, and its many manifestations can be explained using the mathematical structure provided by the nilpotent formalism. Significantly, vacuum is not nothing, but nothing minus fermion. In principle, this is the continuous gravitational vacuum, the zero-point energy which is the only serious candidate for the carrier of nonlocality, but it has three discrete components, created by the quaternion operators used to define the weak, strong and electric charges. These weak, strong and electric vacua have distinctive characteristics and physical manifestations related to those of the corresponding forces. The link between the charges and their respective vacua can be made through the Casimir effect, which can be variously interpreted as either a vacuum effect or a direct result of the interactions of discrete charges.

12.1 Physics and Observables

There is a long-standing belief among some physicists that physics is only about observables, but there is at least one law which denies this. One way of stating the second law of thermodynamics is to say that physics cannot be structured purely on observables. Observable 'time', used in classical mechanics, electromagnetic theory and relativity, is reversible, but 'thermodynamic' time is not. We know from the second law that the absolute order of events cannot be changed. In principle, this involves an infinite number of possible occurrences, though only a finite number will be observable. Every time we observe a finite sequence, it appears to correspond to what could be constructed from an infinite one, but we can only assert this as a supposition; it cannot be observed.[1]

The unchangeable and absolute order of physical events is what Newton meant by 'absolute time', although his words have been distorted ever since to apply to the very different concept of time *measurement* which he specifically distinguished as 'relative time'. However, Einstein used exactly the same concept in defining a concept of *causality*, strictly outside the kinematic definition of special relativity, to determine that the absolute order of events could not be changed by changing one's frame of reference. For Einstein, this was manifested by the inability of transporting 'information' (more specifically, energy) faster than the 'speed of light' (c). In more formal terms, this means defining proper time (τ) as an independent parameter, alongside space and time in the equation

$$r^2 - c^2t^2 = c^2\tau^2.$$

With $c = 1$, $\hbar = 1$, this is equivalent to defining $(\pm ikt \pm \mathbf{i}\mathbf{r} + \mathbf{i}\tau)$ or its conjugate expression, $(\pm ikE \pm \mathbf{i}\mathbf{p} + jm)$, as a nilpotent.

The Newtonian concept of absolute time was based on a universe structured on an instantaneous and universal gravitational action at a distance. The Einsteinian concept can now be seen to depend on the instantaneous correlation required in quantum mechanics, as the relationship between t and $\mathbf{r}$ enshrined in the definition of c, emerges only with the definition of the nilpotent structure also incorporating τ. The instantaneous correlation itself is also a necessary consequence of nilpotent structure as each 'fermionic' term of the form $(\pm ikE \pm \mathbf{i}\mathbf{p} + jm)$ must be defined as instantly distinguishable from all others to preserve a nonzero total wavefunction for any number of fermionic states. Significantly, 'observable' time $(\pm ikt)$ is not pure time (t), while $(\pm ikE)$ is not pure energy (E). In quantum mechanical terms, $(\pm ikt \pm \mathbf{i}\mathbf{r} + \mathbf{i}\tau)$ is only a classical approximation, with the quantum $(\pm ikE \pm \mathbf{i}\mathbf{p} + jm)$ really representing $(\mp ik\partial / \partial t \pm \mathbf{i}\nabla + jm)$, where t is not an observable. Significantly, *CPT* invariance, which is defined to simultaneously preserve relativity *and* causality, is an obvious consequence of the structure $(\pm ikE \pm \mathbf{i}\mathbf{p} + jm)$ or $(\pm ikt \pm i\mathbf{i}\mathbf{r} + \mathbf{i}\tau)$, precisely because each of the E, $\mathbf{p}$, m or t, $\mathbf{p}$, τ terms is preceded by one of the quaternion operators $\boldsymbol{k}, \boldsymbol{i}, \boldsymbol{j}$, which make it nilpotent.

Another way of accommodating instantaneous correlation, either in the Newtonian or quantum mechanical sense, is to describe time and energy in terms of 'continuity', as suggested by the Klein-4 symmetry between space, time, mass and charge, put forward in chapter 2; and another way of describing continuity of either time or energy is to use the concept of 'filled vacuum', at least in a virtual sense. (Interestingly, in particle physics, it is the filled vacuum that leads to the creation of rest mass in discrete particle states, in the same way as the m term, or 'proper energy', is the 'causality' term in the fermionic state vector.) And so, ultimately, the second law of thermodynamics is an expression of the filled vacuum concept; and, while 'pure' relativity, defined in a kinematic sense, is about space and time only, and so does not need a vacuum, as Einstein showed, the extended concept, including proper time and causality, definitely requires one. Nevertheless, this filled vacuum idea is one of the most peculiar aspects of physics, and its acceptance has always been problematical. The reason is that the 'continuity' it evokes is necessarily an indefinite thing. It talks about what we don't know, not what we know. In a sense, when we set out an idea of physics, necessarily based on observables, we unavoidably map out only a part of the total picture.[2] This is what we mean by discreteness, finiteness, or even a defined 'system'. To make this work, however, we have to find some way of specifying 'the rest', the part that cannot be defined. This is what we mean by 'vacuum'. We have to take the undefinable 'rest' (of the universe) into account. But though we cannot define it directly, we can define it indirectly in terms of what it *is not*; and, in many cases, this reflects the structure of the part that we can define.

12.2 Zero-Point Energy

One interpretation of vacuum is as 'the rest of the universe', the 'reaction' half of Newton's third law. This is how we can define it by reference to the 'image' charge or 'reflection' of a discrete source. For the discrete weak, strong and electric vacua, it means that part of the rest of the universe recognized by the appropriate charge, and it is an effective negation of that component. The total vacuum, however, is the *continuous* vacuum produced by the real (gravitational) component, and, for any given fermion, produces a state vector equivalent to $-1(\boldsymbol{k}E + i\mathbf{p} + \boldsymbol{j}m)$, with negative energy. The combination of fermion plus total vacuum then produces a zero totality and zero state vector. 'Continuity', in this context, can only mean the absence of discrete energy levels, and it is this property which gives rise to the infinite virtual energy density and virtual energy of $\frac{1}{2}\hbar\omega$ for every possible mode of vibration, the so-called zero-point energy. The continuous vacuum is thus constituted out of the mirror image states of *all possible* fermion states, and it is this continuous vacuum which makes possible the nonlocal connection required by Pauli exclusion. Each possible state provides a virtual vacuum energy of $\frac{1}{2}\hbar\omega$, like the ground state of a harmonic oscillator – which, of course, is precisely what it is. To create a real fermion state, we excite a virtual vacuum state of $-\frac{1}{2}\hbar\omega$ up to the level $\frac{1}{2}\hbar\omega$, using a total energy quantum of $\hbar\omega$. The vacuum becomes an infinite ensemble of harmonic oscillators, providing a nonzero ground state for the weak interaction. It also constitutes Euclidean 3-D space by providing a line in every possible direction for every possible instantaneous spin (**p**) phase term of the fermionic operator. In this sense the zero-point vacuum and 3-D space are identical. The continuous vacuum, however, can never be observed directly precisely because it *is* continuous, and so the concept of continuity will necessarily remain a 'potential' or virtual one.

12.3 The Weak Vacuum

We can consider the function of charge as 'partitioning' the continuous vacuum, that we never observe directly, in a way which, being discrete, can be observed. Charge thus becomes a kind of vacuum state, associated with the quantum field nature of the state vector, and the different charges are associated with qualitatively different vacuum states through their association with respective pseudoscalar, vector and scalar coefficients. The three discrete vacua describe only that part of the vacuum which the associated type of charge sees, while the three coupling or fine structure constants are a scalar measure of each charge's coupling to its respective vacuum.

The total vacuum which the charges partition originates as an expression of the continuous or noncountable nature of mass-energy. Continuity necessarily makes mass-energy unidimensional and unipolar, and, because it is also real, restricts it to a single mathematical sign, which is usually taken as positive. We can interpret this as implying a non-symmetric ground state or a filled vacuum, which is that of negative energy or antifermions. Physically, it manifests itself in the spin 0 Higgs field, which breaks charge conjugation symmetry for the weak interaction, and gives rest masses to the fermions and weak gauge bosons.[3]

The derivation of the Higgs field (see 10.11) usually begins with the Goldstone theorem, according to which the breaking of a continuous symmetry of a physical system necessarily leads to the appearance of a massless scalar or spin 0 boson state. This theorem is, in fact, a natural consequence of nilpotent structure. Bosonic states can be considered as resulting from the transformation of one fermion into another or, sometimes, into itself. Where a symmetry is *complete*, however, with no degeneracy permitted, there is no mechanism for a fermion to transform into itself. An example occurs in the colour singlet baryon, which cannot transform into itself via the colour force, meaning that no gluon can be formed out of a combination of $R\overline{R}$, $G\overline{G}$ and $B\overline{B}$, with the result that the number of independent $SU(3)$ generators is reduced from 9 to 8. In the nilpotent formalism, this principle ensures that a *free* fermion state (as its own supersymmetric partner) remains unchanged as a result of self-interaction via the vacuum, and so requires no charge or energy renormalization.

An ideal vacuum would have the most complete symmetry possible, with exact and absolute *C*, *P* and *T* symmetries. The maintenance of these symmetries would not, therefore, require the creation of a scalar boson out of a vacuum fermion (with $\pm\, \boldsymbol{ik}E \pm \boldsymbol{i}\mathbf{p}$) and its charge-conjugated antifermionic partner (with $\mp\, \boldsymbol{ik}E \mp \boldsymbol{i}\mathbf{p}$). However, the breaking of charge conjugation symmetry (*C*) and / or either parity ($P = CT$) or time-reversal symmetry ($T = CP$), which would be equivalent (with the alternative option of maintaining *C* by the simultaneous violation of *P* and *T* available only to bosons), would make the existence of such a scalar boson necessary, thus justifying the Goldstone theorem. Of course, in the nilpotent formalism, the spin 0 scalar must necessarily be massive, though, being, again, its own supersymmetric partner, there would be no mechanism for an observable self-interaction, and so no renormalization problem concerning the mass.

We can also see how the Higgs boson gives mass to fermion and boson states through the necessary vacuum connection, if we imagine it as being 'produced' at the vertex of an interaction between a massless antifermion and a massive fermion, or between a massless antifermionic boson component and a massive fermionic boson component. If we suppose a pure (massless) antifermion at the vertex with only right-handed helicity, then the production of a boson whose fermion component also has right-handed helicity (which is forbidden in the pure state) must necessarily require a fermion at the vertex with an element of right-handed helicity. So, the boson 'gives' mass to the fermion. (The reverse, of course, is always true, and we could equally see the existence of a positive mass term in fermion states as breaking charge conjugation symmetry, because no mechanism exists to reverse the sign of *m*.)

Because, it is the $\boldsymbol{k}$ operator which changes fermion to antifermion, the weak vacuum is the one associated with fermion / antifermion annihilation / creation. The pseudoscalar aspect implies that the vacuum or charge state, or potential, may be complex, which is the requirement for *CP* violation. The pseudoscalar representation also naturally implies dipolarity because of the fundamental mathematical duality of $\pm\, i$, and the indistinguishability of the two signs under weak charge conjugation violation. This special property of the weak interaction appears to be the ultimate source of different phases of matter and phase transitions, when the indistinguishability of sign is allowed to effectively eliminate the weak component in fermion-fermion combinations, and so overcome aspects of Pauli exclusion. It is certainly the origin of the nonzero Berry phase where the spin 0 'bosonic' state is such as would be required in a pure weak transition from $-ikE$ to $+ikE$, or its inverse. Because the spin 0 state is necessarily massive, time reversal symmetry (the one applicable to the transition) must be broken in the weak formation or decay of states involving the Berry phase. (A possibility for observing this might be devised using the quantum Hall effect in graphene.)

The use of the ikE term for the weak vacuum ensures that it has to be through the weak vacuum that we express the continuity of mass-energy and the conjugate irreversibility of time. No *physical* state can be defined corresponding to $-E$, although charge-conjugated $-ikE$ states can be defined by reversing the sign of the ik operator. In principle, this leads to weak charge conjugation violation, which means that the weak interaction is indifferent to the *sign* of the weak charge, and can only distinguish between fermion and antifermion. To preserve *CPT* symmetry, either parity or time-reversal symmetry must also be violated.

Assuming that the requirement for continuous vacuum energy (or thermodynamic time) ensures that a physical bias exists in favour of matter over antimatter, means that the vacuum should have a weak dipole moment, exhibited as a single-handedness of rotation, and represented by the $\frac{1}{2}\hbar\omega$ modes of vibration or zero-point energy. In principle this may be seen as the origin of left-handed fermion spin, the fermion being created simultaneously with its vacuum reflection. If the weak vacuum is in a continual state of proclaiming its filled state, by creating weak dipoles which have a dipole moment or specific handedness, we may also expect that 'fluctuations' in this vacuum will be the same thing as the production or annihilation of a weak dipolar fermion-antifermion pair, each of spin ½, via a harmonic oscillator creation-annihilation mechanism. Fluctuations of this kind are responsible for the Casimir or Van der Waals force from the zero point energy, corresponding to the potential for a fluctuating dipole-dipole interaction.

12.4 The Strong Vacuum

The strong interaction, as we know it, is manifested, through a nonlocal gluon sea, with switching of momentum components in terms of both sign and direction to incorporate the six phases. This is exactly what is provided by the $\pm i\mathbf{p}$ term in the state vector. It is notable that baryon structure is essentially affine, dissolving into component gluons and combinations of virtual baryons *ad infinitum*. This is exactly what we would expect from the affine nature of the **p** operator, whose components can be no more separated or fixed than the dimensions of space. The vector nature of the strong operator also means that the strong vacuum is the only one with explicit relative phases. In strongly interacting systems, the phases are associated with the presence or absence of electric and weak charge components. Where the phases associated with these components coincide, there is no way of distinguishing them, and consequently no strong interaction (see chapters 13-14).

12.5 The Electric Vacuum

Fermionic states are ones in which weak charges are present (see chapters 13 and 14). There are, however, two types of fundamental fermionic state: quark and lepton. For quarks, the **p** phases are explicit, and s charges are present; for leptons they are nonexplicit, and s charges are absent. As different types of charge and vacuum exist entirely independently of each other, weak charge must be indifferent to the presence or absence of the strong charge. So weak and electric charge allocations for quarks and fermions must follow the same pattern; hence, the fractional electric charges allocated to quarks are merely an expression of the perfect gauge invariance of the strong interaction – analogous to the process of fractional charge creation in the quantum Hall effect – and are not an intrinsic aspect of quark structure. At the same time, the weak charge must be indifferent to the presence or absence of e.

As discussed previously in 10.9, the fermionic states with and without electric charge are conventionally described as the $SU(2)_L$ states (up / down, neutrino, electron, etc.); these must be made *explicitly* indistinguishable under the weak interaction. Conventionally, we use the third component of weak isospin (t_3), as the quantum number for distinguishing these states. For the two isospin states, $t_3 = \pm \frac{1}{2}$, but only in half the total number of states (the left-handed ones). For free fermions, the quantum number for the electric force takes the value $Q = -1$, where electric charge ($-e$, taken as negative by convention) is present, again in half the number of states (though a different half). If the weak and electric interactions are described by any grand unifying gauge group, then orthogonality and normalization conditions require the mixing ratio, defined as $\sin^2\theta_W$, to be determined by $Tr\,(t_3^2)\,/\,Tr\,(Q^2)$, which in this case is 0.25.

As also previously argued, the ratio cannot apply only to free fermions if the weak interaction is indifferent also to the presence or absence of the strong charge. So exactly the same mixing proportion, with $\sin^2\theta_W = 0.25$, should exist also for quark states, and separately for each 'colour' phase, or momentum direction, so that the weak interaction cannot be detected through 'colour'. Interpreting 'colour' through momentum phases or directions, allows the instantaneous existence of only one quark phase in three. So we find that the charge variation 0 0 $-e$ must be taken against either an empty background or 'electric vacuum' (0 0 0) or a full background (e e e), so that the two states of weak isospin in the three colours become:

e	e	0
0	0	$-e$

The most obvious manifestation of an electric vacuum would, therefore, appear to be in the $SU(2)_L$ for the weak interaction. The weak vacuum, which is full and cannot be reversed, effectively contrasts with the electric vacuum, which can be filled or emptied, or reversed in the case of antifermions. However, while the $SU(3)$ and $U(1)$ structures come directly from the vector **p** and scalar m terms in the Dirac state, the $SU(2)$ structure for the weak interaction is only related to the $SU(2)$ spin structure, associated with the pseudoscalar E, in an indirect way. This is because the E term in the equation doesn't fully express the asymmetry of the physical E. The $SU(2)$ for E is the $SU(2)$ for helicity, and is related to the $SU(2)_L$ for weak isospin only via a matrix (like the CKM matrix) involving rest mass. It is the mass dependence, relating to the filled nature of the vacuum through the Higgs mechanism, which makes $SU(2) \rightarrow SU(2)_L$. It arises because the gravitational / inertial component is not specifically included through a specific term in the compactified Dirac state. We can express the $SU(2)$ / $SU(2)_L$ relationship as follows:

changing	ν / u quark	to	e / d quark
requires	zero $-e$	becoming	nonzero $- e$
	≡ isospin up		≡ isospin down
	100 % weak		partially weak
	left-handed		partially right-handed
	100 % fermionic		partially antifermionic
	say 100 % spin up		partially spin down
	no added m		added m
	unchanged E / **p** sign		partially changed E / **p** sign

The mixing of E and **p** terms, or right-handed and left-handed components, in the nilpotent state vector, is also equivalent to the mixing of e and w charges, or electric and weak vacua, but this mixing cannot affect the weak interaction as such, which has no right-handed component for fermions. So, the weak interaction must be simultaneously left-handed for fermion states and indifferent to the presence or absence of the electric charge, which introduces the right-handed element.

12.6 The Gravitational Vacuum

Three terms in the Dirac 4-spinor for a fermion represent its three discrete vacuum 'reflections'; the fourth (conventionally placed in the first row) represents the particle creation itself. Because the three vacuum reflections are generated by terms which are also charge operators, it is natural to conclude that charge is fundamentally a vacuum generator. At the same time, the mass of the fermion and the related vacuum energy may be assumed to be 'generated' by the 'mass' operator (1). So we can consider gravity, the force generated by mass, as being in some way represented by a vacuum operator of the form $1(\pm \boldsymbol{i}kE \pm \boldsymbol{i}\mathbf{p} + \boldsymbol{j}m)$. However, it is more probable that the gravitational vacuum has the form, $-1(\boldsymbol{i}kE + \boldsymbol{i}\mathbf{p} + \boldsymbol{j}m)$, the term which zeros the Dirac state.

Many people have assumed that gravity is a discrete force. It comes, however, from a continuous vacuum and is the only serious candidate for the carrier of nonlocality, for the instantaneous quantum correlation of Dirac states, and for the source of the infinite zero-point energy spectrum. It is clear also that gravitational 'energy' is not energy in the sense normally used and that it does not participate in energy exchanges as they are usually understood. This is one of the reasons why gravity has never been successfully quantized. In fact, the use of the coefficient 1 may be taken as equivalent to the statement that the gravitational vacuum cannot be quantized directly.[4] One way of representing this would be to define gravitational energy as negative (because of the attractive force) and to refer to the filled vacuum for negative energy states, as proposed in the original positron theory of Dirac, and as appears to be required to explain the absence of antimatter from the universe's ground state. The zero-point energy becomes a kind of 'antigravity'. Energy is not transferred in gravity because the vacuum is already full of 'negative energy' states. The energy is positive, the energy states are negative, and the vacuum is full because there is no negative energy.

There is, nevertheless, a discrete vacuum representation related to mass. This is the inertial component, related to the discrete rest mass, which itself originates in the fermionic or bosonic charge structure. In the Higgs mechanism, it is signalled by a nonlocal finite energy level for the weak vacuum. The inertial component may be seen as a discrete local reaction specified by $1(\pm ikE \pm \mathbf{i}\mathbf{p} + \mathbf{j}m)$ to the continuous nonlocal gravitational energy specified by $-1(\pm ikE \pm \mathbf{i}\mathbf{p} + \mathbf{j}m)$, and this *can* be quantized (cf 18.8). The total zero energy of the 'universe' (its zero wavefunction) could then be said to come from the combination of a positive nilpotent (inertia, sum of charges) with a negative one (gravity).

12.7 The Casimir Effect

The manifestation of the continuous vacuum that we *do* observe, is the well-known Casimir force of attraction between uncharged metal plates of area A and small separation d:

$$F = \frac{\pi \hbar c}{480} \frac{A}{d^4},$$

which has been derived by many authors.[5] Because of the dependence on $1 / d^4$, this force manifests itself over the range 1 μm as a dipole-dipole interaction, of exactly the same kind as the Van der Waals force of cohesion between molecules. This interpretation assumes zero-point fluctuations of virtual photons in the space between the plates or molecules, but it is equally possible to obtain the same result using zero-point fluctuations of the electrons in the metal surfaces. In this case it becomes the London dispersion interaction. Yet another picture (Hellmann-Feynman) sees the quantum charge clouds in the two plates, molecules or other objects becoming deformed as they approach, corresponding to a change in the expectation values of their charge distributions. In this case, the force is identical to that of chemical bonding due to the classical electrostatic force.[6]

The Casimir force is thus not a distinct phenomenon, but an aspect of the classical electromagnetic interaction. While Peterson and Metzger use this as a means of removing such unobservables as quantum fluctuations from the argument, we can turn the argument round so that the ordinary electromagnetic force becomes a vacuum projection. An inverse fourth power Casimir force between objects, which are electrically neutral globally but composed locally of electrostatic dipoles, would then imply an inverse square force between the individual charged particles of which they are composed. And related effects of aggregated matter, such as nuclear forces, could be described in similar terms as Casimir-type manifestations of vacuum fluctuations, either fermionic or bosonic, as much as interactions between discrete charges specified by expectation values. It may well be that the left-handed chirality which is observed in aggregated matter (particularly organic) is due to the weak (Van der Waals force) ultimately producing the states of matter.

If we describe the forces due to discrete charges (electric, strong, weak) as Casimir-type manifestations of the vacuum, then we have a direct physical interpretation for the respective use of the quaternion operators $\boldsymbol{j}$, $\boldsymbol{i}$, $\boldsymbol{k}$, both for these three charges and for the operation of the respective electric, strong, weak vacua via $\boldsymbol{j}(\pm \boldsymbol{i}\boldsymbol{k}E \pm i\mathbf{p} + \boldsymbol{j}m)$, $\boldsymbol{i}(\pm \boldsymbol{i}\boldsymbol{k}E \pm i\mathbf{p} + \boldsymbol{j}m)$, $\boldsymbol{k}(\pm \boldsymbol{i}\boldsymbol{k}E \pm i\mathbf{p} + \boldsymbol{j}m)$. Because the operators are attached respectively to pseudoscalar E, vector $\mathbf{p}$ and scalar m, in the state vector, then their vacua will have different effects, and so the forces will behave differently. However, the key driving mechanism in all Casimir calculations is that they are the result of separating out *discrete* objects from a *continuous* background, and that they only have meaning in the context of object pairs. Creating a discrete object pair at some finite separation generates a force because it creates a discrete space which is shielded from some of the modes of vacuum vibration outside this space. In principle, therefore, all interactions between discrete charged objects, and even the values of the charged coupling constants, can be seen as resulting from the *existence* of the rest of the universe as a vacuum state, exactly in line with renormalization and Mach's principle for the parallel case of inertial mass.

In this interpretation, the Casimir and related effects become the way in which the discrete charged vacua manifest themselves in relation to the continuous total vacuum background; they represent the partitioning of the vacuum through the three types of charge state. The occupation status of the charge states (that is, whether the charges have unit or zero values) is established on the basis of relative phases between the components of the state vector (see chapters 13 and 14). This then determines particle type and possible interactions. The vacuum, however, becomes the mechanism by which this process becomes manifested; the creation of discrete units with non-zero occupation status creates the 'distortions' of vacuum, which we call interactions, in the same way as the presence of discrete sources creates the vacuum response or distortions of simply-connected space which we call the Aharonov-Bohm effect and the nonzero Berry phase. The charges thus act through the Casimir effect in the vacua created by the quaternion operators, and these vacua have the characteristics necessary to make them components of the total gravitational vacuum.

In general terms, vacuum may be thought of as the driving mechanism for assembly / disassembly and self-organization within aggregated matter, and for such things as phase transitions, effectively through the weak charge which defines fermionic matter. The Casimir effect will be attractive for bosons because they are weak dipoles, but repulsive for fermions because they are weak monopoles. The difference in status between the 'real' and image terms in the Dirac 4-spinor for a free particle, even if the 'real' particle is actually a vacuum state, also means that the Casimir effect does not require a broken supersymmetry to be observed, because loop cancellation is only at the level of the 'image' terms.

The creation and annihilation of fermions is, of course, one kind of phase transition, and involves the creation and annihilation of units of weak, and other, charges. The assembly of states of matter at other levels is equally concerned with the effect of the weak charge. Most of the properties of gaseous and condensed matter relate to the harmonic oscillator behaviour of its components, while the dipolar Van der Waals force, which expresses in its most fundamental aspect the nature of the weak vacuum, plays a significant role in all material phases. In addition, the properties of the solid state are determined by the Pauli exclusion principle that invariably accompanies the presence of weak charge, while Bose-Einstein condensation is effectively the elimination of this charge and its dipolarity, through the property of weak charge conjugation violation. Another phase transition of the Van der Waals-type occurs with the creation of interbaryonic, or nuclear, matter through a remnant of the strong forces between quarks, and this can be seen, at least in part, as a Bose-Einstein condensation.

12.8 Berry's Geometric Phase

A final aspect of vacuum phenomena is presented by Berry's geometric phase, which has already been alluded to in earlier sections. In principle, this is a purely geometrical or topological phenomenon, an example of algebraic geometry.[7] If a vector is parallel-transported along any path which returns it to its starting place, it will acquire a gauge invariant phase which represents the angular difference between its final and starting directions. In many cases, in particular when the space enclosed by the path is simply connected and contains no singularities, the phase will be 0 or 360°. However, when the topology is nontrivial and contains singularities, that is, when the space is multiply connected, the phase will be nonzero. In quantum applications, the presence of one singularity generates a phase shift of π or 180°, and so, to return the phase difference to 0 or 360° requires a double circuit of the closed path. So, if we regard a single fermion as a singularity, then we can consider it as defining its own multiply connected space, and imagine that a closed circuit of its own 'path' in its singular state requires a double rotation with respect to the simply connected space defined as lying outside it. In the nilpotent theory, ordinary Euclidean space has its existence only with respect to the structure of the fermions creating it – the fermions are not placed within a pre-existing classical space-time lattice – and the fermion defines itself as a singularity by the arbitrariness of the space which surrounds it. Simply connected (or anything but singular) means nonconserved, variable, uncharacterized, arbitrary, translation and rotation symmetric, and, hence, conserving angular momentum. It is precisely this arbitrariness that is expressed by the concept of spin, the ½ value being required to produce the Berry phase that would define the fermion as a singularity. The fermion becomes part of an entangled system of many fermions when we define a multiply connected space incorporating all the fermion singularities, and each fermion has a (geometric) spin (**p**) phase which is instantaneously different from every other (no two paths being 'completed' at the same moment), but necessarily determined by the presence of all the others, because this decides the nature of the space that has been created external to the fermion singularity itself.

Of course, the fermion itself has four components in what we might describe as a 'vacuum space', 'spin space', or even 'charge space',[8] and each of these creates its own multiply connected topology, as we know from *zitterbewegung*, a pure consequence of the Dirac equation, where we continually switch between all four states (see 7.6). In *zitterbewegung*, of course, the switching is virtual – the state retains its real identity – but we can imagine the switching as being between different parts of the vacuum space, each of which is multiply connected in Euclidean space, and the coexistence of the real and vacuum states provides coexistence of fermion and boson modes, as in section 6.6, with respective Berry phases of 180° and 360° (or π and 2π). Significantly, when we switch from E to $-E$ or **p** to −**p**, even when the transition is purely mathematical, we are switching to *another part of the vacuum space*, as well as making an energy or momentum transition, and the transition can only be validly made by simultaneous switching of the fermion and vacuum states. This is evident for the E to $-E$ transition in the theory of propagators (11.4) but it is also true for the (opposite spins) transition from **p** to −**p**, as in the states for the one-electron atom. The simultaneous switching of all four states is, of course, automatic in the nilpotent formalism.

One aspect of the *zitterbewegung*, and its origin in the topology of the Dirac 4-spinor and its geometric phase is especially important. This is that the process required for the weak interaction – the creation and destruction of fermionic states via bosonic vertices – is built into the *nilpotent (and point-like) structure of the fermion itself* and is not an externally imposed condition (cf 10.13). A fermion, to exist at all, is already weakly interacting. So, if we take the strong interaction as the physical realisation of the vector nature of the **p** operator and the electric (Coulomb) interaction as the minimum condition required for the spherical symmetry of a point source, then all three interactions have their ultimate origin in the intrinsic structure of the fermionic state, and the complete description of this state already incorporates the interactions.[9] The weak interaction then becomes the inevitable consequence of the ambiguity in sign of the iE / **p** ratio, or orientation of its unknown absolute phase.

Chapter 13

Fermion and Boson Structures

With the outline features of the three nongravitational interactions established on the basis of their representation in the Dirac nilpotent state vector, it is possible to proceed to deriving the charge structures of the fundamental fermions which emanate from them, using exactly the same source. On the basis of the connection between conservation of charge type and conservation of angular momentum, as used in chapter 10, we can derive algebraic expressions representing the charge structures of each of the quarks and each of the known leptons, and, in addition, to combine these in a *single algebraic expression* which generates all the structures. Phase diagrams show how the charge-angular momentum conservation actually operates. The formulae can then be used to produce quark and lepton charge structures in the form of four tables, three of which, *A*, *B*, *C* represent quarks, and one, *L*, leptons. Meson and baryon charge structures follow immediately, along with large parts of the Standard Model, including such hitherto unexplained facts as *CP* violation. The quarks are noticeably lepton-like, as would seem to be a basic necessity in a grand unified theory.

13.1 The Charge Structures of Quarks and Leptons

Fermions (that is, quarks and leptons) are states characterized by the presence of the weak charge, and to specify all possible fermion states we simply have to enumerate all particle states *which are indistinguishable from each other in terms of the weak interaction.* Essentially, the weak interaction cannot tell whether a strong or electric action interaction is also operating, and so fermions, for example, with strong charges (that is, quarks) ought to be indistinguishable by this interaction from particles without strong charges (that is, leptons). In terms of the weak interaction, quarks ought to be *lepton-like*. For quarks, also, the weak interaction cannot tell the difference between a filled 'electromagnetic vacuum' (or weak isospin up state) and an empty one (or weak isospin down state). The weak interaction, in addition, is also indifferent to the sign of the weak charge, and responds (via the vacuum) only to the status of fermion or antifermion; this results in mixing between the respective fermion *generations*, defined with $+w$, with $-w$ and P violation, and with $-w$ and T violation. From the sets of equally probable states thus specified (excluding energy considerations), we define all the possible distinctions between fermion / antifermion; quark / lepton; isospin up / isospin down; and the three quark-lepton generations. The distinctions are made in terms of the strong and electric charges, and of mass.

The process can be represented in terms of conservation of angular momentum, which we have already associated with the conservation of each of the charges. Taking $\boldsymbol{\sigma}.\hat{\mathbf{p}}$ (or $-\boldsymbol{\sigma}.\hat{\mathbf{p}}$, using the historically-established sign conventions for charges) as equivalent in unit charge terms to an expression in which $\hat{\mathbf{p}}$ becomes the unit vector components $\hat{\mathbf{p}}_1$, $\hat{\mathbf{p}}_2$, $\hat{\mathbf{p}}_3$, in successive phases of the strong interaction, and applying this to the strong charge quaternion operator $\boldsymbol{i}$, the units of strong charge will become $0\boldsymbol{i}$ or $1\boldsymbol{i}$, depending on the supposed instantaneous direction of the angular momentum vector. Only one component of a baryon will have this unit at any instant. In reality, of course, gauge invariance ensures that all possible phases exist at once, so spin becomes a property of the entire system and not of the component quarks.

The same angular momentum term ($\boldsymbol{\sigma}.\hat{\mathbf{p}}$) carries the information concerning the conservation of the other two charge terms; the three charges are, as we have seen, separately conserved because they represent three different aspects of the angular momentum conservation process. In the case of the weak charges, the random unit vector components $\hat{\mathbf{p}}_1$, $\hat{\mathbf{p}}_2$, $\hat{\mathbf{p}}_3$, are associated respectively with the *sign* of the angular momentum state, and, in the case of the electric interaction, they are associated with the *magnitude*. This occurs through the connections of $\mathbf{p}$ with iE and $\mathbf{p}$ with m. We can, thus, generalise the procedure by applying $\boldsymbol{\sigma}.\hat{\mathbf{p}}_1$, $\boldsymbol{\sigma}.\hat{\mathbf{p}}_2$, $\boldsymbol{\sigma}.\hat{\mathbf{p}}_3$ to the quaternion operators ($\boldsymbol{k}$ and $\boldsymbol{j}$) specifying w and e, but with the sequence of unit vectors determined separately in each case. The various alignments between the sequences of unit vectors or *phases* applied to s, w and e then determine the nature of the fermion state produced.

If we align the unit vectors applied to w and e, we are effectively aligning the E and m phases with each other, and so necessarily with the $\mathbf{p}$ phase (by alignment of the magnitudes), which means that the system has a single phase and so cannot be baryonic. The $\mathbf{p}$ phase is defined with E and m, and there is no strong charge. We thus define a free fermion or lepton. In a baryon system, with strong charges present, the vectors assigned to the weak and electric charges, and hence to E and m, will not be aligned, and, consequently, the **p** phase is not fixed with respect to them.[1]

To complete the representation of all possible fermions, we need to incorporate the effects of weak isospin, and the parity- and time-reversal-violations which will create second and third 'generations'. Reversal of isospin can be accomplished by replacing a term such as $-\boldsymbol{j}\hat{\mathbf{p}}_1$ with $-\boldsymbol{j}(\hat{\mathbf{p}}_1 - \mathbf{1})$, the $\boldsymbol{j}\mathbf{1}$ representing the filled 'electric vacuum' state. Charge conjugation violation may be represented by the non-algebraic symbols z_P and z_T, depending on whether it is accompanied by P or T violation.[2] In using these symbols, we are merely saying that we are treating the $-w$ of the second and third generations as though it were positive in the same way as the w of the first generation. We can now express quark structures in the following form:

down	$-\boldsymbol{\sigma}.\,(-\boldsymbol{j}\hat{\mathbf{p}}_a + \boldsymbol{i}\hat{\mathbf{p}}_b + \boldsymbol{k}\hat{\mathbf{p}}_c)$
up	$-\boldsymbol{\sigma}.\,(-\boldsymbol{j}(\hat{\mathbf{p}}_a - \mathbf{1}) + \boldsymbol{i}\hat{\mathbf{p}}_b + \boldsymbol{k}\hat{\mathbf{p}}_c)$
strange	$-\boldsymbol{\sigma}.\,(-\boldsymbol{j}\hat{\mathbf{p}}_a + \boldsymbol{i}\hat{\mathbf{p}}_b - z_P\boldsymbol{k}\hat{\mathbf{p}}_c)$
charmed	$-\boldsymbol{\sigma}.\,(-\boldsymbol{j}(\hat{\mathbf{p}}_a - \mathbf{1}) + \boldsymbol{i}\hat{\mathbf{p}}_b - z_P\boldsymbol{k}\hat{\mathbf{p}}_c)$
bottom	$-\boldsymbol{\sigma}.\,(-\boldsymbol{j}\hat{\mathbf{p}}_a + \boldsymbol{i}\hat{\mathbf{p}}_b - z_T\boldsymbol{k}\hat{\mathbf{p}}_c)$
top	$-\boldsymbol{\sigma}.\,(-\boldsymbol{j}(\hat{\mathbf{p}}_a - \mathbf{1}) + \boldsymbol{i}\hat{\mathbf{p}}_b - z_T\boldsymbol{k}\hat{\mathbf{p}}_c)$

In this representation, $-\boldsymbol{j}$ stands for electric charge (which is conventionally negative), $\boldsymbol{i}$ for strong, $\boldsymbol{k}$ for weak. a, b, c are *each* randomly 1, 2, 3, except that $b \neq c$. Both $-z_P\boldsymbol{k}$ and $-z_T\boldsymbol{k}$ become equivalent to $\boldsymbol{k}$, for the purposes of the weak interaction. For the corresponding leptons, the components are all in phase ($\hat{\mathbf{p}}_a$), and there is no directional component:

electron	$-\boldsymbol{\sigma}.\,(-\boldsymbol{j}\hat{\mathbf{p}}_a + \boldsymbol{k}\hat{\mathbf{p}}_a)$
e neutrino	$-\boldsymbol{\sigma}.\,(-\boldsymbol{j}(\hat{\mathbf{p}}_a - \mathbf{1}) + \boldsymbol{k}\hat{\mathbf{p}}_a)$
muon	$-\boldsymbol{\sigma}.\,(-\boldsymbol{j}\hat{\mathbf{p}}_a - z_P\boldsymbol{k}\hat{\mathbf{p}}_a)$
μ neutrino	$-\boldsymbol{\sigma}.\,(-\boldsymbol{j}(\hat{\mathbf{p}}_a - \mathbf{1}) - z_P\boldsymbol{k}\hat{\mathbf{p}}_a)$
tau	$-\boldsymbol{\sigma}.\,(-\boldsymbol{j}\hat{\mathbf{p}}_a - z_T\boldsymbol{k}\hat{\mathbf{p}}_a)$
τ neutrino	$-\boldsymbol{\sigma}.\,(-\boldsymbol{j}(\hat{\mathbf{p}}_a - \mathbf{1}) - z_T\boldsymbol{k}\hat{\mathbf{p}}_a)$

Both antiquarks and antileptons simply replace $-\boldsymbol{\sigma}$ with $\boldsymbol{\sigma}$.

13.2 A Unified Representation for Quarks / Leptons

It is possible to incorporate all the information outlined in the previous section into a single unified representation for the entire set of charge structures for quarks and leptons (and their antistates):

$$\boldsymbol{\sigma}_z.(\boldsymbol{i}\,\hat{\mathbf{p}}_a\,(\delta_{bc}-1)+\boldsymbol{j}\,(\hat{\mathbf{p}}_b-\mathbf{1}\delta_{0m})+\boldsymbol{k}\,\hat{\mathbf{p}}_c\,(-1)^{\delta_{1g}}\,g)$$

As previously, the quaternion operators $\boldsymbol{i}, \boldsymbol{j}, \boldsymbol{k}$ are respectively strong, electric and weak charge units; $\boldsymbol{\sigma}_z$ is the spin pseudovector component defined in the z direction (here used as a reference); $\hat{\mathbf{p}}_a$, $\hat{\mathbf{p}}_b$, $\hat{\mathbf{p}}_c$ are each units of quantized angular momentum, selected *randomly* and *independently* from the three orthogonal components $\hat{\mathbf{p}}_x$, $\hat{\mathbf{p}}_y$, $\hat{\mathbf{p}}_z$. $\boldsymbol{\sigma}_z$ and the remaining terms are logical operators representing existence conditions, and defining four fundamental divisions in fermionic states. Each of the operators creates one of these fundamental divisions – fermion / antifermion; quark / lepton (colour); weak up isospin / weak down isospin; and the three generations – which are identified, respectively, by the weak, strong, electromagnetic and gravitational interactions.

(1) $\boldsymbol{\sigma}_z = -\mathbf{1}$ defines left-handed states; $\boldsymbol{\sigma}_z = \mathbf{1}$ defines right-handed. For a filled weak vacuum, left-handed states are predominantly fermionic, right-handed states become antifermionic 'holes' in the vacuum (which is 0 in this representation).

(2) $b = c$ produces leptons; $b \neq c$ produces quarks. If $b \neq c$ we are obliged to take into account the three directions of **p** at once. If $b = c$, we can define a single direction. Taking into account all three directions at once, we define baryons composed of three quarks (and mesons composed of quark and antiquark), in which each of a, b, c cycle through the directions x, y, z.

(3) m is an electromagnetic mass unit, which selects the state of weak isospin. It becomes 1 when present and 0 when absent. So $m = 1$ is the weak isospin up state; and $m = 0$ weak isospin down. The unit condition can be taken as an empty electromagnetic vacuum; the zero condition a filled one.

(4) g represents a conjugation of weak charge units, with $g = -1$ representing maximal conjugation. If conjugation fails maximally, then $g = 1$. g can also be thought of as a composite term, containing a parity element (P) and a time-reversal element (T). So, there are two ways in which the conjugated PT may remain at the unconjugated value (1). $g = -1$ produces the generation u, d, ν_e, e; $g = 1$, with P responsible, produces c, s, ν_μ, μ; $g = 1$, with T responsible, produces t, b, ν_τ, τ.

The weak interaction can only identify (1). This occupies the $\boldsymbol{ik}E$ site in the anticommuting Dirac pentad ($\boldsymbol{ik}E + \boldsymbol{i}\mathbf{p} + \boldsymbol{j}m$), with the i term being responsible for the fermion / antifermion distinction. Because it is attached to a complex operator, the sign of $\boldsymbol{k}$ has two possible values even when those of $\boldsymbol{i}$ and $\boldsymbol{j}$ are fixed; the sign of the weak charge associated with $\boldsymbol{k}$ can therefore only be determined physically by the sign of $\boldsymbol{\sigma}_z$. The filled weak vacuum is an expression of the fact that the 'ground state of the universe' can be specified in terms of positive, but not negative, energy (E), because, physically, this term represents a continuum state.

The strong interaction identifies (2). This occupies the $\boldsymbol{i}\mathbf{p}$ (or $\boldsymbol{i}\boldsymbol{\sigma}.\mathbf{p}$) site and it is the three-dimensional aspect of the $\mathbf{p}$ (or $\boldsymbol{\sigma}.\mathbf{p}$) term which is responsible for the three-dimensionality of quark 'colour'. A separate 'colour' cannot be identified any more successfully than a separate dimension, and the quarks become part of a system, the three parts of which have $\hat{\mathbf{p}}_a$ values taking on one each of the orthogonal components $\hat{\mathbf{p}}_x$, $\hat{\mathbf{p}}_y$, $\hat{\mathbf{p}}_z$. Meson states have corresponding values of $\hat{\mathbf{p}}_a$, $\hat{\mathbf{p}}_b$ and $\hat{\mathbf{p}}_c$ in the fermion and antifermion components, although the logical operators δ_{0m} and $(-1)^{\delta_{1g}} g$ may take up different values for the fermion and the antifermion, and the respective signs of $\boldsymbol{\sigma}_z$ are opposite.

The electromagnetic interaction identifies (3). This occupies the $\boldsymbol{j}m$ site in the Dirac pentad. Respectively, the three interactions ensure that the orientation, direction and magnitude of angular momentum are separately conserved. Gravity (mass), finally, identifies (4).

The charge conjugation from $-w$ to w, in the second and third generations, which is represented in the previous section by z_P or z_T, is brought about, as we have said, by the filled weak vacuum needed to avoid negative energy states. The two weak isospin states are associated with this idea in (3), the $\mathbf{1}$ in ($\hat{\mathbf{p}}_b - \mathbf{1}\delta_{0m}$) being a 'filled' state, with its absence an unfilled state, and the weak interaction acts by annihilating and creating e, either filling the vacuum or emptying it – which is why, unlike the strong interaction, it always involves the equivalent of particle + antiparticle = particle + antiparticle, and involves a massive intermediate boson.[3] We thus create two possible vacuum states to allow variation of the sign of electric charge by weak isospin, and this variation is linked to the filling of the vacuum which occurs in the weak interaction, and could be connected with a mass-related 'bosonic' spin 0 linking of the two isospin states (in addition to the spin 1 gauge bosons involved in the interaction). The weak and electric interactions are linked by this filled vacuum in the $SU(2)_L \times U(1)$ model, as they are in our description of weak isospin, and we can regard these as alternative formalisms for representing the same physical truth. It is significant that the Higgs mechanism for generating masses of intermediate weak bosons and fermions requires the same Higgs vacuum field both for $SU(2)_L$ and for $U(1)$. In addition, the combination of scalar and pseudoscalar phases in the mathematical description of the combined electric and weak interactions clearly relates to the use of a complex scalar field in the conventional derivation of the Higgs mechanism.[4]

The formalism actually explains easily how mass is generated when an element of partial right-handedness is introduced into an intrinsically left-handed system. In principle, anything which alters the signs of the terms in the expression $(\boldsymbol{i}\,\hat{\mathbf{p}}_a\,(\delta_{bc} - 1) + \boldsymbol{j}\,(\hat{\mathbf{p}}_b - \mathbf{1}\delta_{0m}) + \boldsymbol{k}\,\hat{\mathbf{p}}_c\,(-1)^{\delta_{1g}}\,g)$, or reduces any of these terms to zero, is a mass generator, because it is equivalent to introducing the opposite sign of $\boldsymbol{\sigma}_z$ or a partially right-handed state. Thus mass can be produced separately by weak isospin, by quark confinement, and by weak charge conjugation violation. Various calculations of the masses of baryons and mesons, quarks and leptons, using these mechanisms, will be made in chapter 15.

13.3 Conservation of Charge Type and Conservation of Angular Momentum

In generating the particle charge structures, we have established the connection between the conservation of the type of charge and the conservation of angular momentum, which emerges from the fundamental symmetry of space, time, mass and charge. Each type of charge (strong, weak and electromagnetic) is related to a separate aspect of angular momentum conservation (direction, orientation (sign), and magnitude). The angular momentum operator in the nilpotent is also vital to the meaning of quantum mechanics itself, as it effectively defines the classical / quantum transition. Real physical processes involve discrete energy 'transfer' between discrete charged or massive particles. In quantum mechanical terms, discrete energy transfer involves the so-called 'collapse' of the wavefunction. This 'collapse' may be considered as a change which breaks the direct connection between the conservation laws of type of charge and angular momentum. Where the connection is maintained in a coherent way, even on a large scale, we have quantum mechanics. Where it is not, we have decoherence through the vector addition of the noncoherent individual $\mathbf{p}$ components (and different relative 'phases' of $\mathbf{p}_1$, $\mathbf{p}_2$, etc.), and, therefore, energy transfer. (If the $\mathbf{p}$ terms remain aligned, there is no need to alter E.)[5] The same also applies for a change of charge structure or rest mass state. The level of decoherence is measured by the increase in entropy (or the number of noncoherent states), and any process involving an interaction between two fermions will involve some measure of it. In practical terms, making a classical measurement means the application of fields of sufficient size to make the whole system decohere and reduce any quantum mechanical variation in spatial coordinates during a given time interval to the level of the uncertainty principle.

13.4 Phase Diagrams for Charge Conservation

Phase diagrams provide a useful way of picturing lepton, baryon and meson charge structures. In the case of the strong interaction, only one component of angular momentum is well-defined at any moment, and the strong charge appears to act in such a way that the well-defined direction manifests itself by 'privileging' one out of three independent phases making up the complete phase cycle. In a truly gauge invariant system, this can only be accomplished in relative terms. If the weak and electric charges are also related to angular momentum, then the same must apply to them, and the relative 'privileging' of phase can only be defined between the different interactions. We have, here, two options. If the 'privileged' or 'active' phases of E and m (or w and e) coincide with each other, then this also determines the 'privileged' phase of **p**; the result is no 'privileged' relative phase. Since the strong charge is defined only through the directional variation of **p**, via a 'privileged' relative phase, a system in which the phases coincide cannot be strongly bound. If, however, they are different, then this information can only be carried through **p** (or s), and the strong interaction must be present.

We can imagine the arrangements diagrammatically using a rotating vector to represent the 'privileged' direction states for the charges. Each charge has only one 'active' phase out of three at any one time to fix the angular momentum direction; the symbols e, s, and w here refer to these states, not the actual charges. The vectors may be thought of as rotating over a complete spherical surface. In the case of the quark-based states – baryons and mesons – the total information about the angular momentum state is split between three axes, whereas the lepton states carry all the information on a single axis

The axes in Figure 13.1 represent both charge states and angular momentum states for leptons, mesons and baryons. As previously stated, each type of charge carries a different aspect of angular momentum (or helicity) conservation; s carries the directional information (linked to **p**); w carries the sign information (+

or – helicity) (linked to iE); e carries information about magnitude (linked to m). Another way of looking at this is to associate these properties, respectively, with the symmetries of rotation, inversion, and translation, and, ultimately, with the dual processes of dimensionalization, complexification and conjugation, which, through the 3-D representation in section 4.9, can also be mapped onto the dihedral symmetries (or rotations around x, y and z axes) of the fundamental parameter group.

Figure 13.1

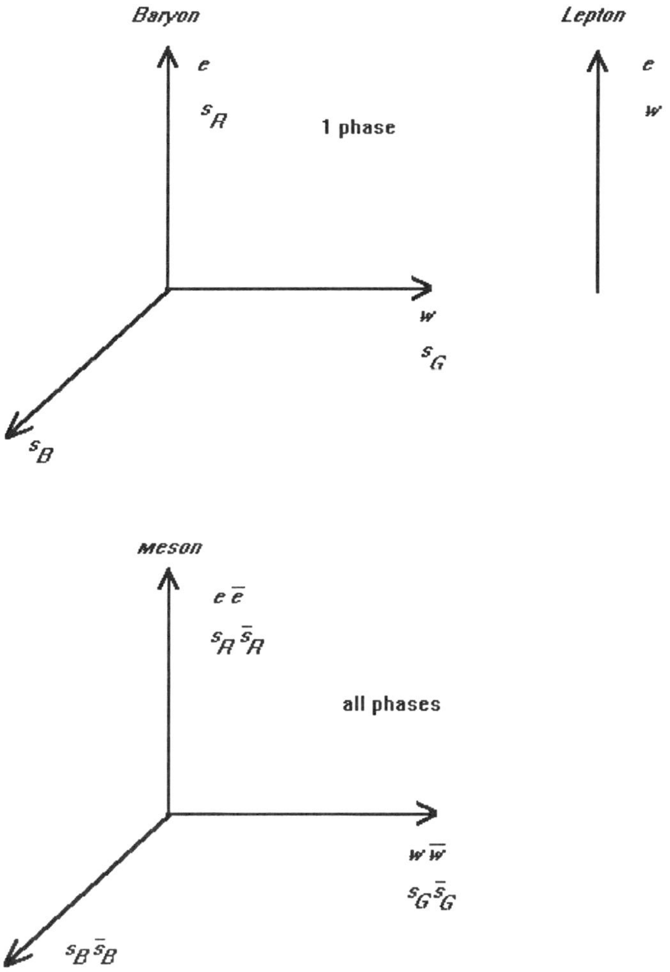

13.5 Quark and Lepton Charge Structures in Tabular Form

From both the separate formulae in section 13.1 and the unified representation in section 13.2, the 0 and 1 charge structures of the fundamental fermions may be expressed in terms of a set of three 'quark' tables, *A*-*C*, with an extra table *L* for the left-handed leptons and antileptons (the unlabelled columns in *L* represent left-handed antineutrinos):

A

		B	G	R
u	+ *e*	1*j*	1*j*	0*i*
	+ *s*	1*i*	0*k*	0*j*
	+ *w*	1*k*	0*i*	0*k*
d	− *e*	0*j*	0*k*	1*j*
	+ *s*	1*i*	0*i*	0*k*
	+ *w*	1*k*	0*j*	0*i*
c	+ *e*	1*j*	1*j*	0*i*
	+ *s*	1*i*	0*k*	0*j*
	− *w*	$z_P k$	0*i*	0*k*
s	− *e*	0*j*	0*k*	1*j*
	+ *s*	1*i*	0*i*	0*k*
	− *w*	$z_P k$	0*j*	0*i*
t	+ *e*	1*j*	1*j*	0*i*
	+ *s*	1*i*	0*k*	0*j*
	− *w*	$z_T k$	0*i*	0*k*
b	− *e*	0*j*	0*k*	1*j*
	+ *s*	1*i*	0*i*	0*k*
	− *w*	$z_T k$	0*j*	0*i*

B

		B	G	R
u	+ *e*	1*j*	1*j*	0*k*
	+ *s*	0*i*	0*k*	1*i*
	+ *w*	1*k*	0*i*	0*j*
d	− *e*	0*i*	0*k*	1*j*
	+ *s*	0*j*	0*i*	1*i*
	+ *w*	1*k*	0*j*	0*k*
c	+ *e*	1*j*	1*j*	0*k*
	+ *s*	0*i*	0*k*	1*i*
	− *w*	$z_P k$	0*i*	0*j*
s	− *e*	0*i*	0*k*	1*j*
	+ *s*	0*j*	0*i*	1*i*
	− *w*	$z_P k$	0*j*	0*k*
t	+ *e*	1*j*	1*j*	0*k*
	+ *s*	0*i*	0*k*	1*i*
	− *w*	$z_T k$	0*i*	0*j*
b	− *e*	0*i*	0*k*	1*j*
	+ *s*	0*j*	0*i*	1*i*
	− *w*	$z_T k$	0*j*	0*k*

C

		B	**G**	**R**
u	$+e$	$1j$	$1j$	$0k$
	$+s$	$0i$	$1i$	$0j$
	$+w$	$1k$	$0k$	$0i$
d	$-e$	$0j$	$0k$	$1j$
	$+s$	$0i$	$1i$	$0k$
	$+w$	$1k$	$0j$	$0i$
c	$+e$	$1j$	$1j$	$0k$
	$+s$	$0i$	$1i$	$0j$
	$-w$	$z_P k$	$0k$	$0i$
s	$-e$	$0j$	$0k$	$1j$
	$+s$	$0i$	$1i$	$0k$
	$-w$	$z_P k$	$0j$	$0i$
t	$+e$	$1j$	$1j$	$0k$
	$+s$	$0i$	$1i$	$0j$
	$-w$	$z_T k$	$0k$	$0i$
b	$-e$	$0j$	$0k$	$1j$
	$+s$	$0i$	$1i$	$0k$
	$-w$	$z_T k$	$0j$	$0i$

L

		$\bar{e}$	$\bar{e}$	ν_e
	$+e$	$1j$	$1j$	$0j$
	$+s$	$0k$	$0i$	$0i$
	$+w$	$0i$	$0k$	$1k$
				e
	$-e$	$0i$	$0k$	$1j$
	$+s$	$0j$	$0i$	$0i$
	$+w$	$0k$	$0j$	$1k$
		$\bar{\mu}$	$\bar{\mu}$	ν_μ
	$+e$	$1j$	$1j$	$0j$
	$+s$	$0k$	$0i$	$0i$
	$-w$	$0i$	$0k$	$z_P k$
				μ
	$-e$	$0i$	$0k$	$1j$
	$+s$	$0j$	$0i$	$0i$
	$-w$	$0k$	$0j$	$z_P k$
		$\bar{\tau}$	$\bar{\tau}$	ν_τ
	$+e$	$1j$	$1j$	$0j$
	$+s$	$0k$	$0i$	$0i$
	$-w$	$0i$	$0k$	$z_T k$
				τ
	$-e$	$0i$	$0k$	$1j$
	$+s$	$0j$	$0i$	$0i$
	$-w$	$0k$	$0j$	$z_T k$

Applying these to the known fermions, *A*-*C* would appear to have all the properties of the coloured quark system, with s (or the $A^{\alpha n}$ term in the covariant derivative) pictured as being 'exchanged' between the three states (although in reality, of course, all the states exist simultaneously), in the same way as the operator **p** in the nilpotent baryon wavefunction. In relation to these tables, we can look on symmetry-breaking, in general, as a consequence of the setting up of the algebraic model for charges. When we map time, space and mass onto the charges w-s-e, to create the anticommuting Dirac pentad, only one charge (s) has the full range of vector options. ‘Fixing’ one of the others (say *e*) for *s* to vary against, gives us only 2 remaining options for *w*, unit on the same colour as *e* or unit on a different one. Putting both *w* and *e* on the same colour denies the necessary three degrees of freedom in the direction of angular momentum, so this is forbidden in a quark system.

In principle, there are also two other quark tables (*D-E*), but, with the application of the exclusion rules defined above, these both reduce to the lepton table *L*. The reduction from *A-E* to *A-C* plus *L* can be thought of as similar to a reduction from the full Dirac pentad to a 4-vector or quaternion representation.

D

		B	G	R
u	+ *e*	1*j*	1*j*	0*i*
	+ *s*	0*k*	1*i*	0*j*
	+ *w*	0*i*	0*k*	1*k*
d	− *e*	0*i*	0*k*	1*j*
	+ *s*	0*j*	1*i*	0*i*
	+ *w*	0*k*	0*j*	1*k*
c	+ *e*	1*j*	1*j*	0*i*
	+ *s*	0*k*	1*i*	0*j*
	− *w*	0*i*	0*k*	$z_P k$
s	− *e*	0*i*	0*k*	1*j*
	+ *s*	0*j*	1*i*	0*i*
	− *w*	0*k*	0*j*	$z_P k$
t	+ *e*	1*j*	1*j*	0*i*
	+ *s*	0*k*	1*i*	0*j*
	− *w*	0*i*	0*k*	$Z_T k$
b	− *e*	0*i*	0*k*	1*j*
	+ *s*	0*j*	1*i*	0*i*
	− *w*	0*k*	0*j*	$Z_T k$

E

		B	G	R
u	+ *e*	1*j*	1*j*	0*j*
	+ *s*	0*k*	0*i*	1*i*
	+ *w*	0*i*	0*k*	1*k*
d	− *e*	0*i*	0*k*	1*j*
	+ *s*	0*j*	0*i*	1*i*
	+ *w*	0*k*	0*j*	1*k*
c	+ *e*	1*j*	1*j*	0*j*
	+ *s*	0*k*	0*i*	1*i*
	− *w*	0*i*	0*k*	$z_P k$
s	− *e*	0*i*	0*k*	1*j*
	+ *s*	0*j*	0*i*	1*i*
	− *w*	0*k*	0*j*	$z_P k$
t	+ *e*	1*j*	1*j*	0*j*
	+ *s*	0*k*	0*i*	1*i*
	− *w*	0*i*	0*k*	$Z_T k$
b	− *e*	0*i*	0*k*	1*j*
	+ *s*	0*j*	0*i*	1*i*
	− *w*	0*k*	0*j*	$Z_T k$

An important aspect of this model is the fact that the charges (*e*, *s*, w) are irrotational, but the quaternion labels (***j***, ***i***, ***k***) are not, and this provides an alternative approach to the primary derivation of the tables. The value of this approach, however, is primarily heuristic. It gives us an idea of the 'physical' meaning involved in the more rigorous algebraic process which produces the symmetry-breaking between the strong, electric and weak interactions in the creation of the Dirac state. Here, we define a unit charge as some combination, ± ***i***s, ± ***j***e, ± ***k***w, where the charge components, *s*, *e* and *w*, may take values 1 or 0. We then suppose that *s*, *e* and *w* are fixed with respect to each other and so may be said to have 'rotational asymmetry', whereas the quaternion components ***i***, ***j*** and ***k*** are not fixed and have rotational symmetry. Thus the charge component *e* could be associated with ***i*** or ***k*** as easily as it could be associated with ***j***.

This becomes physically possible, if, for example, we always associate unit values of *e* with the term ***je*** and zero values of *e* with the terms ***ie*** and ***ke***. Then, as long as physical systems with 1***je*** are indistinguishable from those with 0***ie*** and 0***ke***, a valid model may be constructed. This only becomes possible if unit charges always exist in some form of combination. The individual charges can then be identified but only in such a way as to never be separable. In the case of *e*, for example, charges could be combined in groups of three in such a way as to render systems containing 1***je***, 0***ie*** or 0***ke*** indistinguishable. Alternatively, unit charges could always be combined with unit anti-charges.

As a perfect symmetry between the charges would always collapse to the defined state, ± ***i***s, ± ***j***e, ± ***k***w, which is not allowed, an imperfect or broken symmetry must be expected, in exact parallel to the algebraic combination of the two triads in the Dirac algebra. Using the **B**, **G**, **R** notation, as before, for a three-component system, a pattern like

	B	**G**	**R**
± *e*	1***j***	1***j***	0***i***
± *s*	1***i***	0***k***	0***j***
± *w*	1***k***	0***i***	0***k***

will automatically introduce asymmetry into the status of the three charge components. In this example, the *s* component may take unit value in the **B** colour only, whereas the *e* component may take unit value in either **B** or **G**, or both, and, in fact, we find that in every conceivable arrangement of any workable scheme involving three 'colour' states, there is always at least one charge component that is confined to a unit value in only one of the colours.

Now, there are eight possible combinations of $\pm e$, $\pm s$, $\pm w$. Four of these can be defined as antistates of the others, without, in the first instance, implying any particular physical significance by this distinction. However, the combinations which correspond to antistates may be determined as soon as we decide which of the three components may take a unit value in only one colour. If every combination of three quarks must contain one of each colour, then the unit component in one colour must always be unit in that colour and must always bear the same sign. The choice is arbitrary because *e*, *s* and *w* up to this stage are only labels, but, by comparison with experiment, we choose *s*. So we identify the following

States	Antistates
$+e+s+w$	$-e-s-w$
$-e+s+w$	$+e-s-w$
$+e+s-w$	$-e-s+w$
$-e+s-w$	$+e-s+w$

Such an arrangement is only possible for one of the three charge components, because the four identifiable flavours with + *s* require states which have both positive and negative components in *e* and *w*. However, the division between states and antistates can only be used once, and so we now are obliged to incorporate charges with both + and – sign options for *e* and *w* into both states and antistates. There is only one way of doing this, in fact, and it only works for one charge type. We take, for example, the **B** and **G** colours of all quarks with the positive charge component to have unit value, and the **R** component to have zero; at the same time, the **R** colour of all quarks with negative charge component has unit value, while the **B** and **G** colours have zero. Once again, the choice is arbitrary between *e* and *w*, but experimentally, it is the electromagnetic charge component which is incorporated in this way (i.e. using 'weak isospin'), so that typically we recognize the *u* and *d* flavours as having the following assignments for the *e* and *s* components:

		B	G	R
u	$+e$	1*j*	1*j*	0*k*
	$+s$	0*i*	0*k*	0*j*
	$+w$	*k*	*i*	*k*

		B	G	R
d	$-e$	$0i$	$0k$	$1j$
	$+s$	$0j$	$0j$	$1i$
	$+w$	k	i	k

The final component (w) component cannot be incorporated this way into a structure which preserves the colour singlet nature of the combination with the quaternion operators unspecified. In principle, the $SU(2)$ (or isospin) option, like the $SU(3)$ option, can only be used once. Each of the tables *A*-*E* produces a valid structure for one isospin state, but not for the other. However, w can be incorporated in the same way as s if we assume that, for the w component of charge, *states of + and − are, in some sense, indistinguishable*, with all the expected consequences of P and T symmetry-breaking, which creates the two extra generations in the quark tables A-E. That isospin is a weak interaction, rather than an electric one, is evident from the fact that an interaction is specified by what remains *unchanged*, rather than what is changed; thus, the strong interaction defines all the e and w combinations that change under an unchanging s; while a weak interaction defines all the e combinations that change under an unchanging w. To preserve e unchanged, after taking these changes into account, we are left only with possible changes of phase.

The heuristic approach to the tables is not a substitute for the more rigorous algebraic derivation, but it certainly provides a useful illustration of its physical consequences. Even here, for example, table E appears to be excluded automatically by requiring all three quaternions to be attached to specified charges, thus losing the three required degrees of rotational freedom, and, at the same time, necessarily violating Pauli exclusion.[6] Significantly, also, though the separate and random variation of $\hat{\mathbf{p}}_a$, $\hat{\mathbf{p}}_b$, and $\hat{\mathbf{p}}_c$, in a quark-based system, suggests 27 possible variations in charge structure, the number of different outcomes is reduced by repetitions, to 5, as in the anticommuting pentads of the Dirac algebra. Effectively, we 'privilege' one of $\hat{\mathbf{p}}_a$, $\hat{\mathbf{p}}_b$, $\hat{\mathbf{p}}_c$, by allowing it complete variation with respect to the others; $\hat{\mathbf{p}}_b$ is the one selected, and it corresponds to the 'privileging' of $\mathbf{p}$ as a vector term with full variation in the Dirac anticommuting pentad.

13.6 Mesons and Baryons

The tables *A*-*E* (or *A*-*L*) thus ultimately result from combining the action of two independent 3-dimensional operators in a manner analogous to the Dirac algebra, a conjugate pairing of a nonconserved triad with a conserved one. We can consider it either as the action of thc randomized angular momentum operator on a fixed quaternion structure representing conserved charges, or, as the result of using arbitrary quaternion operators to represent conserved charge states.

Using either algebraic or heuristic methods, we find that we may avoid specific quaternion-charge assignments by always combining unit charges in groups of three in such a way as to never know whether we have, for example, 1*je*, 0*ie* or 0*ke*. The different colour permutations give the same charge structure and at the same time hide the quaternion assignments, unit *je*, for example, being indistinguishable from zero *ie* or *ke*. The obscuring of quaternion assignments or defined phases can also clearly be accomplished by combining unit charges with unit anticharges. The origin of both baryons and mesons thus becomes a natural result of this process, and their charge structures can be represented as follows:

meson octet (0^-)

particle	quarks	charge structures	average structure
π^-	$u\ \bar{d}$	$+e$	
π^+	$d\ \bar{u}$	$-e$	
π^0	$u\ \bar{u}$	0	
η	$d\ \bar{d}$	0	
K^-	$u\ \bar{s}$	$-e$ or $-e-(1+z)\,w$	$-e-(1+z)\,w/3$
K^+	$s\ \bar{u}$	$-e$ or $-e-(1+z)\,w$	$-e-(1+z)\,w/3$
K^0	$d\ \bar{s}$	0 or $+(1+z)\,w$	$+(1+z)\,w/3$
$\bar{K}$	$s\ \bar{d}$	0 or $-(1+z)\,w$	$-(1+z)\,w/3$

baryon octet (½⁺)

particle	quarks	charge structures	average structure
n	udd	$+s+w$	$+s+w$
p	uud	$+e+s+w$	$+e+s+w$
Λ	uds	$+s\pm w$	$+s+w/3$
Σ^-	dds	$-e+s\pm w$	$-e+s+w/3$
Σ^0	uds	$+s\pm w$	$+s+w/3$
Σ^+	uus	$+e+s\pm w$	$+e+s+w/3$
Ξ^-	dss	$-e+s\pm w$	$-e+s-w/3$
Ξ^0	uss	$+s\pm w$	$+s-w/3$

baryon decuplet ($^3/_2{}^+$)

particle	quarks	charge structures	average structure
Δ	ddd	$-e+s+w$	$-e+s+w$
Δ	udd	$+s+w$	$+s+w$
Δ	uud	$+e+s+w$	$+e+s+w$
Δ	uuu	$+2e+s+w$	$+2e+s+w$
Σ^-	dds	$-e+s\pm w$	$-e+s+w/3$
Σ^0	uds	$+s\pm w$	$+s+w/3$
Σ^+	uus	$+e+s\pm w$	$+e+s+w/3$
Ξ^-	dss	$-e+s\pm w$	$-e+s-w/3$
Ξ^0	uss	$+s\pm w$	$+s-w/3$
Ω^-	sss	$-e+s-w$	$-e+s-w$

13.7 The Standard Model

The rules for the Standard Model follow immediately from the charge structures and tables. Both the tables and structures explain many facts related to particle physics, and also make some new predictions. The strong interaction follows an $SU(3)$ symmetry whose exactness means that the exchange bosons or gluons remain massless. All possible 'colour' states in this representation are automatically identical. The electromagnetic charge structures for the quarks in the Standard Model ($2e/3$ for u, $-e/3$ for d, instantaneously averaging out the values for the **R**, **G** and **B** representations) follow immediately from this colour-symmetry and its intrinsic gauge invariance. With coupling constant g, generators of $SU(3)$ $T^{\mu}/2$, and gauge fields $\mathbf{W}^{\mu}$, we are led naturally to a covariant derivative such as:

$$D^{\mu} = \partial^{\mu} + igT^{\mu}.\mathbf{W}^{\mu} / 2.$$

The electric and weak interactions are linked in the $SU(2)_L \times U(1)$ symmetry which acts vertically in the quark tables, as the $SU(3)$ acts horizontally. Here, it is significant that perfect symmetry between s, e and w would be maintained if we could restrict each of these charge-types to one sign only (+ in the case of s and w, −, for purely historical reasons, in the case of e). In this case, the representations A, B, C, D would then allow an $SU(3)$ symmetry to be applied to each of the three interactions: A, B, C for s; B, C, D for e; A, C, D for w. However, the symmetry is broken when the opposite signs of the three charge-types are introduced.

For strong charges, the $SU(3)$ symmetry remains, because all the $-s$ components can be assigned to antiparticles, though this implies that there must be a set of particles with $s = 0$ (leptons). The electromagnetic charge, however, accommodates the opposite sign (+) by, in effect, adding 1 unit (j) to *all possible quark states*. The concept of isospin (I) (not to be confused with weak isospin) groups all particles with identical values of s and w, and the behaviour of the electromagnetic charge can be interpreted as the breaking of isospin symmetry. The concept of hypercharge (Y) then follows as 2 × the average charge of an isospin multiplet.

More significant than these are *weak* isospin (t, with third component t_3), which is an $SU(2)$ symmetry relating all particles with the same value of w but disregarding the value of s, and the corresponding weak hypercharge (y), which deals with the electromagnetic ($U(1)$) component, the relationship between these and the average charge of a weak isomultiplet (Q) being given by:

$$Q = t_3 + y / 2.$$

The specification of the third component of weak isospin is necessary because only one quark carries the weak charge. With coupling constants g and g', Pauli matrices $\boldsymbol{\tau}$ as generators of $SU(2)$, and gauge fields $\mathbf{W}^{\mu}$ and B^{μ} for weak isospin and weak hypercharge, the covariant derivative becomes

$$D^{\mu} = \partial^{\mu} + ig(\tau / 2).\mathbf{W}^{\mu} + i(g' / 2)yB^{\mu}.$$

The $SU(2)$ means that two particle states (with different states of electric charge) are available for every value of w. Also, because of the particular nature of electric charge accommodation, no composite particles depend on the value of *e*; there is no electromagnetic equivalent of 'colour'. Hence, the infinite range of the interaction and the $U(1)$ symmetry with which it is associated.

With the $SU(3)$ and $SU(2)_L \times U(1)$ symmetries established from fundamental principles, it is possible to use the Lagrangians already available for QCD and the Weinberg-Salam electroweak model, and justify them term by term. However, new numerical information can be extracted once the Grand Unified gauge group is identified, while particle masses can be assumed to be the result of the breaking of various symmetry conditions (in particular those involved with the weak interaction) and the suppression of units of each of the charges.

The three generations of weak isospin exist only to accommodate changes of sign in the weak charge. From the point of view of the weak interaction, however, there is only one generation and physical mixing can be expected to occur between all three generations. The mixing ensures that all weak interactions violate parity or time-reversal symmetry. For two generations, this involves the Cabibbo angle and Cabibbo-rotated eigenquarks of weak isospin; for three generations, the mixing needs to incorporate the complex *CP*-violating phase, as originally predicted by Kobayashi and Maskawa.[7]

Now, while baryons have a weak charge structure of $\pm w$, mesons always have a weak charge structure 0, with the exception of states like the *K* mesons, and the other meson states combining a fermion and antifermion of two different generations. For these states, we find alternative weak charge structures of 0 or $\pm (1 + z)w$, where z may be z_P or z_T. (Mesons combining the second and third generations will have structures combining z_P and z_T.) The alternatives depend on the particular colour-anticolour combination, and particular quark representation, we choose. Clearly, 0 and $\pm (1 + z)w$ have to be indistinguishable in weak terms, but we have already broken a symmetry (either *P* or *T*) in creating z, which means that we are also obliged to break another to maintain overall *CPT* invariance. In the case of a bosonic state, charge conjugation must be preserved, so we are obliged to break *CP* (or *T*) as well as *P*, or *CT* (or *P*) as well as *T*. A prediction (now largely fulfilled) may be made that such an additional violation will be found in all states of this kind. It has already been observed in K^0 and $\bar{K}^0$ mixed states, and is now known to occur in K^0 and $\bar{K}^0$, taken separately, as well as being extended to incorporate (*B*) mesons combining first and third generation components. However, it should also be observable in K^+ and K^-, and in the equivalent states in other generations, and it should be observed, additionally, in the weak decays of Bose-Einstein condensates, which again involve a $(1 + z)w$, weak charge structure.[8]

It is assumed in many contexts that parity violation occurs in preference to time-reversal symmetry, but nature knows no such preference. It is impossible to tell, in principle, which violation has actually occurred in any given circumstance. It is simply a matter of convention and convenience that we take parity violation as occurring in the first instance, and time-reversal asymmetry in the second. The real distinction is between one symmetry violation and two. Systems that are assumed to violate time-reversal symmetry, such as the K^o meson, have already been found to violate parity, and time-reversal asymmetry is only invoked because it can no longer be avoided.

We should not, therefore, expect a difference in mass-scale between *individual* P or T violations, but rather a difference between the breaking of one symmetry and the breaking of two. Some evidence of this may perhaps be found in the Cabibbo-Kobayashi-Maskawa mixing between the respective quark generations represented by d, s and b, where the mixing ratio between the first two generations (approximately tan θ_C) appears to be of the same order as that between the electric and symmetry-breaking weak interactions ($\sin^2\theta_W$), while that between second and third generations, with two symmetry-violations, takes this to a second order. The increasing mass-scale for the quarks in the higher generations must be related in some way to this symmetry-breaking.

It has been suggested that *CP* violation may have cosmological significance in explaining the matter-antisymmetry, although the physical quantities involved are many orders of magnitude too small. However, the matter-antisymmetry is fundamental, rather than cosmological in origin. It is the result of a filled vacuum or continuous mass distribution, and has nothing to do with hypothetical conditions in the 'early universe'. In addition, *CP* violation, as so far discovered, concerns only boson states, whereas the distinction between 'matter' and 'antimatter' only has any meaning in the context of fermions and antifermions. Of course, some bosons are antiparticles of others, but, since they are all fermion-antifermion combinations, there is no physical principle by which we can separate out some as particles of 'matter' and others as particles of 'antimatter'.

Other effects associated with the weak charge are strangeness and charm. Strangeness may be defined as $(3n - 3) / 2$, where n is the average number of weak units of charge (or units of $\mathbf{k}$) in a particle. Charm may be defined in a similar way, but it is associated with a quark with $+e$ instead of $-e$, and is also given the opposite sign by convention. Both strangeness and charm are measures of the degree of parity violation necessary to convert $-\mathbf{k}$ to $\mathbf{k}$. Similar measures could be applied in the next generation to the degree of violation of time-reversal.

As strangeness and charm are purely properties of the $\boldsymbol{k}$ component, they are necessarily conserved in strong and electromagnetic interactions.

A strangeness-conserving current may be defined as one which retains the sign of $\boldsymbol{k}$, and a strangeness-changing current one which reverses it. The decay of Λ to $p + e^{-} + \bar{\nu}$ (equivalent to $s + w$ or $\boldsymbol{i} + \boldsymbol{k}$) requires a strangeness-changing current for those states of Λ with $-\boldsymbol{k}$, rather than $\boldsymbol{k}$, and so is slower than the same decay of the neutron, which always has $\boldsymbol{k}$. Another effect is that Ω^{-} is the only member of the spin $^{3}/_{2}$ decuplet which always decays via a weak interaction because it always has a $-\boldsymbol{k}$ weak component, unlike any other state in the octet or decuplet, and can only decay to a state which has some fraction of $+\boldsymbol{k}$.

The charge structures of neutrinos show, at least in the first two generations, the clear distinction between electron and muon neutrinos, which require, respectively, $+w$ and $-w$ or $z_P w$. Mixing, however, occurs in parallel quark states with only the sign of w to distinguish them. There is no obvious way of distinguishing $-w$ or $z_P w$ of the muon neutrino physically from the $-w$ or $z_T w$, representing the tau neutrino. The charge structures suggest that, from the point of view of the weak interaction, they must be mixed or oscillating states.

As previously mentioned in chapter 2, baryon and lepton conservation are obvious consequences of the separate conservation laws for s and w. Baryons are the only particles containing both s and w, and so there is no decay mode available to eliminate the strong charge. Leptons, similarly, are the only particles with w, but no s, and so there is again no decay mode available to create another state.

The distinction between fermions and bosons is another obvious result of the charge structures, in addition to the forms of their state vectors. The presence of a nonzero weak charge separates fermions from bosons. To describe the full range of fermions, we require a charge quaternion structure, whereas bosons require only the specification of the electromagnetic charge (and then only in the form of the added scalar term). The charge quaternion structure for fermions leads on to a requirement for a quaternionic wavefunction and Fermi-Dirac statistics, in contrast to the scalar wavefunction and Bose-Einstein statistics required of bosons. In particular, fermions have $+\boldsymbol{k}$ (either directly or through parity or time-reversal violation); bosons have 0 $\boldsymbol{k}$. The $SU(2)$ spin $\pm \frac{1}{2}$ symmetry maps directly onto the virtually identical symmetry for weak isospin. The derivation of the spin $\frac{1}{2}$ value from the $+\boldsymbol{k}$ component once again relates to the idea of proton spin as a global feature, rather than as one derived from three component $\pm \frac{1}{2}$ spins from the quarks. The properties of bosons and fermions may be summarised as follows:

	Bosons	**Fermions**
	2 charge states	1 charge state
	1 mass state	2 mass states
	2 space states (or parity)	1 space state with 2 signs
	1 time direction with 2 signs	2 time directions
	$C = PT$ preserved	$C = PT$ violated
	→ wavefunctions symmetric	→ wavefunctions antisymmetric
	$P = CT$ preserved	$P = CT$ violated
	$T = CP$ preserved	$T = CP$ preserved
or	$P = CT$ preserved	$P = CT$ preserved
	$T = CP$ preserved	$T = CP$ violated

In the case of quarks, the total fermionic weak charge component of $\pm w$ is allowed to change sign to preserve the quark colour invariance and this is achieved by violation of parity conservation, in the first instance; in terms of the *CPT* theorem, the symmetry of *CP* is preserved at the expense of that of *PT*. (In the second instance, the position is reversed.) Now, in order to introduce the real-imaginary space-time symmetry into the quantum mechanics of the fermion, it is more convenient, in the conventional representation, to introduce space, which now has only one parity state but two mathematically indistinguishable solutions, rather than time, with its uniquely specified direction, as the imaginary component of the relationship (i.e. $i\mathbf{r}$ rather, rather than ikt). By restricting space to one physical state, we now allow *two* physical states to time. By application of the *group* symmetry, this mathematical device should make the mass-energy of fermions appear to be imaginary, with both positive and negative solutions, and charge appear to be the real quantity with a single solution, which is how the conventional Dirac theory predicts 'antimatter': negative mass-energy states exist alongside positive ones, but the equations only recognise a single sign of charge. The theory has to restore the true status of mass and charge and remove the implied negative masses by the *ad hoc* introduction of antimatter or particles with the opposite sign of charge. (The nilpotent representation, of course, uses the negative sign of $\boldsymbol{k}$ for antimatter, rather than that of E.) Bosons, on the other hand, have 0 component of weak charge; this does not require any reversal in the sign of spatial coordinates or in the status of space and time when the imaginary space-time symmetry is introduced into the otherwise asymmetric wave equation;

for bosons, space remains the real and time the imaginary parameter, mass-energy remains positive and the antiparticles are simply a result of *CPT* symmetry; with two equally possible charge states, the *PT* symmetry is preserved. In the production of a fermion and antifermion pair in the *decay* of *K* bosons, of course, the *CP* and *CT* symmetries are violated, but charge conjugation (as we have seen) is preserved.

13.8 A Pentad Structure for Charges and their Transitions

From its ultimate origin in the Dirac equation, the mathematical structure needed to model charges, *and their transitions*, is clearly of the same pentad nature, combining quaternions and vectors, as that for *E*-**p**-*m*. We can illustrate this for the case of the *ddd* combination, which is represented by the scalar product of the terms $-\boldsymbol{\sigma} = \mathbf{i} + \mathbf{j} + \mathbf{k}$ and $-\boldsymbol{j}\hat{\mathbf{p}}_a + \boldsymbol{i}\hat{\mathbf{p}}_b + \boldsymbol{k}\hat{\mathbf{p}}_c$. For a baryon, the quarks are in the *A*, *B* or *C* representations, with $\hat{\mathbf{p}}_a$ and $\hat{\mathbf{p}}_c$ necessarily representing different unit vectors, and $\hat{\mathbf{p}}_b$ cycling through all three possible values via the strong interaction. No generality is lost, therefore, if we take $\hat{\mathbf{p}}_c$ to be **k** and $\hat{\mathbf{p}}_c$ to be **j**. We can now write $(-\boldsymbol{j}\hat{\mathbf{p}}_a + \boldsymbol{i}\hat{\mathbf{p}}_b + \boldsymbol{k}\hat{\mathbf{p}}_c)$ in the form

$$(-\boldsymbol{j}\hat{\mathbf{p}}_a + \boldsymbol{i}1\hat{\mathbf{p}}_b + \boldsymbol{k}\hat{\mathbf{p}}_c) = (\boldsymbol{i}\mathbf{ij}\boldsymbol{j} + \mathbf{ii}\boldsymbol{i}\hat{\mathbf{p}}_b - \boldsymbol{i}\mathbf{ik}\boldsymbol{k}) = -\boldsymbol{i}\mathbf{i}\,(-\mathbf{j}\boldsymbol{j} + \boldsymbol{i}\mathbf{i}\boldsymbol{i}\hat{\mathbf{p}}_b + \mathbf{k}\boldsymbol{k}).$$

The terms $-\mathbf{j}\boldsymbol{j}$, $\boldsymbol{i}\mathbf{i}\boldsymbol{i}$, $\mathbf{k}\boldsymbol{k}$ form a closed cycle of the same form as the quaternion operators $\boldsymbol{k}$, $\boldsymbol{ii}$, $\boldsymbol{ij}$ in the Dirac algebra, although, this time, as the products of vectors and quaternions, they are commutative. With the second term again successively multiplied by the three vector operators **i**, **j**, **k**, we can see that the algebra of charge accommodation has at its core a pentad structure of the same form as that applied to the Dirac equation – *SU*(5) embedding *SU*(3) (see 15.2) – and we can proceed to use this principle to associate specific charge structures connected with particular particle states to specific quaternion wavefunctions.

It is not only the *SU*(3) quark transitions that can be treated in this way in terms of a pentad; the $SU(2)_L \times U(1)$ quark / lepton transitions can also be structured through a pentad of commuting terms, though the pentad will this time be broken, since the terms cannot all be used at once. The pentad can be written in the form: $\boldsymbol{j}\mathbf{i}e$, $\boldsymbol{j}\mathbf{j}e$, $-\boldsymbol{j}\mathbf{k}e$, $\boldsymbol{k}\mathbf{k}w$, $\boldsymbol{k}\mathbf{i}w$. The left-handed leptonic states from *L* (left-handed electron, neutrino, left-handed antielectron) require either $\boldsymbol{j}\mathbf{i}e$, $\boldsymbol{j}\mathbf{j}e$, $\boldsymbol{k}\mathbf{k}w$ or $-\boldsymbol{j}\mathbf{k}e$, $\boldsymbol{k}\mathbf{k}w$. The quark states (from *A*-*C*) (leaving out the *s* component) require either $\boldsymbol{j}\mathbf{i}e$, $\boldsymbol{j}\mathbf{j}e$, $\boldsymbol{k}\mathbf{i}w$ or $-\boldsymbol{j}\mathbf{k}e$, $\boldsymbol{k}\mathbf{i}w$. It may be of interest also that we can simulate all possible left-handed leptonic states with the tetrad: $\boldsymbol{j}\mathbf{i}e$, $-\boldsymbol{j}\mathbf{k}e$, $\boldsymbol{k}\mathbf{k}w$, $\boldsymbol{k}\mathbf{i}w$; and all possible electroweak-interacting quarks with the tetrad: ***j*i*e***, ***j*j*e***, ***–j*k*e***, ***k*i*w***. In all these cases, of course, the *individual* states involved in the $SU(2)_L \times U(1)$ are reduced to 1 or 0 by taking the scalar product with *one* of the unit angular momentum operators: $\hat{\mathbf{p}}_a$, $\hat{\mathbf{p}}_b$, $\hat{\mathbf{p}}_c$. Quark-lepton 'transitions' (*A*/*B*/*C*-*L*) (as in neutron beta decay) (though probably never occurring directly) might be represented using similar tetrads: ***–j*i*e***, ***–j*j*e***, ***k*k*w***, ***k*i*w*** (for down / electron) ***j*i*e***, ***j*j*e***, ***k*k*w***, ***k*i*w*** (for up / neutrino); down / neutrino would be a triad: ***j*i*e***, ***–j*j*e***, ***k*k*w***, while up / electron would involve the full pentad: ***j*i*e***, ***j*j*e***, ***–j*k*e***, ***k*k*w***, ***k*i*w***.

13.9 Lepton-Like Quarks

Both the formulae and the tables suggest that quarks are *fundamentally* (i.e. in symmetry terms) lepton-like objects, with similar intrinsic charge structures. In such a theory, the fractional electric charges observed (indirectly) for quarks (2*e* / 3 for *u*, *c*, *t*, or –*e* / 3 for *d*, *s*, *b*), in experiments such as the ratio of hadron / muon production in electron-positron annihilation events, and the rate of decay of neutral pions to two photons, are attributable to the absolutely unbroken gauge invariance of the strong interaction, which means that individual phases of the interaction will never be observed. In this sense, of course, the electric charges of quarks really *are* fractional, not simply 'observed' as such – an unbroken gauge invariance can mean nothing else. But the symmetries are also important.

Lepton-like quarks have, of course, a long history. They appeared in the original paper by Han and Nambu which introduced the concept of colour to explain the strong interaction.[9] The Han and Nambu quarks had electric charge assignments of the form:

	B	G	R
u	*e*	*e*	0
d	0	0	–*e*

by contrast with those assumed in the first version of the quark theory proposed in the previous year by Gell-Mann and Zweig:

	B	G	R
u	2*e* / 3	2*e* / 3	2*e* / 3
d	–*e* / 3	–*e* / 3	–*e* / 3

The theory had the advantage of making the 'colour' differences a natural result of the existence of different charge structures rather than an arbitrarily added extra property, though 'colour' was later added also to the Gell-Mann-Zweig version. The original implication of Han and Nambu's theory may have been that, at some sufficiently high energy, the integral nature of the charges would become manifest and the colours directly revealed.

However, under conditions of perfect gauge invariance between the coloured phases, and hence perfect infrared slavery (as our model requires), this transition will never occur, and the Han-Nambu model will provide instead a precise way of predicting the observation of fractional charges. There will be no transition regime between implicit and explicit colour, and no finite energy range at which integral charges or colour properties will emerge. Observed charges, measured by quantum electrodynamics, will be exactly fractional at all energies. Frank Close has expressed it in the following way: 'Imagine what would happen if the colour nonsinglets were pushed up to infinite masses. Clearly only colour **1** [singlets] would exist as physically observable states and quarks would in consequence be permanently confined. At any finite energy we would only see the 'average' quark changes and phenomenonologically we could not distinguish this from the Gell-Mann model where the quarks form three identical triplets.'[10] In this picture, however, the observed fractional charges are not even 'averages', but exact values, because they reflect an effectively infinite rate of 'rotation' between the coloured states or phases. They are QED or electroweak eigenstates.

As it happens, an almost exactly parallel phenomenon has been observed in condensed matter, in the fractional quantum Hall effect. Here, ensembles of particles with only exact units of e acquire the characteristics of perfect (odd) fractions of this unit, by becoming associated with the odd number of magnetic flux lines needed to create an overall boson state. Thus, if an electron becomes attached to 3 flux lines, its charge is divided between them in units of e / 3. Laughlin explained this, with reference to the work of Anderson, as 'a low-energy collective effect of huge numbers of particles that cannot be deduced from the microscopic equations of motion in a rigorous way and that disappears completely when the system is taken apart'.[11,12] In his 1998 Nobel Lecture, he even suggested the connection with particle physics: 'The fractional quantum Hall effect is fascinating for a long list of reasons, but it is important in my view primarily for one: It establishes experimentally that both particles carrying an exact fraction of the electron charge e and powerful gauge forces between these particles, two central postulates of the standard model of elementary particles, can arise spontaneously as emergent phenomena. Other important aspects of the standard model, such as free fermions, relativity, renormalizability, spontaneous symmetry breaking, and the Higgs mechanism, already have apt solid-state analogues and in some cases were even modeled after them (Peskin, 1995), but fractional quantum numbers and gauge fields were thought to be fundamental, meaning that one had to postulate them. This is evidently not true.'[13,14]

Of course, in the present quark model, there is no energy regime at which integral charges will appear, and so the 'emergence' of the fractional charges is not in any way energy-dependent; but there are fundamental problems with the assumption that the charges of the Gell-Mann-Zweig formalism are intrinsic, rather than 'emergent', in principle. Thus, either charge is not properly quantized, or quarks and leptons are not truly fundamental, because their charges do not come in fundamental units. We can, of course, choose to redefine the fundamental unit of charge as $e / 3$, rather than e, but this would make a particle with two units of charge, like u, less fundamental in some respect than one with a single unit, such as d; and electrons, which give no evidence of being composite, would need three units of this new fundamental charge. Again, lepton-like quarks would seem to be required by grand unified theories which propose a single overall unification scheme for quarks and leptons, with possible quark-lepton transitions. Different fundamental units of charge, however, would certainly make this difficult to accomplish.

The Han-Nambu proposal has often been seen as an 'alternative' to the Gell-Mann-Zweig fractional theory, which, although never experimentally refuted, has gradually fallen into disuse because of its less direct relationship to QED phenomenology, but we should really see the two models as being different representations of a more fundamental underlying theory. They are, thus, not alternative theories, but different parts of the same one. The Gell-Mann model is really a representation of the strong interaction between quarks, which forces their electric charges to be phenomenologically, and even actually, fractional. The Han-Nambu model seems to suggest the underlying group structure which enables us to propose a scheme of grand unification, and makes sense of the fundamental nature of electric charge. In fact, as we will show, the group theoretical aspects of this representation allow us to propose an exact grand unification of the weak, strong and electric charges at the Planck mass, $(\hbar c / G)^{1/2}$, the energy at which quantum gravity becomes significant. In addition, they resolve an anomaly in the application of the Higgs mechanism, which cannot be resolved in a group representation based on fractional charges.

Chapter 14

A Representation of Strong and Weak Interactions

Sections **14.2** to **14.8** mostly represent collaborative work done with **JOHN CULLERNE.**[1] The purpose of this chapter is to provide an alternative formulation of some of the material in the previous two chapters. In chapters 10 and 11, we showed how the broken symmetry of the Standard Model emerges purely from the structure of the Dirac nilpotent, and how this necessarily also determines the structures of fermions and bosons. Through the angular momentum operator, and the connection between conservation of charge and conservation of angular momentum, we derived algebraic formulae, whose results could also be tabulated in four (or five) different representations. Although the algebra alone shows all the necessary connections, it is possible to use the tables in a more direct way (by representing them as matrices) to construct baryons and mesons within the Standard Model $SU(3)$, and to show how they also specify all aspects of the $SU(2)_L \times U(1)$ electroweak unification. It is additionally possible to represent the properties of the tables directly using nilpotent structures, and to use nilpotent structures to illustrate $SU(2)$ transitions and the Higgs mechanism.

14.1 Charge Occupancy

The four components of the nilpotent Dirac spinor have been identified as the fermion and its three discrete vacuum 'reflections' under transitions which would have the characteristics of the weak, strong and electric interactions. The spinor can thus be considered as containing the full potentiality of what any fermionic state could be transformed into, and the weak, strong and electric interactions as the means of making this transfer. It is significant that the gravitational or inertial interaction is 'passive' in this respect, the vacuum reflection (expressible as 1ψ or scalar $\times$ ψ) leading to the state itself. We can consider a fermion, with creation operator specified by the first term in the spinor, as having the potentiality to be switched by the appropriate interaction into any of the vacuum reflections that it carries with it, and that are specified by the quaternion operators labelled $\boldsymbol{k}$, $\boldsymbol{i}$, $\boldsymbol{j}$, and that might be specified by the respective weak, strong and electric *charges*. In this respect, the action of the weak vacuum force becomes a change of fermion to antifermion, with a corresponding change in helicity; the action of the strong vacuum force becomes a change in helicity, without a change of fermionic status; while the action of the electric vacuum force becomes a change from fermion to antifermion, without a corresponding change in helicity.

There is, however, a significant distinction between the two types of fundamental fermion – quarks and leptons – in that only quarks incorporate the explicit vector behaviour of the momentum operator in their spinor state vectors. We can account for this distinction in terms of vector *phase*. So a baryon state vector might have a form such as

$$\begin{array}{c} \textit{inertial} \\ \textit{strong} \\ \textit{weak} \\ \textit{electric} \end{array} \begin{pmatrix} i\boldsymbol{k}E \pm i\boldsymbol{\sigma}.\mathbf{p}_1 + \boldsymbol{j}m \\ i\boldsymbol{k}E \mp i\boldsymbol{\sigma}.\mathbf{p}_1 + \boldsymbol{j}m \\ -i\boldsymbol{k}E \pm i\boldsymbol{\sigma}.\mathbf{p}_1 + \boldsymbol{j}m \\ -i\boldsymbol{k}E \mp i\boldsymbol{\sigma}.\mathbf{p}_3 + \boldsymbol{j}m \end{pmatrix} \begin{pmatrix} i\boldsymbol{k}E \pm i\boldsymbol{\sigma}.\mathbf{p}_2 + \boldsymbol{j}m \\ i\boldsymbol{k}E \mp i\boldsymbol{\sigma}.\mathbf{p}_2 + \boldsymbol{j}m \\ -i\boldsymbol{k}E \pm i\boldsymbol{\sigma}.\mathbf{p}_3 + \boldsymbol{j}m \\ -i\boldsymbol{k}E \mp i\boldsymbol{\sigma}.\mathbf{p}_2 + \boldsymbol{j}m \end{pmatrix} \begin{pmatrix} i\boldsymbol{k}E \pm i\boldsymbol{\sigma}.\mathbf{p}_3 + \boldsymbol{j}m \\ i\boldsymbol{k}E \mp i\boldsymbol{\sigma}.\mathbf{p}_3 + \boldsymbol{j}m \\ -i\boldsymbol{k}E \pm i\boldsymbol{\sigma}.\mathbf{p}_2 + \boldsymbol{j}m \\ -i\boldsymbol{k}E \mp i\boldsymbol{\sigma}.\mathbf{p}_1 + \boldsymbol{j}m \end{pmatrix}$$

or

$$\begin{array}{c} \textit{inertial} \\ \textit{strong} \\ \textit{weak} \\ \textit{electric} \end{array} \begin{pmatrix} i\boldsymbol{k}E \pm i\boldsymbol{\sigma}.\mathbf{p}_1 + \boldsymbol{j}m \\ i\boldsymbol{k}E \mp i\boldsymbol{\sigma}.\mathbf{p}_1 + \boldsymbol{j}m \\ -i\boldsymbol{k}E \pm i\boldsymbol{\sigma}.\mathbf{p}_1 + \boldsymbol{j}m \\ -i\boldsymbol{k}E \mp i\boldsymbol{\sigma}.\mathbf{p}_3 + \boldsymbol{j}m \end{pmatrix} \begin{pmatrix} i\boldsymbol{k}E \mp i\boldsymbol{\sigma}.\mathbf{p}_2 + \boldsymbol{j}m \\ i\boldsymbol{k}E \pm i\boldsymbol{\sigma}.\mathbf{p}_2 + \boldsymbol{j}m \\ -i\boldsymbol{k}E \mp i\boldsymbol{\sigma}.\mathbf{p}_3 + \boldsymbol{j}m \\ -i\boldsymbol{k}E \pm i\boldsymbol{\sigma}.\mathbf{p}_2 + \boldsymbol{j}m \end{pmatrix} \begin{pmatrix} i\boldsymbol{k}E \pm i\boldsymbol{\sigma}.\mathbf{p}_3 + \boldsymbol{j}m \\ i\boldsymbol{k}E \mp i\boldsymbol{\sigma}.\mathbf{p}_3 + \boldsymbol{j}m \\ -i\boldsymbol{k}E \pm i\boldsymbol{\sigma}.\mathbf{p}_2 + \boldsymbol{j}m \\ -i\boldsymbol{k}E \mp i\boldsymbol{\sigma}.\mathbf{p}_1 + \boldsymbol{j}m \end{pmatrix}$$

However, for any given direction of $\boldsymbol{\sigma}$ (in principle, defining the inertial phase, which is unique to each fermion) only certain $\mathbf{p}$ phases will be 'active'. These representations are similar to those for charges in the particle tables of chapter 13, charge 'occupancy' (1 / 0) being determined by the phases. The strong charge goes through all possible phases, while the weak and electric remain relatively (although not absolutely) fixed on single phases. In a baryon, the weak and electric phases must be on different quarks. The inertial element again goes through all possible phases in fixing the direction of spin $\boldsymbol{\sigma}$. Mesons have the same structure as baryons, except that they are single fermions combined with the corresponding antifermions and the three phases ('colours') should be considered in a purely temporal (i.e. non-spatial) sequence. The weak and electric charges 'switch on / off' as the phase changes through the components 1, 2, 3.

For free fermions (leptons), the phases are purely the inertial phases. Only the direction of vector properties of **p**, of course, define a strong phase – the magnitude is determined by the combination of *E* and *m*. For free fermions, there is no strong charge because no information is carried about direction, and there is no *SU*(3) symmetry. Leptons have weak and electric occupancy on the same phase, with a temporal cycle, 1-2-3, as the structure rotates through the three directions involved in **p**.

$$\begin{matrix} \textit{inertial} \\ \textit{strong} \\ \textit{weak} \\ \textit{electric} \end{matrix} \quad \begin{pmatrix} ikE \pm i\boldsymbol{\sigma}.\mathbf{p}_1 + jm \\ ikE \mp i\boldsymbol{\sigma}.\mathbf{p}_1 + jm \\ -ikE \pm i\boldsymbol{\sigma}.\mathbf{p}_1 + jm \\ -ikE \mp i\boldsymbol{\sigma}.\mathbf{p}_1 + jm \end{pmatrix}$$

We can consider *iE*, $\boldsymbol{\sigma}.\mathbf{p}$ and *m* as the respective coefficients for the weak, strong and electric vacuum terms. There are two pseudoscalar terms $\pm iE$; six vector terms $\pm \boldsymbol{\sigma}.\mathbf{p}_1$, $\pm \boldsymbol{\sigma}.\mathbf{p}_2$, $\pm \boldsymbol{\sigma}.\mathbf{p}_3$; and one scalar term *m*. The weak component switches in such a way as to make *iE* into $-iE$. The strong switches in such a way as to make $\boldsymbol{\sigma}.\mathbf{p}_1$ convert to $-\boldsymbol{\sigma}.\mathbf{p}_1$, and also to $\boldsymbol{\sigma}.\mathbf{p}_2$; $-\boldsymbol{\sigma}.\mathbf{p}_2$; $\boldsymbol{\sigma}.\mathbf{p}_3$; and $-\boldsymbol{\sigma}.\mathbf{p}_3$. The weak transition involves dipolarity. The strong transition requires a constant rate of change of **p**, which is equivalent to a linear potential. The electric component preserves *m*. The respective group structures are *SU*(2), *SU*(3) and *U*(1).

Only *iE* and $\boldsymbol{\sigma}.\mathbf{p}$ vary, and only $\boldsymbol{\sigma}.\mathbf{p}$ varies in magnitude with phase. The *E* terms are, therefore, always global. There are two ways of setting up these transitions, both using covariant derivatives for the operators *iE* and $\boldsymbol{\sigma}.\mathbf{p}$. We can either set up a combination of (pseudo)scalar and vector group generators; or, using the idea that these groups represent the spherical symmetry of a point source, and are covariant, we can replace the scalar and vector parts by scalar potential functions of *r*, associated with *E*. In the first case, the scalar parts are scalar phase (Coulomb) terms; the vector parts are the generators that make the individual interactions have the *SU*(3), *SU*(2) or *U*(1) symmetries associated with the **p**, *E* and *m* operators in the nilpotent, or with the direction, handedness or radial magnitude of the angular momentum. In the second case, we can group all terms related to a single particle under a single representation of *E* as a scalar function of *r*, which applies globally to the entire state. The two *SU*(2) states – filled electric and empty electric background – being global, are automatically set with respect to *E*. That is, the background is incorporated as the potential producing a scalar or $U(1)$ phase in the E term. In the case of spherical symmetry, this becomes a Coulomb potential. It is possible to combine all the information into a single expression by using the fact that Lorentz invariance, in the case of a purely point source with spherical symmetry, allows us to transfer all the information contained in $\boldsymbol{\sigma}.\mathbf{p}$ to the E term by adding a potential function of r which reproduces the specific aspect of spherical symmetry ($SU(3)$, $SU(2)$ or $U(1)$) incorporated in the covariant part of $\boldsymbol{\sigma}.\mathbf{p}$, that is, the part responsible for the interaction. When the frame is chosen such that all this information is transferred to the E term, then all specific phase information is lost. The rotation of vector $\mathbf{p}$ terms ensures that the strong term is a linear function ($\propto r$). The scalar nature of m ensures that the electric term is a scalar phase (Coulombic) ($\propto 1 / r$). The dipolarity of $\pm iE$ ensures that the weak term is a dipolar equivalent of the scalar phase ($\propto 1 / r^3$). These options are also evident as a result of directly applying the condition of spherical symmetry to the fermionic state.

The procedures we have outlined lead to a specification of precise charge structures for all known fermions and bosons, but the Standard Model, as we have stated, also has a mechanism for the generation of fermion and boson masses via the assumed universal Higgs field. In the simplest terms, this is a two-stage process. In the first stage, a degeneracy in the (weak) vacuum requires the creation of massless spin 0 scalars (Goldstone bosons). In the second stage, the imposition of local gauge invariance, acting under the symmetry of a Lie group such as $SU(3)$, $SU(2)$ or $U(1)$, converts the massless scalars into the gauge fields required for the group, together with a massive spin 0 scalar (Higgs boson or Higgs field) which couples to otherwise massless fermions and bosons to give them finite nonzero masses. As we have seen, the nilpotent formalism requires a *free* fermion state (as its own supersymmetric partner) to remain unchanged as a result of self-interaction via the vacuum, and so require no charge or energy renormalization, but it is important to decide what happens if we apply it to *vacuum*, and, in particular, to a vacuum state where we initially suppose that mass does not exist.

For nilpotency, of course, we do not require nonzero mass, only energy and momentum. So we can imagine idealized real fermions generated from an ideal vacuum with structure $\pm \boldsymbol{i}kE \pm \boldsymbol{i}\mathbf{p}$. An ideal vacuum would, of course, have the most complete symmetry possible, with exact and absolute C, P and T symmetries. However, if the vacuum were in any way *degenerate* in some way under charge conjugation (as supposed in the weak interaction), the $\pm \boldsymbol{i}kE \pm \boldsymbol{i}\mathbf{p}$ would be transformable into something with the same E and $\mathbf{p}$, but with properties differing in some way, then one or more of these symmetries would not be exact, and there would have to be a vertex at which the new state could be created out of its degenerate partner. The principal symmetry to be broken would be the one concerned with the whole state, i.e. charge conjugation (C), but, to maintain CPT, either parity ($P = CT$) or time-reversal symmetry ($T = CP$) would also have to be broken, depending on whether the $\boldsymbol{i}kE$ or $\boldsymbol{i}\mathbf{p}$ term was affected.

Now, breaking charge conjugation would mean that $\pm\, \boldsymbol{i}kE \pm \mathbf{i}\mathbf{p}$ (representing creation of the new state) and $\mp\, \boldsymbol{i}kE \mp \mathbf{i}\mathbf{p}$ (representing annihilation of the old state) must become distinguishable, leading to the Goldstone theorem, and requiring the creation of a spin 0 scalar as the vertex of an identifiable interaction: $(\pm\, \boldsymbol{i}kE \pm \boldsymbol{i}\mathbf{p})(\mp\, \boldsymbol{i}kE \mp \mathbf{i}\mathbf{p})$. In the nilpotent formalism, however, spin 0 scalars are necessarily massive, and the process of violating charge conjugation symmetry must, therefore, necessarily create mass in the vacuum state itself. So, for the vacuum to be able to *distinguish* $\pm\, \boldsymbol{i}kE \pm \boldsymbol{i}\mathbf{p}$ and $\mp\, \boldsymbol{i}kE \mp \mathbf{i}\mathbf{p}$, even when the fermions and antifermions are assumed massless, means that there must exist at least one state with the structure $(\pm\, \boldsymbol{i}kE \pm \mathbf{i}\mathbf{p} + \boldsymbol{j}m)(\mp\, \boldsymbol{i}kE \mp \mathbf{i}\mathbf{p} + \boldsymbol{j}m)$, with m necessarily unipolar. Of course, since the spin 0 scalar must be its own supersymmetric partner, there can be no mechanism for an observable self-interaction, and so no renormalization (or hierarchy) problem concerning the mass.

14.2 Symmetries in a Matrix Representation

In chapter 13, we established the charge structures for the spectrum of fermions and bosons on the basis of the gauge symmetries derived in chapter 10. It is also possible, however, to derive the gauge symmetries from the charge structures, and, although it is not absolutely necessary to demonstrate the reciprocal nature of the procedure, it is certainly instructive, as it contributes significantly to a physical interpretation of the fundamental concept of charge. From section 13.5, we learn that 27 orientations of charge structure exist across 5 representations (A, B, C, D and E). These orientations are equivalent and derive from arbitrariness in the labelling of individual charge components or angular momentum phases. A study of the transformations between orientations will reveal the symmetry that spans this 27-dimensional space. To facilitate our discussion, we consider the following matrix for a d quark:

	$\bar{A}$	$\bar{B}$	$\bar{C}$	$\bar{D}$	$\bar{E}$
A	*	*	*		
B	*	*	*		
C	*	*	*		
D					
E					

This matrix represents all the transformations within and between the representations *A*, *B*, *C*, *D* and *E*. The region highlighted with stars represents just the transformations within and between *A*, *B* and *C*. An inspection of charge structure matrices from baryon or meson states within these three representations reveals symmetry involving the transformation of strong charge orientation:

A representation
$$\begin{matrix} 0 & 0 & -\mathbf{j} \\ \mathbf{i} & 0 & 0 \\ \mathbf{k} & 0 & 0 \end{matrix}$$

B representation
$$\begin{matrix} 0 & 0 & -\mathbf{j} \\ 0 & 0 & \mathbf{i} \\ \mathbf{k} & 0 & 0 \end{matrix}$$

C representation
$$\begin{matrix} 0 & 0 & -\mathbf{j} \\ 0 & \mathbf{i} & 0 \\ \mathbf{k} & 0 & 0 \end{matrix} \qquad (14.1)$$

Here, we observe the relative positions of the three components of charge within the three representations. Transformations between *A*, *B* and *C* may be linearly combined into the following set of symmetry generators:

$$|1\rangle = (B\bar{A} + A\bar{B})/\sqrt{2}\,, |2\rangle = -i(B\bar{A} - A\bar{B})/\sqrt{2}\,, |3\rangle = (B\bar{B} - A\bar{A})/\sqrt{2}\,,$$
$$|4\rangle = (B\bar{C} + C\bar{B})/\sqrt{2}\,, |5\rangle = -i(B\bar{C} - C\bar{B})/\sqrt{2}\,, |6\rangle = (A\bar{C} + C\bar{A})/\sqrt{2}\,,$$
$$|7\rangle = -i(A\bar{C} - C\bar{A})/\sqrt{2}\,,, |8\rangle = (A\bar{A} + B\bar{B} - 2C\bar{C})/\sqrt{6}\,, |9\rangle = (A\bar{A} + B\bar{B} + C\bar{C})/\sqrt{3} \qquad (14.2)$$

The representation with the line over it should be read as the one leaving the interaction vertex. If we now introduce the colour labels, *r*, *b* and *g*, to the arrangements in (14.1), the system of operators in (14.2) becomes:

$$|1\rangle = (r\bar{b} + b\bar{r})/\sqrt{2}\,, |2\rangle = -i(r\bar{b} - b\bar{r})/\sqrt{2}\,, |3\rangle = (r\bar{r} - b\bar{b})/\sqrt{2}\,,$$
$$|4\rangle = (r\bar{g} + g\bar{r})/\sqrt{2}\,, |5\rangle = -i(r\bar{g} - g\bar{r})/\sqrt{2}\,, |6\rangle = (b\bar{g} + g\bar{b})/\sqrt{2}\,,$$
$$|7\rangle = -i(b\bar{g} - g\bar{b})/\sqrt{2}\,, |8\rangle = (r\bar{r} + b\bar{b} - 2g\bar{g})/\sqrt{6}\,, |9\rangle = (r\bar{r} + b\bar{b} + g\bar{g})/\sqrt{3}\,.$$

In terms of $SU(3)$ symmetry (on which QCD is based), these nine states constitute a colour octet (transformations 1 to 8) and a colour singlet (transformation 9).

14.3 Constructing a Baryon

We can, in fact, show (using a different set of matrices) that the procedure by which the *s* charge is 'transferred' between tables *A*-*C*, which is clearly analogous to the transfer of the $A^{\alpha n}$ term in the covariant derivative or the **p** term in the nilpotent state vector, has an exact $SU(3)$ symmetry. To do this, we outline the procedure for constructing a neutron (*udd*) in the *A* representation. Within the *A* representation the table for *u* is:

u	**B**	**G**	**R**
$+e$	$1j$	$1j$	$0k$
$+s$	$1i$	$0k$	$0j$
$+w$	$1k$	$0i$	$0i$

If **B**, **G** and **R** correspond to **i**, **j** and **k** respectively in $(-j\hat{\mathbf{p}}_b + i\hat{\mathbf{p}}_a + k\hat{\mathbf{p}}_c)$, then this table is the result if $\hat{\mathbf{p}}_b = \mathbf{k}$, $\hat{\mathbf{p}}_a = \mathbf{i}$, $\hat{\mathbf{p}}_c = \mathbf{i}$, with positive electric charge taking unit values in **B** and **G** and zero in R, and negative electric charge taking unit value in **R** and zero in **B** and **G**. The table for *d* is therefore:

d	**B**	**G**	**R**
$-e$	$1i$	$1k$	$0j$
$+s$	$1i$	$0j$	$0k$
$+w$	$1k$	$0i$	$0i$

To construct a neutron, we first need to combine three different columns from the tables corresponding to the three quarks *u*, *d* and *d*. There are three ways of constructing the total charge for each representation. For the *A* representation the neutron may be constructed by either,

$(\boldsymbol{i} + \boldsymbol{j} + \boldsymbol{k})$ in a colour, $-\boldsymbol{j}$ in another colour, 0 in a third

or, $(\boldsymbol{i} + \boldsymbol{k})$ in a colour, $-\boldsymbol{j}$ in another colour, $\boldsymbol{j}$ in a third

or, $(\boldsymbol{i} + \boldsymbol{k})$ in a colour, 0 in another colour, 0 in a third.

From this we can construct a matrix representation of the neutron, where the columns and rows become the degrees of freedom for the combination of three unit charges; that is,

$$\begin{pmatrix} i+j+k & -j & 0 \\ -j & j & i+k \\ 0 & i+k & 0 \end{pmatrix} \qquad (14.3)$$

Along every column and every row there is a valid charge structure in a given colour for the neutron. The columns and rows are therefore colour labels.

The matrix in (14.3) may be called the charge structure matrix for the neutron in the *A* representation. However, the arbitrariness of the colour labels means that there are six ways of writing this matrix which correspond to the six permutations of the three colours B, G and R. QCD requires that the combination of three unit charges must therefore admit all these combinations with a colour state given by the antisymmetric colour singlet of *SU*(3):

$$\psi \sim (BGR - BRG + GRB - GBR + RBG - RGB). \qquad (14.4)$$

The action of an arbitrary *SU*(3) operation:

$$O(a_1, a_2, \ldots, a_8) = exp\left(i \sum_{k=1}^{8} a_k \lambda_k \right), \qquad (14.5)$$

where a_k are continuous parameters and λ_k are the generators of *SU*(3), would leave ψ invariant. The equivalent transformations of the colour labels in the matrix (14.3), are performed by simply applying the six orthogonal permutations of S_3:

$$\begin{pmatrix} 1 & 0 & 0 \\ 0 & 1 & 0 \\ 0 & 0 & 1 \end{pmatrix}, \begin{pmatrix} 0 & 0 & 1 \\ 1 & 0 & 0 \\ 0 & 1 & 0 \end{pmatrix}, \begin{pmatrix} 0 & 1 & 0 \\ 0 & 0 & 0 \\ 1 & 0 & 1 \end{pmatrix}, \begin{pmatrix} 0 & 1 & 0 \\ 1 & 0 & 0 \\ 0 & 0 & 1 \end{pmatrix}, \begin{pmatrix} 0 & 0 & 1 \\ 0 & 1 & 0 \\ 1 & 0 & 0 \end{pmatrix}, \begin{pmatrix} 1 & 0 & 0 \\ 0 & 0 & 1 \\ 0 & 1 & 0 \end{pmatrix}. \qquad (14.6)$$

The first three are even permutations, with determinant = +1, and the second three are odd permutations, with determinant = −1. Transforming (14.3) by the orthogonal transformations above, leads to the six other ways of writing this matrix which correspond to the six linearly independent terms of ψ.

The symmetry that admits these possibilities into the *A* representation of the neutron, is only a special case of the full symmetry arising through the inclusion of all representations, *A*, *B*, *C*, *D* and *E*. This symmetry would allow the application of the permutations in (14.6) to any of the individual charge components of (14.3). Each charge type in a baryon has three orientations in colour space. Since there are three charge types, the total number of charge structures for a baryon is $3 \times 3 \times 3 = 27$. We could have come to this conclusion by realising that within each of the representations *A*, *B*, *C* and *D*, there are 6 orientations of charge structure. The *E* representation is the odd one out. It does not obey the rules for charge allocation and therefore quaternion-charge assignments collapse to the trivial:

$$\pm \mathbf{i}s \pm \mathbf{j}e \pm \mathbf{k}w. \tag{14.7}$$

The *E* representation cannot have distinct orientations corresponding to the six colour combinations in (14.4). Only three orientations are possible corresponding to the three colours that (14.7) can occur in. The 24 matrices from *A* to *D* and the 3 matrices from *E*, then form the total of 27 charge structures.

The overall *SU*(5)-type or similar structure represented by the tables and the charge-allocation algebra has a clearly defined *SU*(3) component. The *s* charge is the only one which takes the same value in all baryons or combinations of three unit charges and it is the only one which only ever occurs in one unit of charge at any given time. The 'colour' invariance requires that the single *s* component in the combination cannot be specified as belonging to one of the unit charges and a mechanism must exist for a continual exchange of the *s* component between the three quarks in a baryon. Gauge invariance further requires that even the notion of an instantaneous location for *s* must be impossible. According to the tables, transitions between systems *A*, *B* and *C* are equivalent to a transfer of the unit *s* component between the **B**, **R** and **G** quarks, without any change in the values of the *e* and *w* components, and the *s* component is the only one that can be transferred in this way. Consequently, it is the effective exchange of a single strong charge between the three bound quarks which prevents the identification of any one quark by its colour.

We have demonstrated that an exact *SU*(3) symmetry, like the one which has been used to accommodate the strong interaction into quantum field theory, is a precise equivalent of the continuous gauge-invariant transformation between three colour states required by the simultaneous and equally probable existence of the three quark systems *A*, *B* and *C*, and the definition of a quark actually requires it to be necessarily in a strong-interacting state. However, transitions from systems *A*, *B*, *C* to system *L* (or *D* / *E*) cannot be involved in the pure strong interaction because they involve the additional exchange of a weak (or electric) component of charge. The weak interaction (or combined electroweak interaction) may thus be found to occur because the *L* representation of quarks, cannot be obtained in general from any of the others without an exchange in weak or electric, as well as strong, charges between the units. A *pure* weak or electroweak interaction, on the other hand, cannot take place where strong charges are present.

14.4 Constructing a Meson

The $SU(3)$ gauge symmetry can also be shown to apply to the meson states derived from the quark tables. In these structures, we avoid specific quaternion charge assignments by always combining unit charges with unit anticharges in such a way as to make assignments like $1\boldsymbol{je}$, $0\boldsymbol{ie}$ or $0\boldsymbol{ke}$ indistinguishable. We can demonstrate the process by examining the charge structure of the K^0 meson (d quark with $\bar{s}$ antiquark) in the A representation. The charge allocation table for the $\bar{s}$ antiquark is as follows:

$\bar{s}$	$\bar{\mathbf{B}}$	$\bar{\mathbf{G}}$	$\bar{\mathbf{R}}$
$+e$	$0\boldsymbol{i}$	$0\boldsymbol{k}$	$1\boldsymbol{j}$
$-s$	$1\boldsymbol{i}$	$0\boldsymbol{j}$	$0\boldsymbol{k}$
$+w$	$z_P\boldsymbol{k}$	$0\boldsymbol{i}$	$0\boldsymbol{i}$

The K^0 charge structure is constructed by taking a column from the d quark table and combining it with the corresponding column from the $\bar{s}$ antiquark. The resulting matrix has the structure:

$$\begin{pmatrix} 0\boldsymbol{i} + (1 + z_P)\boldsymbol{k} & 0 & 0\boldsymbol{j} \\ 0 & 0\boldsymbol{j} & 0\boldsymbol{i} + (1 + z_P)\boldsymbol{k} \\ 0\boldsymbol{j} & 0\boldsymbol{i} + (1 + z_P)\boldsymbol{k} & 0 \end{pmatrix}$$

which displays the equivalence of unit and zero assignments by possessing six orientations corresponding to the six colour combinations in (14.4). Although mesons and baryons have very different structures, the charge accommodation is achieved in exactly the same way.

The z factor is clearly significant in the case of the K^0 particle. If z were not always -1 for the strange-charm generation, due to parity violation, we would have $+2w$ and 0 as the charge structures for the K^0. This would make it possible to identify the particular colours belonging to particular quarks in a particular combination, in violation of the basic principle of the coloured quark system.

14.5 Lepton Structures

Lepton interactions are subject to an $SU(2)_L \times U(1)$ electroweak gauge symmetry, but not to the $SU(3)$ of the strong interaction. So we should expect the charge structures of leptons to be a specific indication of the nature of the electroweak interaction. Now, the weak interaction provides a mechanism for the transition between the *A-C* and *L* representation. Since leptons are suspected to be the products of weak interactions between these representations, one would expect their structures to be evident within *L*. So we obtain:

Particle	Helicity	Charge Structure
e^-	left-handed	$-\boldsymbol{j}\, e + \boldsymbol{k}\, w$
e^-	right-handed	$-\boldsymbol{j}\, e + 0\, \boldsymbol{k}\, w$
ν_e	left-handed	$\boldsymbol{k}\, w$
μ^-	left-handed	$-\boldsymbol{j}\, e - z_P\, \boldsymbol{k}\, w$
μ^-	right-handed	$-\boldsymbol{j}\, e - 0\, z\, \boldsymbol{k}\, w$
ν_μ	left-handed	$-z_P\, \boldsymbol{k}\, w$
τ^-	left-handed	$-\boldsymbol{j}\, e - z_T\, \boldsymbol{k}\, w$
τ^-	right-handed	$-\boldsymbol{j}\, e - 0\, z_T\, \boldsymbol{k}\, w$
ν_τ	left-handed	$-z_T\, \boldsymbol{k}\, w$

The leptons are unit charges with no *s* component. They behave in all respects like unbound quarks, and we may assume that it is the absence of the strong charge which makes them unbound. Two charge types are now to be accommodated by allocating them to three quaternions. The leptons, however, are not completely free because weak interactions always require them to exist in a weak version of a bound state; that is leptons are always produced as part of a lepton-antilepton pair (or weak dipole), like the quark-antiquark pairing of the meson. Though there is no strong binding involved as there is with quark-antiquark combinations, because of the complete removal of the *s* colour invariant system appropriate for three-component charges, the *w* and *e* charge components of leptons are still associated with quaternion operators or phases of the momentum operator which must remain unspecified.

We can see that the representations for the three generations of quarks collapse to three generations of leptons when there is no strong charge to accommodate. These generations still correspond to the three ways of accommodating the weak charge w; that is, $+w$ is indistinguishable from $-zw$, where z takes on the value of -1 if either P is violated and T is not, or T is violated and P is not; and, in analysing the electroweak interaction, we can use table and matrix representation for leptons, which are identical to the tables and matrices for quarks with the omission of strong charge.

14.6 The Electroweak Interaction Mechanism

The *A-C* and *D-E* (= *L*) representations can be said to occupy two 'subspaces' which differ by the relative positions of the weak and electric charges.

$$\begin{array}{llll} A\text{-}C & 0 & 0 & -\boldsymbol{j} \\ & * & * & * \\ & \mathbf{k} & 0 & 0 \\ & & & \\ D\text{-}E & 0 & 0 & -\boldsymbol{j} \\ & * & * & * \\ & 0 & 0 & \mathbf{k} \end{array} \qquad (14.8)$$

That is, either electric or weak charges are on different colours (*A*, *B* and *C*) or they are on the same colour (*D* and *E*).

If we now apply the condition that the *E* representation is no longer valid, we find that the only valid transitions are those within the *SU*(3) subspace (*A*, *B* and *C*) or those that transform between that subspace and the *D* (= *L*) representation. If the *SU*(3) subspace is represented by *S*, then

$$\begin{array}{ccc} & \overline{S} & \overline{D} \\ S & * & * \\ D & * & * \end{array} \qquad (14.9)$$

is a matrix that represents all the transformations between the subspace *S* and the *D* representation. The symmetry transformations in (14.9) between *S* and *D* correspond to an interaction involving only the electric ($\boldsymbol{j}$) and weak ($\mathbf{k}$) charges. Having factored out the *SU*(3) in the strong charge component, we see that there are still two degrees of freedom which correspond to the two charge orientations in (14.8). But, however we look at it, the two possible states may be taken up by one of the charges while the other charge can only take up one of the states, as in the diagram below.

$$\begin{array}{ccccc} \alpha & \beta & & \alpha & \beta \\ -j & 0 & \leftrightarrow & 0 & -j \\ 0 & \boldsymbol{k} & & 0 & \boldsymbol{k} \end{array} \tag{14.10}$$

In (14.10) we see that the symmetry transformation between the two configurations may be written as though the $\boldsymbol{k}$ stays in the same phase (β) at all times while $-j$ moves between the phases ($\alpha \leftrightarrow \beta$). We could of course have done this with $-j$ staying in the same phase and $\boldsymbol{k}$ moving. But the origin of the problem appears to lie in the way in which we are forced to accommodate the weak charge ($\boldsymbol{k}$), and this seems to be responsible for the parity violation that occurs in the weak interaction. In effect, the weak charge only exists in one of the phases in (14.10); the two discrete phases are therefore states of helicity.[2]

We may write the system of generators from (14.9) as the following linear combinations:

$$\begin{aligned} |a_x\rangle &= \left(S\overline{D} + D\overline{S}\right)/\sqrt{2}\,, \\ |a_y\rangle &= -i\left(S\overline{D} - D\overline{S}\right)/\sqrt{2}\,, \\ |a_z\rangle &= \left(S\overline{S} - D\overline{D}\right)/\sqrt{2}\,, \\ |0\rangle &= \left(S\overline{S} + D\overline{D}\right)/\sqrt{2}\,. \end{aligned} \tag{14.11}$$

The construction of the symmetry transformations here has the purpose of giving it a structure reminiscent of $SU(2)$. From (14.11) we see that:

$$\begin{aligned} D\overline{S} &= \frac{1}{\sqrt{2}}\left(|\sigma_x\rangle - i|\sigma_y\rangle\right) = \begin{pmatrix} 0 & 1 \\ 0 & 0 \end{pmatrix} \\ S\overline{D} &= \frac{1}{\sqrt{2}}\left(|\sigma_x\rangle + i|\sigma_y\rangle\right) = \begin{pmatrix} 0 & 0 \\ 1 & 0 \end{pmatrix} \end{aligned} \tag{14.12}$$

The matrices in (14.12) are in the basis

$$\begin{aligned} |S\rangle &= \begin{pmatrix} 1 \\ 0 \end{pmatrix} \\ |D\rangle &= \begin{pmatrix} 0 \\ 1 \end{pmatrix} \end{aligned} \tag{14.13}$$

This basis is only a mathematical representation of the charge orientations S and D. Pictorially, (14.12) may be expressed as:

$$\begin{array}{cccccccc} & & \alpha & \beta & & \alpha & \beta \\ D\bar{S} & = & 0 & -\boldsymbol{j} & \mapsto & -\boldsymbol{j} & 0 \\ & & 0 & \boldsymbol{k} & & 0 & \boldsymbol{k} \\ \\ & & \alpha & \beta & & \alpha & \beta \\ S\bar{D} & = & -\boldsymbol{j} & 0 & \mapsto & 0 & -\boldsymbol{j} \\ & & 0 & \boldsymbol{k} & & 0 & \boldsymbol{k} \end{array} \tag{14.14}$$

Inspecting the two helicities, we see that the β helicity states undergoes the transformation $-\boldsymbol{j} + \boldsymbol{k} \leftrightarrow \boldsymbol{k}$; that is, the symmetry transformation couples these two states of charge configuration. However, the α helicity states do not seem to have a charge structure to couple to ($-\boldsymbol{j} \leftrightarrow 0$). The quaternion configurations are telling us that there is no coupling for the α helicity states between S and D; i.e., this transformation is helicity sensitive.

The same may be done for

$$\begin{array}{cccccccc} & & \alpha & \beta & & \alpha & \beta \\ S\bar{S} & = & -\boldsymbol{j} & 0 & \mapsto & -\boldsymbol{j} & 0 \\ & & 0 & \boldsymbol{k} & & 0 & \boldsymbol{k} \\ \\ & & \alpha & \beta & & \alpha & \beta \\ D\bar{D} & = & 0 & -\boldsymbol{j} & \mapsto & 0 & -\boldsymbol{j} \\ & & 0 & \boldsymbol{k} & & 0 & \boldsymbol{k} \end{array} \tag{14.15}$$

Here we see that in the first transformation a coupling in the α helicity state is allowed between $-\boldsymbol{j} \leftrightarrow -\boldsymbol{j}$. However, after that, only β helicity states have structured couplings: $\boldsymbol{k} \leftrightarrow \boldsymbol{k}$ in the first, and $-\boldsymbol{j} + \boldsymbol{k} \leftrightarrow -\boldsymbol{j} + \boldsymbol{k}$ in the second transformation.

The above results are a statement of the Standard Model with the following correspondences between this version and that of the Glashow-Weinberg-Salam electroweak theory (GWS):

(1) Transitions between D and S;

$$D\bar{S} \leftrightarrow W^{+}\left(j_{\mu}^{+}\right), \quad S\bar{D} \leftrightarrow W^{-}\left(j_{\mu}^{-}\right).$$

(2) Third component of weak isospin current coupling only left handed (β phase) charge structure;

$$\left|\sigma_{x}\right\rangle = \left(S\bar{S} - D\bar{D}\right)/\sqrt{2}\,.$$

(3) Weak hypercharge current the invariant construct coupling both left and right-handed (both α and β phases).

$$|0\rangle = \left(S\bar{S} + D\bar{D}\right)/\sqrt{2}\,.$$

It is notable that the 1 in 4 proportion which determines the value of $\sin^2\theta_W$ also occurs in the neutral electroweak transitions, the photon having one neutral isospin transition ($0 \rightarrow 0$) and three hypercharge ($-\boldsymbol{j} \rightarrow \boldsymbol{k}$, $-\boldsymbol{j} \rightarrow 0$, $0 \rightarrow \boldsymbol{k}$), while Z^0 has the same proportion in reverse (cf 10.6-10.7).

The four states in (14.9) constitute the generators of $SU(2)_L \times U(1)$, where the L subscript indicates that the $SU(2)$ involves only left-handed (β) states. Since these transformations correspond to relative positions of weak and electric charge in charge structure matrices, we cannot really make a distinction between changes in electric or weak orientations. Therefore changes in orientation due to generators (14.9) can only really be called electroweak.

Globally, then, the transformations between representations A, B, C and D / L, are generated by the symmetry group of the standard model, $SU(3) \times SU(2)_L \times U(1)$, with the $SU(3)$ corresponding to QCD and the $SU(2)_L \times U(1)$ corresponding to the GWS electroweak unification.

14.7 The Production of Leptons

We have seen that the electroweak interaction is represented in this theory as the transformations between the subspace S and the representation D (matrix (14.9)). The matrices in (14.8) reduce these transformations essentially to (14.10); that is, via the placing of the weak and electric charges in the same colour or in different colours. Although the strong charge symmetry has effectively been factored out, we still have these two (helicity) degrees of freedom.

Thus far we have considered only transformations between representations. Close inspection of the quark tables reveals that symmetry with exactly the same form as the standard $SU(2)_L$ is apparent between $u \leftrightarrow d$, $c \leftrightarrow s$, and $t \leftrightarrow b$ in the D (or L) representation. For example,

d quark in *D*

$$\begin{matrix} 0 & 0 & -\boldsymbol{j} \\ * & * & * \\ 0 & 0 & \boldsymbol{k} \end{matrix}$$

$\bar{u}$ antiquark in D

$$\begin{matrix} -j & -j & 0 \\ * & * & * \\ 0 & 0 & -k \end{matrix}$$

Of course, $\boldsymbol{k}$ is required to be unipolar in order to fit the tables. Therefore, through an interaction identical to $S \leftrightarrow D$, the charge structures of the d quark and $\bar{u}$ antiquark are interchangeable as a symmetry operation. When the strong charge is present, these transitions can only occur within a π^- meson charge combination, since this would be the only way to accommodate the components of the two unit charges. However, in the decay of the π^- the strong charge and anticharge in the present combination effectively annihilate. The need for strong charge confinement is relaxed and the two unit charges become 'free'. In the absence of the strong charge, the charge structures of the two free charges will become:

$$\begin{matrix} \alpha & \beta \\ 0 & -j \\ 0 & k \\ & d \end{matrix} \qquad \begin{matrix} \alpha & \beta \\ -j & 0 \\ 0 & k \\ & \bar{u} \end{matrix} \tag{14.16}$$

The two columns represent the helicity phases that we have discussed. The first column represents a free charge state made up of only one electric charge $-\boldsymbol{j}$. The second column represents two free charges, one of structure $-\boldsymbol{j} + \boldsymbol{k}$ and the other of structure $-\boldsymbol{k}$ alone.

When the d quark and $\bar{u}$ antiquark were in the π^- combination, their spins were paired to form a spin singlet state. When they separate as free particles, spin must remain conserved, so the resulting particles must have opposite spin states but essentially the same helicity states. So the second column in (14.16) is telling us that one possibility for the free charges is in states $-\boldsymbol{j} + \boldsymbol{k}$ and $-\boldsymbol{k}$, which in principle have the same helicity. However, the first column essentially tells us that the other spin pairing possibility must be accounted for but that there is only one charge, $-\boldsymbol{j}$, that can take this state up.

This is exactly what is observed in the π^- decay to electron and antineutrino. The pion has a spin 0, so the electron and antineutrino must emerge with opposite spins, and hence equal helicities. The antineutrino ($-\boldsymbol{k}$) is always right-handed, so the electron ($-\boldsymbol{j} + \boldsymbol{k}$) must be right-handed as well. However, the electron is not massless so must admit the other helicity state ($-\boldsymbol{j}$), whereas the antineutrino cannot take up its opposite helicity state, so we obtain $0\ \boldsymbol{k}$.

According to the GWS theory, if the electron were massless, then it would only exist as a left-handed particle. More precisely, the $1 - \gamma^5 = 1 - ij$ in the weak vertex factor would couple only to left-handed electrons, just as it couples only to left-handed neutrinos. If the electron were massless, the decay above would not occur at all. Since the electron is not massless but does have a very small mass, one would expect this decay to be heavily suppressed. This is indeed what is observed. The violation of parity is evident. The requirement that the weak transitions ($S \leftrightarrow D$ type transformations) treat $\pm \boldsymbol{k}$ as if they were the same, leads to a violation of parity in the free charges. This violation of parity is not 'visible' in the π^- combination, but becomes conspicuous as soon as the electron and antineutrino free charges escape the bound state. (In the Standard Model, there are *no* weak-interacting left-handed antineutrino states.)

The fact that the weak interaction cannot tell the difference between $\pm \boldsymbol{k}$ has other important consequences. Transitions $u \leftrightarrow s$, $c \leftrightarrow d$, etc., must be allowed to occur by exactly the same mechanism ($S \leftrightarrow D$ type transformations). The global structure of the quaternionic charge model therefore requires a weak mixing of the Kobayashi-Maskawa type, where the weak interaction may be thought of as a transformation between two subspaces of quark states (d, s and b) $\leftrightarrow$ (u, c and t). The standard model represents these transitions by offsetting the quark generations for the weak interaction:

$$\begin{pmatrix} u \\ d' \end{pmatrix}, \begin{pmatrix} c \\ s' \end{pmatrix}, \begin{pmatrix} t \\ b' \end{pmatrix}$$

where the quarks d', s' and b' are related to the physical quark states d, s and b by the Cabibbo-Kobayashi-Maskawa (CKM) matrix (cf 15.11).

The weak interaction can be thought of as a swapping of $\boldsymbol{k}$ or w from $\boldsymbol{j}$ or e 'on' to e 'off' or vice versa, creating the $SU(2)_L$, but, in fact, there is no mechanism for doing this directly, as there is in the strong interaction, because there is no combined system to do it in. What we *can* do, however, is to annihilate and create, and instead of swapping over w, we annihilate and create e, either filling the vacuum or emptying it. However, we cannot annihilate or create a charge without also annihilating or creating its antistate, and the dipolar weak interaction (unlike the strong) always involves the equivalent of particle + antiparticle = particle + antiparticle, or a double particle interaction going both ways at once. We don't know which it *really* is because the weak interaction (reflecting its origin in a pseudoscalar with indistinguishable + and –solutions) works to *prevent* such knowledge.

It is because of the filling and emptying of the vacuum via the *e* charge that rest mass is involved. W^+ and W^- involve a one-way *e* transition, Z^0 involves a two-way *e* transition, the purely electromagnetic ($U(1)$) γ no *e* transition. This, once again, gives us 0.25 for the electric / weak ratio. The same value also occurs for the weak isospin quantum number squared, with weak isospin representing the process of annihilation and creation, creating a vertical motion in the tables. Sideways motion (swapping over *w*) never happens directly, but can be considered to happen indirectly, and where the signs are those for the antiparticle, this will be taken care of by the particle and antiparticle interactions being always simultaneous, with the weak interaction unable to recognise the difference. Particle + antiparticle also allows for the 'elimination' of *s* in those transitions, like neutron beta decay, that appear to be from quark to lepton. In transitions that appear to be *A*/*B*/*C* to *D*/*E* (neutron decay) the interaction can be represented by two vertical transitions acting in opposite directions.

14.8 Electroweak Mixing

The GWS model (cf 10.6 – 10.8) asserts that the three weak isospin currents couple, with strength g_w, to a weak isospin triplet of intermediate vector bosons, *W*, whereas the weak hypercharge current couples with strength $g'/2$ to an isosinglet intermediate vector boson, *B*:

$$-i\left[g_w J_\mu . W^\mu + \frac{g'}{2} j_\mu^Y B^\mu\right]$$

Within this structure is contained all of electrodynamics and all of the weak interactions.

$$J_\mu . W^\mu = j_\mu^1 W^{\mu 1} + j_\mu^2 W^{\mu 2} + \frac{1}{\sqrt{2}} j_\mu^3 W^{\mu 3}$$

The scalar product can be written out explicitly:

$$J_\mu . W^\mu = \frac{1}{\sqrt{2}} j_\mu^+ W^{\mu +} + \frac{1}{\sqrt{2}} j_\mu^- W^{\mu -} + \frac{1}{\sqrt{2}} j_\mu^3 W^{\mu 3}$$

or, in terms of the charged currents,

$$j_\mu^\pm = j_\mu^1 \pm i j_\mu^2,$$

where
$$W_\mu^\pm = \frac{1}{\sqrt{2}}\left(W_\mu^1 \mp i W_\mu^2\right)$$

are the wavefunctions representing the $W^{\pm}$ particles.

The couplings are easily read off, from the coefficients of the charged W particles. For example, $e^- \to \nu_e + W^-$ is described in GWS as

$$j_\mu^- = \bar{\nu}_L \gamma_\mu e_L = \bar{\nu}\gamma_\mu\left[\left(1-\gamma^5\right)/2\right]e ,$$

giving a term

$$-ig_W\left(1/\sqrt{2}\right)j_\mu^- W^{\mu-} = -\frac{ig_W}{2\sqrt{2}}\left[\bar{\nu}\gamma_\mu\left(1-\gamma^5\right)e\right]W^{\mu-} .$$

The vertex factor is

$$-\frac{ig_W}{2\sqrt{2}}\left[\gamma_\mu\left(1-\gamma^5\right)\right].$$

In terms of global charge structure, the quaternion charge description exhibits exactly this symmetry transformation as

$$S\overline{D} = \frac{1}{\sqrt{2}}\left(\left|\sigma_x\right\rangle + i\left|\sigma_y\right\rangle\right)$$

in the S, D basis.

The underlying $SU(2)_L \times U(1)$ is mixed in GWS theory; that is, the two neutral *states*, W^3 and B, mix, producing one massless linear combination (the photon) and an orthogonal massive combination (the Z^0):

$$A_\mu = B_\mu \cos\theta_W + W_\mu^3 \sin\theta_W$$

$$Z_\mu = -B_\mu \sin\theta_W + W_\mu^3 \cos\theta_W .$$

The quaternion charge description necessarily exhibits this kind of mixing as is evident in (14.16). Here we have evidence of a naturally occurring GWS structure, which arises from the quaternion description, and so we can complete our account of the structures of fermions and bosons which derive from the broken symmetry introduced by the Dirac nilpotent operator.

14.9 *SU*(2) Transitions

A 'weak interaction' is a demonstration that intrinsic left-handedness is an identical phenomenon in all fermionic states, while intrinsic right-handedness is an identical phenomenon in all antifermionic states, irrespective of the composition of the fermion or antifermion. So, the intrinsic handedness is preserved irrespective of any 'transition' between one state and another. All fermionic states, therefore, seek to demonstrate the gauge invariance of one-handedness with respect to all other possible fermionic states with probabilities determined by the energy, momentum and mass terms involved. This is what is meant by a 'transition'. In any such transition, the anti-state to the state to be annihilated and the state which is to be created must exist as a spin 1 bosonic combination. Because of quark confinement, there can be no transition from free fermion to quark, or quark to free fermion – that is, there can be no pure weak transition in which a fermion acquires or loses a 'vector' character. However, fermion states with mass also carry a degree of right-handedness. A non-vector transition from left- to right-handedness, involving only fermionic states (not antifermionic), requires the vacuum which we have described as 'electric'. Only the electric vacuum carries a transition to right-handedness where the vector character is absent, and, to produce a pure transition from left- to right-handedness (and *vice versa*) without a change from fermion to antifermion requires an electroweak combination (***jk***, equivalent to ***i***):

$(\pm ikE \pm \boldsymbol{i}\,\mathbf{p} + \boldsymbol{j}\,m)$ left-handed fermion

$(\mp ikE \pm \boldsymbol{i}\,\mathbf{p} + \boldsymbol{j}\,m)$ weak transition to right-handed antifermion

$(\pm ikE \mp \boldsymbol{i}\,\mathbf{p} + \boldsymbol{j}\,m)$ electric transition to right-handed fermion

Using the concept of electric 'charge' as indicating the presence of right-handedness, we may identify four possible transitions (taking the 'left-handed' / 'right-handed' transition to mean 'the acquisition of a greater degree of right-handedness'), and hence four possible intermediate bosonic states:

Left-handed to left-handed
Left-handed to right-handed
Right-handed to left-handed
Right-handed to right-handed

The left- / right-handed transition clearly has the nature of an $SU(2)_L$ symmetry, with the requirement of three generators, which are necessarily massive, to carry the right-handedness unrecognised by the interaction, and two of which carry electric 'charge' (+ and –), in addition to one which leaves the handedness unchanged. This leaves the fourth transition state or equivalent as an extra generator with a $U(1)$ symmetry. If we assume that massive generators are necessary for a 'weak interaction', and indicate its presence, we can assign the fourth generator to the pure electric interaction. Electric charge, however, is not the sole reason for the massiveness (and hence mixed handedness) of real fermionic states; and the absence of electric charge does not indicate that a weak generator must be massless. So, the two generators without electric charge are assumed mixed, the combination producing two new generators, one of which becomes massless and so carries the pure electric, rather than the weak interaction.

To write the $SU(2)_L$ directly into the nilpotent representation, we can consider a lepton, for example, to be a superposition of states of the form $(\pm\, \boldsymbol{k}E \pm \boldsymbol{i}\, \mathbf{p} + \boldsymbol{j}\, m)$ and $(\pm\, \boldsymbol{k}E \mp \boldsymbol{i}i\, \mathbf{p} + \boldsymbol{i}j\, m)$, of which only the first acts weakly, while the neutrino is more likely to show Majorana behaviour as a superposition of $(\pm\, \boldsymbol{ik}E \pm \boldsymbol{i}\, \mathbf{p} + \boldsymbol{j}\, m)$ and $(\mp\, \boldsymbol{ik}E \pm \boldsymbol{i}\, \mathbf{p} + \boldsymbol{ij}\, m)$. Baryons might be constructed in such a way that *each strong phase* is a superposition similar to that for the lepton. Further splitting of states into superpositions might be needed to fully incorporate the full range of fermionic particles, with successive symmetry violations, within the three quark-lepton generations.

14.10 The Higgs Coupling

The existence of a nilpotent state allows us to understand the basic principle of the Higgs mechanism in a more fundamental way than in 10.11. Imagine a virtual fermionic state with no mass in vacuum, both defined by the nilpotent expression

$$(\boldsymbol{ik}E + \boldsymbol{i}\mathbf{p}).$$

An ideal vacuum would maintain exact and absolute C, P and T symmetries. Under C transformation, $(\boldsymbol{ik}E + \boldsymbol{i}\mathbf{p})$ would become

$$(-\,\boldsymbol{ik}E - \boldsymbol{i}\mathbf{p})$$

with which it would be indistinguishable under normalization. No bosonic state would be required for the transformation. For a vacuum state which is degenerate in some way, however, under charge conjugation (as may be supposed in the weak interaction), then $(\boldsymbol{ik}E + \boldsymbol{i}\mathbf{p})$ will be transformable into a state which can be distinguished from it, and the bosonic state $(\boldsymbol{ik}E + \boldsymbol{i}\mathbf{p})\,(-\boldsymbol{ik}E - \boldsymbol{i}\mathbf{p})$ will necessarily exist. However, this can only be true if the state has nonzero mass and becomes the spin 0 'Higgs boson':

$$(\boldsymbol{ik}E + \boldsymbol{i}\mathbf{p} + \boldsymbol{j}m)\,(-\boldsymbol{ik}E - \boldsymbol{i}\mathbf{p} + \boldsymbol{j}m)$$

The coupling of a massless fermion, say $(\boldsymbol{ik}E_1 + \boldsymbol{i}\mathbf{p}_1)$, to a Higgs boson, say $(\boldsymbol{ik}E + \boldsymbol{i}\mathbf{p} + \boldsymbol{j}m)\,(-\boldsymbol{ik}E - \boldsymbol{i}\mathbf{p} + \boldsymbol{j}m)$, to produce a massive fermion, say $(\boldsymbol{ik}E_2 + \boldsymbol{i}\mathbf{p}_2 + \boldsymbol{j}m_2)$, can be imagined as occurring at a vertex between the created fermion $(\boldsymbol{ik}E_2 + \boldsymbol{i}\mathbf{p}_2 + \boldsymbol{j}m_2)$ and the antistate $(-\boldsymbol{ik}E_1 - \boldsymbol{i}\mathbf{p}_1)$, to the annihilated massless fermion, with subsequent equalization of energy and momentum states. If we imagine a vertex involving a fermion superposing ($i\boldsymbol{k}E + \mathbf{i}\mathbf{p} + \boldsymbol{j}m$) and ($i\boldsymbol{k}E - \mathbf{i}\mathbf{p} + \boldsymbol{j}m$) with an antifermion superposing ($-i\boldsymbol{k}E + \mathbf{i}\mathbf{p} + \boldsymbol{j}m$) and ($-i\boldsymbol{k}E - \mathbf{i}\mathbf{p} + \boldsymbol{j}m$), then there will be a minimum of two spin 1 combinations and two spin 0 combinations, meaning that the vertex will be massive (with Higgs coupling) and carry a non-weak (i.e. electric) charge. So, a process such as

$$(i\boldsymbol{k}E + \mathbf{i}\mathbf{p} + \boldsymbol{j}m) \rightarrow \alpha_1\,(i\boldsymbol{k}E + \mathbf{i}\mathbf{p} + \boldsymbol{j}m) + \alpha_2\,(i\boldsymbol{k}E - \mathbf{i}\mathbf{p} + \boldsymbol{j}m) \tag{14.17}$$

isospin up *isospin down*

requires an additional Higgs boson vertex (spin 0) to accommodate the right-handed part of the isospin down state, when the left-handed part interacts weakly. This is, of course, what we mean when we say that the W and Z bosons have mass. The mass balance is done through separate vertices involving the Higgs boson.

In general, the acquisition of mass in the nilpotent formalism can be related to the capacity for change of sign in the **p** term with respect to that of the E term. In principle, a weak isospin transition can be seen as a change of the form ($i\boldsymbol{k}E + \mathbf{i}\mathbf{p} + \boldsymbol{j}m$) to α_1 ($i\boldsymbol{k}E + \mathbf{i}\mathbf{p} + \boldsymbol{j}m$) + α_2 ($i\boldsymbol{k}E - \mathbf{i}\mathbf{p} + \boldsymbol{j}m$). The down state introduces a degree of right-handedness which is not present in the up state, and which is not weak in origin. Where the strong interaction is not involved, a partial **p** sign transition (involving vacuum operator $\boldsymbol{i}$) can only be accomplished by involving the electric vacuum operator ($\boldsymbol{j}$) as well as the weak one ($\boldsymbol{k}$).

We note here once again that the electroweak interaction (or the weak component of it) is defined always in terms of a 2-component vertex, such as

$$(i\boldsymbol{k}E + \mathbf{i}\mathbf{p} + \boldsymbol{j}m)\,(-i\boldsymbol{k}E + \mathbf{i}\mathbf{p} + \boldsymbol{j}m)$$

Essentially, because of the pseudoscalar nature of the energy term, associated with $\boldsymbol{k}$, i.e. the mathematical indistinguishability of $+i$ and $-i$, the weak interaction is always defined as minimally dipolar, in the same way as the fermion always defines itself as a dipole with respect to vacuum (leading to half-integral spin).

It is possible, of course, that we can take the isospin up state as itself a mixed state, say, this time, of the form $\alpha_1(i\boldsymbol{k}E + \mathbf{i}\mathbf{p} + \boldsymbol{j}m) - \alpha_1(i\boldsymbol{k}E - \mathbf{i}\mathbf{p} + \boldsymbol{j}m)$. In this case, an isospin down of the form $\alpha_1(i\boldsymbol{k}E + \mathbf{i}\mathbf{p} + \boldsymbol{j}m) + \alpha_1(i\boldsymbol{k}E - \mathbf{i}\mathbf{p} + \boldsymbol{j}m)$, would show a relative decrease in **p** with regard to E, and therefore a relative increase in mass. A transition to the second or third generation, with violation of parity or time-reversal symmetry, could be symbolised by the addition of terms of the form $\beta_1(i\boldsymbol{k}E - \mathbf{i}\mathbf{p} + \boldsymbol{j}m)$ and $\beta_2(i\boldsymbol{k}E - \mathbf{i}\mathbf{p} + \boldsymbol{j}m)$, which would again involve a relative decrease in **p** and relative increase in mass.

14.11 The Mass Gap for Any Gauge Group

In our understanding, only nilpotent fermionic states (which are necessarily described on R^4, or 4-D space-time) exist, with bosonic transitions between them. In gauge theories, the bosons will be the generators of the gauge groups, such as $SU(3)$, $SU(2)$, and $SU(2) \times U(1)$. Though we have shown that the gauge group $SU(3)$, in particular, defining the strong interaction interaction between quarks, necessarily generates finite mass (i.e. creates a *mass gap* $\Delta > 0$) (see 6.6), it is possible to show the property more generally for any compact simple gauge group G. Any such group will have a finite number of generators > 1, and consequently a finite number of gauge transformations > 1 between fermionic states. If we suppose that these states are massless, we can write down the transitions between two such states in the form

$$(\pm\, \boldsymbol{i}kE_1 \pm \boldsymbol{i}\mathbf{p}_1)\,(\pm\, \boldsymbol{i}kE_2 \pm \boldsymbol{i}\mathbf{p}_2), \tag{14.18}$$

where we make no prior assumptions about the relative signs and magnitudes of the E and **p** terms. We can use the necessary scalar multiplication to equalise the magnitudes of E_1 and E_2, which will then necessarily equalise the magnitudes of $\mathbf{p}_1$ and $\mathbf{p}_2$ – a local gauge transformation would require this in any case. (14.18) will, then, immediately zero, unless there is some sign variation in *one* of the terms of the second bracket. Which we choose is arbitrary, but it is convenient to make it $\boldsymbol{i}kE_2$. So, now we can write (14.18) as

$$(\pm\, \boldsymbol{i}kE_1 \pm \boldsymbol{i}\mathbf{p}_1)\,(\mp\, \boldsymbol{i}kE_1 \pm \boldsymbol{i}\mathbf{p}_1), \tag{14.19}$$

unless there is also directional variation between $\mathbf{p}_1$ and $\mathbf{p}_2$. In any case, we establish that (14.19) has the properties of a spin 1 boson. In the case of directional variation, a case we have already considered in section 6.8, (14.19) has the properties of a gluon. (Variation in direction only would not be allowed, as it would have to include cases like $(\pm\, \boldsymbol{i}kE_1 \pm \boldsymbol{i}\mathbf{p}_x)\,(\mp\, \boldsymbol{i}kE_1 \pm \boldsymbol{i}\mathbf{p}_x)$ which cannot be nonzero, as required.)

Now, as a generator of the gauge group, (14.19) represents the idea that the state $(\pm\, \boldsymbol{i}kE_1 \mp \boldsymbol{i}\mathbf{p}_x)$ has been annihilated to become $(\mp\, \boldsymbol{i}kE_1 \pm \boldsymbol{i}\mathbf{p}_x)$ (the second bracket) simultaneously with the creation of $(\pm\, \boldsymbol{i}kE_1 \pm \boldsymbol{i}\mathbf{p}_x)$ (the first bracket). In principle, since the interaction is *local*, this means that two fermionic states must coexist with opposite helicities. This can only be true if the states have nonzero mass, thus introducing a mass gap. The only other possibility is that (14.19) has a nonzero mass, thus introducing a mass gap there. In principle, then, all regular gauge bosons of spin 1 must either have mass or introduce mass into the interacting fermions because they require simultaneous existence of opposite helicity states, while those of spin 0 must have mass to avoid zeroing due to the nilpotent structure. (In the more exotic cases of Bose-Einstein condensates and structures involving the Berry phase, where $\pm\, i\mathbf{k}E_1$ and $\pm\, i\mathbf{k}E_1$ are paired, rather than $\pm\, i\mathbf{k}E_1$ and $\mp\, i\mathbf{k}E_1$, the opposite considerations must apply.)

It is significant that all particle-particle (local) interactions, at the fundamental level, require a spin 1 gauge boson, and are parity transformations; other transformations (say T and C) can only be seen in the context of particle-vacuum (nonlocal) interactions (vacuum being the definition of the nonlocal state). If we assume, again, that the outgoing state has the negative E and $\mathbf{p}$ values of the ones it had during existence, then, if the local interaction requires transformation of E, then the nonlocal one requires the transformation of $\mathbf{p}$ and vice versa. So:-

change of spin (parity transformation)

$$(i\mathbf{k}E + \mathbf{i}\mathbf{p} + \mathbf{j}m)\,(i\mathbf{k}E - \mathbf{i}\mathbf{p} + \mathbf{j}m) \quad \text{requires} \quad (i\mathbf{k}E + \mathbf{i}\mathbf{p} + \mathbf{j}m)\,(-\,i\mathbf{k}E + \mathbf{i}\mathbf{p} + \mathbf{j}m),$$

incoming *outgoing* *spin 1*

which is a time reversal transformation in the weak vacuum state; while

change of energy state (time reversal transformation)

$$(i\mathbf{k}E + \mathbf{i}\mathbf{p} + \mathbf{j}m)\,(-i\mathbf{k}E + \mathbf{i}\mathbf{p} + \mathbf{j}m) \quad \text{requires} \quad (i\mathbf{k}E + \mathbf{i}\mathbf{p} + \mathbf{j}m)\,(i\mathbf{k}E - \mathbf{i}\mathbf{p} + \mathbf{j}m),$$

Incoming *outgoing* *Bose-Einstein condensate (spin 0)*

which is a parity transformation in the (strong) vacuum state. However,

antiparticle to particle (charge conjugation)

$$(i\mathbf{k}E + \mathbf{i}\mathbf{p} + \mathbf{j}m)\,(-i\mathbf{k}E - \mathbf{i}\mathbf{p} + \mathbf{j}m) \quad \text{requires} \quad (i\mathbf{k}E + \mathbf{i}\mathbf{p} + \mathbf{j}m)\,(i\mathbf{k}E + \mathbf{i}\mathbf{p} + \mathbf{j}m),$$

incoming *outgoing*

which is impossible, unless we have a superposed state.

Only the pure weak interaction between neutrinos can be imagined to be subject to such a (nonlocal) transition, which would require both Majorana behaviour and a nonzero mass. The picture is completed with the case where no change occurs:

$$\underset{incoming}{(ikE + \mathbf{ip} + jm)}\ \underset{outgoing}{(ikE + \mathbf{ip} + jm)} \quad \text{requires} \quad \underset{spin\ 0}{(ikE + \mathbf{ip} + jm)\,(-ikE - \mathbf{ip} + jm)},$$

which would seem to represent the coupling to the Higgs field, ensuring that the fermion must have a mass to exist at all.

In fact, the final result that emerges from this discussion is an argument that *all* fermions must have nonzero mass. Spin 1 transitions require that fermions (and antifermions) have states with both helicities. This itself suggests that all fermions must be massive. It also emphasizes the role of the **p** operator as the ultimate source of all knowledge about the fermionic state. In this context, if we consider the unique direction for this operator in the quaternionic 'space' provided by the orthogonal components ikE, $\mathbf{ip}$, jm (see section 6.3), then it is clear not only that direction duplication will be avoided only if one of the orthogonal axes (which is, of course, jm) is restricted to a single sign, and if all three axes have nonzero components. Direction duplication immediately rules out massless fermions and antifermions of the same helicity, for example $(ikE + \mathbf{ip})$ and $(-ikE - \mathbf{ip})$, while a nonzero mass term will give a direct measure of the mixture of left- and right-handedness in a fermionic state. However, the necessary equivalence of any nilpotent with form $(ikE_1 + \mathbf{ip}_1)$ with one of form $(ikE_2 + \mathbf{ip}_2)$, shown in the transition from representation (14.18) to representation (14.19), is an even more powerful argument for the necessity of a mass term in the fermionic state, especially as, being massless, these fermions necessarily have the same handedness, and the relative values of E and $\mathbf{p}$ determine the nature of the phase term.

Chapter 15

Grand Unification and Particle Masses

Some sections of this chapter (**15.2-15.3** and **15.5-15.6**) were begun during the active collaboration with **JOHN CULLERNE**. Section **15.5** is a follow up to the discussion, in chapter 4, of the hierarchy of dualities, done with **BRIAN KOBERLEIN**. Deriving the $SU(3)$ strong and $SU(2)_L \times U(1)$ electroweak representations from the Dirac equation leads naturally to the idea of a grand unified theory of all three based on the Dirac nilpotent algebra. Apart from generating a Dirac equation for charge, the Dirac algebra operators can be immediately transformed into a system for producing the 24 generators of $SU(5)$. A 25th generator, suggested by the algebra, appears to be a field quantum or gauge boson which would couple to all other particles in the manner of gravity. Using $\sin^2\theta_W = 0.25$, from chapter 10, and lepton-like quarks, with the standard rescaling ('renormalization') equations for strong, weak and electric fine structure constants, produces exact grand unification between all three forces at the Planck mass, the mass-energy associated with quantum gravity. This suggests that the grand unification is connected with gravity as well and that the 25th generator, indeed, has a gravitational connection. (The discussion in chapter 17 suggests that it is inertial, and spin 1, like the other gauge bosons.) It also suggests that the group structure for grand unification is actually $U(5)$, meaning that all interactions at grand unification become scalar phases of exactly the same type as gravity and the electromagnetic force. The calculation further shows how the number of free parameters in the Standard Model can be drastically reduced. The remainder of the chapter shows that the single expression for generating the charge structures of particles, derived in the previous chapter, is also responsible for the generation of mass via the Higgs mechanism, and can be used directly to calculate many particle masses.

15.1 A Dirac Equation for Charge

The idea that the 5-fold Dirac algebra is responsible for the symmetry breaking which leads to the $SU(3) \times SU(2)_L \times U(1)$ splitting in the interactions between fundamental particles, suggests that Grand Unification may involve the $SU(5)$ group, or even $U(5)$, or something containing $SU(5)$ and extending it to right-handed states, such as $SO(10)$. In principle, we derive five representations of the electric, strong and weak charge states (*A*-*E*), which map onto the charge units (*e*, *s*, *w*), and the five quantities (*m*, **p**, *E*) involved in the Dirac equation. The analogy between the components of the Dirac equation *E*-**p**-*m* and the charge structures of *w*-*s*-*e* is so close that we can represent energy conservation and charge conservation by equations of the same type. Here, it is most convenient to begin with one of the more standard forms of the Dirac equation,

$$(\alpha.\mathbf{p} + \beta m - E)\ \psi = 0,$$

which we expand, using a 4×4 matrix, to

$$(\alpha.\mathbf{p} + \beta m - E)\psi = \begin{pmatrix} -E & 0 & im & -ip \\ 0 & -E & ip & im \\ -im & -ip & -E & 0 \\ ip & -im & 0 & -E \end{pmatrix} \begin{pmatrix} \psi_1 \\ \psi_2 \\ \psi_3 \\ \psi_4 \end{pmatrix} = 0 .$$

The column vector, here, is the usual 4-component spinor, and the terms E and $\mathbf{p}$ represent the quantum differential operators rather than their eigenvalues.

An expression for conserved charge can then be obtained by taking a product of a 4×4 matrix and a 4-component column vector in the same way as for conserved energy in the Dirac equation:

$$\begin{pmatrix} k w & 0 & -j e & -\mathbf{i} s \\ 0 & k w & \mathbf{i} s & j e \\ -j e & \mathbf{i} s & -k w & 0 \\ \mathbf{i} s & j e & 0 & -k w \end{pmatrix} \begin{pmatrix} k w + \mathbf{i} s + j e \\ k w + \mathbf{i} s - j e \\ -k w - \mathbf{i} s - j e \\ -k w - \mathbf{i} s + j e \end{pmatrix} \tag{15.1}$$

The 4×4 matrix used here is almost identical in form to the matrix for the Dirac differential operator, although the + and – signs are in different places. The *s* term effectively takes up the vector-type properties of **p**, and can be represented as a vector with a single well-defined direction. The sign applied to *e* is that of the charge itself, but *e* has the added property of weak isospin, so that the *e*'s on the first and fourth rows of the matrix and on the first and fourth rows of the column vector can be considered as isospin ‘up’ and the others as isospin ‘down’. The opposite states of isospin are not + and – but 1 and 0. So, we should apply to these e terms the matrices:

$$\begin{pmatrix}1\\0\end{pmatrix};\begin{pmatrix}0\\1\end{pmatrix};\begin{pmatrix}-1\\0\end{pmatrix};\begin{pmatrix}0\\-1\end{pmatrix}.$$

The result of this is that all terms involving e disappear on multiplication.

Multiplying out the expression in (15.1) results in a product consisting of a unit column vector times a scalar factor. If we apply the factor i to the s term in the column vector, we derive $w^2 - s^2$, which becomes 0 when $w = s = \pm 1$. We can write this out in the form:

$$\begin{pmatrix} kw & 0 & -\boldsymbol{i}\mathbf{j}e\uparrow & -i\boldsymbol{i}\mathbf{s} \\ 0 & kw & -i\boldsymbol{i}\mathbf{s} & \boldsymbol{i}\mathbf{j}e\downarrow \\ -\boldsymbol{i}\mathbf{j}e\downarrow & i\boldsymbol{i}\mathbf{s} & -kw & 0 \\ i\boldsymbol{i}\mathbf{s} & \boldsymbol{i}\mathbf{j}e\uparrow & 0 & -kw \end{pmatrix} \begin{pmatrix} kw + i\boldsymbol{i}\mathbf{s} + \boldsymbol{i}\mathbf{j}e\uparrow \\ kw + i\boldsymbol{i}\mathbf{s} - \boldsymbol{i}\mathbf{j}e\downarrow \\ -kw - i\boldsymbol{i}\mathbf{s} + \boldsymbol{i}\mathbf{j}e\downarrow \\ -kw - i\boldsymbol{i}\mathbf{s} - \boldsymbol{i}\mathbf{j}e\uparrow \end{pmatrix} = 0. \qquad (15.2)$$

where $\uparrow$ represents isospin up and $\downarrow$ isospin down. If we now create an exponential term $e^{-i(wt - \mathbf{s.r})}$, to produce a state vector for charge, and define $i\partial/\partial t = -iw$ and $-i\nabla = i\mathbf{s}$, we obtain:

$$\begin{pmatrix} ik\partial/\partial t & 0 & -\boldsymbol{i}\mathbf{j}e\uparrow & -\boldsymbol{i}\nabla \\ 0 & ik\partial/\partial t & \boldsymbol{i}\nabla & \boldsymbol{i}\mathbf{j}e\downarrow \\ -\boldsymbol{i}\mathbf{j}e\downarrow & \boldsymbol{i}\nabla & -ik\partial/\partial t & 0 \\ \boldsymbol{i}\nabla & \boldsymbol{i}\mathbf{j}e\uparrow & 0 & -ik\partial/\partial t \end{pmatrix} \begin{pmatrix} kw + i\boldsymbol{i}\mathbf{s} + \boldsymbol{i}\mathbf{j}e\uparrow \\ kw + i\boldsymbol{i}\mathbf{s} - \boldsymbol{i}\mathbf{j}e\downarrow \\ -kw - i\boldsymbol{i}\mathbf{s} + \boldsymbol{i}\mathbf{j}e\downarrow \\ -kw - i\boldsymbol{i}\mathbf{s} - \boldsymbol{i}\mathbf{j}e\uparrow \end{pmatrix} e^{-i(wt - \mathbf{s.r})} = 0. \quad (15.3)$$

The weak isospin terms cancel, suggesting why this becomes the scalar phase term. We can therefore write equation (15.2) as:

$$\begin{pmatrix} kw & 0 & 0 & -i\boldsymbol{i}\mathbf{s} \\ 0 & kw & -i\boldsymbol{i}\mathbf{s} & 0 \\ 0 & i\boldsymbol{i}\mathbf{s} & -kw & 0 \\ i\boldsymbol{i}\mathbf{s} & 0 & 0 & -kw \end{pmatrix} \begin{pmatrix} kw + i\boldsymbol{i}\mathbf{s} \\ kw + i\boldsymbol{i}\mathbf{s} \\ -kw - i\boldsymbol{i}\mathbf{s} \\ -kw - i\boldsymbol{i}\mathbf{s} \end{pmatrix} = 0$$

or

$$\begin{pmatrix} kw & 0 & 0 & -i\boldsymbol{i}\mathbf{s} \\ 0 & kw & -i\boldsymbol{i}\mathbf{s} & 0 \\ 0 & i\boldsymbol{i}\mathbf{s} & -kw & 0 \\ i\boldsymbol{i}\mathbf{s} & 0 & 0 & -kw \end{pmatrix} \begin{pmatrix} 1 \\ 1 \\ -1 \\ -1 \end{pmatrix} (kw + i\boldsymbol{i}\mathbf{s}) = 0.$$

The left-hand side reduces to

$$\begin{pmatrix} kw + ii\mathbf{s} \\ kw + ii\mathbf{s} \\ kw + ii\mathbf{s} \\ kw + ii\mathbf{s} \end{pmatrix} (kw + ii\mathbf{s}) = 0 ,$$

in which each row of the column vector becomes

$$- \; w^2 + s^2 = 0,$$

as in the parallel case of the Dirac equation.

Without the 'phase' terms, equation (15.3) becomes:

$$\begin{pmatrix} ik\partial/\partial t & 0 & 0 & -i\nabla \\ 0 & ik\partial/\partial t & i\nabla & 0 \\ 0 & i\nabla & -ik\partial/\partial t & 0 \\ i\nabla & 0 & 0 & -ik\partial/\partial t \end{pmatrix} \begin{pmatrix} kw + ii\mathbf{s} \\ kw + ii\mathbf{s} \\ -kw - ii\mathbf{s} \\ -kw - ii\mathbf{s} \end{pmatrix} e^{-i(wt - \mathbf{s}.\mathbf{r})} = 0 .$$

Here, each term of the resultant column vector becomes a pseudo-Dirac or Dirac-type equation for charge:

$$(ik\partial/\partial t + i\nabla)(kw + ii\mathbf{s})e^{-i(wt - \mathbf{s}.\mathbf{r})} = 0 ,$$

in the same way as each term of the resultant column matrix becomes a Dirac equation for the E-$\mathbf{p}$-m combination. This equation provides a convenient representation of the parallel between the mathematics for charge allocation, determining particle structures, and that for the Dirac state.

In this interpretation, starting with the real Dirac equation (for E-$\mathbf{p}$-m), we introduce a filled fermion vacuum to create the two-sign degree of freedom required for E. We also define a particular status for antifermions beyond the original requirement that each charge-type has two possible signs. We assume, therefore, that a particular type of charge, say s, can only be unit in one of the three 'colours' needed to make up an observed state. This excludes charges of the opposite sign, so we take the concept of antistates from the Dirac equation, and assign $-s$ to the antifermions. We cannot, however, repeat the same procedure for, say, e, which must have both signs in both states and antistates. So, we preserve the rule that a charge ($-e$ in this case) can be unit in only one of the three 'colours', but make the 'default' position (e, e, e) as opposed to (0, 0, 0) for s, and so produce two signs by creating 'weak isospin', with alternatives (e, e, 0) and (0, 0, $-e$). Subsequently, we find that using 'weak isospin' actually gives us a suitable zero for the matrix equation for charge. Finally, to accommodate two signs of w, we have to refer to the fact that a filled vacuum, with antiparticles nonexistent in the ground state, violates charge conjugation symmetry for the charge (w) which specifies the fermion state.

15.2 *SU*(5) Symmetry

The 5-fold nature of both the charge-allocation system and the Dirac algebra suggests a possible relationship with an *SU*(5) or *SU*(5)-containing symmetry for individual generations, as originally proposed by Georgi and Glashow.[1,2] Here, the 15 left-handed fermions are accommodated into *SU*(5) as follows:

$$\bar{5} = (\bar{3}, 1) + (1, 2)$$
$$= (\bar{d}) + (\nu_e, e)$$
$$10 = (\bar{3}, 1) + (1, 1) + (3, 2)$$
$$= (\bar{u}) + (\bar{e}^+)(u, d)$$

The gauge bosons, in *SU*(5), are obtained from

$$5 \times \bar{5} = 24 + 1$$

$$24 = (8, 1) + (1, 3) + (1, 1) + (3, 2) + (\bar{3}, 2)$$

The first term represents the 8 gluons, essentially the 8 non-singlet combinations of s and $\bar{s}$ states (the ninth being a colour singlet). The next two terms represent the four exchange particles for the combined electroweak interaction, W^-, W^+, Z^0 and γ, which are essentially the four possible combinations of (e, w) and ($\bar{e}$, $\bar{w}$). The last two terms are the six possible combinations of s and ($\bar{e}$, $\bar{w}$). and the six possible combinations of $\bar{s}$ and (e, w), which are considered to be the X and Y bosons, mediating the combined strong-electroweak interaction. Significantly, the weak bosons W^-, W^+, and Z^0 and the X and Y particles are the only gauge bosons which acquire mass, because the weak interaction is the only one which involves changes in the sign of charge.

Exactly the same number of options are available to the terms in the Dirac wavefunction, e, s and w being now replaced by the respective terms m, p and E, and the respective combinations in the $5 \times \bar{5}$ by $p\bar{p}$, (m, E) ($\bar{m}$, $\bar{E}$), p ($\bar{m}$, $\bar{E}$), and $\bar{p}$ (m, E), or the equivalent vector / quaternion operators. The meaning of these terms may be taken as transitions between states represented by different wavefunctions, using annihilation and creation operators, that is the annihilation of one state and the creation of another. A transition like $p_x \rightarrow p_y$ would thus require a gauge boson like the colour transition $\mathbf{R} \rightarrow \mathbf{G}$, or the representation transition $A \rightarrow B$.

There are good reasons for believing that the grand unified gauge group is indeed $SU(5)$, or something containing it, such as $SO(10)$. $SU(5)$ is a group of rank four and so has four invariant Casimir operators, which are nonlinear functions of the generators, and clearly correspond to the four fundamentally conserved quantities w, s, e, M (where M is mass-energy). $SU(5)$ also appears to be the underlying structure of the γ-matrix algebra (especially in the form of the 4-vector-quaternion combination), and its separate allocations to w, s, e and E, $\mathbf{p}$, m, in addition to the five possible quark representations A, B, C, D, E. We have good evidence for the following mappings of these structures onto each other:

E	p_x	p_y	p_z	m
w	s_G	s_R	s_B	e
D	A	B	C	E
$i\mathbf{k}$	$\mathbf{ii}$	$\mathbf{ij}$	$i\mathbf{k}$	j
γ^0	γ^1	γ^2	γ^3	γ^5

The mapping of the strong terms is always exact, but the electroweak tèrms are so closely linked physically that transposition to equivalent representations may be necessary to reflect the physical manifestations of these interactions. (Interestingly, there are also five main representations of string theory.) The s-$\mathbf{p}$ connection provides a dynamical model of quarks using integral charges because, in all these case, the vector element in the mapping is preserved, irrespective of the sign. This is what makes it possible to write down a baryon wavefunction in terms of a 'rotating' $\mathbf{p}$.

15.3 The Grand Unification Group Generators

The proposal made earlier suggests that, although electric charge-related phenomenology is determined by the fractional charges generated by the perfect gauge invariance of the strong interaction, the *gauge relations between interactions* must reflect the more fundamental underlying lepton-like quark structures producing the observed effects. Grand unification of the three non-gravitational forces is currently believed to occur at an energy of order 10^{15} GeV, which is about four orders below the Planck mass (M_P). Minimal $SU(5)$, however, the main model used to make the prediction, which is based on the simplest available group incorporating $SU(3) \times SU(2)_L \times U(1)$, is known to be seriously flawed. In the first place, it fails to predict an exact convergence of the three interactions. In the second, the assumed electroweak mixing parameter, $\sin^2\theta_W = 0.375$, is very different from the experimental value of 0.231, and has different values for quarks and leptons. Although a 'renormalization' procedure can be adopted to reduce the predicted value (assumed to be that for grand unification) to about 0.21 at the Z boson mass (M_Z), a reapplication of the renormalized value to the equations for the coupling constants leads to a completely contradictory grand unification value of 0.6! Again, the supposed grand unification is imperfect: the weak and strong coupling constants are assumed to be exactly unified, but the electric coupling constant occurs only in a mixed state with the weak one via an assumed group structure.

Here, it is proposed that the true group structure can be found through the direct connection between the charge states and the Dirac nilpotent state vector in which they are incorporated. The five charge units (e, s_G, s_B, s_R, w, taking into account the vector nature of s) map directly onto the five Dirac operators (***ik***; **ii**; **ji**; ***ki***; ***j***), and the five quantities (m, p_x, p_y, p_z, E) involved in the Dirac equation, and generate both an overall $SU(5)$ and its breakdown to $SU(3) \times SU(2)_L \times U(1)$. The 24 $SU(5)$ generators can be represented in terms of any of these units. For example:

	$\bar{s}_G$	$\bar{s}_B$	$\bar{s}_R$	$\bar{w}$	$\bar{e}$
s_G					
s_B		gluons		Y	X
s_R					
w		Y		Z^0, γ	W^-
e		X		W^+	Z^0, γ

or:

	$\bar{p}_x$	$\bar{p}_y$	$\bar{p}_z$	$\bar{E}$	$\bar{m}$
p_x					
p_y		gluons		Y	X
p_z					
E		Y		Z^0, γ	W^-
m		X		W^+	Z^0, γ

Possibly we can also express these in terms of the tables *A*-*E*, which apply only *at* grand unification, when all the interactions have equivalent *SU*(3) representations:

	$\bar{A}$	$\bar{B}$	$\bar{C}$	$\bar{D}$	$\bar{E}$
A					
B		gluons		Y	X
C					
D				Z^0, γ	W^-
E		X		W^+	Z^0, γ

The only unobserved generators here are *X* and *Y*, which earlier *SU*(5) schemes have taken to imply direct proton decay. However, such decay would be forbidden by separate charge conservation rules, as it involves the complete elimination of a strong charge unit, and it also disregards the necessary dipolarity of the weak charge. Here, the *X* and *Y* generators remain linked to the dipolar particle + antiparticle mechanism of the ordinary weak interaction, and it may be that, below grand unification energies, they are connected with nothing more exotic than the ordinary process of beta decay, which links strong and electroweak interactions in the manner required.[3]

SU(5), however, is not the full story. If we had a 25th generator (which the Standard Model disregards on the grounds that it is not observed), the group would become *U*(5), and all the generators would be entirely equivalent to scalar phases. Such a particle, if it existed, would couple to all matter in proportion to the amount, and, as a colour singlet, would be ubiquitous.[4,5] This is a precise description of the force of gravity, and, since we will show that Grand Unification of the electromagnetic, strong and weak forces occurs at the Planck mass (cf 15.6), the energy usually taken as characteristic of quantum gravity, then it is highly probable that $U(5)$, or something containing it, is the true Grand Unification group, and that it also incorporates a gravity generator – though most probably a spin 1 generator for the inertial reaction, rather than a spin 2, or other hypothetical, generator for gravity itself (cf chapter 17). The gravity generator would then link the 2 colourless gluons with the Z^0 and γ, along the diagonal of the group table, suggesting a link between all four interactions. Reduction of the generators to scalar phases would mean that, at grand unification, all interactions would be identical in effect, and all non-Coulombic structure would disappear. The unification would be exact.[6]

15.4 The Dirac Algebra Operators and *SU*(5) Generators

The generators may be seen more directly as resulting from the mathematical structure of the operators used in the Dirac algebra. Essentially, there are nine elements of the Dirac algebra which contain a vector and a quaternion term (and another nine if it is made complex). However, the 32-part algebra requires just five 'primitives' or generators, only three of which are vector quaternions. In principle, the system inevitably privileges one set of three out of the nine – only one quaternion is allowed to have a vector in the primitive set (leading to the behaviour of **p** and *s*). Where the system is complex, there seems also to be a requirement for some of the primitive terms to have a coefficient of imaginary *i* and others to have a real coefficient. We can draw up the following multiplication table for the components of the algebra:

		*i*i	*i*j	*i*k	*j*	*ik*
*i*i		−1	*i*k	*i*j	*k*i	− *ij*i
*i*j		*i*k	−1	− *i*i	*k*j	− *ij*j
*i*k		− *i*j	*i*i	−1	*k*k	− *ij*k
j		− *k*i	− *k*j	− *k*k	−1	*ii*
ik		*ij*i	*ij*j	*ij*k	− *ii*	1

By comparison with the tables for the *SU*(5) generators, ***i*i**, ***i*j**, ***i*k** represents the strong charge in its three-colour form, ***j*** the electric charge, and ***ik*** the weak charge. The terms $\pm\, i\mathbf{i}$, $\pm\, i\mathbf{j}$, $\pm\, i\mathbf{k}$ represent the colour non-singlet gluons; $\pm\, ii$ the weak generators W^+ and W^-; and $\pm\, k\mathbf{i}$, $\pm\, k\mathbf{j}$, $\pm\, k\mathbf{k}$ and $\pm\, ij\mathbf{i}$, $\pm\, ij\mathbf{j}$, $\pm\, ij\mathbf{k}$ the generators X and Y of the Grand Unified theory. All possible parts of the Dirac algebra are represented as either source terms or generators, if we take no account of the complex factor i. (It is notable that terms appear as sources or generators only with this factor or without it – there is no duplication, and the total number of possible outcomes remains 16.) The five zero-charge generators along the diagonal, the two colourless gluons, Z^0, γ, and the gravity operator may be assumed completely mixed at Grand Unification in an overall $U(5)$ structure.

This is what happens if we assume the charge allocation produces a complete one-to-one correspondence with the Dirac algebra. If we now choose to construct a system for charge allocation based on a double quaternion algebra (i.e. replacing the vector set **i**, **j**, **k** with a second, independent, quaternion set, using the same symbols), and, leaving out the i term entirely, we obtain the following:

		$i\mathbf{i}$	$i\mathbf{j}$	$i\mathbf{k}$	j	k
$i\mathbf{i}$		1	$-\mathbf{k}$	$\mathbf{j}$	$k\mathbf{i}$	$-j\mathbf{i}$
$i\mathbf{j}$		$\mathbf{k}$	1	$-\mathbf{i}$	$k\mathbf{j}$	$-j\mathbf{j}$
$i\mathbf{k}$		$-\mathbf{j}$	$\mathbf{i}$	1	$k\mathbf{k}$	$-j\mathbf{k}$
j		$-k\mathbf{i}$	$-k\mathbf{j}$	$-k\mathbf{k}$	-1	i
k		$j\mathbf{i}$	$j\mathbf{j}$	$j\mathbf{k}$	$-i$	-1

This is closer to the charge allocation algebra, which is based on a quaternion-vector system. If we use a quaternion-vector system, as such, we obtain:

		$i\mathbf{i}$	$i\mathbf{j}$	$i\mathbf{k}$	j	k
$i\mathbf{i}$		-1	$-i\mathbf{k}$	$i\mathbf{j}$	$k\mathbf{i}$	$-j\mathbf{i}$
$i\mathbf{j}$		$i\mathbf{k}$	-1	$-i\mathbf{i}$	$k\mathbf{j}$	$-j\mathbf{j}$
$i\mathbf{k}$		$-i\mathbf{j}$	$i\mathbf{i}$	-1	$k\mathbf{k}$	$-j\mathbf{k}$
j		$-k\mathbf{i}$	$-k\mathbf{j}$	$-k\mathbf{k}$	-1	i
k		$j\mathbf{i}$	$j\mathbf{j}$	$j\mathbf{k}$	$-i$	-1

This has the advantage of making all the diagonal terms identical, as we would require. Also, all the pure vector terms become complex, while the pure quaternion terms and the vector quaternion terms are not, which creates a greater degree of uniformity in the algebra. In this case, $\boldsymbol{i}\mathbf{i}$, $\boldsymbol{i}\mathbf{j}$, $\boldsymbol{i}\mathbf{k}$ represents the strong charge in its three-colour form, $\boldsymbol{j}$ the electric charge, and $\boldsymbol{k}$ the weak charge. The terms $\pm\, \boldsymbol{i}\mathbf{i}$, $\pm\, \boldsymbol{i}\mathbf{j}$, $\pm\, \boldsymbol{i}\mathbf{k}$ represent the colour non-singlet gluons; $\pm\, \boldsymbol{i}$ the weak generators W^+ and W^-; and $\pm\, \boldsymbol{k}\mathbf{i}$, $\pm\, \boldsymbol{k}\mathbf{j}$, $\pm\, \boldsymbol{k}\mathbf{k}$ and $\pm\, \boldsymbol{j}\mathbf{i}$, $\pm\, \boldsymbol{j}\mathbf{j}$, $\pm\, \boldsymbol{j}\mathbf{k}$ the generators X and Y of the Grand Unified theory. It is significant now that the strong charge is represented by a quaternion $\boldsymbol{i}$, which is 'privileged', by taking on the vector operators $\mathbf{i}$, $\mathbf{j}$, $\mathbf{k}$, but that the same quaternion, without the vector operators, represents the weak interacting generators, W^+ and W^-; while the colour non-singlet gluons are represented by complex pseudovectors, exactly as they are when represented as carried by the spin angular momentum.

Another possibility is to use a vector quaternion algebra which is close to the charge-allocation form:

		$\boldsymbol{i}\mathbf{i}$	$\boldsymbol{i}\mathbf{j}$	$\boldsymbol{i}\mathbf{k}$	$\boldsymbol{j}\mathbf{j}$	$\boldsymbol{k}\mathbf{k}$
$\boldsymbol{i}\mathbf{i}$		1	$-\mathbf{k}$	$\mathbf{j}$	$\boldsymbol{ik}\mathbf{k}$	$\boldsymbol{ij}\mathbf{j}$
$\boldsymbol{i}\mathbf{j}$		$\mathbf{k}$	1	$-\mathbf{i}$	$\boldsymbol{k}$	$-\boldsymbol{ij}\mathbf{i}$
$\boldsymbol{i}\mathbf{k}$		$-\mathbf{j}$	$\mathbf{i}$	1	$-\boldsymbol{ik}\mathbf{i}$	$-\boldsymbol{j}$
$\boldsymbol{j}\mathbf{j}$		$\boldsymbol{ik}\mathbf{k}$	$-\boldsymbol{k}$	$-\boldsymbol{ik}\mathbf{i}$	1	$\boldsymbol{ii}\mathbf{i}$
$\boldsymbol{k}\mathbf{k}$		$\boldsymbol{ij}\mathbf{j}$	$-\boldsymbol{ij}\mathbf{i}$	$\boldsymbol{j}$	$\boldsymbol{ii}\mathbf{i}$	1

or, alternatively, the equivalent double quaternion algebra:

		$\boldsymbol{i}\mathbf{i}$	$\boldsymbol{i}\mathbf{j}$	$\boldsymbol{i}\mathbf{k}$	$\boldsymbol{j}\mathbf{j}$	$\boldsymbol{k}\mathbf{k}$
$\boldsymbol{i}\mathbf{i}$		1	$-\mathbf{k}$	$\mathbf{j}$	$\boldsymbol{k}\mathbf{k}$	$\boldsymbol{j}\mathbf{j}$
$\boldsymbol{i}\mathbf{j}$		$\mathbf{k}$	1	$-\mathbf{i}$	$-\boldsymbol{k}$	$-\boldsymbol{j}\mathbf{i}$
$\boldsymbol{i}\mathbf{k}$		$-\mathbf{j}$	$\mathbf{i}$	1	$-\boldsymbol{k}\mathbf{i}$	$\boldsymbol{j}$
$\boldsymbol{j}\mathbf{j}$		$\boldsymbol{k}\mathbf{k}$	$\boldsymbol{k}$	$-\boldsymbol{k}\mathbf{i}$	1	$\boldsymbol{i}\mathbf{i}$
$\boldsymbol{k}\mathbf{k}$		$\boldsymbol{j}\mathbf{j}$	$-\boldsymbol{j}\mathbf{i}$	$\boldsymbol{j}$	$\boldsymbol{i}\mathbf{i}$	1

Within the context of the entire 32-part algebra, the group relationships may be shown in a 32 × 32 table, of which the first 12 × 12 products are:

*	1	i	ii	ij	ik	ik	j	ji	jj	jk	ii	k
1	1	i	ii	ij	ik	ik	j	ji	jj	jk	ii	k
i	i	–1	iii	iij	iik	–k	ij	iji	ijj	ijk	–i	ik
ii	ii	iii	–1	–ik	ij	–iji	ki	k	–ikk	–ikj	–ii	–ji
ij	ij	iij	ik	–1	–ii	–ijj	kj	ikk	k	iki	–ij	–jj
ik	ik	iik	–ij	ii	–1	–ijk	kk	ikj	–iki	k	–ik	–jk
ik	ik	–k	iji	ijj	ijk	1	–ii	ii	ij	ik	–j	–i
j	j	ij	–ki	–kj	–kk	ii	–1	–i	–j	–k	–ik	i
ji	ji	iji	–k	–ikk	ikj	–ii	–i	–1	–ik	ij	–iki	ii
jj	jj	ijj	ikk	–k	–iki	–ij	–j	ik	–1	–ii	–ikj	ij
jk	jk	ijk	–ikj	iki	–k	–ik	–k	–ij	ii	–1	–ikk	ik
ii	ii	–i	–ii	–ij	–ik	j	ik	iki	ikj	ikk	1	–ij
k	k	ik	ji	jj	jk	–i	–i	–ii	–ij	–ik	ij	–1

The remaining 20 rows and columns are the products of the four pentads (***k*i**, ***k*j**, ***k*k**, ***ij***, ***i***); (***ii*i**, ***ii*j**, ***ii*k**, ***i*k**, **j**); (***ij*i**, ***ij*j**, ***ij*k**, ***i*i**, **k**); (***ik*i**, ***ik*j**, ***ik*k**, ***i*j**, **i**). The full group has 64 elements, constructed from the positive and negative versions of the 32 algebra units. For convenience, the multiplications of the negative elements are not shown, but the multiplications of two negative elements will generate the same products as the multiplications of two positive elements, while the multiplications of positive with negative elements, in either order, will generate the same products as two positive elements, if subsequently multiplied by –1. From the table it is clear that the 32-part algebra may be constructed from 1, *i*, and six independent Dirac pentads, which double to twelve with the signs reversed. Within each pentad, it is possible to observe the eight $SU(3)$ generators and the four generators for $SU(2)_L \times U(1)$, as outlined in the earlier part of this section.

In the table, the 'canonical' pentad (***i*i**, ***i*j**, ***i*k**, ***ik***, ***j***), obtained by priveleging the ***i*** quaternion operator, representing strong charge, is followed by two further pentads, obtained by cycling the quaternion operators and successively privileging ***j*** and ***k***, the operators representing the electromagnetic and weak charges. This demonstrates that the $SU(5)$ / $U(5)$ arrangement of the 'quark' tables *A*-*E* represents three interlocking $SU(3)$ systems for the three interactions, of which only one may be privileged as the *physical* carrier of the vector aspect of angular momentum conservation. The existence of three further pentads, in which complex quaternion operators replace vectors, and vectors replace quaternions, is a consequence of the fact that our original creation of the Dirac algebra privileged the vectors by allowing them to retain their perfect symmetry at the expense of the quaternion operators. Mathematically, however, it would have been just as valid to privilege the quaternion operators at the expense of the vectors.

We can consider the five component terms of the Dirac pentad (E-$\mathbf{p}$-m) as discrete, having become so in the process of superimposing the charges w, s, e (with quaternion units $\boldsymbol{j}$, $\boldsymbol{i}$, $\boldsymbol{k}$) on the original group parameters time, space and mass. The discreteness may be thought of, in some sense, as being due to complexification. A nilpotent Dirac pentad is necessarily complex, and complexity is also necessary to fermions because the complex term $\boldsymbol{ik}E$ (occupying the same position in the formalism as w) is intrinsic to their definition. Complexity creates a discontinuous $U(5)$ because in the E-$\mathbf{p}$-m pentad we have a real quantized mass and an imaginary quantized E. $U(5)$ is also suggested by the gravity operator ('mass') being automatically included in the process of 'compactification', or reduction of the eight original units (1, $\boldsymbol{i}$, $\boldsymbol{j}$, $\boldsymbol{k}$, i, $\mathbf{i}$, $\mathbf{j}$, $\mathbf{k}$) to the five composite Dirac terms ($i\boldsymbol{k}$, $\boldsymbol{i}\mathbf{i}$, $\boldsymbol{i}\mathbf{j}$, $\boldsymbol{i}\mathbf{k}$, $\boldsymbol{j}$). Complex i introduces discreteness in the square-rooting process which produces iE, and this links up with the fermion / boson distinction.

15.5 Superspace and Higher Symmetries

Having thus established the basis of a theory of Grand Unification between the interactions, it may be worth looking briefly at the kind of higher symmetries that have been suggested as relevant to physics at an even more fundamental level before returning to a calculation of the energy of Grand Unification. We have seen (in chapter 4) that the whole set of groups relevant to the foundations of physics comes from the fundamental concept of duality, which is effectively the same thing as applying a C_2 symmetry to the idea of physical measurement, and which has direct expression in physical equations through the numerical factor 2. The symmetry group of the parameters develops from three C_2 symmetries, and extends to higher symmetries on application of the specific mathematical forms applicable to these dualities. Successive applications of the real / imaginary and discrete / continuous (or unidimensional / multidimensional) divisions between the parameters space, time, mass and charge leads to a fundamental group of order 64 (the Dirac algebra) applicable to the symmetry of an object. At the same time, the application of the conserved / nonconserved properties also produces the concept of rotational symmetry, which, applied to the multidimensionality which derives from the discrete / continuous division, introduces a related set of Lie algebras, specified by a finite number of generators rather than elements, including the subgroups of G_2, which relate to the Standard Model. It may be that we can continue to find meaning by doubling beyond this stage, and using the Freudenthal-Tits Magic Square. It is even possible that the doubling is open-ended, like that in Newton's third law, or the 'supersymmetric' creation in the vacuum of infinite numbers of fermionic and bosonic states. It may even be their actual physical representation.

Using the Magic Square, which includes the 5 unique order 8 abstract groups, or exceptional Lie groups, generated by the octonions, we may extend our analysis to groups such as E_6 and E_8, which have particularly interesting characteristics for physical theory. The M_3^8 or 3×3 Hermitian octonionic matrix representations of the complexified E_6, for instance (or $E_6 \times U(1)$, with $U(1)$ representing the phase term), which Gürsey *et al* have seen as a possible grand unification group uniting quarks and leptons,[7-9] have 27 degrees of freedom, which is comparable with the 27 possible particle tables of the form A-E. The group E_8, on the other hand, with 256 generators, 8 of which are 'timelike', which may be needed to complete the set of octonion-related symmetries, has connections with supersymmetry and string theories, and so it is particularly interesting that it is a natural product of the hierarchies created by space, time, mass and charge. The real 8×8 matrices of E_8 could possibly be isomorphic to the complex 4×4 matrices of the Dirac group.

In supersymmetry theories, the vacuum has zero energy if the symmetry is unbroken; in the present case the symmetry-breaking is due to the Dirac / Higgs mechanism, which privileges $+E$ states over $-E$. It is probably very significant that the superspace needed for supersymmetry postulates four antisymmetric coordinates as superpartners of space-time – these look very like the mass and three charges of the present theory. The eight coordinates together provide a superspace, which is like the nilpotent Dirac algebra or double quaternions. Perhaps significantly, E-$\mathbf{p}$-m (or t-$\mathbf{r}$-τ) and w-s-e together provide a 10-dimensional possibility, the eleventh dimension of string theory being introduced to embed the 10-D required in an observable structure, like the 2-D needed to draw a line in space.

The present model does not need explanation in terms of string theory, but it may explain how such theories are generated, and perhaps produce new results at higher levels. The most popular group representation for the superstrings which are generalizations of supersymmetry is $E_8 \times E_8$. In this representation one E_8 forms the $SU(3) \times SU(2)_L \times U(1)$ symmetry, while the other E_8 set is assumed to be gravitational terms. We can speculate on the possibility of deriving all four forces from one set, with the other being a mirror set, thus connecting the real / virtual particle symmetry outlined in chapter 5 to the formalism of string theory. Perhaps one set may represent space (3), time, mass, charge (3), combined as in t-$\mathbf{r}$-τ, and quantized via the charge input, and the other set Dirac momentum (3), Dirac energy, Dirac rest mass, and Dirac angular momentum (3). (We may also note the connection with space and phase space, and with the ten generators of the Lorentz group.) The second set contains all the conserved quantities related to mass, which may be connected to the association of the second E_8 with gravity – conservation of charge and angular momentum rotational / irrotational properties are certainly related, as we have seen, in the same way as those of space are related to momentum, and those of energy to time. Mass, also, is split between the two sets, the source being in one, and the 'field' being in the other, and this would seem to be in agreement with the spirit of supersymmetry.

15.6 Grand Unification and the Planck Mass

In calculating the energy at which grand unification may be expected to occur in the present theory, it will first be convenient to look at the (not completely successful) grand unification proposed for minimal $SU(5)$. We begin with the formula for the electroweak mixing angle[10]:

$$\sin^2 \theta_W = \frac{Tr\left(t_3^2\right)}{Tr\left(Q^2\right)}. \tag{15.4}$$

Taking a weak component with only left-handed contributions to weak isospin, over 3 colours of u, 3 colours of d, and the leptons e and ν, we obtain:

$$Tr\left(t_3^2\right) = \frac{1}{4} \times 8 = 2.$$

This is, of course, independent of the electric charge structure. For the electromagnetic component, however, the phenomenological and lepton-like structures diverge. Taking the phenomenological values, with both left- and right-handed contributions, would lead to

$$Tr\left(Q^2\right) = 2\times\left(\frac{4}{9}\times 3\times\frac{1}{9}\times 3+1+0\right) = \frac{16}{3},$$

from which $\sin^2\theta_W = 0.375.$

For lepton-like quarks, however, we have

$$Tr\,(Q^2) = 2\times(1+1+0+0+0+1+1+0) = 8,$$

leading to $\sin^2\theta_W = 0.25.$

Weinberg[11] is one of many who have observed that the value 0.375 for $\sin^2\theta_W$ is in 'gross disagreement' with the experimental value of 0.231. 0.25 is, of course, much closer, and second order corrections could account for the relatively small discrepancy, especially if the 0.25 occurs at the vacuum expectation energy (246 GeV) rather than at M_W or M_Z. It is usual, in the standard approaches, to take the equations for the running weak and strong coupling constants, derived from their respective $SU(2)$ and $SU(3)$ structures:

$$\frac{1}{\alpha_2(\mu)} = \frac{1}{\alpha_G} - \frac{5}{6\pi} ln\frac{M_X}{\mu^2} \tag{15.5}$$

and

$$\frac{1}{\alpha_3(\mu)} = \frac{1}{\alpha_G} - \frac{7}{4\pi} ln\frac{M_X}{\mu^2}, \tag{15.6}$$

and assume that a particular grand unified gauge group structure will modify the equivalent $U(1)$ equation for the electromagnetic coupling ($1/\alpha$) to one in which it is mixed with the weak value, based on $SU(2)\times U(1)$. So, now we have

$$\frac{1}{\alpha_1(\mu)} = \frac{1}{\alpha_G} + \frac{1}{6\pi} ln\frac{M_X}{\mu^2}, \tag{15.7}$$

where

$$\frac{5}{3\alpha_1(\mu)} + \frac{1}{\alpha_2} = \frac{1}{\alpha}. \tag{15.8}$$

From equations (15.5), (15.6) and (15.7), we derive a grand unified mass scale (M_X) of order 10^{15} GeV, and proceed to apply (15.5) and

$$\sin^2\theta_W = \frac{\alpha(\mu)}{\alpha_2(\mu)}, \tag{15.9}$$

to give 'renormalized' values of $\sin^2\theta_W$ of order 0.19 to 0.21.

The big disadvantage of this procedure is that it does not achieve a true equalization of the interactions, even at grand unification. The strong and weak interactions achieve exact equalization with each other, but not with the electromagnetic interaction. However, equations (15.7) and (15.8) are not well-established results, like (15.5), (15.6) and (15.9). They are not supported by the experimental evidence, and make assumptions about group structure, such as relying on a particular value for the Clebsch-Gordan coefficient, $C^2 = 5 / 3$, that have, as yet, no experimental or theoretical justification. The fit to the data is also poor, for substitution of the calculated grand unification constants into the equations for the individual couplings (15.5), (15.6), (15.8), produces a result which manifestly fails to converge to a single value for the grand unified coupling (α_G), leading to *ad hoc* suggestions that a supersymmetric model may be the only solution.[12] Even worse than this, however, is the fact that, compensating errors in the combination tend to disguise the massive inconsistencies between the separate equations. In particular, recalculation of the value of $\sin^2\theta_W$ at $\mu = 10^{15}$ GeV gives 0.6 rather than the 0.375 which was initially assumed in setting up the equations!

In the present theory, however, we have an *independent* value for $\sin^2\theta_W$ of the right order. We can, therefore, perform a much simpler calculation for M_X without making assumptions about the group structure, and avoiding, in the first instance, the problematic running coupling constant equation for $1 / \alpha_1$. We avoid the speculative equations (15.6) and (15.7), and combine the well-established (15.4), (15.5) and (15.8) to give:

$$\sin^2\theta_W(\mu) = \alpha(\mu)\left(\frac{1}{\alpha_3(\mu)} + \frac{11}{6\pi} ln \frac{M_X}{\mu}\right) \tag{15.10}$$

Equally significantly, we immediately obtain a remarkable value for the Grand Unified mass scale, M_X. Taking typical values for $\mu = M_Z = 91.1867(21)$ GeV, $\alpha(M_Z^2) = 1 / 128$ (or 1/129), $\alpha_3(M_Z^2) = 0.118$ (or 0.12), and $\sin^2\theta_W = 0.25$, we obtain 2.8×10^{19} GeV for M_X. This is of the order of the Planck mass (1.22×10^{19}), and may well be exactly so, as purely first-order calculations overestimate the value of M_X. Assuming that M_X *is* the Planck mass, we obtain α_G (the Grand Unified value for all interactions) = 1 / 52.4, and $\alpha_2(M_Z^2) = 1 / 31.5$, which is exactly the kind of value we would expect for the weak coupling with $\sin^2\theta_W = 0.25$. We also obtain unit strength for the strong interaction ($\alpha_3 = 1$) at the approximate energy level of baryonic and mesonic structure (that is, in the range $\mu \sim m_\pi$).

Higher order calculations based on the phenomenological quark model, using a two-loop approximation, reduce the value of M_X by a factor of about 0.64 in that theory, while theoretical plots for $\sin^2\theta_W$ against μ^2 show a distinct dip at M_W - M_z, against an overall upward trend, suggesting that the emergence of massive gauge bosons depresses the effective values of $1 / \alpha_2$ and $\underline{\sin}^2\theta_W$ in the energy range M_W - M_z, where they are normally measured.[13,14] (The actual decrease, from about 0.22 to 0.21, represents a possible decrease up to about 0.02 in what would be otherwise expected; while 0.25 is probably the value expected at 246 GeV, rather than M_Z (see 15.10).) Similar calculations, applied to the lepton-like quark model may well yield similar results, or perhaps an even better fit to the data.

Use of lepton-like quarks, however, means that the hypercharge numbers for the $U(1)$ electromagnetic running coupling equation will be no longer identical to those for a quark model based purely on QED phenomenology. In the lepton-like model, $\begin{pmatrix} \mathrm{u} \\ \mathrm{d} \end{pmatrix}_L$ changes from 1 / 6 to 1 / 2, while $(\mathrm{u}^c)_L$ goes from −2 / 3 to −1, −1 or 0, depending on the colour, and $(\mathrm{d}^c)_L$ from 1 / 3 to 0, 0 or 1. The fermionic contribution to vacuum polarization is, conventionally,

$$\frac{4}{3}\times\frac{1}{2}\times\left(\frac{1}{36}\times3+\frac{1}{36}\times3+\frac{1}{9}\times3+\frac{4}{9}\times3+\frac{1}{4}\times1+\frac{1}{4}\times1+1\right)\frac{n_g}{4\pi}=\frac{5}{3\pi},$$

where $n_g = 3$ is the number of fermion generations; but, modifying this for lepton-like quarks, we obtain:

$$\frac{4}{3}\times\frac{1}{2}\times\left(\frac{1}{4}\times3+\frac{1}{4}\times3+1+1+0+0+0+1+\frac{1}{4}\times1+\frac{1}{4}\times1+1\right)\frac{n_g}{4\pi}=\frac{3}{\pi}.$$

(The result corresponds to the change in Clebsch-Gordan coefficient from C^2 = 5 / 3 to $C^2 = 3$, when $\sin^2\theta_W = 1 / (1 + C^2)$ changes from 0.375 to 0.25.)

One of our objections to minimal $SU(5)$ was that the strong, weak, and electric interactions were not unified on an equal basis. This suggests that our grand unification treatment of the electric action should be in terms of the pure electric coupling parameter α, and not of a modified, mixed electroweak parameter, α_1, normalized to fit an overall gauge group, as assumed in most grand unification schemes. By this understanding, 0.25 is specifically the value of $\sin^2\theta_W$ for a *broken* symmetry, produced by asymmetric values of charge, and is the value that would be expected at the mass scale appropriate to the electroweak coupling, that is at $\mu = M_W$ - M_z, the energy scale at which the symmetry-breaking takes place, or, more likely, at 246 GeV. It should *not* be the value expected at grand unification.

Using the new values we have obtained for the hypercharge numbers, the running coupling of the pure electromagnetic interaction, will be:

$$\frac{1}{\alpha(\mu)} = \frac{1}{\alpha_G} + \frac{3}{\pi} ln \frac{M_X}{\mu^2} . \qquad (15.11)$$

Remarkably, when we substitute in the values $M_X = 1.22 \times 10^{19}$ GeV, $\mu = M_Z =$ 91.1867 GeV, and $\alpha_G = 1 / 52.4$, we obtain $1 / \alpha = 128$, which is *exactly the value obtained experimentally* at energies corresponding to M_Z. It would appear, therefore, that the unification which occurs at M_X might well involve a direct numerical equalization of the strengths of the three, or even four, physical force manifestations, without reference to the exact unification structure.

Since this unification apparently also occurs at the Planck mass, the exact energy scale relevant to quantum gravity, we may propose, in addition, that the actual symmetry group, incorporating gravity in some form, is $U(5)$, with the additional generator, coupling to all others, representing the gravitational interaction (at least in numerical terms), and that $SU(5)$ occurs as the first stage of the symmetry breakdown. At grand unification, also, we would have $C^2 = 0$ and $\sin^2\theta_W = 1$, creating an exact symmetry in every respect between weak and electric interactions, as well as between weak and strong. The mixing parameter, $\sin^2\theta_W$, would then be interpretable as the electroweak constant for a specifically broken symmetry, taking the value of 0.25 at the energy range where the symmetry breaking occurs (either M_W - M_z, or 246 GeV), and gradually decreasing from the maximum to this value at intermediate energies.

A $U(5)$ grand unification would have the advantage of making all the generators become pure scalar phases, and identical in form, at the grand unification energy. A likely possibility is that the grand unification energy represents a kind of 'event horizon', or unattainable limit, at which separate conservation laws for charges would have no meaning. In fact, the necessity for separate conservation laws would prohibit its attainment, as it already prohibits direct proton decay.

The Planck mass, which is here identified as the grand unification energy, is also the likely candidate for the cut-off energy which ensures the finite summation of self-energies for interacting fermions required by the nilpotent formulation (cf chapter 11). Through the need for a filled vacuum and the continuous nature of mass-energy, gravity may well be the instantaneous carrier of the state vector correlations involved in nonlocality, and the Planck mass may be taken as the quantum of the (GTR-related) inertial interactions, which are proposed elsewhere as the result of the effect of gravity on the time-delayed nature of nongravitational interactions (cf chapter 17). These in turn might produce the inertial masses associated with charged particles, by a coupling to the Higgs field which fills the vacuum state.

A grand unification at the Planck mass would have important consequences for reducing the number of free parameters in the Standard Model. Essentially, the three fundamental constants G, $\hbar$ and c have no intrinsic meaning. They are simply numbers which relate the arbitrary units which we choose to assign to space, time and mass, but, since these parameters are fundamental, it is meaningless to look for additional significance in the units themselves. Only the numerical values attached to structures, such as the electron, have this kind of intrinsic meaning. Now, if M_X is the Planck mass, then it, too, becomes a fundamental unit, since it is composed entirely from G, $\hbar$ and c. The value of $\sin^2\theta_W$ is also, apparently, known from an exact conceptual argument, In setting up the conditions for grand unification, then, we have four equations with just five unknowns at any particular energy (μ), namely μ, α, α_2, α_3, and α_G. Of course, these equations, as we write them, are merely first-order approximations, but we could, *in principle*, refine them to any degree of exactness. In effect, given any assumed μ, we could have exact predictions for any of the other four constants, with no other empirical input.

To go further, it is quite possible that one of the other constants has a theoretically exact value at some particular specified value of μ. The most likely possibility is that $\alpha_3 = 1$ (that is, $\hbar c$) when μ is, say, m_ec^2 / α, the mass-energy equivalent for a unit charge coupling, if the electron's mass is derived directly from the electromagnetic coupling. That is, we define the energy (mass) scale at which α_3, the only constant representing an unbroken symmetry, has unit value, so defining a strong charge / mass, or any idealized perfect charge / mass equivalent, with perfect phase rotation. At this value of α_3, however, perturbation theory breaks down and the calculation can only be done approximately. Agreement is moderately good but not perfect, as $\alpha_3 = 1$ seems to occur for $\mu = 1.5\ m_e / \alpha$ (the muon mass), but the equations, of course, here are very sensitive to the approximations employed, and the true value may really be closer to the one we expect. The evidence certainly indicates that there is a fundamental value of mass (m_f) appropriate to the zero charges in composite states possibly equal to about 70 MeV (see 15.7). If this is so, then *not even* μ *need be assumed* to derive the four fine structure constants; we have, rather, a fifth fundamental equation to derive m_e / α or m_e itself.[15] (16.11 suggests another possible way of scaling α_3.)

The definition of m_f, and hence of m_e / α, as a fundamental unit of mass, would further give fundamental status to the unit of length known as the 'classical radius' of the electron, which is defined as $r_c = e^2 / 4\pi\varepsilon_0 m_e c^2 = \hbar / m_f c$, and becomes the Compton wavelength for m_f. The ratio of the Planck mass to m_f, which is also the ratio of the 'classical radius to the Planck length, is a significant dimensionless number, of order 1.74×10^{20}. The 'classical radius', of course, has no physical meaning in relation to the point-like electron, but a uniform spherical distribution of charge e would produce a mass equivalent m_e over the radius $r_e = 3\, r_c / 2 = 4.2289 \times 10^{-15}$ m.

The theory proposed here has the particular merit of being eminently testable by a measurement of any of the three interaction strengths at increasing energies, where there will be divergences from values predicted by other models. The most dramatic changes will occur in $1 / \alpha$, which, on this model would be 1/118 at 14 TeV (the maximum energy of the large hadron collider (LHC) at CERN), in comparison with the 1/125 predicted by minimal $SU(5)$ and the quark model with phenomenological electric charges.

15.7 The Generation of Mass

According to our understanding of the Higgs mechanism, mass is generated when an element of partial right-handedness is introduced into an intrinsically left-handed system. Thus, anything which alters the signs of the terms in the expression $(\boldsymbol{i}\, \hat{\mathbf{p}}_a\, (\delta_{bc} - 1) + \boldsymbol{j}\, (\hat{\mathbf{p}}_b - \mathbf{1}\delta_{0m}) + \boldsymbol{k}\, \hat{\mathbf{p}}_c\, (-1)^{\delta_{1g}}\, g)$, or reduces any of the terms to zero, should, in principle, be a mass generator, because it is equivalent to introducing the opposite sign of $\boldsymbol{\sigma}_z$ or a partially right-handed state. The three main terms in this expression can be specified as sources for producing mass. They can be described as weak isospin, quark confinement, and weak charge conjugation violation.

The production of mass by the zeroing of charge is a particularly significant process, which seems to be responsible, at least, for the masses of the Higgs boson, Z^0, and the composite baryons and mesons. That mass and charge are, in some sense, mutually exclusive components of the vacuum (effectively combining to form an invariant in the same manner as space and time), is implied by standard treatments of the $U(1)$ component of the Weinberg-Salam theory, in addition to being required by the quaternionic form of the Dirac 4-spinor. For example, Aitchison and Hey, writing on the hypercharge value of the Higgs field, state that: 'we do not allow the particle physics vacuum to give an electrically charged field a non-zero value. Thus we require that the component of ϕ with non-zero vacuum value has zero charge.'[16] Missing charges can be seen as 'unused' vacuum, and occur where there is a superposition of allowed states.[17]

The two states of weak isospin specified by the term ($\hat{\mathbf{p}}_b$ – $\mathbf{1}\delta_{0m}$) are effectively equivalent to taking an undisturbed system in the form $j\sigma_z.\hat{\mathbf{p}}_b$ and of taking the same system with the added 'right-handed' term $-j\sigma_z.\mathbf{1}$. In the pure lepton or free fermion states, when $b = c \neq$ z, and hence the weak component, $k\sigma_z.\hat{\mathbf{p}}_c = 0$, the equation generates residual right-handed electron / muon / tau states, specified by $-j$, with the equivalent left-handed antistates specified by j. The right-handed terms may be considered as the intrinsically right-handed or non-weak-interacting parts of the fermions, generated by the presence of nonzero rest mass. As we have seen, the mixing of the left- and right-handed terms illustrates the fact that the electromagnetic interaction cannot identify the presence or absence of a weakly interacting component. The quarks follow the same procedure as leptons in generating the two states of weak isospin, but there are no separate representations of 'right-handed' quarks, as two out of any three quarks in any baryon system will always require $c \neq$ z and $k\sigma_z.\hat{\mathbf{p}}_c = 0$.

Mass is again generated by quark confinement, because each baryonic system requires quarks in which one or more of $i\sigma_z.\hat{\mathbf{p}}_a$, $j\sigma_z.\hat{\mathbf{p}}_b$, or $k\sigma_z.\hat{\mathbf{p}}_c$ is zero. Zero charges represent complete coupling to the Higgs field; nonzero charges represent a reduction of the vacuum state to less vacuum. This mechanism is more likely to be relevant to composite and superposed states, such as mesons and baryons, than to 'pure' ones, such as quarks and leptons. In these cases, the mass equivalent for a zero charge would appear to be that of a fundamental unit m_f, defined for unit coupling ($\hbar c$), from which we *derive* the electron mass, via the electromagnetic coupling α, as $m_e = \alpha m_f$. Hence, $m_f = m_e / \alpha$. The use of a fundamental mass unit for zero charges irrespective of origin appears to derive from the fact that these 'missing' charges are a result of a perfectly random rotation of the momentum states $\hat{\mathbf{p}}_a$, $\hat{\mathbf{p}}_b$, or $\hat{\mathbf{p}}_c$, in exactly the same manner as applies in the strong interaction to produce its linear potential; $\hat{\mathbf{p}}_a$ is, of course, actually an expression of this interaction, but $\hat{\mathbf{p}}_b$ and $\hat{\mathbf{p}}_c$ follow the identical pattern of variation.

The third mechanism for mass generation arises from the fact that the sign of the intrinsically complex k term is not specified with those of the i and j terms. Physically, however, a filled weak vacuum requires that the weak interaction recognizes only one sign for the k term when the sign of σ_z is specified. Hence, negative values of $k\sigma_z.\hat{\mathbf{p}}_c$ must act, in terms of the weak interaction, as though they were positive. Reversal of a sign is equivalent to introducing opposite handedness or mass. So, the two intrinsic signs of the $\boldsymbol{k\sigma_z.\hat{p}_c}$ term become the source of a mass splitting between a first generation, involving no sign reversal, and a second generation in which the reversal is accomplished by charge conjugation violation. However, since charge conjugation violation may be accomplished in two different ways – either by violating parity or time reversal symmetry – there are actually two further mass generations instead of one. In addition, because the weak interaction cannot distinguish between them, the three generations represented by the quarks d, s and b, are mixed, like the left-handed and right-handed states of e, μ and τ, in some proportion related to the quark masses.

Since all of these mechanisms relate to the idea of a filled vacuum, and, since we believe that the zero point energy of the vacuum is another expression of the same principle, it is worth considering whether the two processes can be related. It is relatively straightforward, in fact, to see how rest mass could arise out of the zero point energy of the vacuum. It is well-known that the spectral density of the zero point field is given by:

$$\frac{dU}{d\omega} = \frac{\omega^2}{\pi^2 c^3}\frac{\hbar\omega}{2} = \frac{\hbar\omega^3}{2\pi^2 c^3},$$

where U is the energy density, since $\omega^2 / \pi^2 c^3$ is the number of modes per unit volume, and $\hbar\omega / 2$ is the energy per unit mode. Suppose that we now introduce a cut-off value for ω, say ω_m. Then the energy density integrates to

$$U = \int_{\omega_m}^{0} \frac{\hbar\omega^3}{2\pi^2 c^3} = \frac{\hbar\omega_m^3}{8\pi^2 c^3}.$$

To relate this to a mass-energy density, we need only assume that we can relate the space 'occupied' by a particle (at least in approximate terms) to the reciprocal of its mass-energy mc^2. Suppose, for example, we assume that the particle is 'confined' within a sphere of Compton radius $\hbar / mc$. Then

$$U = \frac{\hbar\omega_m^3}{8\pi^2 c^3} = \frac{3m^4 c^5}{4\pi\hbar^3}$$

and
$$m = \left(\frac{1}{6\pi}\right)^{1/4} \frac{\hbar\omega_m}{c^2} = 0.48\frac{\hbar\omega_m}{c^2} \approx \frac{1}{2}\frac{\hbar\omega_m}{c^2},$$

which is effectively the zero point energy at ω_m. (It would be increased for an accelerating particle.) This mechanism based on the zero point field may be related to that of the Higgs mechanism, in which the Higgs field, treated as a plasma, acts as a high-pass filter, with the low frequencies transmitted away, and producing no stable bound states. The high-pass filter mechanism of the field allows particles to acquire mass determined by the cut-off frequency. The Higgs boson provides the minimum frequency for the plasma to have a collective mode. In principle, then, the Higgs mechanism requires a cut off in the zero point energy, and a cut off frequency is sufficient to generate mass of the right order.[18]

15.8 The Higgs Model for Fermions

The advantages of the version of quark structure generated by the angular momentum formalism are that it is already lepton-like, allowing for an easier unification of the two types of fermion; that it makes $\sin^2\theta_W = 0.25$ for both leptons *and* quarks (which is close to the observed value); that it predicts an exact Grand Unification of all four forces at the Planck mass; that it avoids difficulties of the definition of a charge unit if both 2/3 and 1/3 are fundamental units; and that it produces *two* values for weak hypercharge, as required, when the Higgs mechanism is applied to the creation of quark masses, and they are also the correct values (±1). A model based on 'real' (i.e. primary) fractional charges does none of these, and it fails also to explain how a fermion can 'know' whether it is a quark or a lepton by any means other than the strong interaction. The hypercharge problem is a serious one for the fractional representation, although it is seldom emphasized.

According to a standard textbook by Halzen and Martin: 'An attractive feature of the standard model is that the same Higgs doublet which generates W and Z masses is also sufficient to give masses to the leptons and quarks.'[9] After application of this to electrons, the authors state: 'The quark masses are generated in the same way. The only novel feature is that to generate a mass for the upper member of a quark doublet, we must construct a new Higgs doublet from ϕ.' 'Due to the special properties of $SU(2)$, ϕ_c transforms identically to ϕ, (but has opposite weak hypercharge to ϕ, namely $Y = -1$). It can therefore be used to construct a gauge invariant contribution to the Lagrangian.' Significantly, the hypercharge of $(u_L, d_L) = -1$ in the lepton-like quark model, when the charge structure actually matches that of the leptons, and $\boldsymbol{\sigma}_z.\hat{\mathbf{p}}_b = -1$; but it becomes 1 when $\boldsymbol{\sigma}_z.\hat{\mathbf{p}}_b = 0$, and the electric charge component is provided purely by the filled electromagnetic vacuum.

However, this is only true for lepton-like quarks. For quarks with phenomenological electric charges, the hypercharge is an invariable 1/3 and there is no negative term: the phenomenological charge values allow only one hypercharge state, though the mechanism requires two. The necessary asymmetry introduced by the lepton-like model is lost. The only way round this problem is by the invention of an arbitrary and unphysical linear combination, relating the Higgs terms to the u and d quark masses. And, of course, the reason why the hypercharge must be reversed in the lepton-like model is that the transition involves a reversal of the 'electromagnetic vacuum' or background condition, from empty to full. The nilpotent Higgs mechanism seems to make perfect sense of this procedure, where it is just a mathematical operation that 'works' with appropriate (unexplained) adjustments in the conventional view.

In the lepton-like model, the Higgs Lagrangian for the mass of e directly transfers from the usual covariant derivative Lagrangian – it is virtually a direct copy now applied to the Higgs doublet. The fermion mass Lagrangian for d when the charge structure matches that of the leptons is then a direct copy of that for e, while the fermion mass Lagrangian for d when the charge structure is not lepton-like is a direct copy of that with reversed hypercharge. In the model based on primary fractional charges, the hypercharge for quark mass is different from the hypercharge for quarks; here it is the same. The use of the Higgs mechanism with lepton-like quarks requires no extra modelling at all, and it stems from a charge 'vacuum' or the absence of charges.

15.9 The Masses of Baryons and Bosons

The Higgs mechanism, in the conventional sense, specifies the manner in which bosons and fermions acquire mass, but is unable to predict specific values for the couplings to the Higgs field, and cannot predict the mass of the Higgs boson. The Higgs field and the mechanism of generating specific masses are, however, clear consequences of the present theory, and it should, in principle, be possible to derive numerical results. Of course, the masses of particles originate in the requirements for energy balance between states subject to various conditions of symmetry. Many of these operate at the same time and the prediction of exact masses is a very complex procedure which has not yet been worked out for any single particle; but, where one particular condition is dominant, it would appear to be possible to predict masses to quite good approximations. Difficulties in physical interpretation, however, mean that such results must remain tentative for the present, and the ones presented are suggestive rather than definitive.

The most probable meaning for the Higgs boson is that it represents complete vacuum or a zeroing of all possible bosonic states, that is, a zeroing of all the charges in the complete range of fermion-antifermion combinations within the *A*-*C* and *L* representations. Assuming 6 flavours, 6 anti-flavours, 3 colours (or equivalent states), 3 charge types for each quark / antiquark, and 2 for each quark-antiquark pairing, over 4 representations (*A*, *B*, *C*, *L*), gives a total of 2592 zeros.[19] Assigning a mass-energy of $m_f c^2$ or $m_e c^2/\alpha$ (the unit coupling value), to each zeroed charge gives an approximate total mass of 182 GeV, which, though only a guess, and higher than current expectations would seem to indicate, is, interestingly, within the range of phenomenological values for m_H (approximately 170-180 GeV) which would exclude the production of explicit supersymmetric particles.[20-22]

Similar procedures may apply to the electroweak bosons, which require only the calculation of the mass of Z^0, because the W mass then follows from $M_W = M_Z \cos\theta_W$. The masses of particles are determined by the strength of their coupling to the Higgs field. Z^0 is completely coupled, γ does not couple. Complete coupling implies full strength of vacuum, i.e. zero charges. If Z^0, γ is derived from the reduced *A*/*B*/*C* - *L* representations for the pure electroweak case (*A*/*B*/*C* being indistinguishable from each other, with no *s*, and no electroweak recognition of colour), then complete summation of the zeros over 2 representations produces 91 GeV. If m_H really is 182 GeV then the most favoured decay mode for the Higgs boson would be via the two Z^0 or four lepton route.

It is of interest also that the total mass of the twelve known fermions (Σm) again appears to be about 182 GeV. Now, according to the standard treatment of the Higgs mechanism,[8] fermion masses m are generated by (harmonic oscillator) couplings g_f to the Higgs field of the form

$$g_f = \frac{e}{\sqrt{2}\sin\theta_W}\frac{m}{M_W},$$

with maximal weak coupling expected to occur in the third generation. Suppose the total fermion mass $\Sigma m = M_H = 2\,M_Z$. Taking $M_W = M_Z \cos\,\theta_w$, $\sin\,\theta_w$ at $M_Z = 0.5$, and the weak coupling constant, $g = e / \sin\,\theta_w$, we obtain, for the total weak coupling

$$\Sigma g_f = \frac{g}{\sqrt{2}}\frac{2}{\cos\theta_W} = \sqrt{\frac{8}{3}}g,$$

so making the total coupling producing the fermion masses *directly determined* by the weak coupling. The masses can then be related to M_H in the ratio of the Higgs coupling to the weak coupling:

$$\frac{m}{m_H} = \frac{g_f}{g}\sqrt{\frac{3}{8}}\,.$$

If the vacuum energy is distributed or partitioned in this way between the possible fermion states, it is noticeable that the three quark *generations* are separated from each other by a factor of the order of α, effectively the separating factor between strong, or exactly unit, and electroweak couplings.

Since the vacuum charge state determines the inertial mass value, we can imagine that the mass deriving from the filled electric vacuum is related to the fundamental mass m_f by the factor α, the electric coupling, and so is given by αm_f. Since the electron appears to be the purest example of mass being acquired by this mechanism, we may equate αm_f with m_e, and so $m_f = m_e / \alpha$. According to this argument, the mass of the electron comes from the electric vacuum state being empty, whereas it is filled in the case of the neutrino.

The vacuum expectation value for the Higgs field (f) is clearly another important parameter which ought to be calculable, in some sense, from the zeroing of charge. Phenomenologically, f can be calculated from the Fermi constant (which gives the phenomenological measure of weak coupling) at ~ 246 GeV, which appears to be, at least approximately, 3 M_W (3 × 80.45 ~ 241 GeV).[8,13] There is, in fact, a fundamental reason for a connection with M_W, as, using

$$(ig\frac{1}{2}\boldsymbol{\tau}.\mathbf{W}_\mu)^\dagger\,(ig\frac{1}{2}\boldsymbol{\tau}.\mathbf{W}_\mu) = \frac{g^2v^2}{8}[(W_\mu^{\,1})^2 + (W_\mu^{\,2})^2 + (W_\mu^{\,3})^2]$$

for a complex electrically charged field W^+ (from 9.8), f is given by $M_W = gf / 2$, with g the weak coupling constant expressed in charge units. Now, since $g = e / \sin\theta_W$, with e the electrical coupling, then $g = 2e$ if $\sin^2\theta_W = 0.25$. This might indicate that the vacuum field value, in its most idealised form, is determined by that expected for a three-phase system with the charge divided between three phases. In addition, combining equations (15.6), (15.10) and (15.11) at the energy 246 GeV, with M_P as the energy for grand unification, gives $sin^2\theta_W = 0.25$ at f, compared with 0.246 at M_Z, suggesting that f, rather than M_Z, is the energy scale to be associated with electroweak symmetry breaking.

The *dynamic* masses of the low-lying baryon and meson states are more definitely predictable. These composite particles generate their masses through the term $\boldsymbol{j}\,(\hat{\mathbf{p}}_b - \mathbf{1}\delta_{0m})$ and the strong interaction mechanism, which is, of course, electromagnetic charge independent. Here, we need to consider the global symmetries such as $SU(3)_f$ which group together particles which are indistinguishable by the strong interaction in various isospin multiplets, which effectively represent a single particle in different states of electromagnetic charge. All states of one multiplicity exist simultaneously, and so the zero charge components of all states must be accommodated in determining the mass of the particle. The global $SU(3)_f$ symmetry shows the spin $^3/_2$ baryons as a decuplet, with four Δ states, three Σ states, two Ξ states and one Ω state. The zero charge components of the Δ particles are simply those of all four states added together, but the four Δ states, when excited, have to be averaged between three Σ states, so each Σ state represents an average of 4/3 states, and the average number of charge components has to be multiplied by 4/3. The four Δ states are eventually excited to one Ω state, so each Ω state represents an average 4 states. If M_0 is the highest multiplicity in a particular baryon octet or decuplet, and n_0 is the total number of zero charges in the components of a multiplet of multiplicity M, then the minimum mass of the components of the multiplet is given by

$$\text{mass} = \frac{n_0 M_0}{M} \frac{m_e}{\alpha}.$$

For example, the multiplet Σ in the spin $^3/_2$ baryon decuplet has $M = 3$, and $M_0 = 4$ (the multiplicity of the Δ particles), while n_0, the total number of zero charges, computed from the quark tables *A*-*C*, in the combinations *dds*, *uds* and *uus*, is 15, 17 or 19. For the ground state,

$$\text{mass of } \Sigma \text{ multiplet} = \frac{15 \times 4}{3} \frac{m_e}{\alpha} = 20 \frac{m_e}{\alpha},$$

which may be compared with the experimental value of 19.8 m_e / α or 1385 MeV. The derivation of the masses for the entire spin $^3/_2$ decuplet may be set out in the following table:

	quark structure	n_0	M_0	M	predicted m	measured m
					$\times\, m_e / \alpha$	$\times\, m_e / \alpha$
Δ	*ddd,udd,uud,uuu*	20,22,24	4	4	20	≈ 17.6 - 19.6
Σ	*dds,uds,uus*	15,17,19	4	3	20	19.8
Ξ	*dss,uss*	11,13	4	2	22	21.9
Ω	*sss*	6	4	1	24	23.9

The masses are all calculated using the ground state values for n_0. The Δ particle is unusual in showing a large spread of measured mass, because, in this case, the energy width (approximately 120 MeV at half-maximum) makes a significant contribution, in addition to the rest mass; this energy width is much greater than that for any other member of the decuplet and explains the particle's instability and very rapid decay. The rest mass value (17.6 m_e / α) preserves the difference of 2 m_e / α (140 MeV) between each multiplet which occurs due to successive transitions of one *d* quark to *s* with the net loss of two *w* charges (in line with an *s* mass of 80-155 MeV). (The values also fit with those calculated from the hyperfine interactions due to the magnetic component of the colour charge, if we assume an effective *u* or *d* mass of 363 MeV and an effective *s* mass of 538 MeV. We can assume that these determine the 'rest' value of the Δ mass from the wide range allowed by the half-maximum energy width.[23]) The increasing accuracy of the predictions, from Δ through to Ω, may be related to the fact that the heavier particles represent fewer alternative states.

It is debatable whether similar principles can be applied to an extended $SU(4)_f$ multiplet including the fourth quark (*c*). The very existence of such multiplets depends on the idea that the quarks in the second and third generation have sufficiently high masses to remove the degeneracy between different sets of three-quark and quark-antiquark states. So the existence of higher $SU(n)_f$ symmetries depends crucially on the assumption that $m_b \gg m_t \gg m_c \gg m_s$. However, if the idea can be extended in this way, we would have a count of 50 zero charges (from 10 base states) for *ccc*, and a mass of 50 $m_e / \alpha = 3.5$ GeV, in line with the assumed mass for the *c* quark of between 1.0 and 1.4 GeV. The *b* and *t* quarks may be assumed to be too massive for this mechanism to be the main factor in determining the masses of baryons and mesons incorporating these as components. Extending to $SU(5)_f$, with 20 base states would produce 120 zero charges (8.4 GeV) for *bbb*, while $SU(6)_f$, with 35 base states, would produce 175 zero charges (12.25 GeV) for *ttt*, if such hadronization could be conceived.

A zero-charge analysis may be applied, however, to the spin ½ baryon octet, generated from *u*, *d* and *s*, though here the value for n_0 is taken at the ground state for the *N* multiplet (*n*, *p*), which contains no *s* quark component, and the other mass values are assumed to be of mixed states determined from within the predicted range by their accommodation within the Gell-Mann-Okubo formula required for $SU(3)_f$

$$\frac{1}{2}\left(m_N + m_\Xi\right) = \frac{3}{4}m_\Lambda + \frac{1}{4}m_\Sigma,$$

and the colour-magnetic hyperfine splitting, which applies exactly as in the decuplet:[22]

	quark structure	n_0	M_0	M	predicted m	measured m
					$\times\, m_e / \alpha$	$\times\, m_e / \alpha$
N	*udd,uud*	9,11,13	3	2	13.5 m_e / α	13.4 m_e / α
Λ	*uds*	5,7	3	1	15 - 21	15.9
Σ	*dds,uds,uus*	15,17,19	3	3	15 - 19	17
Ξ	*dss,uss*	11,13	3	2	16.5 - 19.5	18.9

The meson octets do not represent the regular progression of excited states from the lowest member which we observe in the baryon octet and decuplet, and which ultimately derive from $d \rightarrow s$ quark transitions. The multiplets are, in this sense, independent, with mass determined by $n_0\, m_e / \alpha$, where n_0 is the number of zero charge components in the multiplet. For the pseudoscalar 0^- meson octet, the ground state value of n_0 is once again chosen for the lowest lying member of the octet (π) – which again contains no symmetry-breaking s quark component – and the values for K and η selected from within the predicted range, again to fit a Gell-Mann-Okubo formula for $SU(3)_f$:

$$m_K^2 = \frac{1}{4} m_\pi^2 + \frac{3}{4} m_\eta^2 .$$

n_0 is 2, 6, 8, 10, 12, 14, or 16 for π, so the ground state value is 2; the predicted mass is therefore 2 m_e / α, which is exactly the observed value. For K, n_0 takes values 3, 5, 7, 9 or 11, and so the predicted mass is between 3 and 11 m_e / α, compared with the observed mass of 7.1 m_e / α; while η (which is additionally mixed with a singlet state) has n_0 values of 4, 6, 8, 10, 12, leading to a predicted mass between 4 and 12 m_e / α, compared with an observed mass of 7.8 m_e / α. Here, again, for both pseudoscalar and vector meson octets, a colour-magnetic hyperfine splitting appears to apply, though the effective values for u or d and s masses appear to be slightly lower than for baryons. The large mass difference (140 / 776 MeV) between the pseudoscalar π and the vector ρ, which are the respective singlet and triplet combinations of the u and d quarks and antiquarks, can, in particular, be directly ascribed to the hyperfine interaction.

Regge trajectories may provide observational evidence for the use of charge counting in determining the masses of strongly-bound composite particles. If the strong interaction is carried with the angular momentum operator **p**, the covariant derivative introduces the term $q\sigma r$ or $ig_s\lambda^\alpha \mathbf{A} / 2$, which incorporates a quantity equivalent to the strong coupling or the strong charge squared. In principle, therefore, increasing the angular momentum value assigned to any particular state should also increase the effective value of the strong charge squared in the same proportion. If the masses of the strongly-bound composite particles are determined on the basis of strong charge equivalents, then a change in angular momentum should produce a proportional change in mass squared. This, of course, would only be true if the rate of change of momentum, or linear energy density (κ), remained constant, at all distances, as effectively assumed in the semi-classical 'string' or 'gluon flux tube' explanation of the trajectories. In this case, if a quark-antiquark pair are connected by a flux tube of length $2R$, then the total mass-energy of the string becomes $m = \pi\kappa R$ and the angular momentum $J = \pi\kappa R^2 / 2$, leading to the relationship, $J = m^2 / 2\pi\kappa$, with $2\pi\kappa$ determined phenomenologically at ~ 0.9 GeV^2.

Erik Trell derives particle masses by using a homomorphism between the straight line and spherical curvature.[24-31] Here, QED is taken to be volume-preserving while QCD is (spheroidally) symmetry-preserving. Trell considers the volume-preserving ellipsoidal and automorphic subgroup transformations of $SO(3)$, using the two A_2 Lie algebra root space diagrams that result from the canonical real form $SO(3) \times O(5)$ involutive automorphism of $SU(3)$. These diagrams are identical with those of the original eightfold way of Gell-Mann and Ne'eman (that was the precursor to the quark theory), but in a 3-D, rather than a 2-D lattice, this being the actual 3-D representation space of the structures of physical particles. (Trell calls it the 'eightfold eightfold way'.)

$SO(3) \times O(5)$ provides a canonical geometrical decomposition of $SU(3)$ by projecting the A_2 root space diagrams diagonally in R^3, that is by mapping the algebra onto a unit sphere. This is applied to baryon and meson spectroscopy with possible channels determined by orbits following the root vector geodesics. The only way of preserving $SO(3)$ in this canonical coset decomposition is by representing t isospin transitions such as those involving the mesons π^+, π^0 and π^- at the respective orientations of 60°, 0° and 120° to the horizontal plane in the A_2 root space. Assuming that the primary shape of the nucleon (i.e. that of the proton) is a unit sphere, transitions of length Δx defined in the plane perpendicular to the line of electromagnetic flux (that is, in the minor semiaxis of the ellipsoids) produce particle masses which can be related to that of the proton mass via the quark pressure formula $\Delta p = h / \Delta x$.

The result is a whole spectrum of the masses of baryons, mesons, and their resonances, effectively determined from the possible transitions as a succession of available states, similar to a set of energy levels. The masses of known particle states can be measured up against these, and numerous papers by Trell show the very close correspondence. A typical example is that of the lowest energy hyperon (Λ^0), which is produced by decay of the proton and emission of π^-, or decay of the neutron and emission of π^0. In such transitions, the length from the centre of the nucleon to the new t isospin end point is $\sqrt{2}$. To preserve volume, while retaining the length of the ellipsoid, the minor semiaxis must take up the length $\sqrt[4]{1/2}$. Taking the proton as a unit sphere, the mass of Λ^0 will be 1 / $\sqrt[4]{1/2}$ times the proton mass (938.28 MeV), or 1115.8 MeV, corresponding with the measured value of 1115.8 MeV.

Lattice gauge QCD is now able to predict the masses of baryon and meson states, with some accuracy, by calculation of gluon coupling from assumed bare quark masses. As both this approach and Trell's, as well as the one presented earlier in 15.9, are based on the same theory (QCD), then we should expect a correspondence. As it happens, the three approaches supply different, and perhaps complementary information, and each makes different assumptions. The calculations earlier in 15.9 are based on quark structures and zero charges. They supply absolute dynamic masses for composite particle multiplets, if we assume these are produced by $SU(3)_f$. Trell produces more exact *relative* masses by outlining the mass states which are possible, but without direct relation to quark structure or exact correspondence with particular particle states. The lattice gauge approach supplies the bare masses of quarks which make the composite particle masses possible in QCD, and so lead us in the direction of direct fermion masses.

15.10 The Masses of Fermions

Fermion masses, like those of bosons, may be assumed to be due to a coupling between originally massless fermion fields and the nonzero background Higgs field. For a weak $SU(2)$ transformation, acting only on the left-handed component of a fermion state vector, the free particle equation contains a mass term of the form $m\psi_L + m\psi_R$ and so cannot be locally phase invariant because ψ_L and ψ_R are transformed differently under $SU(2)$. The symmetry is only preserved if the fermions are initially massless and acquire their observed masses by interaction with the Higgs field. The coupling strength, however, varies with the mass of the individual fermion and cannot be predicted independently of the known masses. Some other input is needed, as the present account has suggested.

Though the masses for the composite baryons and mesons give indications for limits on the masses of the heavier quarks, deriving exact masses, by direct methods, for the twelve known fermions is a particularly difficult problem, especially as the concept of quark mass seems to be somewhat ill-defined, with the masses 'running', like the values of the coupling constants, with the energy of interaction. However, the mass of the t quark, at least, seems to be obtainable from first principles, on the assumption that it represents maximal coupling to the Higgs field. The mass of t (~ 174 GeV) seemingly represents the maximum possible energy for a state $f / \sqrt{2}$, where f is the vacuum expectation value. It may also be possible to make some tentative approaches to calculating some of the other quark masses. In particular, the masses may be expected to come from some 'partitioning' of the vacuum related to that produced by the charges, or, equivalently, their fine-structure constants, α, α_2 and α_3.

If the fermion masses are generated by the Higgs mechanism, and the ultimate origin of mass is in the introduction of the electric charge to overcome symmetry violation in the weak interaction, then it is conceivable that the mass scales of the three generations of fermions are related by successive applications of the scaling factor α, as noted in the previous section.[32] In a related way, the Cabibbo mixing between the first and second quark generations seems to be determined (as we might expect) by the same factor as the electroweak mixing (0.23 – 0.25), and the additional mixing produced with the third generation involves terms which are the square of this factor ($\approx$ 0.06).

From both the Higgs mechanism, and our own representation, the weak isospin up state of the quarks u, c, t represents a filled electromagnetic vacuum. We may therefore expect the separation of the generation masses to be determined by the electromagnetic factor α (at some suitable energy). (The second and third generations, where this factor might be assumed to apply, notably reverse the mass ordering to isospin 'up' > isospin 'down'.) The electromagnetic connection is also obvious from the origin of this mass in the term $\boldsymbol{j}\ (\hat{\mathbf{p}}_b - \mathbf{1}\delta_{0m})$. So the mass of c is α times that of t, and the mass of u is α times that of c. Possibly this applies to the quark *generations* ($u + d$, $c + s$, $t + b$), or even quark-lepton generations, rather than the individual particles, giving 179, 1.3, and 9.5×10^{-3} GeV. The masses due to the weak isospin 'up' states, as is evident from the general formula for fermions, do not come from the perfectly random rotation, which determines the masses of all other states.

A fundamental fermion mass (probably m_e, via $m_f = m_e / \alpha$) seems definitely derivable from the relations between α, α_2 and α_3, *without any empirical input*, but the perturbation calculations are too approximate at this stage to yield the exact value. The value produced for first order calculations using a 'unit' charge ($\alpha_3 = 1$) seems to be about 0.112 GeV (slightly above the muon mass). It is quite possible that a calculation with higher order corrections might lead to the fundamental 'unit mass' ($m_f = m_e / \alpha = 0.070$ GeV) involved in the zero-charge $SU(3)_f$ procedure (or perhaps $m_\pi = 2m_e / \alpha = 0.14$ GeV). The unit nature of the strong fine structure constant at the proposed 'unit mass' would be a natural result of the strong interaction being a completely unbroken symmetry connected with an unvarying principle of 3-D rotation – an expression of 'perfect' randomness.

Other approaches to the fermion masses are more phenomenological. The masses of the *d*, *s*, *b* quarks (which may be approximately increased by a factor ~ $1 / \alpha_2$ at each generation) certainly run as a result of the QCD coupling of the strong interaction and it is generally believed that they would become identical to the respective masses of *e*, μ, τ at the energy of grand unification (M_X, which we have fixed at the Planck mass, 1.22×10^{19} GeV). More specific predictions become highly model-dependent, and none has yet produced a completely self-consistent set of results. They are unlikely, I believe, to produce the *fundamental* explanations for quark masses, though they will be significant in determining their running values. One set of calculations, for instance,[12] suggests that, at some unspecified energy (μ), a relationship of the form

$$\frac{m_b(\mu)}{m_\tau(\mu)} = \alpha_3(\mu)^{12/23} \alpha_3(m_t)^{8/161} \alpha_3(m_X)^{-4/7} \left(\frac{\alpha(\mu)}{\alpha(m_W)}\right)^{10/41}$$

should hold. (Here, for convenience, we replace the algebraic indices with their numerical equivalents – the originals are not numerical.) If $m_t = 173.8$ GeV, $\mu = 182$ GeV, $M_X = 1.22 \times 10^{19}$ GeV, we obtain $\alpha_3(\mu) = 0.10827$; $\alpha_3(m_t) = 0.1088$; $\alpha_3(M_X) = 0.01908$. Also $1/\alpha(\mu) = 126.40$, $1/\alpha_2(\mu) = 31.846$, $1/\alpha_1(\mu) = 31.517$; $1/\alpha(M_W) = 127.9$, $1/\alpha_2(M_W) = 31.846$; $1/\alpha_1(M_W) = 32.018$. So

$$\left(\frac{\alpha_1(\mu)}{\alpha_1(m_W)}\right)^{10/41} \approx \left(\frac{\alpha(\mu)}{\alpha(m_W)}\right)^{10/41} \approx 1.003 .$$

From these, we derive

$$\frac{m_b(\mu)}{m_\tau(\mu)} = 2.705 ,$$

and if $m_\tau = 1.770$ GeV, then $m_b = 4.79$ GeV. Adapting this to $m_s(\mu) / m_\mu(\mu)$, with

$\alpha_3(m_c)$ replacing $\alpha_3(m_t)$, we obtain $\alpha_3(m_c) = 1/3.64$ if $m_c \approx 1.2$ GeV. Hence,

$$\frac{m_b(\mu)}{m_\tau(\mu)} = 2.832\,,$$

and, for $m_\mu = 0.10566$ GeV, $m_s \approx 0.299$ GeV. The results are reasonable, if slightly high (and interestingly close to what would result from a combination of the bare lepton mass and a contribution from $SU(3)_f$), but any decrease in μ would make them higher still. Also, for $m_d(\mu) / m_e(\mu)$, the perturbation expansion for $\alpha_3(m_d)$ becomes impossible if $m_d \approx 6 \times 10^{-3}$ GeV, as α_3 then increases uncontrollably. A value of $\alpha_3(m_d) \approx 10^{12}$ would be required to generate the approximate ratio 6 / 0.511, which appears to apply!

15.11 The CKM Mixing

The Cabibbo-Kobayashi-Maskawa mixing between generations is, of course, a significant aspect of the fermion mass problem, and it is produced by the term $\boldsymbol{k}\ \hat{\mathbf{p}}_c\ (-1)^{\delta_1 g}\ g$ in the expression for quark-lepton generation. Using the Wolfenstein parameterization, the mixing is written in the form of the matrix:

$$\begin{pmatrix} 1-\lambda^2/2 & \lambda & \lambda^3 A(\rho - i\eta) \\ -\lambda & 1-\lambda^2/2 & \lambda^2 A \\ \lambda^3 A(1-\rho-i\eta) & -\lambda^2 A & 1 \end{pmatrix}$$

λ, A, ρ and η are defined, principally, as experimental parameters, but λ is the Cabibbo parameter for the first and second generation mixing, and η defines the *CP* violating phase. The matrix in this formulation is largely empirical, but presumably has some basis in the electroweak splitting, which, according to our previous arguments, has an idealised 1 in 4 ratio. We may imagine as a working hypothesis that, ideally, the Cabibbo mixing is 1/4 for the first and second generations (λ) and 1/16 for the second and third (λ^2), though a more complicated picture would result from including mixing between the first and third generations, and *CP* violation.

The CKM matrix was originally produced to derive the weak eigenstates of quarks from the mass eigenstates, but, in a fully unified theory, with parity between quarks and leptons, it is difficult to believe that it does not apply equally to leptons, especially as the electroweak mixing, the mechanism actually producing the mass, is blind to the presence or absence of the strong charge. Of course, as Halzen and Martin write of the quark matrix, 'a more involved mixing in both the u, c and d, s sectors can be used but it can always be simplified (by appropriately choosing the phases of the quark states) to the one parameter form'.[9] They also ask (prior to the discovery of neutrino oscillations): 'Why is there no Cabbibo-like angle in the leptonic sector?' And answer: 'The reason is that if ν_e and ν_μ are massless, then lepton mixing is unobservable. Any Cabibbo-like rotation still leaves us with neutrino mass eigenstates.'

Lepton masses, of course, unlike quark masses, are fixed, with no 'running' aspect, and so, if the CKM matrix applies to leptons, we might expect to find it in a 'purer' form, its values approaching more closely to the idealised ones. Let us suppose, therefore, that a hypothetical 'pure' matrix acts upon a set of lepton mass eigenstates e, μ, τ to produce a mixed set of weak eigenstates e', μ', τ'. That is, we assume that, though there is no compulsion or mechanism for leptons to be mixed in the same way as quarks, the symmetry determining the masses of e, μ, τ requires a set of mixed states e', μ', τ', such that

$$\begin{pmatrix} 1-\lambda^2/2 & \lambda & \lambda^3 A(\rho - i\eta) \\ -\lambda & 1-\lambda^2/2 & \lambda^2 A \\ \lambda^3 A(1-\rho-i\eta) & -\lambda^2 A & 1 \end{pmatrix} \begin{pmatrix} e \\ \mu \\ \tau \end{pmatrix} = \begin{pmatrix} e' \\ \mu' \\ \tau' \end{pmatrix}$$

Applying the principle that the fermion masses are generated through the perfectly random rotation of $\hat{\mathbf{p}}_a$, $\hat{\mathbf{p}}_b$, and $\hat{\mathbf{p}}_c$, we might expect that the intrinsic masses of the fermions are related in some way to the constant α_3, which provides the 'unit' mass under ideal conditions, and which is approximately 1 / 8 at the energy of the electroweak splitting represented in the CKM matrix. Using the accepted values for the respective masses of e, μ and τ at 0.511×10^{-3}, 0.10566 and 1.770 GeV, we obtain an approximate correlation with the numerical value of $1 / \alpha_3$ for the mass ratios of τ' / μ' and μ' / e'. The correlation becomes more exact as $A \to 1$, $\rho \to 0$, and $\lambda \to 0.25$.

So, continuing the parallel between the lepton and quark sets, we imagine that the separation between the mass values for e' and μ' may be determined by the 'strong' factor α_3 (at the energy of M_W - M_Z), with the first generation mass being $\alpha_3 \approx 1 / 8$ times that of the second, and that the same applies to the separation between the mass values for μ' and τ'. Again, the connection with α_3 (if present) must occur through the relation between the strong interaction potential and the perfectly random rotation of the angular momentum operators, rather than due to the necessary presence of any strong charge; so, perfect randomness applied to lepton angular momentum operators has the same structure as that applied to those defined for the quarks in baryons and mesons. In principle, it is the perfectly random rotation of the angular momentum states, $\hat{\mathbf{p}}_a$, $\hat{\mathbf{p}}_b$, and $\hat{\mathbf{p}}_c$, which *determines the behaviour of the strong interaction*, with its linear potential and asymptotic freedom, and the value of its fine structure constant, α_3, and associated unit mass; and not the strong interaction which determines the rotation of the angular momentum states.

The result of the CKM calculations seems to suggest that the masses of e', μ', τ' might be determined as though in a quark mixing, though there is no actual mixing between e, μ and τ. If, then, as is highly probable, the mass of e is determined uniquely in the form of m_e / α for 'unit charge', then the masses of e, μ and τ could be, in principle, determined absolutely.[33]

Identical considerations should apply to the quarks d, s, b and their CKM-rotated equivalents d', s', b', as to the leptons e, μ and τ. At grand unification their masses would be the same as those of the free fermions or leptons. (For example, d, s and b quark masses of something like 6×10^{-3}, 0.25 and 4 GeV would fit the same pattern.) However, at other energies, the mass values associated with d, s, b and d', s', b' would become variable, along with the fine structure constants, and, presumably, the mixing parameters. The exact CKM parameters would be similar to the idealized ones but would diverge from them according to the necessity of fulfilling such conditions, from the rescaling of α_3, as the quark masses at measurable energies being approximately 3 times the lepton masses; and of fixing the sum of fermion masses at 182 GeV.

Neutrino mixing is currently a major topic in particle physics, and has a distinct bearing upon the concept of neutrino mass, and the status of the neutrino as a 'Dirac' or 'Majorana' particle. Ideally, of course, we might expect weak mixing or oscillation between neutrino states whose charge structures are either $+w$, $-z_P\mathbf{k}w$, or $-z_T\mathbf{k}w$. The last two structures, in particular, look virtually identical if we realize that there is no observable difference between parity- and time reversal-violation if no other type of measurement can be made. The structures of both neutrinos and antineutrinos as composed purely of $+w$ and $-w$ values which are indistinguishable via the weak interaction might point to Majorana-type behaviour for massive neutrinos, with left-handed neutrinos mixing with right-handed antineutrinos (and possibly an embedding of the $SU(5)$ grand unified gauge group into something like $SO(10)$). In addition, with only weak charges present, the parity and time-reversal violations required to distinguish between $-z_P\mathbf{k}w$ and $-z_T\mathbf{k}w$ are, in themselves, indistinguishable and suggest maximal mixing of the muon and tau neutrino states. However, neutrino observations cannot, at present, be made outside of their interactions with the other leptons; and the issue of their mixings and oscillations cannot be considered separately from the possibility of mixings between these other lepton states, and the parallels they suggest with the already-observed mixings between the quarks.

Though the observation of neutrino mixing might suggest, at first sight, the existence of 'physics beyond the Standard Model', it needs to be looked at in connection with the parallel mixing in the quark sector, where the u', c', t' eigenstates are effectively 'gauged' away in the production of d', s', b'. This process is represented as a convention in standard theory, as of course it is, but there may be a reason for the convention if we attribute the introduction of mass to the presence of e alongside w. The fact that we need only one isospin state to be mixed must reflect the observation that only one such state has a nonzero value of e for any lepton or colour of quark. If one argues that the neutrinos are mixed, then one should also argue that, by comparison with the quark sector, the other leptons should also be mixed, and that, by symmetry with the quarks, we may transform away any mixing in one of the isospin states for the leptons, and, again by symmetry, this should be that of the neutrino states, which parallel u, c, t. Presumably, we could not tell physically whether it is neutrinos that are 'really' mixed or the other leptons. In this sense, there seems to be no reason why we could not use the mixing matrix to produce e', μ', τ' eigenstates, rather than neutrino eigenstates (i.e. use 'lepton gauge' rather than 'neutrino gauge'), and so restore the mathematical basis of the Standard Model.

15.12 A Summary of the Mass Calculations

Though the suggestions for mass calculations given in these final sections cannot be claimed to have equal status with most of the qualitative results that precede them, or even to have equal status with each other, taken together, they do provide a strategy for calculating particle masses which can be put forward as a working hypothesis, and conveniently presented as a unified approach, though the different arguments used largely stand or fall by themselves. Some of the arguments, I believe, are relatively certain – definitely the grand unification calculation, probably the masses of the composite baryons and mesons; others are strongly suggestive – the ones relating to the Higgs and weak gauge bosons and the total mass of the twelve known fermions. Some other arguments are necessarily more tentative, but most are available to testing by experiment. So it will be useful to collect the arguments in a brief summary.

The electric, weak and strong couplings (α, α_2, α_3) unify at the Planck mass, a quantity which is in effect a pure number, formed from the purely dimensional constants, $\hbar$, c and G. The electric and weak couplings are related by a factor, $\sin^2\theta_W = 0.25$, calculable from first principles, at the expectation energy of the vacuum, with a slightly reduced value at the energy of the intermediate bosons, Z and W (though the actual production of such massive bosons may be thought to reduce the value even further as measured). The running of the couplings is well known from equations derived from their respective $U(1)$, $SU(2)$ and $SU(3)$ symmetries. Together with the value for $\sin^2\theta_W$, they produce absolute values for each of the couplings at any given mass-energy, with no other empirical input. The strong coupling represents a perfect gauge-invariant rotation between its three phases, and its value may be thought to become unit ($\hbar c = 1$) at some fundamental value ($m_f c^2$) of mass-energy appropriate to a 'pure' unit charge. Given such a value, the mass of the electron (m_e), which is assumed to be determined solely from the electric charge, may be supposed to represent a reduction to that determined by the ratio of electrical to unit coupling ($m_f\alpha$). That is, $m_f = m_e / \alpha$. The perturbation calculation from α_3 produces a value of the right order but is not yet exact.

The masses of particles connected ultimately with vacuum states may be assumed to come from the removal of units of charge (the reverse process to creation). Imagining a bosonic state created from the zeroing of all possible charge structures (as specified in the tables A, B, C, L) suggests a possible Higgs mass of $2592\, m_e / \alpha = 182$ GeV, exactly as would be necessary if supersymmetry were implicit, as in this theory, rather than explicit. Eliminating the strong states (i.e. making $A = B = C$) would also indicate a purely electroweak interacting boson with maximal coupling (Z) with a mass (from A / L) of half this value, i.e. 91 GeV. Electroweak theory then equates M_W with $M_Z\cos\theta_W$, which, using the effective value of $\sin^2\theta_W$, makes $M_W \sim 80.4$ GeV. From the standard theory, the expectation value of the vacuum Higgs field (f) is determined by $M_W = gf / 2$, with g the weak coupling constant expressed in charge units. Taking the charge unit as the one expected for a quark-type system with the charge divided between three phases, we obtain $f \sim 3\, M_W$ or 241 GeV, which compares well with the 246 GeV obtained empirically from the Fermi constant. In generating fermion mass by coupling the Higgs boson to the twelve known fermion states, we assume that the Higgs mass is partitioned proportionately via weak coupling so that its value is also the total value of the masses of the fermion states.

The electroweak splitting with $\sin^2\theta_W = 0.25$ is also assumed to operate, ideally, in the weak mixing between quark and lepton generations which converts mass eigenstates to weak eigenstates. So the same ratio determines the idealised Cabibbo angle in a postulated idealised CKM matrix applied to the leptons, e, μ, τ, with neutrino mixing transformed away mathematically in the same way as u, c, t mixing for quarks. Starting with the electron mass, and the assumption that the weak eigenstates for the three generations are approximately separated from each other by the factor which indicates pure coupling (α_3), and mass generation through completely random rotation of the angular momentum operators (as assumed in making m_f the fundamental mass unit), we can, in principle, derive mass eigenstates for μ and τ, though precise calculations will require more exact knowledge of *CP* violation.

The $SU(2)$ symmetry creates paired states of weak isospin. The general formula derived for fermion charge states, however, suggests that the masses relating to the splitting of the isospin states do not come from the perfectly random rotation which determines the masses of all other states, but from the creation of a full or empty 'electromagnetic vacuum'. We therefore separate the generation masses determined this way by the electromagnetic factor α, making the quark, or, more probably the quark-lepton, generations, partition the total Higgs boson mass in this way. Taking into account values found for the lepton states e, μ, τ, we obtain 179, 1.3, and 9.5×10^{-3} GeV for the masses of the three quark generations.

The t quark, at least is taken to be that required for maximal coupling to the Higgs field, $f / \sqrt{2} \sim 174$ GeV, which fixes m_b in the region of 4 GeV. We may also apply $SU(n)_f$ symmetries to the zero charge values produced by the less massive composite baryon and meson states, assuming that the masses of the heavier quarks are of sufficient size to create mass degeneracies within the three-quark and quark-antiquark sets of states. If $SU(4)_f$ is valid in this context, as well as $SU(3)_f$, then the second generation quark masses may be partitioned so that $m_c \sim 1.2$ GeV and $m_s \sim 0.1$ GeV. Model-dependent theories also give us information about the 'running' values of the masses of m_b and m_s which relate them to the values of m_τ and m_μ at certain energies. After putting all this information together, we are left with the problem of $\sim 9.5 \times 10^{-3}$ GeV for the total mass of u and d, with no immediate method of partitioning, except to note that, empirically, $m_d \sim 2m_u$, in the reverse proportion to their electric charges. The neutrinos are here assumed here to have relatively negligible masses, and no direct suggestions can be offered as yet on finding more exact values beyond the speculative proposals already made by theorists.

Chapter 16

The Factor 2 and Duality

While the previous chapters have been concerned with specific consequences of the overall theory, this one brings together many aspects of physics at the fundamental level, as the concept of duality produced by conjugation, complexification and dimensionalization, and emerging from the zero totality with which we began, is shown to be responsible for the appearance of the numerical factor 2 in many seemingly unconnected physical phenomena. Multiple (and equally valid) explanations of the same phenomena show the interchangeability of the processes, and their ubiquity across the whole range of physics. It is apparent, in particular, that the factor is neither quantum nor relativistic in origin, and its significance is as much mathematical as physical.

16.1 Duality and Physics

At the heart of both physics and mathematics is a fundamental principle of duality. Essentially, this is a way of creating 'something from nothing'. If the ultimate thing that we wish to describe is really 'nothing', then we can only create 'something' as part of a dual pair, in which each thing is opposed by another thing which negates it. A simple physical example occurs in the conservation of momentum, where a system such as a gun and bullet has the same zero momentum after firing as it had before, because gun and bullet acquire equal momentum in opposite directions. Of course, this is merely an example of a very general physical principle, but such principles themselves derive from the even more fundamental principle of duality, which preserves the zero total state.

We can describe this mathematically in terms of the simplest known symmetry group (C_2), which is essentially equivalent to an object and its mirror image (or 'dual'), whose components are the positive and negative versions of a quantity which may be left undefined. Of course, duality does not always imply equal status, and may incorporate *chirality*, as in the different status of + and –

units in binary numbering. Duality, in addition, is not a single operation, and the process requires indefinite extension, in the form $C_2 \times C_2 \times C_2 \times \ldots$. If we begin with a unit, there will be an infinite series of 'duals' to this unit, via a process which must be carried out with respect to all previous duals (that is, that the entire set of characters generated becomes the new 'unit') and the total result must be zero at every stage.

In simple terms, we can't define something without defining also what it is not, and we can't characterize 'nature' or 'reality', even to the extent of saying whether it has an independent existence (is ontological) or is a product of our perception (is epistemological). A 'theory of everything', including physics and mathematics, needs first to be a 'theory of nothing'. We start from 'nothing' and we end with 'nothing', and duality is there to ensure that when we introduce 'something', we still end with nothing. But this does not mean that we cannot determine its structure. The fundamental duality operates in the most simple way possible.

Duality has an astonishingly simple manifestation in physics through the appearance of the factor 2 everywhere where it becomes significant. Though there are many competing physical explanations for the appearance of this factor, they can all be shown to be versions of one of the more fundamental dualities, in the same way as such physical phenomena as wave-particle duality, and the dualities between electric and magnetic forces, and between space and time in relativity. None of these is intrinsically mysterious. All have an explanation in terms of duality at the fundamental level, and each at some stage invokes the factor 2. Essentially, each process that doubles the options available also produces a doubling of the physical effect which can be reduced to simple numerical terms, though, at the same time, this is often balanced by a halving of the options in another direction.

16.2 Kinematics and the Virial Theorem

There is a purely geometric factor 2 in the formula for the area of a triangle, ½ × length of base × perpendicular height. In the right-angled triangle, it is created by bisecting a rectangle along a diagonal. This can be applied to kinematics if we represent a motion under uniform acceleration (a) by a straight-line graph of velocity (v) against time (t). The area under the graph now becomes the distance travelled, ½ vt. If the motion had been under uniform velocity, the distance would have been represented as the area of a rectangle, vt. Here, the factor 2, distinguishes between steady conditions and steadily *changing* conditions.

We can develop the idea further to produce the well-known kinematic equations for uniform acceleration. Starting from an initial velocity u, and supposing the same uniform acceleration, we obtain the 'mean speed theorem', in which the total distance travelled under uniform acceleration equals the product of the mean speed and the time: $s = ½ (u + v) t$. If we additionally define uniform acceleration as $a = (v - u) / t$, we obtain the well-known equation for uniformly accelerated motion: $v^2 = u^2 + 2as$, which becomes $v^2 = 2as$, when $u = 0$. If we now apply this to a body of mass m, acted on by a uniform force $F = ma$, we find the work done over distance s is equal to the kinetic energy gained

$$Fs = mas = \frac{mv^2}{2} - \frac{mu^2}{2},$$

which reduces to $½\, mv^2$ if we start at zero speed. Using $p = mv$ to represent momentum, it is convenient also to express this formula in the form $p^2 / 2m$. It is easy, of course, to show that this formula applies additionally to the case of nonuniformly accelerated motion, using a simple integration of force (dp / dt) over displacement:

$$\int \frac{dp}{dt} ds = \int mv\, dv = \frac{m_0 v^2}{2}.$$

In principle, however, we see that a steady increase of velocity from 0 to v requires an averaging out which halves the values of significant dynamical quantities obtained under steady-state conditions.

The same factor makes its appearance, in precisely the same way, in molecular thermodynamics, quantum theory and relativity. Its significance here is that it relates the continuous aspect of physics to the discrete, and, since these aspects are required in the description of any physical system, the factor acquires a universal relevance. An obvious classical manifestation is the fact that *two* types of conservation of energy equation are commonly used in physics. The kinetic energy formula applies when a system is undergoing change. A classic example is that of a body of mass m escaping from a gravitational field, where

$$\frac{mv^2}{2} = \frac{GMm}{r}.$$

But if a system is in steady-state, with no overall change in the energy distribution, as in a classical circular gravitational orbit, the force equation, which here is

$$\frac{mv^2}{r} = -\frac{GMm}{r^2},$$

leads to an equivalent potential energy relation

$$mv^2 = \frac{GMm}{r},$$

in which the potential energy has twice the value of kinetic.

Though the potential and kinetic energy equations may, at first sight, appear to be contradictory expressions of the general principle of the conservation of energy, they can be easily reconciled if we consider the kinetic energy relation to be concerned with the action side of Newton's third law, while the potential energy relation concerns both action and reaction. We can give many physical illustrations. An old proof of Newton's of the mv^2 / r law for centripetal force, and hence of the formula mv^2 for orbital potential energy, had the satellite object being 'reflected' off the circle of the orbit, in a polygon with an increasing number of sides, which, in the limiting case, becomes a circle. The imagined physical reflection, by doubling the momentum through action and reaction, then produces the potential, rather than kinetic, energy formula.

Precisely the same principle applies in the derivation of Boyle's law, from what we often call the 'kinetic theory of gases'. Here, a real reflection of the ideal gas molecules off the walls of the container produces the momentum doubling, which indicates steady-state conditions, though it is immediately removed by the fact that we have to calculate the average time between collisions ($t = 2a / v$) as the time taken to travel *twice* the length of the container (a). The average force then becomes the momentum change / time = $2\,mv / t = mv^2 / a$, and the pressure due to one molecule in a cubical container of side a becomes mv^2 / a^3, or mv^2 / V (volume), leading, for n molecules, to the direct pressure-density relationship, which we call Boyle's law ($P = \rho\,\bar{c}^2 / 3$, where $\bar{c}$ is the root mean square velocity). The kinetic behaviour of the ideal gas molecules is actually irrelevant to the derivation of $P \propto \rho$, since the system describes a steady-state dynamics with positions of molecules constant on a time-average (though we can, of course, observe it using Brownian motion). Taking into account the three dimensions between which the velocity is distributed, the ratio of pressure and density (P / ρ) is derived from the *potential energy* term mv^2 for each molecule and is equal to one third of the average of the squared velocity, or $\bar{c}^2 / 3$. So, the relationship could have been derived (as was done by Newton) using a mathematical model in which the positions of the molecules remained fixed.

What is surprising, however, is that photons, which, unlike material particles, are relativistic objects, behave in exactly the same way in a 'photon gas', producing a radiation pressure of the form $P = \rho c^2 / 3$, with the relativistic energy $E = mc^2$ behaving exactly like a classical potential energy term, and with no mysterious 'relativistic factor' at work. We can consider the photons as being reflected off the walls of the container in exactly the same way as the molecules of materials although the real process obviously also involves absorption and re-emission.

The potential / kinetic energy ratio of 2 is, of course, only a special case of the virial theorem, according to which, in a conservative system governed by force terms inversely proportional to power n of the distance, or potential energy terms inversely proportional to power $n - 1$, the time-averaged kinetic and potential energies, $\overline{T}$ and $\overline{V}$, are related by the formula:

$$\overline{T} = \frac{(1-n)}{2} \overline{V} .$$

For the two special cases, of constant force and inverse-square-law force, $\overline{V}$ is numerically equal to $2\overline{T}$. Such forces, in fact, are overwhelmingly predominant in nature, because they are a natural consequence of three-dimensional space, and this may well be related to the geometric origin of the factor 2 in such formulae as that for the area of a triangle.

In the case of material gases, the kinetic behaviour only becomes significant when we introduce temperature as a measure of the average kinetic energy of the molecules. There is no 'derivation' involved here, because temperature is not defined independently of the kinetic energy, and we are obliged to provide the definition by an *explicit* use of the virial theorem, to find the otherwise *unknown* average kinetic energy from the *known* potential energy. Assuming that the potential energy of each ideal gas molecule is kT for each degree of freedom, and, in total, $3kT$, and taking the pressure law as equivalent to a dynamical system involving a constant force, we apply the virial theorem, for conditions equivalent to a constant force, to obtain the kinetic energy expression ($3kT / 2$) for each of these molecules.

16. 3 Relativity

The whole reason for Einstein's introduction of $E = mc^2$ to represent the total energy of both photons and material particles was to preserve the *classical* laws of conservation of mass and conservation of energy. The total energy equation, unlike the *change of energy* formula $\Delta E = \Delta mc^2$, cannot be derived, by deductive means, from the postulates of relativity; it depends entirely on the choice of an integration constant in the relativistic expression for rate of energy change:

$$\frac{dT}{dt} = \mathbf{F.v}\,.$$

No problem arises if we recognise that mc^2 has a classical, as well as relativistic, meaning. Like many other significant results (the Schwarzschild radius, the equations for the expanding universe, the gravitational redshift, the spin of the electron), the expression does not arise from the theory of relativity itself but is a more fundamental truth which that theory has uncovered.

The doubling of the energy term in $E = mc^2$, by comparison with classical potential energy, is sometimes described as 'relativistic', but relativistic factors tend to be of the form $\gamma = (1 - v^2 / c^2)^{-1/2}$, suggesting some *gradual* change when $v \to c$, rather than an abrupt transition involving a discrete integer. $\Delta E = \Delta mc^2$, used for material particles with rest mass m, is a relativistic equation because it incorporates the γ factor in the Δm term, but $E = mc^2$ is simply a requirement needed by Einstein to reconcile special relativity with classical energy-conservation laws, and, since it is determined solely by an integration constant, it cannot be derived directly from relativity itself. In the case of photons, which have no rest mass and no kinematics, there is no distinction between a 'relativistic' approach and one based on classical potential energy, and it is perfectly possible to do classical calculations for photons, entirely independent of any concept of relativity.

Examples of this can be found in calculations based on the classical corpuscular theory of light dating back to the seventeenth, eighteenth and nineteenth centuries, and some are still used for practical purposes at the present day in textbooks on spherical astronomy.[1] We may mention, for example, Newton's calculation of atmospheric refraction in 1694, and his application of the formula for the velocity of waves in a medium to an optical aether in Query 21 of the *Opticks*.[2] Essentially, Newton's formula,

$$c = (E / \rho)^{1/2}$$

where E is elasticity or pressure and ρ density, is an expression of the fact that the potential energy of the system of light corpuscles, or the aether that acts upon them, is equal to the work done at constant pressure as a product of pressure and volume. Newton's elasticity of the aether is essentially the same as the modern energy density of radiation (ρc^2), which is related by Maxwell's classical formula of 1873 to the radiation pressure. Light necessarily gives a 'correct' result for such a calculation when travelling through a vacuum, because there is no source of dissipation, and the virial relation takes on its ideal form.

In addition, even though free photons have no kinematics, it is also perfectly possible to treat photons acting under the constraint of certain forces as though they have. One example occurs in a plasma, where the photons acquire an effective 'rest mass' – this has often been used as an analogy to the Higgs mechanism for acquiring mass in particle physics. Another occurs in a gravitational field, where the light 'slows down', and behaves exactly as if it had kinetic energy in the field. This is why it is possible to use the standard Newtonian escape velocity equation

$$\frac{mv^2}{2} = \frac{GMm}{r}$$

to derive the Schwarzschild limit for a black hole, as was done as early as the eighteenth century, by assuming $v \to c$, with no transition to a 'relativistic' value. In effect, although light in free space has velocity c, and, therefore, no rest mass or kinetic energy, *as soon as you apply a gravitational field*, the light 'slows down', and, at least *behaves* as though it can be treated as a particle with kinetic energy *in the field.*

Less obviously, but equally correctly, we can derive the full double gravitational bending of light, using the kinetic equation,

$$\frac{mc^2}{2} = \frac{GMm(e-1)}{r},$$

for an orbit which may be assumed to be hyperbolic with eccentricity e. We do *not* use the potential energy equation

$$mc^2 = \frac{GMm(e-1)}{r},$$

which requires steady-state conditions, which do not apply when an orbit is in the process of creation, as here (cf 17.7). From the kinetic energy equation, with $1 \ll e$, the full angle deflection (in and out of the gravitational field) is easily derived as

$$\frac{2}{e} = \frac{4GM}{c^2 r}.$$

Contrary to popular opinion, Soldner, who attempted a calculation in 1801, did, in fact, use the correct kinetic energy, and not the incorrect potential energy equation, and only obtained an incorrect final result because he integrated over the single, rather than double, angle. Soldner rightly saw the procedure as being a kind of reverse analogy of Laplace's black hole calculation, though using a hyperbolic rather than a circular orbit, the significant fact being that the photon's speed outside the gravitational field is not determined by it.

That a purely classical calculation of the full deflection is possible should not surprise us. Energy, in relativity, is, after all, *defined* to be consistent with its classical value in the case of a particle with no material component; and so relativity theory should not produce different energy equations to classical physics for light photons; it merely corrects our naïve understanding of what are steady-state and what are changing conditions. Of course, in the case of photons, we never see a material kinetic energy directly, but the total energy balance means that it must be possible to treat it as though it does exist when the particle is 'slowed down' by a field.[3]

However, the successful use of a classical argument doesn't mean that we *can't* use a special or general relativistic argument to derive the effect. The work of many authors has shown that we can. What it does mean is that the cause of the effect itself is independent of the particular version of physics we use to calculate it. Something more profound is involved. This seems to be the fact that, in every case where a 'relativistic' correction (either special or general) seems to 'cause' the doubling of a physical effect, the relativistic aspect, like classical kinetic energy, is providing a way of incorporating the effect of *changing conditions* if we begin with the potential, rather than the kinetic, energy equation.

It isn't necessarily important what particular physical phenomena we invoke to support the calculation, and the many disagreements over the 'cause' of the double deflection bear witness to this. Authors have generally agreed that there are two separate physical components involved, but not on what they are. In principle, it would seem that the potential energy equation is responsible for gravitational redshift, or time dilation, which gives half the effect, while relativity adds the corresponding length contraction.[4] Now, this may be interpreted as redshift being 'Newtonian' while the length-contraction or 'space-warping' is relativistic,[5] but it may equally be postulated that the length contraction is Newtonian while the redshift is relativistic.[6] It has also been argued that the 'Newtonian' effect has to be added to the Einstein equivalence principle calculation of 1911 (which again gives half the effect),[7] but a counter-argument suggests that these two effects are the same, and need supplementing with a 'true' relativistic effect, like the Thomas precession.[5] Amazingly, *all* of these arguments are correct! They are by no means mutually exclusive. None, however, is fundamental. In reality, it depends on the choice of classical energy equation. If we use the potential energy equation where the kinetic energy equation is appropriate, then we can find correct physical reasons for almost *any* additional term which doubles the effect predicted. Even special relativity is only an alternative approach to a calculation that must also be valid classically.[8]

16.4 Spin and the Anomalous Magnetic Moment

Almost exactly the same reasoning can be shown to apply to the anomalous magnetic moment or, equivalently, the gyromagnetic ratio, of a Bohr electron acquiring energy in a magnetic field. According to 'classical' reasoning, we are told, the energy acquired by an electron changing its angular frequency from ω_0 to ω in a magnetic field **B** will be of the form

$$m(\omega^2 - \omega_0^2) = e\omega_0 rB,$$

leading, after factorization of $(\omega^2 - \omega_0^2)$, to an angular frequency change $\Delta\omega = eB / 2mr$. But a relativistic effect (the Thomas precession, again!) replaces the classical $e\omega_0 rB$ by $2e\omega_0 rB$, doubling the value of $\Delta\omega$. All we need to do, however, to obtain the correct value of $\Delta\omega$ is to realise that we must use the *kinetic* energy equation when we have changing conditions, as, for example, *at the instant we 'switch on' the field.* Then, we automatically write

$$\tfrac{1}{2} m(\omega^2 - \omega_0^2) = e\omega_0 rB,$$

which is nothing more than a version of the kinematic equation $v^2 - u^2 = 2as$. The Thomas precession is only needed as a 'relativistic' correction if we begin with the potential energy equation applicable to a steady state.

George Uhlenbeck, one of the discoverers of electron spin, remarked about the introduction of the Thomas precession in this context: 'I remember that, when I first heard about it, it seemed unbelievable that a relativistic effect could give a factor of 2 instead of something of order v / c. ... Even the cognoscenti of the relativity theory (Einstein included!) were quite surprised.'[9] As we have seen, however, there is nothing exclusively relativistic about the factor, and its origin is much more fundamental.

In showing that the gyromagnetic ratio of a Bohr electron is not truly 'anomalous' or relativistic in origin, but perfectly capable of a classical explanation, we are also showing that the origin of the factor 2 in the electron spin term is not of a fundamentally quantum origin either. Traditionally, of course, electron spin is derived from the relativistic Dirac equation by consideration of the commutator

$$[\hat{\sigma}, \mathcal{H}] = [\hat{\sigma}, i\gamma_0\boldsymbol{\gamma}.\mathbf{p} + \gamma_0 m].$$

Purely formal reasoning reduces this to $2\gamma_0\ \boldsymbol{\gamma} \times \mathbf{p}$, which, in the multivariate vector terminology used in this book (equivalent to Pauli matrices), becomes $2ij\ \mathbf{1} \times \mathbf{p}$. The significant thing here is that the factor 2 emerges from the anticommuting properties of the vector operators in an equation such as

$$[\hat{\sigma}, \mathcal{H}] = 2j\ (\mathbf{ij}p_2 + \mathbf{ik}p_3 + \mathbf{ji}p_1 + \mathbf{jk}p_3 + \mathbf{ki}p_1 + \mathbf{kj}p_2).$$

Ultimately, this leads to

$$[\mathbf{L} + \hat{\sigma} / 2, \mathcal{H}] = 0,$$

with the total angular momentum $(\mathbf{L} + \hat{\sigma} / 2)$, including the spin term $\hat{\sigma} / 2$.

Originally, with its automatic derivation from the Dirac equation, this term was thought to be related to the relativistic nature of this equation. However, the same result (or, more specifically, its manifestation in the presence of a magnetic field) can be derived from the nonrelativistic Schrödinger equation, if we use a *multivariate* momentum operator, as we do automatically in the Dirac equation. Significantly for our purposes, the standard derivation of the Schrödinger equation (which can easily be shown to be a nonrelativistic limit to the bispinor form of the Dirac equation) proceeds by quantizing the classical expression for kinetic energy:

$$T = \left(E - V\right) = \frac{p^2}{2m}$$

using the operator substitions $E = i\ \partial / \partial t$ and $\mathbf{p} = (-i\nabla + e\mathbf{A})$, in the presence of a magnetic field determined by vector potential $\mathbf{A}$. Normally, the right-hand side of this equation is interpreted as $-\nabla^2\psi / 2m$, using the scalar product $(-i\nabla + e\mathbf{A}).(-i\nabla + e\mathbf{A})$. However, various authors have shown that, using a multivariate operator for $\mathbf{p} = -i\nabla + e\mathbf{A}$, we obtain:

$$2mE\psi = (-i\nabla + e\mathbf{A})\ (-i\nabla + e\mathbf{A})\ \psi$$

which leads ultimately to

$$2mE\psi = (-i\nabla + e\mathbf{A}).(-i\nabla + e\mathbf{A})\ \psi + 2m\ \boldsymbol{\mu}.\mathbf{B},$$

which we recognize as the form of the Schrödinger equation in a magnetic field, with a spin state supplied by the *ad hoc* addition of Pauli matrices (cf 7.2). It becomes an automatic component of our equation because we define a *full product* between multivariate vectors (or, equivalently, Pauli matrices or complex quaternions) **a** and **b** of the form **ab** = **a.b** + i **a** × **b**.

If real vectors, such as those representing space and momentum, are intrinsically multivariate, then a spin term will automatically result from taking the full product, and the ½-integral value of fermionic spin (incorporated here in the $2m$**μ.B** term) becomes a consequence of the vectors' anticommuting properties. Relativity doesn't come into it; it is a vector, not a 4-vector effect, as we can also see from the well-known fact that the 4π rotation involved in spin is purely a property of the rotation group. At the same time, the ½ in the Schrödinger equation itself clearly comes from the equation's initial derivation from the expression for classical kinetic energy. The factor is both introduced with the transition in the Schrödinger equation from the classical kinetic energy term, and, at the same time, produced by the anticommuting nature of the momentum operator. It is precisely because the Schrödinger equation is derived via a kinetic energy term that this factor enters into the expression for the spin, and this process is essentially the same as the process which, through the anticommuting quantities of the Dirac equation, makes ($\mathbf{L} + \hat{\boldsymbol{\sigma}} / 2$) a constant of the motion.

16.5 The Linear Harmonic Oscillator

A similar situation occurs when the Schrödinger equation is solved for the case of the quantum harmonic oscillator, with a varying potential energy term, ½ $m\omega^2 x^2$, taken directly from the classical kinetic energy term ½mv^2, added to the Hamiltonian. The ½ in this expression then leads by direct derivation to the ½ in the expression for the ground state or 'zero-point' energy of the system. The factor 2 in the quantum harmonic oscillator is clearly derived from the fact that the *varying* potential energy term is taken from a classical term of the $mv^2 / 2$ type. So, the Schrödinger equation for the eigenfunction $u_n(x)$ and eigenvalue E_n, with the $\hbar^2$ explicitly included and the spatial dimensions reduced to the linear x, becomes:

$$\left(-\frac{\hbar^2}{2m} \frac{\partial^2 u_n(x)}{\partial x^2} + \frac{m\omega^2 x^2}{2} \right) u_n(x) = E_n u_n(x) .$$

This equation, as solved in standard texts on quantum mechanics, produces a ground state energy $\hbar\omega / 2$, with the factor 2 originating in the $2m$ in the original equation. We define the new variables

$$y = \left(\frac{m\omega}{\hbar}\right)^{1/2} x \qquad \text{and} \qquad \varepsilon_n = E_n / \hbar\omega,$$

and the equation now becomes:

$$\left(\frac{\partial^2}{\partial y^2} - y^2\right) u_n(y) = \left(\frac{\partial}{\partial y} + y\right)\left(\frac{\partial}{\partial y} - y\right) u_n(y) + u_n(y) \tag{16.1}$$

$$= \left(\frac{\partial}{\partial y} - y\right)\left(\frac{\partial}{\partial y} + y\right) u_n(y) - u_n(y) = -2\varepsilon_n u_n(y). \tag{16.2}$$

From this we derive

$$\left(\frac{\partial}{\partial y} + y\right)\left(\frac{\partial}{\partial y} - y\right) u_n(y) = (-2\varepsilon_n - 1)\left(\frac{\partial}{\partial y} - y\right) u_n(y) \tag{16.3}$$

and $$\left(\frac{\partial}{\partial y} + y\right)\left(\frac{\partial}{\partial y} - y\right)\left(\frac{\partial}{\partial y} + y\right) u_n(y) = (-2\varepsilon_n + 1)\left(\frac{\partial}{\partial y} + y\right) u_n(y). \tag{16.4}$$

From (16.3), we may derive either $(\partial / \partial y - y)\, u_n(y) = 0$, which produces a divergent solution, or $(\partial / \partial y - y)\, u_n(y) = u_{n+1}(y)$ (say), which means that

$$\left(\frac{\partial}{\partial y} - y\right)\left(\frac{\partial}{\partial y} + y\right) u_{n+1}(y) = (-2(\varepsilon_n + 1) - 1) u_{n+1}(y),$$

which is (16.2) for u_{n+1} if $\varepsilon_n + 1 = \varepsilon_{n+1}$.

From (16.4), we may obtain either $(\partial / \partial y + y)\, u_n(y) = 0$, which gives us the ground state eigenfunction, $u_0(y) = \exp(-y^2 / 2)$; or $(\partial / \partial y - y)\, u_n(y) = u_{n-1}(y)$ (say). In the latter case, (16.4) becomes

$$\left(\frac{\partial}{\partial y} + y\right)\left(\frac{\partial}{\partial y} - y\right) u_{n-1}(y) = (-2(\varepsilon_n - 1) - 1) u_{n-1}(y)$$

if $$\varepsilon_{n-1} = \varepsilon_n - 1,$$

which gives us a discrete series of energies E_n at $n\hbar\omega$ above the ground state. From the ground state eigenfunction and (16.1), we obtain

$$2\varepsilon_0 - 1 = 0,$$

which gives us the ground state or 'zero-point' energy

$$E_0 = \frac{\hbar\omega}{2}.$$

Here, we can derive the factor 2 in E_0 directly from the introduction into the Schrödinger equation of the classical term $m\omega^2 x^2/2$, which is equivalent to $mv^2/2$.

16.6 The Heisenberg Uncertainty Principle

The zero-point energy term deriving from the harmonic oscillator can be related directly to the $\hbar\ /\ 2$ in the Heisenberg uncertainty relation, but the formal derivation of this relation also shows that the factor ½ is generated by anticommutativity in the same way as it is for electron spin. We assume a state represented by a state vector ψ which is an eigenvector of the operator P. In this case, the expectation value of the variable p^2 becomes

$$<p^2> = \psi^* P^2 \psi$$

and the mean squared variance

$$(\Delta p)^2 = \psi^* \{P - <p>I\}^2 \psi = \psi^* P'^2 \psi$$

if $P' = P - <p>I$ and I is a unit matrix. Similarly, for operator Q,

$$(\Delta q)^2 = \psi^* \{Q - <q>I\}^2 \psi = \psi^* Q'^2 \psi .$$

Since $P'\psi$ and $Q'\psi$ are vectors,

$$\begin{aligned}(\Delta p)^2\,(\Delta q)^2 &= (\psi^* P'^2 \psi)\,(\psi Q'^2 \psi) \geq (\psi^* P'Q'\psi)\,(\psi^* Q'P'\psi)\\ &\geq \left|(1/2)\,(\psi^* P'Q'\psi - \psi^* Q'P'\psi)\right|^2\\ &\geq (1/4)\,\left|(\psi^*(P'Q' - Q'P')\psi\right|^2\\ &\geq (1/4)\,[P, Q]^2\end{aligned}$$

Hence

$$\begin{aligned}(\Delta p)\,(\Delta q) &\geq (1/2)\,[P, Q]\\ &\geq \hbar\ /\ 2\end{aligned}$$

if P and Q do not commute. The significant aspect of this proof is that the factor 2 in the expression $\hbar\ /\ 2$ comes from the noncommutation of the p operator.

16.7 Fermions and Bosons

The factor 2 in spin states establishes the distinction between bosons and fermions, with bosons occupying integral spin states and fermions half-integral ones. Ultimately, this factor can be shown to originate in the virial relation between kinetic and potential energies. To do this, it is most convenient to use the nilpotent formulation of the Dirac equation for fermions A boson state vector in this formalism is a nonzero scalar, formed as a product of two nilpotents (each not nilpotent to the other). So a spin 1 boson quaternion becomes the sum of the terms

$$(\boldsymbol{k}E + \boldsymbol{i}\,\mathbf{p} + \boldsymbol{j}\,m_o)\,(-\boldsymbol{k}E + \boldsymbol{i}\,\mathbf{p} + \boldsymbol{j}\,m_o)$$
$$(\boldsymbol{k}E - \boldsymbol{i}\,\mathbf{p} + \boldsymbol{j}\,m_o)\,(-\boldsymbol{k}E - \boldsymbol{i}\,\mathbf{p} + \boldsymbol{j}\,m_o)$$
$$(-\boldsymbol{k}E + \boldsymbol{i}\,\mathbf{p} + \boldsymbol{j}\,m_o)\,(\boldsymbol{k}E + \boldsymbol{i}\,\mathbf{p} + \boldsymbol{j}\,m_o)$$
$$(-\boldsymbol{k}E - \boldsymbol{i}\,\mathbf{p} + \boldsymbol{j}\,m_o)\,(\boldsymbol{k}E - \boldsymbol{i}\,\mathbf{p} + \boldsymbol{j}\,m_o),$$

with the components of the fermion state arranged in a row vector (represented, for convenience, as a column), and the components of the antifermion state in a column vector.

While the Dirac and Schrödinger equations, which are ultimately concerned with kinetic energy states, produce fermions with half-integral spins, the Klein-Gordon equation, which applies to bosons, is a potential energy equation, based on $E = mc^2$, with m the 'relativistic', rather than rest mass m_o, and bosons derive their integral spin values from the fact that the energy term in this equation incorporates unit values of the mass m. The Klein-Gordon equation can be seen as a direct quantization of the classical relativistic energy-momentum equation in the form:

$$\frac{\partial^2 \psi}{\partial t^2} - \nabla^2 \psi = m_0 \psi ,$$

and necessarily applies to fermions as well as bosons. This is related to the fact that a fermion cannot be seen as isolated from its 'environment', and so effectively always acts as a member of a composite bosonic state.

Kinetic energy is always associated with rest mass, and cannot be defined without it; photons 'slowing down' in a gravitational field or condensed matter effectively acquire the equivalent of a rest mass. Potential energy, on the other hand, is associated with 'relativistic' mass because this term is actually *defined* through a potential energy-type expression ($E = mc^2$). Light in free space provides the extreme case, with no kinetic energy or rest mass, and 100 % potential energy or relativistic mass. In both classical and quantum physics, we use the kinetic energy relation when we consider a particle as an object in itself, described by a rest mass m_0, undergoing a continuous change; and the potential energy relation when we consider a particle within its 'environment', with 'relativistic mass', in an equilibrium state requiring a discrete transition for any change.

The particle and its 'environment' can be considered as two 'halves' of a more complete whole. This is evident, in the case of a material particle, when we expand its relativistic mass-energy term (mc^2) to find its kinetic energy (½ $m_o v^2$). In effect, we either take the relativistic energy conservation equation

$$E - mc^2 = E^2 - p^2c^2 - m_o{}^2c^4 = 0.$$

as a 'relativistic' mass or potential energy equation, incorporating the particle and its interaction with its environment, and then quantize to a Klein-Gordon equation, with integral spin; or, we separate out the kinetic energy term using the rest mass m_o, by taking the square root of

$$E^2 = m_0^2 c^4 \left(1 - \frac{v^2}{c^2}\right)^{-1}$$

to obtain

$$E = m_0 c^2 + \frac{m_0 v^2}{2} + \ldots$$

and, if we choose, quantize to the Schrödinger equation, and spin ½. The ½ occurs in the act of square-rooting, or the splitting of 0 into two nilpotents in the Dirac equation; the ½ in the nonrelativistic Schrödinger approximation is a manifestation of this which we can trace through the ½ in the relativistic binomial approximation. If we go directly to the Dirac equation to obtain the spin ½ term, we see that the same result emerges from the behaviour of the anticommuting terms; the anticommuting property is a direct result of taking the quaternion state vector as a nilpotent. So the anticommuting and binomial factors have precisely the same origin.

In addition, the connection between spin and statistics becomes obvious: square-rooting the scalar (and, so, commutative) operator, associated with an integral spin state, to produce two ½-spin states requires the introduction of quaternion operators which are necessarily anticommutative. So particles with integral spins (bosons) follow the Bose-Einstein statistics associated with commutative wavefunctions, while particles with ½-integral spins (fermions) follow the Fermi-Dirac statistics associated with anticommutative ones.

16.8 Radiation Reaction

One way of looking at the factor 2 is that it links the continuous with the discontinuous. Expressions involving half units of $\hbar$, representing an average or integrated increase from 0 to $\hbar$, are characteristic of continuous aspects of physics, while those involving integral ones are characteristic of discontinuous aspects. The Schrödinger theory is an example of a continuous option, while the Heisenberg theory is discontinuous. Stochastic electrodynamics (SED), which is based on the existence of zero-point energy of value $\hbar\omega / 2$, is another completely continuous theory, which has developed as a rival to the purely discrete theory of the quantum with energy $\hbar\omega$.

Again, the Klein-Gordon equation is effectively a 'discrete' one, in the space-time sense, whereas the Dirac, like the Schrödinger equation, is effectively 'continuous'. Now, as a result of the distinction between potential and kinetic energies, it is a characteristic of classical physics that 'discrete' (or steady state) energy equations employ terms which are twice the size of those describing 'continuous' (or changing) conditions; and the distinction is transferred into quantum mechanics with the quantum energy equations which are based on classical ones (though, in fundamental terms, of course, where measurement is not the primary concept, the transfer goes the other way). While the 'continuous' Schrödinger equation, for example, is based on a classical kinetic energy equation, the 'discrete' Klein-Gordon equation derives from a relationship involving potential energy. The choice between the factors 1 and ½ for spin, and other related quantities, seems to be made at the same time as that between timelike and spacelike equations, and between discrete and continuous physics.

It is important that we recognize that alternative options like QM and SED, or Heisenberg and Schrödinger, do not represent different systems; they are different ways of interpreting the *same* system, and both are required for a complete explanation (cf 7.4). Each has to incorporate the alternative option in some way. Thus, the Schrödinger approach is a continuous one, based on $\frac{1}{2}\hbar$, but incorporates discreteness (based on $\hbar$) in the process of measurement – the so-called collapse of the wavefunction. The Heisenberg approach, by contrast, assumes a discrete system, based on $\hbar$, but incorporates continuity (and $\frac{1}{2}\hbar$) in the process of measurement – via the uncertainty principle and zero-point energy. There seems always to be a route by which $\frac{1}{2}\hbar\omega$ in one context can become $\hbar\omega$ in another. A characteristic example is black-body radiation, where the spontaneous emission of energy of value $\hbar\omega$ combines the effects of $\frac{1}{2}\hbar\omega$ units of energy provided by both oscillators and zero-point field. In terms of fundamental particles, we see that a fermionic object on its own shows changing behaviour, requiring an integration which generates a factor ½ in the kinetic energy term, and a sign change when it rotates through 2π, while a conservative 'system' of object plus environment shows unchanging behaviour, requiring a potential energy term which is twice the kinetic energy.

The ½$h\nu$ or ½$\hbar\omega$ for black body radiation appears in both the theories of Planck, of 1911, and of Einstein and Stern, of 1913. (It also occurs in the magnetic flux quantum term, $\hbar / 2e$, which appears in the time-dependent voltage of the Josephson effect, $U(t) = (\hbar / 2e)\, \partial\phi / \partial t$, produced by a *changing* phase difference, ϕ.) In the Schrödinger version of quantum mechanics, as we have seen, the zero-point energy term is derived from the harmonic oscillator solution of the Schrödinger equation, showing the kinetic origins of the factor ½, while, in the Heisenberg version, it comes from the ½$\hbar$ term involved in the uncertainty principle, suggesting an origin in continuum physics. The ½$\hbar\omega \rightarrow \hbar\omega$ transition for black body radiation can also be explained in terms of radiation reaction, which is connected again with the distinction between the relativistic and rest masses of an object. Rest mass effectively defines an isolated object, with *kinetic* energy. *Relativistic* mass, on the other hand, already incorporates the effects of the environment. For a photon, which has no rest mass, and only a relativistic mass, the energy mc^2 behaves exactly like a classical potential energy term, as when a photon gas produces the radiation pressure $\rho c^2 / 3$. We take into account both action and reaction because the doubling of the value of the energy term comes from doubling the momentum when the photons rebound from the walls of the container, or, alternatively, are absorbed and re-emitted. Exactly, the same thing happens with radiation reaction, thus explaining an otherwise 'mysterious' doubling of energy from ½$h\nu$ to $h\nu$. In a more classical context, Feynman and Wheeler find a doubling of the contribution of the retarded wave in electromagnetic theory, at the expense of the advanced wave, by assuming that the vacuum behaves as a perfect absorber and reradiator of radiation. (This is effectively the same as privileging matter over vacuum, or one direction of time.)

From our analysis, it would seem that by incorporating radiation reaction in the process we are also incorporating the effect of Newton's third law, as in the case of many other processes. However, as in the parallel case of the anomalous magnetic moment of the electron, many of the same results are also explained by special relativity. C. K. Whitney[10] has argued that the correct magnetic moment for the electron is obtained, without relativity, by treating the transmission of light as a two-step process involving absorption and emission. In our terms, this is equivalent to incorporating action and reaction, and, as we have seen, the same result follows classically by taking the energy value at the moment when the field is switched on, which then becomes the new potential energy value when the system is in steady state. If, however, we use a one-step process (or kinetic energy), we also need relativity, because, once we introduce rest mass, we can no longer use classical equations. ('Relativistic mass' is, of course, specifically *designed* to preserve classical energy conservation!) The two-step process is analogous to the use of radiation reaction, so it follows, in principle, that a radiation reaction is equivalent to adding a relativistic 'correction' (such as the Thomas precession).

Whitney also argues that the two-step process removes those special relativistic paradoxes which involve apparent reciprocity, and we could say that special relativity, by including only one side of the calculation, effectively removes reciprocity, and so leads to such things as asymmetric ageing in the twin paradox. Very similar arguments also apply to the idea that the problem lies in attempting to define a one-way speed of light that cannot be measured, because a two-way measurement of the speed of light also requires a two-step process.[11] An argument by Morris[12] that the complete reciprocity involves a universal reference frame can be related to the notion here that reciprocity or reaction is the 'environmental' contribution as opposed to that of the particle.[13]

16.9 Supersymmetry and the Berry Phase

Taking 'environment' to apply to either a material or vacuum contribution, we can make sense, not only of the boson / fermion distinction and the spin 1 / ½ division in a fundamental way, but also of such related concepts as supersymmetry, vacuum polarization, pair production, renormalization, *zitterbewegung*, and so on, because the halving of energy in 'isolating' the fermion from its vacuum or material 'environment' is the same process as mathematically square-rooting the quantum operator via the Dirac equation. Taking this further, we can propose that energy principles determine that all fermions, in whatever circumstances, may be regarded either as isolated spin ½ objects or as spin 1 objects in conjunction with some particular material or vacuum environment, or, indeed, the 'rest of the universe'; and fermions with spin ½ automatically become spin 1 particles when taken in conjunction with their environment, whatever that may be. Characteristic examples of this occur when integral spins are produced automatically from half-integral spin electrons using the Berry phase, and, by generalizing this kind of result to all possible environments, we extend the principle in the direction of supersymmetry.

In the most general terms, we can consider that a relationship exists between any fermion and 'the rest of the universe', such that the *total* wavefunction representing fermion plus 'rest of the universe' is necessarily single-valued, automatically introducing a nonzero Berry phase. The Jahn-Teller effect and Aharonov-Bohm effect are examples of the action of this phase. Treated semi-classically, the Jahn-Teller effect couples the factors associated with the motions of the electronic and nuclear coordinates so that different parts of the total wavefunction change sign in a coordinated manner to preserve the single-valuedness of the total wavefunction. In the Aharonov-Bohm effect, electron interference fringes, produced by a Young's slit arrangement, are shifted by half a wavelength in the presence of a solenoid whose magnetic field, being internal, does not interact with the electron but whose vector potential does. The half-wavelength shift turns out to be a feature of the topology of the space surrounding the discrete flux-lines of the solenoid, which is not *simply-connected*, and cannot be deformed continuously down to a point. Effectively, the half-wavelength shift, or equivalent acquisition by the electron of a half-wavelength Berry phase, implies that an electron path between source and slit, round the solenoid, involves a *double-circuit* of the flux line (to achieve the same phase), and a path that goes round a circuit twice cannot be continuously deformed into a path which goes round once (as would be the case in a space without flux-lines). The change of phase or topology is, of course, equivalent to a weak interaction and a *CP*-violating one at that; the prediction of *zitterbewegung* in the solution of the Dirac equation indicates that the weak interaction is inherent in the 4-component Dirac spinor structure, in the same way as the electric interaction is fundamental in defining the spherical symmetry of a point-particle and the strong interaction is a necessary consequence of the vector nature of the momentum term.

The presence of the flux line in the Aharonov-Bohm effect is equivalent, as in the quantum Hall effect and fractional quantum Hall effect, to the extra fermionic ½-spin which is provided by the electron acting in step with the nucleus in the Jahn-Teller effect and makes the potential function single-valued, and the circuit for the complete system a single loop. It is particularly significant that the $U(1)$ (electromagnetic) group responsible for the fact that the vacuum space is not simply connected is isomorphic to the integers under addition, for this is the group that ensures spherical symmetry for a point source. In effect, the spin-½, ½-wavelength-inducing nature of the fermionic state (in the case of either the electron or the flux line) is a product of discreteness in both the fermion (and its charge) and the space in which it acts. In principle, the very act of creating a discrete particle requires a splitting of the continuum vacuum into *two* discrete halves (as with the bisecting of the rectangular figure with which we started, or, in another context, the Dedekind cut, which defines the real numbers), or (relating the concept of discreteness to that of dimensionality) two square roots of 0. Mathematically, the identification of 1 as separate from 0 also implies that 1 + 1 = 2, reflecting the fact that physics and mathematics have a common origin in the process that creates counting.

This duality between the fermion and its environment occurs with the actual creation of the fermion state. Splitting away a fermion from a 'system' (or 'the universe'), we have to introduce a coupling as a mathematical description of the splitting. The converse effect must also exist, with bosons of spin 0 or 1 coupling to an 'environment' to produce fermion-like states. Both fermions and bosons, it would seem, always produce a 'reaction' within their environment, which couples them to the appropriate wavefunction-changing term, so that the potential / kinetic energy relation can be maintained at the same time as its opposite. The Higgs mechanism may be conceived as occurring this way, but, more immediately, we can conceive of the phenomenon causing the coupling of gluons to a quark-gluon plasma and so delivering the total spin of ½ or $^3/_2$ to a baryon. It is possible to show that the whole process of renormalization depends on an infinite chain of such couplings through the vacuum. The coupling of the vacuum to fermions generates 'boson-images' and vice versa.

According to the process described in chapter 6, a fermion generates an infinite series of interacting terms of the form:

$(i\boldsymbol{k}E + \boldsymbol{i}\mathbf{p} + \boldsymbol{j}m)$

$(i\boldsymbol{k}E + \boldsymbol{i}\mathbf{p} + \boldsymbol{j}m)\ (-i\boldsymbol{k}E + \boldsymbol{i}\mathbf{p} + \boldsymbol{j}m)$

$(i\boldsymbol{k}E + \boldsymbol{i}\mathbf{p} + \boldsymbol{j}m)\ (-i\boldsymbol{k}E + \boldsymbol{i}\mathbf{p} + \boldsymbol{j}m)\ (i\boldsymbol{k}E + \boldsymbol{i}\mathbf{p} + \boldsymbol{j}m)$

$(i\boldsymbol{k}E + \boldsymbol{i}\mathbf{p} + \boldsymbol{j}m)\ (-i\boldsymbol{k}E + \boldsymbol{i}\mathbf{p} + \boldsymbol{j}m)\ (i\boldsymbol{k}E + \boldsymbol{i}\mathbf{p} + \boldsymbol{j}m)\ (-i\boldsymbol{k}E + \boldsymbol{i}\mathbf{p} + \boldsymbol{j}m)$, etc.

where $(i\boldsymbol{k}E + \boldsymbol{i}\mathbf{p} + \boldsymbol{j}m)$ (abbreviated from the 4-component vector) represents a fermion state and $(-i\boldsymbol{k}E + \boldsymbol{i}\mathbf{p} + \boldsymbol{j}m)$ an antifermion. The $(i\boldsymbol{k}E + \boldsymbol{i}\mathbf{p} + \boldsymbol{j}m)$ and $(-i\boldsymbol{k}E + \boldsymbol{i}\mathbf{p} + \boldsymbol{j}m)$ vectors are also an expression of the behaviour of the vacuum state, which acts like a 'mirror image' to the respective antifermion / fermion. An expression such as

$$(i\boldsymbol{k}E + \boldsymbol{i}\mathbf{p} + \boldsymbol{j}m)\ \boldsymbol{k}\ (i\boldsymbol{k}E + \boldsymbol{i}\mathbf{p} + \boldsymbol{j}m)$$

for a fermion creation operator is part of an infinite regression of images of the form

$$(i\boldsymbol{k}E + \boldsymbol{i}\mathbf{p} + \boldsymbol{j}m)\ \boldsymbol{k}\ (i\boldsymbol{k}E + \boldsymbol{i}\mathbf{p} + \boldsymbol{j}m)\ \boldsymbol{k}\ (i\boldsymbol{k}E + \boldsymbol{i}\mathbf{p} + \boldsymbol{j}m)\ \boldsymbol{k}\ (i\boldsymbol{k}E + \boldsymbol{i}\mathbf{p} + \boldsymbol{j}m) \ldots$$

where the vacuum state depends on the operator that acts upon it, the vacuum state of $(i\boldsymbol{k}E + \mathbf{i}\mathbf{p} + \boldsymbol{j}m)$, for example, becoming $\boldsymbol{k}\,(i\boldsymbol{k}E + \mathbf{i}\mathbf{p} + \boldsymbol{j}m)$. In each case, the action simply reproduces the original state (after normalization). In addition,

$$(i\boldsymbol{k}E + \mathbf{i}\mathbf{p} + \boldsymbol{j}m)\,\boldsymbol{k}\,(i\boldsymbol{k}E + \mathbf{i}\mathbf{p} + \boldsymbol{j}m)\,\boldsymbol{k}\,(i\boldsymbol{k}E + \mathbf{i}\mathbf{p} + \boldsymbol{j}m)\,\boldsymbol{k}\,(i\boldsymbol{k}E + \mathbf{i}\mathbf{p} + \boldsymbol{j}m)\ \ldots$$

is the same as

$$(i\boldsymbol{k}E + \mathbf{i}\mathbf{p} + \boldsymbol{j}m)\,(-i\boldsymbol{k}E + \mathbf{i}\mathbf{p} + \boldsymbol{j}m)\,(i\boldsymbol{k}E + \mathbf{i}\mathbf{p} + \boldsymbol{j}m)\,(-i\boldsymbol{k}E + \mathbf{i}\mathbf{p} + \boldsymbol{j}m)\ \ldots .$$

It thus appears that the infinite series of creation acts by a fermion / antifermion on vacuum is the mechanism for creating an infinite series of alternating boson and fermion / antifermion states as required for supersymmetry and renormalization. The nilpotent operators defined as quaternion state vectors for fermions and antifermions are also supersymmetry operators, which produce the supersymmetric partner in the particle itself. The Q generator for supersymmetry is simply the term $(i\boldsymbol{k}E + \mathbf{i}\mathbf{p} + \boldsymbol{j}m)$, and its Hermitian conjugate $Q\dagger$ is $(-i\boldsymbol{k}E + \mathbf{i}\mathbf{p} + \boldsymbol{j}m)$. Multiplying by $(i\boldsymbol{k}E + \mathbf{i}\mathbf{p} + \boldsymbol{j}m)$ converts bosons to fermions, or antifermions to bosons. Multiplying by $(-i\boldsymbol{k}E + \mathbf{i}\mathbf{p} + \boldsymbol{j}m)$ produces the reverse conversion of bosons to antifermions, or fermions to bosons. The supersymmetric partners, however, are not so much realisable particles, as the couplings of the fermions and bosons to vacuum states. The 'mirror imaging' process thus implies an infinite range of virtual E values in vacuum adding up to a single finite value, exactly as in renormalisation, with equal numbers of boson and fermion loops cancelling through their opposite signs. That is, if the supersymmetric virtual partners are merely vacuum images of the original particles, their mass values will be *identical*, so the infinite sum of boson masses added to the vacuum will be identical to the infinite sum of fermion masses subtracted, so cancelling out exactly without requiring special assumptions about the nature of the vacuum or the masses of the supersymmetric states.[14] (The 'mirror image' nature of the vacuum states will also be meaningful in the context of the Feynman-Wheeler mechanism discussed in 16.8.)

The existence of such 'supersymmetric' partners seemingly comes from the duality represented by the choice of fermion or fermion plus environment. The isolated fermion represents the action half of Newton's third law, characterized by kinetic energy, continuous variation, and spin in half-integral units, while in the case of the fermion interacting with its environment, it is the action and reaction pair, characterized by potential energy, a stable state, and spin in integral units. The combination then represents either a real boson with two nilpotents (which are not nilpotent to each other), or a bosonic-type state produced by a fermion interacting with its material environment or vacuum, and, as a consequence, manifesting Berry phase, the Aharonov-Bohm or Jahn-Teller effect, Thomas precession, relativistic correction, radiation reaction, the quantum Hall effect, Cooper pairing, *zitterbewegung*, or whatever else is needed to produce the 'conjugate' state in the fermion's 'environment'.[15]

16.10 Physics and Duality

From the foregoing discussion, it would appear that the factor 2 may be seen as a result of action and reaction (A); commutation relations (C); absorption and emission (E); object and environment (O); relativity (R); the virial relation (V); or continuity and discontinuity (X). Many of these explanations, however, overlap in the case of individual phenomena, suggesting that they are really all part of some more general overall process:

Kinematics						V	X
Gases	A					V	
Orbits	A					V	X
Radiation pressure	A		E			V	
Gravitational light deflection					R	V	
Fermion / boson spin		C		O	R	V	
Zero-point energy	A	C				V	X
Radiation reaction	A		E		R	V	
SR paradoxes	A		E				

In addition, the complete description of the system tends to lead to the overall elimination of the factor. The use of the factor 2 is a two-way process, halving in one direction and doubling in another, and the system can only be described in complete terms by taking both into account. Physical phenomena involving the factor tend to incorporate, in some form, the opposing sets of characteristics. Kinetic energy variation, for example, is continuous, but it starts from a discrete state; potential energy variation, on the other hand, is discrete, but starts from a continuous state. Neither kinetic nor potential exists independently: each creates the other, just like action and reaction. But kinetic energy also relates to a changing state, while potential energy is usually related to a fixed one.

In the most general terms, the factor 2 is an expression of the fundamental duality in the whole concept of 'nature' – in physics, mathematics, and even ontology and epistemology – the duality that is the result of trying to create something from nothing, and, in principle, the attempt is only possible if 0 is also the end result. Fundamentally, physics does this when it sets up a probe to investigate an intrinsically uncharacterizable nature. Nature responds with symmetrical opposites to the characterization assumed by the probe, which, in its simplest form, is constituted by a discrete point in space. This, we have demonstrated, generates a symmetrical group of fundamental parameters (space – the original probe – time, mass and charge – the combined response), which are defined by properties which split the parameters into three C_2 groupings, depending on whether they are conserved or nonconserved, real (orderable) or imaginary (nonorderable), continuous or discrete. Each of these divisions may be held responsible for a factor 2, for duality seems to be the necessary result of any attempt to create singularity.

The three distinct mathematical processes involved are identified as *conjugation*, *complexification* and *dimensionalization*, which are manifested, respectively, through opposite signs (or equivalent), the distinction between real and imaginary components, and the introduction of cyclic dimensionality. Conjugation is equivalent to conservation, so a positive charge or positive source of mass-energy cannot be created without also creating a negative one. It is also equivalent to the unchanging state (as implied in potential energy equations) as opposed to the changing or unconjugated (kinetic) state. As recognized earlier, only the (3-)dimensional quantities, space and charge, are countable.

To establish every type of dualling, as described in chapters 2 and 3, requires a group of 64 elements. This is the Dirac algebra, the formalism necessary for the parameterization of the whole of physics. In addition the Dirac algebra includes a dualling of each of the dualling processes within itself, the eight groups of objects involved producing every possible combination of + / – × real / complex × nondimensional / dimensional.

The three processes have many physical manifestations. The first duality (conserved / nonconserved, conjugated / nonconjugated, + / –) manifests itself in the use of pairs of *conjugate variables* to define a system, in both classical and in quantum physics. In each case, a conserved quantity is paired with a nonconserved one. So momentum is paired with space, and energy with time. The (vector) momentum is said to constitute an alternative *phase space*, and the techniques of Fourier analysis can be used to transform a representation in real space into one in phase space and vice versa. In quantum mechanics, the conjugate variables are the ones limited by the Heisenberg uncertainty relations: too much precision in one leads to a loss of precision in the other. In the concept of 'virtual' particles, we are able to consider states in which we can temporarily 'exchange' information about space for information about momentum, and information about time for information about energy. Now, the parameter group suggests that the *fundamental* conjugation of space should be with charge rather than momentum, and that the *fundamental* conjugation of time should be with mass rather than energy, but particle physics shows that there are deep connections between the concepts of charge and momentum (or angular momentum), and the relativistic connection between mass and energy has long been established.

Many examples show the factor 2 occurring in the duality between conjugated and nonconjugated quantities. A classic case in classical physics, already discussed, is that of action and reaction. Conjugation, or conservation, implies the simultaneous application of + and – conditions. Thus, when we describe a physical process using constant, rather than changing, terms, we are effectively using both sides of the + / – duality at once. This is the case when we use potential, rather than kinetic energy equations, or both action and reaction sides of Newton's third law, or even relativistic, rather than rest, mass. Relativity itself does not introduce the factor 2, but relativistic equations can often be used as classic examples of changing conditions. The most controversial instance in historical terms is the double bending of light rays in a gravitational field, which can be seen, as we have shown, as an example of the use of a kinetic, rather than potential, energy equation. Of coursing, halving in one respect may lead to doubling in another. So halving the energy, by using a kinetic term, produces a doubling of the angular deflection; but it is also possible to produce the doubling directly by taking both space- and time-related effects into account.

The real / complex duality is the relativistic one. This allows transformations to be made, for example, between space and time representations; and it also manifests itself in the well-known duality between electric and magnetic fields which is apparent in Maxwell's equations. A more subtle form of it occurs in the creation of massive particle states at the expense of components of charge. The use of both space- and time-related effects, rather than kinetic energy, in gravitational light deflection could be seen as an application of the process of complexification (the adding of a complex term to a real one), in this second route to duality. It is significant that the group relationship between the physical parameters is so integrated that such apparently alternative explanations emerge without any fundamental contradiction. Both explanations are equally true, and neither has precedence over the other.

The third duality can be represented in terms of the discrete / continuous, or the dimensional / nondimensional options. A classic case of the first representation is the well-known wave-particle duality, where waves represent continuous options and particles discrete ones. The reason why it is an option is that the theories apply continuity or discreteness to the *entire system*, instead of only to those components which are fundamentally continuous or discrete; and the balance has to be restored by allowing the possibility of the alternative option. The same applies to the Heisenberg and Schrödinger theories of quantum mechanics, which are, respectively, discrete and continuous, but which each have to incorporate some aspect of the excluded property when applied to a real physical system. The factor 2 is also an expression of the discreteness of both material particles (or charges) and the spaces between them, as opposed to the continuity of the vacuum in terms of energy. The same discreteness further implies, though more subtly, the concept of dimensionality, which is responsible for the noncommutativity of the momentum operator, as well as the discreteness of the division of rectangles into triangles.

Careful study of the factor 2 reveals that it is either the link between the continuous and discrete physical domains, or between the changing and the fixed, or the real and imaginary (orderable and nonorderable), the three dualities of the group, and, in every physical instance, between more than one of these. While the continuous or discrete duality is obvious from the distinction between potential and kinetic energies, this distinction also incorporates the duality between conserved and nonconserved quantities, or fixed and changing conditions. The duality may also be expressed in terms of the distinction between space-like and time-like theories (for example, those of Heisenberg and Schrödinger, or of quantum mechanics and stochastic electrodynamics), which are not only distinguished by being discrete and continuous, but also by being real and imaginary. Though a single duality separates such theories, it is open to more than one interpretation because each pair of parameters is always separated by two distinct dualities.

From the construction of dualities in terms of successive C_2 applications, it is possible to see why, in general, the constant terms produce effects which are 2 × the changing terms, the real produce ones which are 2 × the imaginary, and the discrete produces ones which are 2 × the continuous: the multiplication occurs in the direction which doubles the options. The first combines + and – cases where it remains constant; the second involves squaring imaginary parameters to produce real ones; and the third combines dimensionality and noncommutativity with discreteness, and so doubles the elements. Examples of the first include action + reaction, absorption + emission, radiation + reaction, potential v. kinetic energy, relativistic v. rest mass, uniform v. uniformly accelerated motion, and rectangles v. triangles. Examples of the second include bosons v. fermions, and space-like v. time-like systems. Examples of the third include fermion + 'environment' (Aharonov-Bohm, Berry phase, Jahn-Teller, etc.), space-like v. time-like systems, particles v. waves, Heisenberg v. Schrödinger / the harmonic oscillator, quantum mechanics v stochastic electrodynamics / zero point energy; 4π v. 2π rotation, and all cases in which physical dimensionality or noncommutativity is involved.

16.11 The Factor 2 and Electroweak Mixing

One particular case of the factor 2 resulting from the fundamental dualistic processes discussed here has a special significance. This is in the value of the idealised electroweak mixing parameter $\sin^2\theta_W$ calculated from $Tr\ (t_3^2)\ /\ Tr\ (Q^2)$ (cf 10.9, 15.6). At 0.25, this becomes equivalent to making the weak charge value twice that of the electric charge at the energy at which the mixing takes place. The ultimate reason for this is the fact that the quantum number for weak charge (t_3) is half that for electric charge (Q), because of the $SU(2)$ nature or dipolarity introduced by the complexifying factor i, while the compensating factors of weak isospin and single-handedness contrive to halve the respective numbers of electric and weak states simultaneously – which is, of course, no coincidence, because both are aspects of weak dipolarity. Ultimately, then, the weak charge has, ideally, twice the magnitude of the electric charge at the energy at which they are mixed because the weak charge is dipolar, and the weak charge is dipolar because it is complexified. An even more intriguing fact, however, is that the magnitude of the *strong* charge, at the same energy of interaction, appears to be *twice that of the weak charge*. It is as though there were a further doubling effect due to dimensionalization, exactly as in the fundamental algebra. If this apparent doubling effect could be established on the same deductive basis as the doubling effect due to complexification, then an enormous simplification in Grand Unified theory would become possible.

16.12 Alternative Dualities

In terms of the mathematical structure proposed throughout this work, then, it would be possible to classify the physical phenomena involving the factor 2 as resulting from the following processes:

action and reaction	conjugation
commutation relations	dimensionalization
absorption and emission	conjugation
object and environment	conjugation
relativity	complexification
electroweak mixing	complexification
the virial relation	conjugation
continuity and discontinuity	conjugation / dimensionalization

However, overlap is possible in most, or even all, of these cases, and physical systems which apply a doubling through one route will involve a halving through another. For example, gravitational light deflection, if treated as relativistic, doubles its value because of complexification, both space and time being considered. However, the double deflection can also be derived from the use of a kinetic energy term which is half the total (potential) energy, because it represents the unconjugated rather than the conjugated case. This is why there are so many physical phenomena involving the factor 2 with alternative explanations. The factor appears when we look at a process from a one-sided point of view. Another example I have seen is a calculation of the special relativistic addition law of velocities, which is normally possible only in one dimension, done via a 2-dimensional diagram but without using time: here the duality of dimensionalization replaces that of complexification.[16] (In fact, the common duality between dimensionalization and complexification is the reason why a 2-dimensional diagram can be used as a representation of complex numbers.) Again, though a single duality separates alternative theories, such as Heisenberg and Schrödinger, or quantum mechanics and stochastic electrodynamics, it is invariably open to more than one interpretation because each pair of parameters is always separated by two distinct dualities, and the separate interpretations ultimately act together when we consider a phenomenon in relation to its place in the overall 'environment' of the physical universe.

The ½ ratio between the respective spins of fermions and bosons, previously considered, provides a classic instance in which there are alternative explanations using *any one* of the three fundamental dualities. If we obtain the spin ½ value of the electron indirectly by finding its moment in a magnetic field, the kinetic energy equation (valid at the moment of 'switch-on') becomes the ultimate origin of the factor. Effectively, we are using the nonconserved or nonconjugated half of the duality, or one side of + / –. If we use the potential energy equation, we then need to apply relativity (in the form of the Thomas precession) to obtain the correction factor. At first sight, this suggests we are using the conserved or conjugated aspect and should double the value obtained by the first method, but, by relativistically incorporating the imaginary parameter time at the same level as the real parameter space, we are, in effect, doubling up one of the *divisors* involved in calculating the moment, and so we obtain the same result as previously.

These, of course, are indirect methods. To obtain the spin directly we use the relativistic Dirac equation. However, in this case we find that it is not relativity, but the three-dimensionality of the angular momentum term that is responsible for halving the spin, because a divisor of 2 is introduced through the anticommutativity of the vector operators (in principle, $\mathbf{ij} - \mathbf{ji} = 2\mathbf{ij}$, etc.). This is equivalent to the third process of duality: dimensionalization. (The same applies to using the anticommutativity of multivariate vectors in the Schrödinger equation.) Remarkably, then, it is possible to use *any* of the three dualities involved in the parameter group to explain the appearance of the same numerical factor in the spin term for the electron, showing that, in this respect, they have an exactly equivalent effect. This is summarized in the table below, where the words in bold type apply to the calculation procedure for the electron spin.

duality	**method**
conserved / nonconserved	**potential energy** / kinetic energy
real / complex	**nonrelativistic** / relativistic
nondimensional / dimensional	**commutative** / anticommutative

What the simultaneous application of, say, the kinetic energy halving and the anticommutative doubling indicates is that the symmetrical structure applied to physics is organized in such a way that *both these interpretations of the dualling process apply simultaneously*. In effect, this hidden balancing act also operates in yet other, more subtle ways because the virial relation between potential and kinetic energies is specifically one of doubling only when the force laws which apply are those characteristic of 3-dimensional space; and the action and reaction mechanisms which produce the doubled value for potential energy rely on applying vectorial (or dimensional) considerations to the kinetic energy term.

16.13 Mathematical Doubling and the Self-Duality of the Dirac Nilpotent

That the doubling mechanism also applies in purely mathematical, as well as in physical, contexts is evident from the topological explanation of the Aharonov-Bohm effect, though the physical and mathematical applications must ultimately have the same origin. The very concept of duality also implies that the actual processes of counting and generating numbers are created at the same time as the concepts of discreteness, nonconservation, and orderability are separated from those of continuity, conservation, and nonorderability. The mathematical processes of addition and squaring are, in effect, 'created' at the same time as the physical quantities to which they apply, while all the other fundamental mathematical concepts and processes (e.g. the Dedekind cut) are, in some way, defined by dualling. The factor 2 thus expresses dualities which are fundamental to the creation of both mathematics and physics, and duality provides a philosophy on which both can be based.

There are also other possible mathematical connections. It is tempting, for instance, to believe that the uniqueness of the value ½ as the real part of the zero-solutions of the Riemann zeta function, defined by the summation

$$\zeta = \sum_{n=1}^{\infty} \frac{1}{n^z},$$

where n are positive integers and z is a complex number, $a + ib$, with a and b real, has a significance which is physical as well as mathematical, and that, as Hilbert originally conjectured, the solutions represent the eigenvalues and energy levels of an Hermitian operator, which is the Hamiltonian of a quantum mechanical system – Berry has further conjectured that they represent states of quantum chaos.[17] (See also chapter 20.) It is conceivable that the ½ is related to the zero-point energy term of a series of fermionic harmonic oscillators (cf 10.12), and that the random imaginary component is related to the random 'phase' associated with **p**. It is certainly true that, solving the Dirac nilpotent equation for any spherically-symmetric potential other than a linear or Coulomb one (i.e. under harmonic oscillator conditions) requires a Coulomb or phase term with numerical coefficient ½, which is of the opposite complexity to the rest of the potential, and which can be associated with the zero-point energy or (equivalently) the random directionality of the fermion spin. There may also be some physical significance in the fact that integers, like the fundamental parameters, only add directly to produce other integers or in the form of squares to produce squares of integers, but do so in an infinite progression. Both of these mathematical results suggest the possibility of further fundamental significance in the factor 2.[18]

Square-rooting and halving have an intimate relationship, which is manifested physically in the relation between vector spin terms of bosons and fermions and their respective uses of double or single nilpotent operators, in addition to the halving approximation used to find the kinetic energy term in the binomial expansion for relativistic mass. This relationship is determined entirely by the fact that 3-D Pythagorean addition is a dualistic process, with a numerical doubling arising from noncommutativity, and this applies to both the vector operators used for space and momentum, and the quaternion operators used in the Dirac nilpotent. When we express the parameters mass, time, space, and charge in terms of the respective scalar, pseudoscalar, vector and quaternion units (1, i, $\mathbf{i}$, $\mathbf{j}$, $\mathbf{k}$, $\boldsymbol{i}$, $\boldsymbol{j}$, $\boldsymbol{k}$), which their combined properties require, it becomes evident that the combination of the four parameters in the Dirac equation produces the complete self-dualling which we require for a universal parameterization of nature.

The Dirac state vector is, in fact, the most striking of all the instances of mathematical or physical duality, with space, time, mass and charge combining to produce an object which is self-dual, and so producing the desired total nothingness by acting directly on itself. ($\pm\, \boldsymbol{k}E \pm \boldsymbol{i}\, \mathbf{p} + \boldsymbol{j}\, \mathrm{m}$), as a square root of zero, or its own dual, can be regarded as either a classical object or, in relativistic quantum mechanics, as the state vector for a fermion as specified by the Dirac equation. In the latter case, the infinite imaging of the fermion state in the vacuum and the nonlocal connection between all the state vectors in the entire universe, or infinite entanglement of all nilpotent fermion states, described mathematically in terms of Hilbert space, provides an extension of the dualling process to infinity. This is the perfect way of producing something from nothing. Since the universe is believed to be composed entirely of fermions or fermion-antifermion combinations (bosons), the Dirac equation is, in the final analysis, the most significant way of incorporating the foundational basis for the whole of physics into a single structure, and it would appear that it is itself founded entirely on the principle of duality. Ultimately, it would seem, duality is not merely a 'component' of physics but an expression of the fundamental nature of physics itself.

Chapter 17

Gravity and Inertia

While gravity appears to play a role in the grand unification process discussed in chapter 15, its special characteristics, deriving from the continuity of its source, require separate consideration, and this will be the theme of both this chapter and chapter 21. The continuity of mass-energy must determine our understanding of gravity. The speed of gravity, in particular, is constrained to be infinite and to provide the source of the nonlocal correlations required by quantum mechanics. However, the speed of the gravitational force is *not* the same thing as the speed of gravitational waves or the speed of changes in gravitational potential, which is the thing most likely to be measured as the 'speed of gravity'. This has profound implications for the interpretation of general relativity, as does the related fact that quantum mechanics shows that the concept of *measurement* cannot be central to the definition of the abstract mathematical *system*, as supposed in naïve interpretations of GR. The structure of the Dirac nilpotent further shows that the naïve interpretation, with its assumption of an observable definition of time, is untenable. However, it is shown here that GR can be understood fully, even within a quantum context, if gravity is seen to be intimately connected with *inertia*. In principle, if gravity is truly instantaneous, the space-time of *measurement* will not be appropriate to its description, and introducing it will also produce inertial effects due to the effect of the finite speed of transmission of the measurement information, which can be easily calculated even though the speed itself will be gravitationally affected. Such 'aberration' effects would be entirely analogous to those produced by 'inertial forces' in classical physics, and with this interpretation we can choose either to regard the GR field equations as the full gravity plus inertia package, and work, by successive approximations, in the simplest physical system, of a spherically symmetric point source, towards the Schwarzschild solution, *or* we can begin with the classical gravitational potential and then add inertial effects, as required, by minimally relativistic corrections. The latter approach yields easy derivations of gravitational redshift, the gravitational deflection and time-delay of electromagnetic radiation, perihelion and periastron precession, gravomagnetic effects and gravitational waves. The gravomagnetic equations turn out to be exact analogues of Maxwell's classical electromagnetic equations, but with the vector potential effects multiplied by a factor 4, which originate in the gravitational effect on the speed of transmission of inertial information. Overall, the procedure gives us a linear interpretation of the gravitational field, which means that (once the numerical factor has been taken into account) the system can be quantized in the same way as electrodynamics.

17.1 The Continuity of Mass-Energy

The unbroken symmetry of the parameter group requires that both mass-energy and time are absolute continua. In the context of cosmology this would seem to imply that the universe is infinite both in time and in extent, with a single absolute reference frame, presumably coincident with that of the microwave background radiation. Exactly such a continuous mass-energy is required for the weak filled vacuum responsible, via the Higgs mechanism, for supplying inertial mass to fermionic and bosonic matter in the Standard Model of particle physics. This weak filled vacuum, or 'Higgs field', is of exactly the same kind as the vacuum Dirac originally postulated to explain the negative energy solutions of his fermionic equation as antifermionic states, and it is significant that, in the Dirac nilpotent formulation, the negative energy operator is also the weak charge operator. Negative fermionic energy states are never directly observed in nature because the ground state of the universe has positive fermionic energy. The predominance of matter over antimatter in the universe is, therefore, a pure result of the continuum nature of mass and is not a product of some cosmological evolution. It is exactly parallel to the irreversibility of time. In the context of the universal rewrite alphabet, it is equivalent to the privileging of + over – in the first act of 'creation'.

The universality of the gravitational source also gives us a perfectly good *physical* reason why the gravitational force should be instantaneous and not describable in terms of a localised field quantum. Uniformity in the medium means that gravity cannot convey localised information. If the gravitational source is an *absolute continuum*, as is supposed in both classical aether theories and the quantum mechanical vacuum, then we might well *expect* its interactions to be instantaneous, by contrast with interactions whose sources are discrete and pointlike and necessarily localised; and an absolutely continuous mass would give a simple explanation for the fact that gravitational sources always have the same sign.

The fact that mass-energy is an absolute continuum, besides being the obvious explanation of mass's unipolarity, has major implications for the concept of vacuum and its role in a theory of gravity. Quantum mechanics, for example, requires instantaneous correlation of particle wavefunctions, with no obvious suggestion of a mechanism by which this can be accomplished. Unmediated instantaneous action at a distance seems to be ruled out for forces involving discrete charges of any type, which are all thought to be transmitted with time-delayed action involving 'the velocity of light' (c). Gravity, however, remains unlocalised, and the universality of the interaction and the absolute continuity of the vacuum suggest that it may indeed be instantaneous, and, therefore, the preferred candidate for the carrier of wavefunction correlations.

It seems that there are actually four components to the vacuum, which are equivalent respectively to 1, $\boldsymbol{i}$, $\boldsymbol{j}$, and $\boldsymbol{k}$ times the nilpotent state vector, and identifiable, respectively, as the gravitational, strong, electric and weak vacua. The state vector acts on all four at once to produce the four terms of the Dirac 4-spinor or the four 'solutions' of the Dirac equation. In the derivation of the Dirac equation from a quaternionic matrix differential operator, each row or column has a zero term, and it is this term which determines the nature of the vacuum, because the operator which is missing becomes the one which is applied in the quaternionic row or column to leave the three remaining terms as multiples of $\boldsymbol{i}, \boldsymbol{j}$, and $\boldsymbol{k}$, as in the state vector. The operator which leaves the state vector unchanged is, of course, the one which creates the gravitational vacuum, and, uniquely, in this case, application of the state vector to the gravitational vacuum would reduce it to zero. In other words, the state vector and the gravitational vacuum are the same thing. That the state vector is the carrier of the gravitational force is, therefore, entirely plausible.

Of course, instantaneous transmission of gravity does not require instantaneous transmission of other forces, and the nature of their sources as discrete or localised units suggests that the electromagnetic, strong and weak interactions cannot be transmitted instantaneously, and their energy of communication is likewise discrete. But, though the mathematical combination of mass and charge in the quaternion representation suggests that a quasi-localised type of mass, momentum and energy might be associated with particles or unit charges, the discreteness is only apparent, for the price of making this assumption in quantum mechanics is the introduction of nonlocality. Particles with mass, momentum and energy are gravitational sources which are subject to instantaneous gravitational interaction at a distance and hence necessarily 'communicate' through the infinite energy continuum which we call the vacuum.

It is probable that energy or mass is never transferred from place to place. It is just immovable vacuum. But the way it is realised, as rest mass or energy of some particular kind, or as hidden vacuum energy, changes as the charge structure in any place changes. Both mass and the various types of charge are a kind of vacuum state; and mass and charge are like mutually exclusive vacuum components – a kind of quaternion invariant, like space and time in a 4-vector. For any particle, we can have a charge-like manifestation or mass-like manifestation of the vacuum, or a mixture of both; and rest mass, like charge, is a kind of 'vacuum realisation', which is localised in the same sort of way as charge.

Gravity, in the Pope-Osborne synthesis, previously mentioned, is not a force but an expression of the global correlation of angular momentum states.[1] It is certainly of a different nature to the forces produced by discrete charges and it is clearly necessary that it must be instantaneous. In the theoretical framework described here, it is necessarily instantaneous because its source is continuous, and it is clearly responsible for the instantaneous correlation of the infinitely many Dirac nilpotent states that make up the 'universe'. It is certainly not of the same kind as the forces between discrete sources, though it is characterized by a $U(1)$ symmetry, and at Grand Unification its magnitude will be the same.

The fermionic nilpotent, or quaternion state vector, is, in some sense, an angular momentum operator, although it has other characteristics as well, and much of the information concerning it, especially with respect to the charge states, is carried by angular momentum terms. This is because the nilpotent has been structured on the basis of unit charges, the source of quantization, and the conjugate parameter to the *type* of charge is angular momentum. In the sense that this makes the Dirac nilpotent an angular momentum operator, with gravity as the correlation mechanism for Dirac nilpotents, then angular momentum information is also carried by, or carries, gravity. So, the Pope-Osborne process of using angular momentum to convey information conventionally conveyed by gravity and the electric force is of considerable interest and relevance to the present discussion. However, because of the complicated structure of the sources involved, the angular momentum information carried by a Dirac nilpotent is of a more subtle kind than the $U(1)$-type absorption or emission process of a single quantum of magnitude $\hbar$ would suggest; and the process is also, so far, defined exactly only for an isolated quantum system, or for the type of classical system that can be modelled in the same way.[2] Complications may be expected in respect of the quantum / classical transition, where systems can no longer be defined in an isolated way.

17.2 The Speed of Gravity

What is the speed of gravity? This is a vital question, and the answers we find will determine both our understanding of gravity and of much that is also fundamentally significant in terms of cosmology and the interpretation of astronomical observations. Aspect and others have established the truth of quantum nonlocality in experiments involving photon polarization states. This would seem to be a problem, for both relativity and gauge theories imply that 'communication of information' can take place only at the speed of light. However, there is no compelling reason for assuming that luminal or subluminal communication is applicable to gravity; both the experimental evidence and the logic seem to point in the opposite direction. By 'the speed of gravity', here, we mean the speed of the gravitational force, or the speed with which the gravitational field is established, *not* the speed of gravitational waves or any speed associated with gravitational potential. The distinction, as we will see, is vital, and our fundamental considerations allow us to give an immediate answer. The speed of the gravitational force is infinite. The force of gravity is instantaneous. A filled vacuum, a continuous (and unchanging) energy distribution, and instant correlation between particle wavefunctions, all demand this. The structure of general relativity, for all its prediction of gravitational waves, also demands it, as much as the classical Newtonian theory. And there is also observational evidence.

The only way in which we could establish the existence of light-speed gravity, in fact, would be by purely gravitational measurement, which is almost impossible by definition. (Claims that this has been achieved seem to based on fundamental misinterpretations of the difference between the speed of gravitational waves or changes in gravitational potential and the speed at which the force is transmitted.[3]) Yet evidence certainly exists to the contrary. The absence of an aberration effect suggests that the speed of gravity is at least 10^{10} times that of light. For Van Flandern, for example: 'the Sun's gravity emanates from its instantaneous true position, as opposed to the direction from which its light seems to come.' 'If gravity propagated at the speed of light, it would accelerate the orbital speeds of bodies.'[4] The Poynting-Robertson effect, in which an orbiting body is decelerated by the greater pressure of sunlight on its leading over its trailing hemisphere, 'happens precisely because the action of light is much slower than gravity'. And Van Flandern concludes: 'No relativist has yet, to my knowledge, devised a theory to explain how it can be that the direction of the Sun's gravitational force on the Earth and the direction of the photons arriving from the Sun are not parallel.'

Continuity can mean nothing else than instantaneous interaction. Any action propagated in time must be between localised or discrete units like charge, and not the parts of a continuum like mass; we cannot introduce a concept like 'absolute continuity' as a fundamental proposition, unless we are prepared to follow the logic of our own assumptions. The whole structure of physics requires a duality between discreteness and continuity. Quantum mechanics would be impossible without it. Incorporating the opposing ideas of continuity and discontinuity into our abstract conceptual scheme, and accepting the logical consequences without hesitation, will lead us out of the area of problem and paradox in which physics seems to present a mass of contradictions between fundamental theories.

17.3 What is General Relativity About?

In most accounts, general relativity is described as 'the modern theory of gravity' or 'the best theory of gravity we have', replacing the classical Newtonian theory on the basis of a superior fit to experimental data. However, when we look at the way the theory is structured, we find that it is not a theory of gravity at all, in the sense that it actually provides no explanation or mechanism for the action of the gravitational force – it is rather an expression of the way that gravity is measured. One way of looking at the theory is to say that it *replaces* the idea of force *and* mass (the traditional source of the force) with a space-time which is no longer flat and Euclidean, but arbitrarily curved locally, in a Riemannian manner, according to the local strength of the gravitational field. However, it does not replace the Newtonian theory, in a mathematical sense, but *adds* to it, by requiring that the weak field limit of the gravitational potential becomes the Newtonian value, which must be put in by hand when we 'solve' the gravitational field equations. Tests of the principle of equivalence or the constancy of G, which are sometimes described as tests of general relativity, are really only tests of this Newtonian content.

The rationale of the general theory was derived from three physical principles, none of which is actually true and none of which is preserved in the final theory: the principle of equivalence, that 'a gravitational field of force is precisely equivalent to an artificial field of force'; the principle of general covariance, which Einstein thought expressed the relativity of all motion; and Mach's principle, that 'the inertial field is to be determined only by the distribution of mass-energy'. The general theory rose as a result of Einstein's attempts, from about 1911 to 1915, to embrace all these principles in a single mathematical structure. In deriving the field equations, Einstein set about finding the relativistic equivalents of Poisson's equation:

$$\nabla^2 \phi = -4\pi\rho G.$$

Influenced by his work in special relativity, and by the Minkowski formulation of space and time as a single mathematical and physical entity, he looked for a general covariance, including the Lorentz covariance as a special case. Working initially with Grossmann, he showed (in Strandberg's convenient summary[5]) 'that in a metric geometry the square of a line element of a minimum length path, a geodesic, in a curved space-time manifold can be written as a quadratic form of an arbitrary reference space', using the ten elements of the symmetric metric tensor g_{pq}. In 1912 he realised that all the covariants of $g_{pq}dx_p dx_q$, which replaced the polar invariant line-element

$$ds^2 = -dr^2 - r^2 d\theta^2 - r^2 \sin^2\theta \, d\phi^2 + c^2 dt^2$$

of special relativity, could be derived from the Riemann-Christoffel tensor $K^{\,s}_{pqr}$. Using Riemann's nineteenth-century theory of curvature, he set up generally covariant equations for the 10-component g_{pq} (which had replaced the single gravitational potential of Newtonian theory) to meet all the physical requirements set by his assumption of the principle of equivalence.

In the process, he applied Mach's principle – i.e. all ten potentials g_{pq} representing the gravitational field could be determined solely by the masses or energies of the bodies involved. He had Minkowski's energy-momentum tensor T_{pq} from special relativity to represent mass and replace the ρ term in Poisson's equation; he also had the Ricci tensor K_{pq}, contracted from the Riemann tensor, to replace the left hand side of the Poisson equation $\nabla^2\phi$. But, because the divergence of T_{pq} is zero, while that of K_{pq} is not, he found that the right hand side could not be a simple multiple of T_{pq}. After three years of intensive struggle, and many false starts, he finally proposed the equation

$$K_{pq} = -\kappa(T_{pq} - \tfrac{1}{2} g_{pq} T),$$

in which κ is a constant depending on the Newtonian constant of gravitation (actually $8\pi G / c^2$), and

$$T = g_{pq}T_{pq}.$$

By a fairly straightforward manipulation, this equation could be changed to

$$K_{pq} - \tfrac{1}{2}\, g_{pq}K = -\, \kappa T_{pq},$$

which produces ten equations for the ten unknown values of g_{pq}.

In most accounts, 'general relativity' is assumed to mean both the mathematical formalism and the physical assumptions which Einstein used in deriving it. But, in principle, the theory, as represented primarily by the formalism, is empty of physical content; it is neither more nor less than a mathematical theory of space-time curvature described by an equation which does not correspond to any known physical principle. Since it assumes that gravity is simply an expression of the local curvature of Lorentzian space-time, then there is nothing physical to 'cause' this curvature. The Riemann-Christoffel tensor, which expresses the curvature, does not in itself specify the source of it. It is only when we choose to identify the highly approximate metric equations for very simple and specialised configurations of the gravitational field, based on the (arbitrary) limiting assumption of a spherically symmetric Newtonian potential as 'solutions' of the general relativistic field equations, that we are actually able to attribute physical meaning to the theory and link the curvature tensor with the effects of the parameter mass. But even there the gulf between the equations and the 'solutions' is so wide that we cannot realistically use the 'solutions' to derive the fundamental physical meaning of the original equations. The only justification for privileging these equations, in fact, remains mathematical simplicity and elegance, for there are undoubtedly many other sets of possible equations which would yield the same experimental consequences. And, by privileging this non-deducible mathematical structure with respect to any possible physical explanation, we seemingly reduce our options to a kind of 'archaeology' of space-time: gravity operates in this way because space-time *happens* to be curved at certain points.

The structure of general relativity, in fact, suggests that it is something other than a pure theory of gravity. The field equations do not seem to be describing gravity at all. The indicators point to the fact that it is a theory of *gravity plus inertia*. Gravity, according to the theory, curves space-time, but it is not necessary to assume further that curved space-time produces gravity. That is, there is nothing in the equations of general relativity which demands that the gravitational field must be nonlinear. The 'field equations', in fact, are a purely mathematical description of space-time curvature, with no intrinsic relation to gravity at all. The only connection of the curvature with gravity occurs when the classical Newtonian potential is inserted by hand into the drastically simplified equation for the radial field around a point source, which is described as the 'Schwarzschild solution' of the field equations.

So, general relativity, though it concerns gravity, is not really a theory *of* gravity. It is a theory of the inertial effects of gravity. Gravity is only put in at the stage of solution, and, at that point, effectively introduced in the form of the Newtonian theory. What the field equations are describing is the effect of gravity on the *measurement* of space and time, in other words, the gravitational aberration of space, or the inertial force correction. General relativity is a description of the *epistemology* of the gravitational field, not its *ontology*. And this is entirely compatible with a purely classical approach to gravity, where the idealised equations are valid only in an inertial frame of reference, and require correction by inertial forces where this is not applicable. So the Newtonian classical equations and Einsteinian field equations become ways of approaching the same problem from opposite ends: either inertia has to be added to gravity, or gravity to inertia to solve the full problem. In principle, there is no difference, but, in calculating effects within the range of current measurement techniques, it is much easier to make the calculations using the former rather than the latter. In this case, aspects of relativity are introduced *minimally* as required, rather than wholesale at the beginning, to be necessarily restricted by making drastic assumptions at the point where we want to make a measurable prediction. So, while we retain the option of using the general relativistic field equations as the most general description of the effects of gravity on observable space-time, we are not committed to assuming, in advance of the necessary experimental corroboration, that they are the only option.

17.4 General Relativity and Quantum Mechanics

As is well known, general relativity has major problems in accommodating quantum theory. One of the most serious difficulties occurs with the parameter time. In quantum mechanics, as in the fundamental parameter group, time is not an observable, but GR, in adopting the Minkowski formalism, defines time in operationalist terms as a measured quantity virtually indistinguishable from space. And, in the Dirac nilpotent structure, which we have seen as fundamental to the whole structure of physics, even the Minkowski formalism has no canonical status. Space and time, like energy and momentum, are not a *pure* 4-vector; they are associated with *different* quaternionic charge operators, though these disappear in a scalar product, and the relationship of space and time with each other has no more fundamental status than the relationship of each of these quantities with mass. We can see that the attempt to merge space, time and mass in a mathematical superstructure, which does not pay attention to their fundamentally distinct mathematical characters, prevents us from explaining the very real differences that distinguish these basic quantities from each other.

But this is not the only problem, for the attempt at providing a quantum description of gravity seems to be beset with enormous difficulties of many kinds; and there is no sign of a successful theory. The source of all these difficulties, however, is obvious. The assumption has frequently been made that gravity is similar to the electrical and other fundamental forces in being transmitted at the speed of light, but there is a fundamental distinction between these forces, in that gravity is caused by mass, while the other forces cause mass. The problem lies in the assumption that gravity does the same. As a result of this, the quantum theory of gravity seems to produce *unrenormalisable infinities*. The 'relativistic' mass, associated with the gravitational quantum, or graviton, unlike that of the photon, or quantum of the electromagnetic field, is itself a source of the field which produces it; and so the field equations for quantum gravity become nonlinear. In quantum field terms, the graviton becomes a particle with spin 2, unlike the spin 1 of the photon.

We do not require the sophisticated quantum concepts of the twentieth century to see the source of the problem. Already, in a manuscript of the 1690s, Newton, the architect of the classical theory of gravity, had grasped the major problem posed by quantum gravity and nonlinear field theories in general. If the gravitational force is conveyed by particles, these particles themselves must have gravity, because all matter has gravity. The gravitational interaction between these particles must produce further particles and so on to infinity. 'If anyone should ... admit some matter with no gravity by which the gravity of perceptible matter may be explained; it is necessary for him to assert two kinds of solid particles which cannot be transmuted into one another: the one of denser which are heavy in proportion to the quantity of matter, and out of which all matter with gravity and consequently the whole perceptible world is compounded, and the other of less dense particles which have to be the cause of the denser ones but themselves have no gravity, lest their gravity might have to be explained by a third kind and that so on to infinity. Therefore one must altogether determine that the denser particles cannot be changed into the less dense ones, and thereupon that there are two kinds of particles, and that these cannot pass into one another.'[6]

We can, of course, conceive of exchange particles for electric and magnetic forces, which are not electric or magnetic because electricity and magnetism are not universal; and, even where they are present in matter, they are not cumulative, like gravity, because there is overall cancellation on a large scale between positive and negative charges or forces of attraction and repulsion, and so they can be renormalised to finite values. To take another example, the strong field in particle physics is, indeed, nonlinear, but it is differently structured to the hypothetical field of quantum gravity, because the source of the field, unlike that of gravity, is not universal to matter. Gravity presents a unique problem for a quantum interpretation because its sources are universal to all matter and never negative.

There is, however, a solution to the dilemma. In a gravitational system, unlike any other, we make mass terms independent of charge terms. Mass-charge, when linked, is described by a quaternion. By symmetry, we link space-time in a 4-vector. Does this mean that when mass and charge are separated, so are space and time? This might well be implied if these relationships are determined by fundamental formal symmetries. A formal separation of space and time would lead to the further formal separations within all 4-vector quantities. Thus, we would also need to separate energy and momentum, vector and scalar potentials, and so on. In a nonlinked system of this kind, the general relationship between energy and mass, $E = mc^2$, would hold, but there would no longer be any 'localised' elements of mass corresponding to transferred units of energy, the strictly 'relativistic' equation, $\Delta E = \Delta mc^2$, being relevant only in the case of Lorentz covariance. Conservation of mass and conservation of energy would hold overall, but we would not need to invoke the concept of 'identity' of energy applicable to localised units. And, of course, a gravitational field which was no longer a source of localised units of mass would no longer be subject to the property of nonlinearity.

There is thus no particular reason to suppose that the symmetry of the velocity of light 'mechanism', which is the ultimate source of the supposed nonlinearity and the spin 2 graviton, is more important than that of inverse square laws and three-dimensional space, which is valid only for a classical theory of gravity, with no relativistic space-time connection. One must certainly be abandoned, but there are several perfectly good reasons for assuming that it should be the former rather than the latter. The first is the complete failure of quantum gravity, as we know it. The second is the fact that three-dimensional spaces or four-dimensional space-times are simple and natural, being based on the existing algebra of the quaternions and fitting in perfectly with inverse square laws, while arbitrarily curved nonlinear space-times are not. The third is that very important physical differences between masses and charges suggest that they may be different in the mechanisms of their interactions, while retaining the inverse square law dictated by the nature of three-dimensional space; in particular, masses are elements in continuous fields, while charges are singularities. The difficulties can only be resolved when we have a true understanding of the concept of *inertia*.

17.5 The Schwarzschild Solution

As stated previously, we can deal with the gravity-inertia problem in one of two ways: we can assume the Einstein field equations (inertia) and then bring in the Newtonian potential (gravity) as we seek solutions of the equations with physical meaning; or we can start with the Newtonian description of classical gravity and then incorporate the appropriate gravitationally-affected 'relativistic' or inertial corrections. The second option is, in fact, much easier, as it avoids including the 'measurement' process in the description of gravity itself. Using this option, we can predict all immediately testable consequences using a minimally relativistic theory, which is, in effect, only a version of special relativity.

In these terms, we can reduce the potentially observable consequences to the result of two fundamental facts: that light (or energy, generally) has mass and is acted on by gravitational forces, and that gravity *appears* to travel at the speed of light, a speed which may be gravitationally affected because the carrier has mass. The first requires only classical gravitational theory, while the second, in all observational consequences to date, requires only special relativity. *Theoretically*, the second might appear to suggest nonlinearity, but this would only be true if the transmission of gravity at light speed was real, rather than apparent, and, since all measurement processes make the assumption of finite speed of transmission, *measured* effects would incorporate the linear aspects of finite transmission, whether or not they were real. In any case, the effects, *as tested*, require only classical theory and special relativity, whatever they may be said to *imply*.

The equation which predicts all the test results so far accomplished (apart from gravomagnetic effects) is that for the line element in the Schwarzschild solution of Einstein's field equations for the spherically symmetric field round a point source or a uniformly dense solid sphere:

$$ds^2 = -\gamma^{-2}dr^2 - r^2d\theta^2 - r^2\sin^2\theta\, d\phi^2 + \gamma^2c^2dt^2$$

where
$$\gamma = \left(1 - \frac{2GM}{rc^2}\right)^{-1/2}.$$

It would be useful to remind ourselves how far from the field equations this 'solution' actually is. In arriving at the Schwarzschild solution, we start with the Riemann-Christoffel tensor, defining space-time curvature. This tensor has 256 components, but, because of various identities that may be established, only 20 are linearly independent. To reduce the number of equations further, we arbitrarily choose $r = s$ and replace K^s_{pqr} with the contracted or Ricci tensor K_{pq}, which has sixteen components. We then make the assumption that $K_{pq} = 0$ for all p, q. From the equations for K_{11}, K_{22}, K_{33} and K_{44}, we derive equations for g_{11}, g_{22}, g_{33} and g_{44}, where we assume the limiting case of the Lorentzian metric in which $g_{pq} = 0$ for all $p \neq q$. In the case of a spherically symmetrical mass distribution (with polar coordinates), g_{22} and g_{33} are unchanged from the Lorentzian values, while the assumed Lorentzian limit suggests that g_{11} and g_{44} respectively approach -1 and c^2 as r approaches infinity. From this we obtain the metric in which

$$\gamma = \left(1 - \frac{2k}{rc^2}\right)^{-1/2}.$$

Our final assumption is the Newtonian limit, which sets the constant $2k$ at $2GM / c^2$.

It would seem that, in 'deriving' the Schwarzschild solution from the general relativistic field equations, we have virtually approximated the uniquely general relativistic element out of existence. By assuming the approximation to Lorentzian and classical conditions, we have ended up, to a large extent, with nothing more than Newtonian gravity combined with Lorentzian space-time. Consequently, any tests based on the Schwarzschild solution are tests of the Lorentzian and Newtonian conditions which we have put in by hand, but not specifically of general relativity and its particular mathematical formalism.

In view of the means by which the field equations were originally found, and, in view of the series of assumptions and approximations needed to derive the Schwarzschild solution from them, it can hardly be claimed that either process is a 'natural' or 'obvious' development, however successful they may be in explaining the results of experimental observations. There is, however, a perfectly 'natural' way of obtaining the Schwarzschild metric using a minimally relativistic Newtonian approach. This proceeds via the derivation of gravitational redshift to that of gravitational light deflection, and, finally, encompasses the relativistic correction for gravitational orbits.

17.6 Gravitational Redshift

The gravitational redshift of electromagnetic radiation is, as is well known, but not always stated, purely Newtonian. There is no relativistic or inertial force component of any kind, outside of assuming that relativistic kinematics is structured so as to preserve classical conditions of conservation of energy. A photon's energy $h\nu$ (equivalent to dynamic mass $h\nu / c^2$) is modified by the gravitational potential $-GMh\nu / c^2 r$ due to its being situated a distance r from the centre of an object of mass M. This leads to the simplest of formulae for the modified energy:

$$h\nu\left(1-\frac{GM}{rc^2}\right).$$

Many people have explained the redshift along these lines, some producing formulae which slightly modify the effect at higher orders; it has been usual, however, to couch explanations in terms of the principle of equivalence, although this is by no means necessary. The reason for this is that Einstein himself used the principle in his derivation of 1907. But it is a remarkable fact that it is only the historical circumstances surrounding the Einstein derivation that have led to the incorporation of redshift as a 'test' in textbooks on relativity, for, although he was himself a convinced believer in the corpuscular nature of light, Einstein was forced to avoid using this then-controversial theory in his derivation, and a non-corpuscular derivation is necessarily less simple. But for this special historical circumstance, the gravitational redshift would have been accepted as a routine consequence of Newtonian gravity.

In fact, the equivalent calculation to Einstein's had already been done for classical light corpuscles in the eighteenth century. This was John Michell's prediction that the refractive index of light from massive stars would be changed as a result of their gravity.[7] Michell, of course, knew nothing about line spectra, and assumed that the particle (not wave) velocity was responsible for the change in optical refractive index, so he effectively obtained a blue, rather than a red, shift; but, by applying the usual inversion of velocities to obtain the phase velocity relation, a redshift of the correct value is easily obtained.

An earlier calculation by Michell had used Newtonian mechanics to estimate the slowing down of a particle of light emitted from the Sun as a result of the strong attraction produced by the solar gravitational field.[8] This calculation is easily expressed in modern notation. Beginning with the standard Newtonian equation for the acceleration of a particle in a gravitational field g, due to an object of mass M,

$$\frac{d^2r}{dt^2} = -\frac{GM}{r^2} = g\,,$$

we find the velocity acquired over distance r is given by $c(r)$ where

$$c(r)^2 = \left(\frac{dr}{dt}\right)^2 = \frac{2GM}{r} + \text{constant}\,.$$

If we consider a particle of light emitted from the surface of a body with radius R to have velocity $c(R)$, then

$$c(r)^2 - c(R)^2 = \frac{2GM}{r} - \frac{2GM}{R}$$

and

$$c(r) = \sqrt{c(R)^2 - \frac{2GM}{R} + \frac{2GM}{r}}\,.$$

When $c(R)^2 \gg 2GM / R$, the value of $c(r)$ at $r = \infty$, when the particle has completely escaped from the gravitational field, will be

$$c(\infty) = \sqrt{c(R)^2 - \frac{2GM}{R}}\,.$$

It was from this kind of calculation that Michell subsequently predicted his change in refractive index, and also predicted the existence of stars, of radius $2GM / c^2$, which would be too massive to allow light to escape from their surfaces, while Laplace, in the same century, produced a very similar calculation, from standard Newtonian dynamics, using the kinetic energy equation, of this 'Schwarzschild radius' for a 'black hole'.

Using the fundamental duality between wave- and particle- descriptions of light, relativity interprets gravitational redshift as an observed decrease in the frequency of electromagnetic radiation emitted by a source within a gravitational field, and hence as a dilation in the time as measured by a standard clock. The dilation, by a factor $(1 - GM / rc^2)^{-1}$, is taken as being equivalent to a conventional dilation of the form $(1 - v^2 / c^2)^{-1/2}$.

17.7 The Gravitational Deflection of Electromagnetic Radiation

The gravitational deflection or light bending effect, though providing the original evidence on which general relativity was first established, has since succumbed to special relativistic analysis. This is not surprising, for light is a special relativistic system in itself; but the test does not appear to require anything further than the application of *Newtonian* gravity to this already-special relativistic system. It certainly does not imply that *gravity* is special relativistic. Essentially, for any special relativistic system, any effect, such as gravitational redshift, which leads to a relativistic dilation in measured values of time, must necessarily produce a contraction in measured values of length by the same factor. Applying special relativity to light as a system means that gravity can only be applied *simultaneously* to the space and time coordinates, where an effect depends on both, and not to either separately. The original attempt at predicting gravitational light deflection, by Einstein, according to the equivalence principle, failed to take into account this necessary reciprocity and so only calculated half of the true deflection.

The first special relativistic explanation, by Schiff,[9] used a cumbersome apparatus of observers, measuring rods and clock synchronisations, but far more elegant versions have been put forward by Strandberg,[5] by Ghosal and Chakraborty,[10] and others. However, since energy relations in a system without rest mass are *specifically designed* to coincide with purely classical ones, it ought to be possible to derive the correct light-deflection classically if the problem can be expressed purely in terms of conservation of energy. As discussed in chapter 16, exactly such a solution was first proposed in the late eighteenth and early nineteenth centuries by authors who assumed that a light particle approaching the Sun would be travelling so fast that it would be drawn into a hyperbolic orbit, of eccentricity e. To find e, by this method, we can assume that the light particle has the velocity c at the distance of closest approach, when ϕ, the angle made with the horizontal axis, is 0.

It is essential, here, however, that we realize that the effect is one of *changing conditions*, of a gravitational orbit in the process of formation, the reverse process to that of a photon escaping from a gravitational field, which leads, in the extreme case to the Schwarzschild black hole radius. Then

$$\tfrac{1}{2}c^2 = \frac{2GM(1+e)}{R},$$

$$e = \frac{2Rc^2}{GM} - 1,$$

and
$$\sin\delta = \frac{2GM}{Rc^2 - GM}.$$

Alternatively, we can take the velocity of the particle to be c at infinity, when $\cos\phi = -1/e$. Then

$$\tfrac{1}{2}c^2 = \frac{2GM(e-1)}{R},$$

$$e = \frac{2Rc^2}{GM} + 1$$

and
$$\sin\delta = \frac{2GM}{Rc^2 + GM}.$$

In fact, e is so large compared to 1 that it is immaterial, from an observational point of view which we choose. In either case, with $Rc^2/GM \gg 1$,

$$\delta \approx \frac{1}{e} \approx \frac{2GM}{Rc^2}$$

and the full deflection (in and out of the gravitational field) becomes

$$2\delta \approx \frac{4GM}{Rc^2},$$

exactly as measured.

Although light in free space has no equivalent of rest mass or kinetic energy, in the presence of a gravitational field or a material medium, where it is slowed down, a classical calculation would, perfectly legitimately, assume that it has. This is precisely what happens in a plasma, where photons are assumed to acquire an effective mass, and a kinetic, rather than potential, energy equation is perfectly routine for an orbit *in the process of formation*, that is, for a light particle being dragged into a hyperbolic orbit from a straight line path. We can imagine that a photon coming from a star, with a trajectory bending at the edge of the Sun, had, for most of its journey, no idea that it was going to end up in a hyperbolic orbit, the straight-line velocity from the distant source in the direction of the gravitational field being, of course, a purely random result of the radiation spreading uniformly in all directions from the source. (The light would have had this 'velocity' whether or not a gravitational field was present.) The slowing-down effect can also be expressed, as was understood at an early date, in terms of a 'refractive index' formula:

$$\frac{c'}{c} = \left(1 - \frac{2V}{c^2}\right)^{1/2} = \left(1 - \frac{2GM}{Rc^2}\right)^{1/2}.$$

The two known classical calculations which preceded the relativistic one were, as it happens, either incomplete and unpublished (in the case of Cavendish[11]) or incorrectly applied to only half the angle (in the case of Soldner[12]), but both authors (and, certainly, Soldner) seem to have begun, as far as we know, from (correct) reversals of the respective 'black hole' calculations of Michell and Laplace. Stanley Jaki, in introducing a translation of Soldner's calculation, which was originally published in an astronomical *Jahrbuch* for 1801, misinterpreted this in the light of subsequent (incorrect) attempts at classical calculations, assuming an *existing* orbit: 'Soldner must have, of course, known (and the same holds true for Bode, editor of the *Jahrbuch*) that the gravitational potential is proportional to g and not to $2g$. Therefore one must perhaps assume that behind Soldner's use of $2g$ was his realisation that the bending of light round a celestial body would be 2ω, that is, ω of the diagram and its mirror image.'[13] In fact, Soldner's use of $2g$ was perfectly correct, and, if he had realised that the double angle had to be used as well, he would have found the currently accepted value for the effect.

Comparing classical and relativistic calculations, we can see that a fixed orbit provides the time dilation component only, like Einstein's principle of equivalence calculation of 1911, while the distance of approach (R) remains constant. A nonfixed orbit, on the other hand, provides both time dilation and length contraction. Because the calculation is only concerned with the behaviour of light photons, and not with material particles, the kinematic approach becomes totally unnecessary in classical terms, and, in this instance, the classical calculation is not only easier than the relativistic one (whether special or general), but actually trivial.

17.8 The Gravitational Time-Delay of Electromagnetic Radiation

The time-delay effect has been rather less discussed than the deflection, partly because it is a direct consequence of the same physics as produces light bending, and partly because it was put forward as a possible result of relativity much later than the classic 'tests'. Essentially, derivation of the time delay begins with the relation

$$c^2dt^2 = \left(1 - \frac{2GM}{Rc^2}\right)^2 = dr^2 + r^2 d\phi^2 ,$$

from which we obtain an expression for dt and integrate to find the total time for the forward and return journey of a radar pulse from Earth to the selected planet.

We can, however, start with the refractive index equation[14]

$$c' = c\,(1 - 2GM / rc^2)^{1/2}$$

and the one for gravitational time dilation

$$dt' = dt\,(1 - GM / rc^2).$$

Then, using

$$dt' = ds / c$$

and

$$ds^2 = dr^2 + r^2\,d\phi^2$$

and the standard approximation

$$r \approx \frac{d}{\cos\phi},$$

with d the distance of closest approach, we obtain

$$dt = \left(1 + \frac{GM}{rc^2}\right)\left(\frac{dr^2 + r^2 d\phi^2}{c'}\right)^{1/2}$$

and integrate to obtain the total delay.

With $(r_E, -\theta_E)$ the coordinates of Earth, and (r_P, θ_P) those of the planet, the total time for the radar beam on its outward and return journeys becomes

$$t = \frac{2d}{c}\left(\tan\theta_P + \tan\theta_E\right) + \frac{4GM}{c^3}\ln\frac{\tan(\pi/4 + /\theta_P 2)}{\tan(\pi/4 - /\theta_P 2)}.$$

This is more conveniently expressed by making

$$X_E = d \tan\theta_E$$

$$X_P = d \tan\theta_P$$

$$r_E = d / \cos\theta_E$$

$$r_P = d / \cos\theta_P$$

when
$$t=\frac{2}{c}\left(X_P+X_E\right)+\frac{4GM}{c^3}\ln\frac{r_P+X_P}{r_E-X_E},$$

which is the same as the standard relativistic expression obtained by Ross and Schiff in 1966.[15]

17.9 Perihelion and Periastron Precession

Classical calculations of gravitational light bending and time delay become relatively easy once we have established a special relativistic explanation, as relativistic energy equations for systems without rest mass (and, therefore, without kinematics) are specifically designed to coincide with classical ones. The case is, of course, different for material particles, where kinematics becomes significant. Here, for the first time, we really need to introduce the minimal relativity conditions. The relativistic correction for planetary and pulsar orbits is one of the most significant results derived from the Schwarzschild solution, but as this latter is really only a special relativistic correction to classical gravity, it ought to be possible to derive the non-classical orbital variations in exactly the same way.

The main relativistic effect so far found is an extra degree of precession of the perihelia of planetary orbits, and of the periastra of those of binary pulsars (which is, of course, identical in mathematical terms). While Schiff thought that the effect was resistant to special relativistic analysis as it incorporated nonlinearity and relied on terms of the second order,[9] this cannot be true of any phenomenon dependent on the purely linear, and first order, Schwarzschild equation, and Strandberg has, in fact, provided a special relativistic explanation of the perihelion effect, based on a combination of time dilation and the corresponding effect on the radial length of a spherically symmetric orbit.[5]

Like all gravitationally-induced effects, the perihelion precession is, of course, not strictly special relativistic in that it deals with frames that are not inertial, but it is 'special relativistic' (or, at least, Lorentzian) in making time dilations and length contractions reciprocal effects (which is the only meaning of the term from a fundamental point of view, and the only one relevant, in any case, to the general relativistic extension), with the non-inertial aspect deriving purely from classical gravity. There is nothing needed outside of these two fundamental concepts. Ghosal and Chakraborty, however, have produced what they consider to be a 'purer theory', based on a covariant Lagrangian formulation, to derive a fully (special) relativistic version of Newtonian gravity, at the price of explaining gravitational redshift by 'changes in the atomic (or nuclear) processes themselves' rather than 'as the result of gravitational retardation of clocks'.[10]

Essentially, however, the preservation of the assumptions of the conventional version of special relativity is irrelevant in the context either of general relativity (which abandons them) or a minimally relativistic theory (which has a different physical interpretation). (The same assumptions are equally irrelevant to such examples of 'special relativistic' quantum mechanics as the Dirac solution of the hydrogen atom.) The ultimate aim is to produce a theory which is valid at the fundamental level, not to structure a fundamental theory on the basis of 'measurement' or 'observation', based on what we know to be approximations which will not be preserved at the fundamental level. In principle, therefore, we are only concerned with the Lorentz transformations as part of a more fundamental mathematical structure, which may (as is evident from general relativity) have many explanations other than the purely special relativistic one, and where gravitational transformations really do occur. Within this framework, both the Strandberg and Ghosal-Chakraborty approaches remain acceptable, and compatible, options, but there is yet another approach, which is much more simple.

At the end of the nineteenth century, Gerber[16] developed an argument that a finite speed of transmission would reduce the potential energy between masses m_1 and m_2, and the distance of approach between them, altering the static equation for gravitational potential to

$$V\left(1-\frac{1}{v}\frac{dr}{dt}\right)=\frac{Gm_1m_2}{r\left(1-\frac{1}{v}\frac{dr}{dt}\right)}$$

Substitution of the altered V into the Lagrangian force equation then yielded a force law with the characteristic additional component of the form

$$-\frac{2GM}{r^2}\frac{3v^2}{c^2},$$

to give a result which explained certain observed anomalies in gravitational orbits, particularly that in the planet Mercury. Gerber has been criticized[17] because he used Galilean, rather than Lorentz transformations for space and time coordinates. However, this simply means that the velocity c in his theory becomes a variable, while that in relativity is not – a purely optional choice, as we saw in chapter 9. Hence, the Gerber potential

$$V = \frac{GM}{r(1 - v / c)^2}$$

is different from that used in relativity, but, on differentiation (because of the different status of c), the same force law applies. Gerber has also been criticized because his kinematically-derived Lagrangian yields the wrong result for gravitational light-bending, but this is, of course, irrelevant because the kinematics of massive and massless particles are fundamentally different. It is only in energy terms that we can treat light as a classical particle of the same kind as a massive planet.

Gerber's analysis is complicated and restricted. However, it is possible to do the calculation in a much simpler and more convincing way by using the kinematic assumptions of special relativity. We take the potential

$$V = -\frac{GM / \gamma}{r\gamma} = -\frac{GM}{r(1 - v / c)^2},$$

with 'relativistic' mass and 'contracted' distance. It is important, here, to recognise that any contraction due to gravity will be in rods placed *radially* (in the direction of the field) and not those placed *tangentially* (in the direction of the planet's motion). For an orbit in which the angular momentum per unit mass ($vr = \alpha$) is a constant, this provides an 'inverse-fourth' law force correction. Substituting for v^2 in the expression for V, with the appropriate binomial approximation, we now obtain

$$V = -\frac{GM}{r} - \frac{GM\alpha^2}{r^3 c^2}.$$

Applying the condition that the gravitational field strength or force per unit mass F is $- dV / dr$, we differentiate this expression to find

$$V = -\frac{GM}{r^2} - \frac{3GM\alpha^2}{r^4 c^2}. \quad (17.1)$$

This, of course, is the exact equation for force which emerges from the Schwarzschild solution.

From this we can even derive the exact precession effect by direct substitution into an equation provided in Newton's *Principia* to accommodate such planetary perturbations! Solutions of (17.1) can be involved, requiring transformation into polar coordinates, substitutions, special multiplication factors and further differentiation, but the problem of orbits perturbed by additional force components is a well established one in celestial dynamics, and Newton had effectively solved it as early as 1687. So, we can use his solution to save ourselves further trouble. In Book I, Proposition 45 of his *Principia*,[18] Newton stated that 'if the centrifugal [i.e. centripetal] force be as

$$\left(\frac{bA^m - cA^n}{A^3}\right)$$

the angle between the apsides will be found equal to

$$180^0\sqrt{\frac{b-c}{mb-nc}},$$

Here, A is the distance term and all the other terms are constants. The 'angle between the apsides' is that between successive apses and the angle between two successive perihelia is twice this amount. We can see immediately that, for a pure inverse square law, with $m = 1$ and $c = 0$, the angle between successive perihelia is 360^0 or 2π radian, exactly what we would predict for a closed orbit.

In the present case, leaving out the common term $-GM$, $b = 1$, $c = -3\alpha^2 / c^2r^2$, $m = 1$, and $n = -1$; and the Newtonian precession per orbit becomes

$$2\pi\sqrt{\frac{1+3\alpha^2/c^2r^2}{1-3\alpha^2/c^2r^2}} \approx 2\pi\sqrt{1+\frac{6\alpha^2}{c^2r^2}},$$

which, using another binomial approximation, is an advance of

$$\frac{6\pi\alpha^2}{c^2r^2} = \frac{6\pi v^2}{c^2} = \frac{6\pi GM}{rc^2}$$

radians for a near circular orbit. To take into account the ellipticity of Mercury's orbit, we can replace r by $a(1 - e^2)$, where the a is the length of the semi-major axis = 5.79×10^3 m, and e is the eccentricity = 0.2056. Using the values 1.5475×10^3 m for GM / c^2, and 0.24084 years for the sidereal period, we obtain the required 43 arcseconds per century.

So, in principle, the term which produces the additional perihelion precession is equivalent to a relativistic correction of the gravitational potential by a factor $(1 + v^2 / c^2)$, or, for a relatively low orbital speed v, by a factor $1 / (1 - v^2 / c^2)$. A potential of the form

$$V = -\frac{2GM}{r}\left(1 + \frac{v^2}{c^2}\right)$$

was, in fact, derived by Harvey in 1978 in a discussion of the 'Newtonian limit for a geodesic in a Schwarzschild field'.[19] Though the general relativistic planetary precession effect has been subjected to various physical 'explanations', the real physical reason behind the whole phenomenon is, it would seem, extremely simple: the apparent transmission of the gravitational force at the finite velocity of light. This can be seen from the fact that exactly the same potential energy and force corrections apply in the case of gravitational light bending, although they are there added onto initially zero values. In this case the gravitational effect causes a potential energy and a force to appear where none previously existed and it has no other cause than the finite velocity of light.

17.10 The Inertial Correction

As Ghosal and Chakraborty remark: 'from the pedagogical standpoint one may reasonably ask if it were absolutely necessary to deal with curved spacetime in order to understand gravitation and whether it was completely impossible for certain to extend Newtonian gravity (NG) in the framework of inertial frames to suit the requirements of special relativity. In other words one may enquire whether historically all options had been explored in this regard before one was forced to discard such endeavours.'[10] They note that 'GR is very elegant and rich in its structure and is endowed with interesting philosophical import', but argue that 'it suffers from a serious drawback' in that its 'experimental support' is 'very limited' and, even for that, 'one has to make use of a rather cavalier approximation of GR'.

We have seen how the more straightforward general relativistic effects can be derived from the most minimal addition of special relativistic corrections to the Newtonian potential, which is assumed as the limiting case in all solutions of the field equations of GR. But the question of the *meaning* of these questions remains, if we assume (as both GR and Newtonian gravity effectively do) that the gravitational field – as opposed to changes in the gravitational potential – applies instantaneously. How does the 'velocity of light', the limiting speed of physical measurement, actually enter into the equations at all?

They key concept here is *measurement.* Fundamentally, this requires discrete sources (such as charges) and has little meaning in the context of a universal energy continuum. However, it does have meaning in the context of rest mass and radiant energy. Although these are certainly manifestations of the fundamentally continuous energy state, we can apply the concept of measurement to them, *as long we find a way of making the truly continuous nature of the energy explicit*; and the only way of doing this is to make the instantaneity of the interaction between energy elements explicit. Our system must be a combination of instantaneous interaction and time-delayed measurement.

In such a context, instantaneous transmission of gravity would not lead to Newtonian field equations, as the space-time familiar from measurement would no longer be appropriate. Lorentz invariance is already *assumed* within our system of measurement; but Lorentzian, or 4-vector, space-time is not constructed to describe instantaneous interactions. Instantaneous interaction would be described by Newtonian dynamic equations only if the space-time used was *not* Lorentz-invariant. In respect to the space-time of measurement, then, we would not expect to find Newtonian *equations* even when we were describing a Newtonian system.

To find equations appropriate to instantaneous interaction, we would need to find that system in the space-time of measurement which would be equivalent to a straight line in Newtonian space-time – the 'default' position, as it were, or geodesic, in a Lorentzian space-time, and, more specifically, in a gravitationally-affected Lorentzian space-time, because Lorentzian space-time changes under the influence of gravity by deflecting even 'massless' particles. For this, we need look no further than the gravitationally deflected electromagnetic or Lorentz-invariant path from a straight line of a light ray or any 'massless' field quantum.

Now, the equation, from the Schwarzschild solution, describing the bending of light can be written as

$$\left(\frac{dr}{dt}\right)^2 + \left(1 - \frac{2GM}{rc^2}\right) r^2 \left(\frac{d\phi}{dt}\right)^2 = 0$$

which, in the absence of the gravitational field, reduces to

$$\left(\frac{dr}{dt}\right)^2 + r^2 \left(\frac{d\phi}{dt}\right)^2 = 0 .$$

The 'straight-line' position for a dynamics of instantaneous interaction in the space-time of ordinary measurement is, therefore, described by the first equation rather than the second, which would apply in a *Newtonian* space-time.

The 'default' position describes a system with no dynamic energy. Hence, the application of a Newtonian potential $-GM / r$ to deflect a body of unit mass from straight line motion, in a system of total mechanical energy E, would require a dynamic equation of the form:

$$\left(\frac{dr}{dt}\right)^2 + \left(1 - \frac{2GM}{rc^2}\right) r^2 \left(\frac{d\phi}{dt}\right)^2 = \frac{GM}{r} - E ,$$

compared with the

$$\left(\frac{dr}{dt}\right)^2 + r^2 \left(\frac{d\phi}{dt}\right)^2 = \frac{GM}{r} - E$$

which we would expect for the same system described in a Newtonian space-time. We see at once that a gravitational system involving instantaneous interaction would require dynamic equations in the space-time of measurement of exactly the form required to predict the relativistic effect of planetary perihelion precession.

In this context, we may describe the perihelion (or periastron) precession effect as a result of time and distance changes, produced by the effect of a gravitational potential on Lorentzian space-time, rotating the local coordinate system by an amount which adds the term $-(2GM / rc^2)(d\phi / dt)^2$ to the equation of motion. Naturally, no such 'rotation' would occur if we could employ some 'absolute' system of coordinates separate from the process of measurement. In alternative words, the term $-(2GM / rc)^2 (d\phi / dt)^2$ is the 'inertial force' term expressing the fact that the ordinary process of measurement provides a noninertial frame for a gravitational system. It can even be seen in terms of a curvature of space and time, though as an effect, rather than source, of the gravitational field.

Though gravity, by the adoption of this hypothesis, would be an ordinary vector, rather than a 4-vector system, the space-time of our measurements would remain 4-vector. All measurements in gravitational systems would, therefore, be noninertial and correction terms would need to be applied to make them compatible with other measurements. The size of these correction terms could be imagined as being equivalent to that which provides the bending of light in a gravitational field since the effect is essentially the same. Applying this 'inertial force' term to the energy equation for planetary orbits gives the precise correction which causes the planetary perihelion precession.

It would appear, from this analysis, that we can treat the purely linear effects predicted by the Schwarzschild solution as inertial effects due to the use of the gravitationally-affected metric of measurement for a system which does not require it. Such inertial effects are a standard component of classical Newtonian mechanics, and reflect the fact that a noninertial frame has been selected for our observations; use of a 'wrong' metric would be equivalent to using a noninertial frame, producing the effect which might be described as the 'aberration of space'. The fact that measurement takes place at the velocity of light, or some gravitationally-affected version of it, does not prove that gravity requires the Lorentzian-Schwarzschild metric for its transmission. It is completely consistent with all known facts and mathematical theories to describe the GR 'curvature' equations as a convenient way of describing the inertial effects produced by choosing a metric determined by measurement, in the most general form, rather than as an intrinsic description of the relationship between space and time required by gravity itself.

17.11 The Aberration of Space

If gravity acts instantaneously at a distance, then Lorentzian space-time will not be the appropriate one to describe its effects. However, *measurement* will still take place in this metric, with the assumption that all information (except gravitational interaction) is transmitted at the speed of light, and any energy-carrying information (for example, electromagnetic information), such as is used in the process of measurement, will be subjected to the same gravitational effects as would occur with a beam of light – redshift, deflection, time delay, etc. These effects are derived from the known properties of light (either as classical photons or as special relativistic waves), combined with the action of a purely classical Newtonian potential – there is nothing 'relativistic' about the gravitational component. They can be expressed simply in terms of their effects on the Lorentzian metric, and will have other consequences (such as perihelion precession and gravomagnetic effects) which are immediately calculable from this metric.

It is important that we realise that, while Newtonian theory assumes instant transmission of the gravitational action, *all our processes of measurement* use Lorentzian space-time, with transmission of information at the finite velocity c. A pure Newtonian theory of gravity, then, as we have said, would not be described by Newtonian equations if they were constructed using the space-time of measurement. While theory might demand instantaneous transmission, observation would provide us with transmission at a finite c, and a gravitationally-affected c, at that.

Such a situation was already encountered for astronomical measurements in the early days of Newtonian physics. Roemer's discovery of the finite velocity of light required corrections to the positions of the moons of Jupiter, while Bradley's discovery of aberration required corrections to the measured positions of the stars. In the present case, however, the effect of measuring with a finite c (the transmission speed for all nongravitational information) would be more subtle: the frame would become noninertial. The effect, nevertheless, could be described, by analogy, as the 'aberration of space'.

With this principle established, we could create a method for explaining the precession effect *within Newtonian theory itself.* It is often forgotten that Newtonian theory has a way of dealing with mismatches between theory and observation; but such a system is necessary for the theory to be used at all, for, although Newtonian theory is defined only for inertial frames, *all* experimental observations use frames which are noninertial. The way round this, of course, is to assume that the noninertial frames are actually inertial, and then add on purely fictitious centrifugal and Coriolis forces to accommodate the inertial effects. In the present case, then, there is no reason why, if we *assume* Newtonian gravitational theory to be correct, we cannot simply add on the inertial force terms needed to incorporate the aberration of space.

It is, actually, the universality of mass that makes it necessary to imply only one type of space-time for all systems, whether the forces involved are instantaneous or otherwise; this is because the same masses are present whether we define them through nongravitational fields (as effect) or gravitational field (as sources). So, we have no option but to use the 4-vector space-time of measurement when we are concerned with discrete systems, even when we are describing purely gravitational effects. However, we have also no option but to recognize that 4-vector space-time introduces 4-vector effects in gravitational systems which are *extrinsic to gravity itself*; and, therefore, do not make the gravitational field nonlinear.

From the gravitational point of view, the effects are 'fictitious' effects, resulting from the fact that a 4-vector space-time defines a noninertial frame for a gravitational system, and we attribute them to 'fictitious' *inertial forces*, of exactly the same kind as the centrifugal and Coriolis forces, already mentioned, of classical dynamics. The inertial force correction terms will be, typically, of magnitude $-\ 3GMv^2\ /\ r^2c^2$ or $-\ 3GM\alpha^2\ /\ r^4c^2$, and they will be identical, for obvious reasons, to the terms representing the gravitational effect on energy transmitted at the velocity of light. The light-ray equation will, in effect, represent the 'straight-line' position for a material object which will be altered in the planetary motion equation by the orbital energy term.

In principle, then, in our calculations, the inertial force effect is defined to be exactly that which we would expect to obtain from a relativistic calculation. So, if, to a light-ray geodesic defined in the absence of a gravitational field, we introduce inertial forces exactly sufficient to cover the effects produced by a gravitational potential $-GM / r$, we will, *by definition*, obtain the standard new geodesic equation expected under these conditions, with the extra terms now interpreted as being equivalent to the rotation of the space-time coordinates. In principle, we would expect in all cases the same gravitational curvature of the Lorentzian space-time of measurement as we would obtain from the field equations of general relativity, or equivalent, and those equations could, in fact, be regarded as the most general expression of that curvature.

Under the most general conditions, we would recover the full theory of general relativity, though with the emphasis shifted from ontology (the nature of the gravitational field) to epistemology (the nature of measurement in a gravitational field). The general relativistic field equations, as an aspect of a theory of measurement, would apply in all measured space-time, and predict all observable relativistic effects, including waves, but they would not determine what the unmeasurable 'force' of gravity did in absolute terms. Shifting the ontology to that of a nonlocal force, however, would have positive benefits in terms of gravitational theory, because there would now be no singularities and no nonrenormalizable infinities. Quantum aspects would be available within the epistemological representation, but would not be defined for the ontology of the nonlocal gravitational field, and a field quantum defined by local inertial repulsion would become spin 1, rather than spin 2.

17.12 Gravomagnetic Effects

The aim of the preceding sections has been to suggest that, if it is possible to approach observable general relativistic effects either by starting with the relativity and then supplying the classical approximation, or by starting with the classical theory and then supplying the relativistic correction, it makes sense to look for the option which provides the easiest method of calculation, and the one closest to conditions of observation. For the phenomena considered so far this appears to be the second method, but does this also apply to the more complicated effects which are now within the range of physical observation?

Most such effects can be associated with the existence of a 'gravomagnetic' field analogous to the magnetic field of classical electromagnetic theory, though, according to our assumptions, they would be inertial or 'aberration' effects, rather than directly gravitational. Being purely linear, it should be possible to duplicate such effects by using a 'gravity first' analysis. Kolbenstvedt has, in fact, derived the equations for the gravomagnetic field from special relativity, using a kinematical argument, but has allowed only the effect of time dilation as special relativistic because it requires only the principle of equivalence and the Doppler effect.[20] He considers that the contraction of measuring rods requires general relativity or the 'curvature of space'. Nevertheless, as we have already seen, the length-contraction, or some other doubling effect, is a necessary component of a full special relativistic or even classical treatment.

Allowing, then, for this modification, and following Kolbentsvedt's procedure, we consider an object of mass M, moving with velocity $\mathbf{u}$ in the positive x-direction in the frame of the laboratory, and a particle of mass m moving with velocity $\mathbf{v}$ under its gravitational influence. In the rest frame of mass M, the Lagrangian $\mathcal{L}_0$ of the mass m particle can be found from the variational principle:

$$\delta\int(-mds) = \delta\int \mathcal{L}_0 dt_0 = 0,$$

where $$ds_0^2 = \gamma^{-2}dt_0^2 - \gamma^2 dr_0^2$$

is the line element in the rest frame, and

$$\gamma^2 = 1 - \frac{2GM}{rc^2} = 1 - \frac{2\phi}{c^2}.$$

Integrating for the rest frame,

$$\mathcal{L}_0 dt_0 = -m\,(\gamma^{-2}c^2dt_0^2 - \gamma^2 dr_0^2)^{1/2}.$$

Then, transforming to the laboratory frame, and neglecting higher order terms, we obtain

$$\mathcal{L}dt = -m\,[(c^2 + 2\phi)\,(dt - udx)^2 - (1 - 2\phi)\,(dx - udt)^2 - dy^2 - dz^2]^{1/2}.$$

Once again neglecting the higher order terms, and dividing by dt, we obtain the Lagrangian

$$\mathcal{L} = -\left(c^2 - v^2 + 2\phi - 8\phi\frac{\mathbf{u.v}}{c^2}\right)^{1/2}.$$

When $u \ll v \ll c$, and the rest energy term mc^2 and higher order corrections are neglected, the series expansion approximates to

$$\mathcal{L} = \tfrac{1}{2}mv^2 - m\phi + 4m\phi\frac{\mathbf{u.v}}{c^2}.$$

This expression may be compared with the standard Lagrangian for a particle of mass m and charge q moving with speed $v \ll c$ in an electromagnetic field determined by scalar potential ϕ and vector potential **A**:

$$\mathcal{L} = \tfrac{1}{2}mv^2 - q\phi + q\frac{\mathbf{A.v}}{c}.$$

The equations are clearly analogous, with $4\phi\mathbf{u} / c$ being the gravitational equivalent of the electromagnetic vector potential **A**.

If we extend the argument, the analogue of the magnetic field term $\mathbf{B} = \nabla \times \mathbf{A}$ then becomes the 'gravomagnetic' field

$$\omega c = \nabla \times \left(4\phi\frac{\mathbf{u}}{c}\right) = \left(4\phi\frac{\mathbf{u}}{c}\right) \times (\nabla\phi)$$

$$= 4\frac{\mathbf{u}}{c} \times \mathbf{g}$$

which leads to a series of equations of the Maxwell type:

$$\nabla.\mathbf{g} = 4\pi G\rho$$

$$\nabla.\omega c = 0$$

$$\nabla \times \mathbf{g} = -c\frac{\partial \omega}{\partial t}$$

$$\nabla \times \omega c = 4\pi G\rho\mathbf{v} + c\frac{\partial \mathbf{g}}{\partial t}.$$

Here, $G\rho\mathbf{v}$ takes the place of $\mathbf{j}$; and $4\pi G\rho$ that of ρ / ε_0. In empty space $\mathbf{g}$ and ωc will satisfy the equations for $\nabla \times \mathbf{g}$ and $\nabla \times \omega c$:

$$\Box\mathbf{g} = 0$$

and $$\Box\omega c = 0,$$

with the analogous mass-density terms being added where sources are present.

Other standard results follow immediately, as they do in electromagnetic theory:

$$\text{Poynting vector (energy flux in an element of solid angle)} = \frac{c^2}{4\pi}\, \mathbf{g} \times \boldsymbol{\omega} c$$

$$\text{energy density of fields} = -\frac{g^2 + \omega^2 c^2}{8\pi}$$

$$\text{Lorentz force} = m\,(\mathbf{g} + \mathbf{v} \times \boldsymbol{\omega} c)$$

$$\text{Larmor precession frequency} = \tfrac{1}{2}\omega$$

$$\text{equation of motion of a particle: } \frac{d\mathbf{v}}{dt} = \mathbf{g} + \nabla \times \boldsymbol{\omega} c\,.$$

The question that remains to be answered is: what do these equations mean? One possible interpretation is that they are 'inertial' equations, and that the relativistic correction to gravity is really inertial (although gravitationally-affected – as evidenced by the factor 4 in the expression for $\boldsymbol{\omega} c$), rather than gravitational in direct terms. Inertial forces, significantly, are repulsive, unlike gravitational; this makes the rotational term $\boldsymbol{\omega} c$ positive, unlike the gravitational field, and so, for comparison with Maxwell's equations, in which **E** and **B** are both positive, we take **g**, a 'static' component of inertial repulsion, in place of the gravitational field $-\mathbf{g}$ in the equations from which the wave solutions are obtained. The form of the equation $\nabla.\mathbf{g} = 4\pi G\rho$ is, of course, insensitive to the velocity at which the interaction is transmitted, so a real gravitational attraction *of* the mass density ρ transmitted at infinite velocity would be formally indistinguishable from a fictitious static inertial repulsion *by* the mass density ρ transmitted at the velocity of light. In chapter 21, we will show that the static component of inertial repulsion, which is assumed by the principle of equivalence to be numerically equal to the attractive force of gravity, has a very important cosmological significance.

A whole series of observable effects follow from such formulations, for according to Kolbenstvedt, the linearised gravomagnetic field may be used to derive such effects as 'geodesic deviation of spinning particles, precession of gyroscopes orbiting the Earth, and 'dragging' of inertial frames by rotating masses, by leaning on well-known effects from classical electromagnetism and atomic physics involving spin-orbit and spin-spin coupling'. In 1916 de Sitter calculated that the axis of the Earth-Moon system's orbit would precess at a rate of about 0.02 arcseconds per year due to the curvature of space-time.[21] Eddington pointed out that this was the same geodetic effect as would be expected from the spin axis of a gyroscope no longer pointing in the same direction as it travelled as it would necessarily do in a flat Euclidean space. A gyroscope in a high geocentric orbit might be made to precess at a rate of 6 arcseconds per year.[22] In addition to the geodetic effect, there will be a dragging of inertial frames around a rotating body, that is a Larmor precession in the gravomagnetic field. This effect was first postulated in 1918 by Lense and Thirring,[23] who applied it to the planetary orbits round the rotating Sun. Although much less than the geodetic effect, the Lense-Thirring frame dragging is potentially more important, for it has been regarded since Eddington's suggestion of 1923 as a potential proof of the relative nature of rotational motion.[24] If, however, the effects are due to aberration, then their fictitious nature would preclude their use as a test of the relative aspect.

In a parallel development, various authors have extended the Maxwell formalism to reobtain the results of general relativity for gravitational waves up to the factor 4 which, as we have seen, comes from a combination of the effects of length contraction and time dilation, though Blandford attributes it to 'the spin-2 nature of the gravitational field'.[25-27] In electromagnetic theory, of course, closed systems with constant charge-to-mass ratio have no dipole radiation. So, by analogy, a system whose 'charge' (or gravitational mass) is identical to its inertial mass (and always of the same sign) will emit no dipole radiation. The lowest order 'gravitational' radiation, then, will be quadrupole, and the formula may again be derived by analogy with the well-established results of electromagnetic theory.

When $\mathbf{R}$ is the radius vector from the centre of mass of the system to the point where the field is measured, the equivalent unit vector is $\mathbf{n}$, and $\mathbf{r_i}$ is the radius vector from the centre of mass to the mass m_i, then the gravitational equivalent of the vector potential:

$$\mathbf{A} = 4\phi\frac{\mathbf{u}}{c}$$

becomes, for the quadrupole,

$$\mathbf{A} = -\frac{2}{Rc^2}\frac{\partial^2}{\partial t^2}\sum m_i\mathbf{r}_i(\mathbf{n}_i.\mathbf{r}_i)$$

$$= -\frac{2}{3}\frac{1}{Rc^2}\frac{\partial}{\partial t}\sum_{\mathbf{r}_i} m_i\left[3\mathbf{r}_i(\mathbf{n}_i.\mathbf{r}_i) - \mathbf{n}r_i^2\right]$$

$$= -\frac{2}{3}\frac{1}{Rc^2}\mathbf{D},$$

where **D** is the gravitational quadrupole moment. The components of the vector **D** are $D_{\alpha\beta}n^{\beta}$, where

$$D_{\alpha\beta} = \sum m_i \left[3x_{i\alpha}x_{i\beta} - \delta_{\alpha\beta}r_i^2\right]$$

is the mass quadrupole tensor. The gravomagnetic field intensity is then

$$\omega c = \nabla \times \mathbf{A} = -\frac{2}{3}\frac{1}{Rc^2}\mathbf{D}\times\mathbf{n},$$

and, using the expression for the Poynting flux, we obtain the average energy output over all directions per unit time:

$$-\frac{dE}{dt} = \frac{G}{45c^2}D_{\alpha\beta}^2,$$

which is identical to the expression derived directly from the field equations. Using this, we find the energy loss for a system of two bodies of mass m_1 and m_2, separated by distance r, in orbit round each other with period T, is

$$-\frac{dE}{dt} = \frac{32G}{5c^5}\left(\frac{m_1 m_2}{m_1 + m_2}\right)^2 r^4 \left(\frac{2\pi}{T}\right)^2,$$

as in the experimentally-investigated case of the binary pulsar.

17.13 A Linear Interpretation of the Gravitational Field

It is a widely held assumption that nonlinearity of the gravitational field (or a field which acts as a source of itself) is an inevitable result of the choice of a gravitationally-affected Lorentz-invariant space-time for measurement. The gravity plus inertia interpretation of the general relativistic field equations, however, allows us to propose a purely linear interpretation, and the enormous simplification which results from linearisation suggests that there may be fundamental reasons why this interpretation must be true. The advantages posed by a valid theory of this kind are only too obvious, for nonlinearity has never been accommodated happily to the gravitational field. The idea of nonlinearity, in fact, is not grounded in experiment, and it has no direct presence in the GR field equations. The only place where the characteristic nonlinearity might be displayed – in the discovery of a black hole *singularity* (as opposed to a classical black hole) – is merely a self-fulfilling prophecy. The assumption that an object of a certain size and mass would, *according to the nonlinear interpretation of GR*, have collapsed into a singularity, does not prove that a singularity is concealed by the discovery of such an object, and there is every reason for disputing it on fundamental grounds.[28] Nonlinear theories of the gravitational field proclaim their own fallibility by producing unrenormalisable infinities. They are so constructed that they cannot give a complete description of gravity, even in principle; nor could any conceivable modification of them. If we replace a fundamentally linear theory with a fully nonlinear description of gravity, we rule out all hope of a simple and completely unified theory of physics. We deny, in effect, that the laws of nature are ultimately simple.

Many problems have been associated with the idea that GR must necessarily be such a theory. While the theory interpreted in this way introduces a concept of nonlinearity, it certainly cannot handle it; it similarly requires a quantum formulation but cannot renormalise it; it suggests a unified field theory which turns out to be a hopeless failure; it predicts curvature but cannot explain why the universe is flat. Widely recognised to be incompatible with the highly successful quantum theory, it denies the nonlocality successfully established by Aspect's experiment of 1982, and should predict the immediate closure of a universe with an energy density equivalent to that of the zero-point energy of quantum fields. It turns out to be too difficult to apply in cosmology and the physics of black holes (whose mathematical descriptions seem, additionally, to contradict the theory's physical requirements). It destroys the basis of several symmetries that would otherwise exist, for example the $U(1)$ symmetry between the inverse-square laws for gravity and electromagnetic forces, and the perfect relationship between inverse-square laws and three-dimensional space. In addition, the symmetry between gravomagnetic and Maxwell's equations is true only to first order, and has no fundamental meaning.

From the way it is defined as the first stage in an unending succession of best-fit models, it excludes the possibility of a truly unified theory which would incorporate it. In effect, its inherent structure virtually denies the possibility of any significant progression. In methodological terms, it has led to the idea that sophisticated mathematics is needed at a *fundamental* level – surely a contradiction, in terms – and the widely-held, but erroneous, belief that it is normal to explain simple effects on the basis of complex assumptions – surely a violation of Ockham's razor. Its assumptions also seem to gloss over the very important distinction between the source of gravity, mass, which is a continuous field, and the sources of other interactions, which are centred on discrete particles.

The fundamental reason for developing the original nonlinear interpretation of gravity was to retain the symmetry of the 'velocity of light' mechanism for all four interactions, but the cost of retaining this symmetry was the loss of the alternative symmetry of the inverse square law, and the perfect relation, noted by Kant, which seems to exist between the inverse square law and three-dimensional Euclidean space. The original suggestion of a nonlinear theory, of course, stemmed from the kind of thinking which stressed the primacy of 'light' as a signal, but this was never anything more than a metaphysical conception. 'Light', which really has no *special* status in physics, is nothing other than the action of 4-vector or quantum-based forces, and the introduction of the Minkowski formalism makes any conception of deriving the fundamental properties of space and time by limitations on 'signalling' involved in light's 'physical' action effectively only a heurism. The Minkowski approach showed, as early as 1907, that the properties of space and time did not need to be considered in any 'physical' context at all, but could be simply derived from a formal extension of ordinary vectors into 4-vectors; and fundamental algebra shows us that 4-vectors are a completely natural analogue to the algebraically-determined quaternion system which seems to apply to mass and charge, while the Dirac equation, of course, which is even more fundamental than the Minkowski formalism, further subsumes both 4-vector and quaternion into an overall nilpotent structure.

With the gravitational field assumed linear, like the electromagnetic field, the free-space equations involving ωc, the rotational term representing these inertial forces, become the direct analogues of Maxwell's equations for electromagnetic fields, so an exact formal symmetry can be applied between the equations for gravitational and electromagnetic systems. In the gravitational case, the curl fields (or 'gravomagnetic' fields) become *inertial* fields, equivalent to a rotation of the coordinate system. The inertial effects mean that gravity, expressed in terms of measured space-time, will behave in many respects like an ordinary 4-vector system, though it will be a 'fictitious' 4-vector.

If gravity really is instantaneous, then the Lorentzian (or gravitationally-affected Lorentzian) space-time of measurement, based on discrete sources, will not supply an inertial frame for a gravitational system. In this sense, Newtonian gravitational equations can only be written in a Newtonian space-time, and using a Lorentzian space-time will have the same effect as using a noninertial frame in conventional Newtonian mechanics – the introduction of fictitious inertial forces and rotation of the local coordinate system. The *gravitational-inertial* waves observed will be time-delayed and c-dependent, but will give no information about the transmission of the gravitational force.

In addition, we will not be required to quantize the attractive force of gravity with a spin 2 gauge boson causing nonlinearity, but rather the fictitious repulsive force of inertia, with a QED-like spin 1 boson (whose energy quantum is the Planck mass), thus preserving the theory's linearity with no singularities and full renormalizability. We can imagine a theory (possibly to be named 'quantum gravitational dynamics' or QGD, in the manner of Bell *et al*,[29] or even 'quantum inertial dynamics', or QID), in which (with $G^{1/2}$ normalized to 1) the gravitational / inertial 'charge' would be determined by the relativistic mass of the interacting particles, and in which 'fermion' and 'boson' propagators would have the same structure as in QED, with the effects of space-time curvature absorbed into the value of the vector potential. We could also imagine a Coulomb-like solution of the Dirac equation for a fermion in the radial gravitational field emanating from a point source, and incorporate the $U(1)$ inertial component into the A / r term in the general strong-electroweak solution of section 10.14 to produce a strong-electroweak-gravitational / inertial solution. At a Planck mass Grand Unification scale, then, only the A / r term would remain, and all four coupling constants would become equal in value.[30]

Of course, as quantum field theory predicts, spin 2 intermediaries would not be possible for a *repulsive* force between like particles. As soon as we drop the necessity of the spin 2 intermediary, we ensure that our quantum theory becomes renormalisable in the same way as QED. For such a system, again, representing a purely linear field, there will be no singularities in which normal conservation laws will not apply; black holes (if they exist at all) will be purely classical, and gravitational attractions between discrete sources will only be detectable through the inertia.

General relativity and quantum theory, it is frequently said, cannot both be true. This, in fact, is not so, if we take general relativity for what it really is – a mathematical theory of the apparent effects of curvature of Lorentzian space-time produced by gravity. There is no need, then, for any other theory of the gravitational component *itself* than Newton's original one (as, in any case, is the *de facto* position in calculating physical effects from the GR equations). The inverse-square law of gravity, deriving ultimately from the three-dimensionality of space, will thus remain exactly parallel with the same law in electrostatics, and exactly as general relativity has to assume in solving real problems. Interaction between elements of mass, like that between elements of charge, will remain, as originally assumed, a binary operation produced by the group structure relating the four fundamental parameters to each other. Since mass is intrinsically continuous, no physical agent will be necessary unless we describe the vacuum as the source of continuity.

It is a relatively straightforward procedure to show that the effects of such inertial forces as gravity produces will be of exactly the same kind as the 'non-Newtonian' effects predicted by general relativity. However, the theory will no longer require a nonlinear interpretation, and 'general relativity' will become a theory of inertia, or the *effects of gravity*, rather than of gravity itself. The equivalence principle will have the interpretation that gravity causes inertia, but inertia will not be seen as the cause of gravity. Effects will not be equal to causes. Space-time curvature as a *measure* of mass, and indicator of its presence, will not be identical to mass itself.

In such an interpretation, mass remains an independent parameter affecting Lorentzian space-time but not being *equal* to its effects, and the limiting value of the Newtonian potential is no longer an arbitrary assumption. Because of the 'fictitious' nature of the relativistic correction, both the general relativistic mathematical theory *and* the Newtonian physical theory become simultaneously valid! General relativity will be a combination of classical gravity and relativistic inertia, its field equations directly incorporating inertial effects while preserving the classical structure of gravity itself. Gravity, in fact, unlike the other forces, will not be relativistic at all, in intrinsic terms. General relativity, in this form, it is true, loses some of its cultural significance, as the major plank of Kuhn and Popper's theories of 'scientific revolution' and 'falsification', but it also becomes more perfect by losing the nonlinearity which makes it self-destruct in describing strong gravitational fields. For quantum mechanics, we have the immediate explanation for the 'mysterious' process of nonlocality: gravity really does act instantaneously at a distance.

We can even use the principle (established in chapter 2) that 3-dimensionality is the sole source for discreteness (and finite dimensionality) in physics, to develop a more formal mathematical theory of quantum gravitational inertia, in which the key structure becomes the *3-dimensional* nilpotent structure, represented by $ikt + i\mathbf{r} + j\tau$, or its phase-space equivalent, $ikE + i\mathbf{p} + jm$, and in which the classical 4-D space-time has no fundamental validity (see chapter 18). In the context of gravitational / inertial interactions involving individual fermionic states at the quantum level, there is, of course, an effective reduction or compactification of the spatial dimensions from 3 to 1, since only the **r** parameter is significant. (This is true, of course, also for special relativity.) Significantly, such a 2 + 1 theory of gravity (or, in this case, gravitational inertia) would be both quantizable and renormalizable.[31]

The effect of gravity on the inertial 3 × 3 'quantum metric', based on ***ikt***, ***i*r**, ***jτ***, could be seen in terms of off-diagonal 'curvature' terms, which would be entirely equivalent to the effects of potentials on the phase space or Dirac state equivalent, *ikE*, ***i*p**, *jm*, and these would, in turn, determine the nature of the phase term required to make the Fourier transformation between the quantum metric and the Dirac state (as in the 'solutions' to the Dirac equation presented in chapter 10). Since the Dirac state directly determines the nature of the vacuum which responds to it, this process could be considered as equivalent to the Davies-Unruh effect, where a nonaccelerating system sees a plane-wave version of the zero-point field but an accelerating system sees a distorted one, observable, for proper acceleration α, through the black body temperature $T = \hbar\alpha / 2\pi ck$. The special case of states at a Schwarzschild boundary, where $\alpha = g = GM / r_s^2$, and the metric is curved, with off-diagonal terms, would lead to Hawking radiation.[32]

In fact, the full expression for the spectral density of the zero point field seen by an accelerating source,[33]

$$\frac{dU}{d\omega} = \frac{\omega^2}{\pi^2 c^3}\left(1 + \frac{\alpha^2}{\omega^2 c^2}\right)^2\left(\frac{\hbar\omega}{2} + \frac{\hbar\omega}{exp(2\pi c\omega / \alpha) - 1}\right)^2,$$

which is determined from the space-time properties of an accelerating reference frame, requires two extra components compared to that for the non-accelerating source in a purely Lorentzian metric, given in 15.7, with a correction factor for the density of states in addition to the black-body correction to the zero-point energy; and Puthoff[34] has proposed that the first correction may be used, together with a cut-off in the zero-point field (presumably at the Planck mass), to generate a long-range 'Van der Waals'-type force with the inverse-square characteristics of gravity. It would seem, from the argument used in 15.7, that this force is, in fact, the static inertial repulsion for discrete sources, which is numerically equal to the gravitational interaction, but is generated by a cut-off in the zero-point field, as gravity itself is not.

Following on from this, Haisch and Rueda[35-38] have shown that the Davies-Unruh anisotropy in the distribution of vacuum fluctuations is equivalent to a non-vanishing Poynting vector, which leads to a resistance to acceleration which may be interpreted as inertia or inertial mass, according to Newton's second law, $F = ma$. Haisch and Rueda also use a zero-point cut-off, and their employment of the Poynting vector may be considered as equivalent in principle to the use of a 'gravomagnetic' inertial field to generate the inertial mass within the discrete event horizon defined by the limiting velocity c, which will be presented in section 21.1. This argument depends on the linearity of the gravitational field being determined by the fact that the zero-point cut-off applies only to the discrete inertial repulsion, and not to the continuous gravitational attraction.[39]

The proof that a linear version of the theory can explain all 'non-Newtonian' effects frees it from its less satisfactory physical manifestations. Without the self-inflicted problems brought about by its assumed nonlinearity and spin 2 quantum nature, the many dilemmas seemingly introduced by the acceptance of the general theory rapidly disappear. General relativity also becomes something of a natural development of the fundamental theories of special relativity and Newtonian gravity, rather than a complex and *ad hoc* theory of which they are arbitrary results on approximation, and one which has not yet been derived by a deductive argument from fundamental principles. We are then free to explore whether its mathematical formalism is the best way of describing the effects of gravity on nongravitational energy.

Chapter 18

Dimensionality, Strings and Quantum Gravity

3-dimensionality is identified as one of the most profound and fundamental concepts in physics. With its origin in ideas of anticommutativity, which are antecedent to the concept of number, it is responsible for all discreteness in physical systems, and in particular for quantization. It is responsible for symmetry breaking between the forces, for many significant aspects of particle structure, and for most of the manifestations of the number 3 that are considered fundamental in physics. It is responsible for the selection of the fundamental parameters that we use in the most basic physical explanations, and for their special properties, and the Dirac equation is specially structured to accommodate it. No other dimensionality, not even that of '4-dimensional' space-time, has any fundamental physical significance, a fact which has extremely profound consequences for a unified theory. In Clifford algebras, of course, 'dimensionality' can be defined in a multiplicity of ways for any given algebraic structure, and the nilpotent formalism can be easily accommodated to twistor and string representations, with respective mathematical dimensionalities of 8 and 10 (and even, by analogy, to the topology of the 3-D immersion of the Klein bottle). These, however, are abstract representations and do not require interpretation in terms of any *physical* model. Even within these representations, however, the nilpotent structure shows its fundamental *3*-dimensionality, and it is this inherent 3-dimensionality which allows us to develop a fully renormalizable formal theory of quantum gravitational inertia.

18.1 Discreteness and Dimensionality

String theorists regularly talk of 10 and 11 dimensions. Special relativity uses a 4-dimensional space-time. Kaluza-Klein theory introduces a fifth dimension to general relativity to account for the electromagnetic field. All this is done on the basis that the origin and meaning of dimensionality in nature are matters still to be decided, that, *a priori*, no particular number of dimensions is more likely than any other, that no number associated with dimensionality can be privileged, and that the actual number of dimensions is still negotiable.

Yet 3-dimensionality is very special. It has a mathematical as well as physical validity, which should make us wary of cavalierly expanding the number of dimensions in our system to meet the immediate needs of defining a physically inclusive theory. And the number 3 appears everywhere in fundamental physical contexts – 3 dimensions of space, 3 nongravitational interactions, 3 fundamental symmetries (*C*, *P* and *T*), 3 conserved dynamical quantities (momentum, angular momentum and energy), 3 quarks in a baryon, 3 generations of fermions. Could these be in any way related, and could the explanation of their common 'threeness' somehow lead to a deeper explanation of physical 'reality' than is apparent from the more complex attempts at explanation represented by string theory? What could be special about the number 3 which could unite these apparently disparate manifestations of its application?

As we have seen, space and time, and the sources of the four fundamental interactions, namely mass (or mass-energy) and three types of 'charge' (electromagnetic, strong and weak), which it is convenient to describe as orthogonal dimensions in a 'charge space' or 'vacuum space', are parameters which are symmetric according to a Klein-4 scheme, with exactly opposed properties and 'antiproperties'.

A remarkable aspect of this symmetry lies in the last column, where there are *two* properties and *two* antiproperties, which, if the symmetry is exact, must be linked. We are obliged, it would seem, to suppose that discreteness and dimensionality are intrinsically related properties. In addition, though it is apparent that discreteness and continuity can be considered as a genuinely opposite pairing of property and antiproperty, it is not quite so obvious that the same applies to (3)-dimensionality and nondimensionality, or, as it is sometimes called, 1-dimensionality.

However, '1-dimensional' quantities are not really dimensional at all, and it is relatively easy to see why an absolutely continuous quantity cannot be dimensional (cf chapter 2). Dimensionality requires an origin, a cross-over point or zero position, that is a distinct discontinuity of some kind, which is of course incompatible with the kind of absolute continuity which makes time irreversible and mass-energy unipolar and ubiquitous in the form of the vacuum or Higgs field.

If the linkage here seems relatively obvious, it is far from obvious that discrete quantities must be dimensional, and specifically 3-dimensional at that. However, each of the two known dimensional quantities seems to supply half of the required explanation. Thus, the discreteness of space is associated with its use as the unique channel of physical measurement because the *nonconserved* nature of space means that its discreteness can be endlessly restructured. Measurement would, of course, be impossible in one dimension. A continuous line would offer no possibility of measurement unless it could be drawn in a 2-dimensional space with the other dimension providing the marking off of the zero points or origins. At the same time, the imaginary nature of charge would appear to imply that any dimensionality associated with this quantity must be 3-dimensionality, as required by the algebra of quaternions.

The link between discreteness and 3-dimensionality would appear to be a result of the other properties associated with the parameters which we have defined as discrete, and there appears to be no direct route to be found leading from discreteness itself to dimensionality. However, this is not the case with the reverse process, and it is in fact possible to show that *dimensionality is really the primary property*, and that all discreteness in nature results from dimensionality, and that *only 3-dimensional quantities* are discrete.

18.2 Dimensionality and Chirality

Any 3-dimensional structure which has individually identifiable components is, in principle, a broken or chiral symmetry, and it is always broken in the same way. If we take, say, a quaternion system and identify $\boldsymbol{j}$ (the label is arbitrary, but this choice will be convenient), then we have, typically, a magnitude (scale) or a level of complexification. If, but only if, we bring in a second term, say $\boldsymbol{i}$, we will introduce dimensionalization, and it will necessarily be 3-dimensionalization, automatically generating $\boldsymbol{k}$. This will mean that the $\boldsymbol{k}$ term now has nothing left to do, except determine + or – values, or right- or left-handed axes. In a sense, $\boldsymbol{k}$ has been made redundant, except for 'book-keeping'. Of course, where we don't identify the axes, as for example in the usual description of space rotation, the perfect symmetry is preserved, and it appears that the symmetry-breaking has a close association with the use of a concept of conservation or conjugation in connection with the axes, the 'book-keeping' term being specifically concerned with this, and being of the opposite complexity to the rest to ensure the zeroing of the squared nilpotent quantity, while keeping open the two possible sign options.[1]

The separate roles for the three axes in a 3-dimensional system with identifiable components has a remarkable similarity with the processes involved in creating the infinite algebra of the rewrite system. The role of $\boldsymbol{j}$ is essentially that of complexification, the beginning of a new and as yet incomplete new quaternion system. The role of $\boldsymbol{i}$ is to introduce dimensionalization, while $\boldsymbol{k}$ is restricted to the 'book-keeping' role of conjugation or conservation. These also run parallel to the roles of scalar, vector and pseudoscalar quantities (which an extra i factor has transformed from the sequence pseudoscalar, quaternion, scalar). This is not, in fact, a coincidence, because the key properties of the fundamental parameters have been chosen, by a process of physical 'natural selection' of what can be made to 'work', to reflect the 3-dimensionality which makes it possible to define them at all. The same also applies to the parameter sequence mass, space, time, whose algebraic structures effectively reflect those attributable to the components of charge, the only fundamental 3-dimensional system with identifiable, i.e. independently conserved, components. It is this parallelism which makes it possible to create a closed parameter system with zero totality and in-built repetition.

Table 18.1: 3-dimensional systems with identifiable components

pseudoscalar	quaternion	scalar	(1)
scalar	vector	pseudoscalar	(2)
mass	space	time	
m	$\mathbf{p}$	E	
τ	$\mathbf{r}$	t	
e	s	w	
C	P	T	
$\boldsymbol{j}$	$\boldsymbol{i}$	$\boldsymbol{k}$	
magnitude	direction	orientation	
complexification	dimensionalization	conjugation	
complexification	dimensionalization	conservation	

Here, in Table 18.1, the 'dimensional' term is in the second column and the 'book-keeping' term in the third. (1) = (2) × i and (2) = (1) × i. It may be that we can also include momentum-angular momentum-energy and space translation-space rotation-time translation. The last row refers to the properties of the parameter group, whose fundamental 3-dimensionality is displayed in the diagrams included in chapter 4.[2]

18.3 '4-Dimensional' Space-Time

Minkowski famously said about space and time, after his introduction of 4-vectors (1909): 'From now on, space by itself, and time by itself, are destined to sink into shadows, and only a kind of union of both to retain an independent existence'.[3] Of course, space and time still *are* connected, but the connection is not privileged as Minkowski believed it to be. The connection between space and time is no more significant than that between space and mass and mass and time, or all these parameters and charge. And *there is no fundamental 4-dimensional link between space and time.* There is, however, a 3-dimensional one!

The space-time 4-vector has always run into the problem that one component is physically different from all the others, and it is essentially on account of these differences that the problem of wave-particle duality developed. Wave theories made space timelike (i.e. continuous) while particle theories made time spacelike (i.e. discrete) to fit the two parameters into a single physical model or dimensional structure.[4] The dichotomy even manifested itself in nonrelativistic quantum mechanics, with Schrödinger's timelike theory opposed to Heisenberg's spacelike one. However, the problem, in fundamental terms, is that it cannot be done. The true picture is restored in the Dirac nilpotent theory which is neither spacelike nor timelike, but incorporates elements from both Schrödinger and Heisenberg.

What this theory tells us is that space and time are not a 4-vector. We do not add a pseudoscalar to a pure vector, because each term is premultiplied by a gamma factor or a quaternion before addition. Space and time are actually two dimensions of a 3-dimensional structure, whose third dimension is a mass-related term, the 'proper time', which is of course premultiplied by the remaining quaternion. The 'proper time' is not a time term; it is not a pseudoscalar. It gets its name simply from the fact that it becomes *numerically* equal to the time variable if we equate the space component to zero. We could just as easily describe the actual time variable as the 'proper space' in investigating systems, such as photons, in which the proper time (or rest mass) becomes zero.

Of course, when we take a scalar product, as we invariably do in classical special relativity, the quaternion terms disappear and time acts, to all intents and purposes, as an imaginary fourth dimension of space, fulfilling the role of pseudoscalar needed to complete a mathematical vector theory. Coordinate space, unlike quaternionic charge, is not a pure 3-D system, but already contains, in a hidden form, part of another incomplete 3-D system, which manifests itself as a pseudoscalar under vector multiplication (see 1.6). However, it is important that time is not exactly that pseudoscalar, and no physical quantity exists which can fill this role. The algebraic structure which we have created as a representation of physical 'reality' has no place for 4-dimensional physical quantities. It forces us over and over again into a 3-dimensional pattern.[5] Our quantized, i.e. 3-dimensional, picture denies us the opportunity of representing time as a fourth dimension, denying it status as a physical observable. In a *3-dimensional* theory, time occupies the place of the 'book-keeper', as energy does in the Dirac state, the quantity which preserves conservation or conjugation, but adds only the information of + or –. We only know the direction of the sequence that preserves causality, not a *measure* of time in the same sense as we measure space, in the same way as energy only tells us whether the system is a fermion or antifermion. This fact is well known as a stumbling block to proponents of a quantum theory of gravity, which automatically incorporates time as a physical fourth dimension. It is likely to prove equally damaging to string theories in which a spatio-temporal dimensionality is automatically assumed to be possible.

The fact that the number system we use in mathematics has a 3-dimensional origin is of profound significance. It means that we can't arbitrarily choose the number of dimensions we apply to quantities like space and time without contradicting the principles on which these concepts, and related ones, such as quantization and conservation, were founded. The number of dimensions is not negotiable once we have decided to use the fundamental parameter group and the number system which emerges from 3-dimensionality. Only at the level of classical approximation can we even contemplate any interference in the number of dimensions which nature appears to have thrust upon us.

18.4 Proper Time and Causality

Special relativity is, of course, fundamentally, only a classical approximation to the Dirac equation. In defining its concept of *simultaneity*, it assumes a quantum process (the absorption and emission of light photons) which cannot happen in the quantum world (quantum events requiring a unique birth-ordering), and this is the source of most of the 'paradoxes' which the theory appears to generate. The answer to these paradoxes, of course, is that one can always uniquely determine the true order of events by causality (i.e. via the proper time τ), and that, therefore, they are a result of the approximation introduced by defining simultaneity as a real phenomenon. (Incidentally, as mentioned in 8.3, the 'paradox' that events seen as simultaneous by one observer may not seem simultaneous to another, or that events may appear in the opposite chronological order to two different observers, is not unique to special relativity, but is an obvious consequence of the Galilean transformation. Such occurrences always involve some element of visual parallax, exactly as is invoked to preserve relativity when we observe quasars flying apart at visual speeds much greater than *c*.)

The Einstein-Minkowski formalism for relativity, strikingly leaves out the proper time, and, hence, causality, from its physical picture (cf. 8.4). Einstein's classical theory dispenses with the 'aether' or vacuum precisely because he leaves the 'aether' term out of his equation! Of course, by doing this, he was able to reduce the space-time connection in his classical approximation to kinematics, and, indirectly, to pinpoint the significance of the (excluded) proper time. However, quantum mechanics has been obliged to restore the term directly as an integral component of the Dirac nilpotent, and as the ultimate manifestation of the Higgs mechanism and the filled vacuum, through the observable property of mass. Vacuum, and, in particular, the concept of a filled vacuum, is an expression of the nonlocality inherent in quantum mechanics, which is, of course, antithetical to the classical picture of relativity put forward by Einstein, but which is essential to any fundamental representation of the quantum state.

18.5 The Klein Bottle Analogy

The creation of the fermion state, or fermion-vacuum interface, presents an interesting analogy with the *Klein bottle*. The Klein bottle – a single-sided bottle with no boundary, made from two Möbius strips or single-sided pieces of paper – is a topological structure which, using a twist, effectively reduces a 4-D space to a 2-D space. It is a 2-D manifold which can only exist in 4-D. Visual representation can only be accomplished in a 3-D *immersion*, which introduces a singularity or hole which would not be present in a true Klein bottle structure. Such an immersion, we could say, reduces a 3 + 1 space-time structure to a 2 + 1 structure. Now, the nilpotent structure (ikE + $i\mathbf{p}$ + jm) is a 2 + 1 structure in charge (quaternion) space, and could be considered a reduction from a 3 + 1 mass-charge space. If you derive it from quaternionic matrices, with coefficients 1, $\boldsymbol{i}$, $\boldsymbol{j}$, $\boldsymbol{k}$ (as in 5.7), you zero one of them for each solution, ending up with the elimination of the real term with coefficient 1, which is the continuous vacuum (mass). In conventional theory, each of the 4-component column vector free-particle amplitudes has a zero in a different position in the column. Here, the 4-component spinor space maps onto 4-D Lorentzian space-time, because there is no charge space available. This means that the spinor (with its symmetrical 2 particle-antiparticle × 2 spin up-spin down states) distorts the 4-D space-time into the same 2 + 2 structure, with signature + + – –, instead of the true + + + – (in the nilpotent formalism, this happens to the charge space, causing symmetry breaking), but, by zeroing one component, it always ends as 2 + 1 (or, equivalently, 1 + 2). The process of making the 2 + 1 is that of making the fermion a singularity, and, in the nilpotent formalism, *the rest of the universe as the fermion sees it* has the same structure. This can be considered a 'creation' event. The Klein bottling effect can be considered as being manifested through the spin (the twist) and the creation of a singularity. Infinity – the infinite vacuum – we could say, becomes zero – the fermion singularity – at this point.[6]

18.6 A String Theory Without Strings

Superstring and membrane models have been claimed as offering the best candidate for a unified theory of physics, though it is also believed that the unifying theory will probably not be any of the five classes of string theory currently known (Types I, IIA, IIB, and $SO(32)$ and $E_8 \times E_8$ heterotic), but a more fundamental, unifying theory of which these are model-dependent approximations under specific assumptions. An ideal string or membrane theory would, then, appear to be one which removes the model-dependent assumptions, in effect a string theory without strings.

10 space-time dimensions are apparently required to construct a quantum field theory of superstrings in which all anomalies cancel out, while an eleventh dimension is required to extend to supermembranes embedding all the different classes of string theory. (There are, of course, 10 generators in the Poincaré group and 10 components in the metric tensor.) The nilpotent Dirac theory provides exactly these requirements. Each nilpotent represents 10 conserved quantities and so could be constructed in a 10-dimensional phase space:

energy	3 components of momentum	rest mass
weak charge	3 components of strong charge	electric charge

This set of 10 'dimensions' incorporates a fundamental duality involving vacuum, combining a 'real' space with a 'vacuum' or 'charge' space. All particles are dual with the vacuum, and only exist in relation to it (*zitterbewegung* being the dynamic manifestation of this), and so we require ten pieces of information at the same time for a full specification of a particle state. The energy and the charge components appear as mutually exclusive occupiers of the vacuum and material aspects. In principle, one set of five components represents the particle and the other set the dual vacuum state, or one set represents amplitude and the other phase. All ten, however, are needed to specify the state, and, to convert from phase-space to 'real' space, we would simply use the conjugate metric ($\pm\, \boldsymbol{i}kt \pm \boldsymbol{i}\mathbf{r} + \boldsymbol{j}\tau$). It is significant that six 'dimensions' (all except E and $\mathbf{p}$) are fixed or compactified, exactly as required in string theory, and also that they are constrained by symmetries that are spherical in origin, like the $U(1)$ symmetry in Kaluza-Klein theory, which corresponds here to the special case of electric charge. In addition, the nilpotents are embedded in a Grassmann-Hilbert space, which acts like an 'eleventh dimension' (although it is, of course, infinitely dimensional in itself), and carries the gravitational connection.

'Dimensionality', of course, is definable in many ways within Clifford algebras; and the nilpotent operator can also be seen according to different criteria as 1-, 2-, 3-, 4-, 5-, 6-, 8- or even higher-dimensional, and a variety of different geometrical algebras can be used to create the 64 unit structures needed for the gamma matrices. $G(4,1)$, $G(3,2)$, $G(3,3)$ and $G(6,0)$, for example, are all equally possible. The multiplicity of dimensionalities is provided by the fact the basic units contain two independent 3-dimensional systems.

Other features of string theory are immediately apparent from this representation. S-duality and T-duality are fundamental symmetries of the parameter group, in which each parameter is its own inverse (see chapter 2). These dualities represent those between the respective nonconserved and conserved quantities. It is already known that the superstring theories known as Type IIA and Type IIB are related by T-duality, as are the $SO(32)$ and $E_8 \times E_8$ heterotic theories, while S-duality in 10 dimensions relates heterotic string theories and those known as Type I. On the other hand, Type IIB theories are self S-dual.

Some of these theories have more specific connections with the Dirac nilpotent. $E_8 \times E_8$ can be derived from an octonionic representation of the 4-vector-quaternion formalism for space-time-mass-charge. In the Dirac nilpotent, however, the octonion symmetry is broken, and this gives an indication of the point at which $E_8 \times E_8$ deviates from the more fundamental theory. (See 15.5.)

Other string models also appear to use specific aspects of the Dirac formalism. Thus *SO*(32), which is used in two of the theories, invokes the 32 components of the Dirac algebra or Dirac state, while the 26 dimensions needed for boson strings to provide nilpotent BRST operators may well be related to the 25 bosonic generators plus Higgs needed for grand unification in this formalism.

The eleventh or 'membrane' dimension, in the present representation, is the commutative Hilbert space or Grassmann algebra linking all the nilpotents, which, significantly, is connected to gravity and instantaneous correlation. We should, nevertheless, recognize that both 10- and 11-dimensional models are, really, manifestations of a more fundamental 3-dimensionality, in being determined by the three quaternion operators $\boldsymbol{k}$, $\boldsymbol{i}$, $\boldsymbol{j}$. A quantum nilpotent structure can always be given a 3-dimensional representation, through the affine nature of the **p** or *s* operator. Only one direction for spin is well defined and only one state for the colour charge. This understanding, as we will see in 18.5 and 18.6, allows us to develop a renormalizable theory of quantum gravity or quantum gravitational inertia.

String theories are, by definition, also supersymmetric. It is, of course, the unbroken supersymmetry of the Dirac nilpotent which allows us to define energy and charge states simultaneously. In principle, an unbroken supersymmetry requires the vacuum to have zero total energy, which is what we expect if the total energy due to the matter is exactly cancelled by negative gravitational energy. Spontaneous symmetry-breaking, in this interpretation, is not, then, due to the overall state of the vacuum, but to the discrete weak vacuum, which, via the Dirac / Higgs mechanism, privileges $+E$ states over $-E$ for discrete matter. Significantly, as mentioned in 15.5, the superspace needed for supersymmetry postulates four antisymmetric coordinates as superpartners of space-time; here they become mass and the three charges. The eight coordinates together constitute the superspace, which, in this formalism, takes on the character of the nilpotent Dirac algebra.

18.7 Twistor Representations

Because of its origin in the eight fundamental units of multivariate vectors plus quaternions, there is also an easy mapping from the nilpotent formalism to either an octonion or a twistor representation. The octonion representation is straightforward and leads to the symmetry groups of fundamental interest in particle physics, though not to any specific identification of how those symmetries are actually applied. The twistor (or 4-dimensional complex) representation is more subtle. We may draw attention here to Witten's proposal to represent perturbative gauge theory in terms of a string theory in twistor space.[7] Here, the twistor space introduces the space of the four spinor components, which are mapped by Fourier transform onto the ordinary 4-dimensional momentum space. In quaternionic forms of the Dirac equation, the four quaternion units ***j***, ***k***, ***i***, 1 map quite naturally onto the four components of the Dirac spinor when the structure of the gamma matrix representation is preserved. For example, one way of transforming the conventional Dirac equation to the nilpotent one is through applying the quaternion coefficients to the 4 components of the column vector (see 5.7). A typical, though not unique, representation associates $E \pm \mathbf{p}$ with ***j***, ***k***, and $-E \pm \mathbf{p}$ with ***i***, 1, the transition to the opposite helicity state being a parity transition accomplished through multiplication by ***i***. This maps onto a gamma matrix structure in which one of the momentum components is paired with E in having complex matrix elements, while the other two momentum components have matrix elements that are real. The 4-spin state is identified exactly with a 4-momentum state, at the price of splitting it into two parts with a signature $+ + - -$. The full nilpotent formalism, however, introduces subtle connections which overcome this apparent bifurcation, while the compactification in the new structure ensures that the quaternion units have multiple meanings as symmetry operators, vacuum creators and charge generators, while also representing the Klein-4 group incorporating charge, space, time, and mass.

The twistor aspect emerges naturally from the nilpotent formalism because two 4-D spaces, with opposite signatures (exactly equivalent to a complex 4-D space), are being combined explicitly in a maximally-compactified format.[8] In the Witten representation, the twistor space is an additional mapping, to be derived from the conventional gamma matrices, like the partial mapping onto quaternion units of the 4 components of the spinor space. Infrared divergencies emerge as the twistor space is mapped back onto momentum space; and this occurs because of the conventional bispinor splitting of positive and negative energy states. Despite this, the representation has the power to remove massive amounts of redundancy from QCD calculations by reducing the number of diagrams required. The reason for this appears to be the fact that information referring to positive and negative helicity states is duplicated many times over in calculations which treat them separately. In the nilpotent case, of course, duplication is further reduced by avoiding the remaining split between positive and negative energies, and the removal of the infrared divergence is an added bonus. There is no need to map from one space to another as the momentum space and spinor space are essentially identical and inextricably linked; the spinor space is simply an automatic ('drone') extension of what is already conveyed through the quaternion coefficients in the momentum space. It is significant that the twistor representation derives some of its special power from its use of scalar propagators and its specific confinement (like string theory) to mass-shell fermions. These are, of course, already well-established aspects of nilpotency.

At the very least, making the Witten theory nilpotent would almost certainly enhance its already strong calculating power, but it would also establish the fundamental nature of the nilpotent condition, for the route from the twistor to the 10-dimensional string-like space needed to complete the intended connection is provided exactly via the nilpotent condition. The Witten version of twistor space is created as the square root of a squared invariant mass. Incorporating this term by taking a square root of a zero totality extends the basic structure from a double 4-dimensionality with incomplete links to a double 5-dimensionality with complete links. 'Dimensionality', of course, is definable in many ways, as we have shown in the previous section. However, the 10-dimensional representation is of the exact kind required by the proponents of string theory as the abstract link between the more inexact physical realisations incorporated in the five basic models. Also, the requirement for a perfect string theory is that self-duality in phase space determines vacuum selection. The nilpotent operator is self-dual, expressed in terms of phase space, and completely determines vacuum selection. Here, we have that requirement fulfilled exactly, without any need for an intermediate 'physical' model, and an embedding eleventh 'dimension' is, as stated, provided by the Hilbert space within which the state vectors operate.

18.8 Quantum Gravitational Inertia

The principle that 3-dimensionality is the sole source for discreteness in physics, and that no other dimensionality exists at the fundamental level has consequences for the development of a mathematical theory of quantum gravity, or, in more accurate terms, a mathematical theory of quantum gravitational inertia. According to the argument presented here, there is no fundamental 4-D, and, though there is a mathematical object called a 4-vector, there is no physical realisation of it, except in the classical approximation. The key structure then becomes the *3-dimensional* nilpotent structure, variously represented by $\boldsymbol{ik}E + \boldsymbol{i}\mathbf{p} + \boldsymbol{j}m$ and $\boldsymbol{ik}t + \boldsymbol{i}\mathbf{r} + \boldsymbol{j}\tau$, which is both fully quantum and fully relativistic, and the 3-dimensionality of the structure is essential to its complete quantization.

There is no true 5-dimensionality in the structure, as we might at first think, because the nonconserved 3-dimensionality of the **p** and **r** terms is of a different nature to the conserved 3-dimensionality of $\boldsymbol{k}$, $\boldsymbol{i}$ and $\boldsymbol{j}$, though we can, if we choose, relate the nilpotent information in $\boldsymbol{ik}E + \boldsymbol{i}\mathbf{p} + \boldsymbol{j}m$ and $\boldsymbol{ik}t + \boldsymbol{i}\mathbf{r} + \boldsymbol{j}\tau$, as defining the ten 'degrees of freedom' concerning any fermionic state which lie at the basis of the 10- and 11-dimensional superstring and supermembrane theories. However, although it is worth showing that this is possible, it is not worth pursuing in detail, as there is no point in developing a more limited superstructure, whose ultimate purpose is to provide a route to a more fundamental basis, when that basis is already available. Thus, although various larger algebraic structures, for example octonions and even classical Minkowski space-time, have been shown to produce some of the results that are required in a fundamental theory, this is always at the price of producing others which are invalid, and it would seem that the 3-dimensional pattern is the one that nature prefers, and that in identifying this as the true fundamental context we are likely to discover more universally valid results.

We can now, for example, immediately relate $\boldsymbol{ik}t + \boldsymbol{i}\mathbf{r} + \boldsymbol{j}\tau$ to the discrete gravity theory presented by Koberlein,[9] which is based on the fact that a single object (particle or field) at two points in Minkowski space-time (represented by the 4-vector x) must satisfy the causality constraint $\Delta\tau^2 + \Delta x^2 = \Delta(\boldsymbol{ik}t + \boldsymbol{i}\mathbf{r} + \boldsymbol{j}\tau)^2 = 0$, which defines a hypercone for the object. 'Extended causality' then applies when we shift τ and x by infinitesimal steps $d\tau$ and dx. Using Koberlein's procedure, we can then apply a massless scalar field to obtain a discrete field equation, and a field source represented by a scalar charge to generate a 'graviton'-like object (a 'pseudo-boson') and a metric for a discrete gravitational field. It is clear that if we apply the quantum $\boldsymbol{ik}t + \boldsymbol{i}\mathbf{r} + \boldsymbol{j}\tau$ for a Dirac particle in the appropriate places in place of (x, τ), then we can produce a fully quantum version of this discrete gravity, with the discreteness referring to an interaction between fermions as discrete particles defined by a 3-D Dirac nilpotent. Significantly, the discrete theory also dispenses with the transverse directions, to create a 1 + 1 space-time, paralleling the fact that a quantum Dirac particle, with conserved charge (the kind of object to which quantum gravity or quantum gravitational inertia will apply), requires only **r**, and a single well-defined direction of spin, rather than the classical x, y, z.

It is already apparent, from previous quantum gravity theories, that any attempt at quantizing 4-D space-time is a lost cause, because time is not an observable in quantum mechanics as it is in classical relativity theory; it merely plays the 'book-keeper' role of specifying the direction which preserves causality. This means that, for a fully quantized theory, we need a metric other than the 4×4 representation using x, y, z, t, with added curvature, which is used in classical general relativity. The obvious one that suggests itself is a 3×3 representation, with diagonal terms $\boldsymbol{ik}t$, $\boldsymbol{i}\mathbf{r}$, $\boldsymbol{j}\tau$, in the absence of the curvature resulting from a gravitational field. This formalism would have the distinct advantage of being a natural $2 + 1$ theory of gravity (the 2 representing the 'real' terms $\mathbf{r}$ and τ, and the 1 the imaginary term it), and such theories are already known to be renormalizable, unlike those with a higher number of dimensions. A preliminary investigation of the method suggests that it works exactly as expected.

Now, Bell *et al* have presented a preliminary approach to a QED-like quantum gravity[10] by using a quaternionic mapping of the four solutions of the Dirac equation onto a space which, without curvature, is equivalent to that provided by the usual 4×4 representation of the Lorentzian metric. The natural result of this mapping is the production of the Bohr-Sommerfeld orbitals for the electron in a scalar electrostatic potential in a purely classical way, thus providing a 'natural' generation of space-time curvature, which can be extended when gravitational curvature terms are directly applied to the metric of the four Dirac solutions.[11]

In terms of the theory presented here, of course, any version of the 4×4 Lorentzian metric will be neither fully quantized nor fully relativistic, but the 3×3 'quantum metric', based on $\boldsymbol{ik}t$, $\boldsymbol{i}\mathbf{r}, \boldsymbol{j}\tau$, will fulfil both these criteria, and, in the spirit of Bell *et al* (though using a different set of dimensional quantities), can be mapped onto a 'phase space' metric based on $\boldsymbol{ik}E$, $\boldsymbol{i}\mathbf{p}, \boldsymbol{j}m$, which gives the full information about the Dirac state, and produces the full Dirac 'atom' solution and $U(1)$ QED-type behaviour, with a corresponding photon-like mediating boson, merely on the assumption of spherical symmetry and the multivariate vector nature of the spin term $\mathbf{p}$ (or, equivalently, conserved charge). A purely 'Lorentzian' metric would not, of course, automatically include spin, unless the vector term was assumed to be multivariate, but, more seriously, would exclude the fundamental nilpotent relations between the parameters space, time, mass and charge which are responsible for both quantization and relativity.

The phase space metric has the direct advantage that it can be obtained directly from the 'quantum metric' (and vice versa) via a Fourier transform, and we can thus imagine the quantum metric as being generated by and carried along with the state which defines it. Evidence for 'curvature' (i.e. the effect of a gravitational field on the inertial metric) can then be seen in the phase through which this transformation occurs – which will be the usual complex exponential for a free particle, but distorted in the presence of a field or 'curvature'. Since the Dirac state directly determines the nature of the vacuum which responds to it, this process will be equivalent to the Davies-Unruh effect, where a nonaccelerating system sees a plane-wave version of the zero-point field but an accelerating system sees a distorted one.

In the case of phase space, the reduction of the metric to 3×3 reflects the fact that, in the nilpotent formulation, the specification of four separate solutions becomes redundant information in the Dirac spinor, because knowledge of the signs of $\boldsymbol{ik}E$ and $\boldsymbol{i}\mathbf{p}$ in the first term automatically gives us the entire pattern which follows – this is equivalent to separate specification of x, y and z being redundant in the quantum context. In addition, if the basic metric is 3×3, rather than 4×4, the mediator responsible for any curvature terms becomes spin 1 (as Bell *et al* require for a renormalizable theory), rather than spin 2.

The need for a spin 1 mediator and QED-like theory in 'quantum gravity' has been discussed in many previous publications. There, it has been suggested that the continuity of mass-energy, the filled vacuum, the Higgs field, and the need for instantaneous correlation between Dirac states, together with the fact that energy does not actually move (as opposed to the form of its realisation in connection with a discrete state), require an instantaneous gravitational force, which is undetectable by direct observation, and only ever observed through the c-dependent inertial reaction on discrete fermionic or bosonic states. Being repulsive, this force requires a mediator of spin 1. In this context, we may note that the nilpotent representation significantly makes the Dirac state identical to its own gravitational vacuum, at $1(\boldsymbol{ik}E + \boldsymbol{i}\mathbf{p} + \boldsymbol{j}m)$, whereas the vacuum responses to the weak, strong and electric charges can be represented, respectively, by $\boldsymbol{k}(\boldsymbol{ik}E + \boldsymbol{i}\mathbf{p} + \boldsymbol{j}m)$, $\boldsymbol{i}(\boldsymbol{ik}E + \boldsymbol{i}\mathbf{p} + \boldsymbol{j}m)$, and $\boldsymbol{j}(\boldsymbol{ik}E + \boldsymbol{i}\mathbf{p} + \boldsymbol{j}m)$.

The standard mathematical representation of the gravitational force incorporates no information relating to speed, but the description of gravity as an undetectable property of the vacuum would *require* it to be instantaneous. The c-dependence of the inertial reaction, however, determines that, though linear and renormalizable, this force will itself be affected by gravity, giving rise to the 'curvature' terms in the metric tensor, as in general relativity. It is, however, 'curvature' of the metric for inertia, not for gravity, which has no metric.[12] The work in chapter 21 will show that, if we equate the inertial reaction numerically with the undetectable gravitational attraction (so defining an equivalence principle), we justify a form of Mach's principle, and obtain gravomagnetic effects, redshift, acceleration of the redshift, and perhaps even the cosmic microwave background radiation. In the simplest case of 'curvature', provided by a point source, we will generate the Schwarzschild metric and a factor 4 in the gravomagnetic equations by comparison with those for QED. This factor (incorporating 2 for space 'contraction' and 2 for time 'dilation', if we adopt the usual convention of making c constant), is evident in the factors of 2 which appear in the mass and field terms in the Schwarzschild solution presented by Bell *et al.*[10]

18.9 Calculation of Quantized Gravitational Inertia

If we assume that 3-dimensionality is the sole source for discreteness in physics, the mathematical object called a 4-vector will have no physical realisation at the quantum level. Instead, we use the *3-dimensional* nilpotent structure, represented by the terms $\pm ikt \pm i\mathbf{r} + j\tau$ and $\pm ikE \pm i\mathbf{p} + jm$. The first term may be described as the 'quantum metric'; the second is the realisation of the Dirac state, and may be regarded as the phase space version of the first. No other fundamental structure is both fully quantum and fully relativistic, and the 3-dimensionality of the structure is essential to both of these conditions.

The interpretation of inertia as the result of the interaction between discrete matter and the continuous gravitational vacuum suggests that it is gravitational inertia rather than gravity which is subject to quantization. On the basis that 3-dimensionality is indeed the sole source for discreteness in physics, we can develop a more formal mathematical theory of quantum gravitational inertia, in which the key structure becomes the 3-dimensional nilpotent structure, represented by $\pm ikt \pm i\mathbf{r} + j\tau$, or its phase-space equivalent, $\pm ikE \pm i\mathbf{p} + jm$. The 3-dimensionality comes from the fact that, for gravitational inertial interactions involving individual fermionic states at the quantum level, there is, of course, an effective reduction or compactification of the spatial dimensions, to a single well-defined parameter ($\mathbf{r}$). In this case, we can construct a 2 + 1 theory of gravitational inertia, based on a 3 × 3 'quantum metric' (with $\mathbf{r}$ and τ representing the 'real' parts and it the imaginary), which would be both quantizable and renormalizable.

As indicated in the previous section, one way of quantizing gravitational inertia in this way is via the discrete gravity theory based on the idea of *extended causality*, which has been presented by Koberlein. Koberlein's presentation can be translated (or even transcribed) almost immediately into nilpotent terms and, in principle, quantized, without changing any of the significant details, because it uses explicit proper time and a single direction for **r**. In this formulation, a single object (particle or field) at two points in Minkowski space-time (represented by the 4-vector x) must satisfy the causality constraint

$$\Delta \tau^2 + \Delta x^2 = \Delta(ikt)^2 + \Delta(i\mathbf{r})^2 + \Delta(j\tau)^2 = 0, \qquad (18.1)$$

which defines a hypercone for the object. Extended causality then applies when we shift τ and x by infinitesimal steps $d\tau$ and dx. We then obtain

$$(\Delta \tau + d\tau)^2 + (\Delta x + dx)^2 = (ik\Delta t + ikdt)^2 + (i\Delta \mathbf{r} + id\mathbf{r})^2 + (j\Delta \tau + jd\tau)^2 = 0. \quad (18.2)$$

Defining f as a fibre in the 'spacetime' $x \equiv (i\mathbf{r}, ikt)$, where $f_\mu = dx_\mu / d\tau = id\mathbf{r} / jd\tau + ikdt / jd\tau$ and $f^\mu = -d\tau / dx^\mu = -kd\tau / d\mathbf{r} + iid\tau / dt$, we may combine (18.1) and (18.2) to obtain:

$$\Delta \tau + f \cdot \Delta x = j\Delta \tau + (id\mathbf{r} / jd\tau + ikdt / jd\tau)(i\Delta \mathbf{r} + ik\Delta t) = 0,$$

while extended causality now requires

$$(\tau - \tau_o) + f_\mu (x - x_o)^\mu = j(\tau - \tau_o) + (id\mathbf{r} / jd\tau + ikdt / jd\tau)(i\mathbf{r} + ikt - i\mathbf{r}_o - ikt_o) = 0. \qquad (18.3)$$

Suppose now that we introduce a massless scalar field $\phi_f(x, \tau) \equiv \phi_f(i\mathbf{r}, ikt, j\tau)$, the extended causality in (3) will constrain it to the hypercone generator. Equation (18.3) will also induce a direction to the field derivatives, so that

$$\partial_\mu \phi_f = (\partial_\mu - f_\mu \partial_\tau) \equiv \nabla_\mu \phi_f.$$

With this expression we can now derive a discrete field equation. If χ is the coupling constant and $\rho(x, \tau) \equiv \rho(i\mathbf{r}, ikt, j\tau)$ the discrete field source, then, using standard methods, the action of the field is given by

$$S_f = \int i \, d\mathbf{r} \, dt \, d\tau \, \{½ \, \eta^{\mu\nu} \nabla_\mu \phi_f \nabla_\nu \phi_f - \chi \phi_f \rho(i\mathbf{r}, ikt, j\tau)\}.$$

The field equation then becomes

$$\eta^{\mu\nu} \nabla_\mu \nabla_\nu \phi_f(i\mathbf{r}, ikt, j\tau) = \rho(i\mathbf{r}, ikt, j\tau),$$

with energy tensor

$$T^{\mu\nu}_f = \nabla^\mu \phi_f \nabla^\nu \phi_f - ½ \, \eta^{\mu\nu} \nabla^\alpha \phi_f \nabla_\nu \phi_f.$$

The solution can be conveniently expressed in terms of a Green's function. Here we write:

$$\phi_f(\mathbf{ir}, ikt, \boldsymbol{j}\tau) = \int id\mathbf{r}\, dt\, d\tau(\mathbf{ir}' + ikt')\, G_f(\mathbf{ir} + ikt + \boldsymbol{j}\tau - \mathbf{ir}' - ikt' - \boldsymbol{j}\tau')$$

$$\times\, \rho(\mathbf{ir}', ikt', \boldsymbol{j}\tau')$$

and $$\eta^{\mu\nu}\nabla^{\mu}\phi_f\nabla^{\nu}\phi_f\, G_f(\mathbf{ir} + ikt + \boldsymbol{j}\tau) = i\delta\mathbf{r}\,\delta t\,\delta\tau.$$

Using the Heaviside step function, Θ, with $b = \pm\, 1$, the Green's function is then

$$G_f(x, \tau) = \tfrac{1}{2}\, \Theta(bf^4 t)\, \Theta(b\tau)\, \delta(\tau + f\, .\, x)$$

or $$G_f(\mathbf{ir} + ikt + \boldsymbol{j}\tau) \tag{18.4}$$

$$= \tfrac{1}{2}\, \Theta(b(id\mathbf{r} / \boldsymbol{j}d\tau + ikdt / \boldsymbol{j}d\tau)ikt)\, \Theta(b\boldsymbol{j}\tau)\delta(\boldsymbol{j}\tau + (id\mathbf{r} / \boldsymbol{j}d\tau + ikdt / \boldsymbol{j}d\tau)(\mathbf{ir} + ikt)).$$

Even in classical, discrete form, this equation is independent of the transverse components of the field, paralleling the quantum reduction to a single well-defined direction of spin.

If we now take a single scalar charge $q(\tau)$, with world line $z(\tau)$, as a field source, then:

$$\rho(x, t_x = t_z) = q(\tau_z)\, \delta^{(3)}(x - z(\tau_z))\, \delta(\tau_x - \tau_z),$$

and the solution for the emitted field becomes:

$$\phi_f(x, t_x) = \chi \int d\tau_y\, \Theta(t_x - t_y)\, \Theta(\tau_x - \tau_y)\, \delta[t_x - t_y + f\, .\, (x - y)]\, q(\tau_z),$$

which reduces to

$$\phi_f(\mathbf{ir}, ikt, \boldsymbol{j}\tau) = \chi\, q(\tau)\, \Theta(t)\, \Theta(\tau)\, |_f,$$

or $$\phi_f(\mathbf{ir}, ikt, \boldsymbol{j}\tau) = \chi\, q(\tau)\, |_f,$$

when $\tau \geq 0$ and $t > 0$. In the discrete model, the emission or absorption of a field causes a discrete change in q, and we can apply the standard techniques appropriate to quantum field theory, and, in particular, QED, to develop the formalism for interactions at higher orders, the 2 + 1 nature of the theory ensuring its renormalizability.

Chapter 19

Nature's Code

This chapter is coauthored with **VANESSA HILL**. Mathematical structures apparently underlying different aspects of physics and biology are examined in relation to their possible common origin in a universal system of process applicable to Nature, which we may describe as 'Nature's code'. We begin with a revision of mathematical and physical concepts essential to the rewrite procedure as developed in chapters 1 and 2. This is aimed at showing the significance to fundamental processes of such concepts as 4 basic units, 64- and 20-unit structures, symmetry-breaking and 5-fold symmetry, chirality, double 3-dimensionality, the double helix, the Van der Waals force and the harmonic oscillator mechanism, with an explanation of how they necessarily lead to self-aggregation, complexity and emergence in higher-order systems. Biological concepts, such as translation, transcription, replication, the genetic code and the grouping of amino acids are shown to be driven by fundamental processes of this kind. The role of the Platonic solids, pentagonal symmetry and Fibonacci numbers in organizing 'Nature's code' is explored in detail, with special reference to DNA, RNA and the genetic code.

19.1 The Dirac Nilpotent as the Origin of Symmetry-Breaking

Biological systems, though operating at the edge of chaos, are extremely ordered, whereas the tendency for nature is to become more disordered. Biology is, in effect, a race between order and entropy with the odds stacked in favour of entropy. So biological systems must create order, i.e. process information, with as much efficiency as possible. The work in this chapter is aimed at showing that the efficient processing of information requires certain algebraic and geometric structures, which are also found in systems organized at other scales, in particular, physics.

There are some tantalizing hints that a few simple mathematical structures play a significant rolc in biology, as well as physics, when taken at the fundamental level. A key connecting idea is 3-dimensionality. Using the concept of a universal computer rewrite system and starting from the idea that zero totality is necessary at all time, we showed, in chapter 1, that we could devise an algebraic structure in which 3-dimensionality, through the linked property of anticommutativity, generates the entire system of discrete numbering on which mathematics is founded. At the same time, replicating biological systems seem to be generated in a way that conforms to similar mathematical processes, and to be constrained by the necessity of fitting into a context determined by the 3-dimensionality of space. It is, of course, one thing to hypothesize that this physics / biology link exists, another to prove it; but the link, if it exists, would be so significant that is worth examining a number of possible connections suggested by the underlying mathematics.

A significant question in the creation of all systems with large-scale order is how complexity can arise from simplicity. Related to this is the question of how asymmetry can arise from symmetry. The universal rewrite system suggests how this can occur in physics, and it will be interesting to see if the same applies in biology. In the rewrite system, anticommutativity is seen as the only true source of discreteness in nature. If **a** and **b** are anticommutative with each other, then neither can be anticommutative with anything else, except **ab**. So **a**, **b** and **ab**, or $\boldsymbol{i}, \boldsymbol{j}$ and $\boldsymbol{ij}$, form a closed or discrete set. The rewrite process generates an infinite succession of otherwise identical closed sets of this kind, which thereby generates a system of numbering. It also generates a series of Clifford-type algebras:

real scalar	units	± 1
complex (real + imaginary) scalar	units	$\pm 1, \pm i$
quaternion	units	$\pm \boldsymbol{i}, \pm \boldsymbol{j}, \pm \boldsymbol{k}$
multivariate vector (complex quaternion)	units	$\pm \mathbf{i}, \pm \mathbf{j}, \pm \mathbf{k}$, etc.

Remarkably, the algebras required for these four orders are respectively those required for the four fundamental physical parameters: mass (real scalar), time (pseudoscalar = imaginary scalar), charge (quaternion) and space (multivariate vector).

time	space	mass	charge
pseudoscalar	vector	scalar	quaternion
i	**i j k**	1	***i j k***

The combination of these requires an algebra of 64 units (including + and – signs). This is made up of

4 complex scalars	$(\pm 1, \pm i)$
12 complex vectors	$(\pm 1, \pm i) \times (\mathbf{i}, \mathbf{j}, \mathbf{k})$
12 complex quaternions	$(\pm 1, \pm i) \times (\boldsymbol{i}, \boldsymbol{j}, \boldsymbol{k})$
36 complex vector quaternions	$(\pm 1, \pm i) \times (\mathbf{i}, \mathbf{j}, \mathbf{k}) \times (\boldsymbol{i}, \boldsymbol{j}, \boldsymbol{k})$

This is, of course, also known as the Dirac algebra because it is the algebra which occurs the Dirac equation for the electron, although there it is often, more conventionally, expressed in terms of the five γ matrices. The significant thing here is that the algebra does not need 8 primitive units, and + and – signs, to generate these 64 parts. *It needs only 5 composite ones.* In effect, the most efficient structure is not the most primitive one, and is also not the most symmetrical. The 1 + 3 / 1 + 3 symmetry that can be observed in the 8 primitive units is completely broken, when we write down the 5 composite units that most efficiently produce the 64-component algebra, because this can only be done by taking one of the 3-dimensional quantities (space or charge) and superimposing its units on the others. If we take charge (the conserved quantity), we obtain:

$$i\boldsymbol{k} \qquad \mathbf{i}\boldsymbol{i} \quad \mathbf{j}\boldsymbol{i} \quad \mathbf{k}\boldsymbol{i} \qquad 1$$

and, as we have seen, these five combined or composite units also acquire a composite physical character, derived from their components, respectively, energy (E), three components of momentum ($\mathbf{p}$), and rest mass (m):

$$E \qquad \mathbf{i}p_x \, \mathbf{j}p_y \, \mathbf{k}p_z \qquad m$$

$$= \mathbf{p} \text{ (when components combined)}$$

That is, the charges produce new quantities that are quantized and conserved (like themselves) but also retain the respective pseudoscalar, vector and scalar aspects of time, space and mass. In complete form, we have:

$$i\boldsymbol{k}E \qquad \mathbf{i}\boldsymbol{i}p_x \, \mathbf{j}\boldsymbol{i}p_y \, \mathbf{k}\boldsymbol{i}p_z \qquad 1\boldsymbol{j}m$$

As we have already established in chapter 3, a system made of these components is the most efficient packaging of the information concerning the four fundamental parameters, space, time, mass and charge, that can exist in nature. In fact, no other type of physical system exists. We describe it as a fermion (or antifermion, depending on the sign of $\boldsymbol{ik}E$), and it turns out that the fermion / antifermion described by the combination $(\pm\, \boldsymbol{ik}E \pm \boldsymbol{i}\mathbf{p} + \boldsymbol{1j}m)$ is the fundamental unit of physics. The space, time, mass and charge components become compactified into a single unit. Very significantly, every packaging of this kind is a nilpotent or square root of zero. This means that it squares to zero with the 'vacuum state' $-(\pm\, \boldsymbol{ik}E \pm \boldsymbol{i}\mathbf{p} + \boldsymbol{1j}m)$ created simultaneously when we extract it from a zero totality superposition. Another way of expressing this (which is derivable from certain mathematical properties of the nilpotent) is to say that it has a 'spin' describable in half-integer values, which becomes 'single valued' (full integer) when combined with its vacuum state or equivalent. (Significantly, no new information would be gained by allowing a sign variation in $1m$ in a fermion state, as opposed to vacuum.)

While the symmetry-breaking creation of the concepts of energy (E), momentum ($\mathbf{p}$) and rest mass (m) is the immediate result of the fermionic packaging, the process, as we have seen, also works simultaneously in the opposite direction by breaking the symmetry between the three charges ($\boldsymbol{i}, \boldsymbol{j}, \boldsymbol{k}$) as they take on the respective pseudoscalar, vector and scalar aspects of time, space and mass. So we now have, for charge:

weak	strong	electric
w	$s_R\ s_G\ s_B$	e
	(3 colours)	
$\boldsymbol{ik}$	$\mathbf{i}\boldsymbol{i}\ \ \mathbf{j}\boldsymbol{i}\ \ \mathbf{k}\boldsymbol{i}$	$1\boldsymbol{j}$
pseudoscalar	vector	scalar

19.2 The Significance of the Pseudoscalar Term

Two very significant facts are associated with defining $(\pm\, \boldsymbol{ik}E \pm \boldsymbol{i}\mathbf{p} + \boldsymbol{1j}m)$ as a nilpotent. The first is that, to ensure nilpotency, one term at least must be pseudoscalar, that is, contain the factor i. Otherwise the squared terms will not add up to zero. The pseudoscalar is the term that becomes the 'book-keeper' (energy, weak charge), the creator and destroyer, the aggregator and disperser of matter in all its forms. In the universal rewrite system, this factor occurs as part of an incomplete quaternion set – there are no intrinsic pseudoscalars – and it is in removing this incompleteness that all physical interactions are ultimately situated. Here, the pseudoscalar i occurs in the position occupied by the parameters time, energy and weak charge, and the incompleteness manifests itself in the properties of all these quantities. The second significant fact is that a pseudoscalar quantity is, by definition, mathematically ambiguous. We cannot distinguish between + and – signs, *but both must be present*. This is what allows us to generate antifermions ($-E$) or ($-w$) and time reversal symmetry ($-t$). The conditions are, of course, connected: antifermions do not have negative energy in ordinary time; they only have negative energy in reversed time. The result of the sign ambiguity in the pseudoscalar term is that the fermion and vacuum are always a dual combination.

As previously stated, though the Dirac nilpotent is structured as ($\pm ikE \pm i\mathbf{p} + jm$), it is also often convenient to multiply throughout by i and write it as ($\pm kE \pm ii\mathbf{p} + ijm$). The bracketed expressions are really column (or row) vectors whose four components take up the four sign variations of $\pm ikE \pm i\mathbf{p}$. The four components represent the fermion state itself and three vacuum 'reflections' representing the effects of the three charges. In other words, the nilpotent, when structured in the column vector form ($\pm ikE \pm i\mathbf{p} + jm$), incorporates both energy and charge information. Information about charge 'occupancy' is determined by relative phases. The three vacuum 'reflections' are structured as $\boldsymbol{k}$ ($\pm ikE \pm i\mathbf{p} + jm$), $\boldsymbol{i}$ ($\pm ikE \pm i\mathbf{p} + jm$), $\boldsymbol{j}$ ($\pm ikE \pm i\mathbf{p} + jm$), which represent respective the weak, strong and electric vacua. If we multiply ($\pm ikE \pm i\mathbf{p} + jm$) by any of these any number of times it has no effect. We can consider the three things as 'partitions' of a total, or gravitational, vacuum, which would be 1 ($\pm ikE \pm i\mathbf{p} + jm$). Multiplying ($\pm ikE \pm i\mathbf{p} + jm$) by this would make it disappear (Pauli exclusion). (The reason why there is no independent m sign change, of course, is that there are only three charges. This has very important consequences in ensuring that there are only positive values for mass and ultimately energy, as also for proper time.)

The whole meaning of nilpotency is that the object squares itself to zero, and, from the rewrite mechanism which generates it, it would appear that the reason is that a fermion can have no existence outside of the total vacuum or 'rest of the universe'. In this sense, the total vacuum is a continuum (not partitioned), whereas the three 'partitions' are discrete. We create a fermion only with its other half. The total effect remains zero. A fermion always acts in such a way as it is trying to find this other half. However, the only perfect partner for a fermion is the completely delocalised vacuum – or rest of the universe. Fermions can be pictured as incomplete systems which are forever seeking the ideal partners needed to reduce them immediately to zero. This is the reason why we have aggregated matter. We can consider this by analogy with Newton's third law of motion, where body A does not act on body B but on the rest of the universe, and the same applies for body B supposedly acting on body A. However, body B can become, *in effect*, the rest of the universe if it is, say, very close to body A. (Thermodynamics, as we have previously implied, is relevant here, and the distinction between closed and open systems; the nilpotent fermionic state is unique in being an open system, but defined to conserve energy.)

In the same way, we can create local aggregates of fermionic matter by a fermion, say $(\pm\, \boldsymbol{i}kE_1 \pm\, \boldsymbol{i}\mathbf{p}_1 + \boldsymbol{j}m_1)$, effectively finding its other half in another fermion, say $(\pm\, \boldsymbol{i}kE_2 \pm \boldsymbol{i}\mathbf{p}_2 + \boldsymbol{j}m_2)$. (Of course, 'fermion' is a generic term here: either or both of these states could be antifermionic.) The equivalent of the squaring operation here is the 'vertex' $(\pm\, \boldsymbol{i}kE_1 \pm \boldsymbol{i}\mathbf{p}_1 + \boldsymbol{j}m_1)\ (\pm\, \boldsymbol{i}kE_2 \pm \boldsymbol{i}\mathbf{p}_2 + \boldsymbol{j}m_2)$, which eliminates all quaternionic components when all four product terms are combined; and, if we make the specification that all products of fermions are non-quaternionic, then we only need to specify the lead term in an expression such as $(\pm\, \boldsymbol{i}kE_1 \pm \boldsymbol{i}\mathbf{p}_1 + \boldsymbol{j}m_1)$, say $(\boldsymbol{i}kE_1 + \boldsymbol{i}\mathbf{p}_1 + \boldsymbol{j}m_1)$; the other three terms become automatic.

An interaction occurs where the creation of a vertex leads to changes in the E and / or $\mathbf{p}$ terms of the components; and, where the E and / or $\mathbf{p}$ become equalized, the vertex becomes a new combined, bosonic, state, with a purely scalar value. Such a state can only exist at all, however, if the two halves have different signs for either $\boldsymbol{i}kE$ or $\boldsymbol{i}\mathbf{p}$, or both, which the combining force (the weak interaction) will not recognize. The three possibilities create spin 1 bosons (reverse E in the second fermion), spin 0 bosons (reverse E and $\mathbf{p}$ in the second fermion), and 'Bose-Einstein condensates' (reverse $\mathbf{p}$ in the second fermion). Bose-Einstein condensation is, of course, a particular case of the very general physical concept called the Berry phase, with many distinct manifestations – other examples are the Aharonov-Bohm effect, the quantum Hall effect and the Jahn-Teller effect. Fermions, like electrons, self-aggregate, for example, with something like a nucleus or a magnetic field line to create a system with single-valued spin (not multiples of ½). It is probable that nonzero Berry phase is manifested in some way with all fermions – we just don't see it in most cases, because the system with which it is connected in this way is too dispersed. It is through this kind of phenomenon that matter self-aggregates over long distances.

Ultimately, quantum mechanics is about relating the packaged and conserved energy and momentum terms in the nilpotent expression $(\boldsymbol{i}kE + \boldsymbol{i}\mathbf{p} + \boldsymbol{j}m)$ to the nonconserved parameters time and space, from which they were originally derived; and to do this we make $(ikE + \mathbf{ip} + jm)$ into a differential operator in which E and $\mathbf{p}$ represent the quantum operators $\partial / \partial t$ and ∇, expressing the variation in time and space. In the case of the system not being free or isolated (which is, of course, always true in reality), we also add on respective scalar and vector potential energy terms to these operators, which reflect the particular fields (i.e. other fermionic states) to which the fermion is subject. (Again, in reality, this will be an infinite number, though in practice one or a few may be dominant.) What happens now is that the operator requires a *unique* phase term on which to operate – or unique within the symmetry constraints of the fields involved. The result of the operation will then be an amplitude, which will always square to zero. So, in the case of a free particle, the operator $(-ik\partial / \partial t + \mathbf{i}\nabla + jm)$ generates the uniquely defined phase term $\exp -i(Et - \mathbf{p}.\mathbf{r})$, which, when differentiated by the operator, produces $(ikE + \mathbf{ip} + jm)$, which squares to zero.

In more complicated cases, where the operator iE is not simply $\partial / \partial t$, we get a completely different phase term. A standard example is the hydrogen atom, where the inverse quadratic force (Coulomb, $\propto 1/ r^2$) requires $iE \rightarrow \partial / \partial t + A / r$, and the phase term has a real, not imaginary, exponential. For a discrete point particle, with spherical symmetry in space, it can be shown that there will be no solution unless there is a Coulomb term in iE. We can also show that there are only three solutions with spherical symmetry: inverse quadratic force (Coulomb); inverse quadratic plus constant force; and inverse quadratic plus anything else. The first gives the characteristic electric force solution; the second produces confinement, with the characteristics of the strong interaction; the third is a harmonic oscillator, irrespective of the actual nature of the force law. This is a creator and destroyer, with the characteristics of the weak interaction and the van der Waals forces responsible for aggregated matter. It is exactly what we would expect for dipolar and multipolar forces of any kind.

Ultimately, $(ikE + \mathbf{ip} + jm)$, in whatever form it is, uniquely determines the phase to be associated with it, and also the amplitude. The phase can be seen as corresponding in many ways to what we call 'vacuum', while the amplitude corresponds to the particle. In a sense, the phase is the connection of the particle to the rest of the universe, the carrier of nonlocality, the origin of the idea of phase conjugation (which determines locality) and the holographic principle (see chapter 20). It is like a signal sent out by a fermion asking other fermions, etc., to organize themselves with respect to it, and to aggregate. It is also a fundamental part of the intrinsic duality of the nilpotent fermion. We have amplitude on the one hand, and operator plus phase on the other, and they must be identical.

19.3 Spin and Aggregation of Matter

The mechanism which results in the creation or destruction of (combined) bosonic from or into (uncombined) fermionic states is described as the harmonic oscillator. The harmonic oscillator is a classic indication of aggregation or complexity in a system. It is a statement that no system is ever 'closed'. All fermions interact with each other via discrete quantum transitions; E and $\mathbf{p}$ are never fixed. This lack of closure is an expression of the second law of thermodynamics, and is the driver for all processes. No process is reversible. Time is unidirectional.

To arrive at the standard quantum mechanical harmonic oscillator, means realising that spin ½ is intimately connected with the weak interaction, and results from a particular aspect of it (its incompleteness). Spin ½ in physics comes from a fermion only being created simultaneously with its vacuum, and the spin is defined with respect to vacuum. A fermion cannot be defined otherwise. As we have already stated, it is like an interaction in the case of Newton's third law – not body A on body B, but body A on the rest of the universe (which is mostly B), and body B on the rest of the universe (which is mostly body A). Vacuum is the rest of the universe. It is a manifestation of the nonlocal aspect of quantum mechanics (and is probably the same thing as 'gravity' – again relating to Newton's third law). Now we can imagine the spin of the whole system as a kind of 'helical' motion. Interestingly, it is *double helical*, because the vacuum 'spins' ½ at the same time as the fermion. The total is spin 1 (or 0), so the combination produces an ordinary 2π rotation, though each half rotates through 4π. It may be significant, from the point of view of the universal applicability of the nilpotent rewrite system that a double helical 'DNA-type' nebula, about 80 light years long, and indicative of 'a high degree of order', has now been detected about 300 light-years from the centre of the Milky Way.[1] Here, the requirement of a strong magnetic field acting on the rotating body suggests that the best analogy is with a fermion acting in a boson-like manner in a magnetic field.

Another interesting aspect of spin is that it is generated from the vacuum via the weak interaction. The weak interaction is dipolar (a dual ± source), because of the sign ambiguity in the pseudoscalar parent quantity, and the basic force law for dipole-dipole is inverse quartic ($\propto r^{-4}$), just as the basic unipole-unipole law is Coulomb or inverse quadratic ($\propto r^{-2}$). In physics, *any* third body (or 'pole') introduced into a system where two bodies (or 'poles') have a $1 / r^2$ attraction /

repulsion, will produce an additional $1 / r^4$ dependence. In other words, the $1 / r^4$ dependence is a natural result / expression of aggregation.

The most famous example of inverse quadratic plus inverse quartic in physics is the analogous perihelion precession of the planets produced by gravity. That is, while the inverse quadratic term produces the orbit, the inverse quartic term makes the orbit spin. So, we can consider the weak term as generating the spin. In effect, the ½-integral 'spin' can be seen as the weak interaction operating a continual switching between the $+E$ and $-E$ or $+t$ and $-t$ states (*zitterbewegung*), or fermion and vacuum. In the planetary case, the main reason for the inverse quartic term is the disturbing effect of other planets. That is, the orbit of Mercury round the Sun is disturbed by all the other planets in the solar system. Another way of expressing this is to say that the tendency to aggregation within the Sun-Mercury combination is supplemented by the tendency to aggregation between this combination and other aggregated bodies of matter. Any tendency to such aggregation creates a dipolarity.

The weak term is indeed the crucial one for the collective behaviour of matter. Matter becomes collective only when it overcomes the weak Pauli exclusion (the uniqueness of each fermionic state) by creating *physical* dipoles, rather than matter-vacuum dipoles. This can be seen as a kind of localization. It has been suggested in chapter 10 that spin is a manifestation of a weak dipole moment. That is, it shows a bias to one sign over the other. This does not happen with strong and electric interactions. Particle physicists attribute the left-handedness of the weak interaction for fermions and right-handedness for antifermions to the idea that there is a filled weak vacuum. It is notable that the weak charge and energy are located in the same place in the Dirac nilpotent. We note here also that collective matter has a left-handed bias in the same way as the weak interaction: All proteins are composed of L-amino acids, all sugars are the D form and all nucleic acids in RNA and DNA are D form and it is tempting to conclude that this must have the same origin.

Continuous vacuum energy, such as we would expect from a 'filled' vacuum, is what we mean by nonlocality. It is the continuing connectedness, through the vacuum, of apparently discrete fermionic states, and it is required to maintain Pauli exclusion. Rest mass is a localization and therefore discretization of the continuous total vacuum energy. The continuity of vacuum energy is the reason for the left-handed bias of fermionic states, but, in discretizing it as rest mass, we also allow an element of right-handedness to emerge (as also E and $\mathbf{p}$).

19.4 Self-Organization of Matter

All processes in the entire universe can be considered as corresponding to the elimination of those aspects of the structure of the fermion which keep it separate from the rest of the universe: weak charge; ½-integral spin; a single energy state. In the first instance, fermions aggregate to become bosons or boson-like states, but this then proceeds to higher levels. In general, physical systems self-aggregate through the dipole-dipole van der Waals force, a classic expression of the action of vacuum, and, as we have seen, the concept of aggregation – in effect, complexity – originates, in physics, in the harmonic oscillator. Physically, the 'oscillator' aspect is seen in the behaviour of the molecules responsible for the various physical states of matter – gases, liquids and solids – which also represent different manifestations of the van der Waals force.

We think of this force as being electric or electromagnetic, because the components of matter are electrically charged, and so there are significant electrical forces involved; but the reason for the aggregation in the first place concerns the weak force. Any tendency for matter to aggregate is all about overcoming the weak force of Pauli exclusion or ½-valued spin. Any spin ½ object, or any object with spin ½ components, has a tendency to try to effect a physical realization of the 'rest of the universe'. In terms of weak charges, this is like trying to cancel them. So $+w$ cancels $-w$, or, since the sign of the weak charge is ambiguous, $+w$ cancels $+w$. But this is only a tendency – it can never be satisfied, because a 'real' partner can never cancel out a state completely.

Hydrogen bonding is one of the classic dipole-dipole forces (sometimes described as 'van der Waals' though some authors restrict this term to the intermolecular attraction), and, of course, it is precisely this which keeps the bases together in the two strands of DNA. The strands of DNA can be thought of as like the fermion and its partner (and they are equally subject to harmonic oscillation, as the links continually break and reform). All phase transitions necessarily involve a van der Waals-type force, because what is happening in a phase transition is that seemingly 'independent' systems are being more closely connected, apparently isolated systems are being realised as being connected with other apparently isolated systems. Anything which disturbs an apparently isolated (canonical, energy-conserving system) manifests itself as dipolarity, leading to a van der Waals-type force, and a physical manifestation of helicity.

The formation of a nucleus is a classic phase transition. Even though the nucleus is held together by a van der Waals-type version of the strong force (a multipolar remnant of the quark-quark forces within the individual protons and neutrons), the real thing which makes it possible is the cancelling out of the fermionic nature of its components, or their weak charges. A phase transition occurs when there is a significant change in the number of independent states of energy and momentum in a system. Ultimately, phase transitions which decrease the number of independent states of energy and momentum (e.g. the formation of real bosons by the equalization of energies of fermion and antifermion) increase the amount of order in the system and decrease the entropy (and complexity); those which increase the number of independent states of energy and momentum decrease the order and increase the entropy. Natural processes will always favour the latter because the tendency is to multiply connections. We might suppose also that the emergence of higher structures at this stage follows a fundamental mathematical pattern guaranteeing maximum efficiency. A Bose-Einstein condensation is another classic example of a phase transition. Here the number of independent states of energy and momentum is decreased by the fact that many 'bosonic' states can be aggregated into a single state of energy and momentum because bosonic wavefunctions are scalar and so are not subject to Pauli exclusion.

DNA and RNA appear to behave in a way pre-determined by the mathematical structure required for nilpotency, and for a totality which always remains zero. The nilpotent fermion replicates itself in the vacuum, effectively through its phase, and it may be that a similar thing is involved in DNA. There is no reason to believe that the chemistry is unique (and, of course, T in DNA is replaced by U in RNA), but it may well be the case that, as soon as nature hit on a mode of replication, it took off and replicated. So aggregations happened until a particular replicating one developed. A universal phase effect must certainly have been involved.

It is even possible that some biological form of the Pauli exclusion principle may operate in these circumstances. It is a widely known fact that the enzymes responsible for DNA replication and repair within bacteria (the DNA polymerases) all have an error rate generally between 1×10^3 and 1×10^7. However, these error rates are values given for *in vitro* situations and cannot be directly applied *in vivo*. Different bacterial species have differing numbers of these polymerases each responsible for slightly different processes and each has an error rate that will be cumulative as a whole. These error rates increase when the system is under stress or within suboptimal conditions, and it may be that this error rate is part of a system for adaptive evolution. When the actual error rates are considered as a whole it is unlikely that any one bacterial cell is in fact a true identical clone of another – a situation that may remind us of the Pauli exclusion principle.

19.5 The Filled Weak Vacuum and the One-Handed Bias in Nature

It is important to recognize that the force responsible for the emergence of higher-level structures in Nature is a strictly nonlocal or vacuum one. In fact, the organizing factor in all aggregation of matter is the weak force between fermion and vacuum. The fermion effectively has an unpaired weak charge which can always interact weakly with vacuum, because the weak vacuum is *filled.* In principle, this is precisely what weak filled vacuum means: there is always something with which the fermion interacts, and it is nonlocal. It is the filled vacuum that is responsible for spin ½, chirality, *zitterbewegung* (vacuum fluctuation), van der Waals dipolarity and rest mass. Its manifestations very likely include the prevalence of matter over antimatter, thermodynamics, and the arrow of time, and its origin comes from the fact that mass-energy and time are both continuous concepts, even in the discrete fermionic state.

As we have seen, the fermion's whole objective is to cease to exist, but this cannot happen because it would require the delocalised whole of the rest of the universe to become localised at the same point. However, the thing that makes it want to do this – the unpaired weak charge – can do this by combining with a fermionic / antifermionic partner in the bosonic state. However, many bosonic states are not a true union, and the structure remains open; for example, in the hydrogen atom – the parts are still distinct and try to combine with other fermions / bosons – e.g. the H atom becomes H_2 molecule, etc. The van der Waals force which drives the process originates in the fermion's weak interaction with vacuum. Though the ordinary weak interaction between two real particles or weak charges is very short range, the interaction with vacuum has unlimited range because the weak vacuum is continuous, and we normally describe the van der Waals force as arising out of the fluctuations in vacuum (which are a result of the weak charge / vacuum dipolarity), and calculate it via the Casimir effect, which is an expression of a continuous vacuum. So this is a real link between aggregated structures at all levels. In a sense, this is the 'physical' side of 'process', and there should be a set of fundamental symmetries associated with this which operate at all levels.

An example of this may be the chirality or 'one-handedness' in both physics and biology, already mentioned, where many people have long sought a link. Thus, the spin term which emerges from nilpotent fermionic structure has an inherent bias towards left-handedness in its weak interactions. This is an intrinsic property of the nilpotent fermion operator, and it automatically implies a filled vacuum (or bias towards $+E$ and $+t$). The computer version of the rewrite system suggests that it is an aspect of defining the concept of 'negative' (see 6.11 and Appendix B). Biology also has the same one-handed bias. Thus, amino acids can chemically exist as Dextro or Laevo (light) rotatory forms. However, biological systems only use L-forms, D-forms being toxic in many cases. (Glycine is an exception being a symmetrical molecule.) In addition, it would seem that D-forms of sugars take precedence within Nature, and D-form nucleic acids in RNA and DNA. Perhaps, this also has a mathematical origin.

19.6 The Idea of 3-Dimensionality

Studies of the universal rewrite system, etc., have shown that there are two ways of looking at 3-dimensionality: nonconserved and conserved. In the nonconserved version, as in ordinary space, the dimensions are indistinguishable. But the conserved case is a higher-order, more composite, concept, in which the 3 dimensions are different. This is what happens with the Dirac nilpotent. The $\boldsymbol{k}$, $\boldsymbol{i}$, $\boldsymbol{j}$ are different, and always in the same specific way. Always one dimension represents scale (say m), another dimensionality itself (say $\mathbf{p}$), and the remaining one the book-keeping (e.g. handedness) (say E). The weak interaction is the classic book-keeper. This is its only function. Biology may be using conserved 3-dimensionality and mapping it onto the nonconserved 3-dimensionality of space; and this is why, we will suggest, the *tetrahedron* can reproduce not only the other Platonic solids but also biological structures, at least approximately.

Biology, in fact, seems to require a double 3-dimensionality in the same way as physics does for quantum mechanics and the structure of particles. Illert, for example, has proved that the minimum correct representation of the growth of a sea-shell requires a doubling of conventional 3-D space, and Santilli has applied his mathematical formalisms to accommodate this.[2] It seems to be required to describe development *in time*, which, in biology, is also connected with spirals and handedness. This would make sense if we need the 3-D of $\boldsymbol{k}$, $\boldsymbol{i}$ and $\boldsymbol{j}$ to relate space and time in ($\pm\, \boldsymbol{i}\boldsymbol{k}t \pm \boldsymbol{i}\mathbf{r} + 1\boldsymbol{j}\tau$), or momentum and energy in its Fourier transform, ($\pm\, \boldsymbol{i}\boldsymbol{k}E \pm \boldsymbol{i}\mathbf{p} + 1\boldsymbol{j}m$).

However, the structure of this object suggests that we need an additional pseudoscalar term, which according to the rewrite system is really the manifestation of an incomplete additional 3-D system. This provides the classic pattern of the 'conserved 3-dimensionality' associated with the creation of ($\pm ikE \pm i\mathbf{p} + 1jm$) and ($\pm ikt \pm i\mathbf{r} + 1j\tau$). In addition to the ordinary space-like 3-D of **p** or **r**, there is also the 3-D 'charge space' or 'angular momentum space' of k, i and j, and the incomplete 3-D of iE and it.

From a fundamental point of view, we see that the development of increasing complexity is related to the existence of the fifth, symmetry-breaking, term in the fermionic state, or, more fundamentally, in the rewrite process as applied to natural phenomena.[3] Though the mathematical structures go on to infinity, only 2.5 quaternion series appear to be needed for infinite replication in a 'physical' universe.

At the same time, replicating biological systems seem to be generated in a way that conforms to similar mathematical processes, and to be constrained by the necessity of fitting into a context determined by the 3-dimensionality of space. There appear to be similarities in that the 3-dimensionality has to be applied, strictly, 2.5 times.

19.7 Application to Biology: DNA and RNA Structure

Several important aspects seem to relate fundamental physical structures to those of biology, as though the process was somehow universal and independent of level:

The concept of double helix
The significance of 64 units
The significance of 20 units
Chirality
The relevance of Platonic solids (tetrahedra, cubes, etc.)
The 5-fold broken symmetry (Fibonacci numbers)

We have already made some mention of the double helix and chirality, but all of these relationships are particularly relevant to DNA, RNA, and their action in coding to produce amino acids for proteins. Both DNA and RNA are macromolecules and they are polymers, the monomers of which are called nucleotides. A single nucleotide consists of a 5-carbon sugar, either deoxyribose (DNA) or ribose (RNA), one or more phosphate groups, and a nitrogen containing base. Four different nucleotides are found in DNA differing only in the nitrogenous base. The nucleotides are given one letter abbreviations as shorthand for the four bases; adenine (A) guanine (G) cytosine (C) and thymine (T). RNA is also comprised of 4 different bases and the same convention is followed for nomenclature, however in RNA the base thymine (T) is replaced by the base uracil and a single letter U is employed as shorthand.

Within a cell DNA is present as a supercoiled double-stranded macromolecule; two polynucleotide chains held together by weak thermodynamic forces (hydrogen bonds) to form a double stranded right-handed helical spiral. The two polynucleotide chains 'run' in opposite directions with A always pairing with T and G with C. For a given protein one of the two strands is referred to as the sense (coding) strand and the second the antisense (non-coding) strand. Some idea of the complexity of this system may be reached by reflecting upon the fact that the total amount of DNA in a single human cell is approximately 1.8 m long and 4 nm wide.

Proteins are made up of amino acids. The precise number, and sequence, of amino acids makes up the primary structure of a polypeptide chain. A functional protein may consist of a single, or several polypeptide chains. DNA must therefore carry genetic information that determines not only the number and types of amino acids that appear in a polypeptide, but also their exact sequence in the chain. The code for this primary structure cannot be carried within the sugar-phosphate 'backbone' of DNA since this part of the structure is identical in all DNA, variation only occurs in the base sequence. Therefore the sequence of bases in DNA determines the sequence of amino acids in a polypeptide chain. However it is apparent that, if DNA is to use the sequence of 4 different bases to code for 20 different amino acids, the code cannot be as simple as 1 base coding for 1 amino acid. In fact each amino acid is coded for by a sequence of 3 nucleotide bases, this allows 64 variants. A triplet of bases is called a codon.

19.8 Transcription

To access the genetic information incorporated in DNA it must be converted into a usable molecular 'format'. This is the so-called messenger RNA (mRNA). The process by which this achieved is called *Transcription.* During transcription a single strand of mRNA is synthesised using a double stranded DNA molecule as a template. The two strands of the DNA molecule are separated from one another, exposing the nitrogenous bases. Only one strand is actively used as a template in the transcription process. The RNA sequence that is made is a direct copy of the nitrogenous bases in the DNA sense strand *but* it is the complementary sequence. If a guanine (G) base is part of the sequence on the sense DNA strand, then the RNA molecule has a cytosine (C) base added to its sequence at that point In the RNA molecule uracil (U) substitutes for thymine (T). In this way the process of transcription constructs a small (relatively), mobile mRNA molecule comprised of a nucleotide sequence which is complementary to a coding sequence in the DNA molecule.

19.9 Translation and Triplet Codons

Translation is the process by which the genetic information in mRNA directs the synthesis of a polypeptide by controlling the order of insertion of amino acids into the growing polypeptide. This is achieved by means of a second much smaller form of RNA namely transfer RNA (tRNA). A cell contains about 60 different types of tRNA. It is mainly single stranded, 70-90 nucleotides long, and some portions of the molecule are double stranded giving the whole molecule a cloverleaf shape. Each tRNA molecule possesses two important features:

1. An anti codon site, which consists of a triplet of unpaired bases. The sequence of bases in this site varies from molecule to molecule and there is an anti codon sequence that is complementary to each codon sequence found on mRNA.

2. An amino acid binding site at the free end of the molecule that can bind a specific amino acid.

The particular amino acid that binds to each tRNA molecule is determined by the anti codon sequence. The actual amino acid is that which would be specified by the nucleotide sequence complementary to the anti codon, i.e. the codon on mRNA. For example the mRNA codon UCU specifies the amino acid serine. Thus the tRNA molecule that could recognise and bind serine would carry the anticodon AGA. Consider the whole process:- The DNA triplet codon for the amino acid serine is AGA; this is transcribed into the complementary mRNA triplet codon UCU. This codon is recognised by a tRNA molecule carrying the codon AGA and this tRNA is attached to the amino acid serine.

There are 64 (4^3) different possible triplets that can be obtained from the four DNA bases A, T, G and C which theoretically could code for 64 different protein building blocks or amino acids, but Nature (generally) selects only 20 amino acids which can be coded for by 1-4 triplets as shown in Table 19.1.

Table 19.1 The 64 triplets, 20 amino acids, and stop / start* codons of the genetic code

2nd position

1st position 5' end

3rd position 3′ end

	U		C		A		G	
	UUU	Phe	UCU	Ser	UAU	Tyr	UGU	Cys
U	UUC	Phe	UCC	Ser	UAC	Tyr	UGC	Cys
	UUA	Leu	UCA	Ser	UAA	STOP	UGA	STOP
	UUG	Leu*	UCG	Ser	UAG	STOP	UGG	Trp
	CUU	Leu	CCU	Pro	CAU	His	CGU	Arg
C	CUC	Leu	CCC	Pro	CAC	His	CGC	Arg
	CUA	Leu	CCA	Pro	CAA	Gln	CGA	Arg
	CUG	Leu*	CCG	Pro	CAG	Gln	CGG	Arg
	AUU	Ile	ACU	Thr	AAU	Asn	AGU	Ser
A	AUC	Ile	ACC	Thr	AAC	Asn	AGC	Ser
	AUA	Ile*	ACA	Thr	AAA	Lys	AGA	Arg
	AUG	Met*	ACG	Thr	AAG	Lys	AGG	Arg
	GUU	Val	GCU	Ala	GAU	Asp	GGU	Gly
G	GUC	Val	GCC	Ala	GAC	Asp	GGC	Gly
	GUA	Val	GCA	Ala	GAA	Glu	GGA	Gly
	GUG	Val*	GCG	Ala	GAG	Glu	GGG	Gly

19.10 Triplet Codons and the Dirac Algebra

Is there a link between the 64 amino acid triplets and the 64 units of the Dirac algebra? Are the biological structures determined by the same kind of algebra as the physical ones? In principle, the physical structure is an example of 'conserved 3-dimensionality'. This means that it follows the pattern of a quantity such as angular momentum, with 3 non-symmetric 'dimensions' representing, say, magnitude, dimensionality itself, and handedness; or scale, dimensionality and 'book-keeping'. In the Dirac nilpotent, these concepts are represented by the respective terms jm, $\mathbf{ip}$ and ikE. The structure always requires the interlocking of two complete quaternionic sets and an incomplete one (which manifests itself as the pseudoscalar term); and it seems to occur in every case in which a 3-dimensional quantity becomes a conserved one.

Such an algebraic structure always requires 64 terms, and it is worth examining the structure of the units of the genetic code to see if they can be related to the known algebraic structure of the Dirac fermionic state. In the case of the amino acid triplets, the actual structure in physical space may be considered the conserved object, and the 3-dimensionality is that of space itself. Classifying the codons, we find that there are

$$4 = 4 \times 1 \text{ with 3 letters the same}$$
$$24 = 4 \times 3 \times 2 \text{ with 3 letters different}$$
$$36 = 4 \times 3 \times 3 \text{ with 2 letters the same}$$

The last is made up of 4 possible letters to be duplicated, 3 letters other than the first chosen, and 3 possible ways of arranging the 3 letters (that is 3 positions for the nonduplicated one). Parallel to this, the Dirac algebra has

(A) 4 complex numbers $(+ / -)\ (1, i)$

(B) $24 = 12 + 12$ complex 3-D objects $(+ / -)\ (1, i) \times (\boldsymbol{i}, \boldsymbol{j}, \boldsymbol{k})$ quaternions
$+ (+ / -)\ (1, i) \times (\mathbf{i}, \mathbf{j}, \mathbf{k})$ vector (single 3-D)

(C) 36 complex 3-D × 3-D numbers (complex vector quaternions)
$(+ / -)\ (1, i) \times (\boldsymbol{i}, \boldsymbol{j}, \boldsymbol{k}) \times (\mathbf{i}, \mathbf{j}, \mathbf{k})$ (double 3-D)

To apply this to the genetic structures we could take the $(+ / -)\ (1, i)$ to represent the 4 options for bases (or stop codons). A and U or T could be say +1 and −1; and G and C $+i$ and $-i$. In each case above, this is where the 4 comes from. The groups of 3, etc., would then be to do with the amount of variation allowed. They are certainly connected with dimensionality. It is interesting to see if the 20 that produce the amino acids can be written out to fit into this pattern. Any given representation of the Dirac nilpotent really only uses 20 units of the 64-part algebra, say:

$i\boldsymbol{k}$	$\boldsymbol{i}\mathbf{i}$	$\boldsymbol{i}\mathbf{j}$	$\boldsymbol{i}\mathbf{k}$	$\boldsymbol{j}$
$\boldsymbol{k}$	$i\boldsymbol{i}\mathbf{i}$	$i\boldsymbol{i}\mathbf{j}$	$i\boldsymbol{i}\mathbf{k}$	$i\boldsymbol{j}$
$-\boldsymbol{k}$	$-i\boldsymbol{i}\mathbf{i}$	$-i\boldsymbol{i}\mathbf{j}$	$-i\boldsymbol{i}\mathbf{k}$	$-i\boldsymbol{j}$
$-i\boldsymbol{k}$	$-\boldsymbol{i}\mathbf{i}$	$-\boldsymbol{i}\mathbf{j}$	$-\boldsymbol{i}\mathbf{k}$	$-\boldsymbol{j}$

That is, it uses: none of (A)
8 of (B) $(+ / -)\ (1, i) \times (\boldsymbol{j}, \boldsymbol{k})$
12 of (C) $(+ / -)\ (1, i) \times (\boldsymbol{i}) \times (\mathbf{i}, \mathbf{j}, \mathbf{k})$

So, in a nilpotent fermionic structure such as $(\pm\, i\boldsymbol{k}E \pm \boldsymbol{i}\mathbf{i}p_x \pm \boldsymbol{i}\mathbf{j}p_y \pm \boldsymbol{i}\mathbf{k}p_z + \boldsymbol{j}m)$, there will be just 20 algebraic terms from the 64 that are important, if we want to represent particle and antiparticle, and also vacuum. (Here, we need to take into account multiplying by –1 and or ± i.) We could guess that these match up to the amino acid-producing codons using:

8 with 3 letters different	$\pm \boldsymbol{k}$	$\pm \boldsymbol{ij}$
	$\pm \boldsymbol{ik}$	$\pm \boldsymbol{j}$
12 with 2 letters the same	$\pm i\boldsymbol{i}\mathbf{i}$	$\pm \boldsymbol{i}\mathbf{i}$
	$\pm i\boldsymbol{i}\mathbf{j}$	$\pm \boldsymbol{i}\mathbf{j}$
	$\pm i\boldsymbol{i}\mathbf{k}$	$\pm \boldsymbol{i}\mathbf{k}$

(We note here that, in the biological case, there is a preference in each species for specific codons for each amino acid, i.e. there is a greater percentage of one type of tRNA for each set.)

We require a third of all possible codons with 3 letters different (8 out of 24), e.g. those in which a codon has its own + / – partner only on the outside.

A G U	A C U
U G A	U C A
G A C	G U C
C A G	C U G

Again, we only want one third of the 36 codons with two letters the same (12 out of 36), e.g. those again with their own + / – partners. Effectively, after we have excluded those with three identical letters (equivalent to (+ / –) (1, i)), we want a symmetrical third of the remaining codons. Of course, biological molecules will not be as precise as single physical particles, so we will have variation (and perhaps 'mimicry'). But we might expect an outline pattern of some kind to emerge which reflects this algebraic patterning, and preliminary analysis of the codons producing the amino acids in various species suggests that the 12 / 8 split has at least approximate validity.

For a codon with three identical bases, we should imagine a version of the 20 significant Dirac units as multiplied throughout by quaternion $\boldsymbol{j}$. This still produces the same quaternionic units for the top four pairs in each of the two columns set out above, but the last two pairs would become + / – i and + / 1, which are the four terms that correspond with the four codons with three bases the same. The strong showing of these codons is particularly interesting because it goes against absolute randomness over the whole 64 being the reason for the 12 / 8 split. The randomness, such as it is, is only between the quaternion elements. We could imagine that, at the most primitive level, the variation is not random, but has a highly structured pattern, though higher species might diverge increasingly from the original blueprint, and such an analysis might prove to be a way of measuring evolution. What seems most likely to be the case is that the initial life-forming process follows the mathematical rule because of the demands of conserved 3-dimensionality, and those of self-assembly related to fermionic structures, but, of course, in very complicated life forms the pattern provides only a very general constraint.

Even the breakdown of the two sets of 64 show very similar mathematical patterns, along with the 20 units in each case, which are required:

AUG $1\boldsymbol{k}i$; AUA $1\mathbf{i}\boldsymbol{i}$; ACA $1\mathbf{j}\boldsymbol{i}$; AAA $1\mathbf{k}\boldsymbol{i}$; AAC $1\boldsymbol{j}$;
UGG $-1\boldsymbol{k}i$; UGC $-1\mathbf{i}\boldsymbol{i}$; UAC $-1\mathbf{j}\boldsymbol{i}$; UCC $-1\mathbf{k}\boldsymbol{i}$; UUU $-1\boldsymbol{j}$;
GAC $i\boldsymbol{k}1$; GAA $i\mathbf{i}\boldsymbol{i}$; GUA $i\mathbf{j}\boldsymbol{i}$; GCA $i\mathbf{k}\boldsymbol{i}$; GGG $i\boldsymbol{j}$;
CAC $-\ i\boldsymbol{k}1$; CAA $-\ i\mathbf{i}\boldsymbol{i}$; CGA $-\ i\mathbf{j}\boldsymbol{i}$; CUA $-\ i\mathbf{k}\boldsymbol{i}$; CCG $-\ i\boldsymbol{j}$.

Here, the codons are grouped into four pentads, with the first base determining whether the first coefficient is 1, as –1, *i* or –*i*. The second base in the three central codons of each pentad is represented by a vector term, corresponding to a different base in each; while the quaternion labels correspond to the final bases, which are different for the pseudoscalar, vector and scalar terms. In this representation, we might imagine the stop codons taking algebraic forms such as –1, *i* and –*i*, though a more systematic representation of the 20 units might privilege vectors rather than complexified quaternions in the third and fourth pentads (as in the table in 15.4):

AUG $1\boldsymbol{k}i$; AUA $1\mathbf{i}\boldsymbol{i}$; ACA $1\mathbf{j}\boldsymbol{i}$; AAA $1\mathbf{k}\boldsymbol{i}$; AAC $1\boldsymbol{j}$;
UGG $-1\boldsymbol{k}i$; UGC $-1\mathbf{i}\boldsymbol{i}$; UAC $-1\mathbf{j}\boldsymbol{i}$; UCC $-1\mathbf{k}\boldsymbol{i}$; UUU $-1\boldsymbol{j}$;
GAC $\mathbf{k}1$; GAA $i\boldsymbol{i}\mathbf{i}$; GUA $i\boldsymbol{j}\mathbf{i}$; GCA $i\boldsymbol{k}\mathbf{i}$; GGG $\mathbf{j}$;
CAC $-\ \mathbf{k}1$; CAA $-\ i\boldsymbol{i}\mathbf{i}$; CGA $-\ i\boldsymbol{j}\mathbf{i}$; CUA $-\ i\boldsymbol{k}\mathbf{i}$; CCG $-\ \mathbf{j}$.

By comparison, the 4×3 pentads of particle physics are the fermionic states of the Standard Model, which can be imagined as being divided into fermions and antifermions (corresponding to, say, 1, –1), isospin up and isospin down (1, *i*, or quaternion, vector) and 3 generations (successively privileging $\boldsymbol{i}$, $\boldsymbol{j}$ and $\boldsymbol{k}$ or $\mathbf{i}$, $\mathbf{j}$ and $\mathbf{k}$), each of which repeats the characteristics of the others.

For many amino acids, the third bale in the codon is partially redundant. Nearly all amino acids are predominantly coded by the first two bases, which remain the same as well as unique to that acid – only serine, leucine and arginine, with 6 triplet codons each, are exceptions. In all these cases, one alternative has the complete range of options for the third base, while the other has a choice of two. In serine, the alternatives are UC and AG; in leucine they are UU and CU; and, in arginine, CG and AG. In all cases, where the first two bases make six bonds (in the conventional arrangement), the third base is entirely redundant, with all four options for the third base (A, U, G, C) being available. This is true also in three of the cases where they make five bonds; in nine other cases, there is a choice between two options for the third base, and in one case, there is just a single option. Where the first two bases make only four bonds, there are three options for the third base in one case, two options in five cases, and a single option in one case. There is a general tendency, therefore, for a decreasing number of options for the third base, where there are more bonds made by the first two bases, as we might expect. In those (seven) cases where two different amino acids share the same two first bases, the third base divides into U / C or A / G in nearly every case; only isoleucine (U / C / A) and methionine (G) provide a slight exception to the pattern. In the case of bacterial start codons, it is not at all unreasonable that the most significant bases are the *last* two, which are invariably UG, with the first base entirely redundant (A / U / G / C); whilc stop codons correspondingly *begin* with the same base (U) with limited options for the final two (AA, GA and AG).

An analysis of grouping of amino acids according to their specific triplet codons also yields interesting relationships. A standard method of grouping is shown in Fig. 19.1 and is dependent upon such factors / properties as size, polarity, charge, hydrophobicity, etc. However, if we attempt to group the amino acids according to the positioning of the nucleotide type within the triplet codon a different picture arises. When we attempt amino acid grouping using the A, T / U, G, C in the first position of the triplet codon, a completely random result is obtained and a similar lack of grouping is given by the position of the base within the third position – usually termed as the 'redundant base'. This is not surprising, in that this third position placement allows for variation in coding for the same amino acid. However, when the group is defined by the middle base (i.e. as in the columns in Table 19.1) we find there is a definite pattern (Fig. 19.2) which gives a similar group profile to the standard system of grouping by chemical properties and yet because there are four bases a fourth group is defined. Table 19.2 lists the amino acids of each new group.

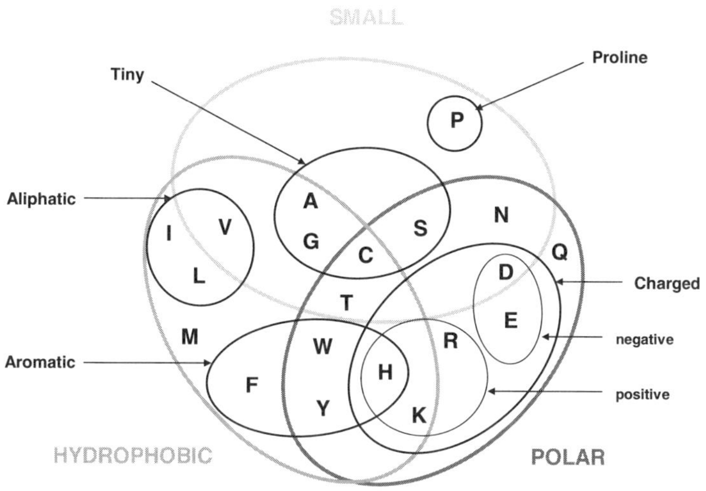

Figure 19.1 Standard 'properties' grouping of amino acids

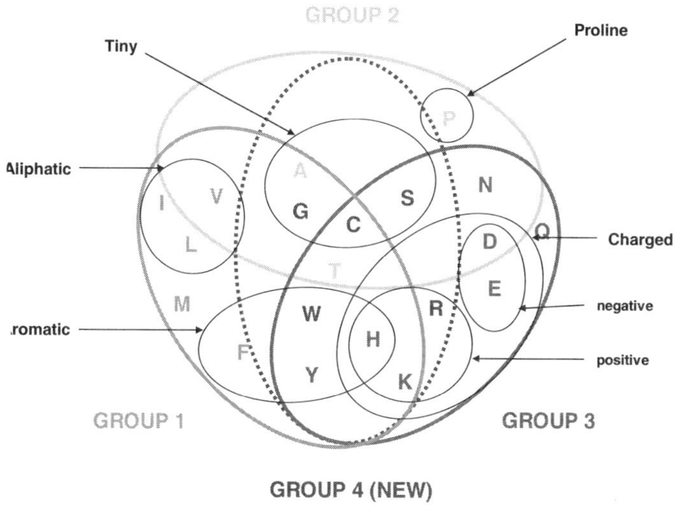

Figure 19.2 Gouping of amino acids using the second base of the triplet:
Group 1: T / U; Group 2: C; Group 3: A; Group 4: G.

When the properties of these four groups (Table 19.2) are looked at closer we see that group 1 now carries all the recognised bacterial start codons; ATG, TTG, CTG, GTG and group 4 contains the amino acids which display the 'extremes' of a certain property (Table 19.3); e.g. glycine is the smallest most flexible amino acid that acts as 'structure breaker' within proteins; tryptophan is the least used amino acid within proteins, it is the largest, is aromatic and absorbs UV light; arginine is the most basic with the most extensive delocalised charge, generally present in protein-nucleic acid interactions; cysteine is involved in disulphide bridge formation and is one of only two sulphur containing amino acids, the other being methionine with its triplet codon acting as the start codon.

Table 19.2 The new amino acid groups

Group 1	Group 2	Group 3	Group 4
Isoleucine: I	Proline: P	Glutamine: E	Glycine: G
Leucine: L	Alanine: A	Asparagine: A	Cysteine: C
Valine: V	Serine: S	Tyrosine: T	Tryptophan: W
Methionine: M	Threonine: T	Histidine: H	Argenine: R
Phenylalanine: F		Lysine: K	(Serine: S)
		Aspartate: D	
		Glutamate: Q	

Table 19.3 Properties of Group 4 amino acids.
Extremes: smallest, largest, most basic, S containing.

Glycine	Smallest, most flexible, structure breaker, achiral.
Tryptophan	Largest, aromatic, rarest, absorbs uv light.
Arginine	Most basic, extensive delocalised charge, present in protein-nucleic acid interactions.
Cysteine	Disulphide bridge formation, typically extracellular, 1 of 2 S containing amino acids (the other being methionine, the start codon).

The algebraic structure can be considered as composed of 32 + terms and 32 – terms. It is interesting to see if such a split can be seen within the triplet codon set, and also to see if the full set of algebraic terms can be allocated to the 64 codons, in addition to the allocation of the 20 algebraic units to the corresponding 20 amino acids.

The system that revealed the previously described group structure within the chemical properties of the amino acids was by division into groups depending upon the bases A, T, G and C as the middle base of each triplet codon. The codon table (Table 19.1) can also be split into two groups dependent upon the type of middle base (pyrimidine or purine) within the triplet. The purines and pyrimidines hydrogen bond to each other upon opposite strands of the DNA helix and can be considered as opposites upon the + and – ve sense DNA strands. Splitting the triplet codons into these two groups dependent upon the middle base of the triplets does indeed reveal another level of order. The group with a pyrimidine base (U / T, C) as the middle codon (Group A) reveals a trend for triplet codons that code for amino acids of predominantly nonpolar / hydrophobic nature and those with a purine (A, G) as the middle base (Group B) as those coding for amino acids of a polar / hydrophilic nature (Fig. 19 3).

If we divide each of these two groups of 32, further into 16 codons with their respective middle bases A, U / T, G and C (Fig. 19.4), we see that the Group 1, containing U / T as the middle base code for amino acids that are all distinctly hydrophobic, nonpolar and neutral; Group 2, containing C as the middle base, code for amino acids that are all neutral and have two amino acids that are hydrophilic and two that are hydrophobic. Group 3, containing A as the middle base, are all polar and 6 out of 7 are hydrophilic, 4 are charged (2 + and 2 –) and 2 neutral. Group 4, with G as the middle base of the triplet codons, gives the most mixed group of amino acids with 3 out of 5 being polar, 1 charged +ve and 4 neutral, 2 hydrophilic and 2 hydrophobic. It appears that Groups 1 and 3 have the most distinctive clustering of amino acid properties based upon nonpolarity / polarity, and Groups 2 and 4 those of less well defined grouped properties. It may be that, for the less well defined groups, there are other properties not yet considered that have greater applicability.

Pyramidine Middle Base | Purine Middle Base

U		C		A		G	
UUU	Phe F	UCU	Ser S	UAU	Tyr Y	UGU	Cys C
UUC	Phe F	UCC	Ser S	UAC	Tyr Y	UGC	Cys C
UUA	Leu L	UCA	Ser S	UAA	STOP	UGA	STOP
UUG	Leu*L	UCG	Ser S	UAG	STOP	UGG	Trp W
CUU	Leu L	CCU	Pro P	CAU	His H	CGU	Arg R
CUC	Leu L	CCC	Pro P	CAC	His H	CGC	Arg R
CUA	Leu L	CCA	Pro P	CAA	Gln Q	CGA	Arg R
CUG	Leu*L	CCG	Pro P	CAG	Gln Q	CGG	Arg R
AUU	Ile I	ACU	Thr T	AAU	Asn N	AGU	Ser S
AUC	Ile I	ACC	Thr T	AAC	Asn N	AGC	Ser S
AUA	Ile*I	ACA	Thr T	AAA	Lys K	AGA	Arg R
AUG	Met*M	ACG	Thr T	AAG	Lys K	AGG	Arg R
GUU	Val V	GCU	Ala A	GAU	Asp D	GGU	Gly G
GUC	Val V	GCC	Ala A	GAC	Asp D	GGC	Gly G
GUA	Val V	GCA	Ala A	GAA	Glu E	GGA	Gly G
GUG	Val* V	GCG	Ala A	GAG	Glu E	GGG	Gly G

T(U)=A
C≡G

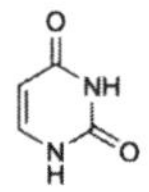
Uracil

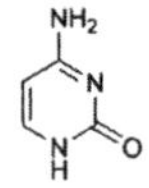
Cytosine

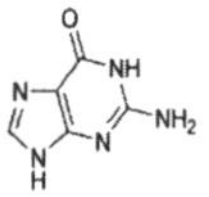
Guanine

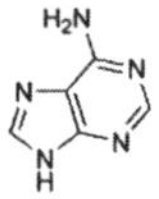
Adenine

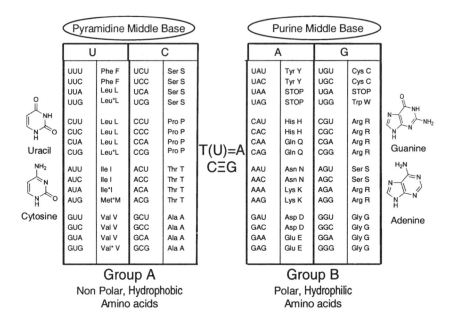

Group A
Non Polar, Hydrophobic
Amino acids

Group B
Polar, Hydrophilic
Amino acids

Figure 19.3 The 32+ and 32– split of the 64 triplet codons

Pyramidine Middle Base | Purine Middle Base

U (Group 1)		C (Group 2)		A (Group 3)		G (Group 4)	
UUU	Phe F: ar,hb,n	UCU	Ser S: p,hl,n	UAU	Tyr Y: p,ar,hb	UGU	Cys C: p,hb,n
UUC	Phe F: ar,hb,n	UCC	Ser S: p,hl,n	UAC	Tyr Y: p,ar,hb	UGC	Cys C: p,hb,n
UUA	Leu L: al,hb,n	UCA	Ser S: p,hl,n	UAA	STOP	UGA	STOP
UUG	Leu*L: al,hb,n	UCG	Ser S: p,hl,n	UAG	STOP	UGG	Trp W: ar,hb,n
CUU	Leu L: al,hb,n	CCU	Pro P: hb,n,*	CAU	His H: p,ar,hl,c+	CGU	Arg R: p,hl,c+
CUC	Leu L: al,hb,n	CCC	Pro P: hb,n,*	CAC	His H: p,ar,hl,c+	CGC	Arg R: p,hl,c+
CUA	Leu L: al,hb,n	CCA	Pro P: hb,n,*	CAA	Gln Q: p,hl,n	CGA	Arg R: p,hl,c+
CUG	Leu L: al,hb,n	CCG	Pro P: hb,n,*	CAG	Gln Q: p,hl,n	CGG	Arg R: p,hl,c+
AUU	Ile I : al,hb,n	ACU	Thr T: p,hl,n	AAU	Asn N: p,hl,n	AGU	Ser S: p,hl,n
AUC	Ile I : al,hb,n	ACC	Thr T: p,hl,n	AAC	Asn N: p,hl,n	AGC	Ser S: p,hl,n
AUA	Ile I : al,hb,n	ACA	Thr T: p,hl,n	AAA	Lys K: p,hl,c+	AGA	Arg R: p,hl,c+
AUG	Met M: al,hb,n	ACG	Thr T: p,hl,n	AAG	Lys K: p,hl,c+	AGG	Arg R: p,hl,c+
GUU	Val V: al,hb,n	GCU	Ala A: al,hb,n	GAU	Asp D: p,hl,c-	GGU	Gly G: al,n,*
GUC	Val V: al,hb,n	GCC	Ala A: al,hb,n	GAC	Asp D: p,hl,c-	GGC	Gly G: al,n,*
GUA	Val V: al,hb,n	GCA	Ala A: al,hb,n	GAA	Glu E: p,hl,c-	GGA	Gly G: al,n,*
GUG	Val V: al,hb,n	GCG	Ala A: al,hb,n	GAG	Glu E: p,hl,c-	GGG	Gly G: al,n,*

Group A | Group B

Key:-
ar=aromatic al=aliphatic p=polar hb=hydrophobic hl=hydrophilic n=neutral charge c+=positively charged c-=negatively charged

Figure 19.4 The chemical properties of the 20 amino acids

19.11 The Five Platonic Solids

At an even more fundamental level, the structure of DNA and the genetic code are related to the more basic ideas of Platonic solids and the Fibonacci sequence. A significant aspect of 3-dimensional space is that there are exactly five Platonic solids, or convex polyhedra with equivalent faces constructed of congruent convex regular polygons. These are the cube, dodecahedron, icosahedron, octahedron and tetrahedron (Fig. 19.5) which are five recognised resonance states of the sphere.

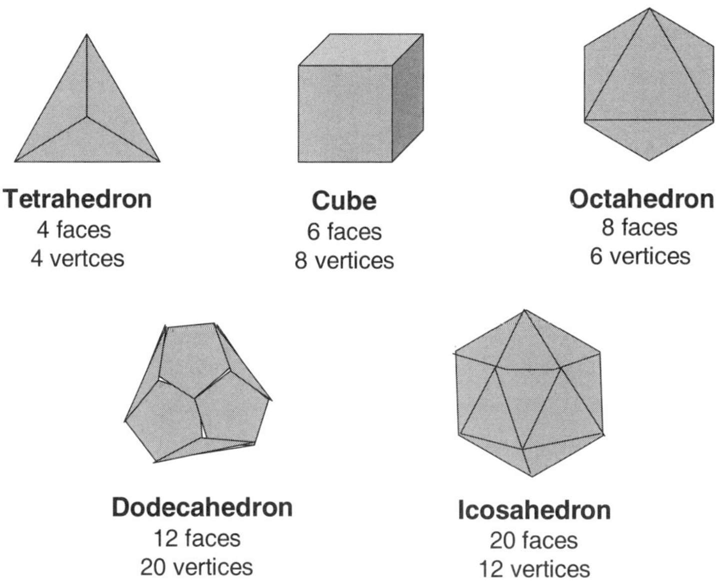

Figure 19.5 The 5 Platonic solids

Significantly, the tetrahedron is a mathematical reciprocal of itself, the octahedron is a reciprocal of the cube and the dodecahedron is a reciprocal of the icosahedron and vice versa. These solids nest one within the other as shown in Fig. 19.6.

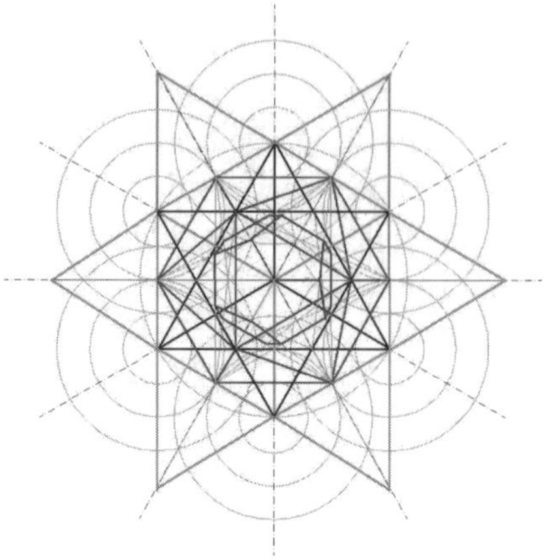

Figure 19.6 Nesting of the 5 Platonic solids

The 4 fundamental parameters involved in the fermionic state, which are fundamental components of the rewrite system, can be easily represented by either a cube or tetrahedron, or, if we include the vacuum (or dual) state as well, a star tetrahedron (see chapter 4). The rewrite system shows that 3-D forces the parameters into a 3-fold separation of dual properties / antiproperties: real / imaginary, conserved / nonconserved, discrete / continuous (= 3-dimensional / nondimensional). As the tetrahedron is the reciprocal of itself, we can also use a tetrahedral representation in which the faces / vertices represent fermion / vacuum states, and are totally reciprocal. We can also use a star tetrahedron to represent this, or relate it to the cubical structure. The cube or star tetrahedron can additionally be used to represent the 8 fundamental units of the algebra required by the fermionic state, as can the octahedron (reciprocal of the cube).

In a Platonic solid, if m polygons meet at a vertex, and each polygon has n vertices, then the number of faces $f = 2 + e - v$, where the number of edges $e = nf / 2$ and the number of vertices $v = nf / m$, each of these being determined uniquely. (Significantly m and n can only be 3, 4 or 5 in 3-D space. These are the obvious 'dimensionalities' to result from the group plus dual, and the 8 primitive algebraic units of space, time, mass, charge.) The tetrahedron shows the set and the dual set, using the six edges coloured to represent the properties and antiproperties; and the vertices or faces to show *either* the set or the dual set. The next tetrahedron filling space would be opposite, like fermion / vacuum (which is what the set / dual set is really all about). If we flatten the tetrahedral representation onto a plane we obtain the kites / darts needed for 5-fold Penrose tiling.

In general terms, the Platonic solids produce pentagonal symmetries which are intrinsic to them as a result of their structuring within a conserved 3-dimensionality (2.5 × 3-D). The fermionic symmetry-breaking, which does exactly this, is clearly related to that of the geometry. And Penrose tiling is not repeatable, exactly like fermionic nilpotents. Symmetry-breaking, in both cases, is related to the creation of a 5-fold structure.

As we have seen, a significant aspect of the Dirac algebra is that it does not need 8 primitive units, and + and – signs, to generate the 64 parts. *It needs only 5 composite ones.* In effect, the most efficient structure is not the most primitive one, and is also not the most symmetrical. The 1 + 3 / 1 + 3 symmetry that can be observed in the 8 primitive units is completely broken, when we write down the 5 composite units that most efficiently produce the 64-component algebra.

19.12 Fibonacci Numbers

5-fold structures, such as Penrose tiling, also introduce the Fibonacci series and the Golden Section, which is so important in biological growth, the sunflower being a characteristic expression of its operation. Fibonacci numbers are an integer sequence in which each new term is defined simply as the sum of the two previous integers: 0 1 1 2 3 5 8 13 21 34 …; the ratios of successive integers: 3/2, 5/3, 8/5, 13/8, 21/13, 34/21 … converge towards the Golden Section value, $\Phi = (1 + \sqrt{5}) / 2 = 1.618 \ldots$, while the inverse ratios tend towards $\Phi' = (1 - \sqrt{5}) / 2 = -0.618 \ldots$, so that $\Phi + \Phi' = 1$ and $\Phi\Phi' = -1$, where the two numbers are the roots of the equation $x^2 - x - 1 = 0$. The Golden Section is the ratio into which any line segment will be cut if the whole segment has the same ratio to the larger part as the larger part has to the smaller; it is also the one that the relative numbers of the 2 different tile types in Penrose tiling (kites / darts – flattened tetrahedra) tends towards. It occurs, in the Platonic solids, with the pentagonal symmetry of the icosahedron (3 golden rectangles transect) and its reciprocal, the dodecahedron (which has pentagonal faces). In addition, the 4-D image of a tetrahedron projected onto a 2-D plane gives a star pentagon which can be constructed within the dodecahedron.

The 20 of the reciprocal icosahedron is clearly visible in physics in the structure of the nilpotent operator which has 4 groups of 5, and contains the dualities of mass-energy / charge, fermion / vacuum, space / phase space, localised / nonlocalised, etc. The 4 × 3 pentads of 19.10 can also be represented by the 12 pentagonal faces of a dodecahedron which connect at vertices in groups of 3, or by the 20 triangular faces of an icosahedron, which connect at vertices in groups of 5. In each case, there are a total number of 4 groups of 5 × 3 or 3 × 5, that is 60, faces connecting at the vertices but the spherical 3-D symmetry (equivalent to privileging one of $\boldsymbol{i}, \boldsymbol{j}, \boldsymbol{k}$ or $\mathbf{i}$, $\mathbf{j}$, $\mathbf{k}$) reduces this to only 20 that are independent.

The icosahedral structure may also be applicable to the 20 amino acids in that the work described in 19.10 shows there is some indication of amino acid grouping into 4 groups, dependent upon the middle base of the associated triplet codon. If a tetrahedron is placed upon each face of the icosahedron to give a tessellated form, with a total of 60 triangular faces, we can then allocate triplet codons to each tetrahedral, triangular face, that relate (directionally) to the appropriate amino acid. It is interesting that here we have to lose 4 triplet codons of the 64 to give us the required 60 and we do have 3 known stop codons. Evolution may well have resulted in the loss of one stop codon and it is already known that there are variations of which codons code for specific amino acids in the process of 'codon capture'.[4] There are also known stop codon variations; for example, the codons that normally code for arginine, AGA and AGG, code for stops in vertebrate mitochondria,[5] while the stop codon UGA, has been replaced by tryptophan in Mycoplasma species[6] and UGA by selenocysteine and UAG by pyrrolysine in Archaea (recently found rare amino acids) in some archaebacter.[7] In physics, the 64 fundamental algebraic units are made up of 12 (= 3 × 4) sets of 5 generators for the entire group (each with an in-built 3-D property) and the 4 units of ordinary complex algebra ($\pm 1, \pm i$), with no dimensionality.

The use of the Fibonacci series as a means to explain all information processing in nature has been discussed by Stein Johansen, partly in connection with the universal rewrite system.[8] His description of the generating process as '1 step back and take it with you' relates to the dual conserve / create of universal rewrite. His algorithm for information processing also splits the numbers into 5s and 3s, with a bifurcation at 8 (8 + 3 = 11; 8 + 5 = 13). The 5-fold symmetry additionally seems to be responsible for introducing the fractal aspect in which the same patterns repeat themselves at different levels in Nature. Here, we observe that the rewrite system produces, in addition to the repetitive Clifford algebra sequence, in which the 2^n algebraic units are the basic ones, a new type of non-repeating 'unit of uniqueness' which, for the first time, combines the properties of recursive and iterative systems through its symmetry-breaking 5-fold symmetric structure.

A classic case of the Fibonacci sequence is the growth of a spiral shell ending in a point-singularity. D'Arcy Thompson observed that a shell grows in size without changing its shape, leading to growth in a logarithmic or equiangular spiral, with radius $r = a \exp(b\theta)$ from the point source.[9] This shape can be observed, typically, when the long side of a Golden Triangle, an isosceles triangle from within a star pentagon, with apex angle = $180^\circ / 5 = 36^\circ$ (the Golden Attractor), is continually used as a base for a new Golden Triangle. Here, the successive bases have the ratios 1Φ, $1\Phi + 1$, $2\Phi + 1$, $3\Phi + 2$, $5\Phi + 3$, $8\Phi + 5$ … . Illert, whose two 3-D systems are constituted from a fixed reference system (ordinary 3-D space) and a set of moving 3-D coordinates representing growth, has used the analogy of the spiral watchspring acting as a classical harmonic oscillator, obeying Hooke's Law ($F = kx$). Significantly, quasicrystals with icosahedral symmetry show no periodicity in ordinary 3-D space, but become periodic, and lose their Fibonacci character when structured within a 6-D cube, constructed from a parallel real 3-D space, and a perpendicular imaginary 3-D space. It may be that the significant distinction between organic and inorganic forms, in this context, is brought about by the extra half-3-dimensionality representing variation with time (and leading to helicity). In physics, this is the role of the weak interaction, which ordered crystalline structure is designed to suppress.[10]

In fact, in the fundamental rewrite system with its 5-component nilpotent operator, we may see, at once, all the fundamental units of *natural process* that apply to biology as well as physics. Defining the operator simultaneously leads to the creation of point singularity and discreteness; compactification (from 8 units to 5) and chirality (as a result of the loss of some sign degrees of freedom in the compactification); symmetry-breaking (between the 5 units) and spontaneous symmetry-breaking (because of the chirality); (double) helicity and angular momentum (with its double 3-D nature); irreversibility (because of the chirality of the time and energy operators); 5-fold symmetry (cubical $\rightarrow$ spiral) and the Fibonacci sequence; and a harmonic oscillator-based tendency to aggregation and complexity (because of the pseudoscalar nature of the time / energy term needed for nilpotency).

In physics, the act of 'creation' is that of the fermionic state (and requires the simultaneous creation of relativity and quantum mechanics); but the structure can repeat at higher levels because it is a recurring pattern. An example from chemistry is the phenomenon of spontaneous chiral symmetry breaking in chiral autocatalytic systems in which continuous stirring of a solution of a salt such as

$NaClO_3$ (a nonequilibirum system in thermodynamic terms) can cause it to go through a symmetry-breaking transition in which it crystallizes almost entirely into either laevo- or dextrorotatory forms.[11] The significance of the creation of a dominant angular momentum state (through stirring) is apparent.

If universal rewrite is valid in the domain represented by Illert's work on the sea-shell, then the 5-fold Fibonacci aspect will require an additional incomplete 3-D (for time). The doubling of space is the creation of a *conserved* 3-D (angular momentum), while time and energy become the sources of chirality. If a tetrahedral arrangement is, say, the most efficient packing of 3-dimensional space, then this structure retains its identity when it is aggregated within a larger one, itself requiring 3-D spiral packing, and double helicity for stability.

19.13 Application of Geometrical Structures to DNA and Genetic Coding

The four bases of DNA – A, T, G and C – can be placed upon the four vertices of a tetrahedron (Fig. 19.7) such that the tetrahedron can now be considered to contain, upon an information level, all the possible 64 (4^3) triplets defined by single stranded (sense) DNA or mRNA (U replacing T). Double stranded DNA can now be represented by interlocking a second tetrahedron to produce a star tetrahedron such that both the sense and antisense strands are combined with the correct base pairing of A to T and G to C that occur within the double helix (Fig. 19.8). The corners of a cube would also serve well here.

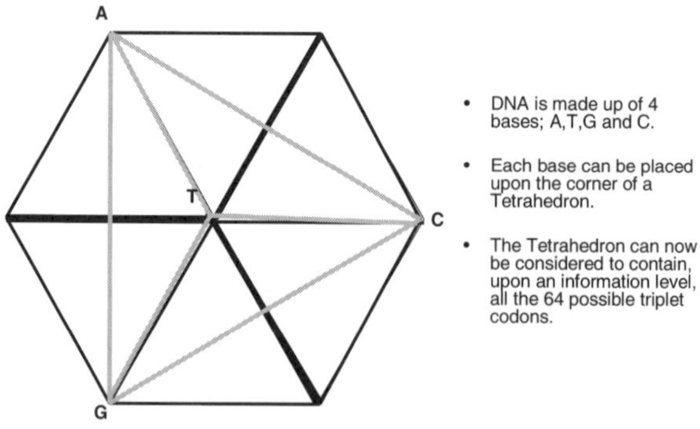

A
T
C
G
• DNA is made up of 4 bases; A,T,G and C.
• Each base can be placed upon the corner of a Tetrahedron.
• The Tetrahedron can now be considered to contain, upon an information level, all the 64 possible triplet codons.

Figure 19.7 Single stranded DNA: tetrahedron

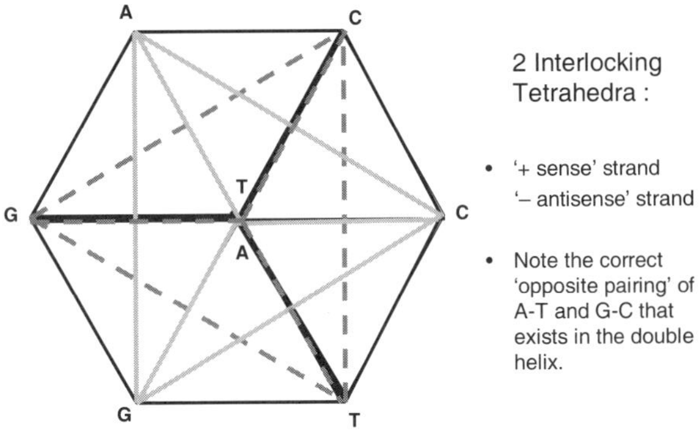

Figure 19.8 Double stranded DNA: the tetrahedron within the cube

As previously mentioned, there are 64 (4^3) different possible triplets that can be obtained from four bases. These theoretically, could code for 64 different protein building blocks (amino acids) but generally Nature selects only 20 amino acids which can be coded for by 1 to 6 different triplets as shown in Table 19.1. If we now look at different higher order levels of tetrahedra (Fig. 19.9) it can be seen that the second order is composed of one octahedron and four tetrahedra and the third order is composed of four octahedra and ten tetrahedra. If each triangular octahedral face is considered to represent a single triplet then each octahedron would have eight possible triplets and if each tetrahedron is considered to represent one amino acid we would have:-

for a second order level tetrahedron : 8 triplet codons and 4 amino acids
for a third order level tetrahedron : 32 triplet codons and 10 amino acids.

Introduction of a second interlocking tetrahedral form (Fig. 19.10 and 19.11) to produce a star tetrahedron would now double these values to:-

second order level star tetrahedron : 16 triplet codons and 8 amino acids
third order level star tetrahedron : 64 triplet codons and 20 amino acids.

The third order level star tetrahedron now meets the requisite numbers of triplets possible from our 4 bases and also the number of amino acid used by Nature to construct proteins.

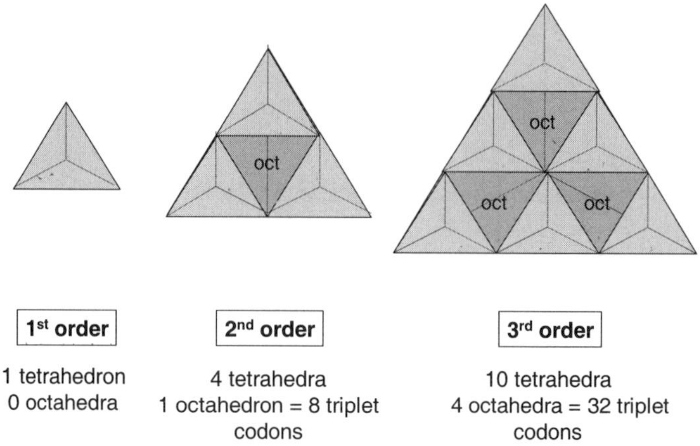

Figure 19.9 Higher order level tetrahedra

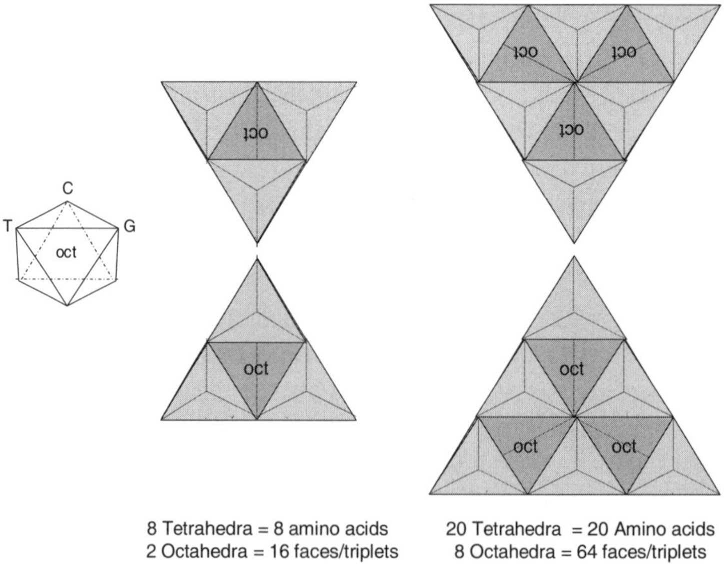

Figure 19.10 The star tetrahedron

Figures 19.11-14 give a deeper insight into how the tetrahedra and octahedra pack within this star. Interestingly, there is a fractal nature to these diagrams, highlighting the reiteration of the star / octahedron / cube. A fractal nature was also previously observed within the higher order tetrahedra where one is reminded of the Sierpinsky fractal triangle.

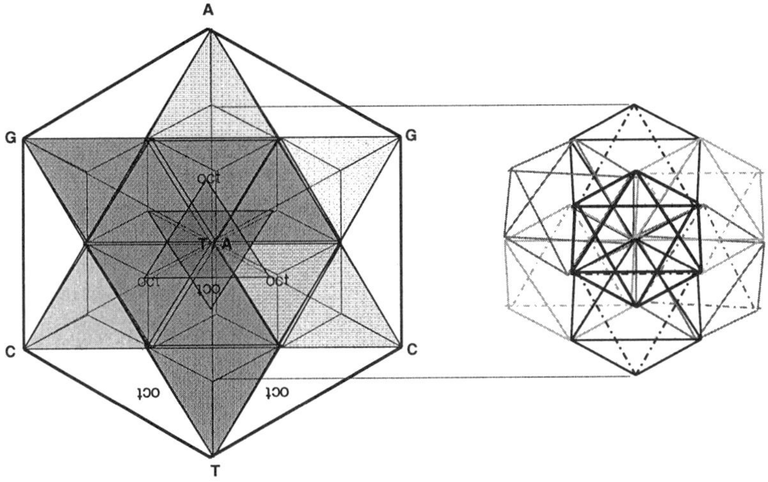

Figure 19.11 Interlocking the two third order level tetrahedra

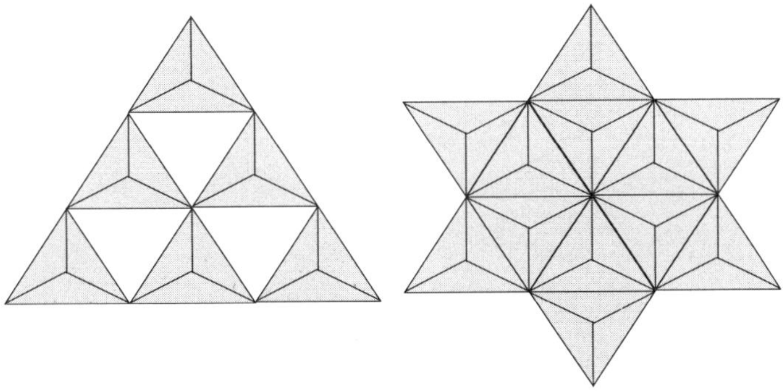

Figure 19.12 Packing of the 20 tetrahedra within the star tetrahedron

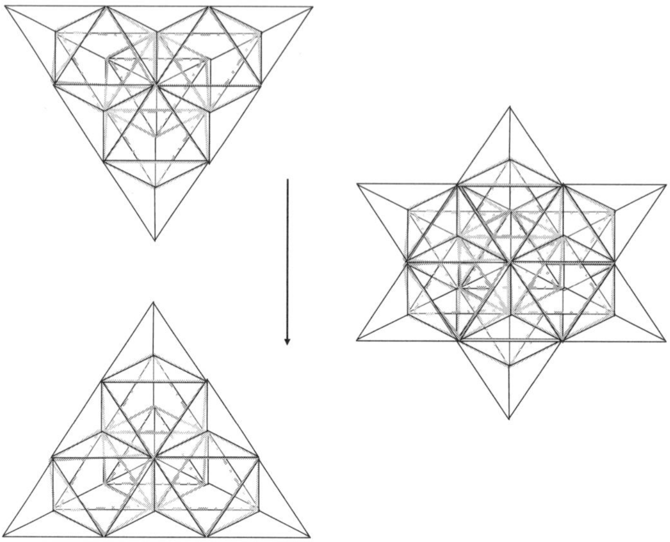

Figure 19.13 Completing the star in full 3D

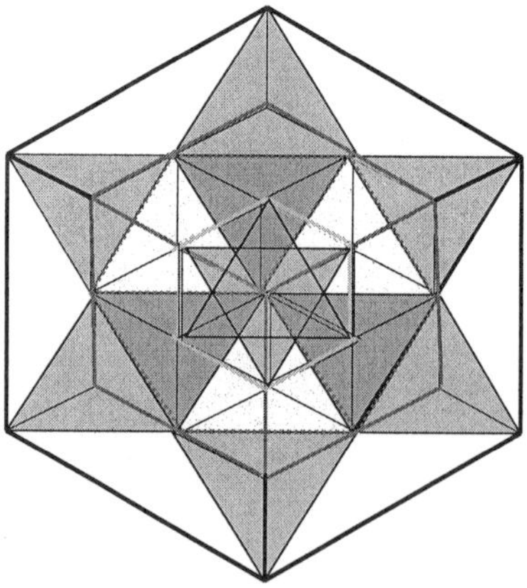

Figure 19.14 Cube reiteration

The direct connection with the rewrite structure is now apparent, with the key information being the number of nested 3-D systems, as we progress from, say, 2 and 4 bases (orders 2 and 4), to a base pairing (order 8) and a pairing bonding in strand formation (order 16), before ending at single-strand RNA / DNA (order 32) and double-strand DNA (order 64):

order 2	$(1, -1)$	$0 \times$ 3-D
order 4	$(1, -1) \times (1, \boldsymbol{i}_1)$	$0.5 \times$ 3-D
order 8	$(1, -1) \times (1, \boldsymbol{i}_1) \times (1, \boldsymbol{j}_1)$	3-D
order 16	$(1, -1) \times (1, \boldsymbol{i}_1) \times (1, \boldsymbol{j}_1) \times (1, \boldsymbol{i}_2)$	$1.5 \times$ 3-D
order 32	$(1, -1) \times (1, \boldsymbol{i}_1) \times (1, \boldsymbol{j}_1) \times (1, \boldsymbol{i}_2) \times (1, \boldsymbol{j}_2)$	$2 \times$ 3-D
order 64	$(1, -1) \times (1, \boldsymbol{i}_1) \times (1, \boldsymbol{j}_1) \times (1, \boldsymbol{i}_2) \times (1, \boldsymbol{j}_2) \times (1, \boldsymbol{i}_3)$	$2.5 \times$ 3-D

Order 8 here corresponds to the second order level tetrahedron with 8 triplet codons in an octahedron and 4 tetrahedral amino acids. Order 16 doubles this to a second order level star tetrahedron with 16 triplet codons in two octahedra and 8 tetrahedral amino acids. Order 32 produces a third order level tetrahedron with 32 triplet codons in 4 octahedra and 10 tetrahedral amino acids, while order 64 produces a third order level star tetrahedron with 64 triplet codons in 8 octahedra and 20 tetrahedral amino acids.

The stages at order 8 and order 32 are key 'phase transitions', producing, respectively the first octahedron and then the first 3-D (tetrahedral) arrangement of 4 octahedra. Orders 16 and 64 produce the direct doubling that is characteristic of timelike transition between spatial states characterized by the hidden time component in the multivariate vector system, due to the additional pseudoscalar $0.5 \times$ 3-D. The cube reiteration in Figure 19.14 shows the exact parallel between the $1.5 \times$ 3-D and $2.5 \times$ 3-D structures when one complete 3-D system is mapped exactly onto another. Pentad structures notably occur only at orders 32 and 64.

The rewrite system requires a double 3-D because an object is dual with the rest of the universe (or vacuum) in that the two combine to a zero totality. In particle physics, this means that a fermion has interactions with all other particles in the universe, and that these determine its final state. They also cause its changes. In a sense the vacuum is what the fermion will become, and we can picture it like two spaces interpenetrating each other in a way that cannot be visualised in 3-D, but can be in higher dimensions. So we have 'static' dimensions and changing ones, just as we have conserved (mass and charge) and nonconserved ones (space and time), or, alternatively, amplitude and phase, or fermion and vacuum. Because the 'phase' part includes the idea of change, it is like Illert's set of moving biological coordinates, but we also need a rest frame of fixed coordinates which express what remains fixed, and as a reference. Significantly, the interaction with the rest of the universe in particle physics is through charge, which provides the second 3-D. Biology, with its self-replicating mechanisms, is even more obviously organised in this holistic way.

In a sense, all our perceptions require the connection between the isolated part and the whole but, through the nilpotent algebra and the universal rewrite system, we know that this is a zeroing. We can picture this in any 3-D structure as the point at which the new shape is realised at any stage in the process – the fixed point – with the moving 3-D system giving us the new point of becoming. The point of preception is the point at which these coincide and at that point we observe only one 3-D. But the fact that we will then go onto a new perception is what justifies us using two lots of 3-D. Our perceptions, however, simply mirror the structure which is inherent in all natural systems. They operate, like everything else in the universe, according to Nature's code.

The task now is to assign the correct placement of the triplets to give the appropriate amino acid represented by each tetrahedron. Figure 19.15 shows the beginnings of this procedure.

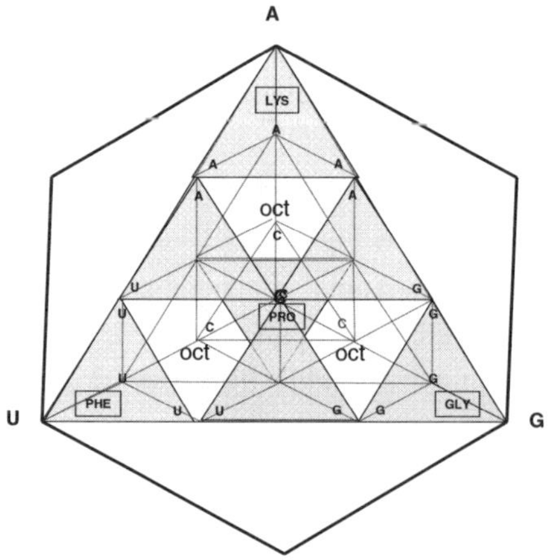

Figure 19.15 Beginnings of placement of the triplets and amino acids

Here, the corner tetrahedra can be nominated as the amino acid that relates to the triplet codon mirrored from the triangular face of an octahedron, e.g.

UUU = phenylalanine (PHE)	GGG = glycine (GLY)
AAA = lysine (LYS)	CCC = proline (PRO)

The predominant start codon AUG, that codes for the amino acid methionine and the three common stop codons (UAG, UGA and UAA) which do not translate into any amino acids, can now all be defined by one tetrahedron as shown in Fig. 19.16. If we consider that each face rests upon that of an octahedral 'codon' face then this tetrahedron is likely to be the one in the very centre of the third order level tetrahedron which is completely surrounded by octahedra.

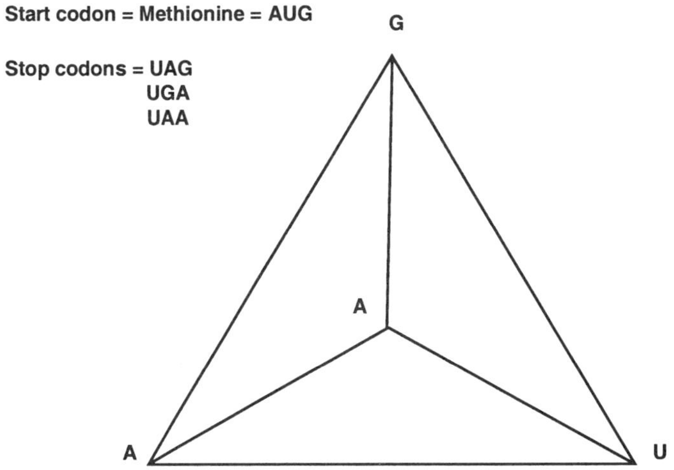

Figure 19.16 The start and stop codons

19.14 Pentagonal Symmetry Within DNA

In the previous section we have been mainly concerned with the 3 Platonic solids of 4-fold symmetry and all these can be defined by constructing a cube with a saltire cross (X) drawn across each cuboidal face as shown in Figure 19.17a. Five-fold symmetry is brought in by the icosahedron and its reciprocal the dodecahedron and hence the golden section proportion Phi ($\Phi = 1.618$) comes into play.

Nature expresses herself using the golden section: Phi and 5 now becomes important. Humans find beauty in both hearing and seeing this relationship and for this reason Phi was used in sacred wall paintings in ancient Egypt and vigorously applied in Renaissance art. The first 3 Platonic solids do not reveal this relationship but the icosahedron and dodecahedron do. Whereas the octahedron is transected by 3 squares the icosahedron has 3 golden rectangles in each of the 3 planes and the dodecahedron interestingly reveals another level of order in that in reality this shape is constructed of 5 interlocking cubes. The icosahedron is often used by viruses and bacteriophages as it gives the greatest volume with stability, e.g. the polio virus, hepatitis A virus and T phages. Phi is very prevalent within Nature and numerous examples of its use exist, but the plant that displays this proportion in every way possible is the sunflower.

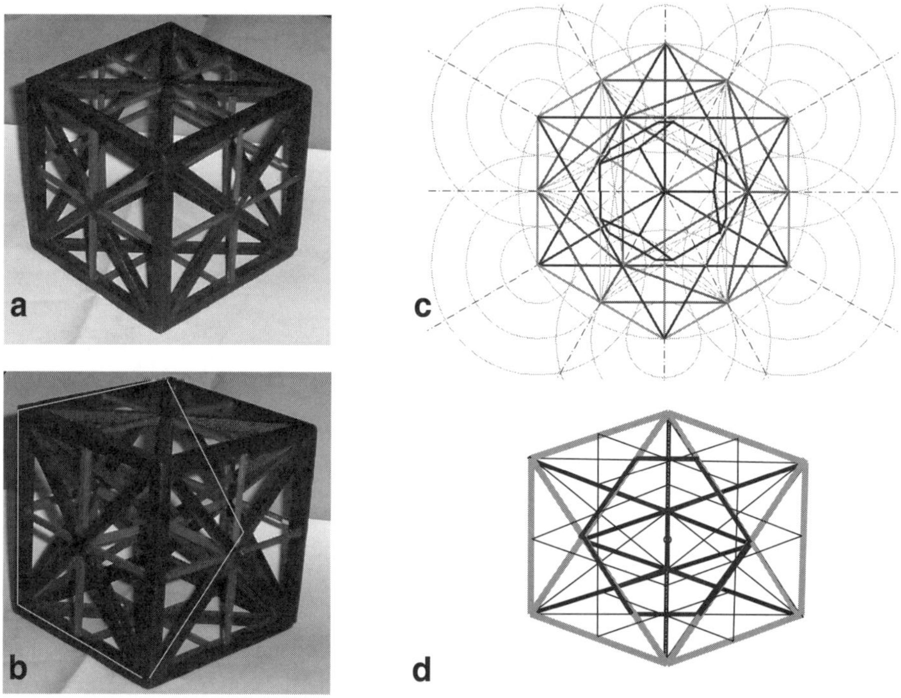

Figure 19.17 The cube: introducing Phi by perspective change

Phi can be seen within our cube by drawing the second set of cross (+) lines on each cuboidal face and allowing a change in the cube's orientation such that the front corner appears to sit over a newly constructed cross point of two lines as shown in Fig. 19.17b. This can be difficult to visualise but Fig. 19.17c is a 2-D representation of the cube in the first orientation and Fig. 19.17d and Fig. 19.18 shows the 2-D outline of the cube in the second perspective. The latter now also gives a 2-D outline of the icosahedron using exactly the same outline of this new perspective of the cube.

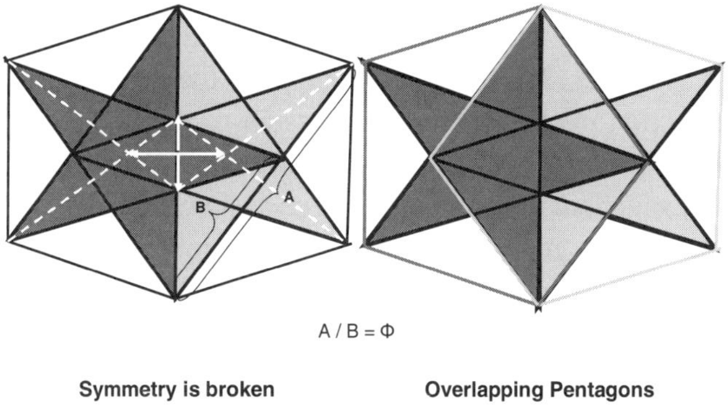

Figure 19.18 Perspective 2: Showing pentagons within the cube. Polarity is introduced and Φ.

This change in cuboidal perspective appears to introduce polarity where the central point of the first perspective now opens up from a single point and expands north-south and east-west into a plane. Two overlapping pentagons with inscribed 5 pointed stars can now be defined (Fig. 19.18). This whole figure is governed by Phi and this image is also a correct 2-D, 'see-through' image of an icosahedron with its reciprocal dodecahedron inside (Fig. 19.19).

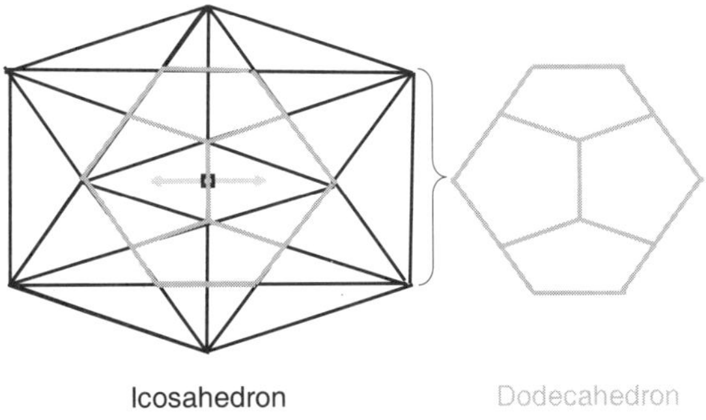

Figure 19.19 Icosahedron with internal reciprocal dodecahedron

Within the algebra of symmetry breaking there is the reduction from the 8 to the 5-fold symmetry which we can represent by packing 5 tetrahedra into a disk (Fig. 20). There is a discrepancy here of 7° 12′, but this has significance for the creation of a spiral structure, as will be described later on. The structure revealed in Figure 19.20 can also be generated by a projection onto 3-space of the self-dual pentatope, which is the 4-dimensional analogue of the tetrahedron. The pentatope, with 5 vertices, 10 edges and 10 faces, is the simplest regular figure in 4-D.[12] The connection may well be an expression of the fact that space, though fundamentally 3-dimensional, is part of a larger structure of nested 3-dimensionalities, which, in the universal rewrite system, is created after the scalar mass, pseudoscalar time, and 3-D conserved quaternion charge; and some of the profoundest insights into physical space's 3-D structure may come from the properties of the larger structure within which it is embedded. In this sense, physical space can only exist as the 3-D spatial projection of the higher dimensionality incorporated in the nilpotent structure, but, in view of the fact that the 'fifth' dimension of the nilpotent structure (the proper time) is an invariant, and therefore essentially redundant, it is significant that the regular Platonic-type figures reach their maximal extent (6) in 4-D; at higher dimensionalities they reduce to 3. The manifestation of the extra dimensionality in biology may then be related to a chaotic variation with respect to time.

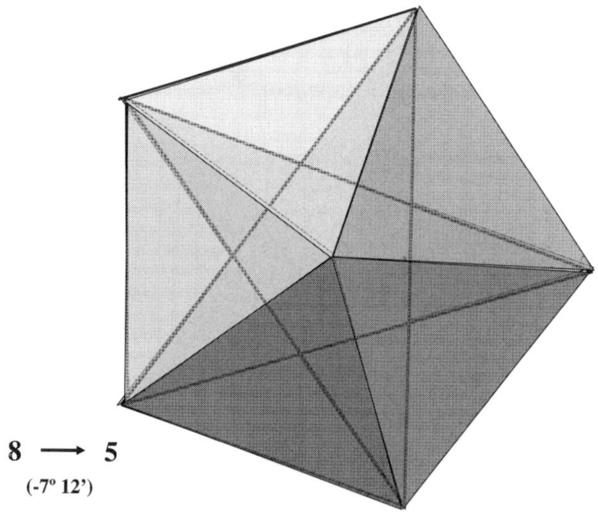

Figure 19.20 Breaking symmetry: 5 tetrahedra make one pentagonal disk

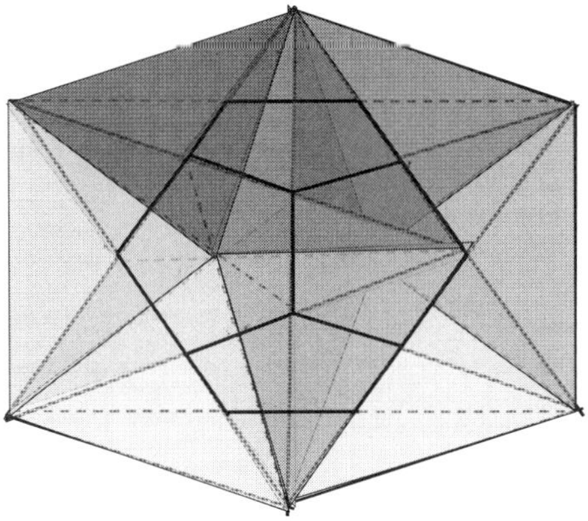

Figure 19.21 2 pentagonal disks overlaid to produce an icosahedron (2D) plus internal dodecahedron

If we now overlay two pentagonal disks a 3-D representation of the icosahedron can again be visualised in 2-D as shown in Fig. 19.21. Here the reciprocal internal dodecahedron has been highlighted which, as we shall see, has certain interesting consequences.

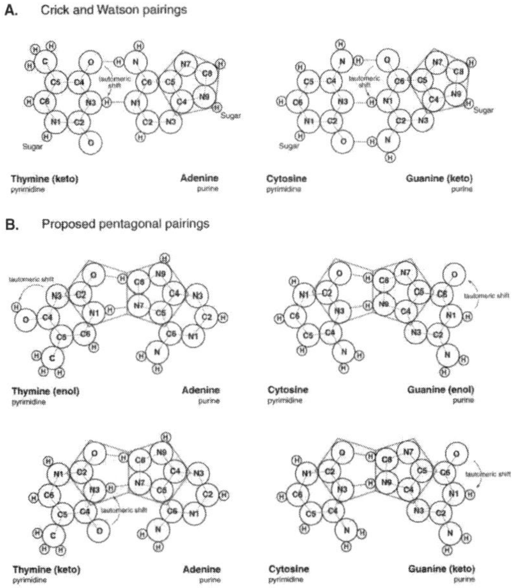

Figure 19.22 Nucleotide pairing within DNA
a. Watson and Crick pairings; b. Curtis pairings (without placement of ribose sugar groups).

The Watson and Crick model for DNA structure[13] is well known but in May

1998 a second possible model was proposed by the artist Mark Curtis who designed one of the stamps for the millennium.[14] Curtis wanted an in depth understanding of this structure and believed that something was amiss when he attempted to build models according to the Watson and Crick theory.

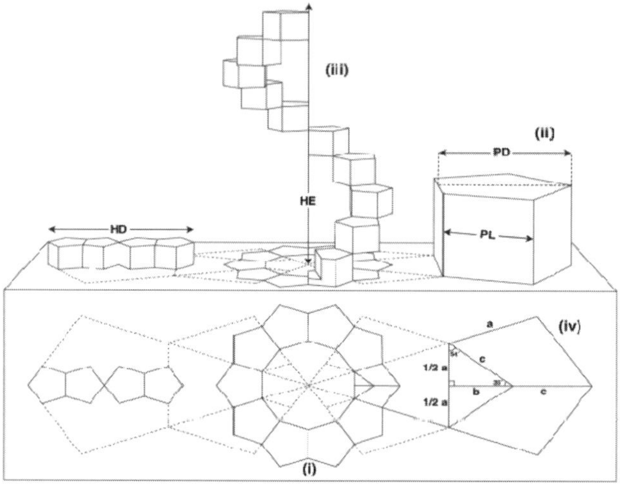

Figure 19.23 Curtis DNA model 1998

After studying the original X-ray crystallography data and conferring with a chemist, it became apparent that there may be a different way to pair the nucleotides within the double stranded DNA that would give the same data. Hoogsteen[15] also proposed different base pairings which have subsequently been implicated to be involved in base pairing within tRNA and triple DNA helices. The Curtis model allows for the inclusion of both possible forms of the dNTPs (or units of base, sugar and phosphate) namely the keto and enol molecular forms of thymine and guanine whereas the Watson and Crick model only allowed for one, the keto form (Fig. 19.22a). However, the Curtis model does not indicate where the ribose plus phosphate group that makes up the backbone of the helix should be positioned but like Hoogsteen pairings there may be examples within nature for these pairings (Fig. 19.22b). Fig. 19.23 shows Curtis's resultant helix as a stack of 10 pentagonal blocks representing the 10 nucleotides predicted by the X-ray crystallography data and below a view looking down the spiral of a ring of 10 pentagonal blocks / nucleotides. The stamp design Curtis finally produced was available to the public in 1998.

It can be seen that by overlaying the icosahedral 2-D outlines or the pentagonal disks over Curtis's drawings of the paired nucleotides, two of the pentagonal faces of the internal dodecahedron align with these paired nucleotides (Fig. 19.24). If we continue this procedure further we build the same ring of pentagons displayed by Curtis for viewing the double stranded DNA spiral head on (Fig. 19.25). The pentagons of the inner dodecahedron in Fig. 19.25 have been highlighted with pentangles and can be seen to match the Curtis's ring which he represented by pentagonal blocks.

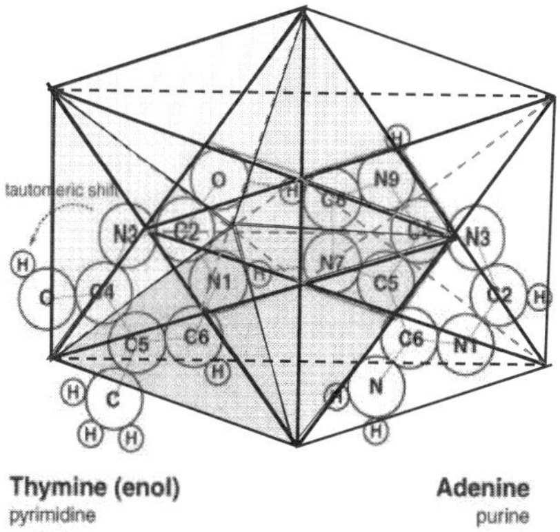

Figure 19.24 Enlargement of Curtis DNA pairings overlaid with cube (perspective 2) and pentagonal disk

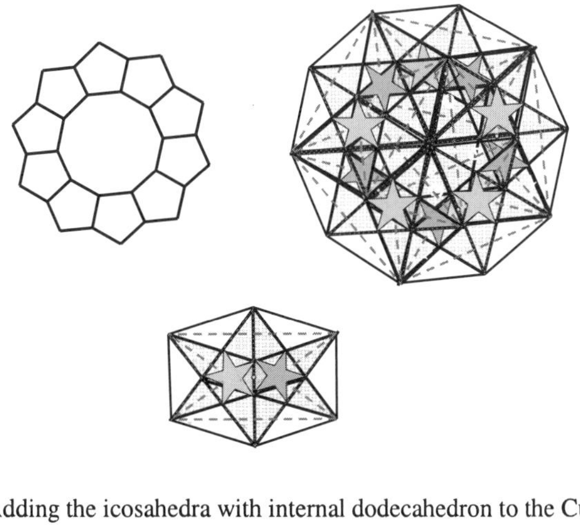

Figure 19.25 Adding the icosahedra with internal dodecahedron to the Curtis model

We can of course overlay single pentagonal disks (of 5 tetrahedra) to give the same result (Fig. 19.26). It is interesting to note that one pentagonal disk now contains 3 nucleotides of one strand of DNA. This could represent an analogy to the triplet codons of 3 nucleotides that are then translated into one amino acid protein building block when 'in frame', i.e. the information of the 3 is contained within the pentagonal disk of 5 tetrahedra. This is reminiscent of the algebra where the 8 is resolved into 5 when the 3 is 'overlaid onto 5', i.e.:

time	space	mass	charge
pseudoscalar	vector	scalar	quaternion
i	**i j k**	1	***i j k***

being reduced from eight simple to five combined or composite units, i.e.:

ik	**i***i* **j***i* **k***i*	1***j***

Considering the original eight, the octahedron would serve as a good candidate for this analogy or the 2 × 4 corner tetrahedra within the second order star tetrahedron. The octahedron is also the reciprocal of the cube which is here the 2-D outline of our icosahedron.

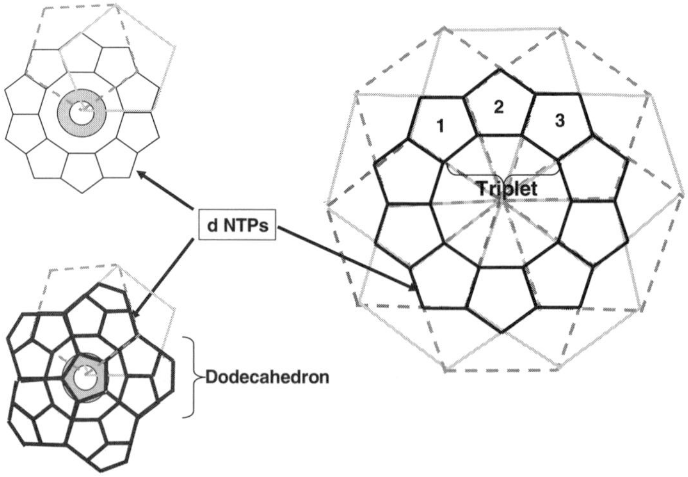

Figure 19.26 Highlighting the nucleotides – one triplet per pentagon

When this pentagonal scheme is applied to the Watson and Crick model of DNA we find that it also fits well as shown in Fig. 19.27. The insert shows a stick model of the helix looking down the top of the spiral and it can be seen that it is the pentose sugar rings of the dNTPs that now relate to the pentagonal faces of the internal dodecahedra.

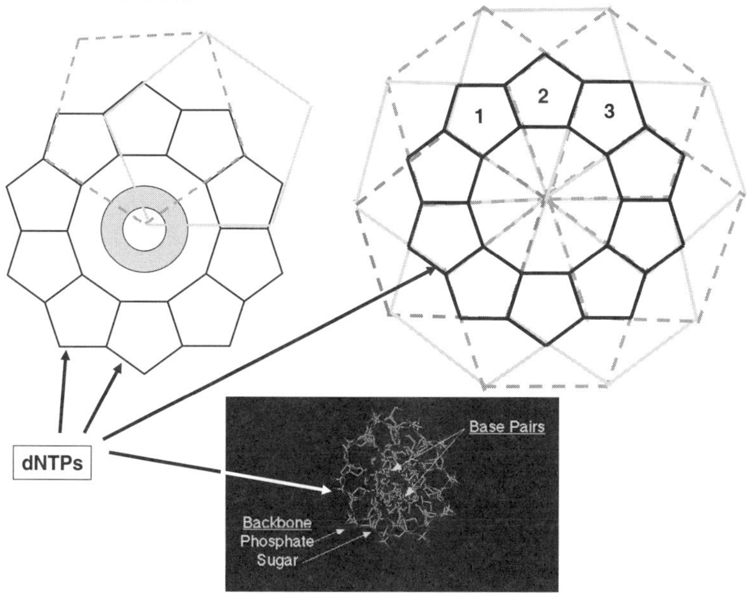

Figure 19.27 Watson and Crick: pentagonal geometry

A computer generated scattergraph of DNA generated by Robert Langridge was analysed by Reginald Brookes[16] who suggested a different pentagonal geometry composed of concentric double pentagons as shown in Fig. 19.28 but our proposed geometry also fits well upon this scattergraph (Fig. 19.28). Again, the scattergraph shows a remarkable similarity to a 3-D projection of a regular solid in 4-D. This time it is the 120-cell or hyperdodecahedron,[17] and again this seems to make sense with relation to the higher dimensionality when the rewrite structure's most fundamental unit – incorporating a time-varying, even chaotic, sequence, which is not present in more ordered structures, such as those of inorganic crystalline materials – is projected onto a 3-dimensionality. Significantly, the dodecahedron and icosahedron, with their pentagonal symmetries, are represented only in 3-D and 4-D – other dimensionalities only have cubes, tetrahedra and octahedra as Platonic solids.

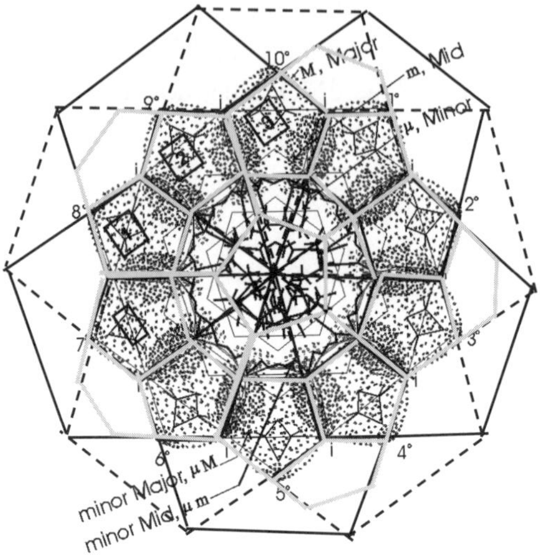

Figure 19.28 Computer-generated scattergraph of DNA: pentagonal geometry

The stacking of the pentagonal disks defining the DNA helix is shown in Figure 19.29. Figure 19.30 shows pictures of models of the stacking diagram made from tetrahedral dice. The model building process clarifies that 2 spirals of 5 stacked pentagonal disks carry the 10 nucleotides and that 2 strands of these fit together to produce the double stranded helix. It is interesting to note that there is a groove running around the molecule that implies space for a third strand of disks – could this be for the RNA or the stabilising water molecules? Double stranded DNA is known to unzip to allow a copy of RNA to be made which is then sent elsewhere to be translated into proteins. After the RNA copy has been completed it needs to unzip itself to release the RNA and the DNA strands are then rezipped together. It is also in the stacking procedure, where the missing 7° 12′ becomes significant. Taken over 2 × 5 dNTPs, this is equivalent to a shift of exactly 2 dNTPs for the double strand (= 2 × 36°), which means that it is of the exact value to ensure that the disks in successive twists of the spiral stack directly over each other. That is, the 'missing' angle not only necessitates the helical structure, but, through its basis in the geometry of the Platonic solids, ensures that it is a regular helix rather than a logarithmic spiral, as in shell structure. (Perhaps, also, the 'missing' angle causes the necessary instability for the zip / unzip process by giving a spring effect and fulfils the *zitterbewegung* effect that we see in physics.)

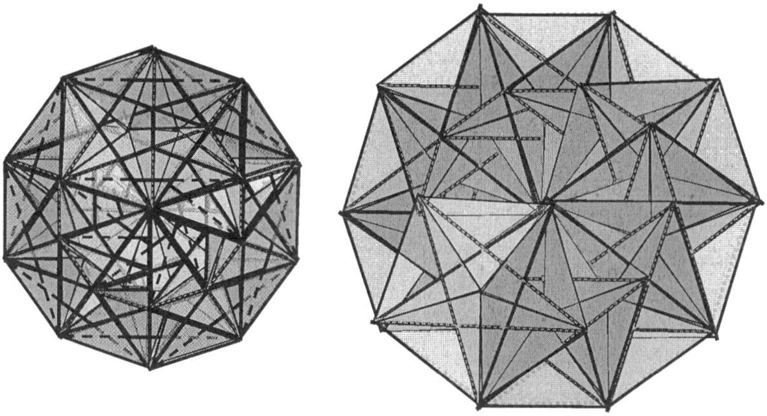

A. Overlaying icosahedra plus reciprocal dodecahedra

B. Overlaying pentagonal disks

Figure 19.29 The DNA helix defined by stacked pentagonal disks

**Single Stranded DNA:
Two Spiral Strands of Pentagonal Disks : 5 per Turn**

**Double Stranded DNA:
2 Spiral Strands of Pentagonal Disks with a Total of 10 Disks per Turn.**

Figure 19.30 3D model of DNA with pentagonal disks of tetrahedra

The need for 2 × 5 dNTPs for every twist of DNA is interesting also for other reasons. Each twist of a single strand contains 1 codon and 2 other dNTPs (cf 1 × 3-D + 2 others in the nilpotent). You can only guarantee a codon if there are 5 dNTPs. Now, if we start with a codon in the twist, it takes 3 sets of 5 dNTPs, i.e. 3 twists, to complete a 'cycle' and start with a codon again. Here, we may think of the Fibonacci series in the way that Johansen does. Also, the twisting in DNA is continuous – it doesn't begin and end anywhere – which is reflected in the fact that the coded part of the sequence is entirely coded, with no gaps.

Two further aspects of the double helix may be mentioned here, in connection with its appearance in both biological and physical contexts. One is the first observation of the structure on the large scale (80 light years = 7.6×10^{17} m) in the nebula near the centre of the Milky Way, already referred to. The other is the phenomenon of spectrin repeats, found in several proteins involved in cytoskeletal structure, for example, spectrin, alpha-actinin and dystrophin. These repeats create a triple helical bundle, which we may liken to the DNA / RNA connection or to vacuum process of fermion becoming virtual boson and then virtual fermion again; or the real process by which all fermionic states absorb and emit gauge bosons (equivalent to a combination of fermion and antifermion) in all discrete interactions.

19.15 The Cube and the Harmonic Oscillator

As previously discussed, the packing of pentagonal disks may have relevance to the spiral which is so important in both DNA and the spin of the fermion with its vacuum. However, as has been described, two disks can also be viewed in 2-D as a cube. The switching between the two perspectives of the cube (Fig. 19.17 and Fig. 19.31a, b) is interesting in that it can also be considered as a harmonic oscillator (Fig. 19.31d). One can visualize the cube spinning and the fermion and vacuum switching places (*zitterbewegung*) when the front flips to the back and the back face then holds precedence. The cycle can then complete when the front face flips forward again. A second element this relates to is the Klein-4 group and dual group (which can be seen as relating to particle and vacuum), where the complementary colours say red and cyan, used for the group / dual group, point to opposite corners of the cube and could be considered as switching. Rotation of the cube would, of course, produce the same effect.

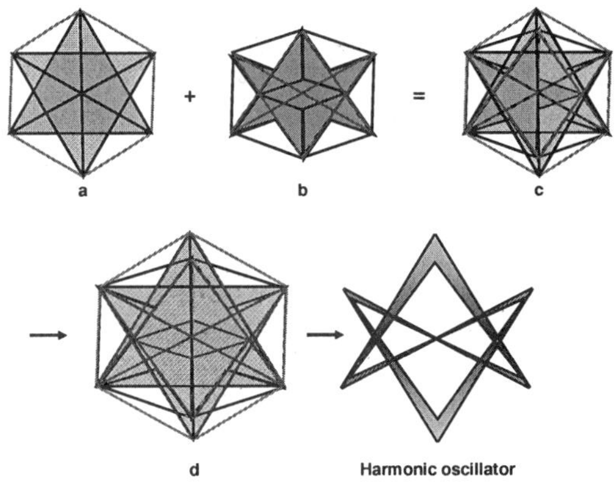

Figure 19.31 The spinning cube as the harmonic oscillator

Extending the dimensions, a 4-D cube with an internal star tetrahedron in rotation can be seen creating and annihilating two disks of five tetrahedra as if it were going through a phase-switching transition, exactly as in *zitterbewegung*, the fourth dimension being identified with $\pm t$, $\pm E$ and $\pm w$, and being the 'vacuum space' in which the real 3-D structures are created.[18]

A particular example of a cubical symmetry with a nilpotent physical, but not yet a biological, application (though it is likely that one is ready to be found), is provided by the Rubik cube structure, which is a regular cube, divided into 27 smaller cubes or cubelets, with each face of the large cube divided into 9 cubelet faces. The 54 cubelet faces are divided equally between 6 colours (or 3 colours and 3 anticolours), so that there are 9 of each colour, and each of the cubelets has the same relative ordering of colours on its faces. The structure, however, can be rotated internally, so that any colour shows on any of the cubelet faces of the larger cube. Corner cubelets have been viewed as giving the correct $\pm$ 1/3, $\pm$ 2/3 twists corresponding to the electric charge values observed for the quark components of mesons and baryons. The 3 colours (primary) and 3 anticolours (secondary), however, could also be seen as representing the three directions of the momentum operator in the nilpotent structure when the cube is in its most organized state $(\boldsymbol{i}kE + \boldsymbol{i}\mathbf{i}p_x + \boldsymbol{j}m)$ $(\boldsymbol{i}kE + \boldsymbol{i}\mathbf{j}p_y + \boldsymbol{j}m)$ $(\boldsymbol{i}kE + \boldsymbol{i}\mathbf{k}p_z + \boldsymbol{j}m)$. The six faces then represent the six possible phases, i.e. when the whole momentum $\mathbf{p} = + \mathbf{i}p_x$, $- \mathbf{i}p_x$, $+ \mathbf{j}p_y$, $- \mathbf{j}p_y$, $+ \mathbf{k}p_z$, $- \mathbf{k}p_z$. Anything else then represents a degree of mixing between two or more phases. When each face has 3 colours and 3 anticolours, the mixing is perfect. Of course, the quantum mechanical picture of the proton requires perfect mixing between the 6 phases ($SU(3)$ symmetry). (This would require a 6 × 6 × 6 cube with faces containing equal numbers of each coloured cubelet.) Another way of looking at it is to rotate the perfectly organized cube (each face one colour), allowing each face to be seen by the viewer for the same amount of time. The 8 agents that produce all possible colour changes (combinations of colour and anticolour) represent the 8 gluons. There are only 8, not 9, because only 2 combinations of red / antired, green / antigreen, blue / antiblue are considered to be independent. (The standard 3 × 3 × 3 becomes relevant if we consider colour / anticolour as occupying the same cube and we consider 3 colours alone. A 4 × 4 × 4 cube would give us the convenient total of 64 cubelets.)

19.16 The Rewrite Process as Nature's Code

If the existence of four bases reflects the basic structure of a universal rewrite system, then we would expect to find a similar set of dualities to those observed for space, time, mass and charge, and we would expect to find some natural process in which they could emerge in a relatively uncomplicated way. If the rewrite process gives us Nature's code, it is not obvious that, apart from existing at the most fundamental level, it will also reappear in similar form in particular structures higher up the natural hierarchy. Such reproductions of features of small-scale systems in larger ones, however, certainly exist in nature, and generally reflect some concept of *coherence* between the small-scale elements, organized by some powerful driving mechanism. The double helical structure and five-fold symmetry of DNA would suggest an application of 'Nature's code' by such a mechanism even if there were no other indications in the number and characteristics of the base elements, and the fact that the component structures are based on the relatively basic chemistry of only a few light elements suggests that the organizing principle is more significant to the process than the specific characteristics of the chemistry.

The nitrogenous bases found in nucleotides are, in fact, relatively simple structures, with a common feature: a heterocyclic ring derived from the parent compounds pyrimidine and purine. The ring structure is composed of both nitrogen (position 1 and 3) and carbon (positions 2, 4, 5 and 6). The pyrimidines T and C, and the purines A and G, are relatively minor variations on this pattern; and purine can be considered as a derivative of pyrimidine as it consists of a pyrimidine ring with an imidazole ring (a five-component ring, again with two nitrogen atoms) fused together. Here, we have one duality. Significantly, also, each of the purines has a pyrimidine *partner* (which makes it 'dual', in another sense), and the bonding atoms of these complementary atoms will line up oppositely on a single strand of DNA. So, we have O-N[-OH] and N-NH[-NH] bonds linking A-T reversing the order for the same bonds used in C-G. If we want a closer analogy with the physical case, we can say that the more structured A and G might be considered to correspond to the more structured physical parameters space and charge, which each have a specific partner in time and mass. Also, the A-T partnership might be considered the 'driver' or source of variation in the same way as the nonconserved space-time partnership is in physics. Thymine certainly initiates change with its replacement by uracil in RNA, while adenine triphosphate (ATP) is the main source of energy transfer in living systems. This suggests the approximate correspondence: A / space; T / time; C / mass; G / charge.

The question remains how such bases could emerge from some natural process which is not purely random (in the sense that, though the chemical structures may emerge randomly, they will be quickly 'selected' for their relevance to the overall scheme). Very possibly, special conditions (for example, high temperature, electricity, magnetism, liquid or gaseous environment) will be required to generate the structures in the right proportions – and so the generation of 'life' is not an *inevitable* result of chemical chaos; it is an inevitable consequence only of the conditions being available for the universal rewrite mechanism to operate on a scale higher than the most fundamental as a result of the creation of some necessary condition of coherence. Given the right conditions, however, positive feedback mechanisms might be expected to take place to enhance a process that would need to evolve only once. Fossil evidence from prokaryotes, which resemble present-day bacteria but may be as much as 3.3 billion years old, suggest that the genetic code developed at a single time at a relatively early date in the Earth's history. There was no multiple evolution.

Proteins are, of course, the essential basis of life, and include a number of enzymes which play a significant role in the transfer of genetic information; but proteins are made of twenty amino acids, selected only from those that follow the coded sequence presented by the four bases in DNA, and there is no other driving mechanism to link them in the seemingly random polypeptide chains, with their multiplicity of folded shapes. Though amino acids are relatively simple chemicals, which are easy to reproduce from basic elements under extreme conditions, there is no mechanism for linking them in protein structures which would then develop a coding system to reproduce them. It is virtually impossible to believe that the more complicated system logically preceded the simpler one, though one could see DNA, RNA and protein structures evolving together through a process in which the evolution of one aided and was aided by the evolution of the others. There are some arguments for believing that RNA preceded DNA, but DNA, with its double helix, is much the more stable structure, and it is certainly difficult to believe that the bases A / T and C / G should form their pairings by accident, rather than as part of the molecule's original design. Although tRNAs also form double stranded conformations they involve the inclusion of over 30 rare bases, with up to 10% of the total number of bases. It is apparent from this, that these are more complicated molecules than DNA. An argument for the temporal precedence of RNA would have to conclude that the extra bases were discarded when the more 'perfect' base pairings of A, T, C and G made the creation of DNA possible.

The structure of the bases as variations on a single basic design almost suggests a modification process to produce the appropriate levels of 'duality', and the creation of DNA through a combination of single types of the elements with each of the valences 1 to 5 (namely, H, O, N, C, P, three of which are gaseous in the free state) suggests something like a minimalist approach. Carbon is, of course, the only basis for creating the molecules that we call 'organic', because its valence 4 gives it the perfect bonding for coherently structured large molecules; and the hexagonal benzene ring is one of the most stable (though dynamic), simple organic structures, easily formed.[19] The most efficient way of driving an evolving process involving small atoms or molecules to break the sterile uniformity produced by the benzene ring itself might be through interaction with a gaseous atmosphere or gas molecules dissolved in liquids (a kind of biochemical equivalent of the physical 'vacuum'); and, from the point of view of stability, bonding, and the kind of 'closure' produced by the base pairing, the two valence 3 nitrogen substitutions found on the rings in each of the bases would seem to provide definite advantages.

The significant role of ATP then suggests the potential importance of the combination of bases and triphosphate as a precursor of DNA. The pentavalence of the phosphorus atoms provides significant opportunities for bigger linkages. Here, we have to imagine the driving process as creating helical structure via a harmonic oscillator mechanism. To create an extended helical structure we need pentagonal symmetry. To create pentagonal symmetry, we need a combination of the two hexagonal structures produced by the A-T and C-G linkages (the hexagonal structure being a double tetrahedron in 2-D), and a pentose sugar (another small 5 ring structure with a gaseous element substitution) interposing between the base and the phosphate. Significantly, proteins, deriving from DNA coding, show elements of helical structure, but much more disrupted than that of DNA itself.

19.17 The Unification of Physics and Biology

We have explored several significant examples in which the mathematics of physics and biology seem to show similar underlying structures relating to more fundamental processes, and to the idea of 'process' itself (Fig. 19.32). In addition to this, we have shown that complexity arises from simplicity in all aspects of nature because the fundamental units of nature (fermions) are symmetrical only when taken in conjunction with the rest of the universe (vacuum). All types of aggregation in matter, at all levels, physical chemical, biological, and all phase transitions, are concerned with an attempt to overcome this asymmetry. The force involved in this process is the weak force (the fifth or asymmetric term) and it is ultimately through this force that we will find a deep link between physics and biology.

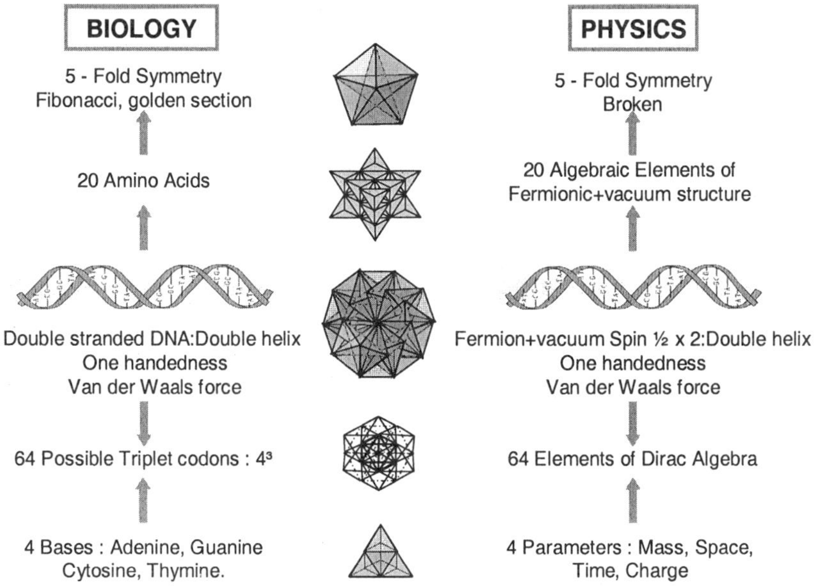

Figure 19.32 The universal rewrite system: Unification of physics and biology

Chapter 20

Nature's Rules

This chapter is written with **PETER MARCER**. It is based, with some significant additions, on four co-authored papers.[1] One of the papers had, as additional co-author, **BERNARD DIAZ**; and another was additionally co-authored with **EDGAR MITCHELL** and **WALTER SCHEMPP**. This chapter concerns the processes by which a sentient being may make sense of a universe structured according to a nilpotent universal computational rewrite system (NUCRS) with an infinite universal alphabet, such as is proposed in chapter 1. Such an alphabet will define the semantics of quantum mechanics in terms of a universal grammar, such that the nipotent generalization of Dirac's quantum mechanical equation is the computational machine order code. In this chapter, various implications of the universal nilpotent computational rewrite system, outlined in chapter 1, are described. Evidence is presented that this discovery not only provides a new semantic fundamental foundation for universal quantum computation, but is the keystone of a fundamental computational foundation for mathematics, quantum physics, the genetic code / molecular biology, neuroscience and cosmology ('Nature's rules'). Models, among other things, of the genetic code and the human brain leave little doubt that 'the Universe, life, consciousness and everything' may derive its entire substance from the difference between classical and quantum mechanical thermodynamic machines. The evidence that the structure of the cosmos, the genetic code, the human brain, and human language corresponds to quantum mechanics as determined by the generalized nilpotent Dirac equation, and to the complementary semantic theory of quantum holographic pattern recognition specified by the corresponding 3-dimensional nilpotent Heisenberg Lie Group, becomes a well determined testable scientific hypothesis. Further, these two nilpotent representations correspond to the required division of the nilpotent quantum mechanical state space into its Clifford / fermionic and Lie / bosonic parts. The power of the NUCRS is that it is able to draw in such significant ideas as the surreal numbers, Wheeler's meaning circuit, the holographic principle, quantum holography, anticipatory computation and quantum thermodynamics, while showing the significance of the Heaviside operator, the bra and ket notation, and the Riemann zeta function, to arrive at an Evolutionary 'Anthropic' Semantic Principle, which defines the rules by which a sentient being is able to comprehend the rules by which Nature is apparently structured.

20.1 A Semantic Model of Computation

Success in science is measured by agreement between the predictions of the particular mathematical language description used and experiment, and the language's interdisciplinary breadth. Syntactic correctness of the language is thus a necessary but not necessarily sufficient condition. It may only guarantee a combinatorial explosion of possible correct solutions, as now appears to be the case in string theory according to the admission of one of its originators, Leonard Susskind.[2] (Peter Woit goes so far as to say that, in its inability to make genuine predictions of testable results, string theory is, in Pauli's famous phrase, 'not even wrong'.[3]) Correct syntax must be accompanied by correct semantics, as in a semantic language model of computation, such as the rewrite systems synonymous with computing / information processing discussed in chapter 1, which specify a grammar as well as a syntax appropriate to the problem needing solution.

Rewrite systems concern the languages in which programs are rewritten as symbols *for computing hardware to interpret.* The nilpotent universal computational rewrite system (NUCRS) outlined in chapter 1 is of particular significance, since its subset alphabets emerge in a minimal way and not only have a mathematical interpretation as algebra, but also concern the nilpotent Dirac algebra, with which one can generalize Dirac's well known quantum mechanical equation, so as include not only mass and electric charge, but also the strong and weak charges, and, implicitly, the property of spin. The NUCRS has a universal grammar in the sense that it delivers the entire infinite alphabet of symbols in one step, when presented with zero, in a simplified presentation, as the initial subset alphabet. (In a more technical extended presentation the need to start with anything at all can be dispensed with.) This universal system, we have conjectured, has a minimum of two rewrite rules or productions:

i) a creation operation, delivering a new symbol at each invocation, where this new symbol may be a single character of the alphabet, a subset alphabet, or the entire alphabet, and

ii) a conserve / proofreading operation, which examines all currently existing symbols to ensure that the bringing into existence of a new symbol or subset alphabet, etc., produces no anomalies.

The nilpotent universal computational rewrite system (NUCRS) (see chapter 1), differs from traditional rewrite systems, of computational semantic language description with a fixed or finite alphabet, in that the rewrite rules allow new symbols to be added to the initial alphabet. (Examples of conventional rewrite systems include ones in which the finite alphabet semantics correspond to geometric rules so as to give very lifelike pictures matching those in botany, for example, a sunflower.) As already stated, we start with just one symbol representing 'nothing' and two fundamental rules (which it turns out are the dual aspects of a single rule): *create*, a process which adds new symbols, and *conserve*, a process that examines the effect of any new symbol on those that currently exist, to ensure 'a zero sum' again. In this way at each step a new sub-alphabet of an infinite universal alphabet is created. However the system may also be implemented in an iterative way, so that a sequence of mathematical properties is required of the emerging sub-alphabets; and we have shown that one such sequential iterative path proceeds from nothing (as specified by the mathematical condition nilpotent) through conjugation, complexification, and dimensionalization to a stage in which no fundamentally new symbol is needed. At this point the alphabet is congruent with the nilpotent generalization of Dirac's quantum mechanical equation, showing that this equation defines the quantum mechanical 'machine order code' for all further (universal) computation corresponding to the infinite universal alphabet.

The property of the universal nilpotent rewrite system that a new symbol can stand for itself, a sub-alphabet or the infinite universal alphabet, allows it to rewrite itself, so as to enable it to describe the ontological structure at a higher (hierarchical) level in terms of those at lower levels, beginning with the fundamental level. This rewrite system with its nilpotent bootstrap methodology from 'nothing / its empty state' thus defines the requirement for universal quantum computation to constitute a semantic model of computation with a universal grammar. It is also significant that, though the universal rewrite system generates both number systems and algebras, it is not confined to these systems, and does not depend on the pre-existence of numerical or algebraic concepts, or any of the ideas of set theory, and the zero is not confined to being that of the empty set. The mathematical structure generated derives rationals from reals, not reals from rationals; and shows that complexification logically precedes discrete numbering, a result that cannot be derived from any form of set theory.

Such a nilpotent rewrite system describing both arithmetic and geometric properties must describe what in computer systems is called universal computer construction, i.e. such nilpotent quantum computation will be both computer universal in the sense of arithmetic and constructor universal in the sense of geometry. That is, it includes both universal digital computation as discovered by Turing[4] in the form of the universal Turing Machine model, and universal computer construction or self replication as revealed by Von Neumann.[5] There is every reason to believe, also, that it is the universal meta-pattern required in nature by the anthropologist Gregory Bateson.[6]

20.2 Scientific Perspectives on Computation

From the above it would appear that:

i) both theoretical physics and mathematics, grounded in nilpotent quantum physical process action, are single, possibly equivalent, bodies of human linguistic / semiotic knowledge emergent from the human brain, a quantum physical semantic machine, as the evolutionary result of the semantic natural physical law NUCRS describes; and

ii) that, as proposed by Langlands, mathematics is (in this new computational NUCRS foundation) indeed a single inseparable body of knowledge, where theoretical physics will be the same thing as mathematics, thus explaining what is often referred to as 'the unreasonable effectiveness of mathematics in relation to physics'.[7]

That is, the NUCRS describes the basis for two new foundational disciplines: the computational foundations of physics and mathematics. Thus, this law's semantic mathematical language description would be expected, in addition to correct syntax, to provide the description of such properties as:

i) a measure, metric and Hamiltonian / Lagrangian for each variable, process and system respectively described; and

ii) thermodynamic principles in relation to quantum measurement, where information becomes a physical resource such as entropy production, able to produce, as in magnetic resonance imaging (MRI), real and virtual imagery of 3 + 1 relativistic space-time physical structures, where these images exist independently of the presence of any observer. This is the fact that natural radiation of any kind, incident on any object it illuminates, shows to be the case; for, as is known and can be demonstrated holographically, any such incidence will, quite independently of any observer, effect local changes of the radiation's amplitude and phase, so as to capture the 3 + 1 relativistic space-time image of the illuminated object as appropriate to the nature of the incident radiation (see section 20.10).

Evidence will be provided in this chapter to support the hypothesis that nilpotent quantum mechanical language description constitutes 'Nature's rules' in accordance with the Premise and Mission Statement of the British Computer Society's Cybernetics Machine Group, that 'In science, Nature sets the rules, but it must never be forgotten, that it is only because life has exploited these rules successfully for billions of years to our evolutionary advantage, that human brains are able to understand them. The mission, at the physical foundations of computing / information processing if one accepts the premise, is therefore to identify how these rules were exploited to achieve this end.' That is, the NUCRS turns this premise into an **Evolutionary Anthropic Semantic Principle.**

The evidence, below, that the NUCRS can indeed be identified with the principal stages needed to accomplish this mission, as set out in its premise, strongly advances the claim that the NUCRS generalization of the computational rewrite concept can be taken as a new fundamental computational foundation for both quantum mechanical and mathematical language description, so as to constitute, from these stages, a likely basis for a 'Theory of Everything' in the form of a fundamental organizational principle. For the evidence is, firstly, that the DNA / RNA genetic code, exemplifying life, fits the essential criteria necessary for it to be a nilpotent rewrite system, defining a further progression of the NUCRS infinite alphabet. That is, the genetic code is a semantic code, which specifies the ontological geometric structures of living systems – including those of all molecular biology – as grounded upon the NUCRS's fundamental level structure of the Standard Model elementary particle physics, within 3 + 1 relativistic space-time. And, similarly, the evidence of the human brain's natural semantic language abilities, exemplifying the *modus operandi* for the advancement of human understanding, points strongly to the fact that the brain's molecular neural and glial structures correspond ontologically to a yet further progression of the infinite alphabet realized within the DNA / RNA genetic code itself, so as to permit the human brain by means of NUCRS rewrite mechanisms to input / output natural language, i.e. to hear and to speak, and to process language semantically, i.e. to proofread it / to think. And thus the reasons for the semiotic, sequential nature of all these processes, for which the evidence is natural language's ability to take the form of the written word, and the speed by means of which the human brain is able to carry out all these thought processes, is furnished with an explanation, as is the human brain's stream of conscious perception. For all are grounded, as will be explained in 20.12, in NUCRS behaviour as sequential single parameter thermodynamic quantum mechanical engine action.

Secondly, the evidence already established in chapters 1 and 2, is that the processes of semantic computation as described in those chapters (including the demonstration itself), are fundamentally quantum physical in nature, an accepted conclusion about the nature of universal computation already reached independently by Deutsch and Feynman.[8,9] The third strand of evidence is that digital computation, which Deutsch has shown quantum computation includes, constitutes a universal regime of rules for syntactical but not yet semantically correct computation, so as to explain why the required semantic basis for any digital computation / algorithms must in general be effected through the agency of the human brain. A nilpotent computational system could loosely be thought of as computation using zero (i.e. topological computation) rather than bits in the binary system 0, 1 and it is more general than binary computation as Deutsch's theory shows. This raises the question 'Could semantic rather than digital computation be what we mean when we refer to the human brain as having 'commonsense'?'[10] For, from the known facts of its working, in particular its human language capabilities, the human brain is almost certainly a universal semantic computational machine. The NUCRS thus marks a clear distinction between human and artificial intelligence, and would explain why the architecture of the human brain is so different from its digital counterpart.[11]

There are also other senses in which digital computation is incomplete. For, if described simply in terms of universal logical primitive NAND, it lacks, as Feynman points out, the additional 'physical' primitives, like those of the unit wire and of signal exchange, that such descriptions of digital computation require if they are to be physically implemented and executed. In addition, descriptions of digital algorithms can have no meaning unless, as John Wheeler has pointed out,[12] there exists some actual physical means by which they can be carried out / executed. Semantic computation, in fact, explains why, despite a digital computer's simplicity, there are no naturally evolved species with nervous systems based on digital architectures. This preference of nature can be attributed to the fact that physical trajectories / systems are known to naturally follow geodesics and principles like that of least action (as indeed does quantum mechanics in Feynman's sum over histories formulation), and so will most likely lead to any natural computation / measurement taking place in a minimum number of computational steps, that is, optimally.[13]

A requirement, cited by Deutsch, is that all valid computation must be canonically labelled, which is satisfied in nilpotent quantum mechanics (NQM) as governed by the nilpotent Dirac equation (D(N)), because the Pauli exclusion principle applies to NQM's fermionic states, so as to be in agreement with Wittgenstein's (semantic) principle[14] that there is necessarily only one proposition for each fact that answers to it, and that the sense of a proposition cannot be expressed except by repeating it. The NUCRS actually shows that, contrary to the Platonic assumption, mathematical language description is just another form of natural semantic language capability, which derives its origin from the semantic computational capabilities of NQM. For natural languages have only now made their evolutionary appearance, partly because they necessitate a nervous system and biological brain of the size, power and complexity of the human brain, which the facts show has never existed until the present era; and also because of the enhanced evolutionary advantage that semiotic / semantic language communication and understanding of the world including mathematics now demonstrably offers for the survival of the human species at the present stage of evolution; and where, in the foundation of any natural language, a necessary grammar for semantics, in addition to syntax, provides what is known in human communication as its 'commonsense'. For example, the objective of mathematical language is surely to demonstrate that the human brain's 'commonsense' semantically-arrived conclusions, do indeed follow syntactically from their respective premises.

20.3 The Nilpotent Structure of the Universal Grammar

By generalizing the concept of the computational rewrite system, the novel, first principles, semiotic approach to mathematical language description, demonstrated here, shows that there exists a semantic mathematical language description with a universal grammar, namely nilpotent quantum mechanics (NQM), governed by the nilpotent generalization of Dirac's quantum mechanical equation D(N):

$$(\mp \boldsymbol{k}\partial/\partial t \pm \boldsymbol{i}\nabla + \boldsymbol{j}m)\,(\pm \boldsymbol{ik}E \pm \boldsymbol{i}\mathbf{p} + \boldsymbol{j}m)\; exp\; i(-Et + \mathbf{p}.\mathbf{r}) = 0 \qquad (20.1)$$

where E, $\mathbf{p}$, m, t and $\mathbf{r}$ are respectively energy, momentum, mass, time, space, and the symbols ± 1, $\pm i$, $\pm \boldsymbol{i}$, $\pm \boldsymbol{j}$, $\pm \boldsymbol{k}$, $\pm \mathbf{i}$, $\pm \mathbf{j}$, $\pm \mathbf{k}$, are used to represent the respective units required by the scalar, pseudoscalar, quaternion and multivariate vector groups. That is, D(N) specifies the rewrite system's computational order code. Potentially, this offers an entirely novel mathematical language means of a semantics with a universal grammar, NQM, to bring about the desired agreement between theory and experiment, where computation / quantum measurement consists of an input / quantum preparation followed by an output. We may now propose the hypothesis that the universal grammar for semantic quantum mechanical mathematical language description that constitutes NQM is a candidate for 'alternative (a)' in Leggett's incisive analysis of 'The Quantum Measurement (QM) Problem',[15] which says that 'QM is the complete truth about the physical world (in the sense that it will always give reliable predictions concerning the nature of experiments) at all levels and describes an external reality'.

The extensive work outlined in previous chapters into the properties of NQM governed by equation (19.1), not only shows that the spontaneous symmetry breaking of the nilpotent Dirac equation (19.1) describes the simultaneous complementary emergence of 3 + 1 relativistic space-time and the experimentally validated quantizations of the strong, weak, and electromagnetic force of elementary particle physics (including spin), from 'nothing', their empty state (or the empty set), but also that these quantizations, including both spin and massive particles, are in excellent agreement experimentally with those of Standard Model elementary particle physics, so that they constitute the entire set of sources and sinks of the 3 + 1 relativistic space-time field. Derived from D(N), this field is thus a quantum phenomenon. That is, this mathematical language description of 3 + 1 relativistic space-time emergence, furnishes a quantum theoretical explanation of 3 + 1 relativistic space-time's existence as a fundamental quantum property of the physical world, something that string theory has long been expected to describe but which it has yet to do.

In principle, these fundamental physical quantum mechanical structures not only define level one of the ontological hierarchy corresponding to the NUCRS semantics, but are those from which all subsequent levels of this hierarchy will be reconstituted in 3 + 1 space-time due to the action of the strong, weak, electromagnetic and gravitational forces. This is in accord with the computational rewrite system description where any symbol *S* of its rewrite alphabet, can stand for, itself as single symbol, a sub-alphabet, and its infinite universal alphabet. The quantum measurement process in NQM is therefore the universal semantic computational decision criterion for deciding among QM descriptions what quantum physics is, where the leading question is 'Will further NQM predictions continue to be in good accord with experimentally validated quantum physics, in the future or as has been already established in the past?'

20.4 General Relativity and NQM Semantic Description

From the 4-vector group description of 3 + 1 space time in the nilpotent Dirac equation, which, among other possibilities, governs both 3 + 1 space-time's quantum emergence and its geodesic behaviour in NQM, it follows that Einstein's legacy of both special and general relativity (in the particular sense specified in chapter 17) also holds universally in NQM. In NQM, this almost certainly follows from the fact that, in the nilpotent Dirac equation (D(N)), spontaneous symmetry breaking, 3 + 1 relativistic space-time and Standard Model elementary particle matter emerge as complementary fundamental quantum physical properties; each are therefore the cause and anti-cause of the other, rather than matter and antimatter as is usually taken to be the case, so this D(N) symmetry-breaking would account for the observed asymmetry in the universe between the latter (see 10.7). The prediction is in good agreement with experimentally validated uses of general relativity, and in particular those of the cosmological models, where general relativity is widely used because the corresponding models of quantum gravity are not known. However, since general relativity is compatible with NQM semantic description, where it coexists with Heisenberg uncertainty and quantum coherence, it is necessary to explain how this is possible, when this has not generally been found quantum mechanically to be the case.

This counterexample to the widely held established view that Einstein's general theory of relativity is incompatible with quantum mechanics, results from NUCRS as a consequence of taking the former as expressed in the form of a multivariate 4-vector group representation, and the latter as the generalized Dirac nilpotent representation. It is accomplished thermodynamically and almost certainly follows from the optimum geodesic nature of nilpotent representations. It indicates that space and time are smooth at the smallest scale, and not fuzzy and foaming as current ideas of quantum gravity appear to require.[16] That is, in an empty universe, this nilpotent symmetry-breaking process constitutes the genesis that brings 3 + 1 spacetime and its complementary elementary particles into existence for the very first time, such that they are first born in a 'virgin birth' from nothing, i.e. their empty set. This (empty) universe is therefore an (empty) white (w)hole. And, just as importantly, this is the initial step in a 'birthordering' or birth order process, which can thus be hypothesized to describe this universe's evolution as defined by the quantum Carnot engine in 20.12. Significantly, also, the evolution will occur at all times without loss of information, the nilpotent theory being consistent with the fact that quantum theory, as distinct from classical, does not allow an information loss (a fact now conceded by Stephen Hawking in the case of black holes).

20.5 Analysis over the Surreals

We have, of course, supposed that the universal computational rewrite system provides, not only for the language descriptions of quantum physics, but also for mathematical language itself. That mathematics itself is indeed a semantic language with a grammar, finds support from John Conway's highly prescient 1976 generation of what Knuth calls the surreal numbers,[17] for it provides a NUCRS foundation for mathematics as an alternative to that of the more usually accepted Zermelo-Fraenkel set theory, in the form of a non-standard mathematical analysis over the surreal number fields.[18] The Conway 'process' definition of the surreals generates a restricted nilpotent infinite universal alphabet, i.e. the numbers become symbols themselves, where, significantly, the number symbol $i = \sqrt{-1}$ has to be added arbitrarily to ensure that this generation process is that of the required universally embedding totally ordered (mathematical) field. We say 'significantly', because in a paper, entitled 'D: the infinite square roots of -1', and in Appendix B, we have shown that these infinite roots of $i = \sqrt{-1}$ provide the universal infinite alphabet in relation to the nilpotent computational path to the nilpotent Dirac equation, and the universal grammar of the semantics of quantum physics.[19]

Conway's work shows that the simplest lexicographical universal model N of a theory concerns the alphabet of the two symbols L and R where the usual convention of L signifying left and R right, which he uses to generate all the numbers great and small, is abolished. This model defines the way of turning the class of all ordinal numbers into a complete mathematical field such that each ordinal extends the set of all previous ordinals in the simplest possible way, by regarding sums, products, inverses, algebraic extensions and transcendental extensions (by means of mathematical groups and rings to fields) as successively more complicated concepts. However these extensions may equally be seen lexicographically[17] as defining the form of alphabetic extensions, as is appropriate to a computational rewrite system; or they may be viewed as extensions of Turing's definition of computation by means of the universal Turing machine over the integers, where the integers are not now seen as integers but viewed as a countable set of symbols.[13]

Surreal analysis[18] also shows that these extensions, which necessitate the introduction of $i = \sqrt{-1}$, are also maximal in the sense that they encompass all the properties of the arithmetic continua R and its Euclidean, hyperbolic and elliptic geometric counterparts, and in particular, such universal models have a unique birthorder field automorphism (birthordering). Furthermore, Conway's model of the surreal number field N_0, where L does signify left, and R right (encompassing all the numbers great and small, including the transfinite and the infinitesimals)[17] is also nilpotent in the sense that it is generated from the empty set; and that its first number is defined to be such that the symbol zero has the value 0. This notably implies that the value ½ and those of the half integers play a special role.[18]

The NUCRS model thus demonstrates that the theory specific to Conway's simplest universal model with two basic alphabetic symbols must be quantum mechanics, as represented by the nilpotent Dirac equation, which we show breaks its nilpotent symmetry (or emptiness) by associating the respective charges with vector, scalar and pseudoscalar operators, where the quantizations (including spin) of the families of elementary particles so realized are the familiar ones known and established by experimental particle physics and can be regarded as the sources / sinks of the 3 + 1 space-time quantum field in both its Lorentz and Einstein general relativistic invariant forms. From the discussion in previous chapters, it is evident that the concept of nilpotence, or, in this interpretation, the empty set as the description of the initial mathematical state of a system as used by Conway in relation to universal models, is foundational to physics, for, from the above arguments, the initial nature of the 'empty set' from which 3 + 1 space-time and elementary particle matter emerge, can be interpreted to be 'dark energy' in the form of quantum coherence. (For further discussion of 'dark energy' and its relation to inertia, see chapter 21.)

Surreal numbers have been shown to provide an appropriate classification of optimal extensions of Turing's definition of computability over the integers. These extensions exhibit many deep connections with the physical world as described by classical, relativistic and quantum mechanical theories; but connections with biological control systems can also be demonstrated. The connections confirm the Church-Turing principle[8] that computability is primarily a physical property and only secondarily a mathematical one. These considerations suggest that this class of extensions can furnish a unified theory within which (extremal) conditions concerning a minimum number of computational steps can be seen to govern perception and cognition as generalized dynamical processes. They further provide explanations of the marked differences between control and information processing systems in living organisms and those employed in conventional (von Neumann / Turing) digital computers, and of how the combinatorial explosion has been avoided in natural systems subject to incremental evolution over time. However, the principal claim is that such a model of optimal computation, denoted by Conway as $C(On_2)$, may furnish a theoretical basis for the genetic code in its contemporary evolved form.

20.6 The Heaviside Operator

Using the NUCRS, Conway's generation of the surreals can be reconstituted as a fixed computational rewrite methodology, which, in agreement with the universal method (though not fundamentally required by that method), has two fundamental productions (identified here as L and R) for the concept of order intrinsic to number. This generation begins from the symbol for the empty state of no numbers, assigning a value to each number symbol so as to generate a unique birthordering of all the numbers great and small including the transfinite and the infinitesimal, where this birthordering is the birthorder surreal number field automorphism, and, finally, also treats the case On_2 where the order actions of Left and Right are no longer distinguished, so as to show that this new birthordering is that of the simplest mathematical field of all the ordinal numbers and so admits all the properties of number.

Giving the symbol L the value 1, and the symbol R value 0, further implies that this particular universal model concerns a Heaviside operator P, equivalent to the corresponding singular Green's function (Schwarz distribution) which permits the same description of the physical wave phenomenon by means of an integral formula, and P is defined by the equation $(P-0)(P-1)$ as an idempotent so that $P^2 = P$. In the three spatial dimensions implicit in the nilpotent rewrite system, this Heaviside operator thus corresponds to Dirac's equally famous 'delta function' as it is known in quantum wave mechanics, and the equation for P is correspondingly that for Von Neumann's *quantum mechanical measurement* projection operator, showing P, as (0, 1), to be a general way of introducing (a) 'bit' signal processing via the concept of the 'bit' and (b) statistical mechanics modelled as coin tossing, into quantum physics. More generally, however, if one takes (P – 1) and (P – 0) to correspond to the Clifford / fermion and Lie / boson partitions of the quantum mechanical state space (and the operation linking them as describing quantum coherence, see 20.12), P can be identified with Berry's postulated dynamical quantum system, with time reversal asymmetry (as the thermodynamics of quantum measurement requires, see 20.17), such that the well known zeta function determines the eigenvalues of its Hamiltonian. In this now generally acknowledged quantum dynamical system (discussed in more detail in 20.13), it is seen that all the fermionic states of spin ½ (appropriate to the fermionic / Clifford partition) constitute the line $x = ½$ such that if the Riemann hypothesis is true then the zeros of the zeta function define by means of their imaginary parts on $x = ½$, the gauge invariant geometric phases of this dynamical system (appropriate to the Lie / boson partition).

This system is therefore nilpotent and satisfies the requirements appropriate to quantum holographic signal processing (see 20.10). Moreover the zeta function defines all the properties of number appropriate to arithmetic via the primes, the distribution of which is known to play a game of chance (see 20.13). (The feature of quantum mechanics that the wavefunction is only determined up to an arbitrary fixed phase (cf the unique phase associated with the **p** operator), while relative phases are measurable as gauge invariant geometric phases, is an absolutely essential one, since any measurement must necessarily be made against a fixed measurement standard to be meaningful, and in the case of the cosmology proposed here, this fixed universal measurement standard cannot, of course, itself be measured, because there can be nothing to measure it against – a fact confirmed by such well known effects as the Lamb shift and the Casimir effect.)

Light, for example, as seen in 20.2, is itself a potential holographic carrier signal existing independently of any observer, where the means for the encoding and decoding of the signals is described through the 3-D nilpotent Heisenberg Lie group so as to determine, as Weyl knew in 1928, the Heisenberg uncertainty (and that of a Lie dual, see 20.10). Thus in the case of signals independent of any observer, we must hypothesize that the direction of such normalized signals, expressed in a wavefunction exp $(i\theta)$, corresponds to that of the universal proper time, and, being conserved, takes the form of a fixed amplitude spiral in the quantum vacuum in the direction of the proper time. That is to say, the active property characterizing the empty nilpotent quantum vacuum is its Heisenberg uncertainty described in the form of creation and annihilation operators (cf *zitterbewegung*).

These conclusions are thus also in agreement with Feynman's conceptual use of Huygens' principle of secondary sources to derive his equally famous path integral formulation of quantum mechanics,[20] since such Heaviside operators (as Jessel has shown) are fundamental to the formalization of Huygens' Principle.[21]

20.7 Wheeler's Meaning Circuit

Once we have a universal nilpotent rewrite system, which yields a description of the recognized laws of quantum mechanics in the form of the generalized nilpotent Dirac equation, starting from the symmetry breaking of the 'empty set', we also have, among other things, a means of mathematically formalizing J. A. Wheeler's argument concerning 'The Meaning Circuit',[12] that, while the laws of physics require description in terms of mathematical algorithms, these algorithmic forms will be useless (i.e. have no meaning) unless they can be executed using the laws of physics themselves. In particular Wheeler argues that this could provide a mechanism or bootstrap for deciding the actual form of physical law, without any foreknowledge of what that law might be, a concept he calls 'physical law without law'.

As we have demonstrated, by introducing the notion of nilpotence, and beginning solely with the symbol zero (i.e. without knowing beforehand anything of the nature of physical law or physics itself), the universal rewrite methodology shows how to generate an actual mathematical description of physical law in recognizable quantum mechanical form. So Wheeler's conceptions can be taken to correspond in this case to physical law in the form of the generalized nilpotent Dirac equation, which the rewrite methodology shows is in fact universal. Furthermore the nilpotent Dirac equation, which implicitly includes the boundary condition of zero or the empty set (implied by its nilpotence) should be (and in fact has so far been) able to predict theoretically the values of nearly all the known and the possible physical constants and invariants, which currently can only be known empirically from experiment. That is to say this methodology can generate all the physical constants so that they can be known without empirical determination, in accordance with the statement of Einstein that 'In a sensible theory, there can be no numbers whose values are determinable only empirically. I can, of course, not prove that .... dimensionless constants in the laws of nature, which from a *purely logical point of view* can just as well have other values, should not exist. To me in my 'Gottvertrauen' (faith in God) this seems evident, but there may well be few who have the same opinion.'[22] (our italics)

This would therefore provide a totally exhaustive means of testing this new model's correctness, as one would expect from its description as a 'proof reading' mechanism. Furthermore the birthordering that the rewrite system provides, is, because of its nilpotence, always entirely renormalizable so as produce an entirely finite representation of the quantum mechanical evolution, where such birthordering defines that evolution's proper time ordering in such a way that it cannot be globally reversed. Such an evolution is thus in conformity with the first, second and third laws of thermodynamics, showing that, while quantum mechanics may be dynamically locally time-reversible on all local scales, its global evolution is, by contrast, thermodynamically irreversible, and can never return to its initial (global) state. Such an evolution therefore must concern the continual thermodynamic reconfiguration in 3 + 1 space-time of a finite quantity of elementary particle matter which appears simultaneously with that spacetime at the first moment of their creation.[23] Thus it follows from the formalization of Huygens' principle of secondary sources,[18] that a source of the 'universe' (corresponding to the white (w)hole from which 3 + 1 spacetime and elementary particle matter emerge) must in this case be equivalent to a set of secondary sources, which are in fact local sinks or 'black holes' at which both 3 + 1 space time and elementary particle matter disappear, so as to function as what in computer terminology is a 'garbage collector'.

Wheeler's hypothesis of 'Physical Law without Law'[12,24] suggests that the physical foundations of computation (see 20.1 and 20.2) constitute 'a Meaning Circuit or 'bootstrap' able to determine physical law, perhaps without any prior knowledge of what that law is'. This bootstrap arises from the fact that, while such a law must be describable in algorithmic form, any such description (of the law) can have no meaning unless there exist actual physical processes by means of which to execute it (the algorithm). Wheeler's hypothesis may be paraphrased as saying that such a semantic circuit describes not only that which can be said by words but that too by means of which words are said (or, alternatively, are written / spelt out semantically by the DNA genetic code of a living system, but where DNA is also the quantum physical means ontologically by which this living system is actually physically brought into existence). And it is a hypothesis for which the demonstration of a universal semantic quantum mechanical state space NQM with its universal grammar provides a definitive solution.

20.8 Anticipatory Computation and Other Ideas Supporting the NUCRS

In support of the general conclusions reached so far in this chapter, we may cite, in addition to Conway's work on the surreal numbers, the anticipatory computation of Daniel Dubois, which can now be understood in terms of the methodology of the computational rewrite process.[25] The concept of semantic computation would provide an alternative explanation of why Dubois's concepts of chaotic / fractal computation, incursion, hyperincursion and computing anticipatory systems are so successful in arriving rapidly at sound computational solutions to difficult problems, and as to why recursion in the form of ordinary digital computation may fail to do this, while incursion and hyperincursion can succeed.[26]

There is also the 'New Computing Principle' of Fatmi and Resconi,[27] in which the description of the optimal design for the physical machine already incorporates a description of a Lagrangian; and the 'Theory of the Cybernetic and Intelligent Machine based Lie Commutators', of Resconi, Fatmi, Jessel and Marcer,[28] in which computer input / output is represented by a categorical arrow, so as to describe formally such machines in terms of 'arrows of human mathematical thought'. Both follow from Dennis Gabor's paper 'A Universal Non-linear filter, Predictor and Simulator which optimizes itself by a Learning Process',[29] which is generalized by using Jessel and Resconi's categorical formulation of 'General System Logical Theory',[30] based on Jessel's formalization of Huygens' principle of secondary sources, already mentioned. These might therefore be more appropriately named, respectively (with hindsight), as the New Semantic Computing Principle, the Theory of the Semantic and Intelligent Machine, and General System Semantic Theory. All follow from the description by Mesarovic and Takahara of category theory in the form of arrows $\rightarrow$ where, as above, these concern a computational input and the subsequent computational output so that the arrow describes the computation taking place.[31]

In connection with General System Semantic Theory, Fatmi and Resconi's paper defines this new computing principle, in terms of the topological structures below:

$$G_s G_j \prod_{i=1}^{p} f_{mi}(0) = G_k G_j \prod_{i=1}^{p} f_{mi}(0) + (G_j G_k - G_k G_j)\left(\prod_{i=1}^{p} f_{mi}(0)\right) \quad (20.2)$$

$$
\begin{array}{ccc}
\prod_{i=1}^{p} f_{mi}(0) & \xrightarrow{\quad G_j \quad} & G_j \prod_{i=1}^{p} f_{mi}(0) = \prod_{i=1}^{p} f_{mi}(t) \\
G_k \Big\downarrow & & \Big\downarrow G_s \\
G_k \prod_{i=1}^{p} f_{mi}(0) & \xrightarrow{\quad Gj \quad} & G_s G_j \prod_{i=1}^{p} f_{mi}(0) = \prod_{i=1}^{p} f_{mi}(0)
\end{array}
\qquad (20.3)
$$

i) where the G's are topological Lie groups describing translation, rotation, Euclidian movement, affine, homographic, gauge and other topological transformations, etc., and require an equivalence relation between groups $G_k \equiv G_s$ that is represented in equations (20.2) and (20.3) where G_k and G_s are the two equivalent continuous groups, Gj is the continuous reference group and $f = (f_1 ,\ldots, fm_1 ,\ldots, fm_2 ,\ldots,$ (*n* terms)) are the computer input and output signals of a vectorial field U where $\exists$ operators $O_j(f(t))$ which constitute its Lie algebra of derivations;

ii) in which the optimal design of the machine for a physical system describable by a Lagrangian is already incorporated, as it can easily be shown that the continuous groups involved are functions of this type; and

iii) where the machine's underlying architecture is that of a unified, multiple ordered, parallel, non-linear analogue computer, able to utilized physio-chemical as well as electronic mechanisms, where no quantization of the input field is necessary.

Other ideas which may be reconstituted according to the rewrite paradigm include Spencer-Brown's Laws of Form (see 7.8);[32] the Dirac formalization of quantum mechanics in terms of bra and ket vectors (see 20.11); and the Alternative Natural Philosophy Association's discrete model of quantum and elementary particle physics, called the Combinatorial Hierarchy. In particular, from Kilmister's Brouwerian Foundation of the Combinatorial Hierarchy (CH),[32] based on Conway's generator for the surreals, and the extensive body of ANPA CH research, it can be hypothesized that the CH can be structured as a nilpotent computational rewrite system for quantum physics based on the two symbols 0 and 1, thus corresponding to another of Wheeler's well known prescient conceptions of 'It (the cosmos) from bit' (as is shown to be the case by means of the Von Neumann operator P in 20.6). Bastin's highly intuitive pre-CH conception that there must exist a computational foundation for quantum physics that led ANPA to the CH can therefore be seen to have been completely correct.[34]

20.9 A Boundary Condition and the Holographic Principle

In NQM, the empty state of 3 + 1 relativistic space-time and matter can be taken as a boundary condition – essential to the correct solution of any problem – for the quantum system that D(N)'s computational order code governs. This is in agreement with the requirement of nilpotence, that the quantizations also specify corresponding phases, so that these are the gauge invariant phases of the quantum mechanical state vector of the system. The empty state, taken as the boundary condition for NQM, thus has the full spatial and temporal quantum coherence necessary for holography in confirmation of the fact that a quantum holographic mechanism is specified in NQM by

$$(\pm\, ikE \pm \mathbf{ip} + jm) \text{ which has a Fourier transform } (\pm\, ikt \pm \mathbf{ir} + j\tau) \qquad (20.4)$$

The nilpotent operators here define the action of two sources of equivalent independent information, **p** or **r**, and E or t, equivalent respectively to amplitude and phase, relative respectively to the proper energy / rest mass (m) or equivalently the proper time (τ), either or both of which can be regarded as fixed / fixing a reference frame (and phase). The nilpotent structure thus admits 3-dimensionality, vector **r**, operating as a single unit, as well as other multi-dimensionalities (1-D, 3-D, 4-D, 5-D, 6-D, 8-D, 10-D and 11-D), depending on the perspective applied to the Clifford algebra of the rewrite language description.

However, in one perspective, they become 2-D, so that the minimum determining information, **r** and t in (20.4) is in agreement with the holographic principle and thermodynamics through the famous concept of a bounding 'area' which determines the relation between a system and the rest of the universe as a unit of thermodynamic information. The 'area' realized within the NQM nilpotent structure is that of the complex plane, involving **r** and t, which determines the nilpotent relation between the fermion state and the rest of the universe. It can be projected as a real area, because the fundamental dualities in the rewrite system (e.g. between amplitude and phase) allow the exchange of information about 2 spatial dimensions for 2 dimensions of space and time; and therefore, because **r** can also be considered as a 3-D quantity in its own right, this NQM minimum rewrite description is that of a 4-D boundary to a 5-D quantum system / universe. The match of phase and amplitude in the theory therefore explains the nature of holographic redundancy, while the 'reference phase' term in (20.4), $+jm$ or $j\tau$, is simply a result of nilpotence, and so is an automatic, not a separately imposed condition, meaning that the quantum holographic condition applies automatically to the fermionic state, and to all structures in which the fundamental code of Nature can be said to operate.

20.10 Quantum Holography

Nilpotency has been generally used, in this work, to refer to the case in which a mathematical object, when multiplied by itself, becomes zero. However, the more general meaning of nilpotency applies to any object X, in which a finite, repeated series of operations (not necessarily multiplication) leads to a zero result. An example is the 3-dimensional Heisenberg group (originally introduced by Weyl), which is a group of upper triangular matrices of the form:

$$\begin{pmatrix} 1 & a_1 & c_1 \\ 0 & 1 & b_1 \\ 0 & 0 & 1 \end{pmatrix} \tag{20.5}$$

where a, b, c are, say, integers, real numbers, or members of any (arbitrary) commutative ring. If they are integers, then this defines the nilpotent Heisenberg Lie group, G(N), which itself defines the 'Heisenberg uncertainty', and whose nilpotent Lie algebra Y, $Y \neq 0$; $Y^2 = 0$, as was known to Weyl, is the simplest such Lie algebra. Extensive researches by Walter Schempp,[35-41] concern the bosonic / Lie partition of nilpotent quantum mechanical state space, as governed by a description in terms of this group, which is nilpotent through its *lower central series*:

$$\begin{pmatrix} 0 & a_2 & c_2 \\ 0 & 0 & b_2 \\ 0 & 0 & 0 \end{pmatrix}, \begin{pmatrix} 0 & 0 & c_3 \\ 0 & 0 & 0 \\ 0 & 0 & 0 \end{pmatrix}, \begin{pmatrix} 0 & 0 & 0 \\ 0 & 0 & 0 \\ 0 & 0 & 0 \end{pmatrix}, \tag{20.6}$$

and its application to quantum holography. Here, the group is used to define nuclear magnetic resonance medical imaging (NMRI) as holographic full wavefront reconstruction in an adaptive resonant process. The description of quantum holography remarkably arises from the fact that, in relation to quantum phase, only phase difference is of physical importance, because each quantum state vector is only defined up to an arbitrary constant phase (i.e. is arbitrary up to an isomorphism). That is, the quantum holographic image encoding / decoding procedure must necessarily involve coherent mixing with a quantum reference signal beam, which defines its reference frame, and this is the role which the 3-D

Heisenberg nilpotent Lie group G(N) plays with regard to 3-D space in 3 + 1 relativistic spacc-time.

Nilpotence is exceptional in determining both the amplitude and phase of the quantum state vector, where phase is known to encode geometric information, i.e. that of 3 + 1 space-time as in a quantum hologram, as described here, and as also proposed in some current cosmological theories. As discovered by Schempp, Heisenberg uncertainty, together with the nilpotence, implies there is, respectively, both quantum (coherent) self-interference and the necessary corresponding zero energy reference frame or wave for quantum holography to take place. It cannot be a coincidence therefore that, at the postulated empty cosmological origin of such theories, there would be both the spatial and temporal quantum coherence necessary for holographic full wavefront reconstruction. That is, a cosmological origin of this kind would indeed constitute a quantum hologram, from which the cosmos itself would come into being, by full wavefront reconstruction, as in fact is evidenced by its 3-D spatial dimensionality from scientific measurement. In our terminology it could alternatively apply to simultaneous particle / vacuum creation, which is the ultimate source of all physical processes.

It is no coincidence that both the nilpotent Dirac and Heisenberg representations coexist as the fundamental basis of the nilpotent system, for in its quantum mechanical state space, they can now be related, respectively, to the required division of that nilpotent space into its complementary fermionic / Clifford and Lie / bosonic representations. Remarkably, therefore, in the nilpotent structure (NQM), in contrast to standard quantum mechanics (QM), Heisenberg 'uncertainty' is the actual optimal means by which to compute, and not an obstacle to computation as is usually considered to be the case, for the Lie nature of G(N) guarantees the existence of a dual G′(N) with Lie differentiable exponential mappings with differentiable inverses,[35,36] so as to ensure this encoding / decoding is indeed possible. In terms of well-established technologies, this enables the description, for example, of the control of the measurement process and the parameterization of MRI machines,[38] where holographic 2-D / 3-D image encoding and decoding is effected by fast symplectic Fourier transform action, synthetic aperture radars (SARs), etc.

As is the case with chaotic computation, where chaos can be used to minimize the number of steps to the specified proximity of the desired result, in NQM computation, Heisenberg uncertainty performs the same role, through Lie exponential diffeomorphic language description, or, as experimentally demonstrated, the optimal control of chemical reactions in a chemical soup so as to produce desired chemical output in real time, a process which formally corresponds to the solution in real time of the Schrödinger equation for the chemistry.[42-46] It is not entirely surprising, therefore, that the role that Heisenberg uncertainty performs in NQM in computing the 3 + 1 relativistic space-time trajectories of objects does not only therefore define geodesic behaviours, but also the ones in agreement with those of general relativity as defined in chapter 17. For, in NQM, where the Standard Model of elementary particles constitute the entire sources and sinks of the 3 + 1 relativistic space-time field, no additional elementary particles such as gravitons are necessary to explain 'gravitational' 3 + 1 space-time effects.[47]

Through the connection with G(N) and its dual G′(N), and the Lie / boson and Clifford / fermion partitions, MRI imaging and microscopy,[38-40] SARs,[35] and other established technologies, thus provide experimental support that NQM semantic mathematical language description includes a universal QM theory of holographic 2-D / 3-D image processing,[38-40] that the work on the nilpotent Dirac equation would lead us to expect, while an abundance of published research, beginning with the seminal work of Pribram,[48] in support of quantum bio- and neural computation, much of it based on Schempp's quantum holography governed by the 3-dimensional nilpotent Heisenberg Lie group, can now be appropriated in support of NUCRS / NQM. It would seem that the nilpotent version of quantum mechanics is the basis for a semantic theory of holographic pattern recognition, which can be conceptualized, as in Wheeler's now famous diagram, as a single eye looking at the body of itself. This may be taken together with the experimental support that the basic semantic ontology of the D(N), corresponds to the fundamental elementary particle structures of the 3 + 1 relativistic space-time physical world, from which, as far as is known experimentally, all other more complex physical / chemical / biochemical structures are derived, through the forces which the NUCRS defines. All this evidence points to the testable hypothesis, that 'nilpotent quantum mechanical language description provides, because of its universal grammar, the semantic descriptive basis for a theory of everything in agreement with experiment, such that in this NQM system, the quantum state vector is that of the universe itself, where should any further prediction of NQM semantic language capabilities, at all, fail to correspond with experiment, the hypothesis will be invalidated'.

20.11 The Bra and Ket Notation

Dirac's formalization of quantum mechanics by means of bra and ket vectors, as set out in his famous and foundational book on quantum mechanics,[49] representing once again the two fundamental operators of its description, can also be restructured as a nilpotent rewrite system for describing quantum mechanical computation. It, by implication, therefore describes not just quantum mechanical dynamics, but also includes quantum mechanical measurement and therefore a thermodynamic decoherent evolution or birthordering, in which the creation of 3 + 1 space-time and elementary particle matter is the fundamental first step. Additionally, too, the rewrite system indicates that the bra vector acts fundamentally as a quantum creation operator, and the ket, as a quantum annihilation operator, whereby this restores nilpotency so as to constitute an operation of proofreading. Thus, since the roles of the bra and ket operators may be reversed, the Dirac notation also includes what is called a Bargmann-Fock model for bosons and the harmonic analysis of the 3-dimensional Heisenberg nilpotent Lie group.[36]

In agreement with Perus and Bischof,[50] in the basic general equation of Dirac's bra / ket notation,

$$|\Psi\rangle = |\Psi\rangle\langle\Psi|\Psi\rangle,$$

the arguments in 20.10 show that the rightmost $|\Psi\rangle$ may represent a holographic output, such that the leftmost $|\Psi\rangle$ denotes the holographic input and $|\Psi\rangle\langle\Psi|$ the action of the associative holographic memory. It is seen therefore:

i) that in correspondence to their classical counterparts, quantum holographic procedures may be described with quantum wavefunctions, as is indeed the case in Schempp's quantum holography based on the 3-dimensional Heisenberg nilpotent Lie group;[36] and

ii) that 3-dimensional geometric space and generalized 3-dimensional spatial image processing are essentially ubiquitous to quantum mechanics, as is illustrated by the application of Schempp's quantum holography to the control of MRI systems in medical use worldwide.[38] Moreover quantum logic gates are not needed to engineer such MRI imaging dynamics.

In addition, if one then expands this basic general equation in the most obvious way, as below:

$$|\Psi\rangle = |\Psi\rangle\langle\Psi|\Psi\rangle\langle\Psi|\Psi\rangle\langle\Psi|\Psi\rangle\langle\Psi|\Psi\rangle\langle\Psi|\Psi\rangle\langle\Psi|\Psi\rangle\langle\Psi|\Psi\rangle\langle\Psi|\Psi\rangle, \quad (20.7)$$

it encapsulates the concept of an extended form of quantum holographic memory by four wave mixing, as is shown by the Frobenius-Schur-Godement identity,[37] where

$$\langle H_\nu(\psi, \varphi;.,.) | H_{\nu'}(\varphi', \psi';.,.)\rangle = \langle \psi\otimes\varphi | \psi'\otimes\varphi'\rangle \quad (\nu = \nu' \neq 0)$$
$$= 0 \qquad (0 \neq \nu \neq \nu' \neq 0) \qquad (20.8)$$

This says that the range of frequencies between ν and ν' allow an adaptive resonant coupling, so specifying a spectrum of very narrow spectral windows, where ψ, φ, ψ', φ' are the quantum wave amplitudes belonging to the complex Hilbert space $L^2(R,t)$, and $H_\nu(\psi,\varphi;.,.)$ are the Liouville densities of the corresponding distributions. It follows therefore that there will be little or no cross talk between, for example in the photon case, the asynchronous collective photonic excitation distributions located in the different hologram planes $(R\otimes R,\Omega_\nu)$, $(R\otimes R,\Omega_{\nu'})$, where the four wavelet mixing ψ, φ, ψ', φ' takes place, so as to make subsequent full wavefront holographic reconstruction possible, that is, so as to constitute a quantum holographic memory that can be both written and read.

It is worth pointing out that all interactions between particles (including vacuum ones) have the same form, with incoming fermion and incoming antifermion (or outgoing fermion) creating a bosonic state at the vertex, and that this applies to the equations above, which are parallel to the automatic fermion antifermion fermion antifermion ... structure of the nilpotent operators acting on vacuum. The system thus creates an automatic and exact supersymmetry of the fermion and its (own) representation as a boson (see 6.11).[51]

20.12 The Universe as a Quantum Thermodynamic Engine

In classical thermodynamics, irreversibility is described by defining the Carnot engine (whose cycle consists of alternate isothermal, or constant temperature, and adiabatic, or constant entropy, phases) as the one that is theoretically most efficient; and one of the many statements of the second law of thermodynamics is that no engine can be more efficient than a Carnot engine. To ensure the incorporation of thermodynamic with quantum and classical dynamical laws, we may propose that the universe is treated as a *quantum* Carnot engine (QCE), consisting of a single heat bath in which the ensemble of elementary particles retains a small amount of quantum coherence, so as to constitute new states of matter, called by Scully *et al.* a 'phaseonium',[52] the first of which is its *Zenergy*, the empty ensemble of dark energy. The explanation of how this initial phaseonium fuels the natural structure of the universe, may be shown to be in complete accord with the existence of a 3 + 1 space-time, the known quantizations of the elementary particles, and galactic structure.

Of particular significance to the development of higher order 'Natural Structures', is the idea that the thermodynamic evolution of the universe is the expression of the way that the unique ordering (birthordering) of quantum events is managed, and that this ordering is what we mean by 'absolute (nonobservable quantum mechanical) time' or 'causality'. The nilpotent universe is totally entangled in a quantum mechanical sense, each nilpotent state requiring correlation with the rest of the universe to complete its description of energy-momentum conservation. There is no closed, isolated, system, and no true simultaneity at the quantum level, only an expression of irreversibility in terms of entropy or decoherence. It seems to be a (Fourth) 'Law of Thermodynamics' that an absolute sequencing of 'events' (involving information transfer) is only possible because entropy change is minimised in a universe structured as a quantum Carnot engine. And entropy is minimised because the QCE operates according to 'Nature's code', the most efficient possible information processing system.

As nilpotent quantum mechanics (NQM) behaviours include quantum coherence, NQM systems will behave quantum coherently as quantum mechanical Carnot engines, and so always include a component of thermodynamic behaviour, where this is governed by the phase of the quantum coherence. This factor thus needs consideration when Heisenberg uncertainty is taken into account. It says that thermodynamic considerations in NQM will almost certainly play a role in relation to NQM general relativistic behaviours as a mechanism by means of which such behaviour is quantum mechanically achieved, in contrast again to other QM models which do not usually include quantum thermodynamic considerations. In this nilpotent system, also, it must be constant arbitrary phase which represents quantum coherence, so as to constitute the 'phaseonium' with the potentiality of the infinite degrees of freedom necessary to its quantum Carnot engine (QCE) evolution, which is indeed universal ('constant' here means invariant, i.e. fixed as in a fixed past rather than forever unchanging).

The postulate of the universe as a quantum Carnot engine (QCE) is in conformity with: the existing thermodynamic understanding of the universe (described as all that exists) as a heat bath where black body radiation conforms to Planck's radiation law based on the quantum of action h; and also with the theoretical findings that there exists in relation to the generalized nilpotent Dirac algebra within NQM, a symmetry breaking from which emerge 3 + 1 space-time and matter in the form of the known experimentally validated strong, electromagnetic and weak quantizations of the elementary particles, such that these quantizations are exactly those that ensure the compatibility between quantum mechanics and general relativity. This symmetry breaking may be said to concern that between the 4-vector algebra and the algebra of the quaternions, so as to constitute an elementary particle 'phaseonium', in the sense of Mach's equivalence principle, since, from the mathematical condition of nilpotence of the generalized Dirac algebra, this phaseonium can be inferred to be a direct consequence of the QCE action of the universe on its original dark energy or 'Zenergy'. (This follows because the condition of nilpotence requires an emitter / absorber model described in terms of quantum creation and annihilation operations.)

This action, which is able to account for the initial creation of material structure of the universe, therefore results in a modified globally isotropic phaseonium almost 'perfectly dynamically balanced' between the 3 + 1 space-time field and the complementary ensemble of elementary particles, where the initial phaseonium / Zenergy, is imprinted with the 3 + 1 space-time field and what can now be described as space-time's most elementary / primitive sources and sinks, so as to prepare the universe for further cycles of its postulated QCE action. These cycles can be thus be inferred as ones of further symmetry breaking and creation of structure, or critical phenomena,[53-54] which will again be imprinted on the modified Zenergy, further irreversibly modifying it. They are processes, in conformity with the experimental discovery of the universe's current background radiation,[55-56] and also the second law of thermodynamics.[51] They provide a further experimental testable hypothesis that there should exist a spectrum of 'perfectly dynamically balanced' thermodynamic structure indicative of these breakings of symmetry imprinted on the current Zenergy, sometimes referred to as the energy of the dynamic quantum vacuum, and on the 3 + 1 space-time field in the sense of Einstein's general relativity. Yet while this thermodynamic model is essentially adiabatic, so that entropy production can be both created and destroyed so as to constitutes an information metric,[57-58] the second law of thermodynamics holds,[51] showing that quantum mechanics is not in fact, as is generally believed, globally reversible. The QCE model must therefore describe what Conway calls a unique birthordering[16] in confirmation of the mathematical fact that *any* universal model of a theory (i.e quantum mechanics in this case) in a language of sets, has a unique birth-order field automorphism.[59]

It is important to realise that the thermodynamic quantum action of the QCE relates to the state vector of the quantum mechanical system called a phaseonium, where the ensemble of particle states retain a small amount of quantum coherence, and so to the observables of this state vector which are gauge invariant geometric phases and describe 3 + 1 relativistic geometric structures. That is to say the phaseonium action of the QCE describes the behaviour of the quantum mechanical system it describes, and not simply the behaviour of the sum of its parts, as is the case in statistical mechanics. A well known example of this distinction is the behaviour of the Ising model.

20.13 The Riemann Zeta Function

The universal computational nilpotent rewrite system, and the fact, as shown by Deutsch, that universal computation is now recognized to be fundamentally a physical process, together suggest a novel physical perspective for viewing the Riemann hypothesis. According to this perspective, quantum coherence / nonlocality is the sole origin for the nilpotent quantum cosmology outlined in the previous section, and, as the 4-vector representation indicates, this cosmology is general relativistic. This conclusion would be in strong agreement with all the evidence of experimental cosmological and elementary particle physics, for, while there exists no confirmed experimental evidence of incompatible physics beyond, there is much in favour that the condition of nilpotency provides, such as a zero total energy, which is the unaccounted for stumbling block to current cosmological theory. It may therefore be hypothesized that nilpotence is the key to proving the Riemann hypothesis, through the properties of the nilpotent quantum mechanical state space of the above hypothesized cosmology or physical system.

Now, the Pauli exclusion principle tells us that the complementary Clifford / fermionic state space provides the canonical labelling required such that the quantum holographic image informational processing described in 20.10 constitutes computation. That is, the gauge invariant geometric phases appropriate to the entire nilpotent state space and, more particularly to quantum holography, must, by the exclusion principle, all lie on the line spin ½ and so can be identified with the zeros of the Riemann zeta function. The Pauli exclusion principle, together with the universality of this nilpotent state space, which corresponds to the universal computational nilpotent rewrite system, thus indicate that the Riemann hypothesis must be true!

According to the universal rewrite system, integers emerge at the same time as nilpotents and the Grassmann-Hilbert space which determines their antisymmetric nature. The zeta function could thus be considered *each* nilpotent's coding of the integers, with an arbitrary phase related to their unique specification in terms of antisymmetric wavefunctions and Pauli exclusion. Since this arbitrary phase may well be determined by the E / $\mathbf{p}$ relationship, it would not be totally unexpected for the zeros of the zeta function to correspond to the energy levels for quantum chaos (see 20.15).

Moreover, the universal rewrite nilpotent system confirms the finding that the nilpotent Dirac equation has a proper time. The nilpotent state space, therefore, has global time reversal asymmetry as is implied by its QCE evolution. Thus not only do the distribution of the primes specified by the zeta function play a game of chance representing the optimum strategy for survival in the specified proper time evolution, but the 3-dimensional structures which correspond to each zero of the zeta function, are universal invariants / constants of the entire state space! That is, to say, the zeta function constitutes a standing wave in the space, and this standing wave which concerns a mapping of all the integers, is therefore the projection of universal quantum onto universal digital computation.

Conversely, in any quantum mechanical state space, because of the required division of the space into its Clifford / fermionic and Lie / bosonic parts, the Pauli exclusion principle implies there must always exist unique 'global' fermionic spin states lying on the line spin ½, and that such states, must, if the Riemann hypothesis is true, necessarily define zeros of the zeta function so as to constitute the 'global' nilpotents of that space, where quantum coherence is necessarily non-zero. That is, the 'global' gauge invariant phase[60-61] corresponding to each zero will have quantum holographic properties, and be such that, at this zero, phase conjugation (where the image of the object coincides with itself)[36] can take place. Remarkably, too, while at each of these zeros there is locally time-reversal symmetry, because of the Pauli exclusion principle, they are susceptible to re-arrangement in a definite order, so it can be postulated, they have a proper global time ordering.

A further perspective on the zeta function is that it can be viewed as a generic calculational principle for describing the solution space for problem solving in NQM in relation to the generalized nilpotent Dirac operator of equation (20.1).

For in relation to (20.1) the solution procedure in principle is to write down the mathematical descriptions of E and $\mathbf{p}$ for the particular fermionic system under investigation, where the associated phase exp $-i(Et + \mathbf{p.r})$ must then be chosen in such a way that (20.1) is nilpotent, i.e. such that the phase is a gauge geometric one. In this way the total of contribution (of the particular fermionic system under investigation) to the Hamiltonian of the whole universe is such that $\mathcal{H}$ remains zero, as it must.

20.14 Galactic Structure

It is known from observation that the structure of the universe constitutes an ensemble of galaxies. From the postulate that the universe as a QCE is the source of all natural structure, constituting a relativistic, almost perfectly balanced phaseonium (rapbp), it can be inferred that each galaxy is itself a rapbp and a QCE, and is a secondary source of the whole natural structure making up the universe. This deduction is in complete agreement with the 'phaseonium' sum of histories approach to quantum mechanics, which Feynman conceptualized in terms of Huygens' principle of secondary sources, and with Everett's many worlds interpretation, except that now each of these many worlds is an actual galaxy or secondary source of the whole universe as all that exists. That is to say, the galaxies are the 'parallel universes' many of which are now visible to us all by various types of telescope. Possible evidence for this rapbp deduction is the spiral (i.e. phase) nature of many of them. It is one that can be tested in relation to equation (1) of Scully,[51] since this equation asserts that such model galaxies as QCEs must exhibit a well defined quantization in relation to radiation pressure on their natural global structure. Further, since the source of the universe as QCE is, as deduced above, effectively a 'white hole' from which emerges the ensemble of elementary particles throughout 3 + 1 space-time, it can be inferred that this 'white (inertial / 'dynamic') hole' must be in almost perfectly balanced equilibrium with the 'black (gravitational) holes' of the secondary sources of its ensemble of galaxies as QCEs, as Huygens' principle would require in relation to the Zenergy. That is, black holes are to be considered now as absorbers of both matter and 3 + 1 space-time, complementary to the emergent white whole.

This explanation, therefore, not only describes the nature of Mach's principle in relation to the equivalence of inertial and gravitational mass (for which see chapter 21), it also shows that the QCE phaseonium model could account for the need for the introduction of the so-called cosmological constant into Einstein's equations of general relativity (see 21.1), while saying that the initial Zenergy phaseonium can almost certainly be described as a universal topological pre-space, so as to explain how the evolving universe as a QCE composed of many worlds is connected together quantum mechanically. It also serves to assert that the question of the equivalence of the well known Schrödinger and Heisenberg representations of quantum mechanics must concern the set of topological equivalence classes. It can no longer be assumed, therefore, that the use of one such representation or the other is simply a matter of convenience, for, as the above arguments show, the whole of the universe may derive its entire structure as a consequence of these topological equivalence relations. It says that the SETI search for intelligent life should be undertaken using pre-spatial quantum mechanical communication via quantum teleportation, and confirms the nature of the QCE as a quantum universal computer constructor model.

In addition, it is a model, which, it appears, remarkably combines both that of Hawking in relation to galactic black holes, with that of Hoyle in respect to the 'continuous creation' of matter. That is to say structure (matter and 3 + 1 space time) absorbed in a black hole has the potentiality to re-emerge in the white (w)hole of 3 + 1 space-time. This could, for example, provide an explanation for the unexplained phenomenon of very high energy cosmic rays. It constitutes quantum tunnelling 'on the grand scale' as the explanation of the creation of 3 + 1 space-time and the nature of matter in accordance with the nilpotent model, its eventual destruction in black holes, and its topologically regulated adaptive re-emergence. Thus the universe as a rapbp would allow a finite fixed quantity of matter to be continually recycled in 3 + 1 space-time to produce structures of greater and greater complexity. Huygens' principle also implies, in order to correctly describe such a model of Zenergy without any or only a vestige of 3 + 1 space-time, that, the galaxies as secondary sources must be sited on a closed surface S described in terms of 3 + 1 space-time, in agreement with the well-known conceptual illustration of the universe as an expanding 'matter bubble'.[23] Similarly, from observation, it is known that such galaxies are an ensemble of stars, which are themselves quantum mechanical ensembles since they produce their energy as a result of nuclear fusion. Thus the stars can also be inferred to be QCEs which are in almost perfectly balanced equilibrium throughout the majority of their lifetimes.

The overall conclusion that can be drawn, therefore, is, that if the universe is itself a QCE, it is also a hierarchy of QCEs, concerning galaxies, stars, solar systems, and eventually biospheres like our planet Earth, all of which are almost perfectly balanced systems, i.e. rapbps in the sense given above. For example, a solar system like our own will itself constitute a 'phaseonium' of orbiting planets, as Kepler demonstrated in relation to the deduction of his laws from the astronomical observations of Tycho Brahe, or as indeed did Newton subsequently in relation to his law of gravitation, when he was forced to assume instantaneous action at a distance. And the QCE idea may extend further, since, from neuron science, it is known, from the very early work of Ramon y Cajal on the brain, that neurons constitute essentially dynamically 'independent structures'. The hypothesis, therefore, that the brain is a QCE consisting of a phaseonium in which the ensemble of neurons retain a small amount of quantum coherence, fits very well with Cajal's finding and the later work of Eccles, Pribram, Hameroff and Penrose, to name but a few. And since each neuron is a type of living cell, it can be further inferred, that any organism is a QCE ensemble of living cells, which again would be themselves QCEs.

The descriptive vehicle of the QCE thus explains exactly how and in what sense any physical system can be more than the sum of its classical parts, as set out, for example, above in relation to the universe itself as a QCE, that is that the universe as a system functions additionally as a white (w)hole, and that it is this contribution of the white (w)hole that ensures the presence of both matter and 3 + 1 space-time, whereas the corresponding classical universe would be fluid and formless. It also establishes the fundamental importance of the recycling of matter and energy to all QCEs if they are to remain rapbps, i.e. relativistic almost perfectly balanced phaseonia, within the universe as a QCE. In effect, mankind's survival as a QCE is constrained by the universal QCE imperative, from which ultimately there is no escape!

20.15 Quantum Thermodynamics and Evolution

The quantum nilpotent universe's state vector at its initial boundary condition, the empty state of 3 + 1 space time and matter, is defined by a single parameter, the phase ϕ of its quantum coherence, appropriate to its NQM description as pure quantum Carnot engine QCE(N).[52] Such a QCE(N) description differs from that of the well-known classical thermodynamic Carnot engine (CCE), by the possession of quantum thermodynamic behaviour governed solely by ϕ. And ϕ as is well known, is for any quantum system arbitrary up to a fixed phase, but its relative phases are the invariant phases of its state vector (cf section 6.4). This NQM boundary condition thus defines a nilpotent quantum thermodynamic evolutionary cosmos described in terms of a single quantum Carnot engine QCE(N), in good agreement with the properties of the nilpotent computational rewrite system that has been demonstrated, where the initial nilpotent quantum preparation is the empty state of 3 + 1 relativistic space-time and matter and each subsequent thermodynamic measurement cycle of QCE(N) is the preparation for the next.

From the fact that this initial preparation results in emergence of the two basic phenomena, 3 + 1 space-time and elementary particle physics, from their empty state, it can be inferred that both the second and third laws of thermodynamics will hold and that the evolutionary quantum cosmology QCE(N) described is an irreversible process, where the initial arbitary fixed phase ϕ provides the measurement standard for all subsequent measurements, and where the initial and subsequent quantum preparations account for the irreversibility.[52] Such conclusions are in agreement with previous research, including that by Deutsch[8] into universal quantum computation that the second and third laws of thermodynamics hold, and that by Berry[62] that there exists an unknown quantum mechanical system with time reversal asymmetry of which the phase space trajectories are chaotic, such that its self adjoint Hamiltonian energy function has eigenvalues / quantizations corresponding to the imaginary parts of the nontrivial zeros of the Riemann zeta function and all lie on the line $x = ½$, if the Riemann hypothesis is true.

That is to say, there exists a quantum coherent quantum chaotic system consisting of a single fermion state with spin = ½, where all its quantizations correspond to gauge invariant phases of its state vector so that they are imaginary and lie on the line $x = ½$, as do the ground states of those, it can be hypothesized, of the irreversible evolutionary nilpotent cosmology QCE(N), as proposed here. The Riemann zeta function can thus be envisaged as a de Broglie pilot ‘standing’ wave that guides the overall evolution, in line with de Broglie’s vision that the principle of least action and the second law of thermodynamics act in a like manner, where respectively the energy $E = hv$ and the entropy $S = kT$.[28] The zeta function’s known overriding critical role in mathematical number theory, is thus a further confirmation of the claim that the universal nilpotent rewrite system provides a new fundamental foundation NQM for both quantum physics and mathematics in terms of their semantic language description.

The universal nilpotent rewrite system, with its semantic approach to a computational language foundation of both physics and mathematics (thus explaining the undue effectiveness of mathematical language in physics), together with the evidence in support of the QCE(N), appears to furnish the highly likely missing links, including the experimentally well validated classical thermodynamic Carnot engine, needed for an improved testable modelling of physical systems, and their evolution, which is more than sufficient to warrant, we believe, their extended theoretical and experimental investigation. A particular example of such a testable missing link is the QCE(N) model of the biosphere, that could advance a more correct understanding of global warming, and of the critical importance of biodiversity.

The view presented here of the QCE(N) as the single phase source of the universe's cosmological evolution, etc., also requires, in line with Feynman's sum of histories approach to quantum mechanics, that this phase action follows that of Huygens' principle of secondary sources, as formalized by Jessel.[21,60] For, as can be shown, the combined effect of such a source together with its secondary sources (also QCE(N)s) on a 3 + 1 space-time surface, can then be such that the source and its secondary sources cancel each other out, so as to satisfy the nilpotent criterion. An example of such a Huygens' cancellation is the phenomenon of anti-sound in relation to sound fields, by means of which a source can be nullified dynamically throughout an entire 3-dimensional space, by secondary sources of sound on its bounding surface.[63] The evidence for such phase action in the case of the cosmos thus comes from observed galactic structure; for, as seen through the most powerful telescopes, the galaxies, as secondary sources of their cosmic source, will undoubtedly each correspond to a QCE(N) on the cosmic 3 + 1 space-time surface, such that the phase of quantum coherence of the cosmos connects them all together, as in the Everett interpretation of quantum mechanics, but in such a way that they are not hidden from one another, as is usually envisaged for 'parallel universes'. Such a model, based on Huygens' principle, thus says that all QCE(N)s are in fact part of the evolving cosmological 3 + 1 relativistic space-time wavefront, plus subsequent wavefronts, with their secondary sources many times removed; and that as this quantum coherent phase precedes the original emergence of 3 + 1 space-time and matter, it could account for the so-called 'dark energy' effects, observed through the acceleration term imposed on the cosmological redshift (see chapter 21).

That is, dark energy must constitute quantum coherence since at this postulated origin of the universe, there is, implicitly from the nilpotent Dirac equation, both the 3-D spatial, and the temporal quantum coherence sufficient for holography, i.e. for full quantum holographic wavefront reconstruction in some hologram plane. This conclusion follows from the well-known quantum mechanical fact that, although the phase of any quantum wavefunction is arbitrary up to a constant phase factor, the phase difference between two wavefunctions is of physical significance (as the geometric / Berry phase, discovered by Berry,[60] shows). That is, these conditions satisfy the requirement of quantum holographic image encoding / decoding procedures, which need the mixing of a coherent reference signal beam (as occurs for example in Mach-Zehnder interferometry) to incrementally record (in the case of encoding) the phase of the object signal beam in the hologram plane so as to form a hologram: a condition that occurs spontaneously in the above circumstances at the point of phase conjugation. And this would therefore result in phase conjugate adaptive resonance, so as to provide a resonance for creation and subsequent adaptive evolution / birthordering.

Such quantum coherence, therefore, not only assures the basic material composition of the universe upon symmetry-breaking, as described by the nilpotent algebra in terms of 3 + 1 space-time and elementary particle matter as it is seen today, and, as far as is known, as has always been in the case in the past, but requires a truly quantum mechanical system / universe. Furthermore, the spatial and temporal quantum coherence necessary for this full wavefront reconstruction, says that this universe can be considered as constituting a quantum hologram; a conclusion in excellent accord with various findings in regard to string / membrane theory. String theory is however only a quantized classical description, which encompasses the four fundamental properties of mass and charge and their corresponding force fields. It does not provide the basis, as does quantum coherence, as dark energy, for a true quantum thermodynamic description of the universe as a description which is quantum holographic; however, we believe nilpotent theory encompasses a 10-D string theory without strings (see 18.6). Such a thermodynamic description is that of a quantum Carnot engine, which will evolve in ways that a classical thermodynamic description of the universe cannot. In particular, therefore, the work on NQM shows that any classical model of the universe would be empty, that is, have neither 3 + 1 space-time nor elementary particle matter, unless these were independently and separately assumed to exist prior to the creation event. And such a classical universe would therefore remain empty so as to be totally without interest. The nilpotent universe is the only possible description that can explain the origin of the universe that we observe today.

Further evidence of this evolution, through the universal rewrite system's computational ability to rewrite the basic descriptions for D(N), in the form of more complex 3 + 1 space-time structures, composed of Standard Model elementary particle matter, is, confirmed by the publication by Stewart[64] of a galactic-like spiral presentation for the periodic table of the atoms / nuclei of the elements. This presentation is in line with their known experimental chemical properties (see Figure 20.1), which, as far as is known, are explained and completely understood in terms of the four forces and particles of the NQM model. The success of this spiral presentation can thus in NQM be attributed to the above QCE(N) evolution as governed by ϕ, the phase of its quantum coherence. For such an evolution would be expected to concern the simplest unit exhibiting all the NQM forces, namely neutronium / the neutron, from which, in the new spiral conception of the periodic table (Figure 20.1), all the elements are then envisaged as evolving, as described by the nilpotent universal rewrite procedure, where at this level of ontological complexity the rewrite sub-alphabet concerns the two fundamental nuclear processes of fusion and fission as its production rules, and where the gaps in the spiral for larger configurations of neutronium are due to nuclear instability kicking in.

It is interesting to recall, here, the speculation of William Crookes, in his 1886 Presidential Address to the Chemical Section of the British Association,[65] that a spiral representation of the periodic table could be explained in terms of a progressive evolutionary genesis of the elements as a result of two forces, one 'operating uniformly in accordance with a continuous fall of temperature', and the other, showing a sinusoidal variation (simple harmonic oscillator) connected with the electric force, together producing a (double helical) generation of elements of increasing atomic mass, but periodically similar chemical properties. This speculation, which predated the discovery of basic atomic structure, the electron, the proton, and the theory of stellar genesis, but predicted isotopes and accommodated the (undiscovered) inert gases, could be seen now as exemplifying the characteristic actions of Nature when operating according to the (then unknown) universal rewrite system, or most efficient natural information processor.

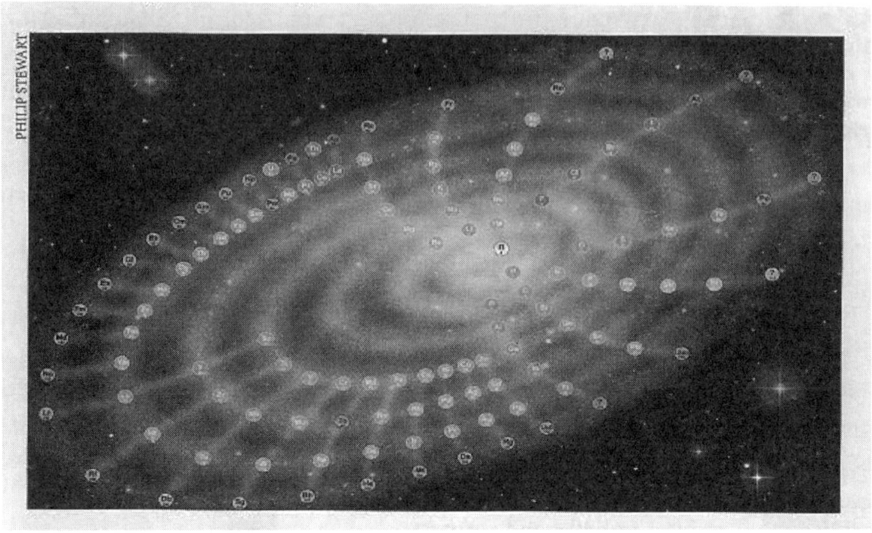

Figure 20.1 This new spiral presentation for the periodic table begins with neutronium, not usually considered an element (but which is cosmically as abundant as oxygen). It situates hydrogen next to carbon which chemically it most resembles. Such a spiral emphasizing the fact that the elements form a continuum, rather than a series of blocks, is in excellent accord with QCE(N) where the quantum phase ϕ follows such a time reversal asymmetric spiral behaviour. (*Illustration, courtesy of Philip Stewart.*)

20.16 DNA as a Rewrite System

The universal nilpotent rewrite system, has implication far beyond physics and cosmology, in such things as living systems, where it applies the most efficient information processing system possible, fractally, at a new level. In particular it follows from its fundamental nature and previous research,[66-68] that it can be hypothesized that the double helix structure of DNA with its base pairings of nucleotides known in their symbolic form as the genetic code (in an explanation relevant also to RNA) is a minimal chemical rewrite system for quantum holographic computation, such that the chemical base pairings constitute its two rules or productions (see also chapter 19). The two fundamental rewrite productions of RNA / DNA, its create and conserve operations, are instantiated by the base-pairings (A = U in RNA which rewrites A = T in DNA) and G ≡ C, respectively, where A, U, T, G and C are the usual biomolecular structures:

adenine, uracil, thymine, guanine and cytosine. That is, the base pairing A = U symbolizes the fact that some initial state must be rewritten as in RNA action where A = T delivers the new symbol, which may be a single character, a subset alphabet or an entire alphabet. That is, it represents the create operation, so that by contrast G ≡ C symbolizes the conserve operation, which examines / proof-reads all symbols currently in existence to ensure no anomalies exist as a consequence of the bringing into existence of the new symbol. This rule therefore verifies / maintains the conservation of 3-dimensional chemical structural stability (such that Dirac nilpotence is maintained !!) in the process of the development of the human embryo via cell division, i.e. it defines natural selection. Thus, at the commencement of human development, A (adenine) may be said to symbolize the human individual to be delivered (i.e. born) as a single stable 3 + 1 dimensional chemical unit or structure, once the entire genetic alphabet of that individual, specified by the subset alphabet of the 46 human chromosomes / symbols has been fully realized, or an anomaly occurs such that the development is aborted.

The DNA / RNA genetic code as an NUCRS would thus explain all living systems (including, of course, the biological human brain) as the product of further intermediate levels of the NUCRS hierarchical semantic ontology, for it is through the relatively fragile structures of these base pairings that the actions of the whole of information transfer throughout the biosphere takes place (including, it can be suggested, that of 3-D holographical imagery[54,66-80]), so that in NQM they are truly the cipher of life. It would also answer the grand unsolved problem of the genetic code, as to how the 3 + 1 space-time structure of its organisms is encoded within it. Just as the periodic table of the elements is the semantic ontology / 'space-time furniture' corresponding to a subalphabet, then the RNA / DNA genetic code for life, and human natural language would both correspond, in the rewrite hierarchy of alphabets, to potentially infinite universal alphabets, which describe rewrites of already existing 'space-time furniture', so that in living systems, the RNA / DNA genetic code would describe 'the biological hierarchy of the space-time furniture of living systems as governed by RNA / DNA, now a semantic / semiotic genetic code, and which would be a generalized computational rewrite methodology in its own right'.[66-69] Living systems as hierarchies of QCEs would retain various degrees of quantum coherence (as does the universe as a QCE), and so would be macroscopic quantum rather than classical machines, where such degrees of quantum coherence constitute the vital difference, as already postulated by many authors.

The replicating structure of DNA, in fact, comes to mind as a 'perfectly dynamically balanced' phaseonium. It is then a natural extension of this hypothesis of DNA as a QCE, the cycles of which enable it to replicate itself, to that of living systems as similar self-replicating QCEs, where the cycles of this self-replication are controlled by their DNA such that, in the case of prokaryote non-nucleated primitive cells, this is a simple replication of the cell itself including its DNA as a QCE, while, in the case of the eukaryote nucleated multi-celled organism, this replication, while still corresponding to the control cycles of its DNA, now concerns the organism's eukaryote celled structure itself as a QCE. From this it may be inferred that each of the nuclei of such cells with its chromatin is there to protect and maintain the perfect dynamic balance of its DNA during the now significantly longer and more complex process of development of the embryo of its organism, so as to ensure the 'gauge invariance' of its DNA throughout the process of replicating of the organism.

These imprinted / encoded phaseonia would indeed fuel the structures of life (in an analogous manner to the way in which 3 + 1 space-time and the quantizations of the elementary particles are fuelled from the Zenergy in the case of the universe). This prediction is in excellent agreement with research[66,68,72,81,82] which shows DNA to be imprinted with the quantum holographic phase information necessary to reconstruct the 3 + 1 space-time structure of the particular dynamic organism for which its genetic code describes the necessary control cycles. That is, these models also provide detailed explanations of the structure of DNA as a phaseonium such that the quantum phase encoding of the holographic images of its organism is sufficient to fuel the 3 + 1 space-time construction of its embryo, since in such quantum holographic models[35-37] it is the 3-dimensional Heisenberg nilpotent Lie group which specifies the 3-dimensional spatial gauge invariance referred to above, and it is this gauge invariance which introduces holographic signal theory into quantum physics.

Thus equations (20.7) and (20.8) of section 20.11 would describe DNA structure in helical form as an 'organized read / write book' of pages corresponding to very narrow spectral windows with no cross talk between them, where these pages contain quantum holograms each encoded on a hologram plane R⊕R, coordinates (x, y), according to reference beam angle $dx \wedge dy$. This provides further confirmation, for, in DNA / RNA structure, the base pairings do indeed occur in planes, where, as phase conjugation would require, each DNA helix acts as the reference read / write beam for the other, and each page would be described as a holographic trace transform implying that $\phi dt = c\phi dt$, where c

denotes a constant relative phase factor appropriate to a gauge invariant geometric relative phase between the different pages, thus showing that DNA does indeed constitute, a read / write quantum holographic memory and filter bank, a phaseonium and a QCE.

The key feature of the nilpotent universal computational semantic rewrite system is that each emergent symbol of its infinite universal alphabet may also stand for its universal alphabet (subject to the nilpotent closure of the previous set of symbols, so that the new symbol corresponds to an empty state). This is to say that, in the resulting hierarchy of single symbols, sub-alphabets, and infinite universal alphabets, as a consequence of the nilpotent rewriting process, each recommencement of the universal alphabet corresponds to a complete repetition (at a higher level of semantic ontological physical structure) of the nilpotent universal computational semantic rewrite system. This repetition is, therefore, in common with the original infinite universal alphabet, realized by a QCE(N), where, in the example already given, of the RNA / DNA genetic code, and in the probable extension to human natural language capabilities (see 20.18), the respective QCE(N)s are those of the prokaryote[72] and eukaryote living cells and of the human brain.[73-80] The further implication is that all living organisms, and indeed many of their subsystems, are QCE(N)s, including human organs, such as the human brain, and indeed, as we will now show, the heart.

An example that should be comparatively easy to demonstrate as a QCE, as there is extensive experimental data and technological diagnosis equipment already available, is the heart-lung system (H-LS) which fuels the body with oxygen, nutrients, and is part of its immune system. The new model should certainly explain those unexplained features of the H-LS in respect of which it differs from that of conventional mechanical fluid pumps, such as, for example, why it has 4 chambers. The now postulated partial quantum coherence of the QCE in view of, for example, the work of Berry in relation to the geometric phase and quantum chaos,[60] and as is evidenced in relation to the Mandelbrot set, which demonstrates a boundary between fractal and wave structures, would imply that such a complex QCE as the heart will show evidence of 'cycles of chaotic working' where this chaos can thus be tuned in relation to the pumping of the blood now to the many rhythms of the body blood fluid complex of organs, veins and arteries in what can be shown to be the most efficient way to maintain the organism in health. It would also say that the blood itself is a phaseonium or active fluid in the sense of possessing a partial quantum coherence, and indeed actual macroscopic quanta, such as its red blood cells, so as to be able to carry the vital supplies of oxygen to those places where at any particular time they are most needed, or similarly for the white cells of the immune system. Again, the frequency organization described in equations (20.7) and (20.8) would provide an appropriate basic model against which to test the above hypothesis, where the model's Liouville densities are the holographic transform and its corresponding trace transform for the fluid wave flows appropriate to the red blood cells and white immune cells, respectively. That is to say, the primary function of the immune system is not just to ensure the optimum oxygenation over the whole body, whatever activity the body is engaged in, but to ensure the integrity of body's geometric structural wholeness, with reference to its holographic geometric ground plan, which would be held throughout the body in the form of its cellular DNA.

In relation to the human fluid circulation system, there can be little doubt that the heart is not just a mechanical amplitude pump for driving the blood, as simply an ordinary fluid, around the body's veins and arteries as fixed pipes, but that it is a 'phase' pump QCE(N) for the whole of the individual human organism, which moves the blood as now a 'living' fluid coherently around veins and arteries with their own living pumping actions, so as to optimize the entire fluid action of the body and brain in relation to whatsoever task the human organism is performing. The overall conclusion that can be reached from the nature of the QCE(N) model is, therefore, that all the living systems above are individual quantum mechanically coherent systems, and that the nilpotent universal semantic rewrite system methodology has the capability to describe quantum computational units, such as QCE(N)s, that are both computer universal and computer constructor universal, so that the latter are able to make replicas of themselves, to produce new generations of the living organism in question.

All the above evidence in relation to the postulate of living systems as QCEs, suggests that QCEs are optimally controlled quantum systems or machines. Since Schempp's quantum holography describes the working of MRI machines[38-41] in widespread use throughout the world for medical diagnosis, such that the process of extracting the desired medical images results in diffraction patterns, which are easily demonstrated to be holographic in nature, it should be comparatively easy to demonstrate experimentally that these machine are indeed QCEs too. Such machines are also known from the extensive research of Dubois[83] to constitute universal fractal, incursive, hyperincursive and anticipatory computation, new conceptions which have now found considerable academic support.

20.17 Brains as Quantum Carnot Engines

A QCE explanation of the system of the human brain as conscious engine may also be given, at the semantic ontological level of its biological space-time neural furniture. Human thought can now be inferred to be quantum measurement, computational input / quantum preparation followed by computational output, so that it is able to function semiotically as a neural computational rewrite system, as is evidenced by its natural semantic language capabilities. So, in the mathematical language of category theory, the arrows of the theory can thus be quite literally taken as formally representing 'such arrows of human thought' so as to describe the rules which govern it. (The fact that QCEs can operate at any constant temperature, i.e. are 'hot running'[52] completely invalidates the 'often invoked' argument that brains as hot thermodynamic systems cannot operate in accordance with quantum mechanical principles.) And yet again it can be seen that equations (20.7) and (20.8) would provide a model for the brain neural information processing in the form of a read / write filter bank and memory which would consist of holograms situated on hologram planes in the form a mathematical chart corresponding to the cerebral cortex, where these holograms are those of its input holographic sensual experience.

Further support for the postulate of the brain as a QCE, in addition to the probability that all DNA can be most appropriately modelled as a quantum Carnot engine, where its structure constitutes a phaseonium, comes, for example, from the AndCorporation,[48] where 3-D holographic face recognition software, based on quantum mechanical phase models, has been realized on conventional digital machines, so as to demonstrate that such phaseonium-based face recognition does work, and has significant advantages over nonquantum mechanical designs, even though the purely quantum mechanical aspects of such computations cannot be realized. It can therefore be postulated, as already argued in general for living systems, that such features of the human brain as the mind and consciousness, absent from classical machines, constitute the difference between the quantum Carnot engine and its classical thermodynamic counterpart. This defines precisely for the first time what constitutes a 'zombie' or robotic machine, as distinct from a conscious machine. That is, the brain's structure constitutes, like the models of the universe and DNA which have been proposed in this chapter, a QCE with a phaseonium working by quantum mechanical base holography so as to perceive the structure of the universe (on a particular scale) in terms of 3 + 1 space-time holographic images, as was first proposed by

Pribram.[48] This is the concept from which the AndCorporation designs have proceeded.

This would then lead to the inference, in this case, that there is again an almost perfect dynamical balance in the quantum Carnot engine that controls our thought processes between, say, the neural workings of a brain, and the corresponding working of the mind (as, say, a field of energy determining its mental arrows of thought). It is therefore a balance, that can be changed at will via the application of a suitable source of energy to this field, so as to produce the required mind-brain interactions in response to, say, data arriving at the senses or as retained in memory. And, since it is to be supposed that this whole brain-mind works now as a single heat bath governed by a single phase parameter, then this would explain why consciousness functions as an essentially serial phenomenon in relation to mind control over the brain. The glial cells[48] or the microtubules[46] have indeed been postulated as the media for the regulation of this controlling energy.

Furthermore, this structure of the brain[84] shows how the human brain is able to assign meaning to human language by providing each name or symbol uniquely with a meaning by means of the object and its properties, with which each stored phase conjugate object image would be associated. And this must be the true power of the human brain that it is able to process meaning, i.e. process words not just syntactically but by their semantics, as known from each human being's actual geometric / holographic experience. Again, although such experience will be subjective in part, since it takes place from the reference frame and viewpoint of that individual, there always remains a fundamental mechanism, the 3-D objects of the real world themselves known through their phase conjugate object images, which provides the common medium for all objective human communication. That is, all parties may reach out to touch, so as to see, hear, or to smell any tangible object to determine the truth about it as stated by the other parties, or nowadays to determine exactly the nature of those properties through common scientific instrumentation and experiment, the process known as science.

20.18 Language and Universal Grammar

The evidence also fits well with the mathematical-linguistic model of Chomsky, that postulates that common principles underlie any language and so produce 'a universal grammar'.[85] On Chomsky's view, such 'universal grammar' is inherent, i.e. it has some genetic determinants. This is an extremely important circumstance, which once again emphasizes the super-genetic relationship of the DNA semiotic structures and human speech structures. To a limited extent this position has already been partially confirmed in the study cited, which shows the similarity of characteristics between DNA and human speech. Chomsky is therefore probably right when he argues that the in-depth syntax constructions which constitute the basis of the language, are passed down from generation to generation, providing each individual with the capacity to learn the language of its ancestors. The fact that a child easily learns any language is then explained through the theory that the grammars of all languages coincide, and the essence of the human language is invariant for all people. It can now be supposed, however, that this invariance extends even more deeply, down to that of the macromolecular semantic ('speech') chromosome structures. Further independent confirmation, in relation to the DNA-wave biocomputer,[68] comes from quantum holographic imagery,[84,86] for here, in 3-dimensional spatial object images, the observations are phase conjugate (pc) so as to coincide with the 3-dimensional objects themselves, the observed. So, it is the 3-dimensional objects themselves that are the symbols that implicitly label all aspects of experience, the observations, in a universal way for all observers, so as to form the basis of communication between all those observers with a common genetic heritage and sensory apparatus. The bases of all languages in this case are therefore shared arbitrary symbols or semiotic labelling of these objects and their properties such that

each (pc) holographic encoding on a 3D object or symbol	maps to and from ⟵⟶	the holographic encoding of the arbitrary label for that object or symbol
right	corpus callosum	left

Such mappings are also unique since no two objects can occupy the same position in 3-dimensional space. This mapping schema could then explain the morphology of the human brain which concerns the two brain hemispheres and the corpus callosum, which joins them. That is, the right hemisphere that realizes the holographic encodings of the real world (concerning the geometric continua), is the artistic brain, and the left, that realizes the arbitrary labellings of these real world objects or symbols and their properties (concerning the arithmetic continua), is the logical brain. For in the latter, an essential element of the mapping of such labelling of objects includes numbers and sets and their logical relationship one to the other, where these must be acquired by learning. This mapping schema can therefore be postulated as the basis of Chomsky's universal grammar or of the universal nilpotent rewrite system in the human brain as a neural system, as fundamentally laid down in the genetic code.

There can be little doubt, in view of natural language, that the human brain is a universal computational semantic machine, and that the NUCRS would provide a natural model and modes by means of which human speech, writing and the hearing of natural language could actually be realized (through a neural NUCRS semantic ontology) so as to allow the human race, as is actually happening in science, to comprehend the evolutionary cosmos in which it exists. The following experiments that anyone can perform provide a partial confirmation of this. If you snap your fingers at some distance away from the head and ask where the hearing senses detect the noise of the acoustic snapping. It is outside yours head exactly coincident with the snap itself. That is, the acoustic object image of the snap and the snap itself coincide, which is the definition of a phase conjugate object image. Similarly, place a glass on a nearby table and reach out and touch it. Again one's senses of sight and touch are such that, in every particular, in 3 spatial dimensions, they coincide with those of the glass itself. That is, both the visual and tactile object images produced by the brain are phase conjugate object images of the glass. And the condition of phase conjugation is a fundamental one, because human survival or indeed the survival of any living system depends upon finding any object where it actually is.

There is even a possibility of support from psychology where the various functions, leading to the actual perception of such things as time and 3-dimensional space, may be coded according to seven combinations of properties or their alternatives in the form A / a, B / b, C / c.[87] If we add the eighth combination (*abc*, perhaps representing a null or neutral state), this has the classic structure of 3-dimensionality (and presumably double 3-dimensionality), expressed as shown above in 20.10, for example, in the nilpotent matrix representations of the 3-dimensional Heisenberg Lie group, and explored in the diagrams in chapter 4, leading to the possibility that a universal structuring of brain functioning, determined by a universal structuring in nature, makes it possible for us to perceive this structure. In principle, a universal semantics determines that we are 'hard wired' in a way that allows us to understand our basic situation.

The discovery of the NUCRS provides a sound scientific basis, by means of which the search for the fundamental physical foundations of computing, as used in brains, can be realized. In principle, therefore, the concept of computational rewrite systems and the remarkable quantum mechanical nilpotent structure appear to define Nature's rules, to provide the desired scientific foundation, so that the conceptual requirement in the Premise and Mission Statement of the British Computer Society Cybernetics Machine Group now becomes the Evolutionary 'Anthropic' Semantic Principle, as previously proposed.

20.19 Nature's Process

Having discussed how the NUCRS may operate in systems at many levels, it may be worth adding a final speculation on how the various contributory components may come together in a complete process of ontological evolution. At all levels, a universal alphabet of symbols, defined by the infinite square roots of –1 (see Appendix B), creates an infinite series of nilpotent zeros in a unique birthordering, each of which is specified by a unique phase term in quaternionic space, which, in the case of fermions, corresponds to the instantaneous vector direction of $\mathbf{p}$, as determined by the 'coordinates' $\boldsymbol{k}E$, $\boldsymbol{i}\mathbf{p}$, $\boldsymbol{j}m$. In relation to each other, the phases constitute 3-D Euclidean space, with a $U(1)$ symmetry (the source of gravity) (see 6.3), which ensures their uniqueness and acts as de Broglie's pilot standing wave, where each zero has a distinct energy $E = h\nu$ and an entropy $S = kT$, which is directly related to the fermionic $\mathbf{p}$. (*Observation*, using Heisenberg uncertainty, effectively assigns the quaternionic phase space term a direction in this 'real' space.) The mathematical function which most closely fulfils this cosmological boundary condition is the Riemann zeta function, and we may speculate that it is this function which characterizes the numbering / ordering of the nilpotent zeros, in a manner which is fractal / quantum chaotic bordering on chaos, as with the interface between the Mandelbrot and Julia sets, so providing maximal transfer of information and minimally entropic (proper time) birthordering.[88] We may further suppose that the totally quantum entangled nilpotent vectors constitute a quantum hologram, determining the QCE evolution of the system's ontology by phase conjugate adaptive resonance, so that the cosmos behaves essentially as an evolving soliton.

Chapter 21

Infinity

Concerned with the large scale and the potentially infinite, this chapter touches on cosmological issues which have been alluded to in chapter 20, but its aim is not to present a theory of cosmology as such, or to comment on existing theories, but to reclaim for physics phenomena which are only usually spoken about in a cosmological context. If anything, it can be described as an *anti*cosmology. In principle, unless one believes in some extreme version of the anthropic principle, the laws of physics, in a *unified theory*, must be true in all places at all epochs, and the concept which drives this chapter is, in fact, purely physical: the linearised gravity-plus-inertia theory and gravomagnetic equations of chapter 18. In the context of this theory of inertia, we can define a version of Mach's principle in which inertial mass is such as would *appear* to be produced by the inertial effect, and, using the principle of equivalence, we can relate the inertial force to that produced by the static gravitational field to derive the equation for cosmological redshift, with acceleration at the deceleration parameter –0.5, exactly as now measured. (A prediction of the effect was made many years prior to the measurement.) The inertial effect, in this interpretation, is equivalent to a loss of information, and the acceleration suggests that the loss of information is equivalent to energy lost through the process of 'radiation'. A further calculation is used to show that this could be the source of the microwave background radiation, meaning that the latter, if derived from an inertial vacuum process, could be conceived of in terms of a universal frame of reference. The calculations, if valid, also demonstrate that parameters such as the Hubble mass, Hubble time, critical density, and background temperature are determined at the same time as the masses of fundamental particles from the vacuum via the Higgs mechanism, and are not independent parameters, while the structure and behaviour of large gravitating objects will depend on such parameters because they are fundamental to inertial processes. The 'cosmic coincidences' assumed in the 'large number hypothesis' would thus be independent of the epoch in which they were measured. At the end of the chapter, a final point is made on the concept of 'creation' as being fundamental to the idea of zero totality and the Dirac nilpotent, irrespective of whether or not it is considered to be the result of a single event in time.

21.1 A Version of Mach's Principle

In a linear gravitational theory, it will not be surprising that the universe is flat, Euclidean, and infinite. There will be no need to invoke an inflationary expansion, an arbitrary cosmological constant or quintessence, or perhaps even any real expansion at all, for an application of a version of Mach's principle can produce the Hubble redshift, with an apparent acceleration, with deceleration parameter $q_o = -0.5$, with no continuous creation of matter. In this instance, the redshift associated with distant astronomical objects becomes, in principle, an 'apparent' effect due to the inertial forces introduced with our system of measurement.

Mach's principle, the idea that the inertia of a body is due to the action of forces produced by all other bodies in the universe, does not fit neatly into the conventional interpretation of relativity, and is generally described as a principle adopted by Einstein only as a stage in the development of his theory, and then subsequently rejected. It is, in fact, probably unsupportable in its original form, but, if we take inertial forces as *fictitious* or apparent effects produced by choice of a noninertial frame for gravitating systems, we can then suppose that the inertia of a body can be such that it will *appear* to be due to the fictitious forces produced by all the other bodies within its event horizon, the event horizon being the limitation produced by the fact that inertial forces are restricted to transmission at the speed of light.

Such a theory may be related to the work of Sciama, who, as early as 1953,[1,2] was investigating the possibility that Mach's principle could be explained in terms of a gravitational analogue of the acceleration-dependent inductive force of electromagnetic theory, which produces radiation or photon emission,

$$F = \frac{Gm_1 m_2 \sin\theta}{c^2 r} \frac{dv}{dt},$$

by analogy with the electromagnetic expression

$$F = \frac{q_1 q_2 \sin\theta}{4\pi\varepsilon_0 c^2 r} \frac{dv}{dt}.$$

The inertial properties of a body, defined by Newton's second law of motion, and relating gravitational to inertial mass, would then, it is assumed, be such as would *appear* to be generated by the inductive inertial field due to an apparent acceleration of the background matter, the apparent acceleration being due to the original choice of a noninertial frame.

To attribute the inertia of a body of mass m to the (fictitious) action of the remaining matter within its event horizon (mass m_u, radius r_u) we require the total sum of all such force terms to be identical with the Newtonian inertial force, mdv / dt, or at least proportional to it. Thus

$$F = \frac{Gm_1 m_2}{c^2 r}\frac{dv}{dt} \propto m_1 \frac{dv}{dt} = Kma \,,$$

so that

$$Gm_u \propto c^2 r \,.$$

This would not only require the components of force on m normal to the direction of acceleration to average to zero, but would also demand, for a universal inertial field, a high degree of uniformity (in terms of isotropy if not homogeneity) in the distribution of matter in the universe as a whole. As we will see $K = ½$, in $F = Kma$, because $Gm_u = ½c^2r$. Effectively, to compare a with an acceleration measured by Newton's laws, we need to divide by a factor 2. On a universal scale, the real attractive force of gravitation reduces a by another factor 2. Another way of looking at this is to say that the inductive / gravomagnetic force is increased by a factor 4 by the general relativistic combination of length contraction and time dilation acting on the gravitational-inertial vector potential.

There is, however, another way of defining inertial force within this Machian paradigm, based on the fact that a continuous and zero-gradient Higgs field must provide a method for defining unit inertial mass nonlocally for the entire universe. We can imagine mass m_u defining a radial inertial field of constant magnitude, acting from the centre of a local coordinate system in such a way that it defines a unit inertial mass for any point in the system, similar to the way that the near-constant **g** can be used to define a unit of mass at the Earth's surface. At the same time, invoking the principle of equivalence (for full comparison with the electromagnetic case), the 'static' component of this force would also be of the same magnitude as the static gravitational gravitational field ($Gm_u \ / \ r_u^2$) which defines the unit of gravitational mass within the same event horizon, but which is independent of the imposition of any local coordinate system. (That is, the inertial effect should appear to act *as though it* confines the nonlocal gravitational field to the event horizon, even though, ontologically, it does not.)

So, adding this further hypothesis, we find that an object of mass m at a distance r from an observer at the centre of a sphere defined by r_u will be seen to experience an inertial force

$$\frac{Gmm_u}{c^2 r}\frac{dv}{dt} = \frac{Gmm_u}{r_u^2},$$

with acceleration
$$a = \frac{dv}{dt} = \frac{c^2 r}{r_u^2}.$$

Writing this in the form
$$v\frac{dv}{dr} = \frac{c^2 r}{r_u^2},$$

and integrating with respect to r, we obtain

$$v = \frac{cr}{r_u},$$

which we recognise as the standard formula for cosmological redshift $v = H_0 r$, with the additional relation,

$$a = \frac{v^2}{r_u} = H_0^2 r.$$

The acceleration term here has the same Machian origin as the supposed expansion itself, and does not require any additional physical principle or adjustable parameter, while very much the same relationship between v and r would emerge from a consideration of the inertial force as a 4-vector correction ($Gmm_u v / r^2 c^2$) to the static gravitational attraction between m and m_u, and that would be the same as applying the 4-vector correction factor $(1 - v^2 / c^2)^{-1/2}$ to each of the masses.

For a universe dominated by relativistic matter, or photon radiation, ρ_u will be replaced by the relativistic $\rho_u + 3P / c^2$, where $P = \rho_u c^2 / 3$ is the pressure of radiation in any direction, and the inertial force becomes twice the nonrelativistic value. In addition, *assuming* a uniform mass density, and taking v as dr / dt in

$$v^2 = \frac{c^2 r^2}{r_u^2},$$

and substituting $8\pi G\rho_u r_u^3 / 3$ for $c^2 r_u$, we obtain (for the relativistic case) the expression

$$\left(\frac{dv}{dt}\right)^2 = \frac{8\pi}{3} G\rho_u r^2 ,$$

from which

$$\left(\frac{dv}{dt}\right)^2 \frac{1}{r^2 c^2} = \frac{8\pi}{3} \frac{G\rho_u}{c^2} ,$$

which is formally similar to the equation for the 'expanding universe', with the terms for 'curvature' and 'cosmic repulsion' automatically equated to zero.

According to the nonlocal interpretation of gravity, with instantaneous transmission, this acceleration is a fictitious ('centrifugal') one, observable through effects on the coordinate system, resulting from the noninertial frame introduced by assuming that the instantaneous gravity resulting from a filled vacuum can be modelled on equations written in a Lorentzian space-time. In such a theory it would not be surprising that the universe is flat, Euclidean, and presumably infinite, that starlight is redshifted, with sources observed as accelerating away from the observer (the so-called 'dark energy'), and that vacuum, as we will now show, provides 2/3 of the mass required to keep the density of the universe at the critical value.

To calculate the vacuum density, we can combine the inertial acceleration with the gravitational term due to total mass m at any distance r, to give:

$$F = \frac{Gm}{r^2} - H_0^2 r = \left(\frac{4\pi G\rho}{3} - H_0^2\right) r ,$$

The equivalent Poisson-Laplace equation becomes:

$$\nabla^2 \phi = 4\pi G\left(\rho - \frac{3H_0^2}{4\pi G}\right) = 4\pi G\left(\rho + \frac{3P}{c^2}\right) = 4\pi G\left(\rho - 3\rho_{vac}\right).$$

From this, we derive a vacuum density ρ_{vac}, equivalent to 'dark energy' density or negative pressure $-P$, and cosmological constant $\lambda = 8\pi G\rho_{vac}$:

$$\rho_{vac} = \frac{H_0^2}{4\pi G} .$$

Hence, with the critical density for a 'flat' universe defined as

$$\rho_{crit} = \frac{3H_0^2}{8\pi G} ,$$

we find that $$\frac{\rho_{vac}}{\rho_{crit}} = \frac{2}{3}.$$

The deceleration parameter, defined in terms of a scale factor R, and incorporating the gravitational acceleration term, will be

$$q_0 = \frac{\ddot{R}R}{\dot{R}^2} = \frac{1}{2}\Omega\left(1 + \frac{3P}{\rho c^2}\right) = -\frac{1}{2},$$

if $\Omega = \rho / \rho_{crit} = 1$, as expected. The 'dark energy' of the universe could thus be the signature of a nonlocal form of gravity, which determines the nature of the Higgs field.

This result (in terms of the acceleration) was predicted long before the discovery of an expansion in the redshift, with deceleration parameter –0.5, in a seemingly flat, Euclidean, and presumably infinite universe.[3-5] The theory is, in fact, totally independent of any specific cosmological model, and depends only on the application of Mach's principle to the theory of inertia, extended to include the equivalence of gravitational and inertial mass. There is no appeal to an arbitrary cosmological constant or a mysterious origin for 'dark energy'. The repulsive force is entirely explicable in terms of similar fictitious forces found throughout physics.

The 'Hubble radius', according to this explanation, defines an event horizon for the inertial reaction of the matter surrounding each individual mass, the inductive inertial field acting from all directions as though the mass were placed at the edge of a sphere of this radius. An observer placed at any point in space would necessarily define a Hubble radius for the inertial force in each direction, and a body placed at that point would be subjected to a resultant inertial force from the total mass within the radius. It may be significant that, in this analysis, no assumption is made that the distribution of matter *has* to be uniform, and it may well be that a relationship in which $m \propto r^2$, as Roscoe has proposed, on Machian foundations (and seemingly with experimental support),[6] is more applicable. It would, for example, be worth exploring the rotation curves of galaxies and galactic clusters to see if there could be an inertial force component in the perceived flattening out of the graphs of v against r after an initial linear increase. This has been almost universally attributed to haloes of a mysterious 'dark matter' surrounding the galaxies and clusters, but the flattening out seems to occur, in the case of galaxies, at the point where v^2 / r becomes of order c^2 / r_u, as though it might be reaching a limit, or observational event horizon, at the value of some universal inertial acceleration for rotating gravitating systems (presumably of the Coriolis type, $2m\omega \times v$, rather than the 'centrifugal' type responsible for redshift[7]). The fact that the ratio of dark / visible matter ratio increases dramatically with the size of object, from stars, to galaxies, to galactic clusters and superclusters, means that it is impossible to explain the dark matter component of the larger objects with the dark matter contribution of the component smaller ones; but such an effect would be totally expected on an explanation connected with an inertial event horizon determining the structures that are gravitationally possible, and a constant maximum value of v^2 / r would imply that the masses of large gravitating systems, calculated as v^2r / G, would show a proportionality to r^2, explaining the 'fractal' structuring of galaxies and clusters observed by Roscoe.

21.2 Gravity and Inertia

The interpretation of Mach's principle given in this analysis is that the value of inertial mass is that which would *appear* to be produced by the interactions of all the other masses in the universe, since the inertial effect is an apparent, rather than a real, one. The 'causation' is thus not a direct physical process, because gravity is not the result of inertia, though the numerical values of the inertial and gravitational quantities have to be equated via the equivalence principle to maintain the undoubted connection. In the universal Machian relation, $c^2r_u \propto Gm_u$ (which, in this case, becomes $\frac{1}{2}c^2r_u = Gm_u$) we assume that the gravitational energy is negative vacuum energy, while the inertial energy is positive, leading to the required zero total. Since the gravitational energy is equivalent here to a vacuum effect we can relate it to the Higgs field, which produces fermionic and bosonic inertial mass, or, equivalently, to the zero-point field using the mass value as a cut-off in the spectral density (a calculation which, as we have seen, yields perfectly reasonable results). Another process which produces inertial mass from the vacuum, is that of renormalization, where we may sum to a finite value of mass for a fermion using a cut-off at the Planck energy; here, the ultimate c-related cut-off value is an indication that the inertial process involves time delay.[8]

Inertia, in fact, is time-delayed, while gravity is instantaneous. However, the expression for the force of static gravitational interaction, like that for the electrostatic, cannot distinguish between instantaneous and time-delayed transmission. The Casimir effect, or attractive inverse fourth-power force between parallel plates due to the excess vacuum pressure on the outside, which is usually cited as an illustration of the reality of a filled vacuum with an infinite virtual zero-point energy density, can, it would seem, be explained alternatively by assuming that electrostatic interaction is instantaneous.[9] According to the reasoning followed here, an instantaneous interaction of some kind is *exactly equivalent* to a filled vacuum. The assumption that it is electrostatic in origin, rather than gravitational, merely transfers the cause. In principle, the difference between the gravitational and electric forces is not determined by whether either or both are transmitted instantaneously, but whether the time-delayed effects associated with them are fictitious or real.

In the same context, it can be argued that the time-delayed aspect observed in the electromagnetic interaction, which results in a finite value for c, the 'velocity' of transmission, is a result of the action on a charge of all the other charges in the universe, each acting instantaneously.[10] Again, there is a valid point in this argument, which can be seen more clearly if we reverse it. Charges of any kind ultimately relate to fermionic state vectors, which are necessarily correlated instantaneously, and the behaviour of any charge is determined by that of all the other charges in the universe. On the other hand, measurable effects involving what is often called 'energy transfer' (but which is really particle or charge transfer) are invariably time-delayed, despite results which may superficially appear to deny this, and we could equally well assume that the behaviour of the rest of the charges in the universe is *determined* by the finite velocity of transmission, but with instantaneous speed of state vector correlation.

It is the inertial process which creates the concept of 'localised' energy. The degree of localisation in the system (inertial rest mass, which is also measured by time delay or by proper time) can be expressed in terms of an apparent metric ('apparent' because the Euclidean metric is the only real metric for space) but the mass is not created by the metric. It is probably incorrect, as we have stated, to say that time-delayed action is an indication of an *actual* 'transfer of energy' from place to place. It is likely that energy itself does not move at all, but defines a complete continuum or 'vacuum' filling space, its value being infinite in terms of density, but fixed at a particular finite level at each point in space (the Higgs field). This level presumably never changes, and its value (which appears to be 246 GeV) must be determined by the fundamental numerical relation between units of charge and units of rest mass. What really changes is the degree of the realization of quantization in the form of the E, $\mathbf{p}$ and m terms in the Dirac nilpotents – exactly as in the field excitations of conventional Quantum Field Theory. It is charge, with 'axes' $\boldsymbol{k}$, $\boldsymbol{i}$, $\boldsymbol{j}$, which creates this quantization, irrespective of the occupation or otherwise of the levels (0 or ± 1) on the axes.

21.3 Cosmology and Physics

There is at present a serious dearth of debate about the fundamentals of cosmology. It is widely assumed that a single model explains all observational facts so smoothly that it is unthinkable to consider any other. The evidence for this model is so compelling, we are told, that, even though it has had to be drastically modified a considerable number of times, it still stands as a monumental and unchallengeable part of the twentieth century's scientific legacy.

Cosmology, however, is not a science of the same degree of exactness as, say, the Standard Model of particle physics, which has had to stand the test of experimental confirmation many times over in a huge variety of forms by groups of physicists in major competition with each other. In cosmology, an unrepeatable 'experiment' has provided two pieces of evidence for a certain interpretation. Nothing that has been done since the discovery of this evidence has provided any further proof of the validity of the interpretation. This does not, of course, provide proof that the interpretation is wrong; but it does mean that we should be much more open to alternative views, especially as the interpretation, as now put forward, is drastically different from, and even contradictory to, the version as first presented.

The two pieces of evidence which count are the cosmological redshift and the microwave background radiation. The first is undoubtedly very strong. The most natural explanation for any such systematic shifting is clearly a Doppler effect produced by a recession of the sources. Anything else will necessarily seem less simple and more contrived. The second piece of evidence, however, is a good deal less powerful. In the first place, it is purely qualitative, despite the fact that the microwave background temperature has an exact value known to considerable precision. There is no theory which predicts this value from first principles, and there is no obvious way of deriving it from the data provided by the redshift. The two phenomena are not associated by a strongly predictive fundamental theory. The link is a good deal more tenuous than is often claimed.

Further evidence is simply not yet forthcoming, while observational data has repeatedly contradicted the assumptions of the particular models put forward to explain the theory at any particular time. This is seldom made clear in the way that theory is presented. Thus, procedures that are merely examples of the fitting of experimental data to the overall structure – for example, the time sequences for various stages of evolution, the abundances of the elements, and the actual temperature of the radiation remaining as a remnant of the original 'fireball explosion' – are frequently spoken of as if they were independent pieces of 'evidence' in support of the theory. The fitting may or may not be correct, but it is effectively only a *descriptive* procedure within the terms of an assumed model. It describes what must have happened if the theory it uses should be true, but it doesn't provide evidence that the theory *is* true. In addition, it has proved remarkably difficult to sustain any particular model of the theory for more than a few years without some drastic modification having to be introduced – for example, inflationary expansion, a positive cosmological constant, cosmic repulsion, dark energy, dark matter (hot or cold), quintessence, infiniteness, flatness, acceleration, etc. We have even had multiple universes, each with their own (completely unpredictable) laws![11]

The promotion strategy for the theory has also found it useful to create the implication that the Big Bang universe is somehow a natural evolution of the culturally significant General Theory of Relativity, with the implication that that theory 'predicts' the current expansionary model through the Friedmann solutions to the general relativistic field equations, when, this again is merely a fitting of observed data *post hoc*, without any predictive element whatsoever. We can only know from observational data whether the universe's density is below, equal to, or above the critical value for continued expansion. In addition, General Relativity and the Friedmann solutions tell us no more about the behaviour of a subcritical, critical or supercritical universe than would classical Newtonian theory (it depends only on the Newtonian potential, however introduced), and the equation used to determine the critical density is identical to the purely classical one.

Reports from astrophysicists have suggested the 'staggering' conclusion that the universe is flat, Euclidean and infinite, with matter (whether visible or dark) at the critical density.[12] Though some of us have never thought otherwise, it seems that the whole picture of the universe, as imagined by the majority of cosmologists not so many years ago, has changed beyond recognition. However, the same theories seem to be assumed to apply, even though the devices now used to explain how we have arrived at this new simple-looking model are becoming increasingly complicated and contrived.[13]

It is possible that current cosmology may eventually be considered to be of the same kind as our understanding of the age of the Earth was before the discovery of radioactivity or the structure of the Earth was before the discovery of plate tectonics. In those instances, alternative models, including the ones now accepted, were dismissed out of hand by the supporters of the dominant theory, but, in each case, the dominant theory was eventually overthrown by the discovery of a new kind of evidence whose existence had been completely unsuspected. What is most significant here is the fact that the too ready assumption of a catch-all cosmology has had the effect of reducing the explanatory power of fundamental physics. So, we look to cosmology to explain the number of neutrinos, or neutrino masses, or the predominance of matter over antimatter, and effectively assume that physics itself is not able to provide the answer. However, it is far more likely that such completely fundamental concepts have as intrinsically a *physical* origin as, say, the conservation of energy, and the same may even be true of the cosmological redshift and the microwave background radiation. That is, it is not necessarily the best possible research strategy to be putting forward model-dependent, essentially untestable, hypotheses of a cosmological nature for fundamental physical phenomena before we have explored the possibility of *physical* origins for them.

21.4 Information Loss and Radiation

The calculation of redshift based on Mach's principle is essentially independent of the cosmological model used to explain the universe. It doesn't *require* an actual expansion to explain the redshift, though it doesn't, as such, exclude one. It is a purely physical argument, and there are many significant physical implications. It implies, for instance, a Machian concept in the 'creation' of both inertial and gravitational masses. The creation is not direct in this theory, but the creation process must be structured so as to make it *appear* direct. Here, also, the inertial process is seen to be a result of using a noninertial frame of reference in the treatment of gravity – a kind of 'aberration' of space-time, to use the terminology already applied. This is, of course, related to the general relativistic idea that inertial mass is the result of local curvature or warping of space-time, and would be identical if that theory were interpreted as being about the inertial effects of gravity, rather than about gravity itself. In some sense, then, the apparent creation of mass must be a vacuum process, related to the actual creation process of the Standard Model, in which inertial mass is generated by an interaction of matter (i.e. sources of strong, weak and electric charge) with the vacuum Higgs field.

However, the theory as an 'aberration' or fictitious force theory is also about loss of 'information', producing a cosmological arrow of time. Light from distant sources is redshifted because we have lost information about them. The energy loss from redshift is thus interpretable also as an information loss. Energy lost from systems in this way (that is, energy lost from the interaction of a system with the 'rest of the universe') usually reappears as thermal energy. Here, the process is systematic on a universal scale, and so may lead to a systematic 'creation' of thermal energy equivalent to the amount lost – though not to any 'heat death', or universal thermal equilibrium, if the universe is infinite. The loss of information, in addition to the fact that the process involves a c-related event horizon, means that we have no 'Olbers' Paradox' problem of a uniformly bright sky, even in an infinite universe. And, where we have acceleration, there is a specific process by which the information loss may happen, because accelerating systems composed of inertial 'matter' radiate energy. The process, here, however, is not concerned with real matter, but is a systematic one applied to space-time as a whole, taking gravity as a vacuum process, and so, just as the acceleration observed in distant sources may be taken as a fictitious one, so may be the accompanying process of radiation, and we may take the charges involved as virtual ones, equivalent to the virtual particles constituting the vacuum in the Standard Model.

21.5 A Numerological Coincidence?

The electron is probably the only particle whose mass is entirely, or almost entirely, electromagnetic in origin. In the Standard Model, it differs from the weak-interacting neutrino only by possessing a non-zero value of this charge, and the Higgs mechanism indicates that this is the origin of the mass. Higher-mass leptons, such as muons or tau particles, have masses which involve additional weak symmetry-breaking, while quarks have masses partly determined by the strong interaction. The electric force is also the only other force besides gravity to be long range in its effect. So, it is natural to think of the vacuum, in the first instance, as being approximated through an infinite number of virtual electron-positron pairs, and as generating inertial mass through the creation of the electronic mass m_e via the electric force.

The real electron appears to be point-like, but it is usual to refer to the so-called classical radius as being significant in relating the values of mass and electric charge:

$$r_c = \frac{e^2}{4\pi\varepsilon_0 m_e c^2}.$$

The classical radius has no specific physical meaning other than as a relation between fundamental units, but, if we assume a sphere of radius r_e with uniformly distributed charge e, then we obtain the relation originally found by Heaviside:

$$\frac{2m_e}{3} = \frac{e^2}{4\pi\varepsilon_0 r_e c^2}.$$

From this, we could say that a uniform spherical distribution of charge e would produce a mass equivalent m_e over the radius:

$$r_e = \frac{3}{2}\frac{e^2}{4\pi\varepsilon_0 m_e c^2} = 4.2289\times10^{-15}\,\text{m},$$

which is 1.5 × the classical' radius. For a uniformly distributed vacuum *virtual* charge density, this may be the more significant value.

Let us suppose, now, that this so-called radius, rather than indicating anything about the size of real electrons, provides us with a fundamental relationship between the value of electric charge and the equivalent inertial mass generated by a vacuum process. In effect, we are saying that it is somehow indicating that the vacuum has the *potential* to create the mass m_e in association with the charge value e if the radius r_e plays some significant part in the process. However, a 'radius' of this kind can mean nothing in terms of a real particle, and can only have significance in terms of a quantity such as charge or energy density. According to the theory of inertia outlined in 21.1, the generation of inertial mass is necessarily connected with the simultaneous creation of gravitational properties. So, the gravitational energy density associated with these values of r_e, m_e and e may be a significant quantity in specifying the vacuum state required. This is easily calculated as:

$$\frac{3Gm_e^2}{4\pi r_e^4} = 4.132\times10^{-14}\,\text{Jm}^{-3}.$$

Remarkably, this is *exactly* the energy density that would be produced by a uniform distribution of black body radiation at a temperature of 2.72 K, and determined by aT^4, where $a = \pi^2 k^4 / 15c^3\hbar^3 = 7.56 \times 10^{-16}$ $\text{Jm}^{-3}\text{K}^{-4}$. This calculation (which arose out of historical work on late nineteenth-century aether theories, and was first published in 1990[14]) looks, at first sight, like a remarkable numerological coincidence, and it is quite possible that that is all it is. Many numerological results of this kind are the result of pure chance, and have no fundamental significance at all. However, there will always be some cases where playing with numbers leads to profound results, the classic one being the derivation of the Balmer series for the hydrogen spectrum in 1885, almost thirty years before it found its explanation in the Bohr theory of the atom. So it is always worth making at least a preliminary investigation of any coincidence, just in case it turns out to be another Balmer series. Could it, perhaps, mean that the microwave background radiation, which, by being isotropic, appears to provide the universal rest frame desired from the vacuum by the physicists of the nineteenth century, is really a fundamental vacuum property?[15] It is certainly worth looking at the possibility of a fundamental *physical* explanation for this phenomenon before we assume that it can only be explained by the necessarily more *ad hoc* models proposed by contemporary cosmologists.

21.6 Vacuum Acceleration and Radiation

Let us imagine a vacuum filled with a uniform distribution of virtual particles of charge e, but assume that the distribution is such as to require the equation of gravitational and inertial forces (as in 21.1). Here, of course, we will not be concerned with real particles, or even with particles at all, but with inertial effects in the vacuum due to the creation of a noninertial frame of measurement, and we suppose that this frame occupies the whole of space (with what we will describe as 'component volumes' of radius r_e), irrespective of whether it contains real particles. The inertial effects, as we have seen, produce an inertial or fictitious acceleration, and accelerating charges radiate, according to the Larmor formula, at a rate

$$\frac{dE}{dt} = \frac{2}{3}\frac{e^2}{4\pi\varepsilon_0}\frac{a^2}{c^3} .$$

As explained in 21.1, a will be reduced from $c^2r \,/\, r_u$, by the general relativistic factor 4 when we include length contraction and time dilation, or by the factor $K = ½$, from the inertial calculation combined with an equal reduction due to the gravitational attraction:

$$\frac{dE}{dt} = \frac{2}{3}\frac{e^2}{4\pi\varepsilon_0 c^3}\frac{c^4 r^2}{16 r_u^4}.$$

Substituting $r = ct$, we obtain

$$\frac{dE}{dt} = \frac{2}{3}\frac{e^2}{4\pi\varepsilon_0}\frac{c^3 t^2}{16 r_u^4}.$$

Then, integrating between limits 0 and E, and 0 and t,

$$\int_0^E dE = \int_0^t \frac{2}{3}\frac{e^2}{4\pi\varepsilon_0 c^3}\frac{c^3 t^2}{16 r_u^4}\,dt\,,$$

from which $$E = \frac{1}{72}\frac{e^2}{4\pi\varepsilon_0 c^3}\frac{r^3}{r_u^4}.$$

This is for one component volume at distance r. For n_e component volumes at uniform number density ρ, consider an element of volume dV:

$$dV = 4\pi r^2 dr$$

$$E\rho dV = \frac{12\pi r^2 \rho}{4\pi r_e^3}\frac{1}{72}\frac{e^2}{4\pi\varepsilon_0}\frac{r^3}{r_u}dr$$

$$= \frac{1}{24}\frac{r^5}{r_u^4 r_e^3}\frac{e^2}{4\pi\varepsilon_0}dr\,.$$

But, for a universe filled with virtual particles, the density of volumes of radius r_e becomes

$$\rho = \frac{3n_e}{4\pi r_u^3}\,,$$

where the number of volumes,

$$n_e = \frac{4\pi r_e^3/3}{4\pi r_e^3/3} = \frac{r_e^3}{r_e^3}$$

and so $$\rho = \frac{3}{4\pi r_e^3}.$$

Hence $$E\rho dV = \frac{1}{24}\frac{r^5}{r_u^4 r_e^3}\frac{e^2}{4\pi\varepsilon_0}dr\,.$$

Integrating to find the energy radiated over the total volume $4\pi r_u^3/3$:

$$E_{TOT} = \int_0^{r_u} \frac{1}{24}\frac{r^5}{r_u^4 r_e^3}\frac{e^2}{4\pi\varepsilon_0 r}\,dr$$

$$= \frac{1}{144}\frac{r_u^6}{r_u^4 r_e^3}\frac{e^2}{4\pi\varepsilon_0}$$

$$= \frac{1}{144}\frac{r_u^2}{r_e^3}\frac{e^2}{4\pi\varepsilon_0}\,.$$

The corresponding energy density is given by

$$\frac{3}{4\pi r_u^3}\frac{1}{144}\frac{r_u^2}{r_e^3}\frac{e^2}{4\pi\varepsilon_0} = \frac{1}{192\pi}\frac{e^2}{4\pi\varepsilon_0 r_e^3 r_u}\,.$$

We could assume that this energy density is the same as that of the microwave background radiation in order to obtain a value for r_u. That is

$$aT^4 = \frac{1}{192\pi}\frac{e^2}{4\pi\varepsilon_0 r_e^3 r_u}$$

and $$r_u = \frac{1}{192\pi}\frac{e^2}{4\pi\varepsilon_0 r_e^3 aT^4}$$

At $T = 2.72$ K, this gives a value for the 'Hubble radius':

$$r_u = 1.22 \times 10^{26}\ \text{m},$$

giving a Hubble time of 4.07×10^{27} s $= 1.29 \times 10^{10}$ yr.

The calculation shows that the model yields a value of the right order. Assuming that we require, at all times, for Mach's principle, a density ρ_u close to the 'critical' value, such that (as both classical theory and general relativity require)

$$\frac{GM}{r} = \tfrac{1}{2}v^2$$

and
$$\frac{4\pi G\rho_u r^3}{3r} = \frac{c^2 r^2}{2r_u^2},$$

we obtain a mass within the radius r_u of

$$m_u = 8.23 \times 10^{52} \text{ kg}$$

and, to a first order approximation, a mass density

$$\rho_u = \frac{3}{4\pi}\frac{m_e}{r_u^3} = 1.08 \times 10^{-26} \text{kgm}^{-3}.$$

This would be equivalent to a baryonic density of 6.47 m^{-3}. Making the assumption that about 4 to 5 % of the mass is of this nature, we obtain a nucleon density of approximately 0.26 to 0.335 m^{-3}. (This value for ρ_u would make the vacuum density = 7.53×10^{-27} kgm^{-3}, and the cosmological constant $\lambda = 8\pi G\rho_{vac}$ = 1.207×10^{-36} s^{-2} = 5.8×10^{-124} $8\pi G\rho_{Planck}$.)

An alternative way of calculating the baryon density would be to assume that the vacuum process which creates the inertial masses of the real electrons in the universe is equatable, at least in approximate energy density terms, to aT^4. That is, if N_e is the particle density of the real electron masses produced by the vacuum process, then

$$N_e m_e c^2 \sim aT^4.$$

This leads to a value of 0.5 m^{-3} for the particle density, which is of the same order as that of the proton density calculated from a baryonic m_u (it would be exactly equal if baryonic matter made up about 8 % of the universe's mass), and we may assume that the numbers of electrons and protons in the universe are at least approximately equal. If the equality is real, and can be discovered by more precise calculations, then we may take the vacuum energy randomly available to any individual proton (which, in thermal units, can be described as kT_p) to be of order $m_e c^2$. With the difference in neutron and proton masses determined at ~ 2.3 $m_e c^2$, the statistical Boltzmann factor exp ($-2.3\ m_e c^2 / m_e c^2$) fixes the neutron / proton ratio at order 0.1 or 10 %, without recourse to any cosmological model. Also, if N_γ is taken as the density of photons of approximate energy kT, then equating $N_e m_e c^2$ and $N_\gamma kT$ fixes the photon / electron ratio at ~ 2×10^9, with the photon / proton ratio at a similar order.

Specifiying the 'creation' of electron mass as a vacuum process further implies that we can define the rate at which the energy $m_e c^2$ becomes available to a proton, converting it with a probability exp (–2.3), as being equivalent to the rate at which the same amount of vacuum energy is converted into thermal energy (in the form kT_p) via redshift, ~ $(8\pi GaT_p^4 / 3c^2)^{1/2}$, and then relate this to the reaction rate for the weak process of proton-to-neutron conversion, ~ $(G_F^2 m_e^5 c^2 / m_p^4 \hbar)$, given the same availability of vacuum energy. Although the calculation can only be done approximately, because the expressions for the process rates are not exact, it suggests an order of magnitude value for the Fermi constant which is in agreement with experimental values (G_F ~ 10^{-5} units of mass^2). This is important because the Fermi constant itself is solely determined by the weak coupling and the mass of the weak interacting boson ($G_F = g^2\sqrt{2} / 8M_W^2 = \pi\alpha_2 / \sqrt{2}M_W^2$), and the value of this mass is in turn fixed by the vacuum expectation energy of the Higgs field – the filled vacuum which provides the instantaneous gravitational action assumed in this analysis. Ultimately, then, it would appear that the classical inertial calculation put forward in the present account has the same fundamental meaning as the calculation of the inertial masses of particles through the Higgs mechanism in the Standard Model.

Some other consequences of the calculations may also be significant. Thus, using the equations

$$aT^4 = \frac{3Gm_e^2}{4\pi r_e^4} = \frac{1}{192\pi}\frac{e^2}{4\pi\varepsilon_0 r_e^3 r_u}$$

and

$$\frac{e^2}{4\pi\varepsilon_0} = \frac{2}{3}m_e r_e c^2,$$

we obtain the relation

$$\frac{m_e}{r_e^2} = \frac{m_u}{108 r_u^2},$$

which could be seen as the origin of the so-called ‘large number coincidences’ discussed by Eddington, Dirac and others. This is because numbers of the order of 10^{80} are derived ultimately from the ratio $m_u / 108\, m_e \sim m_u / m_p$, while numbers of order 10^{40} are derived from r_u / r_e, which equals $(m_u / 108\, m_e)^{1/2}$.

A number of authors have examined the large-scale structures in the universe (planets, stars and galaxies) in terms of the fundamental forces involved (gravity, electromagnetism and the weak interaction) and have concluded that the strengths of these forces, measured by their fine structure constants ($\alpha_G = Gm_p^2 / \hbar c$, for gravity, $\alpha = e^2 / 4\pi\varepsilon_0 \hbar c$, for electromagnetism), determines their optimum masses, radii and lifetimes.[16,17] Planetary radii, for example, are typically of order $(\alpha / \alpha_G)^{1/2} a_o / A$, where a_o is the Bohr radius and A the mass number of the molecular material involved. Stellar radii begin at $(\alpha / \alpha_G)^{1/2} a_o$, with masses in a narrow range around $\alpha_G^{-3/2}m_p$. Lifetimes are, significantly, determined by the product of $\alpha_G^{-1/2}$ and the Compton time for the proton, h / m_pc^2, which, by way of the so-called 'large-number coincidences' (i.e. the gravitational / inertial connection), makes them of the same order as the Hubble time. This does not, therefore, necessarily signify that they have evolved over the same 'lifetime' as the universe itself – only that fundamental gravitational / inertial processes are characterized by numbers involving the Hubble scale. The arguments used are purely physical, and have nothing to do with an evolutionary cosmology.

The evolution of galaxies is less well understood, but it is possible to specify a critical (maximum) radius for their formation, $R = \alpha^4\alpha_G^{-3}(m_p / m_e)^{1/2}a_0$, and a maximum galactic mass, $M = \alpha^5\alpha_G^{-2}(m_p / m_e)^{1/2}m_p$. Lifetimes of galaxies are still unknown, but a gas cloud of these proportions would require a time of $(GM / R^3)^{-1/2}$ to collapse under Newtonian gravity, again producing a time scale within an order of magnitude or so of the Hubble time. It is evident that the Hubble scale is intrinsic to the *physics* of the observable universe, irrespective of the cosmological argument used to explain this. In addition, there is evidence that the same is true of the microwave background radiation. The energy density of starlight, for example, in the Milky Way, is approximately the same as the energy density of the microwave background, and was used, well before the discovery of the latter, to fix the temperature of the 'universe' at approximately 3 K. The same applies to the galactic magnetic field and other indicators, such as cosmic rays. It seems likely that such 'coincidences' will be found in all systems where gravity plays a major role, reflecting in this way the intimate connection between the gravitational interaction and the vacuum 'creation' of inertial mass.

In the approximate set of calculations presented here, of course, assumptions have been made here about uniformity, and so forth, while relativity has not been fully taken into account in the radiation formula. Small numerical factors (resulting from relativity, etc.) may also be ambiguous in a few instances (for example, the GR / inertial factor 4, which seems to be essential to the calculations). No real account has been taken of the actual structure and composition of matter in the universe. So we can hardly expect more than order-of-magnitude agreement with our estimates. However, it is possible that the calculations give an indication that a real, and quantitative, connection between the cosmic microwave background radiation and the cosmological redshift may exist in a purely physical perspective, independent of any cosmological model, and that both radiation and redshift are, in some senses, aspects of the same physical phenomenon. In addition, it would seem to suggest that the inertial and the gravitational energies associated with generating the electron mass from vacuum could be equal to that associated with a microwave background at T = 2.72 K, and that the inertial process could be responsible for generating the background energy.

21.7 The Concept of Creation

Both physics and mathematics require a concept of 'creation'. In the universal alphabet proposed in chapter 2, it is the mechanism by which 'something' is generated, or appears to be generated, from 'nothing'. Of course, nothing still remains and so the idea of 'conservation' is generated simultaneously. We can describe the duality which results in many different ways, for example using iterative and recursive processes in a rewrite alphabet, or variable and conserved quantities in physics. In general, an alphabet can only be described *either* in iterative *or* in recursive terms, but a *universal* alphabet would contain both concepts simultaneously. This is exactly what is provided by the Dirac nilpotent, which, by being a nilpotent structure within an infinite Hilbert space, becomes specified uniquely, and allows us to look at a finite part of what we know must be an infinite whole. At the same time, the Dirac nilpotent packages the whole information needed to paramaterize nature in the way we recognize is physics, and, by being *self-dual*, it does it in such a way that the variable / conserved, iterative / recursive, or even process / object dualities disappear at the point where they reunite to produce the zero totality.

We can imagine the production of a fermionic state as a creation event, which has *within itself* – i.e. by its self-duality – the concept of immediate return to the zero state, and which we realise when we write down the Dirac equation. It is both iterative and recursive, process and object, variable and conserved. It is also both object and environment, in the sense of fermion and vacuum image, and each of these dual systems can be thought of as components in an infinite fractal hierarchy, organized at each stage according to the same fundamental principles. Marcer *et al* have suggested the application of Kenneth Wilson's renormalization group methodology for the determination of critical phenomena to the Dirac nilpotent package proposed by the author and colleagues, based on their earlier ideas of a self-referential cosmology, phase conjugate adaptive resonance and quantum holographic measurement with reference to the cosmological reference frame.[18] (See also chapter 20.) In the terms used in this book, and in earlier publications by the present author, the process of change – which we may describe as 'cosmology' – is driven by the variability which comes packaged within the idea of a Dirac object incorporating the symmetrical concepts of space, time, mass and charge. The 'symmetry' is a set of constraints which determine what is physically and conceptually possible in a universal system which structures itself according to the dynamic which the packaging necessarily contains.

Of course, many people believe it convenient to describe cosmology in terms of a single creation event projected backwards into the past using the Hubble time as a measure of the point of origin. Despite universal confidence at the certainty of such an event, no precise theoretical description of it has held without drastic modification for more than a few years. Beliefs that the universe is curved, enclosed, finite, non-Euclidean, or of subcritical or supercritical density, are now no longer considered to be compatible with experimental evidence, though they were quite firmly held only a relatively few years ago. The apparently neat idea that the creation event was effectively the reverse of the formation of a black hole-type singularity, which could then have been used as the process for the ultimate 'big crunch' in a supercritical universe, has also been abandoned has having the wrong singularity characteristics, and the probable infiniteness of space and the idea that our observed 'universe' is very likely only a local structure has made it increasingly difficult to accept the once widely promoted notion that the 'big bang' event saw the creation of time and space as well as matter. The inflationary universe, introduced in the early 1980s, to explain the relative closeness of the observed density to ρ_{crit}, destroyed the pleasing simplicity of the earlier theory, and made it seem less compelling. In addition, the values of the fundamental constants appear to be finely tuned to produce the structure of the observed universe *as it is now*, not as it would have been under hypothetically different conditions in the past.

Cosmology can solve none of the fundamental problems of physics, because the laws of physics seem to be structured, on fundamental grounds, to be true in any era, despite hypotheses that they have somehow evolved along with the structure of the universe. They cannot depend on the accidents of cosmology unless some extreme form of the anthropic principle is used to argue that the universe and the laws of physics have evolved specifically to allow our conscious observation of it at a particular epoch. On the other hand, the laws of physics, if fundamental in the way suggested, must certainly constrain the cosmological models that are possible. An inertial velocity and acceleration would seem to be required by the Machian argument of 20.1. The cosmology we adopt might seem to depend on how 'real' we take these to be. They are certainly real phenomena and produce real effects, but the forces involved are 'fictitious', in a somewhat novel way, and we could imagine that the physical effects do not manifest themselves directly as effects of motion. The way that physics constrains the cosmological models can be seen directly in the alternative approaches to the flatness of space, which both theory and experiment requires. An evolutionary ('big bang') model, based on real inertial motion and real inertial acceleration develops a 'flatness' problem, which is solved by an inflationary expansion; a non-evolutionary model, based on apparent inertial motion and apparent inertial acceleration, has a built-in flatness, and there is no problem to be solved. In addition, the evolutionary model has a further problem explaining acceleration if the universe is flat, and therefore at the critical density; the non-evolutionary model, however, *predicts* acceleration.

It would certainly be possible to describe the currently-favoured evolutionary cosmology in terms of the kind of Dirac critical phenomenon described above, and the very idea of a nilpotent packaging is suggestive in terms of a big bang event, as creation of a fermionic state also requires a simultaneous recreation of the entire universe. (Such a cosmology has, in fact, been hinted at in parts of chapter 20.) However, the 'creation' concept, which the Dirac nilpotent incorporates, need not necessarily be seen in terms of a singular event in a measurement context. The suggestion here is that the nilpotent carries with it all the information relating to inertial properties, including cosmological redshift and microwave background radiation, which is usually interpreted as time-sequenced cosmology but is considered here in a purely physical context. It is also suggested that the phenomena of particle physics, which are clearly of a fundamental nature, such as the matter-antimatter asymmetry, are also universal, rather than time-evolved, features. Only the large-scale structures of the universe, such as stars, galaxies, and galactic clusters appear to require time-evolution, though it would appear that the Hubble time component in their lifetimes is a characteristic of *any* inertial process. The origin and evolution of galaxies and clusters is not fully understood at present, and there is a great deal yet to be learned about their distribution, and that of other matter, within the observable universe, but nothing so far known compels us to accept that their present state and distribution must *necessarily* be the result of a single evolutionary process, rather than of a series of arbitrarily observed stages in a perpetual cycle, with structures ranging in age from 0 to ~ t_H. Even if we decide, on the evidence of galaxies and clusters, on a single overall process as an evolutionary driving mechanism for their formation, this would still only demonstrate that a large-scale *local* event had taken place, and, even then, there would have to be events which preceded it in time.[19] Most important of all, a single creation event would have to be seen as a *consequence* of processes ultimately necessary in physics, and not as their inexplicable *cause*. Redshift, background radiation, and all the fundamental constants associated with the world as we know it, must be consequences of the unchangeable laws of physics (and, ultimately, of the universal rewrite system) and not accidents of the particular circumstances of creation.

Certain aspects of the whole concept of deriving 'something' from 'nothing' suggest that there are absolutes which must be built into any physical and mathematical description of 'nature' which claims to be complete. Admit any element of compromise and the system ceases to be universal. So, the universe must be infinite in extent and duration because absolute continuity is an essential property of both mass and time. The same property ensures that there must be an absolute rest frame, which is that provided by mass – the one we describe as vacuum. This rest frame must also provide the universal time required by quantum mechanics for nonlocality, and equally by Newtonian physics, and by special and general relativity to preserve causality. Its existence implies that no 'communication' or 'horizon' problem exists with regard to events in parts of the universe at large distances from each other. But the rest frame cannot be the frame used for measurement, which is specifically *local*, and which the Dirac nilpotent proclaims as Lorentzian. An infinite universe also provides the infinite range of Dirac nilpotent values needed to develop that concept of uniqueness which is necessary to defining a universal alphabet; and it is, ultimately, this concept of uniqueness, deriving from an infinite which must exist, which is needed to preserve the zero totality with which we began.

Appendix A

Summary and Predictions

A.1 Summary of the Main Argument

Zero is the beginning and the end of all operations. Physics, mathematics, and epistemology / ontology emerge as attempts to create something are systematically reduced to nothing, which is the only permanent state and uncharacterizable by any process. If we attempt in any way to produce anything else, that is, in effect, to characterize the state, we set in process the means of forcing us back to the uncharacterized zero, and the continual reprocessing of this forces a continual reordering of the zero state.

Mathematics begins by assuming an uncharacterized 'something'; the zeroing process then sequentially introduces conjugation, complexification and dimensionalization, taking the process by repetition to infinity. Conjugation introduces ordinality; dimensionalization (or three-part cyclicity) introduces discrete enumeration. Arithmetic and algebraic processes can then be constructed (notably coming after geometry). Using discrete enumeration, the zeroing process can be interpreted as a repeated application of duality; the stages can be represented using finite groups. The process can be understood in terms of a computer rewrite routine, governed by the rule that a zero-totality alphabet, which reproduces itself by self-action or action on any subset of itself, is continually constrained to extend itself by such self-action to the next zero-totality alphabet.

Conjugation, complexification and dimensionalization can be represented in a symmetric (zeroing) way through the physical parameters space, time, mass and charge, which form a group of order 4, the respective mathematical representations of these parameters being multivariate vector, pseudoscalar, real scalar and quaternion. Fundamental aspects of parameters are derivable directly from their representation in this group. Examples include: the unipolarity of mass, the irreversibility of time, conservation laws of mass and three types of charge, existence of antimatter, gauge invariance, Noether's theorem, the Yang-Mills principle, and universal interactions between all masses and charges. Two of the parameters, space and charge, are dimensional and discrete, while two, time and mass, are nondimensional and continuous. The consequences of this include wave-particle duality and alternative forms of quantum mechanics and mathematical analysis. Dualities within the parameter group lead to a numerical factor 2 in alternative mathematical representations of many physical facts.

Conjugation, complexification, dimensionalization, and repetition, can be encompassed in a 64-part multivariate-4-vector-quaternion group equivalent to the Dirac algebra. Creation of the Dirac algebra from its eight vector, pseudoscalar, real scalar and quaternion elements leads to five composite elements, formed by taking one three-dimensional parameter (most conveniently, the conserved charge) and mapping it onto the other three. The Dirac algebra, constructed in this way, either incorporates or generates all the discrete and continuous groups of interest in fundamental physics, from C_2 to E_8.

The process of mapping the charge dimensions, which we describe as weak, strong and electric, respectively onto time, space and mass, creates the new composite quantities, Dirac energy, Dirac momentum, and rest mass, which are both quantized (dimensionalizing makes discrete) and conserved. At the same time, it breaks the perfect symmetry between the pure charge elements, creating new composite elements, which have the extra mathematical characteristics respectively of pseudoscalar, multivariate vector, and real scalar quantities.

The new five-part structure thus created short-cuts the route back to zero, by being a square root of zero or nilpotent, rather than of a positive or negative quantity. It appears to fulfil the requirements for a parameter space in *topos* theory, and offers a complete method of parameterizing nature in a single package. To demonstrate that it is a conserved quantity, we set it against a term representing the total nonconservation of space and time (in the case of a pure, unentangled, and noninteracting state), and apply a differential operator in space and time, whose eigenvalue is identical to the nilpotent. The use of such a differential operator is equivalent to setting up a parallel nilpotent relationship between space and time, with a third component ('proper time') occupying the position of the fixed quantity (rest mass).

The infinite nature of the dualling process requires the entanglement or superposition of all Dirac or 'fermion' and 'antifermion' states in the universe, in addition to the infinite interaction of any fermion with the vacuum, and the interactions of all charge and mass states. States whose spatial variations are confined by multiple interactions are described by classical approximations. From the parameter group, it is evident that measurement is only possible through the parameter space. If we take space (or possibly space and time) as 'epistemological', it is clear that the other parameters form an 'ontological' component. Neither epistemology nor ontology will fully describe our system.

Nilpotent structures necessarily require the existence of numerical relations between the units of their components. This leads to the existence of fundamental constants, such as $\hbar$, G, c, whose values have no intrinsic meaning. The scalar nature of rest mass and energy mean that the *special relativistic* relations between space and time, and momentum and energy, apply even beyond the case of the pure, unentangled and noninteracting state. The constants also allow us to develop conservation laws, or equivalent mathematical structures (Newton's laws and gravitational theory, Maxwell's equations) under conditions involving multiply interacting states assumed to be classical.

Both classical and quantum laws of momentum conservation lead directly, by a mathematical transformation, to an additional law of angular momentum conservation. By symmetry, it is possible to see that this is equivalent to the concept of the conservation type of charge (weak, strong or electric). This is only true for pure quantum systems; classical systems will break or weaken the connection through decoherence, which occurs through the vector nature of the angular momentum operator. In quantum systems, the conservation of weak, strong and electric charges becomes equivalent to the respective conservation of handedness, direction and magnitude of the angular momentum. Information about the charges associated with a particle is carried by its angular momentum state.

The nilpotent structure for the fermion and antifermion leads to a scalar structure for the boson or combination of the two. A two-fermion 'bosonic'-type state is also possible if the fermions have opposite spins (or opposite momenta for spatially-separated components). The second 'fermion' need not be an actual particle, but could be, for instance, a flux line or a geometric phase. Immediate consequences of nilpotency are the spin-statistics connection, Pauli exclusion, *CPT* symmetry, parities of fundamental particles, geometric phases, no Goldstone bosons but massive Higgs, etc. Analytical calculations of spherically-symmetric bound states yield solutions for the Coulomb potential (electromagnetic interaction), linear potential (the strong interaction), and any other polynomial potential (the spherical harmonic oscillator / weak interaction).

The nilpotent state vectors are automatically second quantized and supersymmetric. The nilpotent treatment of propagators automatically removes the infrared divergency. Fermions and bosons can be shown to have supersymmetric partners in the vacuum, allowing automatic renormalization without requiring the existence of explicit supersymmetric particles. QED, using perturbation theory, can be shown to be valid, in principle, without renormalization, although the usual rescaling results at higher energies. Parallel calculations follow for the weak and strong interactions.

The nondimensionality of the parameters mass and time determines that they are continuous – mass, here, having the meaning of 'mass-energy'. The continuity of mass-energy manifests itself as a filled 'vacuum', which, because of the association of the energy operator with the weak charge quaternion, is identifiable as a filled weak vacuum, or non-zero ground state for the weak interaction. The filled weak vacuum manifests itself through a violation of charge conjugation symmetry in the weak interaction, with a consequent violation of parity or time reversal symmetry. The symmetry violations require a single-handedness in weak-interacting states, which makes fermion states intrinsically left-handed and antifermion states right-handed.

The three-dimensionality of the angular momentum operator allows for the creation of a three-part (i.e. three-'quark') fermionic nilpotent for the baryon. From this we may derive an $SU(3)$ structure for the strong interaction; and, from this, it would appear that the strong charge is 'exchanged' through the angular momentum operator. Spin becomes a property of the baryon as a whole, not of the component quarks. Similar considerations, using the filled weak vacuum, the pseudoscalar and real scalar structure of the composite terms, and their relation to angular momentum conservation, lead to the $SU(2)_L \times U(1)$ electroweak combination. These considerations lead to an idealised mixing ratio of 0.25 between weak and electric interactions, at the stage where the relevant gauge boson (Z) is produced, and to a set of unit and zero charge structures for quarks and leptons which can be expressed in the form of tables (three for quarks, and one for leptons) or even by a single generating formula. From these it is evident that there are six quarks and six leptons, in three generations, each with two states of weak isospin, and that the second and third generations are associated, respectively, with violations of parity and time reversal symmetry. States which violate CP invariance are immediately identifiable. (They include all 'bosonic'-type states which are composed of two fermions, or equivalent, rather than fermion-antifermion.) All the significant results are essentially statements of the fact that weak, strong and electric charges are independent of each other in all respects.

Rest mass is generated when a 'particle' (baryon, boson or free fermion) acquires an element of right-handedness or an antiparticle acquires an element of left-handedness. The Higgs mechanism for generating mass is equivalent to the addition or subtraction of any term from the generating formula which increases this opposite-handed element. Three mechanisms may be identified, one concerned with isospin, one with the quark-lepton generation, and one with the appearance of zero charge units in composite baryon and boson structures. These have been applied to generate the actual masses of the known particles, and the vacuum state (with varying degrees of certainty), in addition to the mixing parameters, and to speculate that the mass of the Higgs boson may be 182 GeV.

Both the tables and the fundamental formula suggest that the electric charges in both quarks and leptons have fundamental integral values, though, because of gauge invariance in the strong interaction, those for the quarks measured by QED at any energy will come out at fractional values. A consequence of these integral values, and of the value 0.25 for the electroweak mixing parameter, is that grand unification will occur at the Planck mass, the energy associated with quantum gravity (and effectively a pure number, determined only by G, $\hbar$ and c). Also, at grand unification, the group structure becomes $U(5)$, incorporating gravity or at least its inertial effect, and all the interactions become identical Coulomb-like phases. There is no electroweak structure at grand unification; the mixing parameter is 1. The value of α is predicted to be 1/118 at 14 TeV, and can also be tested against prediction at much lower energies. Each of the three fine-structure constants is determined absolutely at all energies if one can be fixed at a single value. The best candidate is for α_3 (the constant for the strong interaction) to be 1 at the 'zero-charge' mass used in the Higgs mechanism. This is certainly approximately true (exact perturbative calculations are difficult for the strong interaction at these energies), and, if this, or anything like it is true, the scale for both masses and charges is established on an absolute basis, with no empirical input whatsoever.

Although the grand unification process provides a numerical link with gravity, this force is distinct from the other three in that the source of gravity is continuous, not discrete. This means that the gravitational interaction is instantaneous, and is the ultimate mechanism for the instant correlation or nonlocality required by Pauli exclusion. Since measurement (which becomes possible through entanglement) always involves c-related time delay, then the

Lorentzian space-time of measurement (which is itself affected by gravity) cannot be used to describe the action of gravity without the use of fictitious 'inertial forces', which may well be the origin of the concept of inertia in a Machian sense. The process involved has been described as the 'aberration of space'. Using this gives us a linearised understanding of the GR field equations, with all the usual test results, in addition to a prediction (independent of any cosmological model) of redshift, accelerating with deceleration parameter –0.5. The last was predicted a considerable time before its experimental discovery. The linearised equations provide a theory of inertia, which, as a repulsive force, may be interpreted through a gauge theory with a spin 1 boson, essentially equivalent to QED. The numerical link between gravity and the other forces at grand unification is presumably via its inertial equivalent.

Some preliminary calculations have been made on classical and relatistic calculations of the Larmor radiation associated with the calculated redshift, which may possibly be associated with the thermalized cosmic microwave background radiation. All the cosmological work related to this theory has been based on the idea that the universe is flat and infinite (exactly as is now supposed on the basis of current experimental results), although discrete measurement involves a *c*-related event horizon, whose parameters (Hubble mass and radius) are susceptible to calculation on the basis of Machian assumptions, etc. An infinite universe is a consequence of a continuous definition of mass-energy or vacuum; while continuous time supposes one without beginning or end. In addition, the fundamental necessity for a filled vacuum or non-zero ground state for the universe means that no cosmological origin is necessary for the predominance of matter over antimatter. While the total number of fermions and antifermions is the same, the majority of the latter are vacuum states.

It is assumed in the present theory that 'localized' energy (rest mass) is created by the inertial process, and that it determines the degree of time delay (or localization). It is supposed that energy is not 'transferred' from place to place, and that, though the energy density of the vacuum is infinite, its level has a finite value, calculable from first principles, along with the masses of the particles. The thing that changes is the degree of quantization in the fermion nilpotent wavefunction.

The direct connection between the conservation laws for charge and angular momentum is assumed broken, when a system becomes semi-classical, or no longer isolated, as when a 'measurement' (or multiple entanglement) is made. In this case, the system is no longer purely conservative, but loses some energy to the 'rest of the universe' with which it interacts, and through which it partially decoheres. The first and second laws of thermodynamics are expressions of this process, and are consequently linked to the direction of time.

The structure which apparently provides the foundation for both mathematics and physics appears to represent a universal rewrite system, which is Nature's own most efficient way of information processing. It seems probable that the same kind of information processing structure, based on nilpotency and the algebraic and geometric patterns needed to establish it, will also operate at higher scales of natural organization, in particular, in the creation and reproduction of biological systems. It is remarkable that the genetic code follows precisely the same logical structure as nilpotent quantum mechanics, the culmination at the (maximal) icosahedral symmetry occurring at exactly the same point in the rewrite system as nilpotency emerges in quantum mechanics. Both systems operate from the rewrite structure's process of 'doubling the dimensionalities', using a virtual, as well as 'real', world, to create the logical space for change.

An investigation of both 'Nature's code' and 'Nature's rules', as they are applied in biology, cosmology or even human perception appears to justify the application of a processing system with a fractal structure, or one that shows the same form at all levels. The system which we have described as 'Nature's process' looks like quantum mechanics not only because the physical world is quantum mechanical but also because quantum mechanics has its own origin in the fundamental hierarchy. However, the quantum mechanics needed here is a fermionic one with an infinite semantic logic (based on 0 and ∞), not a finite syntactic logic (based on 0 and 1), which is centred on bosonic states, and we predict that realization of this fact will lead to a new area opening up in AI studies. Ultimately, this is because nilpotency requires a holistic view of Nature, including human perception, in which we can only comprehend the part by instantly connecting to the whole.

A.2 Predictions

Though it is often claimed that physical theories are justified by their predictions, this is seldom the main criterion used for adopting a fundamental theory. It is more usual to adopt theories because of their wide-ranging explanatory power. Successful predictions often rely on model-dependent approximations to the theories rather than directly on the theories themselves, and these developments often follow only after the theory has already been widely accepted. Newtonian gravity, for instance, had to wait seventy years for its first successful prediction, while Maxwell's electromagnetic theory took nearly twenty-five; and in each case the predictions were made by authors other than the theories' discoverers. In more recent times, quantum mechanics only made a few early testable predictions because it was taken up by many researchers immediately on account of its explanatory power, while general relativity was not genuinely tested for many years after its supposed experimental vindication, and then only at a considerable remove from its real fundamental basis. In nearly all these cases, the early predicted results could have been obtained using alternative theories, and in some cases they were. Antimatter, for instance, was long a requirement of aether theorists before it took the form of negative energy states in Dirac's relativistic quantum mechanics. Eventually, of course, successful fundamental theories show their true validity by making predictions of outcomes from many different sets of starting conditions but this only occurs after decades of refinement and original development that the theories' original authors could not have imagined.

While string theory, in particular, has been criticized for its lack of predictions, the more serious objection to that theory is that it does not seem *possible* to imagine making predictions from it that will ever be testable. In addition, the theory – unlike all the others so far mentioned – seems to lack any ability at arriving at results already known; and the number of starting assumptions would seem to outweigh any possible simplification that might result from general applicability. In the case of the theory outlined in this book, the argument for its validity – like the argument for all the fundamental theories now generally accepted – must be its wide-ranging power of explanation, in both conceptual and mathematical terms, of physical facts already known, together with its use of minimal assumptions. That the theory has predictive power should be obvious from its structure, which allows for extensive mathematical development, but I would no more expect to be the author of these predictions than I would have expected Heisenberg or Schrödinger to be the author of all the predictions that have since come out of quantum mechanics. However, the theory *has* made some predictions already, and, although these may be of varied status and viability, collectively they show that the theory can be used in a considerable number of predictive modes.

One unusual aspect of the theory is that it makes predictions of new *mathematical* theorems on the basis of symmetry alone. The most remarkable are those which connect the conservation of angular momentum and charge type, and the conservation of linear momentum and charge magnitude, in an extension of the application of Noether's theorem to the parameter group. These otherwise unexpected connections were predicted as early as 1991, and have now been completely confirmed by the discovery of the *mechanism* by which the conservation of strong, weak and electric charges is associated with the respective $SU(3)$, $SU(2)$ and $U(1)$ Lie groups determining their interactions. This establishes the first theorem and its mathematical basis, while the second theorem follows automatically from the direct connection between linear and angular momentum required by relativistic quantum mechanics. When the results were first predicted, there was no obvious reason why they should be true, except from a reliance on absolute symmetry, and the connections seemed extraordinary and rather counter-intuitive. Such results are, therefore, the best possible test of the theory's validity. Somewhat less remarkable, but still of interest, is the direct prediction of both T duality and S duality (among many other such dual possibilities) from the parameter group as early as 1979. This is because the main application of these dualities has been in string theory, which still awaits canonical status. However, the successful use of the dualities in string theory suggests that there may be a more fundamental underlying reason for their appearance in the physics of fundamental particles.

The grand unification argument in chapter 15, first published in 1999, makes a number of predictions which can be tested experimentally with particle accelerators. Grand unification is predicted to occur at the Planck mass, while the electric, strong and weak fine-structure constants and the weak mixing angle are predicted for all higher energies up to this value. In particular the electric fine-structure constant is predicted to be 1 / 118 at 14 TeV, as opposed to 1 / 125 for minimal $SU(5)$.

More qualitative predictions are also available for particle physics. Hyperentanglement is a fundamental requirement of the nilpotent representation of bosons and baryons (1998) and is now observed. The nondecay of the proton is a fundamental requirement of the conservation of type of charge (1979) and has been maintained, despite arguments to the contrary from many grand unification theorists, and some early experimental results suggesting that it had been observed. This negative prediction, which appears to run counter to all others made so far, has been validated by the most accurate tests to date. It is, however, predicted that, though there is no direct strong-electroweak proton decay, there is a finite amplitude for a strong-electroweak component in ordinary beta decay (1991), though it has not yet been possible to derive a numerical value. Another result expected from the current theory is a weak dipole moment

(2001); it may be that a numerical value can in some way be associated with neutrino mass.

A *CP* violation in the decay of all bosonic states made from mixed generations has been an intrinsic component of the theory since 1979. Experiment has been gradually discovering these, but, while conventional theory may have paralleled this prediction to a considerable extent, the theory discussed in this book has additionally predicted *CP* violation from states involving nonzero Berry phase (1998); the discovery of materials like graphene with interesting versions of the quantum Hall effect suggests that this may be testable in the near future using condensed matter. A very definite prediction of the theory is the nonexistence of supersymmetric particles (1998). This has already been partially vindicated by experiments which find no evidence for squarks, the supposed supersymmetric partners to quarks, up to 400 GeV. While it has been possible to suggest a value for the Higgs mass at 182 GeV (1998), this is higher than other estimates, and, is only a guess based on model-dependent assumptions, but it is the only one I know of based on any kind of first principles.

A number of astronomical predictions have emerged as by-products of the main analysis. The flat and infinite universe required from the beginning (1979), and in opposition to every view then held by mainstream astronomers, has seemingly been vindicated by measurements made since 1998. The prediction of dark energy at 2 / 3 of the value of the energy of the universe, or equivalent, made as early as 1979, could be said to have been comprehensively fulfilled by results since 1999, even though it is frequently stated that dark energy was an 'entirely unexpected' discovery, and 'no-one' predicted it. The 2 / 3 value, of course, is an approximation based on the assumption of absolute uniformity, but it is still very close to the percentage estimated from the astronomical observations. Perhaps less significant is the 2.72 K predicted for the background temperature as early as 1990, as the model it requires is less well-established. However, the prediction was made at a time before experimental results had closed in on values close to this one.

Overall, we can say that, although the theory outlined in this book does not depend for its validity on its ability to make predictions, that ability is clearly apparent in the twenty predictions covered in this appendix, many of which have now received total or partial confirmation. This is already a relatively high total for a new theory, and there is every reason to believe that others would follow if aspects of the theory led to new developments by other researchers.

Appendix B

The Infinite Square Roots of –1

This is a summary of a paper written with **BERNARD DIAZ**. To structure the universal rewrite system in a practically usable form, we propose a computer representation of the infinite square roots of –1, using *D* as a symbol in the universal alphabet. For a universal rewrite system of the type we have suggested there is the need to determine the nature and symbols of the alphabet generated at the complexification stage in the iterative rewrite sequence. Here, we think of *D* as an infinite table of 1's in any representation, e.g. binary or hexadecimal. Any specified column D_i of the table has the property that when multiplied with a row D_i, results in a representation of –1. D_i multiplied with D_j anticommutes as $-(D_j*D_i)$ and produces D_k in a way identical to the quaternions $\boldsymbol{i}, \boldsymbol{j}$, and $\boldsymbol{k}$. With an infinite and uniquely identifiable set of such triad forms, *D* can be considered both a symbol and, because of this behaviour, an alphabet.

We have, of course, a zero sum result at all times. Using the iterative explanation, we use the operators 'create' and 'conserve' to generate, then check for, new symbols in the evolving alphabet. We start with the symbol and an alphabet consisting of just 0. In a first step we 'create' the symbol 1 and to return to the zero sum we generate its conjugate (in the 'conserve' process), the symbol –1. From –1, we can generate the infinite square roots which when considered with their conjugate forms, e.g. $\pm\boldsymbol{i}$, $\pm\boldsymbol{j}$, etc, and, with anticommutativity, drive the identification of further new alphabet symbols. We need now to represent in a computationally tractable form these infinite square roots such that they avoid the requirement for an infinite symbol sequence: $\boldsymbol{i}, \boldsymbol{j}, \boldsymbol{k}, \ldots$, and we suggest that an acceptable solution that captures the notion of an infinite sequence of the square roots of –1, as required, is *D*, where the symbol chosen is arbitrary and represents all the infinite square roots.

We begin by establishing that there does exist an infinite number of square roots of –1.[1] Although several routes to a proof exist, most depend on the observation that the set of quaternions that square to –1 is the infinite set of vectors of absolute value 1.[2] The demonstration of an infinite number of solutions is a corollary of the lemma that the square of any unit (quaternionic) vector is –1.[1] If μ is an arbitrary unit vector defined in terms of a Cartesian representation, its square is given by $\mu = (\boldsymbol{i}x + \boldsymbol{j}y + \boldsymbol{k}z) / b$ where b is $\sqrt{(x^2 + y^2 + z^2)}$ and x, y, z are real, and $\boldsymbol{i}$, $\boldsymbol{j}$, and $\boldsymbol{k}$ are mutually perpendicular unit (quaternionic) vectors that follow the rules defined by Hamilton in 1843.[3] The square is:

$$\mu^2 = (\boldsymbol{i}x + \boldsymbol{j}y + \boldsymbol{k}z) / (x^2 + y^2 + z^2)$$
$$= (\boldsymbol{i}^2x^2 + \boldsymbol{j}^2y^2 + \boldsymbol{k}^2z^2 + \boldsymbol{ij}xy + \boldsymbol{ji}xy + \boldsymbol{ik}xy + \boldsymbol{ki}xz + \boldsymbol{jk}yz + \boldsymbol{kj}yz) / (x^2 + y^2 + z^2)$$

and if we substitute Hamilton's rules $\boldsymbol{i}^2 = \boldsymbol{j}^2 = \boldsymbol{k}^2 = \boldsymbol{ijk} = -1$ we get:

$$\mu^2 = (-x^2 - y^2 - z^2) / (x^2 + y^2 + z^2) = -1.$$

As there are an infinite number of unit vectors there are an infinite number of unique solutions to the equation. All of these will follow the anticommutative Hamilton rules in a cyclical way as for example:

$$\boldsymbol{ij} = -\boldsymbol{ji} \quad \text{and} \quad \boldsymbol{jk} = -\boldsymbol{kj}, \quad \text{etc.}$$

The infinite solutions can also be shown to be a consequence of considering Euler's formula: $e^{ia} = \cos a + i \sin a$, for points on a unit circle where each circle is in the plane of one of the infinite planes that constitute the unit sphere in quaternion space.[2]

The demonstration of an infinity of square roots of –1 leads to a discussion of the computational methods[4] of representing –1 and the observation that this maps through quaternion tesseral addressing to a specific tile, and in the limit, to a point in space.[5] To establish the representation we now borrow notions of infinite series of digits from other number representation domains (e.g. p-adic representation[6]) and create a bracketed notation to simplify our handling of the infinite sequence. Using this and the computational and tesseral representation explanations we will be able to construct a meaning for the *D* symbol to represent the infinite square roots.

In the now standard Booth representation of negative integers in computers,[4] wordlength is fixed and using a twos complement binary form[7] every bit combination of wordlength is used, approximately half yielding positive and half negative integers. In such computational notations –1 is a sequence, wordlength long, of repeated 1's. For general use, we can define an infinite series of 1's to represent –1, which we present as:

| 1 1 1 1 1 1 1 1 1)

where |, is repetition to the left infinitely and) repetition of 1 infinitely to the right. One is reminded of p-adic numbers,[8,9] infinite tesseral addresses,[10] and the bra ket notation.[11] (There is no need in this appendix to invoke the convention that the repeating digit sequences are identified with an overline as only one digit repeats here.)

An alternative method of establishing the same representation based on the description of tesseral quaternions[5] is to consider quaternion space halved at each division by four orthogonal hyperplanes. This process gives rise to an origin, orthogonal axes, and 16 equal divisions of the 4D space. If we label each axis division either side of the origin with a binary 0 and 1, then each division of space will have a 4 digit binary address 0000 to 1111. To distinguish these we can label them using hexadecimal digits 0-F. If we take the hexadecimal 1 division (binary labelled 0001), and repeat the space sub-division process hierarchically we generate a tiling with tesseracts, where one such space (point), in the limit, has the infinite hexadecimal 1 address that can be labelled in a number of ways, including |1111) as above or simply just |1). Provided we retain exactly the same algorithm in dividing the space and numbering sub-divisions, the hexadecimal label |1) will be adjacent to |0) and in the limit will also be adjacent to |2), |3), ..., |F). An arithmetic based on the quaternions can be generated for these addresses where addition corresponds to translation through the 4D space and multiplication to scaling and rotation. (It should be noted that if we restrict the representation to 3D, the labelling is exactly that of the computer space storage structure known as an octtree[5] which is much used in computer graphics and image processing as a spatial data storage structure.[12])

Given the notation established above, and using * to indicate multiplication, we now ask what is the representation of the square root? We seek to find *D* where:

$$D * D = -1$$

or more specifically given the notation already established:

$$D * D = |\,1\,)$$

If we assume that the square of –1 is 1, then an answer might be:

$$|{-1}) * |{-1}) = |1)$$

or in a more expanded representational form:

$$|{-1}\ {-1}\ {-1}\ {-1}\ {-1}\,) * |{-1}\ {-1}\ {-1}\ {-1}\ {-1}\,) = |\,1\ 1\ 1\ 1\ 1\,)$$

where we take each –1 in turn from one representation and multiply it with its corresponding element in the second representation. Although similar to vector and tensor representation, the infinite nature of the process renders it tangibly different.

We can achieve a more intuitive explanation by noting that a simple resolution of a representation for the form |–1) is one in which each –1 in the row representation shown above is written as a |1) in columnar form, that is each –1 in the row representation above is replaced by a column |1), resulting in an infinite table of 1s:

$$\begin{array}{lcl} -\,-\,-\,-\,- & & \\ |\,1\,1\,1\,1\,1\,) & & \_ \\ |\,1\,1\,1\,1\,1\,) & = & |\,1\,) \; = D \\ |\,1\,1\,1\,1\,1\,) & & \cup \\ \cup\cup\cup\cup\cup & & \end{array}$$

which we read as an infinite series of column and row 1's and define generically, simply as D, the closest symbol to just the bounding brackets of the infinite table. In this generic form we can more closely imitate row / column vector behaviour by considering a column of –1's multiplied by a row of –1's each taken one at a time, to yield |1).

Although similar in arrangement to an infinite matrix, or tensor, none of the mathematical properties of these forms should be assumed. This representation is simply tabular, with the table interpreted in terms of the rows and columns. Each row and column can be numbered to identify specific rows and / or columns uniquely in an enumerated form of the representation.

We can define a property of each enumerated row form D_i that it multiplies with a row form D_i in a column vector with row vector way to generate |1) as the result. In practice each D_i, may be simplified for computational purposes to just one (wordlength) column of 1's and one (wordlength) row of 1's, giving rise to the enumerated mathematical form, e.g.:

$$\begin{array}{c} - \\ 1 \\ 1 \\ 1 \\ 1 \\ 1 \\ \cup \end{array} \; * \; |1\,1\,1\,1\,1\,) \; = \; |\,1\,)$$

Although it does not matter which row and column is used, in this enumerated but fixed form there are an infinite set of such identical ways of generating |1). All of these are drawn from the generative formulation D but labeled D_1, D_2, D_3, … D_n. An alternative labelling might be the ***i, j, k***, … (as used by Hamilton), a symbolism we were seeking to avoid, but which we can use to describe the behaviour.

We have seen that D, in its generic form when squared yields –1 which is |1), and that this is exactly the same as the use of complex i to represent $\sqrt{-1}$. A conjugate $-D$ also exists and would be a symbol in the universal alphabet generated by the conserve step. Its square is defined within existing rules; thus we get:

$$-D * -D = (-1 * -1) * (D * D) = -1$$

All other arithmetic associated with D in this generic form follows the rules associated with complex i.

We turn to consider the product of separate enumerated forms, i.e. the result of $D_i * D_j$. In general these cannot be resolved until we know which row and column formulation and wordlengths are used to define D_i and D_j. When these are known, the product collapses to the anticommutative form and a closely related enumerated form, D_k. Thus:

$$D_i * D_j = -(D_j * D_i) = D_k.$$

where also:

$$D_j * D_k = -(D_k * D_j) = D_i$$

$$D_k * D_i = -(D_i * D_k) = D_j$$

These are Hamilton's rules as outlined above and illustrate the anticommutativity required at the outset. Their effect can also be demonstrated using a matrix representation of each form.

Finding conjugates of the enumerated forms in this way identifies one of an infinite number of three-fold groupings (triads) that are closely related. Each of these triads allows us to 'create' distinguishable new symbols in the universal alphabet matched by conjugates which are then reserved and have this defined behaviour.

Other arithmetic properties of the enumerated forms follow Hamilton's quaternion rules with any triad's behaviour reserved in the same way. With this restriction we can also view the arithmetic as the arithmetic of 'tiles' where in the limit each tile is a point in 4D space in exactly the way argued for tesseral quaternions by Bell and Mason.[5] The geometric interpretation of this, for example that a multiplication of address labels using tesseral methods can be generated and can be set to corresponds to rotation and scaling of the tile point set, follows from this observation.[13]

We have proposed a single symbol-based method of extending the alphabet needed for the complexification stage of the universal rewrite mechanism. This complexification stage alphabet now consists of the symbols (0, 1, –1, *D*, and –*D*) with a method for extending the symbol set indefinitely by considering and resolving the nature of the enumerated form triads as and when required by the iterative procedure. Thus *D* and –*D* each constitute alphabet generators as well as symbols in the existing alphabet. We have already shown how this process can be used to generate the nilpotent Dirac equation using the anticommutativity property as the method that determines if a symbol (or indeed an entire sub-alphabet) is 'new'. To program such a process without an infinite number of symbols we need an enumerable method for identifying the infinite square root for –1. This has been achieved here by extending the common Booth method of representing negative numbers but done in two directions to yield a table of 1's, which we call *D*. As any indexed row / column of *D* when squared gives the repeated 1 representation of –1, that is |1), this provides a method for constructing an appropriate rewrite system. An implementation of this approach will need to examine the braiding pattern implied by this symbol creation and use it to uniquely identify each created symbol.

References

Preface

1 S. Wolfram, *A New Kind of Science*, Wolfram Media, 2002.

2 Among the very earliest ideas (c 'Year 8') was an attempted tachyonic solution to special relativity (at $v > c$) which showed the complex relation between space and time, and which also suggested a complex partner to mass. Eventually, Coulomb's law and dimensional analysis pointed towards the concept of charge. The tachyonic process also seemed to be responsible for some kind of creation of discontinuity, discreteness or compactification. The idea of some kind of overall symmetry involving these parameters was already there in embryo. Also very early was a belief that the velocity of light was not a primary concept, and that an abstract approach was more fundamental. 3-dimensionality was a key idea from the beginning; the quaternion algebra, which first surfaced in historical studies a little later, seemed to be the only way of making sense of it. Another important idea was that there must somewhere be a unique series, a unique numbering or a unique mathematical structure, which could not be repeated. An 'algebra of uniqueness' was devised. All these ideas are now fundamental components of the work in this book, except the tachyonic process, which lacked the necessary abstraction.

3 L. Smolin, A crisis in fundamental physics, *Update*, January / February 2006.

3 I had fruitful discussions as a student with Bob Warwicker and Gordon Green, who encouraged my efforts at solving the problem of symmetry-breaking between the interactions. At a later period, four of John Cullerne's students, now pursuing physics-related careers – Philip Muller, Josh Wilcox, Adam Brown and Adam Gamza – worked with me for short periods. Philip Muller and Josh Wilcox devised a computer program to show that the algebra in 13.2 really did uniquely produce the tables in 13.5.

Chapter 1

1 P. Atkins, *Creation Revisited*, Harmondsworth, 1994, p. 23.

2 N. Young, *An Introduction to Hilbert Space*, Cambridge University Press, 1988, p. 59.

3 D. Deutsch, A. Ekert and R. Lupacchini, Machines, logic and quantum physics, arXiv:math.HO/9911150v1, 1999.

4 R. M. Santilli, *Algebras, Groups and Geometries*, **10**, 273, 1993.

5 We may note the argument in M. A. Heather and B. N. Rossiter, The universe as a freely

generated information system, *Conceptions*, 2006 (*Proceedings of XXVII ANPA Conference*, Cambridge, August 2005), 357-388, for a category theory description of the universe terminating at the fourth level; and the four-component 'Combinatorial Hierarchy' described in 20, note 30.

6 D. Hestenes, *Space-time Algebras*, Gordon and Breach, 1966.

7 W. Gough, Mixing scalars and vectors – an elegant view of physics, *Eur. J. Phys.*, **11**, 326-3, 1990.

8 H. von Koch, Une méthode géoméetrique élémentaire pour l'étude de certaines questions de la théorie des courbes planes, *Acta mathematica,* **30,** 145-174, 1905.

9 B. B. Mandelbrot, *The fractal geometry of nature*, W. H. Freeman, 1982.

10 N. Chomsky, Three models for the description of language. *IRE Trans. On Information Theory,* **2(3)**, 113-124, 1956.

11 S. Wolfram, Some recent results and questions about cellular automata, in. J. Demongeot, E. Goles, and M. Tchuente (eds) *Dynamical systems and cellular automata,* Academic Press, 1985, pp 153-167.

12 P. Naur *et. al*, Report on the algorithmic language ALGOL 60, *Communs of the ACM*, **3(5)**, 299-314, 1960.

13 P. Prusinkiewicz and A. Lindenmayer, *The algorithmic beauty of plants*, Springer-Verlag, 1990.

14 Denying the significance of quaternions must be the most obvious example of missing an 'open goal' in the history of physics. It should have been obvious to anyone (as it certainly was to Hamilton) that there was no other route to an explanation of 3-dimensionality. It is certainly within this context that quaternions (as revealed through historical studies) had significance within my own early researches – anything that is 3-dimensional must ultimately be quaternionic.

15 A. Einstein, *Relativity: The Special and the General Theory: A Popular Exposition*, translated by R. W. Lawson, from the German edition of 1916 (London, 1920, fifteenth edition 1954, reprinted London, 1962), Appendix I, 120.

16 A. W. Conway, *Proc. Roy. Irish Acad.*, 29, 1-9, 1911.

17 V. V. Kassandrov (arXiv:gr-qc/0602988 v1) has demonstrated that the biquaternion (i.e. complex quaternion) algebra is the group of physical space-time, with a causality term (i.e. proper time, see 3.8, 8.3 and 8.4) that is always positive definite, and a hidden geometric phase that accompanies any displacement in Minkowski space-time. The geometric phase (see 12.8) can be associated with the origin of spin in our formulation.

Chapter 2

1 The generalised concept of 'charge' (and the general usage of this word to describe it) does not seem to appear early in the literature. I can remember having to explain it in some detail even in lectures of the 1990s, even though grand unified theories were being proposed as early as 1974. My own earliest publication of the idea was in a one-page abstract, dating from 1978, of a paper circulated in 1977. I don't know of any earlier mention of the idea in published sources. By that time, I had managed to justify the idea by finding the double 3-D explanation of symmetry-breaking through using the concept of conservation as implying

different behaviour from that of a quaternionic mathematical structure (see 13.5, etc.). It wasn't really until this concept emerged that the idea of a generalised charge became viable.

2 G. J. Whitrow, *The Natural Philosophy of Time*, Nelson, London, 1961, pp. 135-57.

3 P. Coveney and R. Highfield, *The Arrow of Time*, London, 1990, pp. 28, 143-4, 157.

4 A. Robinson, *Non-Standard Analysis,* Princeton University Press, 1996, original publication, 1966.

5 It is also difficult to imagine how the 2-D metric could be set up without using a third.

6 The need for limits may suggest that the Conway process for generating numbers may not include the solution of Zeno.

7 This is the T duality, which makes space with the topology of a circle in at least one direction transform from radius R to $1 / R$ in the dual theory. So, the equivalence of space and its inverse in the parameter group (predating string theory) could be considered a successful prediction. A similar concept, S duality, maps from one type of coupling or charge (say, weak) to the inverse of another (say, strong) in a dual theory. In the present theory, the parameter group makes charges equivalent in principle and also equivalent to their inverses, and so encompasses this possibility (and also the idea that the individual parameter-inverse dualities may be combined, as they are in string theory's U duality). The dualities of the parameters and their inverses relate, ultimately, to the fundamental duality between zero and infinity, which is an underlying theme of the present book. There are clearly many other dualities which must feature within the structure of the parameter group, and could become aspects of particular model-dependent theories.

8 John Valentine has pointed out that, if we assign 1 and 0 values to the properties and antiproperties, we can represent the relationships between them using a logical function, like XOR. My own representation favours XNOR, but the two functions are effectively equivalent. So, for any two inputs (e.g. time and mass), the output is that of the truth table for this function (in this case, charge).

	space	time	mass	charge
real	1	0	1	0
nonconserved	1	1	0	0
discrete	1	0	0	1

9 I. Newton, *Opticks*, 1706/17, Query 31; Dover, 1952, 401-402.

10 I. Newton, *Opticks*, 1706/17, Query 31; Dover, 1952.

11 Lord Kelvin, *Baltimore lectures on wave theory and molecular dynamics,* 1884, in: R. Kargon and P. Achinstein (eds.), *Kelvin's Baltimore Lectures and Modern Theoretical Physics,* Cambridge/Mass.: MIT Press, 1987.

12 S. T. Coleridge, *The Friend*, in *Collected Works*, ed. B. Rooke, Princeton, 1969, vol. 4, pt. 1, p. 158.

Chapter 3

1 The fundamental distinction between mathematics and physics is that mathematics

constitutes a zero totality system without any time reference, whereas physics sets up a universe of all zero totality systems *now*. To some extent, mathematics provides the infinite, open option (cf commutativity) while physics is structured on the finite, closed one (cf anticommutativity).

2 The creation of a second 3-D operator allows us to structure the rest, that is to structure the alphabets or structure the structures.

3 The dual state could, of course, be based on any of the fundamental dualities, with, for example, coefficients such as 1 and i or $\boldsymbol{i}$ and $\boldsymbol{j}$. In the last case, we could take the vacuum terms as the anticommutative partners of the invisible coefficients, used to separate different fermions in the Grassmann-Hilbert algebra connecting them. Then the fermion and vacuum terms would be zero as antisymmetric wavefunctions as well as nilpotent.

4 The algebra can be defined as a tensor product of the quaternion and multivariate 4-vector components, which are, of course, commutative. In principle, this means that each acts as if the other did not exist. However, according to general relativity, mass and energy terms, which are associated with the quaternion algebra, cause the curvature of the 4-vector space-time, here defined as commutative to it. But the curvature is *external* – it is not perceived internally, where geodesics define the paths that would be equivalent to straight lines in a Euclidean system, and the relation between the two is established by the affine connection.

5 This paragraph is connected to the work done with Brian Koberlein in chapter 4.

6 Several significant facts emerge from the analysis. (1) The packaging process which produces the nilpotent requires (and introduces) discreteness. (2) The principle of relativity, which comes from fixing the E / $\mathbf{p}$ ratio in the nilpotent, only applies to discrete sources and discrete charge-related interactions. (3) Nilpotency is the ultimate reason why we have complex numbers in quantum theory – one term must be imaginary to zero the whole on squaring. Immediately, of course, the i term is introduced with the phase, but the amplitude and phase separation is ultimately related to the separation of $\mathbf{p}$ and E into real and imaginary parts. As the packaging process would suggest, quantum mechanics is ultimately inexplicable without relativity. (4) The commutator relations between $\mathbf{p}$ and $\mathbf{r}$ ultimately lead to Heisenberg uncertainty (formally derived in 16.6), stating that we can't know both exactly simultaneously, meaning that we can't define a fermion in both space and momentum space. Unlike space, time, mass and charge, they are not independent information.

7 Of course, the Clifford algebras $Cl_{4,1}$ and $Cl_{2,3}$, when applied to E and $\mathbf{p}$, become $Cl_{1,4}$ and $Cl_{3,2}$, when applied to $\partial / \partial t$ and ∇.

8 By 'universe', here, is meant the set of interacting fermionic states. If 'universe' is taken to mean fermion plus vacuum, then it *is* zero. E. Bastin and C. W. Kilmister have developed a logic of process, based on the concept of *discrimination*, in which, for example, B will be excluded if, after comparison with A, it is found to be not new. Using a procedure of J. H. Conway's, they define a 'discriminately closed subset', the smallest of which contains A, B, and the discrimination of A against B, which is extended to six members (which could be written $a, -a, b, -b, c, -c$) if the discrimination of B against A is assumed to be different from the discrimination of A against B. With two 'signals' x and y, this group takes on a quaternionic structure. Possible connections are discussed in source 44.

9 The reconciliation between iteration and recursion here parallels that between *reductionism* and *holism* which occurs throughout this book. The nilpotent structure at once requires both

paradigms, because it is simultaneously the simplest possible element (and so introduces reductionism), and only definable in terms of the universe as a whole (and so requires holism). This allows a reductionist form of holism to be constructed, which will be evident, for example, in chapter 20. It could be argued that the most successful form of reductionist thinking is most likely to be performed by those with the most holistic vision, for example, Isaac Newton.

10 D. Matzke, *Quantum Computation using Geometric Algebra*, PhD Thesis, University of Texas at Dallas, 2002, www.utdallas.edu/~cantrell/matzke.pdf.

11 Since the origin of discreteness is seen here as a result of finite (3-)dimensionality, we can conceive the commutative infinite-dimensional Hilbert space as restoring continuity to the system, with the individual nilpotents linked by nonlocal and instantaneous superposition. From the point of view of physical observation, fermionic nilpotents are defined using the units of only three quaternionic systems and have commutative relationships with all other quaternionic systems in the algebra. So, the structure of the units of the form $\boldsymbol{i}_s$, forming the infinite-dimensional Hilbert space through which the nilpotents are linked, is completely arbitrary, as long as they remain commutative with all elements of the nilpotent algebra. In principle, of course, each could represent an independent 'universe' in itself, which is entirely unknowable within the context of ours, though one may suspect that the concept is really meaningless.

12 D. Deutsch, Quantum theory, the Church-Turing Principle and the universal quantum computer, *Proc. Roy. Soc.*, **A 400**, 97-117, 1985.

13 The rewrite system, in fact, provides a route into all mathematics, justifying its use in the remainder of this book. All mathematics that exists must map onto the mathematics of 'reality', because this also includes human thought, although it need not describe any physical thing directly.

14 The fact that a topological space must contain the unions and intersections of its subsets may have significant connections with nilpotent interactions.

15 J.-R. Abrial, *The B-Book: Assigning Programs to Meanings*, Cambridge University Press, 1996, 227-401.

16 A. Turing, On computable numbers with an application to the Entscheidungsproblem. *Proc. Lond. Math. Soc.*, ser. 2, **42**, 230-65, 1936-37.

17 We may note here that quantum computing is a more extensive concept than the usual idea of classical computing with qubits. See also chapter 20.

Chapter 4

1 This formal connection is implicit in any quasi-3-D description of an energy parameter and is explicitly mentioned, for example, in source 40, but there have been many attempts to make time 3-dimensional. J. Carroll, for instance, has a demonstration using an actual three-dimensional time operator, and geometric algebra, to derive the Dirac equation and several other significant results (Relativity with three dimensions of time: space-time vortices, *Proceedings of VIII Conference on Physical Interpretations of Relativity Theory*, British Society for Philosophy of Science, London, September 2002). The significant thing here is that the geometric (essentially multivariate vector) algebra, however introduced, will always

produce correct formal results because it provides a hidden way of introducing the operation of charge.

2 A. W. Conway, Quaternion treatment of the relativistic wave equation. *Proc. Roy. Soc.* **A 162**, 145-54, 1937.

3 J. D. Edmonds, Nature's natural numbers: relativistic quantum theory over the ring of complex quaternions. *Int. J. Theoret. Phys.*, **6**, 205-24, 1972

4 W. Gough, Quaternion quantum mechanical treatment of an electron in a magnetic field. *Eur. J. Phys.*, **8**, 164-70, 1987.

5 W. Gough, A quaternion expression for the quantum mechanical probability and current densities. *Eur. J. Phys.*, **10**, 188-93, 1989.

6 C. W. Kilmister, *Proc. Roy. Soc.* **A 199**, 517 (1949).

7 C. W. Kilmister, *Proc. Roy. Soc.* **A 199**, 517 (1951).

8 D. Hestenes, Vectors, spinors, and complex numbers in classical and quantum physics. *Am. J. Phys.*, **39**, 1013-1027, 1971.

9 J. D. Edmonds, Nature's natural numbers: relativistic quantum theory over the ring of complex quaternions. *Int. J. Theoret. Phys.*, **6**, 205-24, 1972.

10 J. D. Edmonds, Quaternion quantum theory: new physics or number mysticism? *Am. J. Phys.,* **42**, 220-223, 1974.

11 W. Gough, Quaternion quantum mechanical treatment of an electron in a magnetic field. *Eur. J. Phys.*, **8**, 164-70, 1987.

12 S. B. M. Bell, J. P. Cullerne and B. M. Diaz. Classical behaviour of the Dirac bispinor. *Found. Phys*., 30, 35-57 (2000), and references therein.

13 H. Minkowski, *Physikalische Zeitschrift*, 10, 104-11, 1909, lecture on 'Space and Time', Cologne, 21 September 1908, translated in Lorentz, Einstein, Minkowski and Weyl, *The Principle of Relativity*, 1923, 104.

14 The dual group here and in earlier work has a mainly mathematical function, connected with reversals of properties, with an implied connection to vacuum, but John Valentine has investigated a possible application of the dual group to real physical parameters (oral presentation to ANPA, Cambridge, 1998).

15 It is also a consequence of defining spin through a scalar product.

16 The primary (and secondary) colours are such a powerful pseudo-3-dimensionality that we might be tempted to speculate that they are actually connected with 3-dimensionality itself, and that 3-colour vision has conveyed an evolutionary advantage in the perception of 3 dimensions, which has ultimately been the key element in developing human understanding of the natural environment, and hence of the principle of reasoning itself.

17 The Fibonacci series 1 1 2 3 5 8 13 21 … (each number being the sum of the two previous) is recognizable in quasicrystals with 5-fold symmetry, where adjacent short (S) and long (L) 'tiles' may follow the sequence S L LS LSL LSLLS LSLLSLSL LSLLSLSLLSLLS, with the L / S radio tending towards the 'Golden Section' 1.618 … The crystals typically show long-range order but no periodicity.

Chapter 5

1 J. B. Kuipers, *Quaternions and Rotation Sequences*, Princeton University Press, Princeton,

NJ, 1999.

2 E. Witten, arXiv:hep-th/0312171.

3 D. Hestenes, *Space-time Algebras*, Gordon and Breach, New York, 1966.

4 W. Gough, *Eur. J. Phys.*, **11**, 326-33, 1990.

5 For the simple case of the circularly polarized fermion, we can use the nilpotent structure to define Stokes parameters of the form: $I = (\boldsymbol{k}E)\,(-\boldsymbol{k}E) + (\boldsymbol{i}i\mathbf{p} + \boldsymbol{i}jm)\,(\boldsymbol{i}i\mathbf{p} + \boldsymbol{i}jm)$; $Q = (\boldsymbol{k}E)\,(-\boldsymbol{k}E) - (\boldsymbol{i}i\mathbf{p} + \boldsymbol{i}jm)\,(\boldsymbol{i}i\mathbf{p} + \boldsymbol{i}jm)$; $u = (\boldsymbol{k}E)\,(\boldsymbol{i}i\mathbf{p} + \boldsymbol{i}jm) + (-\boldsymbol{k}E)\,(\boldsymbol{i}i\mathbf{p} + \boldsymbol{i}jm)$; $v = i(\boldsymbol{k}E)\,(\boldsymbol{i}i\mathbf{p} + \boldsymbol{i}jm) - i(-\boldsymbol{k}E)\,(\boldsymbol{i}i\mathbf{p} + \boldsymbol{i}jm)$; so that $I^2 = Q^2 + u^2 + v^2 = v^2 = 4E^2$ before normalization.

6 John A. Eisele. *Modern Quantum Mechanics with Applications to Elementary Particle Physics* (John Wiley, 1969), p. 194.

7 Since the information needed to specify the state is reduced from a 16-component matrix × a 4 component wavefunction to a single component operator, the nilpotent formalism effectively makes 98.4 % of the Dirac equation redundant!

8 They are active, not passive components, like 'hypertext'.

Chapter 6

1 The energy and momentum operators have different signs ultimately because the retarded wave is associated with positive energy, the sign difference carrying over from the phase term.

2 J. T. Barraeiro, N. K. Langford, N. A. Peters, and P. G. Kwiat, *Phys. Rev. Lett.*, **95**, 260501, 2005.

3 The angular momentum direction is defined only abstractly relative to the vacuum. (The only conceivable 'physical' interpretation of this would be through the cosmic microwave background.)

4 The use of a real, rather than quaternion, operator distinguishes the 'real' particle from the vacuum 'image' states associated with it, which are induced by the weak, electric and strong elements. This means that the first of the four 'solutions' in the Dirac 4-spinor has a different status to all the others, in the same way as the time coordinate (to which it corresponds) has a different status to the three space coordinates in the conventional Minkowski 4-vector. This status is especially significant in the case of a *free* fermion (or boson), where the vacuum terms make no contribution to the particle's energy, meaning that renormalization is not required (cf chapter 9). The same will apply even if the 'real' or first term in the spinor is actually a vacuum state. In renormalization, only the effects of the 'image' terms cancel, leaving those of the 'real' term unchanged.

5 A single fermion acts as a 1-D particle, quantum waves being plane, not spherical, with classical 3-dimensionality effectively reduced to a single direction with the concept of spin. So, a fermion's four solutions, involving $\pm E$, $\pm \mathbf{p}$, can be considered as following the process of timelike reflection off the vacuum (or universe), related to the Feynman-Wheeler 'removal' of advanced waves; and we would have time-space reflections seeing themselves via weak, strong and electric fields of vacuum and corresponding interactions.

6 Creation and annihilation of fermions is intrinsically connected to the creation and annihilation of a 'one-dimensional' state of spin, which can rotate (in the conventional sense) only into a state of existence or non-existence.

7 R. M. Santilli, *Algebras, Groups and Geometries*, **10**, 273, 1993.
8 R. M. Santilli, *Comm. Theor. Phys.*, **3**, 153, 1993.
9 R. M. Santilli, personal communication.
10 Euclidean space means that the zero point energy exists. This energy, in fact, defines Euclidean space (as does the space of all possible **p** phases).
11 This connection is made in B. Diaz and P. Rowlands, D: The infinite square roots of –1. *International Journal of Computing Anticipatory Systems*, **19**, 229-235, 2006, which is summarised in Appendix B. That duality does not imply equality of status is obvious even from the simple fact that $(-1) \times (-1) = 1 \times 1 = 1$. It is also important, in connection with the concept of the filled vacuum, to realise that the 'probe and response' structure we use for physics cannot be described either in purely epistemological or in purely ontological terms. We should expect the physical structure of the universe to *appear* asymmetric because we are, in effect, looking at it from the 'inside', and cannot access an ontological zero totality, which would only be visible from an 'outside' which does not exist. So, although we must structure our understanding of nature on the concept of zero totality, it will always be an unobservable ideal. So we will have equal amounts of matter and antimatter, but only if the antimatter is the unobservable vacuum response to the matter; and, in the version of Mach's principle discussed in 21.1, we will have equal negative gravitational and positive inertial energies, but only because the negative gravitational energy is not directly observable, and only accessible through the positive inertial reaction.
12 In the Dirac nilpotent theory, we can take the charge operator Q as representing amplitude (Heisenberg) and the energy operator $|\psi\rangle$ as representing phase (Schrödinger), which are, of course, dual in this theory.
13 Like other charge operators, the fermionic nilpotent is only physically meaningful when squared.

Chapter 7

1 W. Gough, Mixing scalars and vectors – an elegant view of physics, *Eur. J. Phys.*, **11**, 326-33, 1990.
2 G. Galeczki, A short essay on closed systems, hiererachy and radiation, chapter 8, 185-198, of N. V. Pope, A. D. Osborne and A. F. T. Winfield (eds.), *Immediate Distant Action and Correlation in Modern Physics The Balanced Universe*, Edwin Mellen Press, 2005.
3 It might even be possible to argue that, if we increase the constraints on one isolated system, we *reduce* the constraints on the rest of the universe. Because it is spread over the whole universe, the effect becomes less structured.
4 B. J. Hiley, From the Heisenberg picture to Bohm: a new perspective on active information and its relation to Shannon Information, presented at Quantum Theory: reconsiderations of foundations, Vexjo University, Sweden, June 2001.
5 Louis H. Kauffman, Non-commutative worlds, *New Journal of Physics*, **6**, 2-46, 2004.
6 Louis H. Kauffman, Differential geometry in non-commutative worlds, in K. G. Bowden (ed.), *Conceptions*, ANPA, London, 2006, 193-213.
7 G. Yu. Bogoslovsky, Lorentz symmetry violation without violation of relativistic symmetry, *Phys. Lett. A*, **350**, 5-10, 2006. If the Berwald-Moor metric of anisotropic Finsler geometry,

$ds = (dx_1dx_2dx_3dx_4)^{1/4}$, is substituted for the isotropic Minkowski metric of Riemannian geometry, $ds = (dt^2 - dr^2)^{1/2}$, the zero interval manifold ($ds = 0$) becomes transformed from the familiar Minkowski light cone to a combination of two pyramids joined at the apex. By introducing exponents into the expression for the metric function, Bogoslovsky finds a geometric phase transition, which could be interpreted as a mass-creating spontaneous-symmetry breaking in the fermion-antifermion consendate. According to Bogoslovsky, the generalised Lorentz transformations responsible for the process lead directly to the Berwald-Moor metric. In the discrete version of the double nilpotent representation of the bosonic state (or 'fermion-antifernion condensate'), no mass term appears in the operator, but the differentials may be replaced by covariant derivatives, and so the opportunity arises to represent the appearance of mass directly in terms of an anisotropic space-time structure. I am grateful to Sergey Siparov for his clarification of Bogoslovsky's ideas.

8 G. Spencer-Brown, *The Laws of Form*, Allen and Unwin, London, 1969. I am grateful to Lou Kauffman for use of his draft book on the Laws of Form.

9 The idempotent wavefunction is also, necessarily, only a partial view of the fermionic state, giving only that fraction of the information relevant to the interaction whose quaternionic operator is used.

10 B. J. Hiley, From the Heisenberg picture to Bohm: a new perspective on active information and its relation to Shannon Information, presented at Quantum Theory: reconsiderations of foundations, Vexjo University, Sweden, June 2001.

11 B. J. Hiley, Towards a dynamics of movements: the role of algebraic deformation and inequivalent vacuum states, K. Bowden (ed.), *Correlations* (May 2002, *Proceedings of XXIII ANPA Conference*, Cambridge, August 2001), 104-34.

12 This was proposed in the *Proceedings* of ANPA, Cambridge 2001 (source 43). Subsequent work by Hiley, based on the metaplectic group, uses a propagator-like term, which, like conventional non-nilpotent propagators, produces a 'pole' or singularity at a particular value of energy-momentum (presented, ANPA, Cambridge, 2002). In Hiley's work – and it is significant that he is working here in a nonrelativistic area – the pole introduces the complex *i* term, and he sees it as an origin for complexity in quantum mechanics, which, of course, is highly relevant to his interpretation of the Bohm-Schrödinger division. In chapter 9, I will argue that the pole can be seen as a result of separating out the positive and negative energy solutions in the Dirac nilpotent. So, in a sense, its origin *is* complexification. In a *fully* relativistic quantum mechanics (in which space, time, mass and charge have equal status) the complexification is packaged into the nilpotent, and its action is absorbed into the overall structure.

13 Or Dirac particle (Heisenberg) within the rest of the universe (Schrödinger).

Chapter 8

1 N. V. Pope and A. D. Osborne, Instantaneous Relativistic Action-at-a-Distance, *Physics Essays*, **5**, 409-421, 1992; Instantaneous Gravitational and Inertial Action-at-a-Distance, *Physics Essays*, **8**, 384-97, 1995. N. V. Pope, A tale of two paradigns, chapter 5, 71-119, of N. V. Pope, A. D. Osborne and A. F. T. Winfield (eds.), *Immediate Distant Action and Correlation in Modern Physics The Balanced Universe*, Edwin Mellen Press, 2005. A. D.

Osborne, The Pope-Osborne angular momentum synthesis, chapter 9, 119-239, *ibid.*

2 A. Aspect, J. Dalibard and G. Roger, *Phys. Rev. Lett.*, **49**, 91 and 1804, 1982.

3 M. Rowan-Robinson, Aether drift detected at last, *Nature*, **270**, 9-10, 1977.

4 R. A. Muller, The Cosmic Background Radiation and the New Aether Drift, *Scientific American*, **238**, 64-74, 1978.

5 Temperature (or, in some respects, chaos) is the only fundamental way of getting round the problem of relativity of motion and of some of the apparent consequences of the second law of thermodynamics. (Chaos appears to be necessary for biology.)

6 We may note here that classical SR is as one-dimensional as quantum mechanics, the velocity addition law being only well-defined in one direction. SR, of course, excludes acceleration, except in those contexts where the single dimension can be radial (and, strictly, this latter should also be excluded).

7 In principle, the creation of the Dirac nilpotent requires an intrinsic connection between relativistic and quantum concepts, which ensures that both are expressions of discreteness or 3-dimensionality. This is why it is possible to use apparently classical, or semiclassical, wave-related (or c-dependent) approaches, such as stochastic electrodynamics, to derive many apparently purely 'quantum' results (cf chapter 14). The nilpotent structure is both a necessary and a sufficient condition for a fully relativistic and a fully quantum theory. It is the origin of both h and c; the concepts are generated simultaneously, and any restricted theory which icorporates part of one structure also necessarily incorporates part of the other. So even the structures of classical mechanics require these constants to have meaning (along with G), whether or not they are explicitly used, while classical electromagnetic theory, in requiring time-delayed transmission of energy between discrete sources, almost demands an extension to QED. The connection is particularly apparent in the QED-like inertial reaction (cf chapter 15), which, while clearly quantum in origin, responds directly to a semi-classical treatment via Maxwell-type equations. In more general terms, c, like h, is *purely a statement of discreteness*. It is created when a discrete state is created, and it has no other basic function. Space-time 'curvature', as produced by a gravitational field in general relativity, merely changes the measure of discreteness (c).

8 Strictly, of course, Newton's third law refers to action and reaction between any defined system and the 'rest of the universe', and so is never *exactly* reciprocal, though, to an extremely close approximation, this can be defined in terms of the action and reaction between systems A and B.

9 One way of interpreting the Laplace equation in empty space is to assume that necessary information, which cannot be made available, must be equated to zero. The field due to each charge and its distribution or variation through space do not depend on the presence of any other charges in space; in terms of the available information represented by the field variation, each charge behaves as though isolated. Yet the field variation cannot be detected, in principle, for an isolated charge.

Chapter 9

1 E. Nelson, Derivation of the Schrödinger equation from Newtonian mechanics, *Phys. Rev.*, **150**, 1079-85, 1966.

2 T. H. Boyer, Derivation of the blackbody spectrum without quantum assumptions, *Phys. Rev.*, **182**, 1374-83, 1970.

3 T. H. Boyer, Third law of thermodynamics and electromagnetic zero-point radiation", *Phys. Rev.* D, **1**, 1526-30, 1970.

4 T. H. Boyer, Concerning zero-point energy of a continuum, *Am. J. Phys.*, **42**, 518-9, 1974.

5 H. E. Puthoff, Ground state of hydrogen as a zero-point-fluctuation-determined state, *Phys. Rev.* D, **35**, 3266-9, 1987.

6 H. E. Puthoff, Sources of vacuum electromagnetic zero-point energy, *Phys. Rev.* A, **40**, 4857-62, 1989.

7 Stochastic electrodynamics, like relativistic quantum mechanics, is structured upon a random phase, and is just a different manifestation of the same concept. Here, we see that randomness and discreteness are entirely different concepts.

8 One way to counter the Einstein statement would be to say that, if God does not play dice, then *we* are God. That is, we can only realise the closed totality zero state, and its unique birthordering, by projecting ourselves outside of it. The final zeroing can only be realised from an 'outside', which, in our terms, does not exist.

9 *Isaac Newton's Mathematical Principles of Natural Philosophy*, translated by A. Motte, revised by F. Cajori, Cambridge 1934, p. 6.

10 M. Wegener, Arguments for the existence of a privileged time, *Proceedings of Conference on Physical Interpretations of Relativity Theory*, British Society for Philosophy of Science, London, September 1990, 317-28, p. 318.

11 R. M. Nugayev, Why did Einstein-Lewis programme force out Lorentz-Poincaré, *Proceedings of Conference on Physical Interpretations of Relativity Theory*, British Society for Philosophy of Science, London, September 1990, 203-6 and 419-36.

12 G. Falk, Entropy, a resurrection of caloric – a look at the history of thermodynamics, *Eur. J. Phys.*, **6**, 108-15, 1985.

13 R. Penrose, *The Emperor's New Mind*, Oxford 1989, chapters 7 and 8.

14 The stationary Platonic universe clearly does not exist in the Lorentzian space-time of observation and measurement, though it could be argued that it exists in the 'absolute space' of the mass frame.

15 See, for example, A. Heslot, Classical mechanics and the electron, *Eur. J. Phys.*, **7**, 35-42, 1983, and Dirac's 'scissors' analogy.

16 The thermodynamic arrow is an expression of the 1-dimensionality of time, and hence of the principle of causality.

17 Thermodynamics only becomes relevant when more than one particle is involved; the particles then combine and the combination ceases to be a pure system. Entropy does not exist for a single particle (nor does time direction); it increases with further interactions and more entanglement. It also needs 3-dimensionality, which again does not exist for one particle.

18 Though we always try to structure our theories using isolated systems, the fact that systems are *never* isolated denies discreteness. This is how thermodynamics relates to continuous time. It expresses the real meaning of time, as a vacuum concept, and not the observable reciprocal of frequency.

Chapter 10

1 The property of perfect gauge invariance will ensure that both baryon and boson structures involving the colour force will be notably 'affine'.

2 T. T. Takahashi, H. Matsufuru, Y. Nemoto, and H. Sugunuma, *Phys. Rev. Lett.*, **86**, 18, 2001.

3 J. Ashman *et al*, A measurement of the spin asymmetry and determination of the structure function $g(1)$ in deep inelastic muon-proton scattering, CERN-EP-87-230, December 1987; *Physics Letters*, **B 206**, 364, 1988.

4 S. J. Brodsky, J. Ellis, and M. Karliner, Chiral symmetry and the spin of the proton, *Physics Letters*, **B 206**, 309-15, 1988.

5 H. Georgi and S. L. Glashow. 'Unity of all elementary-particle forces.' *Phys. Rev. Lett.*, **32**, 438-41, 1974.

6 I. J. R. Aitchison and A. J. G. Hey, *Gauge Theories in Particle Physics*, second edition, Adam Hilger, 1989, 485-6.

7 F. Halzen and A. D. Martin, *Quarks and Leptons*, John Wiley, 1984.

8 We could suppose, here, that the passive Coulomb term in the weak interaction is entirely a manifestation of the creation of the spin ½ state. E and **p** are necessarily mixed in the weak interaction. If we suppose that the creation / annihilation of the harmonic oscillator involves ½ for left-handed states and –½ for right-handed states (associated with the iE term), we can imagine the weak charge w switching signs with the **p** or ∇, which carries the active part of the weak interaction.

9 cf note 13 to 8.6.

10 Here we consider the weak charge as producing both a phase term with two solutions, neither of which is priveleged by the mathematics, and a mechanism for oscillating between them. Physical considerations, however, demand the priveleging of the material condition over the filled weak vacuum, and the reference is provided by the corresponding electromagnetic vacuum, which is either empty or filled in the two available $SU(2)$ states.

Chapter 11

1 In principle, this is the same thing as saying a non-interacting charge has no identifiable value, because it is non-real.

2 If gravity and inertia are linked by the Machian theory described in chapters 15 and 16, then the Planck mass will represent a specifically *inertial* quantum, and an event horizon for the time-delayed inertial force, through which inertial mass must *appear* to be generated. So, the event horizon provides a natural cut-off.

3 Though the nilpotent theory is structured intrinsically on the mass-shell or real particles, with $E^2 = p^2 + m^2$, the propagator method shows that the formalism extends to off mass-shell conditions or virtual particles.

4 J. D. Bjorken and S. D. Drell, *Relativistic Quantum Mechanics*, McGraw-Hill, New York, 1964.

Chapter 12

1 Thermodynamics is not discrete, even when describing a system in a discrete phase space.

2 If the concept of continuity is as 'illusory' as some make out, we have to explain how we can *imagine* the idea at all. The 'real' world has to be defined to include all our thoughts and abstractions as well as 'tangible' things like 'observables'.

3 The Higgs boson has spin 0 because it is not discrete in origin, and so has no angular momentum, thus leading to the gravity of the uniform vacuum.

4 A quantum theory, with nonlocal correlations, cannot be constructed without a continuous vacuum. It is possible to argue that Newtonian gravity is a quantum theory, because it has instantaneous interaction (which, if gravitational 'energy' and 'force' have the special status assumed here, becomes equivalent to nonlocal correlation), whereas 'quantum gravity' is not because it has no continuous vacuum.

5 H. B. G. Casimir, *Proceedings, Koninlijke Akademie van Wetenschappen te Amsterdam*, **51**, 793, 1948.

6 R. Peterson and R. M. Metzger, *Int. J. Chem.*, **7**, 1-4, 2004.

7 We may note the appearance of a geometric phase accompanying any displacement in Minkwoski space-time as a result of its origin in complex quaternion algebra, in V. V. Kassandrov, Algebrodynamics in complex space-time and the complex-quaternionic origin of Minkowski geometry, arXiv:gr-qc/0602988 v1, which we associate here with the spin produced by defining the fermion as a singularity. (See also chapter 1, note 17.)

8 'Particle space' and 'source space' are other alternatives.

9 It is notable that the mass term which must be present in a fermion state (see note 14.3) ensures that there is a term in the creation operator which affects only the magnitude of the angular momentum, and that, consequently, that there must be an interaction mechanism which is neither vector nor pseudoscalar in origin, and so is only concerned with the magnitude of the coupling. This is, of course, the electric interaction, which is specified only by a $U(1)$ scalar phase.

Chapter 13

1 In a lepton we associate all the three angular momentum conservation laws together; in a baryon, we use the vector nature of **p** to separate them.

2 This notation is due to John Cullerne.

3 Among other things, the formalism ensures that the electric and weak sources have to be defined relative to each other to establish whether the state is a quark or lepton. The strong source, of course, has a free variation because of the affine 3-D structure of the strong interaction, transmitted by massless gluons.

4 The option of filled / empty vacuum is, of course, for the electromagnetic interaction; the weak interaction allows only a filled vacuum state. Neutrinos could perhaps be considered as negatively charged holes in a filled electromagnetic vacuum, possibly accounting for their mass via the weak dipole moment.

5 Wavefunction collapse and decoherence may be considered as intrinsic to the creation of the 'classical state', or as tending towards that creation. In the classical state, $\boldsymbol{k}$, $\boldsymbol{i}$, $\boldsymbol{j}$ are not

charges; E and **p** are not fixed or quantised; the conservation of charge is not linked to the conservation of angular momentum. As we separate out the terms associated with i, **i**, **j**, **k**, 1, from the charge structure, we approach classical consitions.

6 The explicit nature of the Pauli exclusion violation in this case was put forward by Adam Brown.

7 M. Kobayashi, and T. Maskawa, *Prog. Theor. Phys.*, **49**, 282, 1972. The complex phase may be seen to result from the complex potential (or, equivalently, complex coupling constant) required for the weak interaction solution of the nilpotent Dirac equation (sections 10.10-10.12). The complex coupling constant or potential introduces complexity into the Lagrangian, and hence into the CKM matrix. An equivalent result is found using a complex vacuum or a complex scalar field, as in the conventional derivation of the Higgs mechanism.

8 Perhaps phase transitions of this kind will be most readily observed in an astrophysical context.

9 M. Y. Han and Y. Nambu. Three-triplet model with double $SU(3)$ symmetry. *Phys. Rev*, **139 B**, 1006-10, 1965.

10 F. E. Close. *An Introduction to Quarks and Partons*, Academic Press, 1979, p. 167.

11 R. B. Laughlin, Anomalous quantum Hall effect: an incompressible field with fractionally charged excitations, *Phys. Rev. Lett.* **50**, 1395-8, 1983.

12 P. W. Anderson, *Science*, **177**, 393, 1972.

13 R. B. Laughlin, *Rev. Mod. Phys.*, **71**, 1395, 1999.

14 M. E. Peskin and D. V. Schroeder, *Introduction to Quantum Field Theory*, Addison-Wesley, 1995.

Chapter 14

1 Historically, 13.5-13.7 preceded 14.3-14.8, which then preceded 13.1-13.4. However, the logical order is as now presented.

2 C. Itzykson and J-B. Zuber, *Quantum Field theory*, McGraw Hill Int., 1980, pp 59, 87, 148, 243.

Chapter 15

1 H. Georgi and S. L. Glashow. 'Unity of all elementary-particle forces.' *Phys. Rev. Lett.*, **32**, 438-41, 1974.

2 H. Georgi, H. R. Quinn, and S. Weinberg, 'Hierarchy of interactions in unified gauge theories'. *Phys. Rev. Lett.*, **33**, 451-4, 1974.

3 Just as in the electroweak process of electron-positron annihilation / creation, there is a small amplitude for the weak interaction, which only becomes measurable at high energies, it may be that β decay and similar quark-lepton processes have a very small amplitude for strong-electroweak decay, which would only appear (if at all) at an extremely high energy, say near grand unification. The conventional notion of proton decay generally fails to take into account the dipolar nature of the weak interaction (fermion + antifermion = fermion + antifermion), and assumes that a baryon particle can decay into a lepton one without the mediation of an antibaryon.

4 E. Fischbach *et al*, *Phys. Rev. Lett.* **56**, 3, 1986.

5 E. Fischbach *et al*, *Phys. Rev. Lett.* **56**, 2423, 1986.

6 The perfect symmetry of the three interlocking $SU(3)$s for the three charge states would also be restored. In addition, charge would acquire a fundamental unit, along with h, c and G, only at this value, as renormalization theory requires. (Charge cannot be a fixed constant because $\boldsymbol{i}, \boldsymbol{j}$ and $\boldsymbol{k}$ can only fix a scale in relation to each other.)

7 M. Günyadin and F. Gürsey, *Phys. Rev.*, **9 D**, 3387, 1974.

8 F. Gürsey, P. Ramond and P. Sikilvie, *Phys. Lett.*, **60 B**, 177, 1976.

9 F. Gürsey, Quaternionic and octonionic structures in physics, in M. G. Doncel *et al* (eds.), *Symmetries in Physics*, 1987, 557 ff.

10 F. Halzen and A. D. Martin, *Quarks and Leptons*, John Wiley, 1984.

11 S. Weinberg, *The Quantum Theory of Fields*, 2 vols., Cambridge University Press, 1996, Vol. II, pp. 327-32.

12 U. Amaldi, W. de Boer, and H. Fürstenau, Comparison of grand unified theories with electroweak and strong coupling constants measured at LEP, *Phys. Lett.* **260 B**, 447-455, 1991.

13 A. Masiero, Introduction to Grand Unified theories, in C. Kounnas, A. Masiero, D. V. Nanopoulos, and K. A. Olive, *Grand Unification with and without Supersymmetry and Cosmological Implication*, World Scientific, 1984, 1-143, pp. 20-25.

14 C. Kounnas. Calculational schemes in GUTs, in C. Kounnas, A. Masiero, D. V. Nanopoulos, and K. A. Olive. *Grand Unification with and without Supersymmetry and Cosmological Implication*, World Scientific, 1984, 145-281, pp. 188-227, pp. 228-9.

15 Of possible relevance here is an argument by F. J. Dyson, in *Phys. Rev.*, **85**, 631, 1952, according to which, taking the perturbative expansion of normalized QED to more than 137 terms in e^2, and assuming that the theory also applies to a theory in $-e^2$, where like charges attract, ensures that a cluster of like charges will become unstable against collapse to infinitely negative energy. Extending the argument, H. P. Noyes and J. Lindesay identify $1 / \alpha \approx 137$ as the number of particle pairs within the Compton wavelength that will produce another pair, or the number of interactions within the Compton wavelength of an electron-positron pair when the Dyson limit is reached. There is a cut-off at this point because a different force takes over, the strong force overtaking the electric.

16 I. J. R. Aitchison and A. J. G. Hey, *Gauge Theories in Particle Physics*, second edition, Adam Hilger, 1989, 434.

17 Possibly, we should consider the hypothetical 'interconversion' of mass and generalised charge as being like a critical phase change.

18 We can consider also a 'Casimir effect' between two rest masses providing a cut-off in the continuous vacuum – that is, in the longer vacuum wavelengths or lower energies. The discreteness of the rest masses thus introduces discreteness (or quantization) in the repulsive force, though this force has the same inverse square form as the continuous vacuum force of attraction. The quantity of rest mass introduced for a discrete object is determined by the vacuum, with m being the quantum of inertial repulsion, just as e is the quantum of electrical repulsion. (Gravity couples to the energy rather than to the rest mass.)

19 Taking the calculation over three representations (136.5 GeV) or five representations (227.5 GeV) seems much less probable.

20 D. T. R. Jones, seminar, University of Liverpool, 2000.

21 J. Ellis and D. Ross, A light Higgs Boson would invite Supersymmetry, arXiv:physics-p/0012067.

22 C. Ford, D. T. R. Jones, P. W. Stephenson, and M. B. Einhorn, The effective potential and the renormalisation group, *Nuclear Physics*, **B 395**, 17-34, 1993.

23 See, e.g., Donald H. Perkins, *Introduction to High Energy Physics*, third edition, Addison-Wesley, 1987, 160-3.

24 E. Trell, Representation of particle masses in hadronic SU(3) transformation diagram, *Acta Physica Austriaca*, 55, 97-110, 1983.

25 E. Trell, Geometrical reproduction of (u,d,s) baryon, meson, and lepton transformation symmetries, mass relations, and channels, *Hadronic Journal*, 13, 277-97, 1990.

26 E. Trell, On rotational symmetry and real geometrical representations of the elementary particles with special reference to the N and Δ series, *Physics Essays*, 4, no. 2, 272-83, 1991.

27 E. Trell, Real forms of the elementary particles with a report of the Σ resonances, *Physics Essays*, 5, no. 3, 362-73, 1992.

28 E. Trell, The eightfold eightfold way. A lateral view on the Standard Model, *PIRT VII, Late Papers*, 263-84, 1998.

29 E. Trell, Marius Sophus Lie's doctoral thesis Over en Classe Geometriske Transformationer, English translation and commentary, *Algebras, Groups and Geometries*, 15, no. 4, 395-445, 1998.

30 E. Trell, The eightfold eightfold way: application of Lie's true *Geometriske Transformationer* to elementary particles, *Algebras, Groups and Geometries*, 15, no. 4, 447-71, 1998.

31 E. Trell, Invariant Aristotelian cosmology: binary phase transition of the universe from the smallest to the largest scales, *Hadronic Journal*, 28, 1-42, 2005.

32 Another possible relation is that between the mass scale of the first generation and $\alpha\alpha_2\alpha_3$ times the vacuum expectation value.

33 It is possible that a muon $g-2$ anomaly which has been claimed by some experimenters, but will probably not be established as a valid observation, may be a result of 'down' weak isospin lepton mixing; if so, then even higher anomalies could be predicted. The mixing requires a higher than observed mass for the mixed state and a value of $g-2$ which is higher than the predicted one.

Chapter 16

1 D. T. Whiteside, discussing this, in *Centaurus*, **24**, 288-315, 1980, 308, 315, cites W. M. Smart, *Text-Book on Spherical Astronomy* (Cambridge, fifth edition, 1971; first edition, 1931), chapter III, 58-73, and E. M. Woollard and G. M. Clemence, *Spherical Astronomy* (New York, 1966), chapter 5, 79-96.

2 I. Newton, *Opticks*, 1706/17, Query 21, Dover 1952.

3 The same is also true, of course, when we consider light has having an 'acceleration' due to a change in direction.

4 R. P. Comer and J. D. Lathrop, *Amer. J. Phys.*, **46**, 801, 1978.

5 W. M. P. Strandberg, *Amer. J. Phys.*, **54**, 321, 1986.

6 J. Gribbin, *In Search of the Edge of Time*, London, 1992, 55.
7 F. R. Tangherlini, *Amer. J. Phys.*, **36**, 1001, 1968.
8 Space-time curvature can itself be seen as doubling, and we can see it leading to the creation of discrete charges (***i***, ***j***, ***k***) in pair production (though not of additional gravity!). (Mass can also be derived from electromagnetism in curved space-time.) The space-time duality also relates to the holographic principle, which is discussed fully in 20.9, where the only information about a system comes from the bounding area. If space and time are fundamentally dual in a nilpotent structure, then, in information terms, doubling can be regarded as equivalent to the Pythagorean squaring involved in zeroing the nilpotent structure. So, space and time become dual because they combine as squared quantities, with an invariant proper time, and the only information is in this squared quantity.
9 G. E. Uhlenbeck, *Physics Today*, **29** (b), 43, 1976.
10 C. K. Whitney, How can paradox happen?, *Proceedings of Conference on Physical Interpretations of Relativity Theory VII*, British Society for Philosophy of Science, London, September 2000, 338-51.
11 Whitney also shows that the classic light-bending and perihelion precession 'tests' of General Relativity can be derived using a two-step process. As we have seen, in 14.3, it is certainly possible to derive the light bending by classical arguments using kinetic energy, which is the same thing as using special relativity, because light has no rest mass, and it is also possible to derive perihelion precession using special relativity, as a number of authors have demonstrated (cf 15.7).
12 T. Morris, 'The unique reference frame – is it superfluous or essential in special relativity?, *Proceedings of Conference on Physical Interpretations of Relativity Theory VII*, British Society for Philosophy of Science, London, September 2000, 267, abstract, plus oral presentation.
13 Another explanation of radiation reaction, by Dirac, uses retarded and advanced potentials, like those of the Wheeler / Feynman theory. This is equivalent to ± *t* or fermion / antifermion.
14 As mentioned previously in chapter 5, four separate vacua can be identified. The 'weak' vacuum is the one described here, but, as mentioned previously in chapter 5, we can also write down expressions for 'strong' and 'electric' vacua, which are respectively ***i*** (***k****E* + *i****i*****p** + *i****j****m*) and ***j*** (***k****E* + *i****i*****p** + *i****j****m*). The first flips the spin states, while the second converts fermion spin up to antifermion spin down, etc. The fourth, or 'gravitational', vacuum, 1(***k****E* + *i****i*****p** + *i****j****m*), is identical to the fermion itself. With a nilpotent structure, there will be no vacuum 'image'. The use of the coefficient 1 may be taken as equivalent to the statement that the gravitational vacuum cannot be quantized (though an inertial one can). One way of looking at this is to define gravitational energy as negative (because of the attractive force) and to refer to the filled vacuum for negative energy states.
15 These processes, together with of those of pair production and annihilation, and boson absorption and emission, can be considered as a necessary consequences of the universality of physical interactions, or the definition of the Dirac nilpotent as itself a unit in a necessarily interconnected infinite-dimensional space.
16 S. H. Talbert, On some intrinsic attributes of motion and propagation, *Proceedings of VIII Conference on Physical Interpretations of Relativity Theory*, British Society for Philosophy

of Science, London, September 2002. The 1 + 1 dimensionality characteristic of special relativity may be seen as a result of its origin in the discrete Dirac quantum state.

17 M. V. Berry, Riemann's zeta function: a model of quantum chaos, in T. H. Seligman and H Nishioka (eds.), *Springer Lecture Notes in Physics*, 263, 1-17, *Quantum Chaos and Statistical Nuclear Physics*, Springer, Berlin, 1986.

18 A special case of doubling occurs with chaotic and fractal systems, which are also, dually, continuous and discrete. 'Doubling' seems to act as though it is increasing the apparent 'dimensionality' of chaotic systems ($\to \infty$) and reducing that of fractals ($\to 1$).

Chapter 17

1 N. V. Pope and A. D. Osborne, Instantaneous Relativistic Action-at-a-Distance, *Physics Essays*, **5**, 409-421, 1992; Instantaneous Gravitational and Inertial Action-at-a-Distance, *Physics Essays*, **8**, 384-97, 1995.

2 The mathematical formalism for the Pope-Osborne theory uses the vitial theorem to convert potential energy information (V) to angular momentum via the kinetic energy term (T). This works well for the gravitational and electromagnetic $U(1)$ symmetries, with inverse-square law forces, where $V = -2T$, and would be valid for the $U(1)$ components of the other forces, with their information about the magnitude of the angular momentum term. However, to accommodate the vector part of the strong force and the pseudoscalar part of the weak force might require some additional mathematical strategy. If the force laws for these components are as specified in this book, then the virial relations would be $V = 2T$ for the additional strong component (linear force); and $3V = -2T$ or $V = -T$ for the additional weak component, depending on whether the force involved was dipole-dipole or monopole-dipole.

3 The claim by S. Kopeikin and E. B. Fomalont (*Bull. Amer. Astr. Soc.*, **34**, 4, 2003) is challenged by H. Asada (*Astrophys. J.*, **574**, L69, 2002) and also by C. M. Will (arXiv.org/abs/astro-ph/0301145).

4 T. Van Flandern, *Dark Matter, Missing Planets and New Comets: Paradoxes Resolved and Origins Illuminated*, North Atlantic Books, Berkeley, California, 1993, pp. 47-52. Van Flandern's interesting argument has received less attention than it merits because of the dubious nature of some of the other material in his work.

5 M. W. P. Strandberg, *Am. J. Phys.*, **54**, 321-31, 1986, 321-2.

6 I. Newton, draft for *Principia*, Book I, Proposition VI, Corollary III, early 1690s; translated by J. E. McGuire, Transmutation and immutability: Newton's doctrine of physical qualities, *Ambix*, **14**, 69-95, 1967, 72-73.

7 J. Priestley, *History and Present State of Discoveries Relating to Vision, Light and Colours*, 1772, 787-90.

8 J. Michell, *Phil. Trans.*, **74**, 35-57, 1784.

9 L. I. Schiff, *Am. J. Phys.*, **28**, 340-3, 1960

10 S. K. Ghosal and P. Chakraborty, *Eur. J. Phys.*, **12**, 260-7, 1991.

11 E. Thorpe (ed.), *The Scientific Papers of the Honourable Henry Cavendish, F.R.S., Volume II: Chemical and Dynamical*, London, 1921, 433-7.

12 J. G. Von Soldner, *Astronomisches Jahrbuch für das Jahr 1804*, Späthen, Berlin, 1801, 161, dated March 1801.

13 S. L. Jaki, *Found. Phys.*, **8**, 927-50, 1978.

14 R. Tian and Z. Li, *Am. J. Phys.*, **58**, 890-2, 1990.

15 D. K. Ross and L. I. Schiff, *Phys. Rev.*, **141**, 1215-8, 1966.

16 P. Gerber, *Z. für Math. u. Phys.*, **43**, 93-104, 1898.

17 N. T. Roseveare, *Mercury's Perihelion from Le Verrier to Einstein*, Oxford 1982, 137-44.

18 I. Newton, *Principia*, Book I, Proposition 45; translated by A. Motte as *Mathematical Principles of Natural Philosophy*, revised by F. Cajori, Cambridge, 1934, 144.

19 A. Harvey, *Am. J. Phys.*, **46**, 928-9, 1978

20 H. Kolbentsvedt, *Amer. J. Phys.*, **56**, 523-4, 1988.

21 W. de Sitter, *Monthly Notices RAS*, **77**, 155-84, 1916.

22 A. S. Eddington, *The Mathematical Theory of Relativity*, Cambridge, 1922, 164.

23 J. Lense and H. Thirring, *Physikalische Zeitschrift*, **19**, 156-63, 1918.

24 A. S. Eddington, *The Mathematical Theory of Relativity*, Cambridge, 1922.

25 D. D. Cattani, *Nuovo Cimento*, **60**, 67-80, 1980.

26 V. B. Braginsky, C. M. Caves and K. S. Thorne, *Phys. Rev.* D, **15**, 2047, 1977.

27 R. D. Blandford, Astrophysical black holes, in S. W. Hawking and W. Israel, *300 Years of Gravitation*, Cambridge, 1987, 277-329, quoting 283.

28 While my own argument against a black hole singularity has always been expressed in physical terms, I am grateful to Jeremy Dunning-Davies for pointing out that Schwarschild's own solution of 1916 (translated by S. Antoci and A. Loinger in arXiv:physics/9905030) avoids any singularity at $r = 2GM / c^2$ by chosing coordinates $x_1 = r^3 / 3$, $x_2 = -\cos\theta$, $x_3 = \phi$, so that the determinant of the transformation is 1. He also cites Einstein (*Annals of Mathematics*, **40**, 922, 1939) and Dirac (*The General Theory of Relativity*, Princeton University Press, 1996), among other authors, in support of the idea that there is no physical 'Schwarzschild singularity'. The only singularity in Schwarzschild's own solution is the one at $r = 0$ which necessarily results from the introduction of polar coordinates.

29 S. B. M. Bell, J. P. Cullerne and B. M. Diaz, A new approach to quantum gravity: an overview, in Amoroso, R. L., Hunter, G., Kafatos, M. and Vigier, J-P. (eds.), *Gravitation and Cosmology: From the Hubble Radius to the Planck Scale*, Kluwer Academic Publishers, 2002, 303-12.

30 In the case of *currently observable* phenomena, the only formal theory required is that used for QED in chapter 9, with the scalar and vector potential terms now interpreted as those determined from the semiclassical derivation of the Maxwell-type equations in 15.12 (cf R11, R24).

31 A formal theory of this kind, also including the concept of discreteness through extended causality, is outlined in 18.9, drawing on work previously presented by Brian Koberlein.

32 Hawking radiation of a particular particle type occurs when the Davies-Unruh $kT > mc^2$ for that particle type. Pair production at a Schwarzschild boundary then leads to the emission of *either* the particle or the antiparticle. Significantly, here, there is once again a manifestation of the factor 2, with a particle-antiparticle pair being created in the curved space-time within the boundary, where the GR doubling effect may be presumed to occur, but only one of the pair being emitted in the Euclidean space-time beyond it. Hawking radiation, also, does not discriminate between particle and antiparticle, because gravity cannot tell the difference: mass is always positive, and the vacuum state = the particle state.

33 T. H. Boyer, *Phys. Rev.*, D, **21**, 2137, 1980.
34 H. E. Puthoff, *Phys. Rev.* A, **39**, 2333, 1989.
35 B. Haisch, A. Rueda and H. E. Puthoff, *Phys. Rev.* A, **49**, 678, 1994.
36 A. Rueda and B. Haisch, *Found. Phys.*, **28**, 1057, 1998.
37 A. Rueda and B. Haisch, *Phys. Lett.* A, **240**, 115, 1998.
38 A. Rueda and B. Haisch, The inertial reaction force and its vacuum origin, in R. L. Amoroso, G. Hunter, M. Kafatos and J-P. Vigier (eds.), *Gravitation and Cosmology: From the Hubble Radius to the Planck Scale*, Kluwer, 2002, 447-58, and Gravity and the quantum vacuum inertia hypothesis, arXiv:gr-qc/050961 v3. H. Sunahata, A. Rueda and B. Haisch, Quantum analysis on contribution to inertial reaction force by the electromagnetic vacuum (submitted for publication, 2007), is a quantum analysis reproducing the results previously obtained by stochastic electrodynamics.
39 The inertial c will, of course, be gravitationally affected, but only from the localised component of gravity from discrete sources. The argument in 15.7 suggests how a *specific* cut-off produces a specific inertial mass, but the actual cut-offs required for individual particle states are produced by the couplings to the infinite energy density (but finite energy level) Higgs field or filled vacuum state generated by the specific violations of (charge) symmetry with which they are associated.

Chapter 18

1 The conserved 3-D symmetries even break the unbroken symmetry of the vector term by privileging one direction, as in spin and special relativity.
2 A quite remarkable illustration of the closed nature of this system is the famous formula $e^{i\pi} = -1$, which connects numbers derived from complexification (i), dimensionalization (π, the expression of spherical symmetry) and (non)conservation (e, derived from the fundamental properties of differentiation, and hence nonconservation), to provide a single dual to unity (–1), in a way that parallels the structure of *CPT* symmetry, and becomes a kind of mathematical equivalent. Significantly, the respective properties associated with e, i and π – nonconserved, imaginary, and dimensional / discrete – are the precise duals or opposites of those associated with mass – conserved, real, and nondimensional / continuous – which is the first parameter to emerge from the rewrite structure, and acts here as the 'unit' to be conjugated.
3 H. Minkowski, *Physikalische Zeitschrift*, 10, 104-11, 1909, lecture on 'Space and Time', Cologne, 21 September 1908, translated in Lorentz, Einstein, Minkowski and Weyl, *The Principle of Relativity*, 1923, 104.
4 There is no discreteness of space-time, because it doesn't exist, except classically.
5 Quantum mechanics has fractal order 2 (Mandelbrot); if we use more than 3 dimensions, we lose charge into the fourth dimension. There would otherwise be no conservation (as with E and $\mathbf{p}$ in the 3-D nilpotent). In the nilpotent theory, of course, there is no greater dimensionality than 3, but there are nested 3-dimensionalities.
6 Though the work in this section is differently conceived, I am aware here of the work of Shiela Morgan and Melanie Purcell who have seen profound significance in the Klein bottle.
7 E. Witten, arXiv:hep-th/0312171.

8 The nilpotent structure combines twistor space and momentum space, making them essentially identical. Though two spaces – real space and momentum – are used in conventional quantum theory (and even in classical physics), they are not commutative, because one is derived from the other; but the quaternion and vector spaces that combine to produce the Dirac nilpotent are independent, and therefore commutative, though neither will commute with momentum space, which is derived from both.

9 B. D. Koberlein, Extended causality as a model of discrete gravity, in K. G. Bowden (ed.), *Implications* (*Proceedings of ANPA 22*), 56-63, 2001, and *Correlations* (*Proceedings of ANPA 23*), 18-23.

10 S. B. M. Bell, J. P. Cullerne and B. M. Diaz, Classical behaviour of the Dirac bispinor, *Found. Phys.*, **30**, 35, 2000.

11 S. B. M. Bell, J. P. Cullerne and B. M. Diaz, A new approach to quantum gravity: an overview, in R. L. Amoroso, G. Hunter, M. Kafatos and J-P. Vigier (eds.), *Gravitation and Cosmology: From the Hubble Radius to the Planck Scale*, Kluwer Academic Publishers, Dordrecht, 303-12, 2000.

12 Curvature of space-time-proper time due to inertial reaction, and not gravity, can thus be quantum. We may note that Riemannian space-time must be curved if it is continuous, but doesn't need curvature if it is discrete.

Chapter 19

1 M. Morris, K. Uchida and Tuan Do, A magnetic torsional wave near the Galactic Centre traced by a 'double helix' nebula, *Nature*, **440**, 308, 16 March 2006.

2 C. Illert, *Foundations of Theoretical Conchology ... from Self-Similarity in Non-Conservative Mechanics*, Hadronic Press, 1992. C. Illert and R. M. Santilli, *Foundations of Conchology*, Hadronic Press, 1996.

3 Remarkably, 'complexity' in mathematics seems to be ultimately responsible also for 'complexity' in systems!

4 D. J. Brooks, J. R. Fresco, A. M. Lesk and M. Singh, Evolution of amino acid frequencies in proteins over deep time. Inferred order of introduction of amino acids into the genetic code. *Mol. Biol. Evol.*, **19**(10), 1645-55, 2002.

5 V. Ivanov, A. Beniaminov, A. Mikheyev and E. Minyat, A mechanism for stop codon recognition by the ribosome: a bioinformatic approach. *RNA*, **7**, 1683-92, 2001.

6 B. N. Chaudhuri and T. O. Yeates, A computational method to predict genetically encoded rare amino acids in proteins. *Genome Biology*, **6**, R79, 2005.

7 B. Cobucci-Ponzano, M. Rossi and M. Moracci, Recording Archaea. *Molecular Microbiology*, **55** (2), 339-48, 2005.

8 S. E. Johansen, Initiation of 'Hadronic Philosophy', the philosophy underlying hadronic mechanics and chemistry, Hadronic Journal, **29**, no. 2, 111-135, 2006.

9 D'Arcy Thompson, *On Growth and Form* (1917). Of course, the Golden Ratio is not uniquely determined by the Fibonacci series, but by any series constructed according to the relationship $f(n + 1) = f(n) + f(n - 1)$, where $f(n)$ is an integer for all n. In the work described in this chapter, the Golden Ratio is the necessary result of the fundamental nature of a 5-fold broken symmetry, and the Fibonacci series becomes one means of expressing it

mathematically.

10 Crystalline form is a way of overcoming the possibly chaotic state representing the weak behaviour (e.g. *zitterbewegung*, weak dipole moment) of the isolated fermionic state (cf. chapter 20 on the Riemann zeta function as states of quantum chaos). The regular order and structure of an inorganic crystal provides a minimum entropy state with a high degree of coherence. It may be that this remains dynamic within organic materials, manifesting itself in the less organized, and more time-dependent form of the spiral.

11 D. K. Kondepudi and K. Asakura, *Acc. Chem. Res.*, **34**, 946-954, 2001 (review article).

12 http://mathworld.wolfram.com/Pentatope.html

13 J. D. Watson and F. H. C. Crick, *Nature,* **171**, 737-738, 1953.

14 M. Curtis, *The geometry of DNA: a structural revision*, The Blue Gallery, London, 1997. M. Curtis and A. Ochert, *New Scientist,* 16 May 1998, 32-35.

15 K. Hoogsteen, *Acta Cryst.*, **16**, 907, 1963.

16 R. Brookes, *Proceedings of the Eighth ICECDGD*, Tokyo, **2**, 378, 1998; http://www.brooksdesign-cg.com/Code/Html/godna2.htm. http://www.brooksdesign-cg.com/Code/Html/scodna2.htm.

17 http://mathworld.wolfram.com/120-Cell.html

18 The dodecahedron in 3D is composed of 5 cubes. If we take a 3D cube and consider each of the four axes through the corners, with a cube rotating upon each, there is a position that is created from the 5 cubes that constructs the dodecahedron. As we know we can construct the star tetrahedron within each cube and if we extend this for the constructed dodocahedron, we will have a total of 10 interlocking tetrahedra. Interestingly, these interlocking tetrahedra can be seen to produce an appearance of two vortices, one going one way and the other the other way in a manner analogous to a double helix.

19 The 'dynamic' 2-D hexagonal (or flattened cube) symmetry of benzene reflects that of the more dynamically interesting parent graphite as opposed to the more perfectly ordered, and therefore more inert, 3-D tetrahedral structure of diamond.

Chapter 20

1 P. Marcer, E. Mitchell, P. Rowlands and W. Schempp, Zenergy: The 'phaseonium' of dark energy that fuels the natural structures of the Universe, *International Journal of Computing Anticipatory Systems*, 16, 189-202, 2005. P. Marcer, P. Rowlands and B. Diaz, Nilpotence: the key to a theory of everything, in K. Bowden (ed.), *Against Bull*, 2005 (*Proceedings of XXVI ANPA Conference*, Cambridge, August 2004), 231-46. P. Marcer and P. Rowlands, The Evolutionary 'Anthropic' Semantic Principle, in K. Bowden (ed.), *Conceptions*, 2006 (*Proceedings of XXVII ANPA Conference*, Cambridge, August 2005), 361-70. P. Marcer and P. Rowlands, A remarkable quantum mechanical discovery, *International Journal of Computing Anticipatory Systems*, **19**, 261-278, 2006.

2 L. Susskind, *The Cosmic Landscape: String Theory and the Illusion of Intelligent Design*, Little, Brown and Company, 2005.

3 P. Woit, *Not Even Wrong The Failure of String Theory and the Continuing Challenge to Unify the Laws of Physics*, Jonathan Cape, 2006.

4 A. M. Turing, On Computable Numbers with reference to the *Entscheidungdproblem*,

Proceedings of the London Mathematical Society, **2**, no. 42, 230-265 and 544-546, 1936.

5 J. von Neumann, Theory of automata: construction, reproduction, homogeneity, Part II of A. W. Burks (ed.), *The Theory of Self-Reproducing Automata*, University of Illinois Press, Urbana, Illinois, 1966.

6 G. Bateson, *Mind and Nature: A Necessary Unity*, Bantam Books, Toronto, 1979. We are grateful to Stein Johansen for this reference.

7 G. Chapline, Is Theoretical Physics The Same Thing as Mathematics? *Elsevier Physics Reports*, **315**, 95-105, 1999.

8 D. Deutsch, The Church-Turing principle, and the universal quantum computer. *Proceedings of the Royal Society of London*, A **400**, 97-117, 1985.

9 R. P. Feynman, Quantum Mechanical Computers. Foundations of Physics, **16**, 6, 507-531, 1986.

10 P. J. Marcer, Commonsense, what is it?, presented at the British Theoretical Computer Science Colloquium, University of Warwick, March 24-26, 1986.

11 It is evident that artificial intelligence (AI) cannot be developed efficiently within the constraints of classical digital architecture.

12 J. A. Wheeler, Physics as Meaning Circuit: Three Problems. G. T. and M. O. Scully (eds.), *Frontiers of Non-equilibrium Statistical Physics*, Plenum Press, New York, 1986.

13 B. E. P. Clement, P. V. Coveney and P. J. Marcer, Surreal Numbers and Optimal Encodings for Universal Computations as a Physical Process Interpretation Of the Genetic Code. *CCAI the Journal for the Integrated Study of AI, Cognitive Science and Applied Epistemology*. **10**, 1/2, pp. 149-163, 1993.

14 L. Wittgenstein, *Philosophical Remarks*, Oxford University Press, 1975.

15 A. J. Leggett, The Quantum Measurement Problem, *Science*, **307**, 871-872, 11 February 2005.

16 See *Science*, **301**, 1169-70, spacetime 'Einstein 1, Quantum Gravity 0', 29 August 2003.

17 J. H. Conway, *On Numbers and Games*, Academic Press, London, 1976.

18 N. Alling, *Foundations of the Analysis over Surreal Numbers*, *Mathematical Studies*, 141, North Holland, Amsterdam, 1988. Conway's assertion (ref. 17) that his definition of the surreals and Zermelo-Fraenkel set theory are 'one and the same' appears not to be true. Once we accept the arbitrary inclusion of *i*, Conway's definition of the surreals ensures that the mathematical language used has both a sound syntax and semantics and so must include a satisfactory approach to the theory of types, whereas the Zermelo-Fraenkel theory often leads to paradox. It may be that both the problems of set theory and the limitations of mathematics defined by Gödel's theorem (and therefore the Halting Problem, too) result from assuming that mathematics follows a syntactic, rather than semantic, logic, and that such problems do not arise when we define mathematics, semantically, in terms of a universal rewrite system.

19 B. Diaz and P. Rowlands, D: The infinite square roots of –1. *International Journal of Computing Anticipatory Systems*, **19**, 2006.

20 R. P. Feynman and B. J. Hibbs, *Quantum Mechanics and Path Integrals*, McGraw-Hill, New York, 1965.

21 M. Jessel, *Compte Rendu Academie des Sciences*, Paris, 239, December, 1599-1601 (in French). This first actual contribution to the theory of Huygens' principle led Jessel to the discovery of anti-waves, and to investigate in particular to those of anti-sound, i.e. the

dynamic annihilation of sound throughout an entire 3-dimensional spatial domain.

22 A. Einstein, quoted in A. Pais, *Subtle is the Lord*, Oxford University Press, 1982, p. 34.

23 A theory of cosmology and universe evolution is presented in this chapter, as an aspect of the discussion of the principle of the quantum Carnot engine. While evolution and unique birth-ordering are fundamental aspects of this cosmology, the concept of singularity is not, except in the case of the fermionic components. So readers are invited to put their own interpretations on such concepts as the origin of sources, sinks, 3 + 1 space-time, etc. The 'creation' metaphor used in this chapter does not necessarily refer to a single 'creation event', as will be explained in 21.7, though that might be considered by some as a convenient explanation.

24 R. Landauer, Computation and Physics: Wheeler's Meaning Circuit? Foundations of Physics, **16**, 551-564, 1986

25 D. Dubois, The Fractal Machine, Presses Universitaires de Liège, 1992. D. Dubois and G. Resconi, *Hyperincursivity: A new mathematical theory*, Presses Universitaires de Liège, 1992. This sets out a new generalization of the computational recursion called by Dubois 'incursivity' and 'hyperincursivity'.

26 Dubois's strong anticipation may be connected with recursion, and weak anticipation with iteration (see Chapter 1). The nilpotent structure includes anticipation through its automatic inclusion of $-\partial / \partial t$.

27 H. A. Fatmi and Resconi, A New Computing Principle. *Il Nuovo Cimento*, **101B**, no.2, pp. 239-242, 1988.

28 G. Resconi and P. J. Marcer, A Novel Representation of Quantum Cybernetics using Lie algebras. *Physics Letters A*, **125**, no.6/7, pp 282-290, 1987.

29 D. Gabor *et al.*, A Universal Non-linear filter, Predictor and Simulator which optimizes itself by a Learning Process. *Proceedings IEE*, **108B**, 422-438, 1960.

30 M. Jessel and G. Resconi, General System' Logical Theory (GSLT). *International Journal of General Systems*, **12**, 155-182, 1986.

31 M. D. Mesarovic and D. Takahara, *General System Theory*, Academic Press, New York, 1975.

32 G. Spencer-Brown, *The Laws of Form*, Allen and Unwin, London, 1969.

33 C. W. Kilmister, Brouwerian Foundations for the Combinatorial Hierarchy, *Proc. 1st Annual Western Regional Meeting Discrete Approaches to Natural Philosophy*, Stanford University, USA, November, 1984. (See also Towards a Process Formalism in Quantum Physics, in A. van der Merve, F. Selleri and G. Tarozzi (eds.), *Microphysical Reality and the Quantum Formalism*, vol. 1, editors, Kluwer, Dordrecht, 1987.) The Combinatorial Hierarchy is based on a numerical hierarchy discovered by F. Parker-Rhodes: 3; $2^3 - 1 = 7$; $2^7 - 1 = 127$; $2^{127} - 1 \sim 1.7 \times 10^{38}$, which bears some resemblance to the relative strengths of the four physical interactions.

34 E. W. Bastin and C. W. Kilmister, The Concept of Order 1. The Space-Time Structure, *Proceedings of the Cambridge Philosophical Society*, **50**, Part 2, 278-286, 1954.

35 W. Schempp, *Harmonic Analysis on the Heisenberg Group with Applications in signal theory. Pitman Notes in Mathematics*, Series 14. Longman Scientific and Technical, London, 1986. (This book discusses the general properties of nilpotent Lie groups and their nilpotent Lie algebras and their application to Synthetic Aperture Radars.)

36 W. Schempp, Quantum Holography and Neurocomputer Architectures. *Journal of Mathematical Imaging and Vision*, 2, 279-326, 1992.

37 W. Schempp, Bohr's Indetermincy Principle in Quantum Holography, Self-adaptive Neural Network Architectures, Cortical Self-Organization, Molecular Computers, Magnetic Resonance Imaging and Solitonic Nanotechnology, *Nanobiology*, **2**, 109-164, 1993.

38 W, Schempp, *Magnetic Resonance Imaging, Mathematical Foundations and Applications*, John Wiley, New York., 1998. See also http://www.civm.duke.edu.

39 E. Binz and W. Schempp, Creating Magnetic Resonance Images, Proceedings of the Third International Conference on Computing Anticipatory Systems, *International Journal of Computing Anticipatory Systems*, **7**, 223-232, 2000.

40 E. Binz and W. Schempp, A unitary parallel filterbank approach to Magnetic Resonance Tomography, Proceedings of the Third International Conference on Computing Anticipatory Systems, ed. D. Dubois, *American Institute of Physics Proceedings*, **517**, 406-416, 2000.

41 E. Binz and W. Schempp, Quantum Hologram and Relativistic Hodogram: Magnetic Resonance Tomography and Gravitational Wavelet Detection, Proceedings of the Third International Conference on Computing Anticipatory Systems, ed. D. Dubois, *American Institute of Physics Proceedings*, **573**, 98-131; section 9, 118-120, 2001.

42 S. A. Rice, New Ideas for Guiding the Evolution of a Quantum System, *Science*, **258**, pp. 412-413, 1992.

43 R. S. Judson and H. Rabitz, Teaching Lasers to Control Molecules. *Physics Review Letters*, **68**, 10, 1500-1503, 1992.

44 M. Dahleh, A. P. Pierce and H. Rabitz, Optimal Control of Uncertain Systems. *Physics Review A*, **42**, 3, 1065-1079, 1990.

45 W. P. Schleich, Sculpting a Wavepacket. *Nature*, **397**, 207-208, 1999.

46 C. Leichtle, W. P. Schliech, I. Sh. Averbukh and M/ Shapiro, Quantum State Holography. *Physics Review Letters*, **80**, 7, 1418-1421, 1998.

47 Because momentum and energy are created in the act of compactification of the fermionic nilpotent out of the original space and time, they necessarily occupy the same logical space, associated with same quaternion operators, and so are anticommutative. This anticommutativity becomes expressed through the Heisenberg uncertainty principle. So, Heisenberg uncertainty is saying directly that a fundamental quantum state, or unit of quantum information, has been created through the association of the space, time, mass and charge operators in creating the generators of the Dirac nilpotent Lie group.

48 K. H. Pribram, *Brain and Perception: Holonomy and Structure in Figural Processing*. Lawerence Eribaum Associates, New Jersey, 1991.

49 P. A. M. Dirac, *The Principles of Quantum Mechanics*, Clarendon Press, Oxford, 1947.

50 M. Perus and H. Bischof, The Most Natural Procedure for Quantum Image Processing, Symposium 10, International Conference for Computing Anticipatory Systems, *International Journal of Computing Anticipatory Systems*, 2003.

51 We may here also cite the covariant and contravariant category theory mappings between the universe and the information system, defined as intensional / extensional (or 'push' and 'pull'), in M. A. Heather and B. N. Rossiter, The universe as a freely generated information system, *Conceptions*, 2006 (*Proceedings of XXVII ANPA Conference*, Cambridge, August 2005), 357-388.

52 M. O. Scully *et al*, Extracting work from a single heat bath via vanishing quantum coherence. *Science*, **299**, 862-864, 2003. 10.1126/science.1078955.

53 P. W. Anderson, The 1982 Nobel Prize in Physics, *Science*, **218**, 763-764, 19 November 1982.

54 K. G. Wilson, The Renormalization Group and critical phenomena, *Reviews of Modern Physics*, **55**, 3, 583-600, July 1983.

55 R. R. Caldwell and M. Kamionkowski, Echoes of the Big Bang, *Scientific American*, January 2001, 28-33.

56 F. Close, *The Cosmic Onion*, Heineman Educational Books, 1982, 117.

57 W. H. Zurek, Algorithmic randomness and Physical Entropy, *Physical Review A*, **40**, 8, 4731-4751, 1998.

58 W. H. Zurek, Thermodynamic cost of computation, algorithmic complexity, and the information metric, *Nature*, **341**, 119-124, 14 September 1989.

59 In H. A. Fatmi et al. 1990, Theory of Cybernetic and Intelligent Machines based on Lie Commutators, *International Journal of General Systems*, **16**, 123-164, based on Fatmi and Resconi's New Computing Principle (ref. 20), and Fatmi and R. W. Young's definition of Intelligence *(Nature*, **228**, 97 1970), the cycles of the machine are also, as was first proposed by Resconi, 'fueled' by energy in the sense of Scully *et al* (ref. 43). This is to say, in relation to the universe as a QCE, that entropy constitutes the basis of its universal information metric in respect of the fuelled natural structure.

60 M. V. Berry, The Geometric Phase, *Scientific American*, December 1989, 26-32.

61 R. Resta, The Geometric Phase, *Europhysics News*, **28**, 19, 1997; see also Manini, http://www.sissa.it/~manini/berryphase.html Berry's geometric phase: a review, 03/05,1-9, 2001.

62 M. V. Berry, Riemann's Zeta function: a model for quantum chaos? T. H. Seligman and H. Nisioka (eds.), *Quantum chaos and statistical nuclear physics*. Springer Lecture Notes no. 263, Springer Berlin, 1986, 1-17.

63 M. Jessel, Secondary Sources and Their Energy Transfer, *Acoustic Letters*, **4**, 9, 174-179, 1981.

64 P. J. Stewart, A Chemical Galaxy. *Today* (Oxford University Magazine), **17**, no 3, Trinity issue, News and Events, 2005.

65 W. Crookes, *Reports BAAS*, **55**, 558-76, 1886. See also *J. Chem. Soc.*, **53**, 487-504, 1888. Crookes was influenced by a Periodic Table, arranged as a spiral for pedagogic purposes, by J. Emerson Reynolds, in Note in a method of illustrating the Periodic Law, *Chem. News*, **54**, 1-4, 1886. Crookes thought that a fundamental substance ('protyle') was responsible for the generation of the elements (which he later associated with the electron, after that particle was discovered). He imagined a double 'cosmic pendulum', in which an electric force, $x = a \sin (mt)$, acted at right angles to a uniform lineer motion, representing the fall in temperature y with time, $y = bt$. Another oscillatory motion ar right angles to each of these two, and at twice the frequency of the other oscillation, involved another force, which was probably electrical, $z = c \sin (2mt)$, and brought into the picture a spatial origin for the process of generation. The combination of these motions produced a 3-dimensional figure-of-eight arrangement of the elements, with elements from the same groups appearing vertically above each other on the same positions in the double helix. Crookes thought that valency and

chemical properties were determined by a fixed quantity of electricity which each element had at the moment of its birth, while the atomic weight was a measure of the degree of cooling which had occurred at that moment. Variations in this quantity would lead to atoms of the element with slight (whole number) variations from the mean value (similar to modern isotopes). He supposed that there was continuous creation in the universe, rather than heat death, with radiant heat from one part of the universe contributing to elementary production in another.

66 P. Gariaev, B. Birstein, A. Iarochenko, K. A. Leonova, P. Marcer, U. Kaempf and G. Tertishy, Fractal Structure in DNA Code and Human Language: Towards a Semiotic of Biogenetic Information, *International Journal of Computing Anticipatory Systems*, **12**, 255-273, 2002.

67 P. Marcer and W. Schempp, A Mathematically Specified Template for DNA and the Genetic Code, in terms of the physically realizable Processes of Quantum Holography, *Proceeding of Greenwich (University) Symposium on Living Computers*, A. Fedorec and P. Marcer (eds.), 45-62, 1986.

68 P. Gariaev, B. Birstein, A. Iarochenko, K. A. Leonova, P. Marcer, U. Kaempf and G. Tertishy, The DNA-wave Biocomputer. Fourth International Conference Computing Anticipatory Systems, *Journal of Computing Anticipatory Systems*, **10**, 290-310, 2001.

69 P. Gariaev, B. Birstein, A. Iarochenko, K. A. Leonova, P. Marcer, U. Kaempf and G. Tertishy, Fractal Structure in DNA Code and Human Language: Towards a Semiotics of Biogenetic Information. *International Journal of Computing Anticipatory Systems*, **12**, 255-273, 2002.

70 P. Marcer and W. Schempp, Model of the Neuron Working by Quantum Holography. *Informatica*, **21**, 519-534, 1997.

71 P. Marcer and W. Schempp, The brain as a conscious system. *International Journal of General Systems*, **27**, 1/3, 231-248, 1998.

72 P. Marcer and W. Schempp, The Model of the Prokaryote Cell as an Anticipatory System Working by Quantum Holography. *International Journal of Computing Anticipatory Systems*, **2**, 307-315, 1998.

73 P. Marcer and E. Mitchell, What is consciousness? Philip Van Loocke (ed.), *The Physical Nature of Consciousness*, Advances in Consciousness Research series, John Benjamins B.V., Amsterdam, 2001, 145-174.

74 W. Schempp, Bohr's Indeterminacy Principle in Quantum Holography, Self-adaptive Neural Network Architectures, Cortical Self-organisation, Molecular Computers, Magnetic Resonance Imaging and Solitonic Nanotechnology. *Nanobiology*, 2, 109-164, 1993.

75 W. C. Hoffman, The Visual Cortex is a Contact Bundle. *Applied Mathematics and Computation* **32**, 137-167, 1989. This work is an excellent summary, with many references to papers published by Hoffman as early as the mid 1960s, beginning with such papers as The Neuron as a Lie group germ and Lie product, Quarterly Journal of Applied Mathematics, **25**, 4, 423-440, 1968.

76 R. Noboli, Schrödinger Wave Holography in the Brain Cortex. *Physical Review A*, **32**, 6, 3618-3626, 1985.

77 R. Noboli, Ionic Waves in Animal Tissue. *Physical Review A*, **35**, 4, 1901-1922, 1987.

78 M. Bauer, M. and Nartienssen, W. (1991) Coupled Circle Maps As a Tool to Model Synchronisation on Neural Networks. Network 2, pp. 345-351.
79 B. E. P. Clement, P. V. Coveney, M. Jessel and P. J. Marcer, The Brain as a Huygens' Machine. *Informatica*, **23**, 387-398, 1999.
80 J. Sutherland, Holographic / Quantum Neural Technology, Systems and Applications. *ISCAS*, 313-334, 1999; also http://www.andcorporation.com.
81 J.-C. Perez, The Eight Codes of Life, etc. (private web communications, 2003); see also www.genum.com.
82 A. Patel, Quantum Algorithms and the Genetic Code, *Proceedings of the Winter Insitute of Quantum Theory and Quantum Optics*, 1-13 January 2000, S.N. Bose National Centre for Basic Sciences, Calcutta, India.
83 D. Dubois, http://www.ulg.ac.be/mathgen/CHAOS
84 Ref. 71, p. 235.
85 N. Chomsky, *Reflections on Language*, New York, 1975.
86 ref. 73, p. 159
87 K. Kingsland, oral presentation, Frontiers Workshop, London, 2004.
88 Maximal information transfer appears to occur at the edge of chaos, by maximising the '3-D doubling' or 'bifurcation' that, according to the NUCRS, allows change to occur.

Chapter 21

1 D. W. Sciama, *Monthly Notices RAS*, **113**, 34, 1953.
2 D. W. Sciama, *The Physical Foundations of General Relativity* (London, 1972), pp. 36 ff.
3 P. de Bernardis *et al*, *Nature*, **405**, 955, 2000.
4 B. P. Schmidt *et al*, *Astrophys. J.*, **507**, 46, 1998.
5 S. Perlmutter *et al*, *Astrophys. J.*, **517**, 565, 1999.
6 D. Roscoe, A perspective on Mach's principle and the consequent discovery of major new phenomenology in spiral discs, chapter 7, 169-184, of N. V. Pope, A. D. Osborne and A. F. T. Winfield (eds.), *Immediate Distant Action and Correlation in Modern Physics The Balanced Universe*, Edwin Mellen Press, 2005. D. Roscoe, *Astron. & Astrophys.*, **343**, 788, 1999; **385**, 431, 2002; *Gen. Rel. Grav.*, **34**, 5, 577, 2002.
7 A serious calculation is not really possible at present, as the result would be heavily dependent on having an exact value for the density of baryonic matter. For a universe with 8 % baryonic matter (which is not entirely inconceivable), a 'Coriolis force' might provide four times the effect of this matter's gravitational energy, and so be responsible for 32 % of the universe's density.
8 A relation between electromagnetism and gravity involving an exponential of the fine structure constant has long been suspected, and, using a nonperturbative calculation of the self-energy of the electron from a zero bare mass (K. Johnson *et al*, *Phys. Rev. Lett.* **11**, 518, 1963; Th. A. J. Maris *et al*, *Phys. Rev. Lett.* **12**, 313, 1964), P. F. Browne, for example, argued for a QED cut-off in which the maximum self-energy of the electron due to the emission and absorption of photons would be equal to the particle's rest energy (m_ec^2) if the maximum intermediate state energy were of order exp ($2\pi / 3\alpha$) in units of the rest energy (*Intnl. J. Theoret. Phys.* **15**, 73, 1976; *Found. Phys.* **7**, 163, 1977). It is possible that an

argument of this kind might produce results with appropriate modification.

9 N. Graneau, personal communication. The Casimir effect is attractive for bosons because they are weak dipoles, but repulsive for fermions because they are weak monopoles. The difference in status between the 'real' and image terms in the Dirac 4-spinor for a free particle, even if the 'real' particle is actually a vacuum state (cf note 5.3), means that the Casimir effect does not require a broken supersymmetry to be observed, because loop cancellation is only at the level of the 'image' terms.

10 N. Graneau and P. Graneau, The evidence and consequences of Newtonian instantaneous forces, chapter 6, 121-168, of N. V. Pope, A. D. Osborne and A. F. T. Winfield (eds.), *Immediate Distant Action and Correlation in Modern Physics The Balanced Universe*, Edwin Mellen Press, 2005.

11 We could argue, in fact, that there is no such thing as a Big Bang theory, only a Big Bang concept.

12 A. G. Riess *et al*, *Astron. J.*, **116**, 1009, 1998.

13 A typical example among many is the 'surprise' result that stars were forming only 200 million years after the Big Bang.

14 It was first published in the historical work, *Oliver Lodge and the Liverpool Physical Society* (1990), and could be said to include a *prediction* of a background temperature of 2.72 K, as the value of this parameter was not then well established.

15 The Davies-Unruh effect provides a similar opportunity for detecting an absolute inertial acceleration against a uniform temperature background, another way of defining a universal reference frame. It is significant, however, that it cannot detect a gravitational one, for, in general relativity, only gravitational acceleration is relative – inertial acceleration requires an absolute space-time.

16 B. J. Carr and M. J. Rees, *Nature*, **278**, 605, 1979.

17 J. Silk, *Nature*, **265**, 710, 1977.

18 P. Marcer, D. M. Dubois, E. Mitchell and W. Schempp, Self-reference, the dimensionality and scale of quantum mechanical effects, critical phenomena, and qualia, in K. Bowden (ed.), *Correlations* (May 2002, *Proceedings of XXIII ANPA Conference*, Cambridge, August 2001), 247-267. Further developments are reported in MR1.

19 Though some discussion of cosmology is included in chapter 21, the subject (though not, of course, such structures as galaxies, clusters, superclusters and even the 'Hubble universe') has been *specifically excluded* from the account in this chapter because it is aimed at producing an argument that the laws of physics should be independent of the specific evolution of the universe, though they could, of course, ultimately determine it. However, some discussion of such a fundamental topic may be expected in a book that goes by the title *Zero to Infinity*. So, here, I will state that I consider it a matter of philosophical interpretation whether the 'accelerating expansion' equations derived in 21.1 compel us to accept whether a 'real' expansion is actually taking place in the way that many cosmologists believe. While there have been some hints, in chapter 20, that a cosmology could be constructed on that basis, I do not believe that the issue is fully resolved at the present time and can only state a philosophical *preference* for a view that the universe is infinite in time, space and matter (although, of course, representing a zero totality), whatever time-evolution may have taken place. I also think that there is as yet no compelling evidence for such evolution on a

universal scale, although localised evolution has certainly happened at various levels. (Cosmologists have been too ready, in my view, to project local evidence onto a universal scale, and describe the apparent point origin of the observed distribution of matter as the 'beginning' of time itself, as well as space and matter – even though this kind of projection of the infinite onto the finite is only a philosophical gloss that can never be realised in physical terms.) Clearly, the so-called 'steady state theory' of Bondi, Gold and Hoyle (which has unfortunately appropriated the name for the only possible alternative to a catastrophic 'creationist' theory) is not compatible with a deceleration parameter of –0.5 (as it requires twice this value), and is in any case incompatible with fundamental conservation laws, but this does not mean that a *true* steady state theory cannot exist on a universal scale. This would be a simpler solution than any other so far devised, and it may be that Ockham's razor will ultimately demand it.

Appendix B

1 T. A. Ell and S. J. Sangwine, Quaternion involutions, arXiv.org/pdf/math.RA/0506190.

2 J. Kuipers, Quaternions and Rotation Sequences: *A Primer With Applications to Orbits, Aerospace, and Virtual Reality*. Princeton University Press, 2002.

3 W. R. Hamilton, *Elements of Quaternions*. Longman, 1899.

4 A. D. Booth, A signed binary multiplication technique, *Quart. Jl. Mech. Appl. Math*, **4**(2), 236-240, 1951.

5 S. B. M. Bell and D. C. Mason, Tesseral quaternions for the octtree. *Computer Journal* **33**(5), 386-397, 1990.

6 N. R. Scott, *Computer Number Systems and Arithmetic*, Prentice-Hall, 1985.

7 *ibid.*, ch. 4.

8 N. Koblitz, *p-Adic Numbers, p-adic Analysis, and Zeta Functions*. Springer-Verlag, 1977.

9 Reference 6, ch. 7.7.

10 B. M. Diaz and S. B. M. Bell, *Spatial Data Processing using Tesseral Methods*. NERC publication, Swindon, 1987.

11 D. J. Griffiths, *Introduction to Quantum Mechanics* (2nd ed.). Prentice Hall, 2004.

12 e.g.. I. Gargantini, Linear octtrees for fast processing of three-dimensional objects. *Comput. Graphics and Image Processing*, **20**, 365-374, 1982; H. Samet, The quadtree and related hierarchical structures, *ACM Computing Surveys*, **16**(2), 187-260, 1984, and *Foundations of Mutidimensional and Metric Data Structures*, Morgan-Kaufman, San Francisco, 2006; I. Navazo, D. Ayala and P. Brunet, A geometric modeller based on the exact Octtree representation of polyhedra. *Computer Graphics Forum*, **5,** 91-104, 1986; etc.

13 S. B. M. Bell, B. M. Diaz, F. C. Holroyd and M. J. J. Jackson, Spatially Referenced Methods of Processing Raster and Vector Data. *Image and Vision Computing*, **1**(4), 211-220, 1983.

Index

Abelian group, 101
Abelian symmetry, 102
aberration, 448, 471
aberration of space, 444, 452, 470, 628
absolute, 7, 8, 32, 38-41, 47-49, 56, 59, 61, 74, 90, 98, 123, 164, 178-9, 195-6, 199-203, 211, 219, 221, 225-8, 231, 233-5, 238, 264, 310-1, 314, 323, 352, 369, 405, 412, 445-6, 449, 469, 472, 485, 520, 579, 622, 627, 631-3
absolute conservation, 39-40, 44, 74, 123
absolute continuity, 47-8, 90, 224-5, 446, 449, 485, 622
absolute frame, 229
absolute identity, 59, 61
absolute knowledge, 228, 234
absolute motion, 200, 233, 235
absolute nonconservation, 41, 74, 123
absolute opposition, 32, 59, 61
absolute order of events, 179, 196, 202, 226-7, 234, 310-1
absolute phase, 41, 221, 323
absolute position, 234-5
absolute potential, 41
absolute reference frame, 445
absolute rest frame, 200, 622
absolute space, 200, 228
absolute symmetry, 38, 56, 222, 631
absolute time, 196, 226, 234, 310-1, 579
absolute uniformity, 632
absolute value of potential, 211
absolute variability, 203, 226
absolute variation, 123, 219
absolute zero of temperature, 238
absoluteness, 32, 41, 56
absorber, 430
absorption, 148, 201, 215, 230, 290, 418, 430, 435, 439-40, 447, 489, 501
abstract, 4, 35-7, 51, 56-61, 86, 96, 104, 184, 227, 235-6, 387, 444, 449, 484, 495
abstract machine, 86
abstract system, 56, 58, 60
abstraction, 35-6, 61, 62
accelerated motion, 35, 416, 439
accelerating reference frame, 482
acceleration, 45, 48, 160, 210, 416, 458, 482, 499, 587, 600-5, 609, 611, 620
acceleration-dependent force, 160, 601
acceleration of the redshift, 499, 587, 600-6, 611, 613-9, 628
Achilles and the Tortoise, 46
achiral, 524
action, 181, 203, 230, 304, 500, 561, 586
action and reaction, 101, 149, 174, 210, 417, 430, 434-5, 437, 440-1
adaptive evolution, 512, 588
adaptive re-emergence, 584
adaptive resonance, 574, 588, 619
adaptive resonant coupling, 578
addition, 23, 84, 432, 442, 635
adenine, 516, 553, 591
adenine triphosphate, 553
adiabatic, 578, 580
adjoint wavefunction, 131-2, 295
advanced wave, 430
aether, 58, 195, 199-201, 225, 228-9, 231-5, 419, 445, 490, 612, 630
aethereal displacement current, 214
affine structure, 40, 154, 184, 316, 493
affine transformations, 572
age of the Earth, 609
aggregated matter, 134, 257, 320

aggregation, 509-11, 513, 531, 555
aggregator, 505
Aharonov-Bohm effect, 79, 143, 158, 320, 432, 435, 439, 442, 507
Aitchison, I. J. R., 394
algebra, 9, 13-14, 16, 18, 23, 26, 28-31, 53, 63, 66, 69-72, 74, 78, 81-2, 86, 98, 102, 104, 113, 116-17, 124, 133, 171, 177, 186-90, 195-6, 240, 260, 272, 280, 285, 309, 345, 349, 357, 374, 379, 382, 384-5, 404, 436, 439, 455, 479, 486-7, 503-4, 518-19, 528-9, 541, 546, 557-8, 574, 580, 624
algebra of process, 186
algebraic extensions of the ordinals, 565
algebraic geometry, 321
algebraic numbers, 10, 48, 66
algebraic pattern, 629
algorithm, 20, 530, 561, 569-70, 635
algorithmic process, 47, 49, 232
alpha-actinin, 550
alphabet, 1, 6-10, 19-20, 22-24, 67, 86, 445, 556-558, 560, 563, 565, 591, 593, 619, 623, 633, 638
amino acid, 502, 510, 514-26, 530, 533, 537-39, 546, 553
Ampère, A. M., 58, 214
Ampère's law, 214
amplitude, 117-8, 121-3, 132-3, 164-6, 168, 172, 181-2, 184, 185, 240, 242-3, 248, 274, 276, 278, 281, 300-2, 492, 508, 537, 560, 568, 573, 575, 594, 631
analogue, 47
analogy, 1, 4, 29, 36, 42, 54, 58-9, 202, 265, 375, 420-1, 471, 476, 484, 490, 507, 509, 531, 546, 553, 601
analytic, 56, 59, 133, 239, 241
AndCorporation, 595
Anderson, P. W., 347
angles, 42, 110, 112, 144-5, 165, 224, 249, 409, 420-1, 459, 461, 466, 475, 531, 548, 592, 631
angular distribution, 301
angular frequency, 194, 422
angular momentum, 37, 39-41, 55, 68, 73, 76, 80, 84, 93, 108, 130, 137, 157, 164, 176, 178-9, 191-2, 196, 203-5, 209, 212, 224, 228, 234, 236, 239, 251-2, 256, 258-60, 265-6, 268, 278, 283-4, 322, 324-5, 327-330, 334, 338, 346, 349, 351, 353, 386, 388, 397, 403, 409, 413, 437, 441, 447, 465, 485, 487, 515, 518, 531-2, 625-6, 628, 631
angular momentum operator, 164, 178, 260, 266, 284, 329, 338, 346, 349, 403, 409, 413, 447, 625-6
angular momentum phases, 353
angular momentum space, 515
anisotropic metric, 184
anisotropy in the cosmic microwave background radiation, 200
anisotropy in the vacuum, 482
annihilation, 40, 91, 137, 142, 160-4, 184, 237, 257, 278- 80, 314-5, 321, 353, 366, 378, 568, 580
annihilation operator, 91, 160-4, 184, 278, 568
anomalous magnetic moment, 173, 422, 430
ANPA, 572
anthropic principle, 600, 620
anticharge, 364
anticipatory computation, 556, 571, 594
anticodon, 517
anticolour, 341, 552
anticommutation relation, 76
anticommutative, 1, 8-10, 13, 23, 31, 68, 74, 76, 143, 184, 426, 428, 441, 484, 503, 633-4, 637-8
anticommutator, 162
anticosmology, 600
antielectron, 91, 345
antifermion, 90-1, 118, 120, 140-2, 144, 146-9, 151, 157-9, 261-3, 266, 268, 270, 278-280, 286, 297, 308-9, 313-5, 317, 325, 327-8, 340-2, 345, 350, 353, 367,-8, 370, 373, 377, 399, 427, 433-4, 443, 489, 505-6, 510, 512, 521, 550, 578, 624-6, 628
antifermion state vector, 149, 157, 159
antigravity, 318
antilepton, 326, 332
antimatter, 262, 315, 318, 342, 344, 445, 513, 564, 610, 624, 628
antineutrino, 279, 332, 364-5, 410
antiparticle, 45, 89-90, 119, 141, 150, 261, 328, 340, 342, 345, 365-6, 372, 378, 381, 491, 519, 627
antiproperties, 51-2, 95, 102, 105-107, 109, 485, 528

antiquark, 153, 239, 252, 305, 326-7, 358-9, 361, 399, 402-4, 413
anti-sound, 587
antistates, 121, 147, 265, 268, 327, 336, 365, 369, 377, 395
antisymmetric, 69, 82, 144, 152, 164-5, 205, 342, 344, 356, 387, 493, 582
antisymmetric coordinates, 387, 493
antisymmetric wavefunction, 69, 82, 144, 152, 164-5, 342, 344, 356, 582
apparent metric, 607
apparent order of events, 199
archaea, 531
'archaeology' of space-time, 451
area, 29-30, 184, 195, 319, 415, 418, 449, 573
Argand diagram, 14, 25
arginine, 521, 524, 530
Aristotelian physics, 35
arithmetic, 9, 48, 66, 83, 559, 566, 568, 597, 635, 637
arithmetic continua, 566, 597
aromatic, 524
arrow of time, 218, 226, 235, 513, 611
arrows (categorical), 571, 595-6
artificial intelligence, 561, 629
Aspect, A., 197, 448, 478
assembly, 174, 321
associative, 11, 28, 31, 43, 45
associative holographic memory, 577
asymmetric vacuum state, 261-3, 266, 270
asymmetric wave equation, 344
asymmetry, 41, 44, 58, 90, 99-100, 317, 335, 342, 398, 503, 555, 564
asymptotic freedom, 252-5, 410
Atkins, P., 3
atmospheric refraction, 419
atom, 178, 248, 323, 497, 513, 613
atomic structure, 589
ATP, 553-4
attraction, 57, 319, 454, 458, 509, 511
attractive force, 318, 475, 480, 602
attractor, 112, 531
autonomous actions, 86
axes, 25, 42, 44, 99, 107-8, 25, 126, 135, 144-5, 197-8, 209, 258-59, 330, 373, 459, 466, 475, 486-7, 607, 635
axis division, 635
axiomatic, 83, 265
axiomatization, 83
b quark, 407, 410
background radiation, 233, 445, 580, 600, 608, 610, 613, 615, 618, 621
background temperature, 600, 608, 632
bacteria, 512-3, 522, 524, 553
bacteriophages, 540
Balmer series, 245, 613
bare charge, 303
bare lepton mass, 408
bare quark mass, 405
Bargmann-Fock model for bosons, 577
baryon, 39, 44, 137, 141, 144, 151-158, 239, 250-1, 255-6, 258, 282, 305, 313, 316, 324-5, 327, 329-30, 333, 338-9, 341, 343, 345, 349, 354, 356-8, 379, 394-5, 400-6, 410-1, 413, 433, 485, 551, 616, 626-7, 631
baryon charge structures, 324
baryon conservation, 39, 258
baryon decay, 258
baryon decuplet, 339, 401
baryon density, 616
baryon number, 44
baryon octet, 339, 401-403
baryon spin, 258
baryon state vector, 250, 251, 350
baryon wavefunction, 152-3, 157, 239, 256, 333, 379
baryonic density, 616
base, organic, 511, 515-7, 519-23, 525, 530-3, 536, 544, 552-555, 590-2, 595
base pairings, 544, 554, 590-2
Bastin, E., 572
Bateson, G., 559
Bell, S. M., 480, 497-9, 638
Bell inequality, 107, 197
benzene ring, 554
Bergson, H., 46
Berry, M. V., 442, 568, 586, 588, 593
Berry phase, 79, 82, 149, 158, 296, 315, 320-3, 372, 431, 432, 435, 439, 507, 588, 632
Bertrand's theorem, 246
beta decay, 346, 366, 381, 631
big bang, 3, 609, 620-1
big crunch, 620
bilinear covariant, 111, 131-2
binary address, 635
binary computation, 561

binary digits, 9
binary form, 99, 634
binary numbers, 9, 11, 24, 415
binary operation, 32, 53, 98, 193, 207, 481
binary pulsar, 463, 477
binary system, 561
binomial approximation, 428, 465-6
binomial combinations, 18, 70
binomial expansion, 443
biochemical, 554, 576
biodiversity, 587
biological systems, 502-3, 514, 515, 629
biology, 502-3, 514-5, 531, 542, 555
biosphere, 587, 591, 629
birthordering, 565-567, 570, 577, 579, 588, 600
birthordering field automorphism, 566-7, 581
Bischof, H., 577
bispinor, 168, 170, 188, 423, 494
bits, 561
bivector, 30
Bjorken, J. D., 302
black body radiation, 229, 430, 579, 612
black body spectrum, 221, 229
black body temperature, 482
black hole, 420-1, 458-9, 461, 477-8, 480, 565, 570, 583-4, 620
Blandford, R. D., 476
Bloch function, 247
Bloch theorem, 246
blood, 593-4
Bode, J. E., 461
Bohmian quantum mechanics, 181, 186, 188, 190
Bohr, N., 60
Bohr atom, 613
Bohr electron, 422
Bohr radius, 244, 248, 617
Bohr-Sommefeld orbitals, 497
Boltzmann constant, 237
Boltzmann factor, 616
book-keeping, 486-7, 514, 518
Boolean encoding, 86
Boolean logic, 9
Booth, A. D., 634, 638
bootstrap, 558, 569-70
Born, M., 176
Bose-Einstein condensate, 82, 133, 143, 158, 296, 341, 372, 507
Bose-Einstein condensation, 238, 321, 507, 512
Bose-Einstein statistics, 229, 343, 428
boson, 55, 58, 89, 121, 132, 137, 141, 143-9, 151, 154-8, 174, 187, 190, 205, 231-2, 259, 270, 279, 286, 291, 296-7, 305-6, 314, 321-2, 342-3, 345, 347, 349, 352-3, 366-7, 370-1, 386, 412, 398, 405, 426-428, 431, 433-5, 439-40, 443, 480, 493, 497, 509, 511-513, 568, 576-578, 617, 625-627, 629, 631
boson propagator, 296, 305-6
boson state vector, 427
boson string, 493
botany, 558
bound meson state, 252
bound solutions, 277
bound state, 123-4, 165, 242, 252, 359, 365, 397, 625
boundary conditions, 57, 83, 600
bounding surface, 587
Boyer, T. H., 229
Boyle's law, 418
BPHZ method of renormalization, 304
bra operator, 128-9, 131, 134, 294, 297, 557, 572, 577, 635
Bradley, J., 471
Brahe, Tycho, 585
braiding pattern, 638
brain, 556, 559-562, 585, 591, 593-598
bremsstrahlung, 304
British Computer Society, 560, 599
Brodsky, S. J., 258
broken octonion, 95, 98
broken supersymmetry, 321
broken symmetry, 76, 100, 284, 335, 349, 367, 391-2, 515
Brookes, R., 547
Brouwerian foundation, 572
Brownian motion, 417
BRST conserved fermionic charge, 167
BRST operators, 493
BRST quantization, 166-7
c quark, 255, 402
Cabibbo angle, 341, 413
Cabibbo-Kobayashi-Maskawa (CKM) matrix, 270,317, 342, 365, 408-9, 413
calculus, 64, 182, 184
canonical labelling, 581

canonical quantization, 167
Cantor, G., 66
Cantorian real numbers, 10, 49, 232
Carathéodory, C., 237
carbon, 515, 552, 590
cardinality, 10
Carnot engine, 578-9, 585, 587, 595
Cartesian representation, 634
Casimir effect, 310, 320-1, 513, 568, 606
Casimir force, 319
Casimir operators, 379
category theory, 571, 595
causality, 176, 179, 196, 201-2, 218, 227-8, 230, 234, 311, 489-90, 496-7, 500, 579, 622
Cavendish, H., 461
celestial dynamics, 466
cell, 513, 516-7, 547, 585, 591-2
cell division, 591
centrifugal force, 466, 471, 604, 606
centripetal force, 417, 466
cerebral cortex, 595
Chakraborty, P., 459, 463-4, 467
changing conditions, 421-2, 437-8, 459
changing phase difference, 430
chaos, 502, 575, 593, 600
chaotic, 542, 547, 571, 575, 586, 593, 600
chaotic computation, 575
chaotic system, 586
character, 12, 20, 24, 63, 67, 80, 86, 202, 235, 260, 264, 368, 493, 504, 531, 557, 591
characterization, 27, 32, 37-8, 53, 56, 59, 61-2, 64, 137, 190, 265-6, 324, 415, 434, 436, 447, 537, 568, 618, 623
charge, 32, 36-45, 49-55, 58-60, 63-8, 71-6, 79-81, 84, 89-94, 96-100, 102-5, 110, 130, 141, 147-149, 151, 157, 159, 166-7, 172, 176-9, 188, 191, 193-6, 202-8, 210, 212, 215, 217-9, 222-6, 231-8, 241, 246, 248, 251, 256, 258-66, 268, 270, 278-80, 283-85, 291, 297, 300-5, 309-22, 324-30, 332-3, 335-48, 350, 352-70, 372, 374-5, 377-81, 383-4, 386-8, 391, 393-400, 402-3, 406-7, 410, 412-3, 432, 436-9, 443, 446-9, 453-5, 467, 474, 476, 479, 481, 485-8, 490, 492-4, 496-7, 501, 503-6, 511, 513, 515, 522, 528-9, 537, 542, 546, 552-3, 566, 588, 607, 611-3, 620, 623-8, 630
charge accommodation, 340, 345, 358
charge allocation, 263, 357, 358, 377, 383
charge allocation algebra, 383
charge assignment, 338, 357-8
charge components, 73, 79, 335-6, 353, 356, 359, 401, 492, 505
charge conjugation, 45, 50, 141, 148, 151, 159, 261-34, 270, 313-4, 328, 341, 345, 352-3, 369, 372, 378, 396, 626
charge conjugation symmetry, 159, 261, 262, 313-4, 353, 378, 626
charge conjugation violation, 264, 396
charge density, 212
charge dimensions, 624
charge distribution, 248, 319
charge generator, 383, 494
charge magnitude, 630
charge operator, 167, 318, 453
charge space, 322, 485, 491, 515
charge state, 89, 148, 278, 314, 320, 330, 338, 344-5, 364, 380, 413, 447, 493
charge structure, 262, 284, 319, 324, 327, 329-30, 332, 337- 9, 341, 343, 345-7, 352-4, 356-9, 362-4, 367, 374-5, 397-8, 410, 412, 447
charge structure matrices, 354, 363
charge transfer, 607
charge type, 80, 279, 284, 336, 356, 359, 399, 631
charge*, 93-4, 96, 193
charge-conjugation, 265, 268
charged particle, 148, 204, 215-6, 228, 238, 305, 319, 393
charged scalar particle, 271
chargelike, 50, 90, 232, 447
charm, 342
charmonium, 255
chemical bonding, 319
chemical chaos, 553
chemical reaction, 576
chemical rewrite system, 590
chemical structural stability, 591
chemistry, 512, 531, 552, 576
chiral autocatalytic systems, 531
chiral symmetry, 486
chirality, 280, 320, 414, 502, 513, 515, 531-2
choice of gauge, 295
Chomsky, N., 18, 596, 598

chromatin, 592
chromosomes, 591
Church-Turing principle, 566
circular orbit, 421, 466
classical approximation, 200, 311, 472, 489-90, 495, 625
classical dynamics, 471
classical electrodynamics, 221
classical field theory, 229
classical harmonic oscillator, 531
classical limit, 168
classical machine, 591, 595
classical measurement, 236-7, 330
classical mechanics, 33, 36, 38, 180, 195, 203, 206-7, 209-210, 212, 215, 217, 232, 310
classical physics, 31, 36, 42, 48, 85, 123, 177-8, 180, 191, 195, 203-4, 206, 207, 212, 217-8, 221-2, 239, 277, 421, 429, 437, 444
classical radius, 394, 611-2
classical system, 80, 205, 221, 448, 625
classical theory, 191, 208, 231, 238, 246, 453--5, 472, 490, 615
classical thermodynamics, 177, 220-1, 578
classical universe, 585, 588
Clebsch-Gordan coefficients, 390-1
Clifford algebra, 5, 28-9, 67, 78, 116, 125, 186, 262, 484, 492, 530, 573
  Cl (1, 3), 28
  Cl (3, 1), 28-9
Clifford / fermionic partition of state space, 556, 568, 576, 581-2
C-linear maps and lifts, 125
clock paradox, 234
Close, F. E., 347, 363
closed set, 101, 503
closed system, 12, 74, 476
closure, 9, 16, 24-5, 28-9, 68, 103, 478, 509, 554, 593
clusters, 605, 621
coded sequence, 553
codification, 5, 83
coding, 109, 515-517, 522, 525, 553, 555, 582
codon, 516-522, 524-526, 530, 533, 537-539, 546, 550
codon capture, 530
cognition, 567
coherence, 552-3, 579, 587-8, 591, 593
cohesion, 319
Coleridge, S. T., 61
collapse of the wavefunction, 50, 93, 172, 175-7, 220-1, 226, 228, 329, 429
colour, 105-9, 266, 551-2
colour (strong interaction), 151-2, 154-5, 251-2, 255-7, 266, 305, 307, 313, 317, 327-8, 333-4, 346-7, 351, 354-60, 363, 377-9, 381-4, 391, 399, 402-3, 411, 493, 505, 551-2
colour charge, 252, 402, 493
colour combinations, 105, 357, 358
colour factor, 305
colour flux tubes, 255
colour force, 313
colour labels, 354, 356
colour nonsinglet, 347
colour octet, 355
colour representation, 105, 106
colour singlet, 257, 313, 337, 355-6, 378, 381
colour space, 357
coloured quark, 152, 333, 358
colour-magnetic hyperfine splitting, 403
column matrix, 377
column vector, 75, 79, 118-120, 128, 132, 146, 240, 242-3, 300, 375-7, 427, 490, 494, 506, 636
combinatorial explosion, 557, 567
Combinatorial Hierarchy, 572
commonsense, 56, 561-2
communication, 87, 446, 448, 562, 596-7, 622
commutation relations, 161, 180, 435, 440
commutative, 1, 8-9, 23, 31, 69, 95-6, 101, 345, 428, 441, 493, 574
commutative ring, 574
commutativity, 13, 25, 28-9
commutator, 231, 423
compact simple gauge group, 371
compactification, 72, 125, 236, 386, 482, 494, 499, 531
compactified, 67, 70, 72, 79, 120, 149, 187, 317, 492, 494, 505
compactify, 261
complex 4-D, 98, 494
complex algebra, 25, 43, 265, 530
complex conjugate, 44
complex coupling, 280

complex double quaternion, 17, 66-7
complex forms, 1, 23
complex nilpotent, 240
complex numbers, 5, 12-14, 25, 28, 70, 101, 116, 440, 519
complex operator, 13, 125, 261, 328
complex phase, 280
complex plane, 114, 573
complex pseudovector, 384
complex quaternion, 15-17, 28, 67, 95, 126, 132, 386, 424, 503-4
complex quaternion conjugate, 132
complex scalar, 17, 67, 240, 270, 272, 329, 504
complex scalar field, 270, 272, 329
complex system, 144, 192, 203, 218
complex terms, 15
complex vector, 70, 504, 519
complex vector quaternion, 70, 504, 519
complexification, 1, 13-14, 16, 24, 63-4, 66, 84, 88, 260, 262, 331, 386, 414, 436-7, 439-40, 486-7, 558, 623-4, 633, 638
complexity (mathematical), 11, 24, 65, 83, 100, 114, 132, 280, 386, 442, 486
complexity (systemic), 19, 34, 87, 178, 502-3, 509, 511-2, 515-6, 531, 555, 562, 584, 589
Compton radius, 396
Compton time, 618
Compton wavelength, 181, 394
computability, 566
computation, 4, 558, 561-563, 566, 570-1, 575, 576, 581, 590
computational input / output, 571, 595
computational language, 586
computer, 1, 4, 18-19, 82, 86, 503, 514, 547, 559, 570-572, 594, 623, 633, 635
computer constructor universal, 594
computer graphics, 635
computer language, 19
computer program, 18
computer space storage, 635
computer universal, 559, 594
computing, 4, 18-19, 63, 82, 86, 557, 560, 571, 576, 599
computing anticipatory system, 571
concatenation, 6, 7
condensed matter, 237, 246, 321, 347, 427, 632
confinement, 261, 278, 495, 508
confining potential, 254, 256
conjugate, 1, 6, 10, 13, 16, 21-24, 68, 71, 76, 79-80, 86, 90, 97, 121, 132, 134, 160, 172, 189, 192, 196, 258, 261, 311, 315, 338, 435-6, 447, 492, 596-598, 633, 637
conjugate charge state, 90, 261
conjugate nilpotent expression, 121
conjugate pairing, 338
conjugate parameter, 76, 258, 447
conjugate symmetry, 172
conjugate triad, 71
conjugate variable, 80, 192, 258, 436
conjugation, 1, 9, 13-14, 16-17, 24, 63-4, 66, 74, 84, 86, 88, 134, 150, 261, 264, 326-7, 331, 353, 396, 414, 436-7, 440, 486-7, 489, 558, 623
conscious observation, 620
consciousness, 556, 595-6
conservation, 32, 34, 37, 39-41, 44, 54-5, 60, 72, 75-6, 79, 80, 84-5, 88, 92, 99, 102, 139, 148, 165, 168, 170-1, 174, 178, 181, 184, 187, 192, 195, 203-207, 209, 212, 218-220, 222, 224, 228, 234, 236-7, 239, 246, 251, 258-60, 264-5, 277, 283-4, 324, 325, 329-30, 343-4, 349, 388, 392, 414, 416-7, 419, 428, 436-7, 442, 454, 457, 459, 480, 486-7, 489, 579, 591, 610, 619, 624-5, 628, 630-1
conservation laws, 34, 37, 39-40, 42, 44, 84, 148, 178, 203, 212, 219-20, 224, 265, 284, 329, 343, 392, 419, 480, 624-5, 628
conservation of angular momentum, 32, 54-5, 72, 76, 139, 165, 205, 209, 236, 239-40, 246, 259-260, 264-5, 277-8, 284, 324-25, 329, 349, 386, 625-6, 630
conservation of charge, 76, 205, 224, 258, 283, 324, 349, 375, 381, 388
conservation of charge type, 32, 54, 224, 239, 258, 324, 631
conservation of electric charge, 55, 204
conservation of energy, 40, 54, 80, 123, 168, 170-1, 174, 181, 205, 301, 375, 416-7, 419, 431, 454, 457, 459, 610
conservation of handedness, 625
conservation of mass, 54, 80, 92, 195, 204-5, 207, 224, 419
conservation of momentum, 205, 209, 414
conservation of probability equation, 181

conservative system, 39, 55, 203, 205, 208-9, 211, 220, 223, 418
conserve (operation), 20-22, 558, 590, 633, 637
constant force, 154, 256, 280, 418, 508
constant potential, 246
continuity, 10, 14, 46-50, 56, 65-8, 84, 93, 99, 101, 103, 105, 145, 147, 151, 159, 161, 166, 174-7, 179, 182, 187-8, 200, 202, 204, 212, 218, 222-235, 239, 247, 261, 310-3, 315, 318-20, 328, 342, 356-7, 387, 392, 416, 427-30, 433-6, 438, 440, 442, 444-9, 455, 468, 478, 481, 483, 485-6, 488, 490, 498-9, 506, 510, 513, 528, 531, 550, 572, 584, 589-90, 602, 607, 624, 626-8
continuity equation, 212
continuous change, 174, 427
continuous creation of matter, 601
continuous field, 455, 479
continuous gravitational vacuum, 310, 499
continuous group, 84, 572, 624
continuous vacuum, 49, 147, 151, 225, 231, 312-3, 315, 318-9, 490, 513
continuous vacuum field, 225, 231
continuous variation, 204, 434
contour integral, 294, 296
control and information processing system, 567
Conway, A. W., 27
Conway, J. H., 10, 18, 161, 565-7, 571, 572, 580
Cooper pair, 143, 158, 435
coordinate system, 40, 111-2, 197, 205, 602, 604
coordinates, 14, 38, 40-42, 55, 74, 85, 92, 96, 111-2, 140, 150, 172, 178, 197, 199, 204-5, 207-209, 219-221, 241, 249-50, 330, 344, 387, 432, 459, 462, 464, 466, 469, 472, 493, 531, 537, 592, 600, 602, 604
Copenhagen interpretation, 60, 172, 222
Coriolis force, 471
corpus callosum, 597
corpuscular theory of light, 419
correlation, 23, 79, 81, 202, 311, 409, 447-8, 579, 607, 627
Correspondence Principle, 177
cosmic coincidences, 600
cosmic microwave background radiation, 200, 499, 618, 622, 628
cosmic rays, 584, 618
cosmic repulsion, 604, 609
cosmic source, 587
cosmological, 342, 445, 475, 564, 575, 581, 583, 587, 600-1, 603-5, 608-11, 616, 618-21, 628
cosmological arrow of time, 611
cosmological boundary condition, 600
cosmological constant, 583, 601, 604-5, 609, 616
cosmological evolution, 445, 587, 600
cosmological redshift, 587, 600, 603, 608, 610, 618, 621-2
cosmological reference frame, 619
cosmology, 62, 445, 448, 478, 556, 568, 581, 586, 590, 600, 608-9, 620-1, 629
cosmos, 556, 572, 575, 586-7, 598, 600
Coulomb component, 251
Coulomb field, 239, 242, 246, 274
Coulomb force, 42
Coulomb gauge, 167
Coulomb interaction, 145, 241, 246, 276, 282
Coulomb potential, 242, 246, 253-4, 277, 298-99, 352, 625
Coulomb term, 241, 246, 252, 254, 256, 283, 508
Coulomb-like phase, 627
Coulomb-like solution, 255, 480
countability, 33, 65, 66, 76, 80, 99, 102
countable, 10, 18, 38, 50, 56, 65-6, 74, 85, 93, 100, 179, 223, 232, 436, 566
countable numbers, 18
countable set, 566
counting, 5, 10, 23, 33, 49, 64, 84, 99, 159, 179, 403, 433, 442
coupling, 157-8, 215, 241, 251, 264, 269-70, 283, 291-2, 304, 309, 313, 320, 340, 362-3, 369, 373, 380, 390-393, 395, 399-400, 405-6, 412-3, 433, 480, 500, 578
coupling constant, 215, 241, 251, 269, 283, 320, 340, 380, 406, 480, 500
covariant, 90, 283, 352, 403, 450, 463
covariant derivative, 122, 154, 173, 184, 241, 250, 256, 267-9, 271-2, 333, 340, 351, 355, 398, 404
covariant form, 168, 274, 280-1,
Coveney, P. D., 46
CP invariance, 141, 626

CP violating phase, 408
CP violation, 280, 314, 324, 342, 408, 413, 631
CPT symmetry, 90, 137, 150, 315, 345, 625
CPT theorem, 38, 344
create (operation), 20-3, 558, 590-1, 633, 637-8
creation, 1, 3, 23, 34-36, 47, 64-5, 73-4, 76, 90-1, 100-1, 137, 142, 146-149, 159-165, 168, 184, 192, 200, 207, 237, 257-8, 263-4, 278-80, 286, 304, 306, 311, 314-6, 320-1, 323, 335, 349, 352-3, 366, 371, 378, 386-7, 397, 412-3, 420, 433-4, 437, 442, 445, 490, 503, 505, 507, 509, 515, 529, 531-2, 542, 553-4, 557, 568, 570, 575, 577, 580, 584, 588, 601, 610-13, 616, 618-22, 626, 629
creation event, 165, 588, 619-20, 622
creation operator, 137, 146, 159-62, 165, 184, 349, 378
critical density, 600, 604, 609, 621
critical phenomena, 580, 619
Crookes, W., 589
crystal, 46, 238, 246
crystal lattice, 246
crystalline materials, 547
crystals, 238
cube, 29, 515, 527-8, 531-33, 535, 537, 539-41, 545-47, 550-1
cubical, 417, 528, 531, 551
Cullerne, J. P., 111, 137, 349, 374
curl field, 479
current, 251, 295, 302, 343, 363, 366
current density, 131, 212
current electricity, 58
current-probability density, 131
Curtis, M., 543, 544, 545
curvature, 183, 450-2, 456, 469-70, 472-3, 475, 478, 480-482, 497-8, 604
curved space-time, 450-1
cut-off, 229, 302-3, 392, 396-7, 482-3, 606
Cybernetics Machine Group, 560, 599
cyclical, 634
cyclicity, 15, 24, 74, 76, 623
cysteine, 524
cytosine, 516-7, 591
cytoskeletal structure, 550
D symbol, 634
Dalembertian operator, 213
dark energy, 566, 579-80, 587-8, 604-5, 609, 632
dark matter, 605, 609
Davies-Unruh effect, 482, 498
de Broglie pilot standing wave, 586, 600
de Sitter, W., 475
decay, 44, 267, 315, 343, 345-6, 364-5, 381, 399, 402, 405, 632
deceleration parameter, 600-1, 605, 628
decoherence, 161, 178-9, 237, 329, 579, 625
Dedekind cut, 433, 442
deduction, 126, 583, 585
deductive, 63-4, 198, 212, 215, 419, 439, 483
deductive rationalism, 64
defragmentation, 116-7, 120, 122, 133, 241
defragmented Dirac equation, 117, 120
degeneracy, 313, 352, 402
degenerate vacua, 271
del operator, 26
delocalised, 175, 229, 231, 506, 513, 524
delocalised charge, 524
delocalised vacuum, 506
delta function, 122, 247, 567
density of states, 482
density of the universe, 604
denumerable, 10, 47, 49
deoxyribose, 515
dependent variable, 48
Descartes, R., 58
determinant, 247, 356
Deutsch, D., 4, 82, 561-2, 581
dextro, 334, 360, 514
diagonal matrix, 146, 300
Diaz, B. M., 1, 63, 556, 633
dice, 222, 548
dichotomy, 488
differentiable, 219, 575
differential, 26, 36, 39, 42, 57, 63, 74-5, 78, 83, 85, 90, 119-124, 126, 137, 141, 145, 161, 166-7, 170, 182, 185, 188, 196, 206-7, 212-3, 240-243, 250, 274, 280-1, 283-4, 287, 296, 446, 508, 624
differential equation, 36, 42, 57, 83, 85, 126, 170
differential operator, 26, 63, 75, 78, 85, 90, 119-124, 137, 141, 145, 161, 166-7, 185, 188, 212-3, 240-243, 250, 274, 281, 283-4, 287, 296, 446, 508, 624
differentiation, 48, 75, 184, 203, 208, 465-6

diffusion equation, 183
digital, 47, 86, 561, 567, 571, 595
digital algorithm, 561
digital computation, 561, 571
digital computer, 561, 567
dihedral, 331
dimension, 32, 40, 42-45, 48, 61, 65, 70, 75, 88, 92, 101, 103, 105, 107, 111, 235, 242, 316, 328, 387, 424, 440, 482, 484-486, 488-9, 491-493, 495, 497, 499, 514, 518, 537, 542, 551, 567, 573, 598
dimensional, 1, 10, 13-14, 16, 23, 25-6, 32, 38, 40, 45, 48-50, 59, 63, 65-6, 68-70, 82-3, 88, 91, 97-8, 101, 103-4, 109-10, 112, 114, 130, 145, 202, 224, 353, 412, 436, 438, 440-2, 455, 481, 484-9, 492-9, 518, 542, 577, 591-2, 597, 624
  1-D, 196, 573
  2-D, 388, 404, 490, 529, 540-1, 543, 545-6, 550, 554, 573, 575-6
  2n-D, 125
  3-D, 9, 13, 15-16, 23, 25, 31, 40, 42, 45, 58, 63, 67-68, 70-3, 88, 97, 99, 101-3, 105-10, 112-3, 115, 133, 145, 165, 196, 202, 216, 251, 258, 277, 313, 328, 331, 338, 404, 407, 417-8, 441, 443, 454, 478-81, 484-90, 492-3, 495-6, 499, 502-4, 514-5, 518-9, 521, 527-532, 536-8, 542-3, 547, 549-51, 556, 568, 573-7, 582, 587, 591-2, 597-8, 600, 624, 626, 635
  3-D Heisenberg Lie group, 598
  3-D holographic face recognition software, 595
  3-D representation, 107-9, 331, 404, 543
  4-D, 125, 262, 371, 481, 488, 490, 494-5, 497, 529, 542, 547, 551, 573, 635, 637
  5-D, 71, 495-6, 573
  6-D, 531, 573
  7-D, 110
  8-D, 573
  10-D, 387-8, 491, 495, 573, 588
  11-D, 493, 496, 573
dimensional analysis, 38
dimensionalization, 1, 11, 13-14, 16, 24, 63-4, 66, 84, 88, 262, 331, 414, 436, 439-441, 486-7, 558, 623-4
dimensions of charge, 32
dipolar, 237, 240, 257, 277-280, 309, 315, 321, 352, 365, 370, 381, 439, 508-9
dipolar field, 240, 280
dipolarity, 238, 257, 314, 321, 351-2, 381, 439, 510-1, 513
dipole, 240, 257, 259, 278-280, 315, 319, 370, 476, 509-511
dipole radiation, 476
dipole-dipole force, 259, 511
dipole-dipole interaction, 278-9, 315, 319
Dirac, P., 73, 78, 89-90, 261, 319, 445, 617, 630
Dirac 4-spinor, 118-120, 125, 148, 279, 318, 321, 323, 394, 446
Dirac algebra, 63, 67, 69-72, 74, 76, 92, 98, 102, 110, 127, 160, 188, 256, 335, 337-8, 345, 374-5, 378, 382, 383, 386-7, 436, 493, 504, 518-20, 529, 557, 580, 624
Dirac atom solution, 497
Dirac angular momentum, 73, 388
Dirac code, 111
Dirac convention, 135
Dirac critical phenomena, 621
Dirac differential operator, 114, 201, 375
Dirac delta function, 567
Dirac energy, 73, 110, 201, 230, 388, 576, 624
Dirac equation, 18, 31, 52, 63, 71, 73, 75, 77-9, 85, 88-92, 94, 111-3, 115, 117-8, 121, 123-25, 127-9, 133, 135, 137, 141, 145, 165-6, 168-70, 173, 177, 185, 192, 195, 217, 221-2, 230, 237, 239-42, 246, 252, 261, 263, 273, 277-9, 284, 286, 291, 298, 322, 345, 374-5, 377, 380, 423-4, 427-9, 431-2, 441, 443, 446, 479-80, 482, 484, 489, 494, 497, 504, 556-8, 562-4, 569, 574, 576, 588, 619
Dirac equation for charge, 374-8
Dirac field, 296
Dirac filled vacuum, 230, 261-2, 445
Dirac formalism, 81, 188, 244, 263, 266, 493
Dirac group, 71, 95, 387
Dirac idempotent, 187
Dirac Lagrangian, 131-2
Dirac matrix, 130
Dirac momentum, 73, 388, 576, 624
Dirac nilpotent, 68-9, 84, 85, 88, 137, 161, 166-8, 179, 187, 191-2, 194, 200, 223, 231, 236, 239, 250, 257, 280, 310, 324, 349, 367, 374, 380, 442-5, 447, 452, 488, 490, 492-3, 496, 506, 510, 514, 518-9,

564, 591, 601, 607, 619, 621-2
Dirac nilpotent operator, 166, 280, 367
Dirac nilpotent state, 69, 161, 250, 257, 324, 380, 447
Dirac nilpotent state vector, 250, 324, 380
Dirac nilpotent theory, 488
Dirac notation, 577
Dirac object, 620
Dirac operator, 167, 241, 276, 280, 380, 582
Dirac particle, 410, 496
Dirac pentad, 71, 80, 328, 333-4, 337, 385-6
Dirac propagator, 298
Dirac quaternion state vector, 159
Dirac rest mass, 73, 388, 576, 624
Dirac solution, 111, 253-4, 263, 279, 464, 497
Dirac spinor, 120, 127, 130, 308, 349, 432, 494, 498
Dirac state, 53-4, 63, 66-9, 71-5, 80, 125, 132, 168, 177, 188, 191, 193-4, 202, 207, 260, 262-3, 317-8, 335, 377, 443, 482, 489, 493, 497-9, 518, 624
Dirac state vector, 74, 78, 91, 380, 443, 572, 577
Dirac theory, 89-90, 93-5, 105, 187, 344, 491, 576, 630
Dirac wavefunction, 78, 118, 125, 130-1, 187, 378
Dirac / Hiigs mechanism, 387, 494
disassembly, 321
discontinuity, 48, 65, 176, 218, 222-5, 227-35, 386, 428, 435, 440, 449, 485
discrete, 1, 10, 14, 48-50, 56, 64-5, 67-8, 72, 76, 79, 83-4, 93, 102-3, 105, 147, 168, 174-178, 182-184, 188, 202, 223-4, 228, 231, 233, 235-6, 310-313, 318, 320, 329, 349, 361, 386-7, 416, 419, 425, 427, 429, 432-3, 435-6, 438, 445-447, 449, 467, 471, 479-80, 482-3, 486, 488, 493, 496, 498-501, 503, 506, 508-510, 513, 528, 550, 558, 572, 623-4, 627-8
discrete calculus, 182-3
discrete differentiation, 168, 184
discrete energy transfer, 329
discrete field, 496, 500
discrete field equation, 496, 500
discrete field source, 500
discrete gravitational field, 496
discrete gravity, 496, 500
discrete group, 14
discrete interaction, 550
discrete particle, 311, 433, 479, 496
discrete source, 79, 178, 202, 236, 312, 320, 447, 467, 479-80, 482
discrete transition, 174, 427
discreteness, 1, 5, 9-10, 14, 23-4, 47-50, 64-5, 76, 82, 93, 99, 101, 103, 176, 184, 187, 202, 206, 228, 312, 386, 429, 432, 438-9, 442, 446, 449, 481, 484-486, 495-6, 499, 503, 531
disorder, 237-8
disperser, 505
displacement, 181, 209, 220, 222, 416
displacement current, 58, 238
distance-related potential, 240
distribution of matter, 602, 605
disulphide bridge, 524
divergences, 290, 297, 306, 309, 394, 450, 494
divergent diagrams, 304
divisibility, 1, 32, 46-7, 49-51, 95, 222
division algebra, 45, 103
DNA, 502, 509-12, 514-7, 525, 527, 532-3, 536, 539, 543-50, 552-4, 560, 570, 590-5, 597
DNA polymerase, 512
DNA-wave biocomputer, 597
dNTPs, 544, 547-8, 550
dodecahedron, 527, 529-30, 539, 541, 543, 545, 547
Doppler effect, 473, 608
double 3-D, 519, 531, 537
double algebra, 187
double helix, 502, 509, 515, 532, 550, 552, 554, 589-90
double nilpotent, 185
doubling of elements, 102
down quark, 270
dragging of inertial frames, 476
Drell, S. D., 302
drone terms, 164, 495
dual, 4, 12, 14-15, 24, 68, 69, 84, 88, 91, 93-96, 99-100, 102-3, 105, 107, 109, 149, 159, 188-90, 240, 259, 262, 280, 331, 414, 443, 492, 495, 506, 509, 528, 530, 537, 550, 553, 558, 575-6, 619, 631
dual group, 93-4, 96, 100, 102, 105, 107, 109, 550

dual state, 68, 190, 259
duality, 4, 6, 7, 11, 13-14, 19, 32, 51, 53, 65-6, 69, 75, 82, 84-5, 92, 99-103, 105, 107, 168, 175, 186-90, 193-4, 218, 222, 232, 258, 261, 314, 386, 414-5, 433-8, 440-3, 449, 458, 491-2, 508, 552, 554, 619, 623
Dubois, D., 571
dynamic mass, 400, 405, 457
dynamism, 49
dystrophin, 550
early universe, 342
Earth, 200, 233, 449, 462, 475, 553, 584, 602
Earth-Moon system, 475
Eccles, J. C., 585
Eddington, A. S., 475, 617
effective mass, 460
eigenfunction, 135, 247-8, 424-5
eigenquarks of weak isospin, 341
eigenstates of helicity, 135
eigenvalue, 75, 85, 119, 123-4, 128, 134, 137, 145, 167-9, 180, 241-3, 274, 284, 294, 375, 424, 442, 568, 586, 624
eightfold eightfold way, 404
eightfold way, 404
Einstein, A., 26-7, 59, 176, 196-8, 311-2, 321, 418-9, 421-2, 430, 450-1, 457, 459, 461, 490, 564, 566, 569, 601
Einstein's field equations, 452, 455-6, 490, 584-5
Einsteinian relativity, 26, 93, 199-201, 227-8, 231-4
elasticity, 47, 419
elasticity of the aether, 419
electric charge, 39, 45, 55, 73, 76, 148, 194, 205, 216, 236, 259-60, 264-5, 280, 284, 303, 305, 310, 316, 318, 325-6, 328, 330, 340, 346, 348, 350-1, 355, 360, 363-4, 369, 379, 382, 384, 388, 394, 397-8, 406, 412-3, 439, 491-2, 498, 551, 557, 610-612, 625, 627, 631
electric charge allocation, 316
electric coupling, 380, 391, 400
electric coupling constant, 380
electric field, 279
electric fine structure constant, 292, 374
electric force, 36, 260, 265, 278, 316, 447, 508, 589, 607, 611
electric interaction, 239, 258, 265-6, 278, 316, 325, 328, 349, 368, 391-2, 432, 510, 626
electric phase, 350
electric transition, 368
electric vacuum, 147, 317, 326, 350-1, 368, 370, 400
electric vacuum force, 350
electric vacuum operator, 370
electrodynamics, 38, 39, 42, 222, 235, 285, 366, 429, 445
electromagnetic coupling, 389, 393, 395
electromagnetic field, 59, 148, 167, 215-6, 230, 291, 453, 474, 479, 484
electromagnetic flux, 404
electromagnetic force, 256, 319, 374, 478, 563
electromagnetic information, 470
electromagnetic interaction, 73, 148, 217, 239, 256, 266, 268, 270, 285, 305, 319, 328, 343, 390, 392, 395, 607, 625
electromagnetic mass unit, 327
electromagnetic radiation, 229, 444, 457, 458
electromagnetic signal, 234
electromagnetic theory, 33, 36, 59, 191, 195, 204, 207, 212, 215, 217, 232, 305, 310, 430, 473-4, 476, 601, 630
electromagnetic wave equation, 204
electron, 16, 31, 60, 91, 135, 143, 148, 157, 162, 173, 175, 194, 224, 229, 248-9, 288-90, 292-3, 295-7, 299-300, 302, 304, 316, 319, 323, 326, 343, 345-8, 364-5, 393-5, 397, 400, 413, 419, 422=3, 426, 430-2, 441, 497, 504, 507, 589, 611-2, 616, 619
electron charge, 347
electron current, 300
electron mass, 302, 304, 395, 400, 412-3, 616, 619
electron orbit, 60
electron propagator, 293, 295-7
electron scattering, 302
electron spin, 16, 224, 422-3, 426, 441
electron-muon scattering, 292
electron-positron annihilation, 346
electrostatic dipole, 319
electrostatic field, 210, 214, 216
electrostatic interaction, 210, 607
electrostatic potential, 55, 205, 210, 270, 497
electroweak boson, 306, 307, 399
electroweak boson propagator, 306-7
electroweak combination, 368, 626

electroweak coupling, 391, 400
electroweak eigenstates, 347
electroweak field, 240
electroweak gauge boson, 306
electroweak interaction, 268, 306, 357, 359-60, 363, 370, 378, 381
electroweak mixing, 240, 268, 380, 388, 406, 408, 439-40, 627
electroweak mixing angle, 388
electroweak mixing parameter, 240, 380, 439, 627
electroweak mixing ratio, 265, 268, 316, 342, 626
electroweak splitting, 408-9, 412
electroweak symmetry breaking, 400
electroweak theory, 305, 362
electroweak unification, 349, 363
elements, 5, 11, 13-16, 19, 21-2, 27, 28, 32, 40, 49-51, 53, 66, 81, 98, 102, 107, 112, 133, 137, 193, 203, 207, 219, 226, 258, 382, 385, 387, 436, 439, 450, 454-5, 468, 481, 488, 494, 520, 552, 553-5, 589-91, 609, 624
ellipsoid, 404, 405
elliptic geometry, 566
ellipticity, 566
Ellis, J., 258
embryo, 591-2
EMC experiment, 256
emergence, 63
emergent, 33, 66, 348, 559, 583, 593
emergent phenomena, 348
emission, 148, 201, 215, 230, 290, 405, 430, 435, 439-40, 447, 489, 501
emitter / absorber, 580
empirical, 2, 137, 393, 406, 408, 412, 569, 627
empiricism, 64
empty set, 558, 563-4, 566, 569
empty state, 558, 563, 567, 573, 585-6, 593
empty universe, 564
energy, 3, 37, 39, 41, 49, 54-5, 58, 65, 68, 74, 76, 78, 80, 84, 88-93, 98, 115-8, 121-3, 126, 130, 137, 139-40, 143-5, 148-9, 153, 158-60, 165-7, 172-3, 176-9, 183, 186-90, 194-5, 200-5, 207-9, 212, 217-8, 220-1, 224-5, 228-34, 236-8, 241-2, 244-9, 255, 257-9, 261, 268-70, 275-7, 279, 283, 290-2, 294-5, 301, 304, 309-15, 318-9, 322, 325, 328-9, 345, 347-8, 352, 368, 370, 372, 375, 379, 382, 386-8, 390-4, 396, 398, 400, 402, 404-7, 409, 412, 416-22, 424-31, 435-40, 442, 445-8, 450, 453-5, 457, 459-60, 465, 467-70, 472, 475, 477-8, 480, 482, 485, 489, 491-3, 498, 500, 504-7, 510-2, 514, 531-2, 553, 562, 566, 575, 579-80, 582, 584-6, 596, 600, 606, 607, 611-2, 615-6, 618-9, 625-8, 632
energy density, 396, 404, 420, 475, 478, 612, 615-6, 618, 628
energy density of radiation, 420
energy density of starlight, 618
energy eigenvalue, 246-7
energy exchange, 231, 233, 236-7, 318
energy level, 178, 244-5, 275-7, 312, 390, 405, 442, 582
energy loss, 477, 611
energy operator, 121, 166, 183, 189, 269, 445, 531, 626
energy output, 477
energy renormalization, 314, 352
energy state, 89, 90, 121, 139, 143, 188, 294-5, 318, 372, 445, 468, 511
energy tensor, 500
energy transfer, 329, 553, 607
energy-momentum, 85, 89, 165, 188, 450, 579
energy-momentum tensor, 450
energy-momentum-mass, 85
entanglement, 60, 79, 85, 144, 151, 154, 161, 177, 179, 322, 579, 600, 624, 627
entropy, 179, 228, 237-8, 330, 502, 512, 559, 578-80, 586, 600
enumerated form, 636-8
enumeration, 11, 23, 76, 79, 623
environment, 79, 86, 99-100, 158, 174, 427-8, 430-1, 433-5, 439-40, 553, 619
enzymes, 512, 553
epistemology, 415, 435, 452, 472, 623, 625
equally probable states, 220-1, 237, 325
equation of motion, 132, 173, 175, 180-1, 211, 215, 347, 469, 475
equiangular spiral, 531
equipartition, 229
escape velocity, 420
Euclidean metric, 607
Euclidean space, 27, 145, 184, 322, 476, 479, 600

Euclidean space-time, 184
eukaryote, 592, 593
Euler, L., 39
Euler angle, 112
Euler's formula, 634
Eulerian representation in dynamics, 120
event horizon, 292, 392, 483, 601-2, 605, 611, 628
Everett, H., 583, 587
evolution, 83, 173, 298-9, 521, 553-4, 562, 565, 567, 570, 577, 579, 582, 586-8, 600, 609, 618, 621
evolution of galaxies, 618, 621
evolutionary cosmology, 618, 621
evolutionary quantum cosmology, 586
evolving universe, 584
exceptional Lie group, 96, 387
exchange boson, 282, 339
exchange particle, 221, 230, 378, 454
exotic states, 155
expanding universe, 419, 604
expectation value, 171, 174, 221, 248-9, 264, 319-20, 412, 426
experiment, 5, 33, 57, 200, 336, 411, 477-8, 557, 563, 569, 576, 596, 608, 621
exponential, 74-5, 120, 122-4, 128, 146, 187, 253, 274, 283, 376, 498, 508, 575
extended causality, 500
external field, 131, 290, 292, 309
extremum principles, 39
factor 2, 15, 100-1, 103, 173, 190, 386, 414-5, 418, 423-4, 426, 428, 435, 437-40, 442-3, 602, 624
factor 4, 445, 475, 476, 499, 602, 613, 618
Fadeev-Popov ghost fields, 167
falsification, 481
Fatmi, H. A., 571
Fermi constant, 400, 412, 617
Fermi-Dirac statistics, 343, 428
fermion, 74, 79, 84, 90, 110, 118, 120-2, 133, 138, 141-9, 151, 153, 157-66, 184, 187, 190, 229, 231-2, 239, 241-2, 246, 257, 259-66, 268, 270, 275, 278-80, 284-6, 291, 296-8, 302, 305-6, 308-10, 312-5, 318, 322-5, 327-8, 341-2, 344, 349-50, 352, 367-70, 373, 378, 386, 391, 397-400, 405, 406-10, 412-3, 427, 431-34, 439, 443, 480, 489-90, 505-14, 528-9, 537, 550, 568, 573, 576, 578, 586, 600, 606, 619, 624 -6, 628-9
fermion creation operator, 146, 433
fermion field, 90
fermion generation, 325, 391
fermion loops, 147, 157, 291, 302, 434
fermion mass, 398, 399, 405-10, 412, 434
fermion mass Lagrangian, 398
fermion operator, 151, 514
fermion propagator, 141, 157, 285, 297, 305-6, 309
fermion spin, 138, 142, 148, 279, 315, 443
fermion state vector, 286, 405
fermion wavefunction, 79, 153, 159, 164
fermion-antifermion condensate, 184
fermion-antifermion field, 91
fermionic state, 79, 81-2, 101, 112, 118, 122, 133, 154, 165, 178, 184, 186, 202, 257, 263, 284, 311, 316, 323, 327, 349, 352, 367, 369, 371, 373, 432, 481, 496, 499, 507-10, 513, 515, 518, 521, 528, 531, 550, 562, 568, 574, 581, 607, 619, 621
fermion-boson, 285
fermion-like state, 158, 433
fermion-vacuum interface, 490
Feynman, R. P., 561, 569, 583
Feynman diagram, 112, 148, 285, 290, 303, 309
Feynman formalism, 293
Feynman gauge, 306-7
Feynman loop diagram, 290
Feynman representation, 121
Feynman rules, 301-2, 305, 307
Feynman sum over histories, 581, 587
Feynman-Wheeler mechanism, 430, 434
Fibonacci numbers, 502, 515, 529
fictitious acceleration, 613
fictitious force, 601, 605, 611
fictitious inertial force, 192, 479
field, 41, 45, 49, 58, 90-1, 93, 111, 122-4, 143, 148-9, 166-7, 172-3, 178, 184, 190, 204, 210-1, 215, 220-1, 226, 228, 239, 241, 246, 250-1, 264, 271, 273, 278, 280, 292, 295, 304, 309, 330, 352, 374, 388, 395-6, 398, 400, 405, 420-3, 429-30, 432, 441, 444-5, 450-6, 458, 461, 465, 467-8, 471-7, 479-82, 496, 498-501, 508-9, 563, 565-7, 572, 576, 580, 588, 596, 602, 606-7, 628
field energy, 90, 309

field equations, 90-1, 444, 450-3, 455-6, 467, 472, 477, 481, 496, 500, 628
field excitation, 607
field intensity, 210-1
field operator, 91, 123
field quantum, 91, 374, 445, 468, 472
field source, 184, 496, 500, 501
field terms, 41, 122-4, 178, 204, 210, 220, 241, 499
filled electric vacuum, 400
filled fermion vacuum, 377
filled vacuum, 49, 89, 159, 261-2, 268, 271-2, 292, 309, 311, 313, 318, 328, 342, 378, 392, 396, 445, 448, 490, 498, 513-4, 604, 607, 617, 628
filled weak vacuum, 239, 279, 306, 327-8, 395, 510, 626
fine structure constant, 303, 313, 393, 410, 617
finite, 6, 9, 11, 13-14, 19, 23, 46, 49, 81, 86, 101, 165, 206, 220, 229, 247, 264, 285, 291, 294, 303-4, 306, 309-10, 320, 347, 352, 371, 387, 392, 434, 444, 454-5, 464, 467, 470-1, 481, 558, 570, 574, 584, 606-7, 619-20, 623, 629, 631
finite dimensionality, 481
finite dimensions, 86
finite groups, 11, 101, 623
finiteness, 19, 48, 312
Finsler geometry, 184
first-order correction, 292
first-order coupling, 288, 290
five-fold symmetry, 109, 502, 530-1, 541
flat and infinite universe, 631
flatness problem, 609, 621
flavour, 305
flux line, 143, 347, 432, 625
force, 1, 36, 38-9, 42-5, 58, 60, 68, 85, 110, 113, 154, 191-2, 201, 203, 206, 208-12, 214-6, 223, 238, 246, 251, 257, 260, 264, 278, 280, 310, 318-21, 348, 374, 381, 388, 392, 397, 416-8, 420, 441, 446-50, 453-4, 464-7, 471-2, 475, 479, 481-2, 484, 489, 498, 507-9, 511, 513, 528, 555, 576, 588-9, 601-2, 605-6, 611, 617, 621, 623, 627-8
force laws, 210, 212, 441
four dimensions, 27
Fourier analysis, 436
Fourier series, 287
Fourier sum, 91
Fourier superposition, 162
Fourier transform, 80, 172, 289, 293, 295, 300, 482, 494, 498, 514, 573
Fourier transformation, 80, 172, 482
fourth dimension, 42, 59, 488, 551
fractal, 110, 530, 534, 571, 593-4, 600, 606, 629
fractal computation, 571
fractional electric charge, 316, 346
fractional quantum Hall effect, 347, 432
fragmentation, 115, 133
frame of reference, 283, 311, 600
free charges, 364-5
free electron, 287, 290
free fall, 35
free fermion, 124, 181, 240, 253-4, 262-3, 265, 266, 314, 316-7, 325, 348, 351-2, 368, 395, 410, 627
free particle, 74-5, 89, 112, 120, 123-4, 165-6, 180, 185, 321, 364, 405, 498, 508
free plane wave, 299
free state, 122, 124, 196, 256, 554
frequency, 60, 181, 194, 230, 236, 397, 458, 475, 594
frequency organization, 594
Freudenthal-Tits Magic Square, 104
Friedmann solutions, 609
Frobenius, G., 25
Frobenius-Schur-Godement identity, 578
full product, 44, 113, 424
fundamental charge, 76, 348
fundamental constant, 53, 73, 193, 393, 620, 622, 625
fundamental mass, 395, 400, 413
fundamental parameter, 34, 38, 53, 55, 57, 65, 91, 100, 103, 105, 130, 175, 187, 199, 203, 219, 224, 331, 436, 443, 452, 481, 484, 487, 489, 504, 528
fundamental properties, 74, 235, 479, 588
Furry's theorem, 290
Gabor, D., 571
galactic black hole, 584
galactic cluster, 605, 621
galactic magnetic field, 618
galaxies, 583-4, 587, 605, 617-8, 621
Galilean transformations, 198
Galileo, 2, 35, 61

gamma factor, 488
gamma matrices, 18, 63, 70-1, 112, 114, 137, 492, 494
gases, 221, 232, 417-8, 511, 554, 618
gauge boson, 147, 202, 306, 307, 328, 372, 374, 378-9, 391, 480, 550, 626
gauge condition, 199, 213, 235
gauge field, 183, 268, 271, 271-3, 340, 348, 352
gauge fixing, 167, 306
gauge force, 347
gauge group, 265, 316, 341, 371, 391
gauge invariance, 41, 152-4, 167, 204-5, 211, 213, 219, 221, 228, 234, 238, 250, 256, 258, 270-2, 280, 290, 305, 316, 321, 325, 330, 340, 346-7, 368, 379, 397, 568, 573, 581-2, 586, 592-3, 624, 627
gauge invariant geometric phase, 568, 581
gauge invariant phase, 321, 573, 582, 586
gauge relations between interactions, 379
gauge symmetry, 149, 358-9
gauge theory, 205, 305, 371, 448, 628
gauge transformation, 122, 154, 271, 371
Gell-Mann, M., 346-7, 348, 404
Gell-Mann-Okubo formula, 402-3
general object symmetry, 95
general relativistic field equations, 451-2, 456, 472, 477, 609
general relativity, 2, 35, 46, 59, 111, 184, 195, 208, 217, 421, 444, 448-9, 451-2, 456, 459, 463-4, 467, 472-3, 476-7, 480-1, 484, 497-8, 564, 566, 576, 579-81, 584, 602, 609-10, 613, 615, 622, 630
generalised coordinates, 216
generality, 35, 60, 127, 169, 271, 345
generations (quark-lepton), 270, 325-8, 337, 341-3, 358-60, 365, 369-70, 378, 391, 396, 399-400, 402, 406-9, 413, 485, 521, 626-7, 632
genesis of the elements, 589
genetic, 502, 516-9, 527, 553, 556, 560, 567, 570, 590-3, 597-8, 629
genetic alphabet, 591
genetic code, 502, 518, 527, 553, 556, 560, 567, 570, 590-3, 598, 629
geno-Dirac, 151
geodesic, 450, 467-8, 564, 576
geodesic deviation of spinning particles, 475
geodesic equation, 472
geodetic effect, 475
geometric continua, 597
geometric pattern, 629
geometric phase, 321, 323, 593, 625
geometrical algebra, 14, 16, 492
  G(3,2), 492
  G(3,3), 492
  G(4,1), 492
  G(6,0), 492
geometry, 529, 547-8, 559, 623
Georgi, H., 378
Gerber, P., 464-5
Ghosal, S. K., 459, 463-4, 467
Gibbs, J. W., 26
Glashow, S. L., 378
Glashow-Weinberg-Salam electroweak theory (GWS), 362-7
glial cell, 596
glial structures, 560
global, 40-1, 71, 205, 343, 351, 365, 367, 400, 447, 570, 582-3
global correlation, 447
global symmetry, 71, 401
global time ordering, 582
global transformation, 40
global warming,, 587
glueball, 155
gluon, 147, 149, 152, 155, 158, 251, 256-7, 291, 305, 307, 313, 316, 339, 371, 378, 380-4, 404-5, 433, 552
gluon coupling, 405
gluon exchange, 257
gluon flux tube, 404
gluon plasma, 147
gluon propagator, 307
gluon sea, 155
gluon-gluon interaction, 305
glycine, 524, 538
God, Einsteinian references to, 222, 570
God, mediaeval conception of, 35, 37,
Gödel, K. 83
Gödel's theorem, 83-4
golden attractor, 531
golden section, 529, 539
golden triangle, 531
Goldstone boson, 142, 271, 273, 306, 352, 625
Goldstone theorem, 313-4, 353
grammar, 557, 562-3, 565, 596

Gram-Schmidt method, 135
grand unification, 256, 348, 374, 380-2, 387-8, 390-3, 400, 407, 410-1, 444, 493, 627, 631
grand unification energy, 392
grand unified gauge group, 379, 389, 410
grand unified mass, 389
grand unified theory, 324, 374
graph theory, 110
graphene, 315, 632
Grassmann algebra, 14, 69, 82, 143-4, 493
gravitational attraction, 475, 480, 483, 499, 613
gravitational collapse, 478, 618
gravitational component, 470, 480
gravitational constant, 192
gravitational energy, 3, 318-9, 493, 606, 612
gravitational energy density, 612
gravitational field, 45, 210, 416, 420-1, 427, 437, 445, 448-52, 454, 458-60, 465, 467-9, 471-2, 475-7, 479, 481, 483, 497-8, 602
gravitational field equations, 449
gravitational field intensity, 210
gravitational field strength, 465
gravitational force, 56, 59, 208, 379, 444-6, 448-9, 453, 455, 467, 480, 498, 563
gravitational inertia, 482, 499-500
gravitational interaction, 282, 327, 392, 447, 453, 470, 482, 606, 618, 627
gravitational light deflection, 437, 440, 457, 459
gravitational mass, 208, 476, 583, 602, 610
gravitational potential, 160, 444, 448-50, 457, 461, 464, 466-7, 469, 472
gravitational potential energy, 160
gravitational quadrupole moment, 477
gravitational redshift, 419, 421, 444, 457-9, 464
gravitational source, 80, 445-7
gravitational system, 454, 469, 471, 479
gravitational time dilation, 462
gravitational vacuum, 318, 321, 446, 498
gravitational waves, 445, 448, 476
gravitational-inertial vector potential, 602
graviton, 453-4, 496
gravity, 50, 58-9, 85, 159-60, 202, 217, 228, 244, 292, 318-9, 374, 381, 383, 386, 388, 392, 444-9, 451-7, 459, 463, 465, 467-73, 475, 477-83, 489, 493, 496-9, 509-10, 600, 604-6, 610-1, 617-8, 627, 630
gravity generator, 382
gravity operator, 383, 386
gravomagnetic effects, 445, 455, 470, 499
gravomagnetic equations, 445, 499, 600
gravomagnetic field, 473, 476-7
gravomagnetic field intensity, 477
gravomagnetic force, 602
Green's function, 285, 289, 293-5, 298, 500-1, 567
Grossmann, M., 450
ground state of the universe, 80, 261, 328, 445
group, 11, 13-17, 19, 32, 38, 41, 51-3, 55, 65-6, 74-6, 79, 81, 88, 94-105, 107, 110, 112, 130, 152, 154, 160, 187-9, 193, 207, 261, 266, 270, 331, 335, 338, 340, 344, 348, 351-2, 363, 371, 374-5, 379-81, 385-8, 390, 392, 401, 414, 432, 436-8, 452, 481, 489, 493, 515, 519, 522, 524-5, 528-30, 543-4, 550, 563-5, 572, 574, 608, 623-4, 627, 631
C_2, 14-16, 51, 53, 65, 69, 84, 94-96, 99, 100-2, 126, 281, 386, 390-2, 414, 436, 438, 624
$C_2 \times C_2$, 14, 15, 69, 94-5
$C_2 \times C_2 \times C_2$, 15, 69, 94-5
$C_2 \times C_2 \times C_2 \times C_2$, 15, 69, 95
$C_2 \times C_2 \times C_2 \times C_2 \times C_2$, 15, 69, 95
$C_2 \times C_2 \times C_2 \times C_2 \times C_2 \times C_2$, 15, 69, 95
C_4, 14-16, 95, 100-2
$C_4 \times D_2$, 95
$C_4 \times Q_4$, 95
C_8,, 101
D_2, 14, 52-3, 94, 96, 99-100, 193, 276, 281, 637
D_4, 101
E_6, 104, 387
$E_6 \times U(1)$, 387
E_7, 104
E_8, 84, 104, 387-8, 491-2, 624
$E_8 \times E_8$, 388, 492
F_4, 104
G_2, 96, 104, 289, 387
Q_4, 15, 16, 94, 95, 101, 102
QQ_{32}, 16
$SO(10)$, 76, 375, 379, 410
$SO(12)$, 104
$SO(3)$, 104, 112, 404

$SO(32)$, 491-3
$Sp(3)$, 104
$SU(2)$, 73, 96, 104, 147, 149, 178, 239, 259-60, 264-66, 268-70, 272, 316-7, 328, 337, 340-1, 343, 345, 349, 351-2, 359, 361, 363, 365, 367-9, 371, 374-5, 380, 385, 388-9, 397, 405, 412-3, 439, 626, 631
$SU(2) \times U(1)$, 389
$SU(3)$, 71, 73, 96, 104, 149, 152-4, 178, 239, 250, 256, 258-9, 266, 313, 317, 337, 339-41, 345, 349, 351-2, 355-60, 363, 371, 374-5, 380-1, 385-6, 388-9, 401-5, 407-8, 412-3, 551, 626, 631
$SU(3)$, 104, 401-3, 405, 407-8, 413
$SU(3) \times SU(2) \times U(1)$, 96, 375, 380
$SU(3) \times SU(3)$, 104
$SU(4)$, 402, 413
$SU(5)$, 76, 98, 345, 357, 374-5, 378-82, 386, 388, 391-2, 394, 402, 410, 631
$SU(6)$, 104, 402
$SU(n)$, 402, 413
$U(1)$, 73, 96, 145, 148, 178, 215, 239, 256, 259-60, 264, 266, 268-72, 280, 305, 317, 328, 340-1, 345, 349, 351-2, 359, 363, 366-8, 371, 374-5, 380, 385, 387-9, 391, 394, 412, 432, 447, 478, 480, 492, 497, 600, 626, 631
$U(5)$, 76, 98, 374, 375, 381, 383, 386, 392, 627
V_{16}, 16
VQ_{64}, 16
group generators, 351
guanine, 516-7, 544, 591
Gupta-Bleuler quantization, 167
Gürsey, F., 387
gyromagnetic ratio, 422
gyroscope, precession of, 475-6
hadron, 346, 394
hadron-muon production, 346
Haisch, B., 482
half-integral spin, 113, 149, 370, 427, 431
Halzen, F., 397, 409
Hameroff, S., 585
Hamilton, W. R., 25-8, 39, 224, 634, 637
Hamilton algebra, 23
Hamilton's equations, 180, 184
Hamiltonian (function), 75, 124, 180, 189-90, 203, 211, 249, 298, 424, 442, 559, 568, 583, 586
Hamiltonian dynamics, 180, 186, 188, 213
Hamiltonian energy function, 586
Hamilton-Jacobi equation, 181
Han, M. Y., 346-8
handedness, 259-60, 278-80, 315, 351, 367-8, 373, 396, 514, 518
Han-Nambu quark theory, 347, 348
hard wired, 598
hardware, 4, 18, 557
harmonic oscillator, 158-9, 174, 221, 230, 237, 240, 246, 257, 260, 275-81, 283, 312, 315, 321, 399, 424, 426, 430, 439, 442, 502, 508-9, 511, 531, 550-1, 554
harmonic oscillator term, 281, 283
Harvey, A., 467
Hawking, S. W., 565, 584
Hawking radiation, 482
He^3, 143, 158
He^4, 158
heart, 62, 64, 414, 593-4
heart-lung system, 593
heat, 221, 236-7, 578-9, 596, 611
heat bath, 578-9, 596
heat death, 611
Heaviside, O., 26, 612
Heaviside operator, 557, 567, 569
Heaviside step function, 501
Heisenberg, W., 60, 630
Heisenberg algebra, 186
Heisenberg equations of motion, 180
Heisenberg's quantum mechanics, 50, 60, 93-4, 100, 103, 168, 172, 174-7, 180, 182, 188-90, 224, 230-1, 233-4, 429-30, 436, 438-40, 488, 575, 584
Heisenberg symplectic spinors, 186
Heisenberg uncertainty principle, 50, 93, 174-5, 177, 220, 222, 224, 226, 229-30, 233-4, 238, 426, 564, 568, 574-6, 579, 599
helical, 509, 516, 548, 550, 554, 592
helicity, 137, 139-41, 143, 148, 154, 252, 260-1, 263-5, 283-4, 314, 317, 330, 350, 361-4, 372-3, 494, 511, 531-2
helium atom, 248-9
helix, 525, 544, 547-9, 592
Hellmann-Feynman picture, 319
Hermitian conjugate, 131, 134, 147, 157, 160, 434

Hermitian matrices, 104
Hermitian operator, 442
Hestenes, D., 16, 28, 113
heterocyclic ring, 552
heurism, 479
hexadecimal label, 635
Hey, A. J. G., 394
hidden variables, 222
hierarchical semantic ontology, 591
hierarchy, 19, 88, 99, 103, 157, 353, 374, 552, 563, 584, 591, 593, 629
hierarchy of alphabets, 591
hierarchy of dualities, 88, 99, 374
hierarchy of QCEs, 584
hierarchy of symmetry, 103
hierarchy problem, 157
Higgs boson, 272-3, 314, 352, 369-70, 394, 397-9, 412-3, 627
Higgs boson vertex, 370
Higgs coupling, 370, 400
Higgs doublet, 397-8
Higgs field, 159, 264, 292, 313, 352, 373, 393-5, 397-400, 405-6, 412-3, 445, 485, 498, 602, 607, 610, 617
Higgs Lagrangian, 398
Higgs mass, 412, 632
Higgs mechanism, 49, 80, 154, 158, 201, 239, 270, 317, 319, 329, 348-9, 369, 374, 387, 394, 397-9, 406, 420, 433, 445, 490, 493, 600, 611, 617, 627
Highfield, R., 46
high-pass filter, 397
Hilbert space, 5, 23, 69, 74, 81, 443, 492-3, 495, 578, 582
Hiley, B. J., 181, 186
Hill, V. J., 502
holism, 629
holographic encoding, 597
holographic principle, 184, 508, 556, 573
holographic redundancy, 573
holographic signal theory, 592
holographic transform, 594
holography, 573-4, 587, 595
homogeneity, 602
Hoogsteen, K., 544
Hooke's law, 531
Hoyle, F., 584
Hubble mass, 600, 628
Hubble radius, 605, 615
Hubble redshift, 601
Hubble time, 600, 615, 618, 620-1
human language, 556, 561, 596-7
human organism, 594
human thought, 83, 595
hydrogen, 513, 590, 613
hydrogen atom, 111, 178, 188, 192, 239, 241-2, 244-5, 248, 464, 508
hydrogen bond, 516, 525
hydrogen-like spectral series, 254
hydrophilic, 525
hydrophobic, 525
hyperbolic geometry, 566
hyperbolic orbit, 459-60
hyperboloid inscribed cones, 185
hypercharge, 340, 363, 391-2, 394, 397-8
hypercharge numbers, 391-2
hypercone, 496, 500
hyper-Dirac, 151
hyperdodecahedron, 547
hyperentanglement, 144, 251
hyperfine levels, 241
hyperfine structure, 244
hypergeometrical algebra, 133
hyperincursion, 571
hyperincursive, 594
hypothesis, 34, 57, 199, 408, 411, 469, 556, 560, 563, 570, 576, 580-1, 585, 592, 594, 600, 603, 610, 620
icosahedron, 527, 529-31, 539-41, 543, 545-7, 629
ideal, 36, 44, 78-9, 82, 86, 111-2, 194, 209, 296, 409, 420, 491, 506
ideal gas molecules, 417
ideal system, 220
ideal vacuum, 314, 352, 369
idempotent, 78-9, 168, 185-90, 567
identity, 11, 40-1, 44, 51-2, 61, 94, 96, 98, 105, 107, 151, 183, 190, 195, 204, 213, 219, 222, 227, 258, 270, 299, 322, 454, 532
identity element, 51, 52, 94, 96, 98, 105
Illert, C., 514, 531-2, 537
image processing, 576, 635
imaginary, 12, 16, 25-30, 32, 42-5, 49, 50-2, 65-6, 70, 75, 80, 88, 92-3, 95-6, 98-103, 105, 110, 115-6, 131, 134, 181, 194, 197, 206-7, 210, 213, 223, 225, 231, 235, 253, 275, 282, 304, 344, 382, 386-7, 436, 438,

441-2, 486, 488, 497, 499, 503, 508, 528, 531, 568, 586
imaginary numbers, 25, 43-5, 49, 103, 223, 225
imaginary vector, 70
imaginary vectors, 70
imidazole ring, 552
immune system, 593
impressed force, 58, 160
incursion, 571
incursive, 594
indefinite divisibility, 47, 49
indefinite elasticity, 47
independent variable, 48, 206
indeterminacy, 218, 221-2, 235
indistinguishability, 6-7, 9, 12, 17, 24, 40, 44, 73, 144, 152, 208, 225, 265, 278, 314, 316, 324, 335, 337-8, 341, 344, 358, 360, 365, 369-70, 399, 401, 410, 452, 475, 514
indivisible, 32, 46, 50-1, 95, 222, 223
induction, 214
inductive, 2, 32, 64, 104
inductive empiricism, 64
inductive force, 601
inductive inertial field, 602, 605
inert gases, 589
inertia, 4, 45, 160, 184, 202, 217, 304, 319, 444, 451-2, 455, 477, 480-1, 483, 499, 566, 600-2, 605-6, 612, 628
inertial 'charge', 480
inertial acceleration, 604, 621
inertial calculation, 613, 617
inertial component, 317, 319, 480
inertial effects, 444, 452, 470-1, 479, 481, 610, 613
inertial energy, 606
inertial field, 450, 479, 483
inertial force, 444, 452, 457, 469, 471-2, 479, 481, 600-3, 605, 613, 628
inertial frame, 121, 233, 452, 467, 471, 475, 479
inertial interaction, 292, 349, 392, 481, 499
inertial mass, 208, 292, 320, 393, 400, 445, 476, 483, 600, 602, 605-6, 610-2, 616-8
inertial phase, 350-1
inertial process, 292, 600, 606-7, 610, 618-9, 621, 628
inertial properties, 602, 621
inertial reaction, 160, 292, 382, 498, 605
inertial reaction force, 160
inertial repulsion, 475, 483
inertial system, 191
inertial vacuum, 600
inertial vacuum process, 600
inertial velocity, 620
inertial waves, 480
inexact differential, 237
infinite, 1, 4, 9-11, 13, 16, 19-23, 31, 36, 40, 46-7, 49, 57-8, 63, 69, 74, 81-2, 84-6, 110, 145, 147-8, 157, 159, 161, 165, 177, 211, 220-1, 224, 229, 237-8, 285-6, 291, 298, 306, 310, 312, 318, 341, 347, 387, 415, 433-4, 443-5, 447-8, 475, 487, 491, 503, 508, 515, 556-8, 560, 563, 565, 579, 593, 600-1, 604-5, 607, 609, 611, 619, 622, 624, 628-9, 633-8
infinite degeneracy, 4, 11
infinite dimensions, 16
infinite divisibility, 46
infinite energy continuum, 447
infinite energy density, 229
infinite entanglement, 84, 443
infinite fractal hierarchy, 619
infinite hexadecimal 1 address, 635
infinite Hilbert space, 81, 619
infinite imaging, 84, 443
infinite mass, 347
infinite matrix, 636
infinite momentum, 238
infinite progression, 443
infinite range, 341, 434, 622
infinite semantic logic, 629
infinite series, 1, 9-11, 13, 16, 22, 46, 49, 110, 177, 221, 286, 291, 415, 433-4, 634, 636
infinite solutions, 634
infinite square roots of −1, 565, 600, 633
infinite succession, 47, 147-8, 157, 161, 503
infinite sums, 285, 306
infinite superposition, 81
infinite tesseral addresses, 634
infinite universal alphabet, 556, 558, 563, 565, 593
infinite universe, 145, 605, 611, 622, 628
infinite vacuum, 491
infinite variation, 69, 224
infinite vector, 31, 63
infinite vector space, 63

infinite virtual energy density, 312
infinite whole, 619
infinite zero-point energy spectrum, 318
infinitesimals, 47-8, 566
infinities, 112, 285, 291, 453, 478
infinity, 9, 13, 16, 20-1, 24, 48, 60, 63, 66, 69, 81, 84, 100, 109, 146, 148, 161, 443, 453, 456, 460, 515, 623, 634
inflationary expansion, 601, 609, 621
inflationary universe, 620
information, 2, 5, 15-16, 34, 38, 51, 53, 56-7, 63-4, 67, 80-1, 83, 84, 104-5, 108, 112, 116, 122, 124, 131-2, 137, 164, 166-7, 175-7, 184, 186, 191, 200, 202, 206-7, 222, 227-8, 233, 239, 258, 261, 284, 309, 311, 325, 327, 330, 341, 351-2, 405, 413, 437, 444-5, 447-8, 470-1, 480, 489, 492, 494, 496-8, 502, 504, 506, 516-7, 530, 532, 536, 546, 553, 557, 559-60, 565, 573, 575, 579-80, 589-92, 600, 611, 619, 621, 629
information loss, 565, 611
information metric, 580
information processing, 530, 557, 560, 579, 590, 629
information processing system, 579, 590
information transfer, 579, 591
infrared divergence, 112, 141, 158, 285, 294, 304, 305, 309, 495
infrared slavery, 252, 254-5, 347
initial state, 19-20, 299-300, 591
instant of time, 46
instantaneous, 144, 160, 165, 292, 311, 313, 317-8, 325, 357, 392, 444-9, 468-71, 479, 493, 498, 585, 600, 604, 606-7, 617, 627
instantaneous correlation, 311, 446-7, 493, 498
instantaneous gravitational force, 498
instantaneous interaction, 449, 468-9, 607
instantaneous interactions, 468
instantaneous quantum correlation, 318
instantaneous transmission, 160, 292, 446, 468, 470, 604
integers, 5, 9-11, 24, 49, 55, 66, 83-4, 230, 245, 419, 432, 442, 505, 529, 566, 574, 582, 634
integral charge, 347-8, 379
integral solution, 289
integral spin, 231, 426-8, 431
interaction, 4, 32, 38-9, 41, 44, 54-5, 58, 76, 81, 84-5, 122, 140, 144-5, 149, 153-5, 157, 161, 165, 178-9, 191-2, 194, 202-3, 206-8, 210, 217, 221, 226, 228, 231, 236, 238-40, 246, 248-9, 251-2, 256-8, 260-1, 264-5, 267-8, 270-1, 278-9, 281, 283, 285, 292, 298, 304-7, 309, 310, 314, 316-8, 320, 323-4, 328, 330, 337, 340-1, 346, 349-54, 357, 359-60, 364-6, 368, 371-2, 374-5, 379-82, 386, 390, 394-5, 401-3, 405-6, 410-1, 428, 432, 439, 445-6, 455, 468, 475, 479, 485, 496, 499, 501, 505, 507, 509-10, 513, 537, 554, 578, 596, 606, 610-1, 624-7, 631
interaction vertex, 144, 268, 354
interference, 161, 230, 432, 575
interference fringes, 432
intermediate boson, 328, 368, 412
internal energy, 220, 236, 237
invariability, 206
invariance, 41, 55, 75, 89, 148, 154, 167, 198, 204-5, 215, 219, 221, 223-4, 226, 228, 233-4, 250, 258, 261-2, 290, 293, 311, 341, 346, 357, 592, 597
invariant, 27, 41, 130, 197-8, 201, 203-4, 215, 221, 256, 271, 280, 301-4, 307, 356-7, 359, 363, 379, 394, 405, 412, 447, 450, 468, 477, 495, 542, 566, 579, 585, 597
invariant amplitude, 301-4, 307
invariant line-element, 450
invariant phase, 585
inverse, 42, 51, 52, 57-9, 76, 192-4, 209, 220, 231, 236, 239, 252, 259, 277, 315, 319, 418, 454, 465-6, 478-80, 482, 492, 508-10, 529, 606
inverse fourth-power force, 606
inverse linear, 240, 252, 259
inverse relationship, 231
inverse-square, 42, 57-9, 209, 277, 418, 478, 480, 482
inverse-square force, 42, 277
inversion, 11, 331, 458
inversion of velocities, 458
irreversibility, 34, 145, 218, 223, 225-8, 235, 315, 445, 531, 578-9, 586, 624
irreversibility paradox, 228
irreversible, 65, 179, 225, 228, 234, 237-8, 485, 570, 586
irreversible process, 586

irrotational, 44, 258, 335, 388
Ising model, 581
isodual, 151
isolated quantum system, 178, 448
isolated system, 178-9, 511
isolation, 218
isoleucine, 522
iso-Minkowski, 11
isospin, 262-3, 265-7, 316-7, 325-8, 337, 340, 363, 366, 370, 376-7, 395, 401, 404-6, 411, 413, 521, 627
isospin multiplet, 340, 401
isothermal, 578
isotopes, 589
isotropic, 200, 233, 580, 613
iteration, 19
iterative, 20-1, 81, 86, 530, 558, 619, 633, 638
Jacobi identity, 152, 184
Jahn-Teller effect, 79, 143, 158, 432, 435, 439, 507
Jaki, S. L., 461
Jessel, M., 569, 571, 587
Johansen, S., 530, 550
Jordan algebra, 104
Josephson effect, 430
Julia set, 600
Jupiter, moons of, 471
K meson, 341
Kaluza-Klein theory, 484, 492
Kant, I., 58, 479
Kauffman, L., 182-4
Kelvin, Lord, 59
Kepler, J., 585
ket operator, 129, 131, 134, 173, 295, 297, 557, 572, 577, 635
keto form, 544
Kilmister, C. W., 572
kinematic, 199-200, 233, 311-2, 416, 422, 461, 465
kinematic equations, 416
kinematics, 59, 201, 415, 419-20, 457, 463, 465, 490
kinetic behaviour, 417-8
kinetic energy, 149, 159, 173, 209, 230-2, 238, 248, 271-2, 416-8, 4204, 427-31, 433-5, 437, 439-41, 443, 458, 460
kinetic theory of gases, 417
Klein bottle, 484, 490
Klein-4 group, 54, 494, 550
Klein-Gordon equation, 89, 121, 166, 195, 230, 295-7, 427-9
Klein-Gordon field, 296
Klein-Gordon operator, 296
Knuth, D., 565
Kobayashi, M., 341, 365
Koberlein, B. D., 88, 168, 374, 496, 500
Kolbenstvedt, H., 473, 475
Kronig-Penney model, 246
Kuhn, T. S., 481
label, 39, 260, 264, 486, 597, 635
labelling, 353, 581, 597-8, 635, 637
laevo, 510, 514, 532
Lagrange, J. L., 39, 60
Lagrange's equations of motion, 211
Lagrangian, 40, 120, 122, 167, 186, 188, 203-4, 211-2, 216, 270-3, 306, 397-8, 463-5, 473-4, 559, 571-2
Lagrangian density, 40, 211
Lagrangian dynamics, 188
Lagrangian force equation, 216, 464
Lagrangian representation, 120
Lamb shift, 568
Langridge, R., 547
language model of computation, 557
Laplace, P. S., 421, 458, 461
Laplace equation, 210, 282, 604
large number coincidences, 617
large-scale structures, 617, 621
large-scale systems, 209
Larmor precession, 215, 475-6
Larmor radiation formula, 215
lattice gauge QCD, 405
Laughlin, R. B., 347
law of calling, 186
law of crossing, 186
law of moduli, 28
laws of nature, 4, 478, 569
laws of physics, 4, 38, 49, 83, 150, 196, 203-4, 219, 223, 225, 569, 600, 620, 622
laws of thermodynamics, 85, 218, 235, 570, 586, 629
zeroth law of thermodynamics, 236
first law of thermodynamics, 179, 237, 570, 629
second law of thermodynamics, 145, 179, 228, 235, 310-1, 509, 570, 578, 580, 586, 629

third law of thermodynamics, 229, 238, 570, 586
fourth law of thermodynamics, 579
laws of transformation, 186
left ideal, 79, 186
left-handed, 91, 139-40, 142, 171, 263, 265, 267-9, 278, 315-8, 320, 327, 329, 332, 345, 359, 363, 365, 367-8, 370, 373, 378, 388, 394-6, 405, 410, 486, 510, 514, 626-7
Legendre polynomial, 248
Leggett, A., 563
Leibniz differentials, 232
length contraction, 200, 421, 461, 463, 476, 602, 613
Lennard-Jones potential, 276
Lense-Thirring effect, 476
lepton, 39, 44, 256, 258, 266-7, 269-70, 284, 292, 316, 324-7, 329-30, 332, 334, 340, 343, 345-6, 348, 350-1, 359, 366, 369, 374, 379-80, 387-9, 391, 395, 397-9, 406, 408-11, 413, 611, 626-7
lepton charge structure, 324
lepton conservation, 44, 343
lepton gauge, 411
lepton mass, 256, 409-10
lepton mixing, 409
lepton-antilepton pair, 359
lepton-like quarks, 348, 374, 389, 391, 397-8
leucine, 521
Levi-Civita affine connection, 184
LHC, 394
Lie algebra, 101, 104, 387, 404, 572, 574
Lie dual, 568
Lie exponential diffeomorphic language description, 575
Lie group, 14, 352, 575, 577, 592, 631
Lie / bosonic partition of state space, 556, 568, 574, 576, 582
life, 521, 553, 556, 560, 584, 591-2
life-forming process, 521
lifetime, 618
light, 59, 195-6, 200, 228, 233, 419, 420-1, 430, 435, 437, 448, 455, 457-63, 465, 467-70, 472, 479, 489, 524,
light-cone, 196
limit, 20, 37, 46-8, 49, 78, 112, 220, 222, 247, 249, 309, 392, 406, 423, 449, 456, 605, 614, 634-5, 637
Lindenmayer, A., 18
line element, 27, 450, 455, 473
line spectra, 457
linear dynamics, 192
linear effects, 469
linear gravitational theory, 601
linear harmonic oscillator, 230
linear potential, 252, 255, 351, 395, 410, 625
linearised gravomagnetic field, 475
Liouville densities, 578, 594
living organism, 567, 593-4
living system, 553, 560, 570, 590-2, 594-5, 598
local, 40-1, 57, 79, 122, 154, 204, 206, 212, 219, 271-2, 319, 352, 371-2, 449, 451, 469, 472, 507, 560, 570, 602, 610, 620, 622
local conservation, 40, 79, 212
local conservation of charge, 212
local coordinate system, 469, 602
local curvature, 451, 610
local gauge invariance, 352
local gauge symmetry, 272
local inertial repulsion, 472
local interaction, 372
local phase invariance, 204
localization, 202, 510, 628
localized fermion, 165
locus, 66
logarithmic spiral, 548
logic, 4, 448-9, 629
logical functions, 186
London, F., 55
London dispersion interaction, 319
loop cancellation, 321
loop diagrams, 148
Lorentz condition, 214
Lorentz covariance, 450, 454
Lorentz force, 215-7, 475
Lorentz gauge, 213
Lorentz group, 388
Lorentz invariance, 75, 167, 196, 223-4, 226, 229, 233, 236, 352, 468, 477
Lorentz transformations, 27, 196, 198-9, 464
Lorentz-Poincaré relativity, 93, 199-200, 231-4, 566
Lorentzian frame, 228, 229
Lorentzian metric, 456, 470, 482, 497
Lorentzian space-time, 451, 456, 468-70,

472, 479-81, 491, 604, 628
Löwenheim-Skolem theorem, 49
lower central series, 574
luminal, 448
lysine, 538
Mach's principle, 4, 320, 450, 499, 583, 600-6, 610, 615
Mach-Zehnder interferometry, 588
Machian, 217, 602-3, 605-6, 610, 621, 628
magnetic field, 58, 173, 216, 437
magnetic field intensity, 214
magnetic field line, 507
magnetic field vector, 213
magnetic flux line, 347
magnetic flux quantum term, 430
magnetic force, 415, 454
magnetic moment, 60, 149, 292, 430
magnetic monopole, 73, 216
magnetic resonance imaging (MRI), 559, 575-7, 594
magnetism, 36, 454, 553
Majorana particle, 268, 369, 372, 410
Mandelbrot, 18, 593
Mandelbrot set, 593, 600
many worlds, 583, 584
mapping, 7-D, 110
Marcer, P. J., 556, 571, 619
Martin, A. D., 397, 409
Maskawa, T., 341, 365
Mason, D. C., 638
mass, 3, 5, 32, 35-45, 49-55, 57-60, 63-7, 69, 72, 74, 76, 78, 80-1, 84, 88-94, 96, 98-100, 102-5, 110, 115-7, 121-2, 130, 140, 145, 148-9, 154-5, 159-61, 167, 170, 176, 179, 183-4, 187, 193-6, 200-4, 206-10, 218-9, 221-6, 228, 230-5, 239, 241, 255, 259, 261, 263-4, 269-71, 282-3, 292, 297, 302-4, 308-9, 311, 313-5, 317-8, 325, 328-9, 333, 342, 344, 352-3, 365, 368-71, 373-4, 378-80, 382, 386-8, 390-413, 416, 419-20, 427-8, 430-1, 434, 436-7, 443-7, 449-51, 453-8, 465, 469, 473-81, 483, 485, 487-8, 490, 492-5, 497-9, 503-, 510, 513, 528-9, 537, 542, 546, 552-3, 557, 562, 588-9, 602-7, 610-2, 616-8, 620, 622-8
mass density, 475, 616
mass eigenstate, 408, 413
mass frame, 218, 228
mass gap, 154, 371
mass generation, 395, 413
mass increase, 200
mass of the universe, 203
mass operator, 88, 98, 121, 183, 241
mass quadrupole tensor, 477
mass scale, 389-91, 406
mass shell, 167, 185, 297
mass state, 89, 121, 170, 344, 405, 625
mass term, 115-6, 122, 183-4, 208, 283, 308, 314, 368, 373, 405, 454
mass*, 93-4, 96
mass-charge, 43-4, 54, 89, 98, 110, 193, 223, 226, 490, 492
mass-charge space, 490
mass-energy, 49, 65, 84, 90, 159, 161, 187, 202, 228, 239, 261, 313, 315, 344, 374, 379, 392-3, 396, 399, 404, 412, 436, 444-6, 450, 485, 498, 513, 529, 626, 628
massive stars, 457
massless antifermion, 314
massless boson, 282, 297
massless fermion, 140, 352, 369, 405
massless quark, 155
massless scalar field, 496, 500
masslike, 36, 50, 76, 90, 232, 447
mathematical analysis, 565, 624
mathematical language description, 557, 560, 562-3
mathematics, 1, 2, 4-6, 9-11, 18, 23-4, 27, 32-3, 37, 43, 47, 53, 62-5, 67, 74, 81-4, 86, 99, 103, 132, 186, 232, 377, 414-5, 433, 435, 442, 478, 489, 503, 555-6, 559, 562, 565, 586, 619, 623, 629
matrices, 18, 31, 70, 78, 89, 112-6, 120, 125-30, 134, 136, 146, 195, 262, 300, 317, 349, 353-8, 360-1, 363, 365, 375-7, 379, 387, 408-9, 411, 424, 494, 505, 598, 637
matrix element, 134, 494
matrix form, 113, 125-6, 129, 136, 146
matrix methods, 112
matrix operator, 136
matrix representation, 112-3, 125, 262, 356, 360, 387, 598, 637
matter, 3, 4, 18, 34, 58-9, 71, 90, 134, 151, 192, 199, 221, 225, 233, 235, 237, 257, 262, 278, 315, 320-1, 342, 381, 430, 445, 453-4, 493, 499, 505, 507-8, 510-1, 513, 555, 564, 566, 570, 573, 577, 580, 583-5,

587-9, 602-3, 605, 609-11, 616, 618, 620-1, 628, 637
matter-antimatter asymmetry, 621
Matzke, D., 81
Maupertuis, P. de, 39
Maxwell, J. C., 26, 58-9
Maxwell's displacement current, 239
Maxwell's equations, 58-9, 213-7, 301, 437, 445, 474-6, 478-9, 625, 630
Maxwell's formula for radiation pressure, 420
mean speed theorem, 416
Meaning Circuit, 569, 570
measurability, 33
measurable, 34, 37, 47, 56, 60, 100, 197, 199-200, 234, 410, 452, 568, 607
measure, 37, 45, 59, 151, 160, 179, 200, 215, 227, 236, 237, 304, 309, 313, 330, 400, 418, 481, 489, 559, 568, 620
measurement, 33-4, 36-8, 45, 47-50, 53, 56-7, 59-60, 62, 64, 66, 83, 85, 89, 93-4, 99-100, 175-9, 191-3, 195-7, 200-1, 203, 210, 218, 220-3, 225-9, 231, 233-4, 236, 311, 386, 394, 410, 429, 431, 444, 448, 452, 455, 464, 467-72, 477, 479, 486, 562, 567-8, 575, 577, 586, 600-1, 613, 619, 621-2, 625, 627-8
Measurement Dual, 99
measuring apparatus, 60, 172, 220, 222, 228
membrane theory, 491, 588
Mercury, 464, 466, 510
Mesarovic, M. D., 571
meson, 252, 256, 305, 327, 329-30, 338, 341-2, 349, 354, 358-9, 364, 394-5, 400, 402-6, 410-1, 413, 551
meson octet, 338, 403
metaplectic group, 189
methionine, 522, 524, 538
metric, 89, 300, 450-1, 456, 470, 482, 491-2, 496-8, 559, 607
metric equations, 451
metric geometry, 450
metric tensor, 300, 450, 491, 498
Michell, J., 457-8, 461
microcanonical ensemble, 178, 237
microcanonical system, 236
microwave, 200, 233, 445, 499, 600, 608, 610, 613, 615, 618, 621, 628
Milky Way, 509, 550, 618
mimicry, 520
mind, 35, 592, 595-6
minimalist, 184, 554
minimally relativistic theory, 455, 463-4
Minkowski, H., 27, 92, 488
Minkowski formalism, 45, 50, 219, 232-4, 450, 452, 479, 490
Minkowski metric, 27
Minkowski space-time, 27, 42-3, 45, 59, 92, 479, 496, 500
mitochondria, 530
Mitchell, E., 556
mixed handedness, 368
mixed state, 262-3, 341, 370, 380, 402, 409
mixing, 115, 265-8, 270, 316-8, 325, 341-2, 367, 392, 395, 406, 408-11, 413, 439, 551, 574, 578, 588, 626-7
mixing ratio, 265, 316, 342, 626
Möbius strip, 490
model-dependent, 62, 137, 407, 491, 610, 629, 631
model-dependent assumptions, 491, 632
molecular biology, 556, 560
molecular forces, 276
molecular theory, 219
molecular thermodynamics, 416
molecule, 178, 221, 319, 417-8, 511, 513-4, 516-7, 520, 548, 554
momentum, 37, 39-42, 54-5, 60, 68, 73-4, 76, 78, 80, 84-5, 88-9, 91, 93, 98, 110, 113, 115-7, 119, 121-3, 125-6, 133, 135-7, 139-40, 143-5, 152-4, 158, 165, 167, 169, 172, 175-6, 179-80, 184, 188-90, 192, 194-5, 200-6, 209-12, 217, 220-2, 224-5, 230-32, 234, 238, 241-2, 251, 256-59, 278, 281, 287-91, 293, 303, 315, 317, 322, 350, 352, 359, 414, 416-7, 423-4, 427, 430, 432, 436-8, 443, 446-7, 450, 4544, 485, 487, 491, 494, 504, 507, 512, 514, 551, 562, 579, 624-5, 630
4-momentum, 300, 494
momentum components, 122, 153, 315, 494
momentum coordinates, 211
momentum direction, 158, 169, 317
momentum eigenfunction, 241
momentum operator, 113, 116, 133, 153-4, 180, 184, 230, 258, 284, 350, 359, 409, 423-4, 438, 447, 551, 626
momentum space, 293, 494
momentum vector, 119, 125, 135, 145

mRNA, 516-7, 532
multidimensionality, 45, 49, 65, 101-3, 387
multiple entanglement, 628
multiple fermion-boson processes, 285
multiple interactions, 166, 625
multiple universes, 609
multiplication of address labels, 638
multiplicity, 83, 401, 484, 492, 553
multipolar, 257, 277, 508, 512
multivariate 4-vector, 17, 67, 564
multivariate vector, 13, 15-18, 31-2, 66-7, 73, 76-8, 95, 113, 117-8, 125, 182, 241, 275, 423-4, 441, 493, 497, 503, 537, 563, 623-4
multivariate vector quaternion, 17, 66-7
multivariate vector space, 16
muon, 307, 326, 343, 346, 393, 395, 407, 410, 611
muon decay, 307
muon mass, 393, 407
muon neutrino, 343
mutual annihilation, 148, 279
Mycoplasma species, 530
Nambu, Y., 346-8
natural language, 560, 562, 591, 593, 598
natural selection, 36, 64, 487, 591
natural world, 2, 3, 37
Nature, 6, 15, 22, 25, 31-7, 41-2, 47, 51, 53, 56-64, 70, 73, 75, 79, 85, 88, 96, 98, 100, 105, 115, 124, 130, 132, 137, 144-5, 148, 152-14, 159, 164-18, 173, 178, 183, 187, 193, 196-7, 199-200, 202, 210, 213, 219, 222, 224, 227, 232, 237-9, 250, 260, 265, 277, 280, 292, 297-8, 313, 316-7, 321-3, 325, 337, 340, 342, 345, 347-8, 352-3, 359, 368, 370, 373, 378, 380, 392, 407, 415, 418, 423-4, 432, 434-5, 439, 443, 445-7, 455, 457, 468, 472, 476, 481-2, 484, 486, 489, 493, 495-8, 501-3, 505, 508, 512, 525, 530-1, 534, 544, 552, 555, 559-61, 563-4, 566, 569, 573, 575, 582-4, 590, 594, 596, 598, 605, 610, 616, 619, 621-2, 624-5, 629, 633, 636, 638
Nature's code, 502, 538, 579, 629
Nature's process, 556, 560, 599, 629
Nature's rules, 599, 629
Naur, P., 18
nebula, 509, 550
negative binary numbers, 159, 634
negative energy, 89-90, 115, 121, 126, 135, 140, 143, 159, 229, 294-5, 306, 312-3, 318, 328, 445, 494, 506, 630
negative integer, 634
negative mass-energy, 344
negative numbers, 4, 638
negative parity, 156
negative pressure, 604
negative vacuum energy, 606
nervous system, 561-2
neural computational rewrite system, 595
neural information processing, 595
neural NUCRS semantic ontology, 598
neural system, 598
neuron, 585
neuroscience, 556
neutrino, 45, 171, 268, 279, 284, 316, 326, 343, 345, 365, 369, 372, 400, 409-11, 413, 610-1, 632
neutrino eigenstates, 411
neutrino gauge, 411
neutrino mass, 409-10, 610, 632
neutrino mixing, 411, 413
neutrino oscillations, 409
neutron, 45, 343, 346, 355-6, 366, 405, 512, 589, 616
neutron beta decay, 346, 366
neutronium, 589-90
New Computing Principle, 571
Newton, I., 35, 39, 56-9, 61, 149, 160, 192, 226-7, 310, 387, 417, 419, 453, 465-6, 480, 585
Newton's gravitational constant, 192, 452
Newton's law of gravitation, 42-3, 57, 208, 212, 585
Newton's laws of motion, 212, 602, 625
  Newton's first law of motion, 208
  Newton's second law of motion, 208, 602
  Newton's third law of motion, 39, 99, 149, 159, 208, 210, 507
Newton-Coulomb interaction, 145
Newtonian concepts, 234, 311
Newtonian escape velocity, 420
Newtonian field equations, 468
Newtonian fluxions, 232
Newtonian gravitational theory, 457, 470, 481
Newtonian inertial force, 602
Newtonian limit, 456, 467

Newtonian mechanics, 36, 191, 212, 229, 458, 470, 479
Newtonian methodology, 36, 56-7
Newtonian physics, 39, 59, 192, 202, 225, 471, 622
Newtonian potential, 451-2, 455, 467, 469-70, 481, 609
Newtonian redshift, 421
Newtonian space-time, 468-9, 479
nilpotent, 21-2, 49, 63, 67-9, 74, 76-9, 81-2, 84-6, 88-92, 95, 100, 118, 120-5, 132-4, 141, 144-6, 148-53, 155, 159-61, 163-8, 170, 173-4, 177, 179, 182-91, 196-7, 201-2, 217, 236-7, 239-42, 246, 248, 251-3, 256, 258-9, 261, 263, 269, 271, 273-4, 276, 279-81, 283, 285, 291-2, 294-8, 304-11, 313-4, 318-9, 322-3, 329, 333, 344, 349, 351-3, 355, 369-73, 386-7, 392, 398, 427-8, 434, 443, 446-7, 479, 481, 484, 486, 490-500, 505-9, 512, 514, 519, 529, 531, 538, 542, 550-1, 556-69, 572-82, 584-90, 592-4, 598, 599-600, 619, 621-2, 624-6, 628-9, 631, 638
nilpotent algebra, 81, 187, 190, 272, 285, 304, 306-7, 309, 538, 588
nilpotent differential operator, 145, 252
nilpotent Dirac equation, 77, 90, 177, 182-3, 556, 562-6, 569, 576, 582, 587, 638
nilpotent equation, 185, 240
nilpotent formalism, 79, 82, 151, 165-6, 184, 197, 201, 217, 239, 273, 294, 296, 298, 310, 314, 323, 352-3, 370, 484, 491, 493-4
nilpotent method, 188, 274, 285, 309
nilpotent operator, 76, 81, 84, 121, 151, 163-4, 167, 246, 248, 309, 434, 443, 492, 495, 529, 531, 573, 578
nilpotent quantum mechanical state, 556, 574, 581
nilpotent quantum mechanics, 185, 562, 579, 629
nilpotent quaternion state vector, 240
nilpotent representation, 125, 150, 153, 291, 297, 307, 344, 369, 498, 556, 564, 631
nilpotent state, 67-9, 79, 123, 125, 149-51, 163, 165, 253, 259, 261, 280, 318, 355, 369, 446, 579, 581-2, 626
nilpotent state space, 581, 582
nilpotent state vector, 67, 123, 125, 149-50, 163, 253, 259, 280, 318, 355, 446, 626
nilpotent structure, 100, 133, 144, 148, 155, 159, 165, 179, 188, 196, 202, 217, 251, 256, 279, 311, 313, 349, 372, 479, 481, 484, 490, 496, 499, 542, 551, 573, 575, 599, 619, 625
nilpotent theory, 246, 322, 565, 588
nilpotent universal computational rewrite system (NUCRS), 556-67, 571, 576, 591, 598, 600
nilpotent universe, 579, 588
nilpotent vertex, 269
nilpotent wavefunction, 49, 79, 144, 145, 150, 174, 295, 628
nitrogen, 515, 552, 554
nitrogenous bases, 516, 552
Noether current, 167
Noether's theorem, 37, 40, 44, 54-5, 167, 180, 258-9 624, 631
non-Abelian group, 101
nonalgorithmic, 47
non-Archimedean geometry, 47
nonassociative, 15, 17, 97
noncommutative, 15, 25, 96, 101-3, 171, 438, 443
nonconservation, 40-2, 47-8, 54, 60, 65, 74, 79, 84-5, 99, 102, 184, 187-8, 192, 195, 203-7, 218-9, 221-2, 226-7, 234, 236, 238, 250, 258, 442, 624
noncountable, 5, 50, 65, 93, 99-100, 102, 159, 313
nondecay of the proton, 631
nondimensional, 45, 48, 50, 65, 98, 224, 436, 438, 441, 528, 624
nonequilibrium thermodynamics, 165, 192, 237
nonexistence of supersymmetric particles, 632
nongravitational energy, 483
nongravitational field, 471
nongravitational force, 193
nongravitational interactions, 38, 217, 239, 292, 324, 393, 485
nonidentifiable, 54
nonidentity, 219
noninertial frame, 192, 469-71, 479, 601-2, 604, 610, 613
noninvariance, 219
nonlinear, 379, 452-5, 463, 471, 478, 479-81,

483
nonlocal, 54, 69, 145-6, 154, 202, 251, 312, 315, 319, 372, 443-4, 472, 509, 513, 602, 604-5
nonlocal connection, 145, 312, 443
nonlocal finite energy, 319
nonlocal force, 472
nonlocal gluon sea, 315
nonlocality, 60, 143, 160, 164-5, 176, 197, 201, 292, 310, 318, 392, 446, 478, 481, 490, 508, 510, 581, 622, 627
nonlogical, 5
non-Newtonian effects, 481, 483
nonorderable, 50, 95, 436, 438
nonpolar, 525
nonquaternionic, 124, 133
nonreal, 206
nonrelativistic physics, 175
nonrelativistic quantum mechanics, 31, 168, 174-5, 186, 190, 488
nonrenormalizable infinities, 472
nonstandard analysis, 49
nonuniform acceleration, 35
nonuniform velocity, 35
nonuniqueness, 50, 81, 219
nonzero, 1, 6, 40, 79, 81, 130, 141, 143, 152, 158, 259, 264, 284, 294, 296, 311-2, 315, 317, 320, 322, 343, 352, 369, 371-3, 395, 405, 411, 427, 432, 507, 632
nonzero charge, 395
nonzero ground state, 312
nonzero mass, 141, 143, 284, 352, 369, 371-3
nonzero total wavefunction, 311
nonzero universe, 81
normalization, 122, 144, 146-7, 152, 245, 285, 297, 316, 369, 434
nothing, 3-4, 6-7, 19, 31, 39, 74, 79, 84, 99, 131, 137, 166, 173, 186, 201, 222, 291, 310, 342, 346, 381, 414-5, 422, 436, 449, 451, 456-7, 463, 470, 479, 486, 558, 563-4, 568, 612, 618-9, 621-3
nothingness, 4, 19, 51, 85, 106, 443
nuclear fission, 589
nuclear forces, 320
nuclear fusion, 584
nuclear instability, 589
nuclear matter, 257
nuclear process, 589
nuclei, 143, 155, 158, 249, 432, 507, 511, 589, 592
nucleic acids, 510, 514
nucleon, 282, 404, 405, 616
nucleon density, 616
nucleotides, 515, 517, 544-6, 548, 552, 590
number, 1, 3, 5, 9-10, 12-13, 15, 19-21, 23-5, 36-7, 40, 42-3, 46, 48-9, 53, 57-8, 63, 66-8, 71, 74, 81, 83-5, 91-2, 95, 99, 102, 114, 130, 158-9, 172, 179, 193, 207, 211, 218, 220, 223, 225, 229, 232, 237, 258, 264-5, 277, 280, 290-1, 310-1, 313, 316, 330, 337, 342, 347, 357, 366, 371, 374, 378, 383, 387, 391, 393-4, 396, 401, 403, 412, 417, 434, 439, 442, 456, 484-5, 489, 494, 497, 503, 506, 508, 512, 516, 519, 522, 528-30, 533, 536, 552-3, 558, 562, 565-9, 575, 586, 598, 600, 608, 610-1, 613-4, 616-8, 627-8, 630-2, 633-5, 637-8
number system, 10, 23, 67, 84, 114, 489, 558
number theory, 586
numbering, 6, 10-11, 67-8, 83, 415, 503, 559, 600, 635
O(3), 259
object, 18-20, 46, 95, 99-100, 103-4, 115, 174, 187, 200, 202, 320, 387, 414, 417, 427, 429-30, 435, 440, 443, 457-8, 472-3, 478, 495-6, 499-500, 506, 511, 514, 519, 537, 559, 574, 582, 588, 596-8, 603, 606, 619
object dual, 619
object symmetry, 96, 103
observability, 33, 178
observable, 46, 56, 60, 171, 175-6, 192, 203, 310-1, 314, 341, 347, 353, 388, 410, 444, 452, 455, 472, 475, 482, 489, 490, 497, 604, 618, 621
observation, 5, 32, 36, 59, 64, 83, 174, 181, 191, 203, 220, 347, 411, 464, 471-2, 498, 550, 583-4, 600, 633-4, 638
Observation Dual, 100
observational data, 608-9
observational physics, 210
observed, 16, 21, 36, 49, 55, 58, 65, 100, 144, 158, 179, 197, 199, 204-5, 218, 224, 228, 234, 266, 310, 313, 320-1, 341, 346-7, 364-5, 377, 379, 381, 389, 397, 403, 405, 411, 445, 458, 464, 480, 498, 504, 529, 531, 535, 551-2, 564, 587, 597, 604, 606-7, 609, 611, 620-1, 631

observer, 100, 181, 196, 199, 490, 559, 568, 603-5
octahedron, 527-8, 533-55, 537-9, 546, 549
octonions, 11, 15, 17, 43, 96-8, 103-4, 110, 387, 492-3, 496
Olbers' paradox, 611
one gluon exchange, 252, 279
one-electron nuclear atom, 244
one-handedness, 368, 513
one-way speed of light, 199-200, 431
ontology, 435, 452, 472, 576, 591,600, 623, 625
open system, 165, 292, 507
opposed properties, 485
opposition, 51, 58-9, 61, 83, 222, 632
optical refractive index, 457
Opticks (Newton), 57, 419
optics, 36
optimal computation, 567
optimal control, 576
optimal design, 571-2
orbit, 192, 404, 416-7, 420, 457, 459-61, 464-6, 475, 477, 510
orbital angular momentum, 138, 258
orbital energy, 472
orbital motion, 191-2
orbital radius, 60
order (of group), 11-17, 50-1, 66-9, 95-7, 100-3, 387, 503, 536-7, 623
order (within systems), 237, 502-3, 510, 512, 531
order of events, 179, 196, 199, 202, 226-7, 234, 310-1, 489-90, 582
order of events, 179, 196, 199, 202, 226-7, 234, 310-1, 489-90, 582
orderable, 42, 45, 50, 95, 436, 438, 442, 547, 580
ordered pairs, 15, 103
ordering, 5, 7, 165, 406, 489, 551, 570, 579, 600
ordinal, 1, 9, 11, 21, 23-4, 84, 565, 567
ordinal numbers, 565, 567
ordinality, 5, 10, 23-4, 623
organic, 320, 531, 554
organizational principle, 560
organizing principle, 552
origin (graphical), 14, 48, 65, 107, 246, 485-6, 635
origin of the universe, 58-88
origins, 2, 4, 32, 34-5, 53, 58, 61, 67-8, 72, 80, 84-5, 103, 111-3, 158, 174, 184, 195, 201, 213, 218, 221-3, 230, 255-7, 265, 294, 304, 315, 323, 338, 342, 345, 361, 365, 370, 395, 406, 414, 418, 422, 428, 430, 433, 441-2, 489, 492-3, 502, 508, 510, 513-4, 562, 575, 581, 587-8, 603, 605, 607, 61-1, 617, 620-1, 628-9
orthogonal axes, 145, 635, 373
orthogonal Clifford algebras, 186
orthogonal hyperplanes, 635
orthogonal permutations, 356
orthogonal projections, 164
orthogonal transformations, 356
orthogonality, 202, 249, 265, 316
orthonormal, 111, 134
Osborne, A. D., 191-2, 202, 447
outer product, 69, 82
oxygen, 590, 593
p-adic representation, 634
pair production, 148, 278, 304, 431
paradigm, 20-1, 572, 602
parallel agent, 86
parallel computation, 86
parallel universes, 69, 583, 587
parameter, 20, 32, 36, 38-9, 50-1, 53-5, 69, 80, 83, 85, 88, 92, 95, 99, 102, 105, 107, 110, 137, 151, 172, 180, 191, 193-4, 199, 203, 217-8, 225, 235, 260, 265, 311, 345, 391-2, 400, 408-9, 437, 441, 445, 451-2, 481-2, 487, 492, 499, 561, 585, 596, 603, 624-5, 627, 631
parameter group, 55, 69, 83, 105, 107, 137, 180, 191, 193, 217-8, 437, 441, 445, 487, 492, 624-5, 631
parameter space, 80, 85, 441, 624-5
parameterization, 137, 408, 436, 443, 575
parameterization of nature, 137, 443
parities of ground-state baryons and bosons, 151
parity, 140-1, 156-7, 205, 259, 261-3, 266, 300, 314-5, 326-7, 341-4, 353, 358, 361, 365, 370, 372, 396, 408, 410, 494, 626
parity operator, 141
parity transformation, 140, 156, 372
parity violation, 342, 358, 361
particle, 36, 38-9, 41, 45, 49-50, 53, 55, 57-8, 64-5, 75, 77, 84-5, 90-1, 93, 95-6, 103, 111, 115, 117, 123, 133, 139, 141-2, 144,

149-50, 159-61, 166, 174-6, 181, 184, 187, 189, 192, 194, 205, 211, 216, 218, 220-1, 223, 226, 230-4, 246, 255-7, 261, 264, 268, 270, 272-3, 284-6, 291-2, 311, 318, 320-1, 324, 328-9, 338-45, 347-8, 350-1, 358, 364-7, 369, 372, 374-5, 377-8, 381, 387, 394, 396-406, 410-2, 418-21, 427-9, 431-4, 437-9, 445-8, 453-4, 457-61, 463, 465, 468, 473-5, 480, 484, 488, 491-3, 496, 500, 508, 514, 519-21, 537, 550, 560, 563-4, 566, 570, 572, 575-8, 580-1, 583, 586, 588-9, 592, 600, 607-8, 611, 613, 616-7, 621, 625, 627-8, 631
particle accelerators, 631
particle charge, 268, 329
particle creation, 318
particle density, 616
particle equation, 90
particle mass, 341, 374, 404-5, 411
particle physics, 36, 38, 41, 58, 65, 115, 133, 149, 311, 339, 347, 394, 410, 420, 437, 445, 454, 493, 521, 537, 560, 563, 566, 572, 581, 586, 608, 621, 631
particle state, 50, 111, 123, 161, 187, 189, 246, 324, 340, 345, 405, 437, 492, 581
particle structure, 55, 187, 256, 284, 377, 484, 576
particle theory, 234
particle wavefunction, 41, 90, 211, 446, 448
particle-like, 176
partitioning of vacuum, 147-8, 313, 320, 400, 406, 506
partitioning of Higgs mass, 412-3
path integral, 167, 569
Pauli, 13, 16-17, 30, 78-9, 81, 113-4, 118, 125, 143-4, 151, 164, 166, 178, 238, 312, 315, 321, 337, 340, 423, 506, 510-2, 557, 562, 581-2, 625, 627
Pauli exclusion, 79, 81, 143-5, 151, 164, 166, 179, 238, 312, 315, 321, 337, 506, 510-2, 562, 581-2, 625, 627
Pauli matrices, 13, 16-17, 30, 113, 118, 125, 340, 423
Penrose, R., 109-10, 145, 228, 528-9, 585
Penrose tiling, 109-10, 145, 529
pentad, 70-2, 110, 124, 337, 345, 385-6, 521, 529
pentagon, 541, 545-7
pentagonal, 502, 529-30, 542-50, 554
pentagonal symmetry, 502, 529, 554
pentaquarks, 155
pentatope, 542
pentose, 547, 554
perception, 64, 415, 538, 561, 567, 598, 629
periastron precession, 445
perihelion precession, 463, 466, 469-70, 510
periodic potential, 246
perturbation calculations, 407
perturbation expansion, 157, 286-7, 290, 309, 408
perturbation series, 288
perturbation theory, 285, 393, 626
perturbative calculation, 627
perturbative gauge theory, 494
perturbative method, 301
Perus, M., 577
Peskin, M. E., 348
phase, 41, 55, 75, 115-7, 119-20, 122-5, 133, 144, 152-5, 163-6, 172, 178-9, 184, 187-9, 204, 219, 221, 224, 228, 236-7, 242, 245, 247-8, 250-4, 256-7, 260, 266, 274-7, 279, 281-3, 313-7, 321, 325-6, 329-30, 337-8, 341, 346-7, 350-2, 359, 361-4, 369, 373, 377, 382, 387-8, 393, 400, 405, 409, 412, 432, 436, 442, 458, 481-2, 491-2, 495, 497-9, 507-8, 511-2, 529, 537, 551, 555, 560, 568, 573-5, 578, 579, 582-3, 585-9, 592, 594-5, 596-8, 600, 619, 625
phase action, 587
phase angle, 165
phase changes, 41, 55, 204, 351
phase conjugate adaptive resonance, 588, 600, 619
phase conjugation, 508, 582, 588, 592, 598
phase cycle, 260, 330
phase difference, 322, 574, 588
phase information, 352
phase of quantum coherence, 587
phase rotation, 393
phase space, 116, 178-9, 388, 436, 482, 491, 495, 497-9, 529, 586, 600
phase space trajectories, 586
phase transformation, 205
phase transition, 155, 184, 236-7, 257, 314, 321, 511-2, 537, 555
phase velocity, 458
phaseonium, 578-81, 583, 585, 592-3, 595
phases of matter, 314

phenomenological, 388, 391, 394, 398, 399, 400, 407
phenomenology, 348, 379, 391
phenylalanine, 538
Phi, 539-41
philosophy, 82, 442
phosphate, 515-6, 544, 555
phosphorus, 554
photon, 144, 148, 201, 215, 228-31, 270, 282, 288-90, 295-7, 302-6, 346, 363, 367, 418-21, 427, 430, 448-9, 453, 457, 459-61, 470, 488-9, 497, 578, 601, 603, 616
photon / electron ratio, 616
photon / proton ratio, 616
photon emission, 601
photon gas, 418, 430
photon momentum, 303
photon polarization, 448
photon propagator, 295-7, 306
photon radiation, 603
physical law, 36, 55, 57, 64, 220, 222, 559, 569-70
physical universe, 74, 79, 82, 86, 440
physical world, 5, 74, 234, 563, 566, 576, 629
physics, 1-5, 9, 11, 14, 24, 27, 32-41, 47, 53, 56-64, 71-7, 79, 82-6, 88, 98, 100-1, 104, 111, 137, 145, 151, 170, 177-8, 181, 183, 187, 190-2, 195-6, 203, 206, 208, 218-9, 225, 231-2, 234-5, 258, 310, 312, 386, 411, 414-6, 421, 429-30, 433, 435-6, 441-3, 449, 453, 462, 475, 478-9, 481, 484, 495, 499, 502-3, 505, 509-11, 513-4, 529-31, 537, 548, 553, 555, 559, 563-6, 569, 572, 581, 586, 590, 600, 605, 610, 618-20, 622, 624, 629, 631
pion, 44, 155, 257, 282, 346, 364
Planck, M., 430
Planck black body spectrum, 229
Planck constant, 55
Planck energy, 606
Planck length, 195, 394
Planck mass, 195, 217, 292, 309, 348, 374, 380, 382, 390, 392-4, 397, 407, 412, 480, 482, 627, 631
Planck radiation law, 581
Planck time, 195
plane wave, 117, 170, 247
plane wave solution, 117, 170
planet, 192, 462, 464-5, 510, 584-5, 617
planetary orbit, 463, 469, 476
planetary perturbations, 465
plasma, 397, 420, 460
plate tectonics, 610
Platonic solids, 502, 514-5, 527-9, 539, 547-8
Poincaré, H., 27
Poincaré group, 75, 491
point particle, 508
point source, 145, 239, 241, 246, 259, 282, 299, 323, 351-2, 432, 444, 452, 456, 480, 499, 531
point-like, 147, 323, 394, 611
Poisson brackets, 80
Poisson equation, 211, 450
polar, 111, 181, 241, 246, 249, 252, 272, 275, 450, 456, 466, 525
polar angles, 249
polar coordinates, 111, 241, 246, 252, 272, 275, 456, 466
polarity, 522, 525, 541
polarization, 291
pole, 294-5, 297, 509
polynomial potential, 277, 625
polypeptide, 516-7, 553
Pope, N. V., 191-2, 202, 447
Popper, K. R., 481
position operator, 175
position vector, 181
positive feedback, 553
positive parity, 157
positron, 44, 136, 162, 290, 294, 302, 318, 611
positrons, 91
potential, 41, 55, 69, 84, 87, 115, 149, 154, 159-60, 172, 174, 205, 208, 211, 216, 230-2, 239, 241-2, 246, 248, 251-2, 254-6, 271-2, 274-5, 276-80, 282, 285, 288, 290-1, 297, 299, 313-5, 351, 417-22, 424, 427-30, 432-42, 460, 464-7, 476, 554, 568, 612
potential energy, 149, 159-60, 172, 174, 208, 216, 230-2, 241-2, 248, 252, 274, 276, 280, 417-22, 424, 427-30, 434-6, 441-2, 464, 467
potentially infinite, 13, 86, 88, 591, 600
Poynting flux, 477
Poynting theorem, 215
Poynting vector, 475, 482

Poynting-Robertson effect, 449
precession of gyroscopes, 475
prediction, 60, 89, 339, 341, 380, 393, 398, 402, 407, 432, 448, 452, 457, 557, 563-4, 576, 592, 600, 627-32
pre-spatial quantum mechanical communication, 584
pressure, 30, 417-9, 435, 449, 603
pressure-density relation, 417
Pribram, K. H., 576, 585, 596
Principia (Newton), 192, 227, 465
principle of equivalence, 208, 449-50, 457, 461, 473, 475, 600, 602
principle of general covariance, 450
principle of least action, 561, 586
principle of relativity, 196, 198
probability, 91, 119, 131-2, 176, 181, 204, 215, 219, 221, 264, 299, 307, 308, 595, 616
probability amplitude, 176
probability density, 119, 132
probe and response, 53, 64, 100, 218, 436
product state, 190
production, 18-20
production system, 18
program, 18, 87, 557, 638
projection operator, 187, 567
prokaryote, 592
proline, 538
proofreading, 558, 570, 577, 591
propagator, 133, 294, 295, 297, 302, 306-7
proper energy, 121, 311, 573
proper space, 488
proper time, 76, 121, 196, 201-2, 311-2, 488-90, 500, 506, 542, 568, 570, 573, 582, 600, 607, 624
protein, 510, 515-7, 524, 533, 546, 548, 550, 553, 555
protein-nucleic acid interactions, 524
proton, 44, 248, 256-7, 292, 343, 381, 392, 404-5, 512, 551, 589, 616, 618, 631
proton decay, 381, 392, 631
proton density, 616
proton mass, 404-5, 616
proton number, 248
proton spin, 343
proton-to-neutron conversion, 617
Prusinkiewicz, P., 18
pseudo-boson, 496
pseudoscalar, 16-18, 29-30, 32, 43, 53, 63, 66-7, 70, 73, 76, 80, 114, 121-2, 132, 140, 183, 201, 239, 257, 260, 265, 280-1, 283, 313-4, 317, 320, 329, 351, 365, 370, 403, 443, 487-8, 503-5, 509, 515, 518, 521, 531, 537, 542, 546, 563, 566, 623-4, 626
pseudoscalar phase, 281, 329
pseudovector, 16, 29-30, 44, 80, 114, 132, 284, 327
psychology, 598
PT symmetry, 345
pulsar orbit, 463
purine, 525, 552
pyrimidine, 525, 552
pyrrolysine, 530
Pythagorean addition, 26, 32, 42-3, 45, 226, 443
quantization, 72-3, 76, 78, 90, 94, 123, 137, 166-7, 189, 194, 202, 225, 231, 236, 427, 447, 484, 489, 496-7, 499, 572, 583, 607, 628
  first quantization, 189
  second quantization, 78, 122, 162-3, 189, 285
quantization of angular momentum, 73
quantization of energy, 194
quantum (coherent) self-interference, 575
quantum annihilation, 577
quantum Carnot engine (QCE), 565, 578-96, 600
quantum causality, 201
quantum chaos, 442, 582, 593, 600
quantum charge clouds, 319
quantum chemistry, 246, 248
quantum chromodynamics (QCD), QCD, 112, 153, 252, 285, 305, 307, 309, 341, 355-6, 363, 404-5, 407, 494
quantum coherence, 564, 566, 568, 573, 575, 578-9, 581-2, 585, 587-9, 591, 593
quantum computation, 84, 86, 559, 561, 594
quantum computer, 82
quantum computing, 5, 24, 82
quantum cosmology, 581
quantum creation, 577, 580
quantum differential operator, 169, 375
quantum duality, 95
quantum electrodynamics (QED), 41, 166, 215, 232-3, 250, 271, 285-6, 290, 292, 296-7, 347, 305-6, 309, 347-8, 391, 404,

480, 497-501, 626-8
quantum event, 201, 489, 579
quantum fermionic state, 244
quantum field, 78, 89-91, 95, 112, 123, 137, 141, 148, 157, 161, 163-4, 166, 172, 187, 189, 202, 211, 215, 229, 291, 306, 313, 357, 453, 478, 480, 491, 501, 566
quantum field integral, 123, 157, 162-3, 291, 306
quantum field theory, 89-91, 112, 123, 141, 162, 167, 172, 202, 215, 229, 357, 480, 491, 501
quantum flavour dynamics (QFD), 285, 305, 307, 309
quantum fluctuation, 3, 319
quantum gravitational dynamics (QGD), 480
quantum gravitational inertia, 481, 484, 493, 495-6, 499
quantum gravity, 196, 217, 292, 348, 374, 382, 392, 453-5, 493, 495-8, 564, 627
quantum Hall effect, 82, 143, 158, 315-6, 347, 432, 435, 507, 632
quantum harmonic oscillator, 424
quantum holographic image encoding / decoding, 574, 588
quantum holographic pattern recognition, 556
quantum holographic phase, 592
quantum holographic signal processing, 568
quantum holography, 556, 574-7, 581, 594, 600
quantum inertial dynamics (QID), 480
quantum information, 82
quantum logic gates, 577
quantum measurement, 559, 563, 568, 595
quantum mechanical computation, 577
quantum mechanics, 2, 23, 31, 33-6, 38, 40-1, 44-6, 48-50, 53, 56, 59, 65, 70, 85, 92, 94, 98, 100, 112, 122-3, 132, 158, 165-6, 168, 172, 176, 180, 182-3, 186, 188-9, 191-2, 201, 207, 218-9, 221-2, 226-7, 230, 232-4, 311, 329, 344, 424, 429-30, 436, 438-40, 444, 446, 452, 464, 481, 490, 497, 507, 509, 514, 531, 556, 561, 564, 566, 568-70, 572, 575-7, 580, 583-4, 587, 622, 624, 629-30
quantum metric, 482, 497-9
quantum nature, 221, 483
quantum nilpotent, 493, 585
quantum nonlocality, 200, 448
quantum of action, 580
quantum operator, 121, 123-4, 128, 132, 172, 431, 508
quantum phase, 574, 590, 592
quantum phase encoding, 592
quantum physical process, 559
quantum physics, 32, 41, 75, 79, 111, 131, 177-8, 217, 232, 427, 436, 556, 564-5, 568, 572, 586, 592
quantum potential, 181
quantum preparation, 563, 586, 595
quantum process, 200, 239, 489
quantum representation, 111
quantum state, 68, 73, 76, 163, 206, 490, 574-6
quantum system, 89, 172, 178, 192, 201, 219-21, 226, 568, 573, 585, 594, 625
quantum teleportation, 584
quantum theory, 69, 191, 224, 416, 452-3, 478, 480, 489, 565
quantum thermodynamics, 557
quantum transition, 179, 509
quantum tunnelling, 584
quantum uncertainty, 218
quantum universal computer constructor model, 584
quantum vacuum, 568, 580
quantum world, 489
quantum-classical boundary, 192, 218
quantum-classical transition, 100, 168, 177-8, 180, 191, 329
quark, 104, 137, 152-4, 158, 238-9, 251-2, 254-6, 258, 260, 262-3, 266-7, 270, 278, 284, 291-2, 305, 316-7, 321, 324-30, 332, 336-48, 350, 353, 355, 357-9, 363-6, 368,-9, 371, 379-80, 386-7, 391, 394-413, 433, 485, 512, 551, 611, 626-7, 632
quark colour invariance, 344
quark confinement, 329, 368, 394-5
quark doublet, 397
quark mass, 256, 396-8, 406-7, 409-10, 413
quark mixing, 410
quark pressure formula, 404
quark propagator, 305
quark representations, 379
quark structures, 326, 379, 405
quark tables, 334, 337, 340, 358, 363, 401
quark wavefunction, 153
quark-antiquark pairing, 359, 399

quark-antiquark potential, 252
quark-gluon interaction, 305
quark-gluon plasma, 158, 433
quark-lepton transitions, 348
quark-quark force, 512
quasars, 199, 490
quasicrystals, 531
quasi-localised, 446
quaternion, 5, 10, 12-13, 15, 17-18, 24-9, 31-2, 43-4, 50, 52-3, 63, 66-7, 70-1, 73-6, 78, 80, 88, 92, 94-8, 101-3, 106-7, 110, 112, 117, 122, 124-39, 146, 151, 158, 160-1, 171, 182, 194, 201, 206, 223-4, 260-2, 264, 291, 310-1, 320, 325, 327, 334-5, 337-8, 343, 345, 350, 357-9, 362-7, 378-9, 382-7, 427-8, 434, 443, 446-7, 454-5, 479, 486-8, 490, 492-4, 503-5, 515, 519-21, 542, 546, 563, 580, 623-4, 626, 633-5, 637-8
quaternion algebra, 26, 28, 71, 76, 95, 130, 383-4
quaternion charge description, 367
quaternion nilpotent algebra, 127
quaternion operator, 15, 44, 53, 63, 71, 75, 80, 88, 92, 94, 117, 125, 128, 130, 138, 160, 194, 201, 262, 310-1, 320, 325, 327, 337-8, 345, 350, 359, 378, 385, 428, 443, 493
quaternion space, 145, 373, 600, 634-5
quaternion state vector, 78, 124-5, 158, 161, 291-2, 428, 434, 447
quaternion tesseral addressing, 634
quaternion wavefunction, 92, 345
quaternionic matrices, 125, 446, 490
qubits, 82
quintessence, 601, 609
R3, 112, 404, 618
R4, 371
radar pulse, 462
radial field, 452
radial gravitational field, 480
radial inertial field, 602
radial source, 111
radiant energy, 468
radiation, 229-33, 235, 238, 304, 418, 420, 429-30, 435, 439, 461, 476, 559, 580, 583, 600-1, 603, 609, 611, 618, 628
radiation field, 231, 238
radiation pressure, 418, 420, 430, 583
radiation reaction, 430, 435
radioactivity, 36, 610
Ramon y Cajal, S., 585
random directionality, 275, 443
random fluctuation, 218, 222
random rotation, 395, 406, 409, 413
random variation, 220-2, 337
randomness, 236, 407, 409, 520
rational numbers, 10
Rayleigh-Jeans radiation law, 229
read / write filter bank and memory, 595
read / write quantum holographic memory, 593
real, 5, 10, 12, 15-17, 23-30, 32, 38, 43-5, 47, 49-51, 60, 63, 65-8, 70, 73-4, 79-82, 85, 93, 95, 98-103, 105, 107, 112, 115-6, 121, 147-9, 157-9, 161, 176, 181, 187, 196, 202, 206, 210, 213, 220-1, 223, 225, 231-3, 242, 254, 257, 272-3, 275, 279-80, 282, 292, 304, 312-3, 321-2, 342, 344, 352, 368, 377, 382, 386-8, 397, 404, 417-8, 424, 433-4, 436-8, 441-2, 453, 455, 467, 475, 480, 489-90, 492, 494, 497, 499, 503, 508, 511-3, 528, 531, 550-1, 559, 573-4, 576, 596-7, 601-2, 606-7, 611-3, 616, 618, 621, 623-4, 626, 629-30, 634
real numbers (reals), 5,-6, 10, 23-5, 47, 49, 66, 69, 79, 83, 161, 210, 232, 433, 558, 574
real particle, 60, 159, 181, 513, 612-3
real particle trajectories, 181
real world, 596, 597
realisation system, 18
reality, 4, 33-4, 46, 56, 59-62, 64, 112-3, 116, 175, 223, 233, 325, 333, 415, 422, 485, 489, 508, 540, 563, 607
reciprocity, 431, 459
recountability, 47
rectangle, 29, 99, 415, 438-9, 529, 539
recursion, 19, 571
recursive, 20-1, 81, 530, 619
red blood cells, 593
redundancy, 112-3, 116, 298, 494
redundancy barrier, 113, 298
re-emission, 418
reference frame, 573-5, 596
reference phase, 573
reference space, 450
reflection, 47-8, 79, 107, 132, 148-9, 205,

220-1, 297, 312, 417
refractive index, 457-8, 461-2
Regge trajectories, 403
relative motion, 197-8
relative nature of rotational motion, 476
relative phase, 316, 320, 330, 506, 568, 585, 593
relative time, 226, 311
relative velocity, 197
relativistic almost perfectly balanced phaseonium (rapbp), 583-4
relativistic Doppler shift, 233
relativistic energy, 76, 418, 427-8, 463
relativistic inertia, 481
relativistic mass, 174, 427-8, 430, 443, 480
relativistic quantum mechanics, 112, 133, 184, 443, 630-1
relativistic time, 201
relativity, 27, 31, 38, 42, 50, 59, 68, 76, 92, 131, 165, 191, 195-6, 199-202, 210, 225, 227, 230, 232-4, 310-2, 348, 415-6, 419, 421-2, 430, 435, 440-1, 448-52, 455, 457-8, 462, 464-5, 472, 480-1, 483-4, 489-90, 497, 531, 564, 601, 618
relativity of knowledge, 234
Renaissance art, 539
renormalization, 112, 137, 147-8, 157, 285, 290, 301, 303-5, 309, 314, 320, 353, 374, 380, 431, 433-4, 606, 619, 626
renormalization group, 304, 619
renormalization group equation, 304
reordering, 623
replication, 502, 512, 515, 559, 592
repulsion, 454, 475, 510
repulsive force, 480, 605, 628
rescaling, 285, 374, 410, 626
Resconi, G., 571
resonance states of the sphere, 527
resonances, 405
rest energy, 474
rest frame, 199, 229, 290, 473, 537, 622
rest mass, 49-50, 54, 68, 73, 75-6, 89, 137, 145, 159, 172, 174, 179, 196, 201-3, 239, 263-4, 311, 313, 317, 319, 329, 366, 395-6, 402, 419-20, 427-8, 430-1, 439, 447, 459-60, 463, 468, 488, 491, 504-5, 510, 513, 573, 607, 624-5, 628
rest of the universe, 68, 74, 145, 149, 165, 179, 236, 312, 320, 431-2, 491, 506-9, 511, 513, 537, 555, 573, 579, 611, 629
retardation, 236, 237, 464
retarded wave, 430
reverse coupling, 158
reversed time, 295, 506
reversibility, 46, 49
reversibility paradox, 49
rewrite algebra, 77
rewrite language, 573
rewrite language description, 573
rewrite rules, 18-20, 22-3, 87, 557, 558
rewrite system, 1, 4, 10, 18-20, 22, 67, 86-7, 101, 159, 487, 503, 509, 514-5, 528, 530-1, 537, 556-60, 562-3, 565-7, 569-70, 572-3, 577, 581-2, 586, 590, 593,-4, 598-9, 629, 633, 638
ribose, 515, 543-4
Ricci tensor, 450, 456
Riemann hypothesis, 568, 581-2, 586
Riemann zeta function, 442, 557, 581, 586, 600
Riemann-Christoffel tensor, 450-1, 456
Riemannian geometry, 184
Riemannian metric, 184
right ideal, 186, 188
right-handed, 140, 143, 154, 171, 205, 263, 265, 268-9, 278, 314, 317-8, 327, 329, 359, 363-4, 367-8, 370, 373, 375, 389, 394-6, 410, 510, 516, 626-7
right-handed spinor, 171
right-product octonions, 97
ring (algebra), 565
ring (chemsitry), 544-5, 547, 552, 554-5
RNA, 502, 510, 512, 514-7, 536, 548, 550, 553-4, 560, 590-593
Robinson, A., 10, 47, 66, 85
robotic machine, 595
Roemer, O., 471
root mean square velocity, 417
root space, 404
root space diagram, 404
root vector geodesic, 404
Roscoe, D., 605
Ross, D. K., 463
rotating vector, 330
rotation, 37, 40-1, 44, 54-5, 72, 80, 85, 95-6, 101-2, 112-3, 133, 139, 145, 187, 197, 209, 215, 219-22, 224, 226, 229-30, 234, 238, 256, 258-9, 315, 322, 330-1, 335,

337, 347, 379, 352, 387-8, 407, 409-10, 412, 424, 439, 469, 472, 475-6, 479-80, 486-7, 509, 551, 572, 605, 635, 638
rotation asymmetry, 41, 44, 219
rotation curves, 605
rotation group, 424
rotation invariance, 226, 229
rotation of the coordinate system, 479
rotation of the local coordinate system, 480
rotation symmetry, 40, 54, 145, 197, 215, 219, 221, 238, 258
rotation symmetry of space, 54, 197, 215, 221, 258
rotational energy, 230
rotational motion, 476
rotational symmetry, 96, 238, 335, 387
row vector, 120, 132, 146-7, 240, 243, 300, 427, 636
Rubik cube, 551
Rueda, A., 483
running coupling constant, 390
running values, 407
Rutherford scattering, 301
Rydberg constant, 245
S duality, 631
s quark, 402-3
Santilli, R. M., 11, 151, 514
saturated potential, 257
scalar, 13, 16-18, 26, 28-30, 32, 41, 43-4, 52-3, 63, 66-7, 70, 73, 76-7, 80, 92, 94, 96, 98, 118-21, 125-6, 131-2, 142-6, 148, 152, 158, 161, 166, 168, 171-3, 178, 183, 185, 194, 201, 204, 210, 213, 216-7, 225, 239, 249, 251-2, 256-7, 259-60, 265-6, 268-9, 271-3, 280, 282, 284, 286-7, 294, 296-7, 300, 305-6, 308-9, 313-4, 317, 320, 329, 343, 345-6, 349, 351-3, 366, 371, 374, 376, 381, 392, 423, 427-8, 443, 453-4, 474, 487-8, 495-7, 501, 503-5, 507-8, 512, 521, 542, 546, 563, 566, 623-6
scalar boson, 156, 314
scalar field, 225
scalar phase, 148, 194, 217, 252, 256, 280, 282, 305, 351-2, 374, 376, 381, 392
scalar phase invariance, 217
scalar phase term, 256, 282, 305, 376
scalar potential, 41, 210, 216, 259, 351, 454, 474
scalar product, 26, 28-30, 44, 120, 126, 132, 142, 201, 249, 296-7, 300, 308-9, 345, 346, 366, 423, 453, 488
scalar propagator, 495
scalar wavefunction, 121, 297, 343
scaling, 183, 191-5, 203, 207, 210, 304, 309, 393, 406, 635, 638
scaling factor, 406
scaling mechanism, 309
scaling parameters, 193
scaling relations, 183, 191, 194-5, 207, 210
scattergraph, 547, 548
scattering, 299-300, 303
scattering matrix, 299
Schempp, W., 556, 574-7, 594
Schiff, L. I., 459, 463
Schrödinger, E., 233, 630
Schrödinger equation, 30-1, 89, 113, 171-4, 176, 181, 183, 189, 195, 222, 229-30, 246, 423-4, 426-430, 441, 576
Schrödinger's wave mechanics, 50, 93-4, 100, 103, 168, 171-7, 180-1, 188-90, 231, 233-4, 429-30, 438-40, 488, 584
Schwarz distribution, 567
Schwarzschild boundary, 482
Schwarzschild limit, 420
Schwarzschild metric, 457, 470, 499
Schwarzschild radius, 419, 458
Schwarzschild solution, 111, 444, 452, 456, 463, 465, 468, 470, 499
Sciama, D. W., 601
science, 2-3, 33, 46, 99, 557, 560, 585, 596, 598, 608
scientific revolution, 481
Scully, M. O., 578, 583
secondary sources, 569-71, 583-4, 587
second-order coupling, 289
self-action, 623
self-adjoint, 586
self-aggregation, 502
self-assembly, 521
self-conjugated, 24
self-dual, 165, 187, 190, 443, 495, 542, 619
self-dynamism, 100
self-energy, 141, 290, 298, 304
self-fulfilling prophecy, 478
self-interaction, 74, 271, 314, 352-3
self-organization, 321
self-referencing, 23-4

self-referential cosmology, 619
self-replication, 537, 592
semantic, 556-65, 570-1, 576, 586, 591, 593-8, 629
semantic code, 560
semantic computation, 561-3, 571
semantic language description, 558, 586
semantic machine, 559, 598
semantic mathematical language description, 559, 562, 576
semantic ontological level, 595
semiotic, 559-60, 562, 591, 597
sentient being, 556
serial, 86, 596
serine, 517, 521
set theory, 558
SETI, 584
shell, growth of, 514, 531-2, 548
Sierpinsky fractal triangle, 535
sign ambiguity, 506, 509
signal exchange, 561
signalling, 200, 228, 236, 479
simple harmonic oscillator, 275, 589
simplicity, 34-5, 61-2, 122, 451, 503, 555, 561, 620
simply-connected space, 320
simulation, 45, 106
simultaneity, 196, 201, 234, 489, 579
single-handed, 278-9, 309, 315, 439, 626
singlet state, 257, 364, 403
singularities, 111-3, 116, 133, 145, 294, 321-2, 436, 455, 472, 477, 480, 490, 531, 620
six gluon interaction, 112
Skolem, 10, 47, 66, 74, 85
Slater determinant, 69, 82, 143-4, 158
software, 4, 18
solar system, 192, 510, 584
Soldner, J. von, 421, 461
solid state, 238, 321
soliton, 600
solution space, 582
something from nothing, 53, 65, 84, 99-100, 414, 436, 443
sound field, 587
space, 2, 14, 16, 25, 27, 32-3, 35-55, 57-69, 72, 74-6, 79-85, 88-100, 102-5, 109-10, 113, 116, 120, 125, 130, 145, 150, 154, 165, 167, 172, 175-6, 178-9, 184, 188-9, 192-206, 210-2, 214, 218-29, 231-2, 234-5, 238, 251, 258-9, 261-2, 277, 283, 299, 311, 313, 316, 319-22, 329, 333, 344, 353, 371, 386-8, 393-4, 396, 404, 415, 418, 420-1, 424, 427, 429, 432, 436, 437-41, 443-4, 447, 449-52, 454, 456, 459-60, 464, 468-99, 503-5, 507-8, 514-5, 519, 527-9, 531-2, 537, 542, 546, 548, 552-3, 556, 559-60, 562-4, 566, 568, 570, 573, 574-7, 579-89, 591-2, 595, 597-8, 600, 605, 607, 610-1, 613, 620-1, 623-5, 628-9, 634-5, 637
space coordinates, 150, 205
space derivative, 89, 154, 283
space reflection, 50
space vector, 210
space*, 93-4, 96, 193
spacelike, 46, 50, 76, 89-90, 93, 116, 232, 429, 438-9, 488, 515
space-time, 27, 35, 43-4, 47, 51, 54, 59, 89, 98, 110, 125, 130, 184, 188, 193, 198-9, 201, 212, 219, 221, 224, 226, 262, 322, 344, 371, 387, 429, 444, 449, 451-2, 454, 456, 468-72, 475, 477, 479-82, 484, 488, 490-3, 496-7, 553, 559-60, 563-4, 566, 570, 573, 575-7, 579-80, 583-9, 591-2, 595, 610-1, 628
space-time anisotropy, 184
space-time curvature, 451-2, 456, 480, 497
space-time manifold, 450
space-time signature, 125, 130, 262
spatial image processing, 577
special relativity, 50, 76, 93, 166, 177, 196, 199-202, 231-3, 235, 311, 419, 422, 430-1, 450, 455, 459, 464-5, 467, 473, 482-3, 488, 490
specific heats, 229
specification, 17, 60, 79, 86, 124, 144, 202, 236, 340, 343, 352, 492, 498, 507, 582
spectral density of the zero point field, 396, 482
spectrin repeats, 550
speculation, 589
speed of changes in gravitational potential, 444
speed of gravitational waves, 444, 448
speed of gravity, 444, 448
speed of light, 196, 201, 228, 311, 431, 448, 453, 455, 470, 601
Spencer-Brown, G., 186, 572

sphere, 18, 396, 456, 603, 605, 612
spherical astronomy, 419
spherical curvature, 404
spherical harmonic oscillator, 274, 625
spherical harmonics, 248-9
spherical symmetry, 239, 241-2, 246, 251-2, 254-5, 259, 274-277, 280, 323, 351-2, 432, 497, 508
spherically symmetric field, 456
spherically symmetric orbit, 463
spherically symmetric potential, 244, 246, 273, 442
spin, 30-1, 44, 55, 78, 82, 91, 94, 101, 113, 118-9, 124, 131, 133, 135, 137-40, 142-5, 147-9, 152-5, 158-9, 162-4, 173, 184, 187-8, 192, 205, 224, 229, 230-2, 241, 251, 257, 259, 262-3, 265-6, 268-9, 271-3, 275, 278-80, 286, 292, 294-6, 307-8, 313-5, 317, 322, 325, 327-8, 343, 350, 352-3, 364, 368-74, 382, 384, 401-2, 419, 423-4, 426, 428-9, 431-5, 441, 443, 453-4, 472, 475-6, 480, 483, 491, 493-4, 496-8, 501, 505, 507, 509-11, 513-4, 550, 557, 563, 566, 568, 581-2, 586, 625, 628
spin 0, 133, 142-3, 147, 155, 158, 266, 271-3, 313-5, 328, 352-3, 364, 369-70, 372-3, 433, 507
spin 0 boson, 133, 142, 313, 507
spin ½, 153, 187, 232, 257, 259, 278-80, 315, 343, 402, 428, 431, 441, 509, 511, 513, 568, 581, 582
spin 1, 133, 142-3, 147, 152, 154, 158, 232, 268, 296, 328, 368, 370-2, 374, 382, 427, 431, 453, 472, 480, 498, 507, 509, 628
spin 1 boson, 133, 142, 154, 296, 368, 371, 427, 480, 507, 628
spin 2, 155, 382, 453-4, 472, 480, 483, 498
spin 3/2, 153, 343, 401
spin $^3/_2$ baryons, 153, 401
spin directions, 145, 158
spin down, 118, 142, 162-4, 173, 263, 317, 491
spin reversal, 148
spin singlet, 364
spin space, 322
spin state, 82, 119, 149, 159, 164, 188, 294-5, 364, 423, 426, 428, 494, 582
spin up, 118, 124, 142, 162-4, 173, 263, 317, 491
spin-averaged probability, 307-8
spin-spin coupling, 475
spin-orbit coupling, 475
spinor, 115, 118-9, 121-2, 124, 126, 128, 130, 133, 139-41, 144, 153-4, 162, 168, 186, 188, 349-50, 375, 491, 494
spinor components, 494
spinor space, 491, 494
spin-spin coupling, 475
spin-statistics, 625
spiral, 109, 516, 531-32, 542, 544-5, 547-8, 550, 568, 583, 589-90
spontaneous chiral symmetry breaking, 531
spontaneous emission, 221, 230, 429
spontaneous symmetry breaking, 95-6, 306, 348, 563, 564
square root of zero, 74, 77, 118, 137, 177, 187, 443, 505, 624
squarks, 632
stable orbit, 246
standard analysis, 10, 47, 66
standard arithmetic, 10, 47
Standard Model, 96, 103-4, 268, 324, 339, 349, 352, 362, 365, 374, 381, 387, 393, 411, 445, 521, 560, 563-4, 576, 589, 608, 610-1, 617
star pentagon, 529, 531
star tetrahedron, 109, 528, 532-537, 546, 551
starlight, 604, 618
stars, 354, 458, 460, 471, 584, 606, 617, 621
start codon, 522, 524, 538
state operator, 190
state vector, 141, 149, 154-5, 161, 164, 179, 266, 343, 350, 443, 495
state vector for charge, 376
states of matter, 320-1, 511, 578
static gravitational attraction, 603
static gravitational field, 600
static gravitational force, 160
static inertial repulsion, 475, 482
stationary states, 242-3
statistical, 237, 616
statistical mechanics, 567, 581
steadily changing conditions, 416
steady-state conditions, 416-7, 420
steady-state dynamics, 417
stellar genesis, 589
Stern, 430
Stewart, P., 590

stimulated emission, 230
stochastic, 19
stochastic electrodynamics, 93, 218, 221, 229-30, 233, 438-40
stop codon, 519, 521-2, 530, 538-9
stopping mechanism, 19
Strandberg, W. M. P., 450, 459, 463-4
strange-charm generation, 358
strangeness, 342-3
strangeness-conserving current, 343
string, 18-20, 404, 484
string representation, 19, 484
string theory, 51, 104, 167, 379, 387-8, 485, 489, 491-5, 557, 563, 588, 629, 630-1
string theory without strings, 491, 588
string-like space, 495
strong binding, 359
strong charge, 44, 55, 205, 251, 256, 258, 260, 266-7, 270, 305, 316-7, 324-6, 330, 340, 343, 350-1, 354, 357-60, 363, 364, 381-2, 384-5, 393, 404, 408-9, 439, 491, 626
strong charge confinement, 364
strong charge orientation, 354
strong coupling, 380, 389, 404, 412
strong coupling constant, 380, 389
strong dipole moment, 278
strong field, 251, 454
strong fine structure constant, 407
strong interaction, 44, 149, 154, 188, 239, 250, 252, 254, 257-8, 274, 277, 279-80, 285, 305, 315-16, 323, 325, 328, 330, 337, 339, 345-6, 348, 357, 359, 365, 370-1, 379, 390, 395, 397, 400, 403, 407, 409, 432, 508, 611, 625-7
strong interaction phase, 254
strong interaction potential, 240, 277, 409
strong interaction solution, 188
strong states, 412
strong transition, 351
strong vacuum, 147, 153, 316, 350
strong vacuum force, 350
strong-electroweak component, 631
strong-electroweak interaction, 378
strong-electroweak solution, 480
strong-electroweak-gravitational / inertial solution, 480
structure of the Earth, 610
subalphabet, 1, 6-7, 9, 11, 19-20, 22, 86-7, 557-8, 563, 589, 591, 593, 638
subcritical density, 609, 620
subluminal communication, 448
successor, 24, 65
sugars, 510, 514-6, 543-4, 547, 554
sum, 91, 131-2, 143, 172, 208, 291-3, 302, 319, 410, 427, 434, 529, 558, 561, 581, 583, 585, 587, 602, 606, 633
Sun, 448, 458-60, 476, 510
sunflower, 529, 540, 558
superclusters, 606
supercritical density, 609, 620
super-duality, 107
supermembranes, 491
superpartners, 387, 493
superposition, 68-9, 81-2, 172, 267, 369, 395, 505, 624
superspace, 387, 493
superstrings, 388, 491
supersymmetry, 95, 137, 141, 147, 161, 163, 174, 190, 285-6, 291, 387-8, 412, 431, 434, 493, 578
supersymmetry operator, 147, 161, 190, 434
supervenient, 22, 86
surreal numbers (surreals), 556, 565, 567, 571-2
Susskind, L., 557
symbol, 9, 19-22, 182, 557-8, 563, 565-7, 569, 591, 593, 596-7, 600, 633, 636-8
symbol creation, 638
symmetric wavefunction, 205
symmetry, 29, 32, 36-7, 40, 43-4, 49-51, 53-6, 59-65, 71-3, 76, 89, 95-6, 98-100, 102-4, 110, 116, 122, 137, 145, 148-9, 151-4, 180, 183, 187-8, 203, 205, 209, 215-20, 225-6, 235, 239, 246, 250, 258-60, 263-4, 267, 270-2, 277, 280, 282-3, 305, 311, 313-4, 329, 333, 335, 337, 339-44, 346, 351-7, 360-4, 367-9, 375, 378, 386-8, 391-3, 398, 401, 403-7, 409, 411, 413-4, 445, 447, 454, 478-9, 484-6, 491-3, 502-5, 508, 515, 529-32, 539, 541-2, 551-2, 554, 564, 566, 569, 580, 588, 620, 624-6, 629-30
symmetry breaking, 71, 375, 392, 484, 491, 541, 569, 580
symmetry generators, 354
symmetry operator, 494
symmetry relations, 204, 217

symmetry transformation, 360-2, 367
symmetry violation, 342, 369, 406, 626
symplectic Fourier transform, 575
symplectic group, 189
symplectic spinor, 186
synchronization, 198
syntactic, 557, 561-2, 596, 629
syntax, 557, 559, 562, 597
syntax construction, 597
synthetic, 32, 56
synthetic aperture radar, 575
system, 1, 4, 6, 9-14, 18-20, 23-6, 29-30, 32-3, 35-9, 41, 43, 48-9, 55-7, 60-1, 63, 66-9, 74, 81, 83-8, 91, 95, 99-101, 105-6, 111-3, 116, 130, 133, 145, 147, 149, 151, 160-1, 165, 173-5, 177-9, 181, 184, 186-7, 190-3, 195, 197-8, 200-9, 211-2, 215, 218-24, 226, 231, 234, 236-8, 242, 244, 255-6, 260, 280, 291, 312-3, 316, 322, 325, 328-30, 333, 335, 337, 347, 354, 357-9, 361, 365, 374, 378, 382-3, 386, 394-5, 400, 412, 414, 416-9, 424, 429, 431-3, 435-6, 438-40, 442, 444, 448, 454, 459, 463, 468-71, 476-7, 479-80, 482, 484-9, 492, 498, 502-4, 506-9, 511-2, 515, 516, 522, 525, 528, 530-2, 536-8, 552-3, 557-9, 561, 566, 568, 570, 572-3, 575-9, 581-3, 585-8, 591, 593-5, 600-2, 605, 607, 611, 618-9, 622, 625, 628-9
System Dual, 99
T duality, 631
t quark, 402, 406, 413
Tait, P. G., 26
Takahara, D., 571
Takahashi, T. T., 256
tau, 326, 343, 395, 410, 611
tau neutrino, 343, 410
temperature, 200, 236-8, 418, 553, 578, 589, 595, 609, 612, 618
tensor, 82, 126-7, 132, 153, 215, 307, 450-1, 456, 636
tensor notation, 215
tensor product, 82, 126-7, 153
tensor representation, 636
tensor terms, 132
ternary encoding, 86
tesseracts, 635
tesseral address, 634
tesseral quaternion, 635, 638
tetrad, 345
tetrahedral representation, 528
tetrahedron, 109, 514, 527-30, 532-3, 537-9, 542, 548, 554
't Hooft gauge, 306
theoretical computation, 82, 86
theoretical physics, 195, 559
theory of everything, 415, 576
theory of nothing, 415
thermal energy, 611, 617
thermal equilibrium, 236, 238
thermodynamic forces, 516
thermodynamic information, 573
thermodynamic structure, 580
thermodynamics, 121, 166, 179, 221, 229, 235, 237-8, 513, 568, 573, 580, 586
third component of weak isospin, 265, 316, 340
Thomas precession, 422, 431, 435, 441
Thompson, D'A., 531
thymine, 516-7, 544, 591
tiling, 110, 165, 529, 634-5, 637
time, 1, 2, 9-10, 16, 26-7, 31-43, 45-57, 59-69, 72, 74-6, 79-81, 83-5, 88-90, 92-4, 96, 98-105, 110, 120-1, 123, 125, 130, 142, 145, 149-50, 153-4, 156, 167, 169, 173-9, 181, 183, 185-9, 193-7, 199-207, 209-11, 213, 218-38, 246, 251, 258-63, 280-1, 283, 292, 294-5, 297-301, 305, 310-1, 314-6, 318, 326-7, 329-30, 333, 336-8, 341-5, 353, 357, 370, 372, 386-8, 393-4, 396, 398, 410, 415-8, 421, 424, 429-30, 433, 436-52, 454, 458-9, 461-4, 468-71, 473, 476-7, 479-81, 485-9, 491-2, 494, 497, 499, 503-7, 509, 513-5, 528, 530-2, 537, 542, 546-7, 552-3, 560, 562-4, 566-8, 570, 573, 576, 579-80, 582-8, 590-2, 594-5, 598, 600, 602, 606-8, 610, 613, 618-29, 632, 636
time derivative, 89, 283
time dilation, 200, 421, 461, 463, 473, 476, 602, 613
time direction, 145, 225, 344
time instant, 221
time reversal, 50, 261-3, 315, 341-2, 353, 370, 372, 396, 410, 506, 568, 582, 586, 590, 626
time reversal asymmetry, 568, 582, 586
time reversal symmetry, 261-3, 314, 315,

341-2, 353, 370, 396, 506, 626
time*, 93-4, 96
time-delayed interaction, 200
time-delayed measurement, 468
time-dependent operator, 174
time-independent, 301
timelike, 50, 76, 89- 90, 93, 116, 232, 387, 429, 438-9, 488, 537
time-reversal symmetry, 582
topological computation, 561
topological equivalence classes, 584
topological Lie group, 572
topological transformation, 572
topology, 47, 88, 110, 321-3, 432, 484
topos theory, 49, 85, 624
total angular momentum, 241, 423
total energy, 75, 159, 172, 189, 230, 248-9, 252, 312, 418, 421, 493
total quantum number, 244
total vacuum, 312-3, 320, 506, 510
trace theorem, 307
transcendental extensions, 565
transcendental numbers, 48
transcription, 502, 516
transfinite, 5, 566-7
transformation, 27, 40-1, 54, 55, 80, 89, 111-2, 122, 141, 150-1, 153-4, 156, 172, 185, 197-9, 205, 212, 215, 230, 258, 271, 313, 353-6, 361-3, 365, 367, 369, 371-2, 404-5, 437, 464-5, 482, 498, 625
transformation of coordinate systems, 111
transition, 21, 22, 66, 76, 80, 85, 125, 168, 175, 177, 183, 220, 237-8, 255, 257, 261, 269, 299-300, 315, 321-2, 329, 347, 359, 363, 366, 368, 370, 372-3, 379, 398, 419-20, 424, 430, 448, 494, 511-2, 532, 537, 551
transition current, 300
transition probability, 299
translation (biology), 502, 517
translation (physics), 37, 40-1, 54-5, 74-5, 85, 160, 204, 221, 234-5, 331, 487, 572, 635
translation asymmetry, 80, 219, 235
translation symmetry, 37, 40, 54-5, 197, 216, 219, 221-2, 234, 238, 258-9, 258, 322
translational invariance, 293
tree level, 290, 291
Trell, E., 404-5
triad, 71, 335, 337-8, 346, 633, 637-8
triangle, 110, 255, 415, 418, 438-9, 531
triplets, 25, 347, 366, 403, 516-9, 522-6, 530, 532-3, 537-8, 546
trivector, 30
tRNA, 517, 520, 544
tryptophan, 524, 530
Turing, A. M., 86, 559, 566
Turing machine, 86, 566
twin paradox, 200, 431
twistor, 98, 112, 484, 493-5
twistor space, 98, 112, 494-5
two-loop approximation, 391
twos complement, 634
two-way speed of light, 199
u quark, 317
Uhlenbeck, G., 422
ultraviolet divergence, 304
unaccelerated motion, 210
unambiguous division, 28
uncertainty, 50, 93, 174-5, 177, 181, 219-20, 222-4, 226, 229-30, 233-4, 236, 238, 330, 426, 429-30, 436, 564, 568, 574-5, 579, 600
unchanging momentum, 210
uncharacterizable, 436, 623
unentangled, 624-5
unidimensionality, 65, 103
unidirectionality of time, 178, 227, 236-7
unified fermion state, 284
unified field theory, 478
unified theory of physics, 2, 478, 491
uniform acceleration, 415-6
uniform mass density, 603
uniform motion, 45, 196, 225
uniform relative motion, 196
uniform velocity, 45, 415
unipolar, 49, 65, 80, 159, 225, 236, 313, 353, 364, 485
unique, 1, 4, 6-7, 9, 18-20, 25, 40-1, 43, 50, 51, 63, 69, 79, 81, 85, 86, 91, 102, 110, 132, 144-5, 165, 177, 179, 186-7, 202, 204, 207, 215, 219, 226, 230, 262, 350, 387, 442, 454, 486, 489, 494, 507-8, 510, 512, 521, 530, 566-8, 579-80, 582, 597, 600, 622, 634
unique birthordering, 567, 580, 600
unique direction for time, 145
unique spin axes, 144
unit charge, 325, 335, 338, 356-9, 364, 393,

410, 412, 446-7
unit inertial mass, 602
unit matrix, 426
unit sphere, 404-5, 634
unit vector, 28, 30, 70, 206, 208, 325, 345, 476, 634
unit wire, 561
unitary gauge, 271
universal, 4, 9, 11, 13, 18-20, 32, 35, 54, 57, 61, 67, 69, 81, 86-7, 90, 99, 101, 145, 177, 179, 204, 206, 210, 213, 231, 296, 311, 352, 416, 431, 443, 445, 454, 467, 502-3, 505, 509, 512, 514-5, 530, 532, 538, 542, 552-3, 555-9, 561-70, 576, 579, 581-2, 584-6, 588-91, 593-4, 596, 598, 600, 602, 605-6, 611, 613, 619-22, 624, 629, 633, 637-8
universal alphabet, 19, 86, 87, 558, 591, 593, 600, 619, 622, 633, 637
universal computation, 561, 565, 581-2, 593, 598
universal computer construction, 559
universal digital computation, 559, 582
universal energy continuum, 468
universal grammar, 556-58, 562-3, 565, 570, 576, 596, 598
universal gravitational interaction, 81
universal inertial acceleration, 605
universal inertial field, 602
universal interaction, 177, 210, 624
universal laws, 57
universal meta-pattern, 559
universal phase effect, 512
universal quantum computation, 556, 558, 586
universal reference frame, 431
universal rest frame, 613
universal rewrite system, 11, 20, 69, 86, 99, 101, 503, 505, 514, 530, 538, 542, 552, 555, 558, 582, 588-9, 629, 633
universal scalar, 213
universal semantic quantum mechanical state space, 570
universal semantics, 598
universal structuring, 598
universal system, 4, 13, 20, 67, 502, 557, 620
universal system of process, 502
universal thermal equilibrium, 611
universal time, 179, 622
universal topological pre-space, 584
universal zero condition, 296
universality of mass, 471
universe, 2-4, 9, 20, 34, 40, 68-9, 73, 79, 81, 91, 149, 159, 161, 179, 222, 231, 262, 311-2, 318-9, 432, 433, 443, 445, 447, 478, 506-7, 509, 511, 515, 537-8, 556, 564, 570, 573, 576, 578-80, 583-5, 587-8, 591-2, 595, 601-7, 609-11, 614, 616-8, 620-2, 624, 628, 632
up quark, 270
upper triangular matrices, 574
uracil, 516-7, 553, 591
vacuum, 58, 68, 74, 84-5, 90-1, 100, 104, 110, 123, 125, 133, 137, 146-9, 153, 155, 157-66, 174, 176, 179, 185, 188, 190, 199-201, 206, 218, 228-30, 233, 239-40, 257, 261-3, 266, 270-1, 273, 278-80, 286, 290-2, 297, 304, 307, 309-22, 324, 327-8, 349, 352-3, 365-6, 368-70, 372, 387, 389, 391, 393-400, 406, 412-3, 420, 430-5, 438, 443, 445-7, 481-2, 485, 490, 492-5, 498, 505-6, 508-13, 519, 528-9, 537, 550-1, 554-5, 568, 575, 578, 600, 604, 606-7, 610-3, 616, 618-9, 622, 624, 626-8
vacuum antifermion, 158
vacuum charge state, 400
vacuum creator, 494
vacuum density, 604, 616
vacuum energy, 318, 400, 447, 510, 616
vacuum expectation energy, 389, 617
vacuum expectation value, 400, 406
vacuum fermion, 314
vacuum field, 206, 230, 329, 400
vacuum fluctuations, 309, 320, 482
vacuum generator, 148, 318
vacuum image, 110, 149, 159, 280, 291, 304, 434, 619
vacuum operator, 146, 185, 291, 297, 318, 370
vacuum phase, 240
vacuum polarization, 290, 391, 431
vacuum pressure, 606
vacuum process, 550, 610-2, 616
vacuum projection, 319
vacuum reflection, 315, 318, 349-50
vacuum selection, 495
vacuum space, 322, 432, 485, 551
vacuum state, 148, 159, 161, 174, 190, 279-

80, 313, 320-2, 328, 352-3, 369, 372, 393, 395, 412, 433-4, 447, 492, 505, 528, 612, 627-8
vacuum virtual charge density, 612
vacuum wavefunction, 146
vacuum wavefunction operator, 146
valence quarks, 256
Van der Waals force, 237, 279, 315, 319-21, 502
Van Flandern, T., 448
variability, 48, 66, 123, 130, 189, 203, 206, 223, 620
variable, 37, 46, 74, 81, 124, 182, 203, 207, 219, 284, 294, 322, 410, 426, 465, 488, 559, 619
variable grain size, 219
variance, 426
variation, 35, 48, 60, 66, 74, 84, 88, 120, 122, 144, 152, 164, 166, 178, 188, 203-4, 212, 220-1, 224, 228, 262, 266, 283, 317, 328, 330, 337, 371, 395, 435, 505, 508, 516, 519-20, 522, 531, 542, 553, 589
variational principle, 473
vector, 5, 13, 15-16, 23, 25-31, 40-1, 43-4, 53, 63, 69-74, 78- 80, 82, 84, 88, 92, 95-6, 98, 101-2, 106-7, 109, 113, 115, 118, 124-5, 127, 130-2, 138-9, 141-2, 144-8, 152, 154, 157, 165, 173, 178, 189, 197-202, 204, 206, 208, 210, 212-3, 223, 230, 241, 249-54, 256-7, 259-61, 264, 266, 268-9, 273, 280, 283-4, 286-7, 291-2, 296, 300, 308, 311-3, 316-7, 320-1, 323, 325, 326, 329-30, 334, 337, 345, 350-2, 366, 368, 375-6, 378-80, 382-4, 386, 392, 403, 423-4, 426, 432-3, 436, 441, 443, 445-6, 454, 469, 471, 474, 476-7, 479-80, 483, 487-8, 497, 504-6, 508, 519, 521, 546, 566, 573-6, 581, 585-6, 600, 603, 607, 624-5, 633-5
4-vector, 27-8, 31, 42-3, 45, 50, 75, 79, 89, 92, 97-8, 110, 115, 127-8, 130-1, 195, 197-8, 200-202, 206, 212-5, 221, 223, 236, 261, 289, 293, 297, 299-300, 334, 379, 424, 447, 453-4, 468-9, 471, 479, 488, 492, 495-6, 499-500, 564, 580-1, 603, 624
4-vector algebra, 580
4-vector potential, 300
vector addition, 115, 329
vector algebra, 26, 28, 102
vector boson, 306, 366
vector calculus, 26
vector component, 300
vector gauge boson, 272
vector meson, 156, 403
vector multiplication, 125, 489
vector operator, 71, 74, 260, 264, 296, 345, 384, 423, 441, 443
vector phase, 350
vector potential, 173, 204, 283, 292, 423, 432, 445, 474, 476, 480, 508
vector potential energy, 508
vector product, 28, 44
vector quaternion, 69-70, 382, 384
vector representation, 581
vector space, 16, 23, 27, 72, 223, 471
vectorial field, 572
velocity, 27, 48, 59, 180, 182, 196, 199, 209-10, 228, 234, 415-7, 419-20, 446, 454, 457-60, 464, 467, 470-3, 475, 479, 483, 607
velocity addition law, 196, 440
velocity of light, 27, 228, 234, 446, 454, 467, 470-2, 475, 479
velocity operator, 180
vertex, 109-10, 144, 148, 154, 267-9, 290-1, 303, 305, 308-9, 314, 323, 353-4, 367, 369-70, 578
vertex factor, 367
vertex graph, 290-1
virial relation, 174, 209, 255, 277, 420, 426, 435, 440-1
virial theorem, 101, 230, 418
virtual, 50, 147, 148-9, 158-9, 161, 175-6, 215, 218, 228, 230, 233-4, 256, 290, 311-2, 316, 319, 322, 369, 388, 434, 437, 550, 559, 607, 611, 613-4, 629
virtual antistate, 147
virtual boson, 147-8, 550
virtual energy, 230, 312
virtual fermion, 147, 369, 550
virtual particle, 147, 159, 388, 611, 613-4
virtual photon, 256, 290, 319
virtual process, 228
virtual quanta, 148, 215
virtual vacuum, 50, 149, 158, 312
virtual vacuum energy, 158, 312
virtual vacuum state, 159, 312
virtual zero-point energy, 607

virus, 540
visual parallax, 490
visual representation, 88, 95, 105, 110
volume, 29, 30, 107, 212, 245, 396, 404-5, 417, 419, 540, 614-5
volume-preserving, 404
von Koch, H., 18
von Neumann, J., 567
wave, 61, 74, 103, 135, 168, 201, 218, 222-4, 228, 230-1, 246, 248, 293, 298, 415, 419, 438-9, 457, 458, 470, 472, 475, 482, 488, 498, 575, 578, 582, 586, 593, 624
wave equations, 59, 90, 213, 214
wave mechanics, 50, 93, 176-7, 232, 233, 567
wave number, 91, 172, 194
wave solution, 475
wave structure, 593
wavefront, 299, 574-5, 578, 587-8
wavefunction, 41, 45, 49-50, 74, 78-9, 81, 85, 90-3, 119-21, 123-4, 126, 128, 130-1, 133, 140-1, 145, 156-7, 166, 168, 170-3, 175-6, 178, 181, 184-7, 202, 204, 220-1, 226, 228, 240-3, 245-6, 288-90, 292, 299, 329, 343-4, 367, 378, 426-9, 432-3, 446, 512, 568, 577, 588
wavefunction collapse, 50, 93, 172, 175-7, 220-1, 226, 228, 329, 429
wavefunction correlation, 292, 446
wavelength, 91, 175, 194, 219, 432
wave-particle duality, 50, 61, 201, 218, 222-4, 415, 438, 488, 624
weak boson, 305-6, 329, 378
weak charge, 45, 148, 205, 217, 237, 258-61, 264, 270, 278-80, 284, 305, 314-6, 321, 324-5, 327-9, 340-4, 360-1, 375, 381, 383-4, 386, 394, 410, 439, 445, 491, 505, 510-3, 557, 626
weak charge conjugation violation, 314-5, 321, 329, 394
weak charge operator, 445
weak charge structure, 341
weak coupling, 270, 390, 399-401, 412, 617
weak coupling constant, 270, 399, 400, 412
weak decay, 341
weak dipole, 260, 278-9, 315, 321, 359, 510, 631
weak dipole moment, 278-9, 315, 510, 631
weak eigenstate, 408-9, 413
weak field, 278, 449
weak gauge boson, 160, 313, 411
weak generators, 383, 384
weak hypercharge, 279, 340, 366, 397
weak hypercharge current, 366
weak interaction, 32, 36, 145, 149, 151, 159, 206, 217, 237, 239, 257, 259-61, 263, 265-70, 272-3, 278-82, 284, 305, 307, 312-8, 323-4, 326, 328-9, 335, 337, 340-3, 352, 357, 359, 361, 365-70, 372, 378, 381, 390, 395, 406, 410, 432, 446, 507-10, 513-4, 531, 617, 625-6
weak isomultiplet, 340
weak isospin, 239, 265-6, 317, 325-9, 336, 340-1, 343, 362, 366, 370, 375-7, 388, 394-5, 406, 413, 439, 626
weak isospin current, 362, 366
weak isospin transition, 370
weak mixing, 365, 410, 413, 631
weak monopole, 321
weak process, 279, 617
weak singlet, 257
weak symmetry-breaking, 611
weak transition, 315, 351, 365, 368
weak vacuum, 147-8, 237, 261, 264, 279, 314-5, 317, 319, 321, 350, 372, 493, 513, 626
weak vacuum force, 350
weak vertex factor, 365
Wegener, M., 227
Weinberg, S., 44, 341, 394
Weinberg-Salam electroweak theory, 44, 341, 362-7, 394
Weyl, H., 568, 574
Wheeler, J. A., 430, 434, 556, 561, 569-70, 572, 576
white (w)hole, 565, 570, 583-5
white blood cells, 594
Whitehead, A. N., 46
Whitney, C. K., 431
Whitrow, G. J., 46
Wilson, K. G., 619
Witten, E., 112, 494-5
Wittgenstein, L., 562
Woit, P., 557
Wolfenstein parameterization, 408
Wolfram, S., 18
wordlength, 634, 636
world, 5, 36, 222, 453, 501, 562, 594, 596-7,

622
world-line, 145
X-ray crystallography, 544
Y boson, 378
Yang-Mills principle, 41, 205, 624
Young, N., 4, 432
Young's slits, 432
Yukawa potential, 282-3
Z boson, 149, 268, 370, 380
Zenergy, 578, 580, 583-4, 592
Zeno's paradoxes, 46-7, 49, 61
Zermelo-Fraenkel set theory, 565
zero, 1, 3-7, 9, 11-12, 14, 16, 18, 20, 22-4, 28, 32-3, 39-40, 42, 49, 51, 63-5, 67-9, 72, 74-5, 78-9, 81, 83-6, 100, 103, 129-32, 135, 137, 151, 153-4, 160-1, 164, 166, 177, 181, 184, 187, 189, 195, 202-4, 206, 208-9, 216, 220-1, 224, 227, 235, 243, 247, 249, 255-6, 259, 304, 317-20, 327, 329, 335-6, 338, 355, 358, 371, 377, 383, 387, 407, 414-6, 424-5, 446, 450, 467, 483, 485, 486-8, 490, 495, 498, 503, 505-6, 508, 512, 537, 557-8, 561, 566, 569, 574-5, 582-3, 600-2, 604, 611, 619, 622-4, 628, 633
zero charge, 264, 393, 395, 399, 401-3, 405, 413, 626-7
zero force, 246
zero mass, 139, 268
zero total energy, 493, 581
zero wavefunction, 319
zeroth-order coupling, 287
zero-point energy, 145, 148, 158, 174, 218, 229, 230, 232-3, 238, 279, 310, 312, 315, 318, 396, 426, 429, 430, 439, 442, 478, 482, 607
zitterbewegung, 159, 181, 186, 221, 230, 257, 279, 322-3, 431-2, 435, 492, 510, 513, 548, 550-1, 569
zombie, 595
Zweig, G., 346, 347, 348

SERIES ON KNOTS AND EVERYTHING

Editor-in-charge: Louis H. Kauffman *(Univ. of Illinois, Chicago)*

The Series on Knots and Everything: is a book series polarized around the theory of knots. Volume 1 in the series is Louis H Kauffman's Knots and Physics.

One purpose of this series is to continue the exploration of many of the themes indicated in Volume 1. These themes reach out beyond knot theory into physics, mathematics, logic, linguistics, philosophy, biology and practical experience. All of these outreaches have relations with knot theory when knot theory is regarded as a pivot or meeting place for apparently separate ideas. Knots act as such a pivotal place. We do not fully understand why this is so. The series represents stages in the exploration of this nexus.

Details of the titles in this series to date give a picture of the enterprise.

Published:

Vol. 1: Knots and Physics (3rd Edition)
by L. H. Kauffman

Vol. 2: How Surfaces Intersect in Space — An Introduction to Topology (2nd Edition)
by J. S. Carter

Vol. 3: Quantum Topology
edited by L. H. Kauffman & R. A. Baadhio

Vol. 4: Gauge Fields, Knots and Gravity
by J. Baez & J. P. Muniain

Vol. 5: Gems, Computers and Attractors for 3-Manifolds
by S. Lins

Vol. 6: Knots and Applications
edited by L. H. Kauffman

Vol. 7: Random Knotting and Linking
edited by K. C. Millett & D. W. Sumners

Vol. 8: Symmetric Bends: How to Join Two Lengths of Cord
by R. E. Miles

Vol. 9: Combinatorial Physics
by T. Bastin & C. W. Kilmister

Vol. 10: Nonstandard Logics and Nonstandard Metrics in Physics
by W. M. Honig

Vol. 11: History and Science of Knots
edited by J. C. Turner & P. van de Griend

Vol. 12: Relativistic Reality: A Modern View
edited by J. D. Edmonds, Jr.

Vol. 13: Entropic Spacetime Theory
by J. Armel

Vol. 14: Diamond — A Paradox Logic
by N. S. Hellerstein

Vol. 15: Lectures at KNOTS '96
by S. Suzuki

Vol. 16: Delta — A Paradox Logic
by N. S. Hellerstein

Vol. 17: Hypercomplex Iterations — Distance Estimation and Higher Dimensional Fractals
by Y. Dang, L. H. Kauffman & D. Sandin

Vol. 19: Ideal Knots
by A. Stasiak, V. Katritch & L. H. Kauffman

Vol. 20: The Mystery of Knots — Computer Programming for Knot Tabulation
by C. N. Aneziris

Vol. 24: Knots in HELLAS '98 — Proceedings of the International Conference on Knot Theory and Its Ramifications
edited by C. McA Gordon, V. F. R. Jones, L. Kauffman, S. Lambropoulou & J. H. Przytycki

Vol. 25: Connections — The Geometric Bridge between Art and Science (2nd Edition)
by J. Kappraff

Vol. 26: Functorial Knot Theory — Categories of Tangles, Coherence, Categorical Deformations, and Topological Invariants
by David N. Yetter

Vol. 27: Bit-String Physics: A Finite and Discrete Approach to Natural Philosophy
by H. Pierre Noyes; edited by J. C. van den Berg

Vol. 28: Beyond Measure: A Guided Tour Through Nature, Myth, and Number
by J. Kappraff

Vol. 29: Quantum Invariants — A Study of Knots, 3-Manifolds, and Their Sets
by T. Ohtsuki

Vol. 30: Symmetry, Ornament and Modularity
by S. V. Jablan

Vol. 31: Mindsteps to the Cosmos
by G. S. Hawkins

Vol. 32: Algebraic Invariants of Links
by J. A. Hillman

Vol. 33: Energy of Knots and Conformal Geometry
by J. O'Hara

Vol. 34: Woods Hole Mathematics — Perspectives in Mathematics and Physics
edited by N. Tongring & R. C. Penner

Vol. 35: BIOS — A Study of Creation
by H. Sabelli

Vol. 36: Physical and Numerical Models in Knot Theory
edited by J. A. Calvo et al.

Vol. 37: Geometry, Language, and Strategy
by G. H. Thomas

Vol. 38: Current Developments in Mathematical Biology
edited by K. Mahdavi, R. Culshaw & J. Boucher

Vol. 39: Topological Library
Part 1: Cobordisms and Their Applications
edited by S. P. Novikov and I. A. Taimanov

Vol. 40: Intelligence of Low Dimensional Topology 2006
edited by J. Scott Carter et al.

Vol. 41: Zero to Infinity: The Fountations of Physics
by P. Rowlands